Metal Phosphonate Chemistry
From Synthesis to Applications

Metal Phosphonate Chemistry
From Synthesis to Applications

Edited by

Abraham Clearfield
Department of Chemistry, Texas A&M University, College Station, Texas, USA

Konstantinos Demadis
Department of Chemistry, University of Crete, Heraklion, Crete, Greece

RSC Publishing

ISBN: 978-1-84973-356-4

A catalogue record for this book is available from the British Library

Published by The Royal Society of Chemistry,
Thomas Graham House, Science Park, Milton Road,
Cambridge CB4 0WF, UK

Registered Charity Number 207890

For further information see our web site at www.rsc.org

Preface

In Chapter 1 it is stated that Giulio Alberti *et al.* were the first to publish on metal phosphonates, by reporting on compounds of the formula $Zr(O_3PR)_2$, where R is an alkyl or aryl group ("Crystalline $Zr(R\text{-}PO_3)_2$ and $Zr(R\text{-}OPO_3)_2$ compounds (R = organic radical): New class of materials having layered structure of zirconium-phosphate type", G. Alberti, U. Costantino, S. Allulli, N. Tomassini, *J. Inorg. Nucl. Chem.*, 1978, **40**, 1113) However, almost simultaneously Cunningham *et al.* published a paper on divalent metal phosphonates ("Divalent metal phenylphosphonates and phenylarsonates", D. Cunningham, P.J.D. Hennelly, T. Deeney, *Inorg. Chim. Acta.*, 1979, **37**, 95). Although, they reported on a number of properties of these compounds their structures were not determined. Due to the very poor X-ray patterns of the zirconium compounds, Alberti did not publish the structures; however, he did draw up a model showing a bilayer of phenyl groups between the ZrO_6P_2 inorganic layers. This model served as a guide for the dozens of subsequent zirconium phosphonates that were synthesized. Sulfonation of the phenyl groups in 1987 ("The preparation and ion-exchange properties of zirconium sulfophosphonates", C. Y. Yang, A. Clearfield, *React. Polym.*, 1987, **5**, 13) and their subsequent exfoliation proved the layered nature of these compounds and years later we were able to solve the crystal structure of the phenyl compound from a highly crystalline powder prepared by the Alberti group ("Determination of crystal structures from limited powder data sets: Crystal structure of zirconium phenylphosphonate", D. M. Poojary, H.-L. Hu, F. L. Campbell, III, A. Clearfield, *Acta Cryst. B*, 1993, **49**, 996). The structure was as predicted except that the phenyl rings are tilted about 30° from the perpendicular. Another milestone occurred in 1988 with the synthesis and structure determination of a number of transition metal phenyl phosphonates by

Metal Phosphonate Chemistry: From Synthesis to Applications
Edited by Abraham Clearfield and Konstantinos Demadis

Published by the Royal Society of Chemistry, www.rsc.org

Mallouk ("Synthesis and structural characterization of a homologous series of divalent-metal phosphonates, $M^{II}(O_3PR)\cdot H_2O$ and $M^{II}(HO_3PR)_2$", G. Cao, H. Lee, V. M. Lynch, T. E. Mallouk, *Inorg. Chem.,* 1988, **27**, 2781). Because of the one to one ratio of phosphonate groups to metal ions there are half as many oxygen atoms to coordinate to the metal. Nevertheless the 6-coordination was accomplished by chelation of the metal by phosphonate groups above and below with donation of electron pairs by the chelating oxygens to adjacent metal atoms. The third phosphonate oxygens bonded across these chains to form layers and the sixth metal orbital is occupied by a water molecules. In Chapter 1, it is also shown that Group I cations from Li^+ to Cs^+ bond to 4,4′-diphosphonic acids with increasing coordination from 4 to 8 with exactly the same composition. All of this demonstrates the versatility of phosphonic acids as ligands.

From these beginnings and many other interesting features of metal phosphonate chemistry published before 1990 it was realized, as some of the early pioneers remarked, "that a vast new field was developing." A huge number of organic moieties can be converted to phosphonic acids, allowing for a vast array of metal phosphonates to be synthesized. Thus, like a rolling snowball descending a mountain the publications grew and grew, as shown in Figure 1.

In 2009 I was contacted by the Royal Society of Chemistry to consider a book on metal phosphonate chemistry. Yes, I thought it was a good idea but the amount of work required was daunting. Shortly after that I attended the 1st International Conference on Multifunctional, Hybrid and Nanomaterials, in Tours, France, and there I met Kostas Demadis for the first time. At lunch with him, I mentioned the book idea. Kostas thought it was great idea and right then and there the decision to go forward with it was made. So here we were a duo of apostates in an ocean of metal-carboxylate chemists plotting heresy. The comfort of having a young enthusiastic co-editor

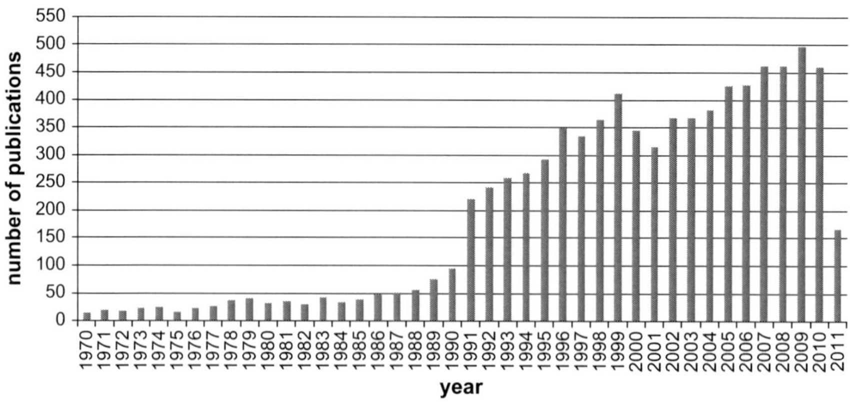

Figure 1 Graph with number of publications per year in the "phosphonate" field (source Thomson ISI, May 5, 2011).

whose English was perfect clinched the deal. We both drew up a list of potential authors, all leaders in their respective expertise and made our contacts. The result is now before you! The quality of the authors and the variety of areas covered make this a very special offering. The book is diverse in the chapter topics (a quick look at the Table of Contents will convince the reader), well referenced and with some amazing illustrations. Kostas and I wish to thank the authors for a job well done and I am indebted to Kostas as he not only edited half the chapters but he picked up the ball when I faltered and assembled all the chapters in the required way and transmitted them to our RSC publishers.

Abraham Clearfield
Texas A&M University
College Station, USA

and

Kostas D. Demadis
University of Crete
Heraklion, Greece

Abraham Clearfield

"I am pleased to dedicate this book to my wife Ruth who urged me to go on for a PhD and has been a steady supporter and life-long help mate. What more could anyone ask for?"

Kostas Demadis

"This book is dedicated to several people who directly or indirectly, knowingly or unknowingly contributed to its completion: my wife Foteini for emotional support, my students who give purpose to our research, Wilhelm Conrad Röntgen for inventing X-rays, and Lemmy Kilmister for inspiration."

Contents

Metal Phosphonate Chemistry: From Synthesis to Applications
Edited by Abraham Clearfield and Konstantinos Demadis

Published by the Royal Society of Chemistry, www.rsc.org

CHAPTER 1

The Early History and Growth of Metal Phosphonate Chemistry

ABRAHAM CLEARFIELD

Department of Chemistry, Texas A&M University, College Station, USA 77843-3255

1.1 Early History of α-Zirconium Phosphonates

It is fitting that we begin with a history of the formation and growth of metal phosphonate chemistry. In 1978 Giulio Alberti *et al.*[1] prepared a number of zirconium phosphonates by the direct reaction of phosphonic acids with Zr^{IV}. In general

$$Zr^{IV} + 2RPO_3H_2 \rightarrow Zr(O_3PR)_2 + 2H^+ \tag{1.1}$$

where R is an alkyl or aryl group. It was recognized that these compounds were layered in nature but no structures were forthcoming because of their very sparingly soluble nature. It was not until 1993 that the structure of the phenylphosphonate $Zr(O_3PC_6H_5)_2$ was solved by Poojary *et al.*,[2] based on a sample prepared in Alberti's laboratory. This synthesis required hydrothermal treatment for 30 days at 190–200 °C in the presence of HF as a solubilizing agent. The crystalline powder yielded unit cell dimensions $a = 9.0985(5)$ Å, $b = 5.4154(3)$ Å, $c = 30.235(2)$ Å, $\beta = 101.333(5)°$, space group C2/c. The structure (Figure 1.1) is indeed layered and similar to that of α-zirconium phosphate, $Zr(O_3POH)_2 \cdot H_2O$.[3] The unit cell dimensions of the phosphate (abbreviated α-ZrP), in space group $P2_1/n$: are $a = 9.060(2)$ Å, $b = 5.297(1)$ Å, $c = 15.414(3)$ Å, $\beta = 101.71(2)°$. The a and b dimensions are very close to the corresponding dimensions of the phenylphosphonate but in this latter structure

Metal Phosphonate Chemistry: From Synthesis to Applications
Edited by Abraham Clearfield and Konstantinos Demadis

Published by the Royal Society of Chemistry, www.rsc.org

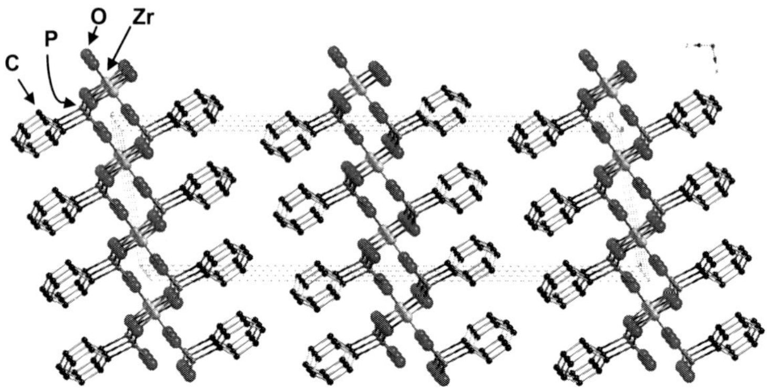

Figure 1.1 Structure of α-zirconium phosphonate as viewed down the *b* axis.

the phenyl rings are tilted 30° relative to the layers (Figure 1.1) whereas the P–OH groups of α-ZrP are almost perpendicular to the layers. This difference arises from the space group requirements. In $P2_1/n$ the Zr^{4+} ion is slightly above and below the mean layer plane but in C2/c the metal ion lies exactly in the plane. The interlayer distances are 7.55 Å and 14.87 Å for the phosphate and phosphonate, respectively.

Subsequent to Alberti's discovery, a large number of similar compounds with many different functional groups were prepared, some of which are listed in Table 1.1. All of them have the α-ZrP structure. (See Chapter 2 for additional details.)

It is possible to prepare layered compounds of group 4 metals with two ligands such as phenyl and methyl groups. The ability to prepare such compounds with two pendant groups has some appealing features. If one group is large, as for example a biphenyl or long chain alkyl group and the other small, then microporosity is built into the structure. However, there are some pitfalls. If the small group is present in excess of 50%, interdigitation may occur removing part or all of the porosity.[4] Further the composition obtained may be different than the starting composition and this may depend upon the synthesis conditions, particularly temperature and choice of solvent and whether HF is present. Another factor depends upon the relative hydrophobic–hydrophilic nature of the two ligands. In such cases the complete mixing may not be possible. However, it may be possible to achieve porosity and functionality, even between hydrophobic and hydrophilic species by choice of the ligands. As an example, mixed derivatives such as $Zr(O_3PC_6H_5)_{2-x}(O_3PCH_2COOH)_x$ introduces ion exchange capability and possible further reactivity at the carbonyl site.

1.1.1 Advent of γ-Zirconium Phosphonates

γ-Zirconium phosphonates are obtained by ester interchange of a phosphonic acid with γ-ZrP. Thus, it is necessary to understand the structure of γ-ZrP before continuing further. This compound was first prepared by refluxing a

Table 1.1 Basal spacing of some zirconium bis(monophosphonates), and organophosphates with α-layered structures.

Compound	*Basal spacing (Å)*
$Zr(O_3PH)_2$	5.58
$Zr(O_3PMe)_2$	8.90
$Zr(O_3PCH_2OH)_2H_2O$	10.10
$Zr(O_3PCH_2OH)_2$	9.10
$Zr(O_3PCH_2Cl)_2$	10.10
$Zr(O_3P(CH_2)_2Cl)_2$	12.30
$Zr(O_3P(CH_2)_2Br)_2$	13.00
$Zr(O_3PCH_2CN)_2$	10.80
$Zr(O_3P(CH_2)_2CN)_2$	11.80
$Zr(O_3PPr^n)_2$	14.00
$Zr(O_3PBu^n)_2$	15.90
$Zr(O_3PC_6H_{13})_2$	19.20
$Zr(O_3PC_{12}H_{25})_2$	32.40
$Zr(O_3PC_{22}H_{45})_2$	56.20
$Zr(O_3PCH{=}CH_2)$	10.60
$Zr(O_3P(CH)_2CH{=}CH)_2$	13.50
$Zr(O_3PPh)_2$	14.70
$Zr(O_3PC_6H_{11})_2$	15.20
$Zr(O_3PCH_2C_6H_5)_2$	16.60
$Zr(O_3PC_6H_4C_8H_{17})_2$	20.80
$Zr(O_3PC_2H_4C_5H_4N)_2$	18.60
$Zr(O_3PCH_2CHOHCO_2H)_2$	14.30
$Zr(O_3PCH_2SO_3H)_2$	15.40
$Zr(O_3P(CH_2)_3SO_3H)_2$	17.30
$Zr[(O_3PO)(CH_2CH_2O)_2PO_3]$	15.30
$Zr[(O_3PO)(CH_2CH_2NH_2)_2]\cdot 2HCl$	14.30
$Zr(O_3PCH_2NHCH_2CH_2NH_2)2HCl\cdot 4H_2O$	17.10
$Zr(O_3P(C_6H_4)_2C_6H_5)$	33.00
$Zr(O_3PC_6H_4NO_2)_2$	15.80
$Zr(O_3P(CH_2)_2COCl)_2$	13.50
$Zr(O_3PCH_2CO_2H)_2$	11.10
$Zr(O_3P(CH_2)_2CO_2H)_2$	13.00
$Zr(O_3P(CH_2)_3CO_2H)_2$	15.00
$Zr(O_3P(CH_2)_4CO_2H)_2$	16.70
$Zr(O_3P(CH_2)_5CO_2H)_2$	18.70
$Zr(O_3POEt)_2$	11.70
$Zr(O_3PMBu^n)_2$	15.90
$Zr(O_3PO(CH_2)_2CH)_2$	13.20

solution of $ZrOCl_2$ in phosphoric acid containing a high concentration of ammonium or sodium phosphates.[5] This procedure yields the half ammonium exchanged or sodium exchanged salt. Treatment with dilute HCl provides the protonated α-ZrP. The composition is $Zr(PO_4)[O_2P(OH)_2]\cdot 2H_2O$. The structure was first solved incompletely from X-ray powder data on the titanium phase[6] and later completely for Zr.[7] The crystals are monoclinic, $a = 5.3825(2)$ Å, $b = 6.6337(1)$ Å, $c = 12.4102(4)$ Å, $\beta = 98.687(2)°$, space group $P2_1$. The layers consist of phosphate groups within the layer bonding through oxygen

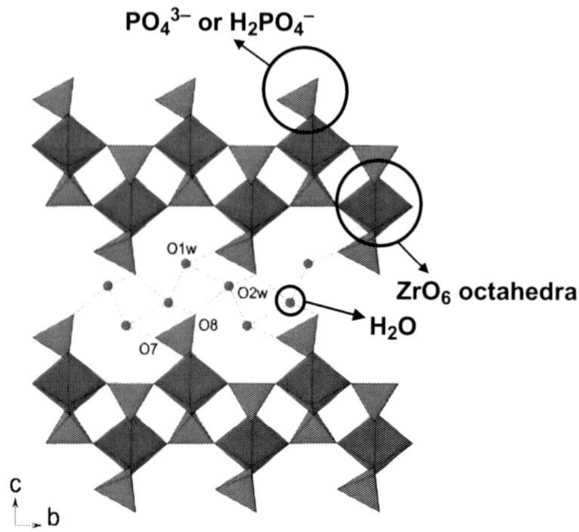

Figure 1.2 Crystal structure of $Zr(PO_4)(H_2PO_4)\cdot 2H_2O$ (γ-ZrP). Adapted from reference 7 with permission.

atoms to four different Zr ions. Four such phosphate tetrahedra bond to a single Zr ion as shown in Figure 1.2. The coordination number of each Zr is six, the remaining two bonds are supplied by the dihydrogen phosphate groups. These groups bridge across two Zr ions in the *a* axis direction, tying together the chains running along the *b* axis direction into layers. The two hydroxyl groups are pendant, pointing into the interlayer space. They hydrogen bond to the water molecules, as shown in Figure 1.2, bringing the layers together. The γ-ZrP has been found to undergo ester interchange reactions by a topotactic mechanism.[8] The group replaced is $O_2P(OH)_2$ and if phenyl phosphonic acid is interchanged, the in-going group is $C_6H_5PO_3{}^{2-}$ (as shown in Figure 1.3). Alberti points out that if the in-going groups are too bulky or of the type $O_2P(C_6H_5)_2$, steric hindrance may occur and only a portion of the sites can be occupied.[8] While the Alberti group has dominated the use of topotactic exchange reactions with γ-ZrP, the first reactions were carried out by Yamanaka and his co-workers.[9–11]

1.1.1.1 *Early Chemistry of Zirconium Diphosphonates*

Martin Dines and his co-workers prepared many zirconium phosphonates,[12] much of which is in the patent literature. The idea behind their study was to prepare catalysts, so many of the phosphonates contained functional groups such as sulfonic acid, carboxylic acid, and other functionalities. However, an important advance was made by the Dines group when they introduced α,ω-diphosphonic acids. These compounds were layered but with interlayer spacings as shown in Table 1.1. These interlayer spacings indicate that the ligands were situated almost perpendicular to the layers and cross-linked them. The

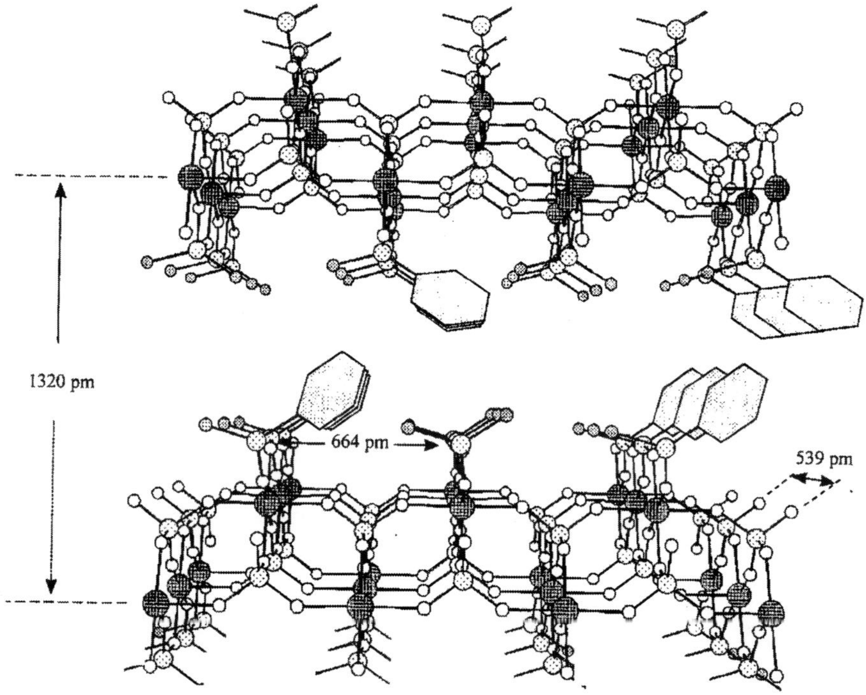

Figure 1.3 Schematic representation of the structure γ-$ZrPO_4(H_2PO_4)_{0.5}(HO_3PC_6H_5)_{0.5}$. Adapted from reference 8 with permission from Pergamon Press.

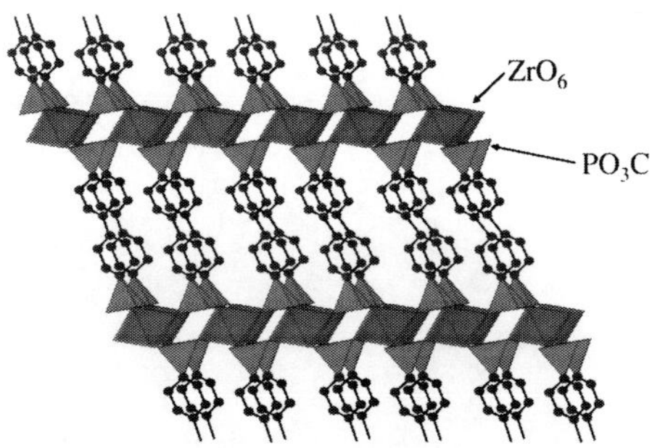

Figure 1.4 An idealized schematic representation of $Zr(O_3PC_{12}H_8PO_3)$ showing the expected non-porous nature of this compound.

distance between these ligands, acting as pillars as in Figure 1.4, is 5.3 Å, leaving no porosity between pillars.[12a] Porosity was obtained by spacing the pillars by including phosphorous acid as a reactant.[12b] However, their data showed that a maximum in surface area was obtained when the mole percent pillar was 50% with a surface area of $390\,m^2\,g^{-1}$. What they did not explain is that the compound with 100% pillars had a surface area of $316\,m^2\,g^{-1}$. In point of fact, the parent compound $Zr(O_3PC_6H_4\text{-}C_6H_4PO_3)$ is porous. Direct determination of the structure of the compound is not possible because the X-ray powder pattern consists of only a few very broad peaks. Thus, the mystery is why the fully pillared compound has such a large surface area. More about this subject will be presented in Section 1.7.

1.2 Divalent Metal Phosphonates

Ten years after Alberti's publication on the zirconium monoaryl and alkyl derivatives, Mallouk *et al.* presented crystal structures of Mn, Zn, Co and Mg phenyl, and alkyl phosphonates.[13] They all have the same layered structure with unit cell dimensions for the zinc compound $a = 5.634(2)$ Å, $b = 14.339(5)$ Å, $c = 4.833(2)$ Å with $Z = 2$ and space group $Pmn2_1$. The composition of these compounds is $M(O_3PR)\cdot H_2O$ where R is the aryl or alkyl group. The metals are six coordinate but there are only three phosphonate oxygens versus six for the four valent layered compounds. The required coordination is obtained by a sharing of oxygen atoms as shown in Figure 1.5. Two of the phosphonate oxygen atoms chelate the metal ion and then both of these oxygen atoms donate to adjacent metal atoms. These oxygens are three coordinate and they

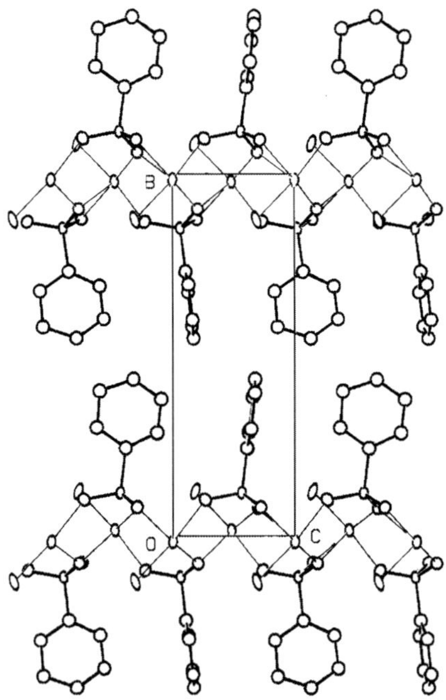

Figure 1.5 Layer structure of $Mn(O_3PC_6H_5)$ viewed down the *a* axis showing the pleated sheet nature of the layers and the arrangement of the phenyl rings. Adapted with permission from reference 13, Figure 2. Copyright © 1988 American Chemical Society.

are situated above and below the mean plane of the layer. A fifth coordination site is filled by the third phosphonate oxygen atom that connects these rows of metal atoms aligned along the *a* axis direction to create the layer. The sixth site is occupied by the water molecule. In the study of the zinc phenylphosphonate structure[14] the phenyl rings were shown to be disordered. Closer inspection of the data revealed weak reflections requiring a doubling of the *c* axis. Refinement in this larger cell showed the phenyl rings in adjacent small cells to be rotated 90° from each other. Removal of the water molecule creates an open coordination site which can be occupied by primary amines.[15] In a similar study only NH_3 was taken in from the gas phase by the zinc phosphonate.[16] After dehydration of the zinc hydrate at 200 °C under vacuum, liquid primary amines were readily intercalated.[17] In the case of the mixed derivatives phosphite phosphonates, $Zn(O_3PH)_x(O_3PC_6H_5)_{1-x}$ a certain amount of branched chain amines were also taken up.[18] Crystal structures of the amine intercalates revealed the mechanism of the intercalation reaction.[19] In the hydrated state the layers have a crenellated or zigzag appearance but in the intercalate the zinc atoms are four coordinate of distorted tetrahedral geometry. The metal bonds to three oxygens from three separate phosphonate groups and an amine nitrogen. In the process the crenellated shape straightens into a rather flat layer. This process is reversible.

Copper phosphonate structures are an exception to the six-coordinate first row transition metal phosphonates as the copper atoms are five coordinate for the phenyl and methyl compounds.[20] Although the composition is the same, $Cu(O_3PR)\cdot H_2O$, the copper coordination is distorted tetragonal pyramidal. The structure of $Cu(O_3PCH_3)$ is layered, one oxygen atom from each phosphonate bonds to two copper ions forming a chain while the remaining two phosphonate oxygen atoms bond to two copper ions in an adjacent chain forming the layers. The base of the pyramid consists of three phosphonate oxygen atoms and the coordinated water molecule. The apical oxygen comes from an adjacent phosphonate oxygen atom and this bond distance is more than 0.3 Å longer than the Cu–O equatorial bonds. The copper phenylphosphonate has a similar layered structure but the phenyl rings within the interlamellar region are tilted to each other at a 98° angle.

Two copper arylenes were prepared in which the 1,4-phenylenebis(phosphonic acid) yielded a layered compound of composition $Cu_2(O_3PC_6H_4\text{-}PO_3)(H_2O)_2$ whereas the compound prepared with 4,4′-biphenylenebis-(phosphonic acid) retained two protons, $Cu(HO_3PC_{12}H_8PO_3H)$. In both cases the copper coordination is square planar.[21] Subsequently we synthesized a series of copper α,ω-alkylenebis(phosphonates) with 2 and 3 carbon chains[22] and with 4 and 5 carbon chains;[23] all four structures were solved utilizing X-ray powder data. The general formula for all the compounds is $Cu_2[(O_3PC_nH_{2n}\text{-}PO_3)(H_2O)_2]$. The copper coordination is distorted square pyramidal with bonding by four phosphonate oxygen atoms and one water molecule. The metal–oxygen bridging interactions form layers that are then connected to each other by the alkylene groups to form 3D structures. The three carbon chains form open rings with the layers and the water molecules protrude into the pores. These pores fill with water that hydrogen bonds to the bonded water molecules. The layer alkylene chain compounds are six coordinate, $CuO_4(OH_2)_2$, forming layers that are cross-linked by the alkyl chains. This pillared arrangement creates tunnels running parallel to the *c* axis direction and contain additional water molecules. These water molecules are readily replaced by ammonia molecules.

Zinc compounds with two and three carbon chains were also prepared.[22] In $Zn_2[(O_3PC_2H_4PO_3)(H_2O)]$ the phosphonate groups chelate the Zn ions and bridge to neighboring zinc ions. The third phosphonate oxygen links these rows to form a layer and a water molecule bonds to zinc forming distorted octahedra. This type of bonding is similar to that in Figure 1.5 but in this case the layers are tied together by the ethylene groups. However, the unit cell extends over two layers as the ethyl groups point in opposite directions in adjacent layers. The C_3 compound is anhydrous, $Zn_2(O_3PC_3H_6PO_3)$. The zinc ions are tetrahedrally coordinated and corner shared with the phosphonate group leaving empty spaces within the layer. The unit cell extends over two layers ($c = 18.865(1)$ Å).

Two additional zinc compounds were synthesized using 1,3-propylenbis-(phosphonic acid).[24] In $Zn(HO_3PC_3H_6PO_3H)$ the Zn ions are tetrahedrally coordinated by four oxygen atoms, two each from two independent phosphonate groups. The remaining third oxygen atom of both groups is protonated. The metal phosphonate interactions lead to double chains that are held together through the organic linkages leading to 2D sheets. These in turn are connected though hydrogen bonding forming a 3D metal phosphonate network. In the second compound, $Zn_3[(HO_3PC_3H_6PO_3)_2]\cdot 2H_2O$, there are two independent zinc ions, one in a general position and one in a special position providing six zinc ions per unit cell. They are both four coordinate but Zn1 is coordinated by four phosphonate oxygen atoms and Zn2 by two water molecules and two phosphonate oxygen atoms. The structure contains large pores; however, no surface areas were obtained for this compound or the $Zn_2(O_3PC_3H_6PO_3)$ compound.

The main group elements such as calcium and magnesium are known to yield multiple phases other than those of the transition elements.[13,15,25] Larger cations such as Pb^{2+} form eight coordinate structures and in acid solution may form 1:2 derivatives such as $M(HO_3PC_6H_5)_2$.[26] In these compounds both phosphonate groups chelate the metal ions and one oxygen of each chelate coordinates to an adjacent metal ion. The remaining oxygen atoms cross-link these chains into layers.

1.3 Trivalent Metal Phosphonates

At this juncture a word about solubility is in order. We have seen that the four-valent aryl and many alkyl phosphonates are noncrystalline. This feature may be attributed to the great insolubility of these compounds. The few crystalline compounds usually contain fluoride ions and are prepared in strong HF solutions. However, inclusion of nitrogen or other hydrophilic groups into the phosphonic acid changes the solubility of the M^{IV} complex to the point that crystallization may occur. Three-valent phosphonates are more soluble, leading to many crystalline compounds. Almost all divalent phosphonates are soluble in slightly acidic media and can be crystallized by slow evaporation of solutions or by addition of a second solvent or layering. All monovalent phosphonates are highly soluble and readily crystallizable. We shall illustrate that in Section 1.11.

1.3.1 Lanthanide Phosphonates

Mallouk and co-workers prepared lanthanide methylphosphonates of the type $LnH(O_3PMe)_2$[27] but the structures were not determined because the crystals were in the form of very thin flakes and twinned. Subsequently, the La phenyl and benzyl derivatives were prepared and their crystal structures solved. The crystals were grown hydrothermally in 0.4 M HCl. The structures are similar to the Ba and Pb phenyl phosphonates[28] in that the ratio of phosphonate to metal is 2:1 in both cases. The La^{3+} is eight coordinate with the same chelation–donation structure of the Ba and Pb layers. However, the phenyl groups are disordered and that the true unit cell requires a doubling of the *c* axis, 5.636(1) Å.

The benzyl derivative is different from the phenyl compound in that it is a dihydrate $LaH(O_3PCH_2Ph)_2 \cdot 2H_2O$.[28] It is orthorhombic with $a = 10.801(2)$ Å, $b = 10.301(2)$ Å, $c = 33.246(8)$ Å, $Z = 8$. The two water molecules are coordinated to La so only six phosphonate oxygens are needed to fulfill the eight coordination of La. This is accomplished by having one phosphonate ligand chelate the metal and one of these oxygen atoms donate to one near neighboring La. The other phosphonate ligand bridges across lanthanum ions and the third oxygen connects the chains to form a layer. One of the oxygen atoms has a P–O distance of 1.57 Å and must be the one bearing the proton. A different stoichiometry, $Ln_2(O_3PC_6H_5)_3$, was obtained in alkaline solution at pH values above 8. Acidification converts this compound into $LaH(O_3PR)_2$ and $LaCl_3$.[28]

1.3.1.1 Iron(III) Phosphonates

$FeH(HO_3PPh)_4$ was prepared in dichloromethane from FeOCl and an excess of phenylphosphonic acid in a sealed tube reaction.[29] The crystals are triclinic, $P\bar{1}$, $a = 14.968(9)$ Å, $b = 5.36(1)$ Å, $c = 8.678(6)$ Å, $\alpha = 88.5(1)°$, $\beta = 86.41(6)°$, $\gamma = 89.6(1)°$ with $Z = 1$. The compound is layered with coplanar Fe^{3+} ions. The iron ions are octahedrally coordinated by oxygen atoms from six different phosphonate groups. Four of the oxygen atoms come from phosphonate groups that bridge across two Fe ions to form chains parallel to the *b* axis. The remaining two coordination sites are filled by two phosphonate groups through only one of their oxygen atoms. The chains are then linked to form layers. These sheets are held together by van der Waals forces. There are five protons on the four phosphonate groups requiring that one phosphonate group retains both of its protons yet still be bonded to the metal ions.

The iron tetraphosphonate forms when the ratio of phosphonic acid to Fe is equal to 4 or more.[30] When this ratio is 2, a series of three compounds, all with the same composition $FeH(O_3PPh)_2 \cdot H_2O$ but different structure, depending upon the length of the reaction time, is formed. These three phases were designated as α, β, and γ. Spectroscopic and thermal analysis led to the conclusion that these phases have α-ZrP type structures. A similar result was obtained for $FeH(O_3PCH_3)_2$.[31]

1.4 Literature Reviews

At this juncture it is well to summarize the early work with reference to a series of review articles. A wealth of information is available in articles by three of the pioneers of metal phosphonate chemistry, Alberti,[8,32] and Clearfield and Costantino.[33] The most comprehensive review to literature up to 1998 is in *Progress in Inorganic Chemistry*.[34] Mallouk's summary includes layer by layer growth of metal phosphonate thin films[35] among other topics. Burwell and Thompson[36] describe a number of mixed ligand compounds such as $Zr(O_3PCH_2COOH)_{2-x}(O_3POH)_x$ and Thompson discusses molecular architecture of zirconium phosphonates and provides a summary of his work on photophysical studies.[37] Two additional sources are by Clearfield in *Current Opinion in Solid State Science*.[38] Some additional topics of interest by the author's research group are macrocyclic leaflets,[39,40] zirconium azacrown ether phosphonates,[40–42] deprotonation of phosphonic acids by divalent cations,[43] 3D hexagonal structures,[44] and novel materials based on self assembly.[45] Other reviews of interest include a summary of early work on fuel cell membranes[47] recently synthesized Cu^{II}, Zn^{II}, Cd^{II} compounds,[46] metal phosphonate open-framework materials by Maeda,[48] alkaline earth metal phosphonates by Demadis,[49] phosphonate and sulfonated metal organic frameworks (Shimizu, *et al.*),[50] hybrid metal phosphonate corrosion inhibiting thin films by Papadaki and Demadis,[51] supramolecular coordination networks employing sulfonate and phosphonate linkers, Shimizu *et al.*,[52] Kumar, *et al.*,[53] and Francis and Martodam[54] on chemical, biochemical, and medicinal properties of the diphosphonates, a comprehensive review on supramolecular chemistry and complexation abilities of diphosphonic acids and finally a review on phosphonates in coordination chemistry.[55]

1.5 Staged Materials

In Section 1.1 we described compounds containing two phosphonate ligands. A variant of such materials may be prepared with two different ligands that are separated into alternating layers. The first recognition of staging in the phosphate–phosphonate family was the synthesis of a phosphate phosphite of composition $Zr(HPO_4)(HPO_3)$.[56] Actually three phases were obtained, one phase yielded an X-ray powder pattern (XRPD) of α-ZrP containing a small amount of phosphite, a second phase had an XRPD of $Zr(O_3PH)_2$ containing a small amount of phosphate, and a third phase that was staged with the composition $Zr(HPO_4)_{0.85}(HPO_3)_{1.15} \cdot 0.5H_2O$. This latter phase had an interlayer spacing of 12.9 Å, which is less than the sum of the interlayer spacings of the individual compounds, $Zr(HPO_3)_2$ (5.58 Å) and $Zr(HPO_4)_2 \cdot H_2O$ (7.6 Å). Because about 15% of phosphite was present in the phosphate layer this interlayer was reduced to ~7.35 Å. However, in the phosphate layer the phosphate groups mainly abut against each other in adjacent layers and should not reduce the interlayer size. The structure of the phosphate–phosphite was further established by X-ray methods.[57]

Subsequently, Alberti *et al.*[58] prepared a mixed phosphate–phosphonate of composition $Zr(HPO_4)_{1.15}(O_3PC_6H_5)_{0.85} \cdot 0.84H_2O$. The interlayer spacing was 24.9 Å. These data were interpreted as a staged compound as represented in a schematic way in Figure 1.6. The interlayer spacing of zirconium

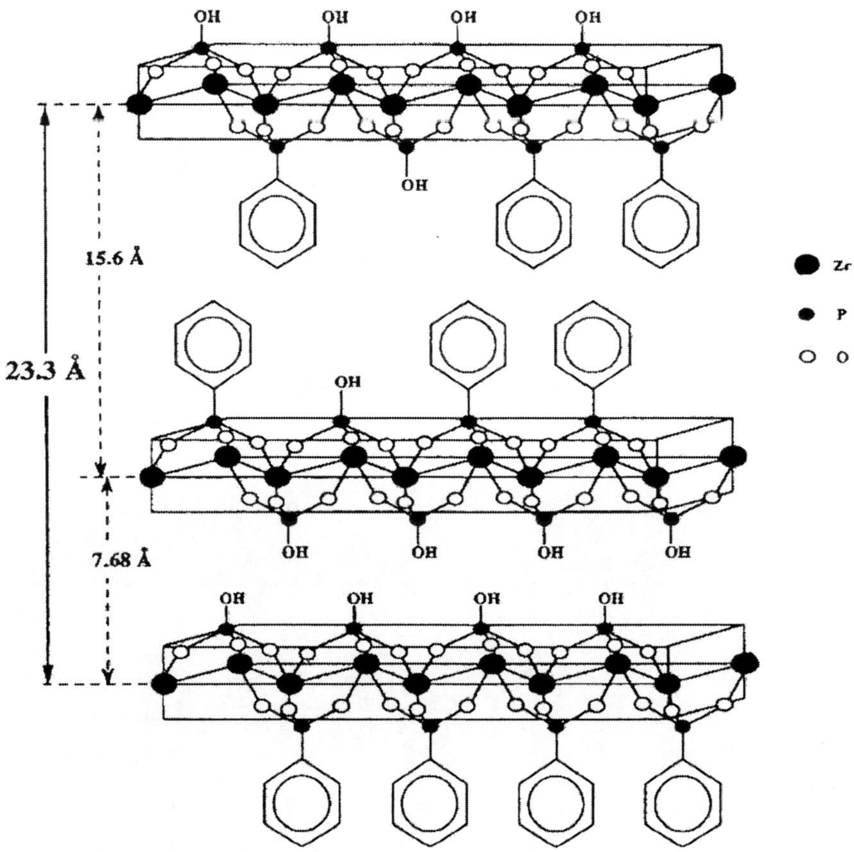

Figure 1.6 Schematic representation of zirconium phosphate phenylphosphonate staged compound. Excess phosphate groups may enter the phosphonate layer but not the reverse.

phenylphosphonate is 14.8 Å and for α-ZrP is 7.6 Å, adding to 22.4 Å. Since there are only half as many layers of phosphate, there is the equivalent of 1.68 moles of water in $Zr(HPO_4)_2$. This extra water accounts for the increased interlayer distance. The structure of this staged material was reasonably well substantiated by a combination of X-ray powder diffraction and spectroscopy. Our studies, following Alberti's procedure, produced a staged product with an interlayer distance of 22.5 Å with a water content of 0.5 mol. On exposure to a moist atmosphere the layers swelled to 24.5 Å. MAS NMR yield two major resonances for ^{31}P, one at –5.3 ppm for the phosphonate and another at –20.2 ppm for the monohydrogen phosphate group.[59] This staged compound intercalated butylamine, increasing the combined interlayer spacing to 34.7 Å. It was also found that the composition of the staged material could be varied over a range of phenyl phosphonate to phosphate of from 1 to 0.7.[59] Also, a staged compound was not obtained when phosphorous acid was used in place of phosphoric acid. Instead the phosphite group is distributed throughout the layers with phenylphosphonate. If the latter ligand exceeds 50%, the interlayer spacing is ~15 Å but at higher phosphite content the interlayer spacing is ~10.6 Å. The smaller interlayer spacing results from the phenyl groups all being opposite phosphite groups within the layer.[59] Immersion in hot sulfuric acid converts the phosphite groups into phosphate. Figure 1.7 (right) is a transmission electron micrograph of zirconium phenylphosphonate prepared at 200 °C for 3 days with added HF. Even under these extreme synthesis conditions the particles are nano-sized. However, measurement of the interlayer spacing shows that it is clearly ~15 Å. Figure 1.7 (left) shows that the staged compound is much less regular, has much fewer layers, and the interlayer spacing is ~25 Å. We will find that this is a feature of zirconium phosphonate chemistry. Unless an excess of HF is used or the ligand has some hydrophilic

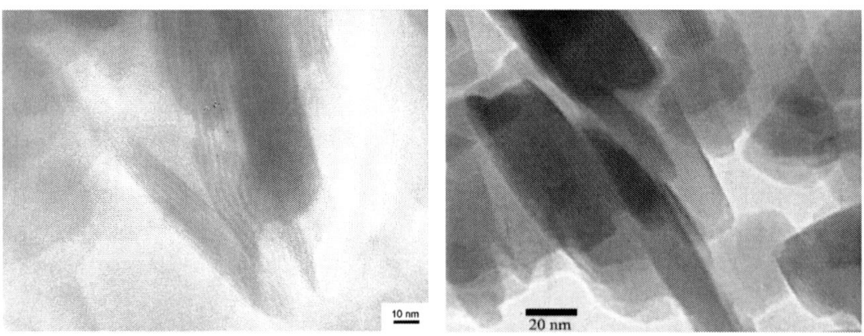

$Zr(PhPO_3)(HOPO_3)$ $Zr(PhPO_3)_2$

Figure 1.7 Transmission electron micrographs of zirconium phenyl phosphonate, $Zr(O_3PC_6H_5)_2$ (right) and the staged compound $Zr(PhPO_3)(HOPO_3)$ (left). Note the wider layer spacing of the staged compound and the smaller number of layers.

atoms or groups present, very poorly crystalline materials are obtained. For example, carboxyl groups impart a measure of solubility yielding crystals in some cases.

More recently, we have prepared staged compounds using alkyl phosphonates with chains of three or more carbons. Thus, it would appear as a general rule that sufficiently hydrophobic ligands in general will separate from very hydrophilic groups such as phosphates, and possibly sulfonates.

We have found that it is possible to immobilize different species in the two layers. The hydrophobic bilayer of phenyl groups does not swell in hydrophobic media such as toluene or benzene. However, if sulfonated using chlorosulfonic acid the layers swell or exfoliate in water. Addition of methylviologen to the exfoliated system results in its incorporation into the phenylsulfonic acid layer. If then the α-ZrP layer is enlarged with tetrabutylammonium chloride, ruthenium bipyridyl, $[Ru(bpy)_3]$, can be inserted therein. We are in the process of determining the rates of electron transfer and the back reaction.[60] This technique is amenable to many variations. The hydrophobic layer may be expanded by using biphenyl or triphenyl groups or alkylaryl groups. It also might be possible to use other hydrophilic groups such as polyetherphosphates.

1.6 Sulfonated Phosphonates

We have indicated that it is possible to sulfonate the phenyl rings in the staged materials but the reaction is more general. DiGiacomo and Dines[12a] prepared $H_2O_3P(CH_2)_3SO_3H$ through a sulfone intermediate and also utilized fuming sulfuric acid to prepare $HO_3S–C_6H_4–CH_2CH_2PO_3H_2$. They sulfonated the ethylphenyl bromide first and then converted it to a phosphonic acid by an Arbuzov reaction. Subsequently it was shown that aromatic phosphonic acid derivatives of zirconium could be sulfonated in fuming sulfuric acid directly.[61] Alternatively the phenylphosphonic acid could be sulfonated and then allowed to react with a soluble zirconium salt.[62] Complete sulfonation was attained in 28% fuming sulfuric acid in 1 h at 70 °C. This sulfonated compound readily exfoliates in water and requires addition of methanol to precipitate. Preparing a mixed derivative that is more hydrophobic by inserting methylphosphonic acid or phosphorous acid groups along with the phenyl groups allows recovery of the sulfonated product by filtration. The dry sulfonated phenylphosphonate has an interlayer spacing of 16.1 Å but if phosphite is present in near equal amounts the value of this spacing is 14.7 Å.

The ease of exfoliation of the sulfonate arises from the fact that in aqueous media a solid electrolyte forms due to hydrolysis of the protons with formation of highly negatively charged layers. Preliminary light scattering studies have shown that ~95% of the dispersed single layers are nano-sized being about 40 Å on a side with about 5% being several times as large. Reference to Figure 1.7 shows that the zirconium phenyl phosphonate particles are 40 nm thick and 80 nm long but the much smaller size of the sulfonate may arise partly from the severe conditions of sulfonation and some breakup resulting from exfoliation. During the last several years we have carried out sulfonations using chlorosulfonic acid. The reaction is much easier to control using this sulfonating agent. Alberti *et al.*[63] sulfonated mixed derivatives such as $Zr(O_3PC_6H_4SO_3H)_{0.74}(O_3PCH_2OH)_{1.26}$ in order to more easily recover the solid. Almost any aryl derivative of Zr, Ti, or Sn, even the 3D porous pillared phosphonates can be sulfonated.[64,65]

It is not possible to determine the crystal structure of $Zr(O_3PC_6H_4SO_3H)_2$ as the X-ray powder pattern of the dry compound consists of a series of (00*l*) reflections.[61] However, the crystal structure of the fluorobenzyl phosphonate and its Ni salt were determined[62] and the sulfonated group was shown to be meta to the phosphonic acid position. This structure was in accord with a computer simulated model of zirconium sulfophenylphosphonate that was based on the 16.1 Å interlayer spacing of the XRPD.[63]

1.6.1 Ion Exchange Properties of the Zirconium Sulfophosphonates

Organic polymer ion exchange resins swell in aqueous media. The swelling may be controlled by the ionic strength of the solvent. Zirconium sulfophenylphosphonate swells to exfoliation in aqueous media but the swelling may be controlled by preparing mixed derivatives such as $Zr(O_3PC_6H_4SO_3H)_{2-x}(O_3PR)_x$[63] or one may limit the percent of the phenyl rings sulfonated. In the exfoliated state it is possible to sequester charged species as the layers are attracted by the cation charge encapsulating the cation. For example, ruthenium bipyridyl, $Ru(bpy)_3{}^{2+}$, was readily incorporated between the layers from an aqueous solution to which the sulfophosphonate was added.[66] Absorption and emission spectra were readily obtainable. In order to limit the expansion of the interlayer distance two mixed derivatives of the type $Zr(O_3PC_6H_4SO_3H)_x(HPO_4)_{2-x}$ were prepared with $x=0.767$ and 0.43. Titration with NaOH showed that the sulfonic acid protons were exchanged first and then the phosphate protons, being weaker, were neutralized at higher pH.[67] However, the K_d values were found to increase remarkably with increase in cation size and charge as shown in Table 1.2. For the exchanger with $x=0.767$ the K_d values ranged from 110 ml g^{-1} for Li to 6500 ml g^{-1} for Cs^+ and from 21 000 ml g^{-1} for Mg^{2+} to 400 000 ml g^{-1} for Ba^{2+}. K_d values for the exchanger with $x=0.43$ were considerably lower, ranging from 9800 ml g^{-1} for Mg^{2+} to 19 000 for Ba^{2+}. Still these K_d values are much higher than those for the strong acid polystyrene sulfonic acid resin AG50W$^-$X8. K_d values are defined by the following equation:

$$K_d = \frac{C_0 - C_e}{C_e}\left(\frac{V}{M}\right) \tag{1.2}$$

in units of ml g^{-1} where C_0 is the initial ion concentration, C_e is the concentration at equilibrium, V is the volume of the solution, and M is the mass of

Table 1.2 Distribution coefficients for alkali and alkaline earth metal ions on exchangers $Zr(O_3PC_6H_4SO_3H)_x(HPO_4)_{2-x}$ (2 samples), amorphous zirconium phosphate, and AG 50W-X8 at 25 °C.

	Kd (mL/g)			
Ion	*MY-IV-95*[a] x = *0.767*	*MY-VI-2*[a] x = *0.43*	*Amorphous ZrP*[a,b]	*AG 50W-X8*[c]
Li^+	110		7	33
Na^+	205		11	54
K^+	1500	650	120	99
Cs^+	6500		1600	148
Mg^{2+}	21 000	9800		790
Ca^{2+}	89 000	37 000		1450
Ba^{2+}	400 000	190 000		5000

[a] K_d at pH = 2.00 and a metal loading of 0.1 meq/g.
[b] Calculated from selectivity coefficients in L. Kullberg, A. Clearfield, *J. Phys. Chem.*, 1981, **85**, 1578.
[c] K_d in 0.1 M HNO_3. Data from F. W. E. Strelow, R. Reth, C. J. C. Bothma, *Anal. Chem.*, 1965, **37**, 106.

the exchanger, respectively. It is evident that the higher the uptake of cation the larger is the K_d value. However, K_d values are not thermodynamic equilibrium constants but vary with choice of experimental conditions. In addition to K_d values, some thermodynamic data for selected ion exchange processes are available.[68] The very large changes in the observed K_d values are indicative that both group separations and separations within a group are possible.

1.6.1.1 *Proton Conduction of Zirconium Sulfophenylphosphonate*

The very weak attraction of the sulfonated group for the proton in aqueous media would lead to the expectations that the zirconium sulfophosphonate is an excellent proton conductor. This is the case. Alberti *et al.*[69] found that for the mixed derivative $Zr(O_3PC_6H_4SO_3H)_{0.97}(O_3PCH_2OH)_{1.03}$ the conductivity was strongly dependent on the relative humidity (RH). At 81% RH and room temperature the conductivity was found to be $\sim 8 \times 10^{-3}\,\Omega^{-1}\,cm^{-1}$. Our own data[65] for the pure Ti and Zr sulfophenylphosphonate at 25 °C and 85% RH were $1.3 \times 10^{-1}\,\Omega^{-1}\,cm^{-1}$ and $2.1 \times 10^{-2}\,\Omega^{-1}\,cm^{-1}$, respectively. These are among the best conductivity values recorded, and qualifies them for consideration to be utilized in fuel cell membranes. The Alberti group has pursued this objective to the present time developing a broad spectrum of membranes and formulations.[47,69–71] The emphasis now is to obtain conductors that work well at 120 °C. The problem at lower temperatures is that sufficient CO is liberated when a carbon-containing fuel is used, resulting in poisoning of the platinum catalyst. At 120 °C this is no longer the case. However, the proton conductors work best at a high relative humidity and this condition is difficult to maintain at 120 °C.

We have prepared two types of sulfonated materials aimed toward maintaining water content at elevated temperatures. At first we attempted to over-sulfonate $Zr(O_3PC_6H_5)_2$. It was found that to sulfonate all the phenyl rings in fuming sulfuric (30% acid) it was necessary to constantly stir the mixture of the solid phosphonate in the viscous acid solution at 70 °C. Samples were removed at 20 min intervals. It was found that the TGA curves for all the samples from 20 min to 120 min were almost identical. Each compound contained from 3.5 to 4.7 mol of water per mol of Zr. The samples were dried at 150 °C under vacuum to constant weight and analyzed for C, H, S. There was no weight loss until 250 °C. Then all the weight loss could be accounted for by a theoretical formula of $Zr(O_3PC_6H_5SO_3)_2$ with 11.38% S. However, the actual samples had somewhat different amounts of carbon, as shown in Table 1.3. This reduction in carbon content arises from cleavage of some of the aryl groups that are replaced by hydroxyl groups. This is shown convincingly by ^{31}P solid state MAS NMR. A new resonance appears at −22 ppm along with the phosphonate resonance at −5 ppm. The O_3POH ^{31}P resonance in α-zirconium phosphate occurs at −21 ppm. To account for the loss of 5% of the 25.58% carbon in the ideal structure the loss of phenyl rings of 0.5 mole is required. Thus the formula is approximately $Zr(O_3PC_6H_5)_{1.5}(HPO_4)_{0.5}(SO_3)_2$, Fwt = 533 g. This composition would be reduced to ZrP_2O_7 at 1000 °C or a weight loss of ~50.0% as compared to an observed value of 48%. Additional evidence is derived from IR and X-ray data, the latter technique showing that the interlayer spacing with water present is 18–19 Å and in dried state is 16.1 Å.

A TGA of the sulfonated zirconium phenyl phosphonate in general[65] has only two weight losses, 12%–15% water to 200 °C and an additional loss of approximately 44% from 200 to 600 °C. The formula that fits this TGA is $Zr(O_3PC_6H_5)_{1.5}(HPO_4)_{0.5}(SO_3)_{1.96}\cdot 4H_2O$, Fwt = 602.7 g. It also agrees well with the analytical data in Table 1.3. This also depends on the amount of P–C bond breakage. What is interesting is that only water is lost until 200 °C, and even at 300 °C only an additional 2%–3% weight loss may occur. Thus the bulk of the sulfonic acid groups are present at this temperature because of the very large high temperature (above 300 °C) weight loss, indicating the possibility of

Table 1.3 Results of sulfonation of zirconium phenylphosphonate in 30% fuming sulfuric acid.

Sample	*Time (70 °C)*	*Interlayer spacing (Å)*	*M. Wt. dry*	*X·H₂O*	*C (%)*	*H (%)*	*S (%)*
SRAM-36A	20 min	15.01/17.20	541.18	4.66	20.81	2.08	11.22
SRAM-36B	40 min	17.26	542.66	4.16	21.11	2.17	11.64
SRAM-36C	60 min	16.71	547.65	3.76	21.52	2.35	11.70
SRAM-36D	90 min	17.55	542.20	3.58	20.20	2.21	11.31
SRAM-36E	120 min	18.80	535.58	3.6	19.82	2.10	11.42
theoretical[a]	–		563.40	4.66	25.58	1.79	11.38

[a]Based upon the composition $Zr(O_3PC_6H_5)_2\cdot 2SO_3$.

using the zirconium sulfophenylphosphonate as a strong Brønsted acid heterogeneous catalyst, as well as a proton conductor.

The second type of sulfonated product is the porous cross-linked diphosphonates. We prepared the biphenyl and terphenyl zirconium derivatives and sulfonated them in fuming sulfuric acid as before.[64] The advantage here is that these compounds can be washed free of excess sulfuric acid without exfoliation. They could then be thoroughly dried for catalysis or on standing in air saturate the pores by water uptake for ion exchange or conductivity studies. A sample TGA for the biphenyl derivative is shown in Figure 1.8. The knowledge that the end product of combustion is always the pyrophosphate, ZrP_2O_7, allows determination of the original formula weight, which in this case is 676 g. The weight loss to 150 °C, 15.49% amounts to 5.81 moles of water. A second weight loss of 11.01% may be a combination of bonded water and SO_3. The final weight loss results from loss of biphenyl groups which, if no C–P bond cleavage occurs, is 22.51%. This leaves 22.77% for the SO_3 and remaining water in the temperature range of 150–400 °C. Thus, significant amounts of water and SO_3 are present above 120 °C. Conductivity and catalysis studies are in order for these materials.

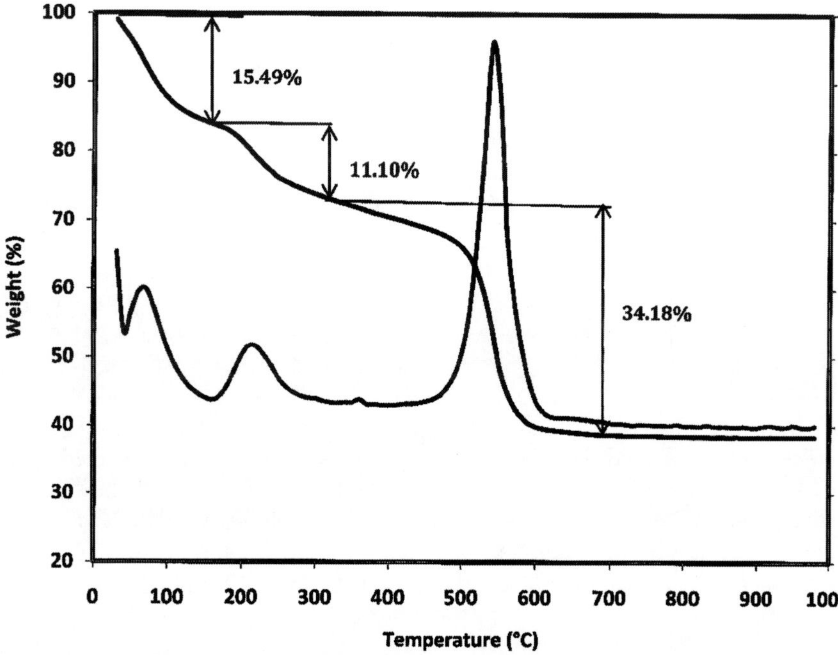

Figure 1.8 TGA curve for highly sulfonated zirconium phenyldiphosphonate with formula weight 676 g for $Zr(O_3PC_{12}H_8PO_3)(SO_3)_{2.1} \cdot 5.8H_2O$. Reproduced from reference 85 with permission.

1.6.1.1.1 Catalytic Properties of Sulfonated Zirconium Arylphosphonates. It was soon after the synthesis of zirconium sulfophenylphosphonate(ZrSPP)[61] that catalytic reactions were carried out. As we shall see this compound is a strong Brønsted acid and as such may be used in organic solvents to effect many reactions. Early experiments were reported in the Chinese literature and were generally inaccessible to a wide audience. In 1991 it was shown that esterification, acetal and ketal formation as well as etherification and preparation of alkenes by dehydration of alcohols were catalyzed by ZrSPP.[72] It was claimed that the use of this catalyst was easy with no waste and the catalyst could be reused repeatedly.

A series of papers using ZrSPP as catalyst were carried out by Curini *et al.* at the University of Perugia. The reactions they reported on included the preparation of cyclic and acyclic β-amino alcohols by addition of amines to epoxides,[73] cyclic ketals, and thioketals.[74] Early papers by the Curini group used a mixed catalyst $Zr(O_3PCH_3)_{1.2}(O_3PC_6H_4SO_3H)_{0.8}$.[75,76] It was found to be an efficient catalyst for the tetrahydropyranylation of alcohols and phenols. Additional related papers in use of such catalysts appeared regularly.[77] It was also shown that the mixed methane-sulfophenyl phosphonate mediated a regioselective synthesis of 2,3-disubstituted tetrahydro-2*H*-indazols. *In vivo* evaluation of these compounds proved the presence of anti-inflammatory activity without any gastric injury.[78] More recent work in China showed that the ZrSPP was an efficient and stable solid acid catalyst for the carbonylation of formaldehyde.[79]

In contrast to the catalytic and ion exchange study described above, we prepared zirconium derivatives of biphenyldiphosphonic acid and terphenyldiphosphonic acid hydrothermally.[64] The temperature of preparation was kept at 80 °C for 3 days with added HF to develop some crystallinity in the products. Surface areas were in the range of 370 to 416 $m^2 g^{-1}$ with 95% internal pores of 8–18 Å diameter. The terphenyl derivatives surprisingly had smaller surface areas than the biphenyl derivative, ranging from 273 to 407 $m^2 g^{-1}$ with ~97% internal pores ranging in size from 11 to 17 Å in diameter. The porous pillared zirconium bisphosphonates were sulfonated in 26–29% fuming sulfuric acid at 150 °C for several days. The sulfur content of a 3-day treated solid was 3.86% or 0.28 SO_3 per ring and a 6-day treatment increased the sulfur content to 8.74% or 0.714 sulfonic acid groups per ring. The surface areas of the sulfonated products were not determined because exposure to air allowed the pores to fill with water. Instead the recovered solids were kept under dry nitrogen. A typical X-ray powder pattern for the biphenyl derivative is shown in Figure 1.9. The interlayer spacing is 13.65 Å and slightly lower for the sulfonated complex.

For the purposes of determining the acid strength of the sulfonated products the compounds prepared in fuming sulfuric acid could not be used. To recover the solid from the fuming sulfuric acid required dilution and neutralization of the SO_3. It was therefore difficult to obtain a totally water-free product. Therefore, we prepared samples for acid strength determination by use of SO_3 gas under pressure.[64] Two biphenyldiphosphonates were prepared: one

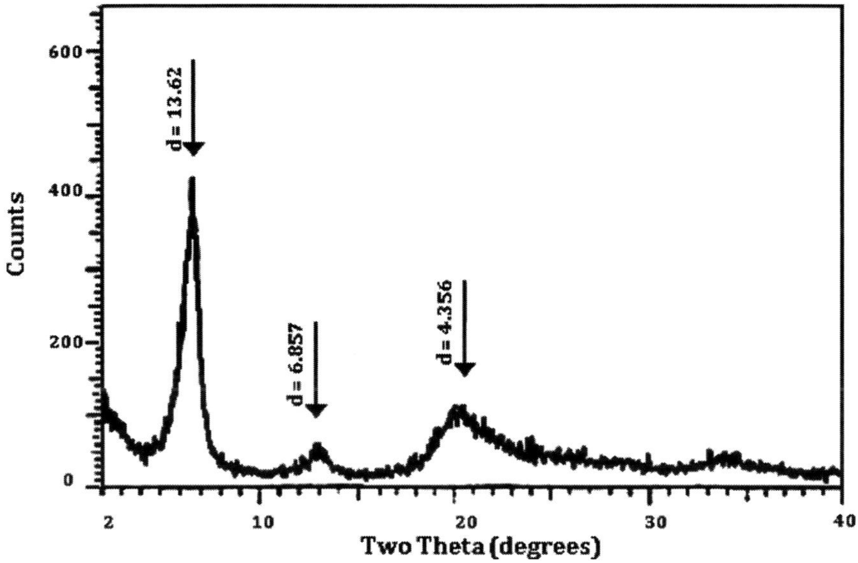

Figure 1.9 X-ray diffraction powder pattern of zirconium bipheyldiphosphonate. Reproduced from reference 85 with permission.

contained 3.86% S and the second 8.74% S, equivalent to 0.27 and 0.714 SO_3 per aromatic ring, respectively. A terphenyl derivative with 15.5% S or 1.32 SO_3 groups per ring was also prepared. To determine the nature of the acid sites in the biphenyl derivative pyridine-^{15}N was used as a probe molecule. An NMR procedure developed by Maciel and Haw[80] showed that Brønsted acids produced a shift in ^{15}N of 200 ppm and below while Lewis acids gave shifts between 226–235 ppm. The zirconium biphenyl derivative with 8.74% S gave a ^{15}N shift of 203 ppm.

A second set of experiments developed by Haw *et al.*[81] for zeolites involved interaction of the acid with a hydrocarbon containing a single atom of enriched ^{13}C. This carbon atom undergoes a considerable chemical shift from its normal resonance position upon interaction with a solid acid. A scale of acid strengths based upon chemical shifts of known strong acids and superacids was developed. For example, the isotropic chemical shift for acetone-2-^{13}C in 100% H_2SO_4 is 244 ppm from an original position of 205 ppm. We determined that the chemical shift induced in acetone by the 8.74% biphenyl derivative was 243 ppm and the resonance peak was extremely sharp, indicative of the uniformity of the acid groups.[64] As a final experiment it was shown that the polymeric forms of acetone, mesityl oxide, phorone, isophorone, and mesitylene were also generated by contact of the probe acetone with the catalyst sulfonic acid groups. In summary it would appear that the sulfonated phenylphosphonate deserves further study especially as it can be made in a porous state by preparation of a mixed derivative with methylphosphonic acid.

1.7 Porous Pillared Layered Zirconium Arylphosphonates

We have seen in Section 1.1.1.1 that zirconium biphenylphosphonate is a layered compound and Martin Dines *et al.*[12] prepared porous materials by interspersing a small group such as $-O_3PH$ between the pillars. However, we pointed out that these materials are naturally porous. Our early preparations using 4,4′-biphenyldiphosphonic acid (BPDP) exhibited variable amounts of internal porosity and a broad spectrum of pores.[82,83] However, by carrying out the reactions in DMSO or a mixture of DMSO-C_2H_5OH at 100–110 °C products with high surface area and a significant portion of micropores were obtained.[64,84] A typical X-ray powder pattern is shown in Figure 1.9. A representative N_2 sorption–desorption isotherm is shown in Figure 1.10 and the pore size distribution curve in Figure 1.11. It was also found that the use of excess Zr helped to produce a higher level of internal porosity and a narrower range of pore sizes. The first peak in the X-ray pattern, 13.62 Å, represents the interlayer spacing and the second peak, at 6.857 Å, may be the 002 reflection. This lack of crystallinity is present in all of the four-valent cross-linked compounds. Solubility tends to increase as the cation charge decreases. We observed that addition of a diphosphonic acid to a soluble Ti, Zr, or Sn^{IV} solution results in immediate precipitation with almost no further Ostwald ripening. Addition of HF produces a somewhat more crystalline product as a

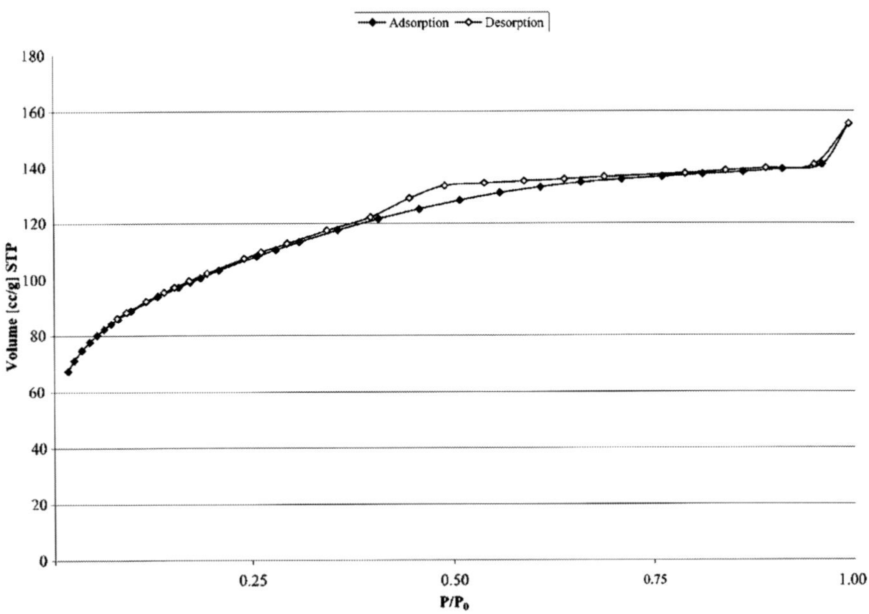

Figure 1.10 A representative N_2 adsorption (◆), desorption (◇) isotherm obtained at 77 K for zirconium biphenyldiphosphonate prepared in DMSO. Reproduced from reference 85 with permission.

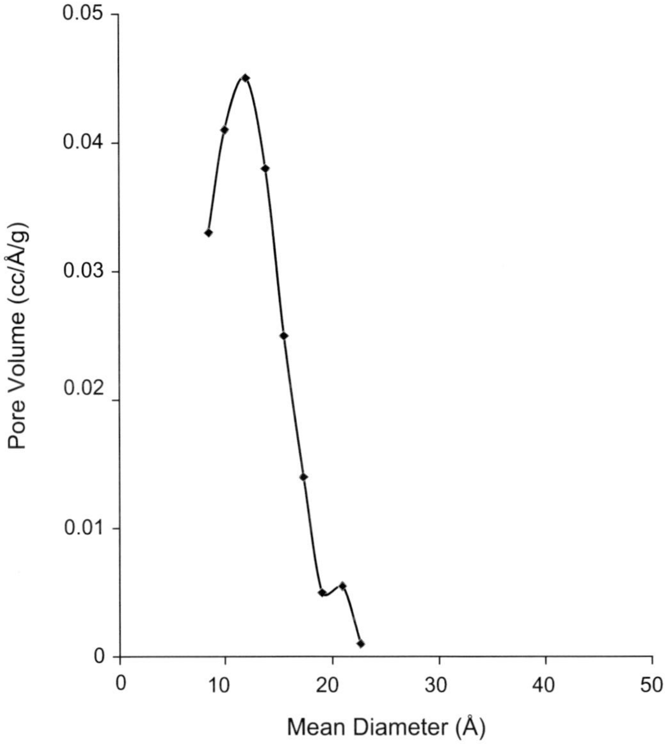

Figure 1.11 The pore size distribution curve as determined by the MP method for the isotherm shown in Figure 1.10. Adapted from reference 85 with permission.

result of ZrF_6^{2-} formation with slow release of Zr^{4+} above 60 °C. Surface areas are of the order of 300–400 $m^2 g^{-1}$ with internal porosity of 95–97%.[85] As already indicated, these compounds can be sulfonated whereupon they may be used as ion exchangers and Brønsted acid catalysts. A number of spacer groups may be used in an effort to spread the pillars farther apart. They include phosphonic acids with small alkyl or aryl groups (phenylphosphonic acid) or those with functional groups such as carboxyl, amino, thiol, phosphite, O_3PH, phosphate, O_3POH, *etc*. It is interesting that preparations with a high level of methyl groups as spacers yield isotherms that are more zeolite-like (Figure 1.12). Also using phenylphosphonic acid as the spacer molecule does not change the interlayer spacing which is 13.6 Å. In zirconium phenylphosphonate the interlayer spacing is 14.8 Å because of the van der Waals gap. This must mean that two phenyl groups cannot occupy positions in adjacent layers in the biphenyl compound unless they are severely tilted, for which there is no room, or some other mechanism holds. This mechanism must also account for the high porosity of the phenyl group as separator and the high internal surface area created by a bulky ligand. For example, a series of such compounds with a ratio of Zr:BPDP:PhP = 1:0.25:0.5 in solvents ranging from pure DMSO to pure ethanol

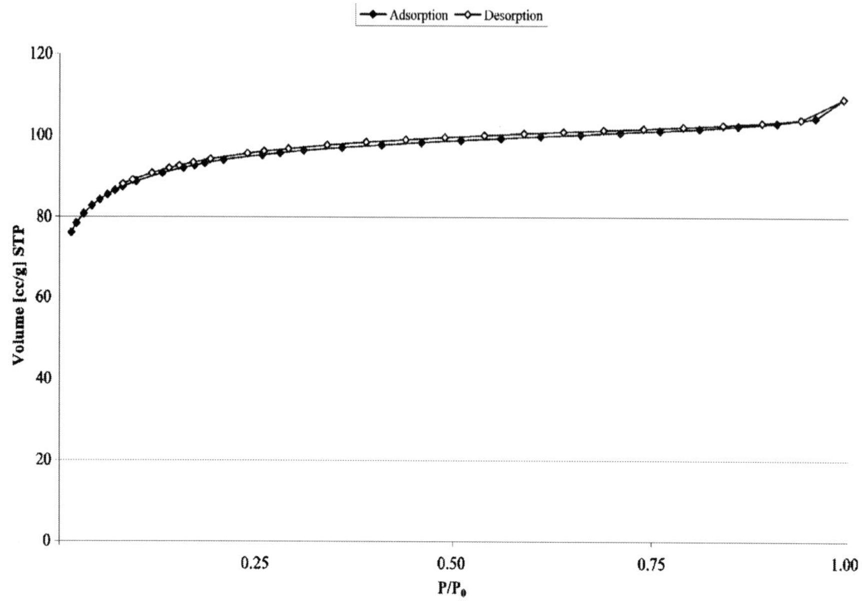

Figure 1.12 The zeolite-like N_2 sorption-desorption isotherm for a mixed derivative $Zr(O_3PC_{12}H_8PO_3)_{0.5}(O_3PCH_3)$. Reproduced from reference 85 with permission.

gave surface areas of 336 to 421 $m^2\,g^{-1}$ and micropore volumes of from 0.175 to 0.238 $cm^3\,g^{-1}$.[85] Similar hybrids using triphenyldiphosphonic acid and phenyl spacers gave a product with 490 $m^2\,g^{-1}$ of which 470 $m^2\,g^{-1}$ is in micropores. The pore size distribution was 8–16 Å with a pore volume of 0.249 $cm^3\,g^{-1}$.

There is no doubt that these zirconium phosphonates have special growth patterns that distinguish them from those that crystallize such that their structures can be determined by X-ray diffraction methods. Crystallization of zirconium α,ω-bisphosphonates requires that the ligand contain a softer element such as nitrogen[86] and possibly a solubilizer such as HF. It has been earlier shown that the particles are nanosized.[64,85]

A workable structure of the pillared porous group IV metal diphosphonates is required to supply answers to these observations: (1) the compounds produce nanosized particles with a minimum of long range order; (2) the pore sizes often exceed the interlayer spacing; (3) choice of solvent can alter the pore size and pore regularity; (4) the use of phenyl rings as spacers does not increase the interlayer distance of the monophenyl or biphenyl cross-linked derivatives; (5) often the ratio of phosphorus to M^{IV} exceeds 2:1. We had earlier proposed a model for these cross-linked zirconium materials based on crystal growth that explained several aspects of their character.[64,85,87] Such a model, a pictorial representation of which is shown in Figure 1.13, applies equally well to the tin cross-linked materials. The central idea is that upon mixing the diphosphonic acid with the dissolved four-valent metal, rapid precipitation of fine particles

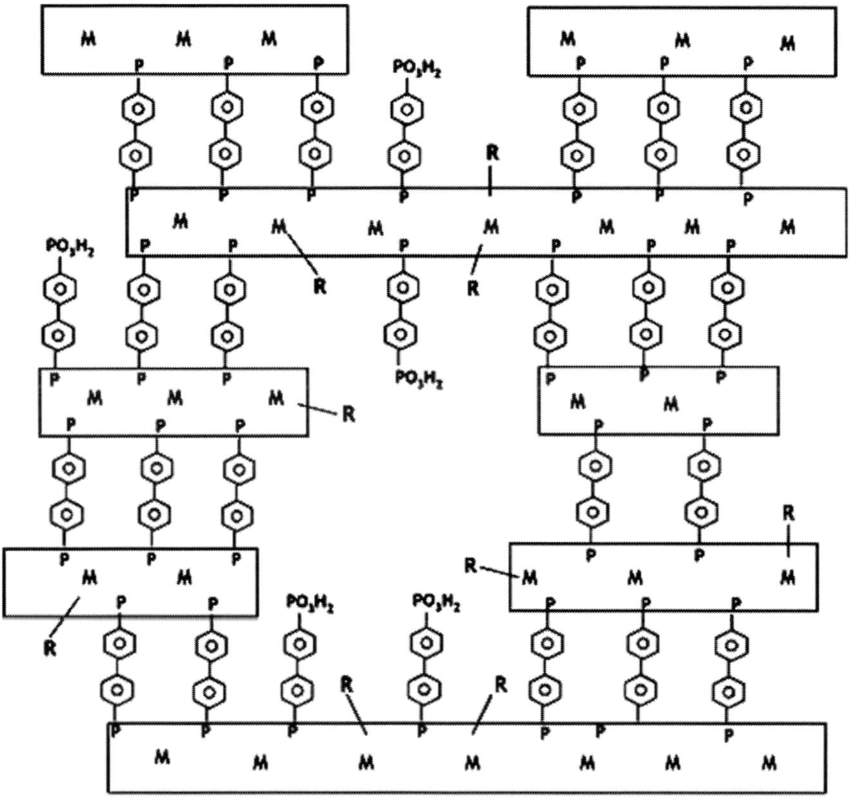

Figure 1.13 A schematic drawing of how micropores may be formed in the metal biphenyldiphosphonates. The horizontal rectangles represent the inorganic layers (M = Zr or Sn) which are cross-linked by the organic moieties. R stands for F^-, if HF is used in the preparation, and the surfaces are capped by M because an excess of metal is used in the syntheses. Reproduced from reference 85 with permission.

occurs. In this process both phosphonic acid groups bond to metals forming part of two layers. Statistically, it is expected that the growth rate of the layers would not be the same because much depends on how the second acid group attaches itself to a nascent layer after the first in the same molecule has become part of an adjacent growing layer. This uneven growth leads to gaps in the structure, as shown in Figure 1.13. It also requires that some of the phosphonic acid groups remain unbonded, as shown. These free acid groups can be determined by titration. Another factor to consider is that if the final particle has a small number of layers, the external surface area should be significant. If the layers end with phosphonic acid groups, the number of free acid groups would be considerable, with resultant high reactivity of the surface. This phenomenon would lead to a phosphorus–zirconium ratio of greater than two, which in fact does occur in many preparations.[63]

1.7.1 Use of 2,2′-bipyridyl-4,4′-diphosphonic Acid as a Pillar

Odobel *et al.*[88] prepared both the α-Zr and γ-Zr phase of the subject pillar (BPyDP) and a mixed phase of composition $Zr(HPO_3)_{0.8}(BPyDP)_{0.6} \cdot 5.3H_2O$. The α-derivative had a mesoporous surface area of 330 $m^2 g^{-1}$ with a pore size of 40 Å and a microporous surface area of 90 $m^2 g^{-1}$. The γ-phase in contrast had only narrow pores of ~5 Å diameter and a surface area of 350 $m^2 g^{-1}$. Both complexes sorbed Fe^{2+} and Cu^{+} from alcohol solutions binding to the bipyridine nitrogen atoms. We prepared a pillared porous zirconium derivative of the type similar to that described in the previous section for the biphenyl derivative using the 2,2′-bipyridyl ligand.[89] Additional samples were prepared using methyl phosphonic acid as the spacer molecule. The compound with a ratio of methyl to bipyridyl of 2:1 yielded the N_2 sorption–desorption isotherm shown in Figure 1.14. The surface area was slightly above 500 $m^2 g^{-1}$ with pore sizes in the range of 8–18 Å.

The zirconium bipyridyl compounds were able to sorb a variety of first row transition metals and to sorb $Ru(bpy)_2{}^{2+}$ to form $Ru(bpy)_3{}^{2+}$ within the pores. In all these complexes the accompanying anions must also be present in the pores. Catalytically active cations, Pd^{2+}, Pt^{2+}, Ru^{2+}, Au^{3+}, and Ag^{+} were also taken up. These latter ions were reduced to their respective metal nanoparticles using 5% H_2 in nitrogen at 70 °C. The particles migrate to the surface of bipyridyl complex as shown for Pd^0 in Figure 1.15. We have now been able to make the particles as small as 1–3 nm. Future studies are directed to determine whether interesting catalysts may be prepared in this fashion.

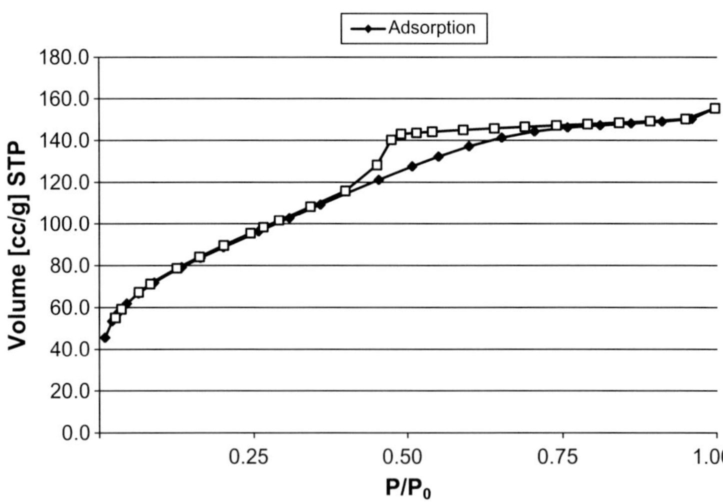

Figure 1.14 A representative N_2 adsorption (◆), desorption (□) isotherm obtained at 77 K for zirconium biphenyldiphosphonate prepared in DMSO. Adapted from reference 85 with permission.

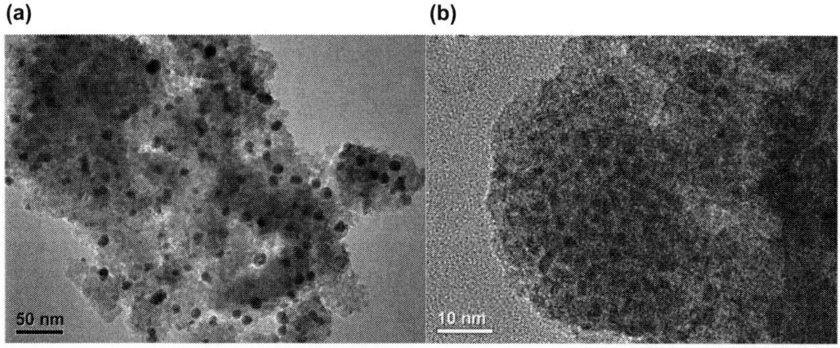

Figure 1.15 Pd^0 nanoparticles on the surface of the porous zirconium 2,2′-bipyridyl-4,4′-diphosphonate, (a) prepared by ethanol reduction and (b) by H_2 reduction. Note the difference in size from 8–12 nm to 1–4 nm.

Figure 1.16 Particle sizes of zirconium 2,2′-bipyridyls with and without the addition of methylphosphonic acid. (a) The pure bipyridyldiphosphonate prepared at 110 °C, 3 days with a small addition of HF. The particles are 60–80 nm. (b) Same preparation, no HF but bpy:me = 1, particles are ~ 20 nm. (c) Same preparation, bpy: me = 0.5. the particles are 8–14 nm.

An interesting feature of these types of cross-linked 3D porous materials was revealed by transmission electron microscopy. The sample containing no spacer groups prepared at 200 °C hydrothermally for 10 days is given in Figure 1.16a. The particles are of the order of 80 nm and have an elongated teardrop shape. As the amount of methylphosphonate included in the structure increases, the particle size decreases, as in Figures 1.16b and 1.16c. At the limit of solid solution of methyl in groups in the bipyridyl mixed solid the particles are 8–14 nm in size. We shall see that this behavior is general.

1.8 Tin(IV) Phosphonates

The thoughtful investigator might expect that the four-valent tin phosphonates have structural and behavioral properties similar to those of their zirconium analogs. In some respects this is true but in others there are

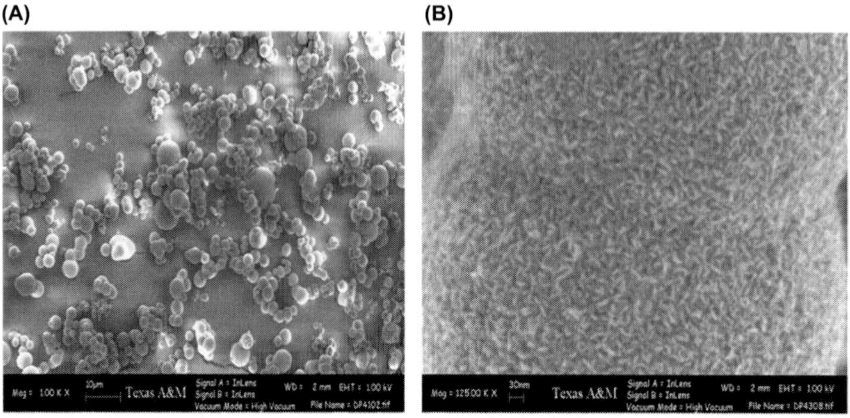

Figure 1.17 A SEM of the spherical shaped globules of $Sn(O_3PC_6H_5)_2$ (A) and (B) an enlargement of one of the gloubles. Reproduced from reference 85 with permission.

differences. Tin(IV) monophenylphosphonate was first prepared by Mal *et al.*[90] who reported the composition as $Sn(O_3PC_6H_5)_2 \cdot H_2O$. In their preparation they used sodium dodecylsulfate (SDS) as the template and prepared a microporous product. In the absence of SDS the tin phenylphosphonate was mesoporous with a broad distribution of pore sizes. Our hydrothermal preparation contained no template and used $SnCl_4 \cdot 5H_2O$ as the source of tin. To our surprise the product was not only porous but scanning electron micrographs (Figure 1.17) showed that the product took the shape of micron-sized spheres. The surface area of this first preparation was $367\,m^2\,g^{-1}$ with the majority of pores in the ~9 Å area and a lesser quantity in the 18–27 Å range.[91] The N_2 sorption–desorption curve and the pore distribution curve determined by the MP method are shown in Figures 1.18a and 1.18b, respectively. The origin of the porosity is revealed by the enlarged micrograph in Figure 1.17b. The particles are small flat layered in shape and aggregate by a "house of cards" mechanism. We theorize that the particles are relatively uniform in size and shape so that the voids between them are mainly within a narrow size range.

Heating tin(IV) methylphosphonate hydrothermally for 30 days produced a powder pattern from which the structure was solved.[92] Indeed the structure is layered with an interlayer distance of 9.74 Å. This very long heating time and high temperature were required to obtain a crystalline product of sufficient quality for structure solution. As an example of how slowly the particles grow, we illustrate the increase in crystallinity of $Sn(O_3PC_6H_5)_2$ in Figure 1.19.[93] Only after hydrothermal treatment at 220 °C for 20 days is a good crystallinity achieved with sharp reflections out to 70° 2θ. A sampling of a variety of tin(IV) phosphonates that are prepared is provided in Table 1.4.[85,92] We note that addition of the methylphosphonate as a spacer does produce higher surface areas.[92] In contrast the functionalized carboxylates and amino derivatives

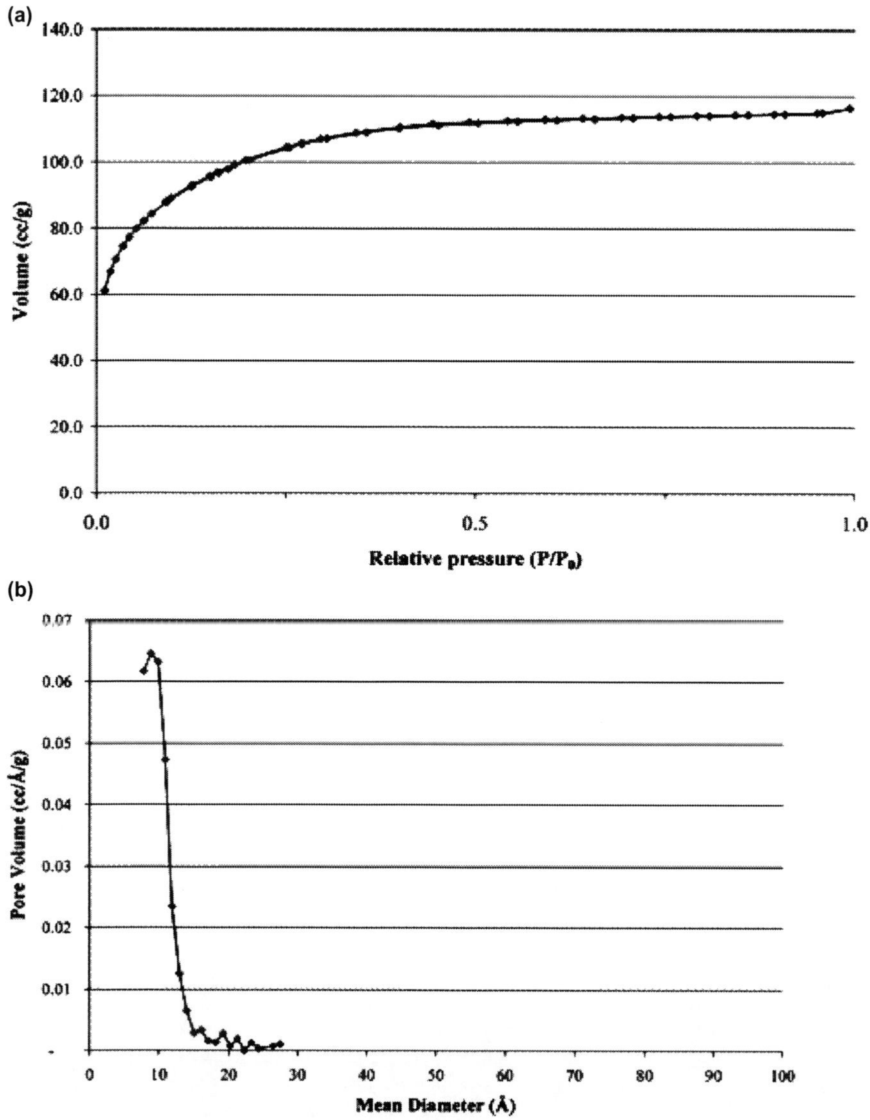

Figure 1.18 (a) N_2 sorption-desorption isotherm for Sn^{IV} monophenylphosphonate and (b) pore size distribution derived from the isotherm, by the MP method. Adapted from reference 91, Figure 1, with permission, American Chemical Society.

exhibited low surface areas and low internal porosities. However, the alkyl derivatives are microporous with a high level of internal micropores. Three bisphosphonates are listed in the table, one with an alkyl chain and two with aryl pillars. These compounds do not form spherical shaped particles but rather have a flat layered shape because they cross-link the layers. Much of what we

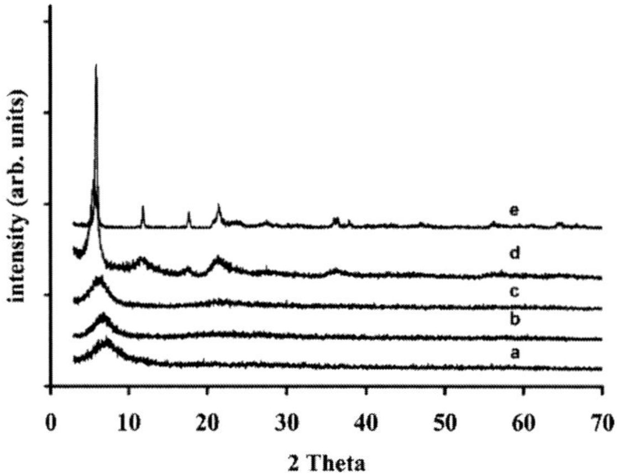

Figure 1.19 Powder XRD patterns of $Sn(O_3PC_6H_5)_2$. Samples were prepared by (a) room-temperature precipitation, (b) reflux for 1 h, (c) reflux for 12 h, (d) hydrothermal treatment at 140 °C for 3 days, (e) hydrothermal treatment at 220 °C for 30 days. Adapted from *Chem. Mater.*, Figure 1 of reference 93, with permission, American Chemical Society.

Table 1.4 Tin(IV) compounds and some of their important characteristics.

Ideal formula	D_{001} *(Å)*	*Surface area (m^2/g)* Total	*Micro*	*Max in pore size distribution curve (Å)*	*Total Wt loss (%)*
$Sn(O_3PC_6H_5)_{1.95}F_{0.05} \cdot H_2O$	15.35	367	350	9	34.09
$Sn(O_3PC_6H_5)_{0.95}(O_3PCH_3)_{0.95}F_{0.10} \cdot 0.4H_2O$	15.22	514	493	11	24.05
$Sn(O_3PCH_3) \cdot 0.4H_2O$	9.74	68	0	–	6.8
$Sn(O_3PC_{10}H_{21})_2$	28.07	264	220	9	47.6
$Sn(O_3PC_6H_{13})_2$	19.8	137	128	11	35.39
$Sn[O_3P(CH_2)_8PO_3] \cdot H_2O$	13.9	361	321	18	29.05
$Sn(O_3PC_6H_4PO_3) \cdot 5.8H_2O$	9.51	368	359	8	36.0
$Sn(O_3PC_{12}H_8PO_3)$	13.8	442	318	18	33.5
$Sn(O_3PCH_2CH_2COOH)_2 \cdot 0.8H_2O$	11.07	224	102	27	33.09
$Sn(O_3PCH_2COOH)_2$	11.08	147	0	–	29.04

said about the porous zirconium bisphosphonates applies to the tin analogs as to preparation, porosity, and other structural features.

My colleagues at the University of Malaga, Spain, synthesized 4(4′-phosphonophenoxy) phenylphosphonic acid and prepared porous pillared tin(IV) phosphonates of this ligand.[94] It was found that using this ligand a variety of pore sizes were obtained dependent on the nature of the solvent. This is shown in Figure 1.20 where alcohols were used as solvent. This figure, published in reference 85, differs slightly from that in reference 94. This discovery would indicate that fine tuning of the pore size may be possible by utilizing

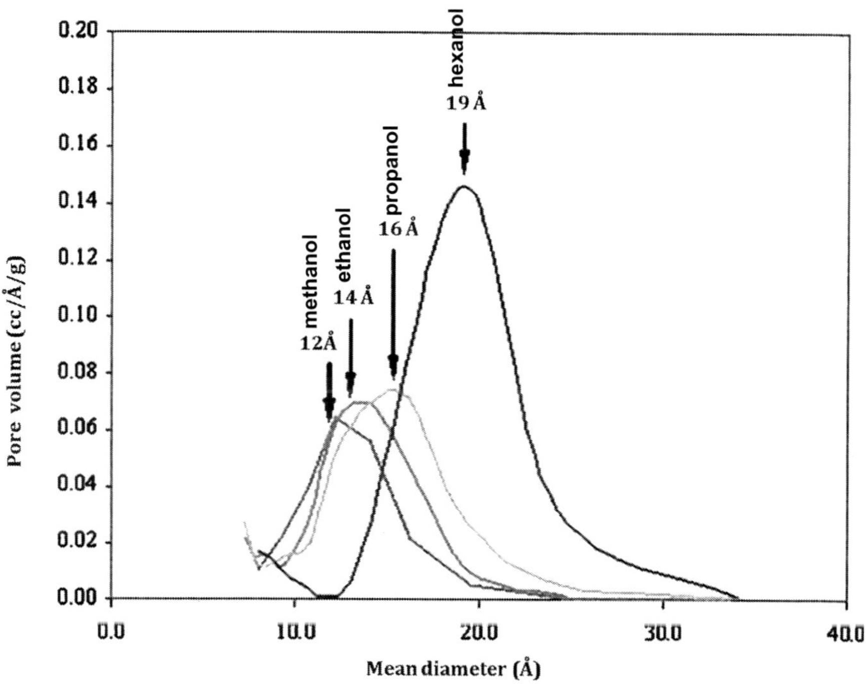

Figure 1.20 Pore size distribution of tin(IV) derivatives of (4,4-phosphonophenoxy)phenylphosphonic acid in alcohol-water mixed solvents (methanol, ethanol, propanol, hexanol). Reproduced from reference 85 with permission.

combinations of solvents based upon size, shape, and hydrophobic/hydrophilic character.

1.8.1 Tin Phosphonates as Baeyer–Villiger Catalysts

We have seen in the previous sections that certain properties such as Brønsted acid catalysis, ion exchange, and separations may be effected by functionalizing the ligands or adding a second ligand with special properties. Thus, we were surprised to discover that tin(IV) phosphonates are active catalysts for Baeyer–Villiger oxidations. BV oxidation reactions involve the oxidative cleavage of a carbon–carbon bond adjacent to a carbonyl that converts ketones to esters and aldehydes to phenols (Scheme 1.1). The reactions generally involve the use of peracids as the oxidizing agent. Peracids are expensive, hazardous, and ecologically undesirable because of the generation of unwanted carboxylic acids. Use of a catalyst that would activate H_2O_2 in water solutions would be cost effective and environmentally sound. Corma, *et al.*[95] have designed a Sn β-zeolite and a Sn-MCM-41 heterogeneous mesopore catalyst for use in BV reactions with H_2O_2 in non-halogenated solvents. We have found that our catalysts, primarily disordered tin phenylphosphonates, can function in

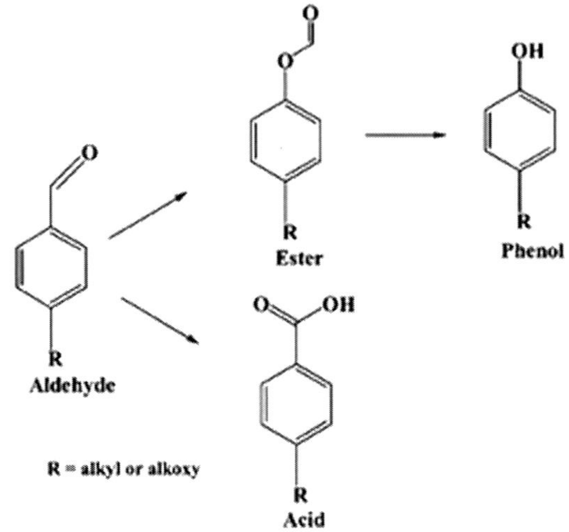

Scheme 1.1 Baeyer–Villiger oxidation of an aromatic aldehyde. Adapted from ref. 96 with permission.

Table 1.5a BV oxidation of different 4-substituted benzaldehydes with Sn^{IV}-phenylphosphonate as catalyst.

Aldehyde	*Conversion (%)*	*Selectivity (%)*
p-ethylbenzaldehyde	46	89
p-tolualdehyde	52	100
p-anisaldehyde	58	100
p-ethoxybenzaldehyde	55	100

Reaction conditions: 80 °C, 60 min, 3.7 mmol of aldehyde and 4.5 mmol of 30% H_2O_2; 0.025 g catalyst.

Table 1.5b Comparison of catalytic activity of different Sn catalysts with 4-methoxybenzaldehyde.

Catalyst	*Sn (wt %)*	*Conversion (%)*	*Selectivity (%)*	*Comments*
Beta Zeolite SnO_2	~2	56	46	acetonitrile solvent[84a]
Sn Beta Zeolite	~2	79	79	acetonitrile /50% H_2O_2[84d]
Sn Phosphonate*	24.5	98	100	30% H_2O_2, no solvent

Reaction conditions: 80 °C, 7 h, 3.7 mmol of aldehyde and 4.5 mmol of H_2O_2; 0.05 g catalyst (*0.025 g).

aqueous 30% H_2O_2 without the need for a solvent.[96] A number of results are listed in Table 1.5 where it is seen that high conversions are obtained in 30% H_2O_2 compared to lower yields obtained with the Sn-zeolite in 50% H_2O_2 in organic solvents. Our hypothesis as to why both the tin phenylphosphonates and bis(biphenylphosphonates) work in these reactions is two-fold; the catalysts have high surface area and are full of defects which allow H_2O_2 to form a complex with Sn at a defect site. Furthermore, these phosphonates contain ~20% by weight of tin as compared to 2% Sn in the zeolites. Only those tin ions at defect sites may be active but given the high level of Sn^{IV} in the compound this may still be a high number. An added benefit to the use of the phosphonate catalyst is the fact that the reaction needs no solvent other than the water supplied by the hydrogen peroxide.

1.9 Use of Porous Disphosphonates of Zr and Sn^{IV} as Ion Exchangers for Nuclear Waste Problems

A number of years ago we prepared compounds of the type $Zr(O_3PC_6H_4PO_3)_x(O_3POH)_{2-2x}$ and $Zr(O_3PC_{12}H_8PO_3)_x(O_3POH)_{2-2x}$.[97] The presence of the phosphate group imparts cation exchange capabilities to these complexes. It was found that their affinity for univalent ions was negligible and for divalent ions moderate. For example K_d values for alkaline earth cations ranged from 320 ml g^{-1} to 420 ml g^{-1} for Mg^{2+} to Ba^{2+} for the monophenyl compound and 207 ml g^{-1} to 687 ml g^{-1} for the biphenyl compound. In contrast, the K_d values for lanthanides were in the tens of thousands for the monophenyl compounds and approximately one third as large for the biphenyl derivatives. These data has remained dormant all these years. However, we have recently reactivated research with these materials for the reasons presented below.

The demands for electrical power are predicted to double in the next 10–15 years. This enormous increase cannot and should not be supplied only from fossil fuels as it would greatly accelerate global warming. The implementation of solar, wind, and other renewable sources needs to be accelerated but will not entirely fill the need. Thus, nuclear power has a major role to play. The advances in reactor design, safety, and lowered cost makes nuclear power even more attractive. However, nuclear energy will only be non-polluting or "green", protecting the environment, if the nuclear waste is dealt with in a sane and safe way, and accepted as such by the public. Therefore the focus now is on the nuclear fuel cycle, that entails reconstituting the spent fuel rods at the 104 operating nuclear reactors in the United States. 95% of the original fuel is recoverable from the spent rods because the fission process generates plutonium. Thus, the remaining U-235 and plutonium can be reused as fuel. Only 5% of the mass is waste.

During operation of the reactors, lanthanides and heavier actinides, americium and curium, are generated. These elements are objectionable as the lanthanides are neutron sorbers and Cm and Am are highly radioactive alpha emitters and need to be separated from the lanthanides and from each other. It occurred to us that our hybrid ion exchangers may be suited to the task because of their selectivity for lanthanides.

In recent work we have been able to improve the uptake of lanthanides by optimizing the pore structure of the zirconium monophenylphosphonate phosphates.[98] The analogous tin derivatives behave in a similar fashion. The

Table 1.6 K_d values in ml/g for selective actinides with hydrid zirconium phosphonate phosphates of general composition $Zr(O_3PC_6H_4\text{-}PO_3)_x(HPO_4)_{2-2x}$.

Sample	NpO_2^+ K_d	PuO_2^{2+} K_d	Am^{3+} K_d	Pu^{4+} K_d
JDB-I-45_4	301	10,405	–	127 449
JDB-I-62	131	2920	2030	17 076
JDB-I-96*	261	18 713	51 041	526 272
JDB-I-98	111	2118	1469	7354

*Protons replaced by Na^+.

maximum selectivity for lanthanides by these compounds occurs at pH 3. In recent work carried out at Los Alamos Carlsbad we were able to show that the affinity for actinides is similar to that of lanthanides but the highest selectivities for the actinides occur at pH values below 2. This feature is shown in Table 1.6. Thus, there is a basis for separation of the lanthanides from the actinides. There may also be instances where lanthanides need to be recovered from non-nuclear processes for which our materials may be useful.

1.10 Porous Pillared Aluminum Phosphonates

Aluminum phosphonate chemistry would appear to be more complex than the four-valent compounds. As an example we prepared multiple aluminum phosphonates with stoichiometry variations. This section is not a comprehensive treatment of aluminum phosphonates as there is further treatment in Chapter 10. Mainly it will deal with the type of porous pillared materials and the more soluble crystallizable compounds are treated elsewhere in this book.

Six aluminum phenylphosphonates were prepared.[99] Their respective compositions are (1) $Al_2(O_3PC_6H_5)_3 \cdot 2H_2O$, (2) $Al_2(O_3PC_6H_5)_3$, (3) α-$Al(HO_3PC_6H_5)(O_3PC_6H_5) \cdot H_2O$, (4) β-$Al(HO_3PC_6H_5)(O_3PC_6H_5) \cdot H_2O$, (5) $Al(HO_3PC_6H_5)_3 \cdot H_2O$, and (6) $Al(OH)(O_3PC_6H_5)$. The structure of the α-phase was solved from synchrotron X-ray powder data. It is layered with six-coordinate Al and pendant phenyl groups. The *b* axis is 29.39 Å, spanning two layers as the phenyl rings point in opposite directions in alternating layers. A phase with the same composition was reported earlier but only a model was deduced from X-ray powder data, elemental analysis, and NMR spectra.[100] The unit cell given for this compound is triclinic as compared to an orthorhombic cell for the α-phase (compound 3) indicating that this compound is the β-phase of the same composition. Also the methods of preparation are quite different; our phase was prepared hydrothermally at 200 °C whereas Hakey, *et al.*[100] refluxed their aqueous mix of reactants for 12 hours. Also a note was published for the synthesis of $Al_2(O_3PC_6H_5)_3 \cdot 4H_2O$.[101] This phase was considered to be layered, with two coordinated water molecules and two interstitial water molecules.

Chaplais *et al.*[102] synthesized three aluminum phosphonates, $Al(OH)(O_3PCH_2C_6H_4Br) \cdot H_2O$, $Al(OH)(O_3PC_6H_5) \cdot H_2O$, and $Al_2(O_3PC_6H_5)_3 \cdot H_2O$.

The structure of the benzyl bromide was solved from its powder pattern. It is layered with the bromophenyl rings tilted so that the P–C–C(phenyl) angle is 107°. The structure of the layers is very similar to the layers in α-Al $(OH)(O_3PCH_3) \cdot H_2O$.[103] The authors also stress the versatility of the synthetic processes that influence the final product that include pH, temperature, hydrothermal conditions, the source of aluminum, and the ratio of Al to phosphorus.

A concerted effort was made to prepare porous aluminum phosphonates. Madea *et al.* were able to prepare two such compounds termed AlMepO-β[104,105] and AlMepO-α.[106] A review paper on the many porous phosphonates prepared up through 2003 has been published.[48] A number of authors developed methods to obtain mesoporous aluminum phosphonates. Kimura *et al.* utilized methyl, ethyl, and propyl diphosphonic acids with aluminum alkoxides and various templates to synthesize a series of mesoporous materials.[107–109] Mesoporous materials were also obtained from aluminum atrane complexes and monophosphonic acids or $H_2O_3P(CH_2)_2PO_3H$ using nitrogenous templates as surfactants.[110,111] Amino triamine and penta (methylene–phosphonic) acid were used to obtain aluminum mesoporous materials with high sorptive ability.[112] A large collaborative group synthesized crystalline Ti and Al compounds of the diphosphonic acid $H_2O_3PCH_2$-NC_4H_8N-$CH_2PO_3H_2$.[113] These compounds contained small pores, 3.5×4.0 Å^2, surface area of ~500 m^2 g^{-1}. A number of additional studies that resulted in crystalline products that were either non-porous or contain small pores utilized one or two carbon alkyldephosphonic acids.[114–117]

Our own research with aluminum phosphonates has consisted in preparing monophenyl and biphenyl derivatives. These products are not as spontaneously porous as the analogous four-valent compounds. There is a high level of external porosity as a fraction of the total porosity. This is shown by reference to Table 1.7. The preparations were carried out in alcohol–water mixtures at 120 °C for 5 days. The XRPD[118] shows a weak peak at 26.56 Å and a strong peak at 13.27 Å. These peaks are undoubtedly the 001, 002 reflections. The interlayer distance is expected to be the smaller value but the presence of a spacing of twice that value indicates a repeat distance of two layers. The complexity is further attested to by the appearance of ^{27}Al SS MAS resonances at +46.15 ppm and –21.37 ppm indicative of the presence of both six-coordinate and four-coordinate Al. There are also four ^{31}P resonances at –1.26,

Table 1.7 Surface areas of aluminum biphenyldiphosphonate prepared in alcohol-water solvents.

Sample no.[a]	*Solvent*[b]	*Total SA* (m^2/g)	*External SA* (m^2/g)	*Micropore SA* (m^2/g)
49B	methanol/water	478.1	250.0	228.1
39A	ethanol/water	439.5	184.0	255.6
50B	butanol/water	343.3	230.2	113.1
50A	pentanol/water	352.8	264.9	87.85

[a] $AlCl_3 \cdot 6H_2O$:BPDPA = 8:3, T = 120 °C, 5d.
[b] 50:50 mix of alcohol-water.

–7.08, –9.50 and –12.77 ppm. The reason for the low porosity of sample MS-II-39A is not apparent except that the peaks in the powder pattern are broad and the 001 reflection is not present.

In order to increase the internal porosity we utilized mixtures of BPDPA and phosphorous acid with the expected small phosphite group replacing some of the biphenyl pillars. The results are collected in Table 1.8. A second set of preparations was carried out in mixed alcohol–water solvents for which the ratios were Al:BPDPA:H_3PO_3 8:1.5:1.5. To our surprise the internal surface area decreased leaving some extraordinary high external surface areas. The particles prepared with phosphorous acid were in the range of 15–25 nm as depicted by the electron microgram of Figure 1.21.

Table 1.8 Surface areas of aluminum biphenyldiphosphonate phosphites, $Zr(O_3PC_{12}H_8PO_3)_x(O_3PH)_{2-2x}$.

Reaction[a] *no.*	*Solvent*[*]	*Reactant ratio Al : BPDP : O_3PH*	*Total SA (m^2/g)*	*External SA (m^2/g) (%)*	*Micropore SA (m^2/g) (%)*
51A	ethanol(:H_2O)	8 : 3 : 1.5	531.2	428.7 (80.7)	102.5 (19.3)
58A	ethanol:water	8 : 3 : 1.5	361.0	263.4 (73.0)	97.63 (27.0)
58A[**]	ethanol:water	8 : 3 : 1.5	410.2	299.3 (73.0)	110.9 (27.0)
56A	methanol:water	8 : 1.5 : 1.5	611.2	571.0 (93.34)	40.25 (6.6)
51B	ethanol:water	8 : 1.5 : 1.5	779.5	657.8 (84.4)	121.6 (15.6)
58B	ethanol:water	8 : 1.5 : 1.5	389.3	331.0 (85.0)	58.28 (15.0)
56B	butanol:water	8 : 1.5 : 1.5	783.9	783.9 (100)	0.000 (0)

[**]second preparation.
[*]3:1 alcohol:water.
[a]Reaction conditions same as in Table 1.7.

Figure 1.21 Electron micrograph of aluminum biphenylphosphonate phosphite. The individual particles average about 25 nm.

My collaborator Ernesto Brunet (who wrote Chapter 9 of this book) sent his graduate student Carlos Cerro to work in my laboratory for a six-month stay. Carlos was intrigued by the aluminum phosphonate work and upon his return to Madrid carried out similar mixed ligand reactions. However, he increased the temperature of synthesis to 185 °C for 7 days and used mixed acetone–water 50:50 as solvent. With these differences the results became more regular as to surface area and pore distribution as shown in Table 1.9 and the N_2 sorption–desorption isotherms mirrored this regularity (Figure 1.22).[118] The lack of hysteresis indicates no mesopores are present and in fact analysis of the pore structure yielded the pore maxima at 1.5 nm and 0.6 nm – a bimodal pore structure. However, the inset in Figure 1.22 shows a unimodal pore distribution centered about a diameter of 0.6 nm. It should be noted that the curves in Figure 1.23 for H_2 uptake end at a pressure of 1 atm, but still have a high upward slope especially for sample 152. This large increase of H_2 uptake relative to the other samples may in part be due to the small pore size and the greater ability to approach the aromatic rings and the metal ion

Table 1.9 Results of the BET analysis of the aluminum biphenylphosphonate phosphites.

	Surface (m^2/g)		
Material UAM	*Microporous*	*External*	*Total*
150	96.0	71.8	167.8
151	187.3	107.1	294.4
152	211.4	153.8	365.2

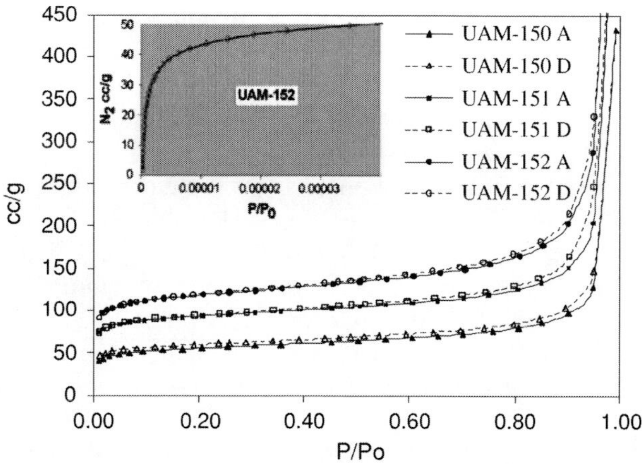

Figure 1.22 Nitrogen adsorption isotherms for the indicated materials (A: adsorption; D: desorption). Inset shows nitrogen isotherm of UAM-152 at very low P/P_0 values. With kind permission from Springer Science + Business Media, adapted from reference 118, Figure 6.

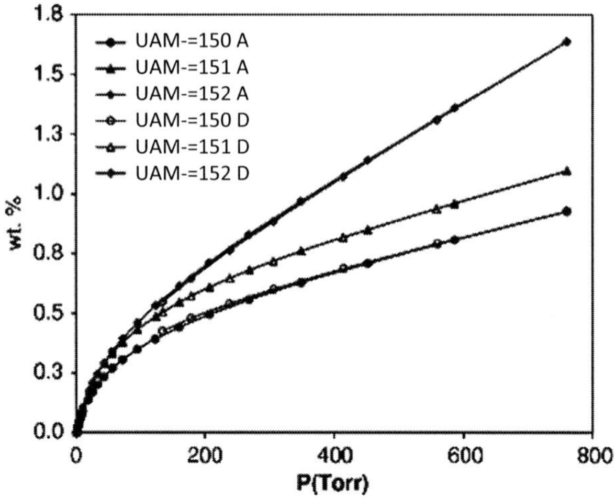

Figure 1.23 Hydrogen adsorption (A), desorption (D) curves at 77 K of compounds 150, 151 and 152. With kind permission from Springer Science + Business Media, adapted from Figure 1, reference 118.

resulting from defects and the large amount of phosphate groups present in the pore wall.

1.11 Monovalent Metal Phosphonates

The literature dealing with metal phosphonates is voluminous for divalent cations, large for four-valent cations and growing for trivalent metal phosphonates. We close our chapter with a brief discourse in the phosphonate chemistry of univalent cations. Only recently did Rao and Vidyasagar[119] report on the phenyl phosphonates of the alkali metal cations ammonium ion and Tl^+. All the compounds are layered except lithium and NH_4^+ which form linear chains. The intriguing aspect of these compounds is the fact that they all have the same composition $M(HO_3PC_6H_5)(H_2O_3PC_6H_5)$, in which the ligands retain three protons. This means that a neutral phosphonate group bonds to a cation in addition to a charge balancing phosphonate group in each of these compounds. Details of the structure were only given for the Li^+, NH_4^+, and Na^+ compounds. The lithium ion compound is tetrahedrally coordinated and connected to two adjacent tetrahedra by sharing corners forming chains running along the *a* axis. The chains are about 8.5 Å apart with the phenyl rings facing each other in the *c* axis direction. The ammonium ion phase is also 1D. The four ammonium ion protons hydrogen bond to four phosphonate oxygens, forming ladder-like chains parallel to the *a* axis. The remaining alkali metals form layered compounds. In the sodium compound the Na^+ is octahedral, utilizing five of the six phosphonate oxygens. These octahedra share an edge to form Na_2O_{10} units which in turn are corner linked to other such units to form the layer. K^+, Rb^+, and Cs^+ are eight coordinate or Cs^+ is nine coordinate if one accepts a bond of 3.52 Å as significant. The sum of the ionic radii for nine-coordinate Cs^+ is 3.14 Å.

Subsequently, the same authors carried out similar reactions using the ligand $H_2O_3P(CH_2)_2PO_3H_2$ with the same univalent ions.[120] The structures were of four different types. The Li^+ are coordinated tetrahedrally by phosphonate oxygens. Each tetrahedron shares two edges with two such tetrahedra to form chains parallel to the *c* axis. These chains are then connected in the *a* and *b* axis directions to form a 3D structure. The sodium ions are octahedrally coordinated but share an edge to create Na_2O_{10} units. These units are further bonded into layers which are in turn are pillared into a 3D structure. The Rb^+ and Cs^+ compounds have structures similar to that of the potassium bis(phosphonate). In the ammonium ion structure the nitrogen atoms are confined approximately to the *ab* planes and are tetrahedrally hydrogen bonded to the diphosphonate oxygen atoms. The connectivity of the NH···O and CPO_3 phosphonate groups result in ladder-like chains parallel to the *a* axis. These chains are similar to those that exist in ammonium phenylphosphonate but in the ethylenediphosphonate the chains are connected to form a 3D structure.

Our interest in these compounds stems from the idea that if the 3D compound could be made porous, then the protons would be accessible for further Brønsted acid reactions within the pore space. We therefore prepared compounds of the alkali metal ions but not ammonium or thallium ions using 4,4′-biphenyldiphosphonic acid as the ligand.[121] Our lithium compound was similar in terms of crystal system (monoclinic, C2/c and unit cell dimensions except for the larger *a* axis, 27.37 Å versus 16.19 Å for the ethylenediphosphonic acid compound). The Na^+ compounds are also similar. However, the remaining alkali metal compounds differ in that our compounds are all triclinic $P\bar{1}$ and the Rao Vidyasager compounds are monoclinic C2/c. The unit cells are different which for the K^+ compounds are $a = 5.879(5)$ Å, $b = 7.778(7)$ Å, $c = 15.134(13)$ Å, $\alpha = 88.10(1)°$, $\beta = 84.50(1)°$, $\gamma = 82.628(7)°$ as compared to the Rao cell $a = 9.237(1)$ Å, $b = 8.909(2)$ Å, $c = 9.484$ Å, $\beta = 92.60(2)$. The pillared nature of the potassium biphenylphosphonate is shown in Figure 1.24.

In this view it is seen that each K^+ ion is coordinated by three oxygens atoms on top and three below to form a trigonal prism. The three oxygens of any one phosphonate group bond to three different K^+ forming a double stranded chain running down the *a* axis direction. The biphenyl groups then connect the chains together forming a 3D structure. However, the biphenyl group on the right becomes the biphenyl group on the left in the adjacent layer and vice versa along the *c* axis, leaving gaps between the pillars. The rubidium and cesium compounds are similar but with increased coordination of seven and eight requiring increased sharing of oxygen atoms. Thus, the stoichiometry does not change but the geometry of alkali metal coordination does. None of our compounds are porous and attempts to insert spacer groups such as methylphosphonic acid all failed. Furthermore, the 3D nature of the compounds prevented any intercalation. However when we included an amine as one of the reactants, totally new complexes were obtained. The structure of the ethylene diamine compound

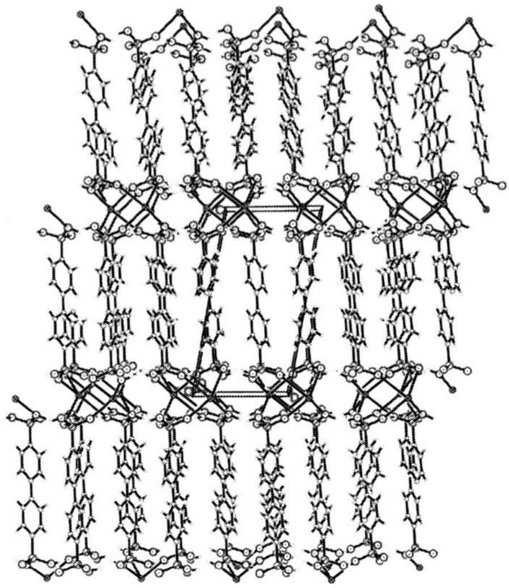

Figure 1.24 The crystal packing of potassium biphenyldiphosphonate as viewed down the *a* axis. The phosphonate groups form K_2O_{12} dimers through bridging links across K^+. The remaining oxygen atoms tie these dimers into chains running along the *a* axis while the biphenyl groups link the chains into a 3D structure.

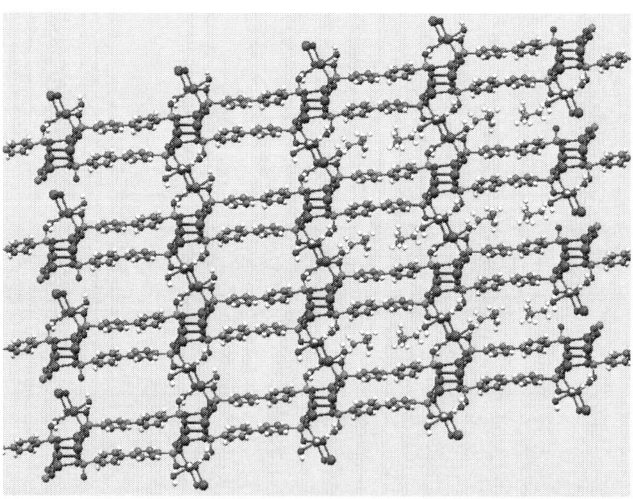

Figure 1.25 A synthesis with Na^+ and biphenyldiphosphonic acid and the addition of ethylenediamine yielded the pictured compound containing alternating narrow and widely spaced biphenyl pillars. The $NH_2CH_2CH_2NH_3^+$ ions are bonded in the wide channels (right-hand half) and omitted on the left-hand half.

$Na_3(O_3PC_{12}H_8PO_3)(H_2NC_2H_4NH_3)$ is shown in Figure 1.25. The cross-linking biphenyl groups are alternately close together and then farther apart. The positively charged ethylene amine groups are bonded within these layer channels. We have now synthesized a number of these type compounds and also found that a variety of interesting structures can be obtained by varying the synthetic procedures.[121] Thus, there appears to be a rich phosphonate chemistry waiting to be exploited with univalent ions including Ag^+.

1.12 The Last Word

In the ensuing chapters the vast amount of progress made in metal phosphonate chemistry will unfold. The end is not in sight and many fruitful years of progress is in store for those with innovative ideas.

Acknowledgements

It is with great measure of gratitude and a sense of humility that I acknowledge the financial support provided for research described in this chapter initially by the U.S. Army Research Office (1984–1988), followed by the R. A. Welch Foundation since 1988 (Grant No. A-0673) and the National Science Foundation grants since 1989. Of course all the work was carried out by my many graduate students, postdoctoral associates, and undergraduates. I can never thank them enough. To my brother Ted, who every month sent me part of his navy pay so I could save for college, my wife Ruth, who encouraged me to go on for a Ph.D. while she worked, and the pride on my immigrant parents' faces when I was hooded – it could never have happened in the old country.

References

1. G. Alberti, U. Costantino, S. Allulli and N. Tomassini, *J. Inorg. Nucl. Chem.*, 1978, **40**, 1113.
2. D. M. Poojary, H.-L. Hu, F. L. Campbell, III and A. Clearfield, *Acta Crystallgr. Sect. B.*, 1993, **49**, 996.
3. (a) A. Clearfield and G. D. Smith, *Inorg. Chem.*, 1969, **8**, 431; (b) J. M. Troup and A. Clearfield, *Inorg. Chem.*, 1977, **16**, 3311.
4. J. D. Wang and A. Clearfield, *Mater. Chem. Phys.*, 1993, **35**, 208.
5. A. Clearfield, R. H. Blessing and J. A. Stynes, *J. Inorg. Nucl. Chem.*, 1968, **30**, 2249.
6. A. N. Christensen, E. K. Andersen, I. G. Andersen, G. Alberti, M. Nielsen and M. S. Lehmann, *Acta Chem. Scand.*, 1990, **44**, 865.
7. D. M. Poojary, B. Shpeizer and A. Clearfield, *J. Chem. Soc. Dalton Trans.*, 1995, 111.
8. G. Alberti, in *Comprehensive Supramolecular Chemistry*, Vol. 7, Eds. G. Alberti, T. Bein, Pergamon, Oxford, UK, 1996, pp. 151–187.
9. S. Yamanaka, *Inorg. Chem.*, 1976, **15**, 2811.

10. S. Yamanka, H. Maeda and H. Tanaka, *J. Inorg. Nucl. Chem.*, 1979, **41**, 1187.
11. S. Yamanka and M. Hattori, *Chem. Lett.*, 1979, **9**, 1073.
12. (a) P. M. DiGiacomo and M. B. Dines, *Polyhedron*, 1982, **1**, 61; (b) M. D. Dines, P. M. DiGiacomo, K. P. Callahan, P. C. Griffith, R. H. Lane and R. E. Cooksey, *ACS Symp. Ser.*, 1982, **12**, 192.
13. G. Cao, H. Lee, V. M. Lynch and T. E. Mallouk, *Inorg. Chem.*, 1988, **27**, 2781.
14. K. Martin, P. J. Squattrito and A. Clearfield, *Inorg. Chim. Acta*, 1989, **155**, 7.
15. G. Cao and T. E. Mallouk, *Inorg. Chem.*, 1991, **30**, 1434.
16. K. J. Frink, R.-C. Wang, J. L. Colon and A. Clearfield, *Inorg. Chem.*, 1991, **30**, 1438.
17. Y. Zhang, K. J. Scott and A. Clearfield, *J. Mater. Chem.*, 1995, **5**, 315.
18. K. J. Scott, Y. Zhang, R.-C. Wang and A. Clearfield, *Chem. Mater.*, 1995, **7**, 1095.
19. D. M. Poojary and A. Clearfield, *J. Am. Chem. Soc.*, 1995, **117**, 11278.
20. Y. Zhang and A. Clearfield, *Inorg. Chem.*, 1992, **31**, 2821.
21. D. M. Poojary, B. Zhang, P. Bellinghausen and A. Clearfield, *Inorg. Chem.*, 1996, **35**, 4942.
22. D. M. Poojary, B. Zhang and A. Clearfield, *J. Am. Chem. Soc.*, 1997, **119**, 12550.
23. D. I. Arnold, X. Ouyang and A. Clearfield, *Chem. Mater.*, 2002, **14**, 2020.
24. D. M. Poojary, B. Zhang and A. Clearfield, *Chem. Mater.*, 1999, **11**, 421.
25. G. Cao, H. Lee, V. M. Lynch and T. E. Mallouk, *Solid State Ionics*, 1988, **26**, 63.
26. D. M. Poojary, B. Zhang, A. Cabeza, M. A. G. Aranda, S. Bruque and A. Clearfield, *J. Mater. Chem.*, 1996, **6**, 639.
27. G. Cao, V. M. Lynch, J. S. Swinea and T. E. Mallouk, *Inorg. Chem.*, 1990, **29**, 2112.
28. R. C. Wang, Y.-P. Zhang, H. Hu, R. R. Frausto and A. Clearfield, *Chem. Mater.*, 1992, **4**, 864.
29. B. Bujoli, P. Palvadeau and J. Rouxel, *Chem. Mater.*, 1990, **2**, 582.
30. P. Palvadeau, M. Quegnec, J. P. Venien, B. Bujoli and J. Villieras, *Mater. Res. Bull.*, 1988, **23**, 1561.
31. M. Riou-Cavellec, C. Serre, J. Robino, M. Nogues, J.-M. Grenech and G. Ferey, *J. Solid State Chem.*, 1999, **147**, 122.
32. G. Alberti, U. Costantino, C. Dionigi, S. Murcis-Mascaro's and R. Vivani, *Supramol. Chem.*, 1995, **6**, 29.
33. A. Clearfield and U. Costantino, in *Comprehensive Supramolecular Chemistry*, Eds. G. Alberti, T. Bein, Pergamon Press, New York, 1996, vol. 7, p. 107.
34. A. Clearfield, *Prog. Inorg. Chem.*, 1998, **47**, 371.
35. G. Cao, H. -G. Hong and T. E. Mallouk, *Acc. Chem. Res.*, 1992, **25**, 420.
36. D. Burwell and M. E. Thompson, *ACS Symp. Ser.*, 1992, **499**, 166.
37. M. Thompson, *Chem. Mater.*, 1994, **6**, 1168.

38. A. Clearfield, *Curr. Opin. Solid State Mater. Sci.*, 1996, **1**, 268; 2002, **6**, 495.
39. C. V. K. Sharma and A. Clearfield, *J. Am. Chem. Soc.*, 2000, **122**, 1558.
40. A. Clearfield, C. V. K. Sharma and B. Zhang, *Chem. Mater.*, 2001, **13**, 3099.
41. A. Clearfield, D. M. Poojary, B. Zhang, B. Zhao and A. Derecskei-Kovacs, *Chem. Mater.*, 2000, **12**, 2745.
42. B. Zhang and A. Clearfield, *J. Am. Chem. Soc.*, 1997, **119**, 2751.
43. C. V. K. Sharma, A. Clearfield, A. Cabeza, M. A. G. Aranda and S. Bruque, *J. Am. Chem. Soc.*, 2001, **123**, 2885.
44. C. V. K. Sharma and A. Clearfield, *J. Am. Chem. Soc.*, 2000, **122**, 4394.
45. C. V. K. Sharma, A. J. Hessheimer and A. Clearfield, *Polyhedron*, 2001, **20**, 2095.
46. V. Chandrsekhar, T. Senapati, A. Day and S. Hossain, *Dalton Trans.*, 2011, **40**, 5394.
47. (a) G. Alberti, M. Casciola, M. Pica and G. Di Cesare, *Annals New York Acad. Sci.*, 2003, **984**, 208; (b) G. Alberti, U. Costantino, R. Narducci, M. Pica and M. Sganappa, *Desalination*, 2006, **199**, 4.
48. K. Maeda, *Micropor. Mesopor. Mater.*, 2004, **73**, 47.
49. K. D. Demadis, in *Solid State Chemistry Research Trends*, Ed. R. W. Buckley, Nova Science Publ., Inc., Hauppauge, NY, 2007, p. 109.
50. G. K. H. Shimizu, R. Vaidhyanathan and J. M. Taylor, *Chem. Soc. Rev.*, 2009, **38**, 1430.
51. M. Papadaki and K. D. Demadis, *Comments Inorg. Chem.*, 2009, **30**, 89.
52. G. K. H. Shimizu, J. J. Taylor and R. Vaidhyanathan, in *Macromolecules Containing Metal and Metal Like Elements: Supramolecular and Self-Assembled Metal-Containing Materials*, Vol. 9, Eds. A. S. Abd-El Aziz, C. E. Carraher, C. U. Pittman, M. Zeldin, John Wiley & Sons, Inc., Hoboken, NJ, 2009, p. 125.
53. C. H. Kumar, A. Bhambhani and N. Hnatiuk, in *Handbook of Layered Materials*, Eds. S. M. Auerbach, K. A. Carrado, P. K. Dutta, Marcel Dekker, Inc., New York, NY, 2004, p. 313.
54. M. D. Francis and R. R. Martodam, in *The Role of Phosphonates in Living Systems*, Ed. R. L. Hilderbrand, CRC Press, Boca Raton, FL, 1983.
55. E. Matazak-Jon and V. Videnova-Adrobinska, *Coord. Chem. Rev.*, 2005, **249**, 2458.
56. G. Alberti, U. Costantino and R. Guilietti, *Gazz. Chim. Ital.*, 1983, **113**, 547.
57. G. Alberti, U. Costantino and G. Perego, *J. Solid State Chem.*, 1986, **63**, 455.
58. G. Alberti, U. Costantino, J. Kornyei and M. L. L. Giovognotti, *React. Polym.*, 1985, **4**, 1.

59. A. Clearfield, J. D. Wang, Y. Tian, E. Stein and C. Bhardwaj, *J. Solid State Chem.*, 1995, **117**, 275.
60. P. Zhang and A. Clearfield, work in progress.
61. C. Y. Yang and A. Clearfield, *React. Polym.*, 1987, **5**, 13.
62. A. F. Benedetto, P. J. Squattrito and F. Adani, *E. Montoneri. Inorg. Chim. Acta*, 1997, **260**, 207.
63. G. Alberti, M. Casciola, U. Costantino, A. Peraio and E. Montoneri, *Solid State Ionics*, 1992, **50**, 315.
64. Z. Wang, J. M. Heising and A. Clearfield, *J. Am. Chem. Soc.*, 2003, **125**, 10375.
65. E. W. Stein, Sr., A. Clearfield and M. A. Subramanian, *Solid State Ionics*, 1996, **83**, 113.
66. J. L. Colon, C.-Y. Yang, A. Clearfield and C. R. Martin, Jr. *J. Phys. Chem.*, 1988, **92**, 5777.
67. L. H. Kullberg and A. Clearfield, *Solv. Extr. Ion Exch.*, 1989, **7**, 527.
68. L. H. Kullberg and A. Clearfield, *Solv. Extr. Ion Exch.*, 1990, **8**, 187.
69. G. Alberti, M. Cosciola, U. Costantino, A. Peraio and E. Montonari, *Solid State Ionics*, 1992, **50**, 315.
70. G. Alberti, M. Casciola, U. Costantino, S. Ferroni, L. Massinelli and P. Staiti, *Solid State Ionics*, 2001, **145**, 249.
71. M. Casciola, G. Alberti, A. Ciarletta, A. Cruccolini, P. Piaggio and M. Pica, *Solid State Ionics*, 2005, **176**, 2985.
72. X. Fu, B. Luo and Q. Lei, *Yingong Huaxue*, 1991, **8**, 6.
73. M. Curini, F. Epifano, M. C. Marcatullio and O. Rosati, *Eur. J. Org. Chem.*, 2001, **21**, 4149.
74. M. Curini, F. Epifano, M. C. Marcotullio and O. Rosati, *Synlett.*, 2001, **7**, 1182.
75. M. Curini, F. Epifano, M. C. Marcotullio, O. Rosati and U. Costantino, *Tetrahedron Lett.*, 1998, **39**, 8159.
76. (a) M. Curini, O. Rosati, E. Pisani and U. Costantino, *Synlett*, 1996, 333; (b) M. Curini, M. C. Marcotullio, O. Rosati, E. Pisani and U. Costantino, *Synlett*, 1997, 769; (c) M. Curini, F. Epifano, M. C. Marcatullio, O. Rosati and U. Costantino, *Synth. Commun.*, 1999, **29**, 541.
77. (a) M. Curini, O. Rosati and U. Costantino, *Curr. Org. Chem.*, 2004, **8**, 591; (b) M. Curini, O. Rosati, V. Compagna, F. Montanari, G. Cravotto and M. Boccalini, *Synlett*, 2005, **19**, 2927.
78. O. Rosati, M. Curini, M. C. Marcotullio, A. Macchiarulo, M. Perfumi, L. Mattioli, F. Rismondo and G. Cravotto, *Bioorg. Med. Chem.*, 2007, **15**, 3463.
79. Y. Sun, Y.-Y. Zhang, Y. Qi, H. Wang, H.-C. Liu, J.-G. Shen and Z.-G Liu, *Cuihua Xuebao*, 2009, **30**, 786 (in Chinese).
80. (a) J. F. Haw, I. S. Chuang, B. L. Hawkins and G. E. Maciel, *J. Am. Chem. Soc.*, 1983, **105**, 7206; (b) G. E. Maciel, J. F. Haw, I. S. Chung, B. L. Hawkins, T. A. Early, D. R. McKay and L. Petrakis, *J. Am. Chem. Soc.*, 1983, **105**, 5529.

81. (a) T. Xu, N. Kob, R. S. Drago, J. B. Nicholas and J. F. Haw, *J. Am. Chem. Soc.*, 1997, **119**, 12231; (b) J. F. Haw, J. B. Nicholas, T. Xu, L. B. Beck and D. B. Ferguson, *Acc. Chem. Res.*, 1994, **29**, 259.
82. (a) A. Clearfield, in *Design of New Materials*, Eds. D. A. Cocke, A. Clearfield, Plenum, New York, 1986, pp. 121–134; (b) A. Clearfield and Z. Wang, *Dalton Trans.*, 2002, 2937.
83. A. Clearfield, *Chem. Mater.*, 1998, **10**, 2801.
84. A. Clearfield, Z. Wang and P. Bellinghausen, *J. Solid State Chem.*, 2002, **167**, 376.
85. A. Clearfield, *Dalton Trans.*, 2008, 6089.
86. R. Vivani, U. Costantino and M. Nocchetti, *Inorg. Chem.*, 2006, **45**, 2388.
87. A. Clearfield and Z. Wang, *J. Chem. Soc., Dalton Trans.*, 2002, 2937.
88. F. Odobel, B. Bujoli and D. Massiot, *Chem. Mater.*, 2001, **13**, 163.
89. H. P. Perry and A. Clearfield, *Microporous Mesoporous Mater.*, in press.
90. (a) N. K. Mal, M. Fugiwara, Y. Yamada, K. Kuraoka and M. Matsukata, *J. Ceram. Soc.*, 2003, **111**, 219; (b) N. K. Mal, M. Fujiwara, Y. Yamada and M. Matsukata, *Chem. Lett.*, 2003, **32**, 292; (c) N. K. Mal, M. Fujiwara, Y. Yamada and M. Matsukata, *Chem. Commun.*, 2005, 5199.
91. A. Subbiah, D. Pyle, A. Rowland, J. Huang, R. A. Narayanan, P. Thiyagaran, J. Zon and A. Clearfield, *J. Am. Chem. Soc.*, 2005, **127**, 10826.
92. S. Kirumakki, J. Huang, A. Subbiah, J. Yao, A. Rowland, B. Smith, A. Mukherjee, S. Samarajeewa and A. Clearfield, *J. Mater. Chem.*, 2009, **19**, 2593.
93. J. Huang, A. Subbiah, D. Pyle, A. Rowland, B. Smith and A. Clearfield, *Chem. Mater.*, 2006, **18**, 5213.
94. M. M. Gomez-Alcantara, A. Cabeza, P. Olivera-Pastor, F. Fernãndez-Moreno, I. Sobrados, J. Sanz, R. E. Morris, A. Clearfield and M. A. G. Aranda, *Dalton Trans.*, 2007, 2394.
95. (a) A. Corma, M. T. Navarro, L. Nemeth and M. Renz, *Chem. Commun.*, 2001, 2190; (b) A. Corma, L. Nemeth, M. Renz and S. Valencia, *Nature*, 2001, **412**, 423; (c) M. Renz, T. Blasco, A. Corma, V. Forn's, R. Jensen and L. Nemeth, *Eur. J. Chem.*, 2002, **8**, 4708; (d) A. Corma, V. Fornés, S. Iborra, M. Mifsud and M. Renz, *J. Catal.*, 2004, **221**, 67.
96. S. Kirumakki, S. Samarajeewa, R. Harwell, A. Mukherjee, R. H. Herber and A. Clearfield, *Chem. Commun.*, 2008, 5556.
97. R. Cahill, B. Shpeizer, G. Z. Peng, L. Bortun and A. Clearfield, in *Separations of Elements*, Eds. K. L. Nash, G. R. Choppin, Plenum, NY, 1995, p. 165.
98. J. D. Burns, D. Reed, A. Clearfield, *Radiochem. Acta*, submitted.
99. A. Cabeza, M. A. G. Aranda, S. Bruque, D. M. Poojary, A. Clearfield and J. Sanz, *Inorg. Chem.*, 1998, **37**, 4168.
100. J. E. Haky, J. B. Brady, N. Dando and D. Weaver, *Mater. Res. Bull.*, 1997, **32**, 297.

101. L. Raki and C. Detallier, *Chem. Commun.*, 1996, 2475.
102. G. Chaplais, J. L. Bideau, D. Leclercq, H. Mutin and A. Vioux, *J. Mater. Chem.*, 2000, **10**, 1593.
103. L.-J. Sawers, V. J. Carter, A. R. Armstrong, P. G. Bruce, P. A. Wright and B. E. Gore, *J. Chem. Soc., Dalton Trans.*, 1996, 3159.
104. K. Maeda, Y. Kiyozumi and F. Hizukami, *Angew. Chem., Int. Ed. Engl.*, 1994, **33**, 2335.
105. K. Maeda, J. Akimoto, Y. Kiyozumi and F. Mizukami, *J. Chem. Soc., Chem. Commun.*, 1995, 1033.
106. K. Maeda, J. Akimoto, Y. Kiyozumi and F. Mizukami, *Angew. Chem., Int. Ed. Engl.*, 1995, **34**, 1199.
107. (a) T. Kimura, *Chem. Mater.*, 2003, **15**, 3742; (b) *Chem. Mater.*, 2005, **17**, 337; (c) *Chem. Mater.*, 2005, **17**, 5521.
108. T. Kimura and K. Kato, *Micropor. Mespor. Mater.*, 2007, **101**, 207.
109. T. Kimura, K. Kato and Y. Yamauchi, *Chem. Commun.*, 2009, 4938.
110. J. El Haskouri, C. Guillem, J. Latorre, A. Beltran and P. Amoros, *Chem. Mater.*, 2004, **16**, 4359.
111. J. El Haskouri, C. Guillem, J. Latorre, A. Beltran and P. Amoros, *Eur. J. Inorg. Chem.*, 2004, 1804.
112. T. Y. Ma, X.-J. Zhang and Z.-Y. Yuan, *J. Phys. Chem.*, 2009, **113**, 12854.
113. C. Serre, J. A. Groves, P. Lightfoot, A. M. Z. Slowin, P. A. Wright, N. Stock, T. Bein, M. Haousas, F. Taulelle and G. Ferey, *Chem. Mater.*, 2006, **18**, 1451.
114. H. G. Harvey and M. P. Attfield, *Chem. Mater.*, 2004, **16**, 199.
115. M. P. Attfield, H. G. Harvey and S. J. Teat, *Solid State Chem.*, 2004, **177**, 2951.
116. R. N. Devi, P. Wormald, P. A. Cox and P. A. Wright, *Chem. Mater.*, 2004, **16**, 2229.
117. H. G. Harvey, S. J. Teat, C. C. Tang, L. M. Cranswick and M. P. Attfield, *Inorg. Chem.*, 2003, **42**, 2428.
118. E. Brunet, C. Cerro, O. Juanes, J. C. Rodriguez-Ubis and A. Clearfield, *J. Mater. Sci.*, 2008, **43**, 1155.
119. K. P. Rao and K. Vidyasagar, *Eur. J. Inorg. Chem.*, 2005, 4936.
120. K. P. Rao and K. Vidyasagar, *Eur. J. Inorg. Chem.*, 2006, **813**.
121. T. Kinnibrugh, A. Clearfield, *J. Solid State Chem.*, Submitted..
122. G. Alberti, U. Costantino, G. Dionigi, S. Murcia-Mascaros and R. Vivani, *Supramolecular Chem.*, 1995, **6**, 29.
123. A. Clearfield, J. D. Wang, Y. Tian, F. L. Campbell III and G.-Z. Peng, in *Materials Synthesis and Characterization*, Ed. D. L. Perry, Plenum, NY, 1997, p. 103.
124. G. Alberti, M. Casciola, U. Costantino, M. Pica and D. Cesare, *Annals N.Y. Acad. Sci.*, 2003, **984**, 208.

CHAPTER 2

Zirconium Phosphonates

RICCARDO VIVANI, FERDINANDO COSTANTINO AND MARCO TADDEI

University of Perugia, Department of Chemistry, Via Elce di Sotto 8, 06123 Perugia, Italy

2.1 Introduction

Zirconium phosphonate chemistry has grown considerably in recent years, due to the increased demand of new functional solids for application in many fields of materials chemistry. The development of these compounds has been essentially made by the modification of well known "base compounds" having different structures, corresponding to different chemical features. Most research in zirconium phosphonate chemistry has been focused on layered derivatives with α- and γ-type structure as "base compounds"[1,2] A variety of functional groups, that can range from superacidic to basic, from lyophilic to highly hydrophilic, can be inserted in them by means of various procedures.[1–5] These derivatives have been investigated for potential application in the fields of photochemistry,[6] molecular and chiral recognition,[7] biotechnologies[8] and catalysis.[9–11]

More recently many new compounds, essentially based on aminophosphonate derivatives, have been discovered. This class of compounds, first studied by Clearfield and co-workers,[3,12–15] has been recently enriched with a large variety of structural types, with different dimensionality, ranging from single or composite linear chains, to layered compounds, to open-framework structures.

Low-dimensional 1D and 2D structures are, for their own nature, easily accessible and able to be functionalised for many application; on the other hand, 3D open-framework compounds are attractive due to their modular structure, that may be tuned in order to create tailor-made galleries and cavities, suitable to host different classes of guest molecules.

Metal Phosphonate Chemistry: From Synthesis to Applications
Edited by Abraham Clearfield and Konstantinos Demadis

Published by the Royal Society of Chemistry, www.rsc.org

Zirconium phosphonates are generally very insoluble compounds. This feature, although attractive for many potential applications, makes it difficult to deduce their structural characteristics. Large crystals prepared from zirconium phosphonates are very rare, and most of the structural knowledge on this field comes from powder diffraction studies. Therefore the recent development of new instrumental apparatus and numerical strategies for the analysis of powder diffraction data[16] has given a new impulse to the study of these compounds. In many cases, these structural studies were crucial for the development of the chemistry of a given compound. For example, zirconium phosphates and phosphonates with a layered structure of γ-type were first prepared in the 1960s, and have been the object of growing interest since 1990,[1] just after the publication of the structure of γ-zirconium phosphate.[17] Different from α and γ-zirconium phosphonates, in which the structure of layers, and therefore the spatial organization of organic groups attached to them, remain essentially unchanged, the subclass of zirconium aminophosphonates has revealed an unexpected structural richness, so that a reliable prediction of the global crystal structure on the basis of stereochemistry of constituting building blocks is still far to be achieved. On the other hand this structural richness represents a vast reservoir of new possibilities, and new chances for the materials chemistry.

This chapter will present a survey of the recent results on this class of compounds and will mainly concern the contribution from our laboratory.

2.2 α-Zirconium Phosphonates

2.2.1 Layered α-Zirconium Phosphonates

The chemistry of layered α-zirconium phosphonates began after 1964, in which Clearfield and Stynes obtained and characterized the first crystalline member of the α-zirconium phosphates with formula $Zr(HPO_4)_2 \cdot H_2O$ (Figure 2.1).[18]

The structure of this archetype compound is layered and it is constituted of planes of zirconium octahedra ZrO_6 and monohydrogen phosphate tetrahedral groups (HPO_4). Each Zr atom is coordinated by six phosphate groups while three oxygens of each phosphate link three different Zr atoms. The fourth oxygen is protonated and points toward the interlayer region that is occupied by one water molecule per mol of compound and linked via a hydrogen bond to the protonated P–O groups.

From a structural point of view, Zr atoms are not placed in the same plane and the phosphate groups are not equivalent. Each phosphate is surrounded by six phosphate groups and bonds three different metal atoms.

Looking at the layer surface, the free area around each phosphate group is about 24 Å^2, while the P–OH groups are spaced at about 5.3 Å, as shown in Figure 2.2.

α-Layered zirconium phosphonates have general formula $Zr(O_3P\text{-}R)_2 \cdot nS$ where R is an organic group (such as Ph, Me, Et, $-C_nH_{2n+1}$, etc.) and S is a polar solvent intercalated in the interlayer region.

The synthetic procedures for obtaining α-Zr phosphonates are similar to those used for α-ZrP, that is, refluxing amorphous or semicrystalline precipitates, or by

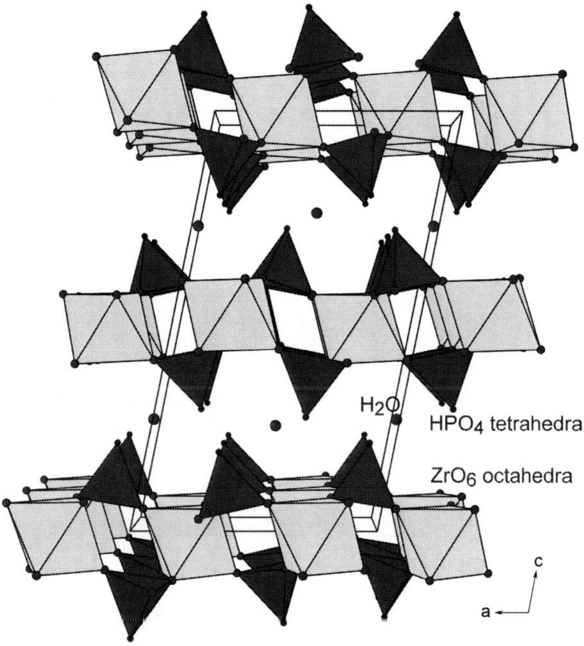

Figure 2.1 Polyhedral representation of the structure of α-ZrP.

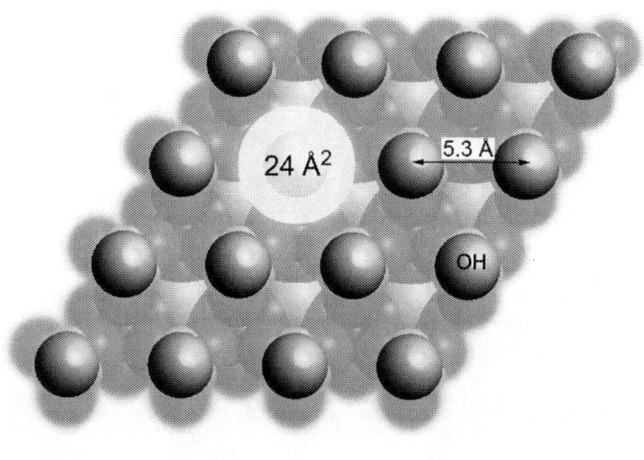

Figure 2.2 Van der Waals representation of the surface of a layer of α-ZrP. The white region indicates the free area associated to each phosphate or phosphonate site. (Reproduced with permission from *Microporous Mesoporous Mater.*, 2008, **107**, 58. Copyright 2008 Elsevier.)

using the so-called HF method[19] that consists in a previous complexation of Zr ions with fluoride. Zr is then slowly released into a solution containing the previously dissolved phosphonic acid by thermal decomposition of the soluble zirconium fluorocomplexes.

From a structural point of view, α-zirconium phosphonates are closely related to each other, because they are made of inorganic layers with a common structure of α-type and the organic groups –R, that in phosphonates substitute the –OH phosphatic groups, placed in the interlayer region.

It is evident that, due to the large choice of organic groups that can be attached to the inorganic layered backbone, a very high number of compounds can be, in principle, obtained. The only steric requirement in order to observe the formation of α-Zr phosphonates, is that the organic groups covalently attached to the layer does not exceed the 24 Å^2 cross section around each R–PO_3 group.

In order to understand the role played by structural invariance among this class, we may recall that in the past, the structural model of several zirconium phosphonates was proposed, on the basis of the structure of α-type layers, the rough molecular geometry of the R groups employed and the experimental distance between two adjacent inorganic layers, this last obtained by X-ray powder diffraction (XRPD) analysis. Some years later, these models have been found to be basically correct. As an example one of the first α-zirconium phosphonates characterized was the phenylphosphonate with formula $Zr(PO_3C_6H_5)_2$, shown in Figure 2.3.

A first structural model of this compound was proposed by Alberti in 1978,[20] whereas the final structure was solved 15 years later by Clearfield from XRPD data.[21] The compound crystallizes in the $C2/c$ space group with cell parameters $a = 9.0985(5)$ Å, $b = 5.4153(3)$ Å, $c = 30.235(8)$ Å, $\beta = 101.33(5)°$. The a and b

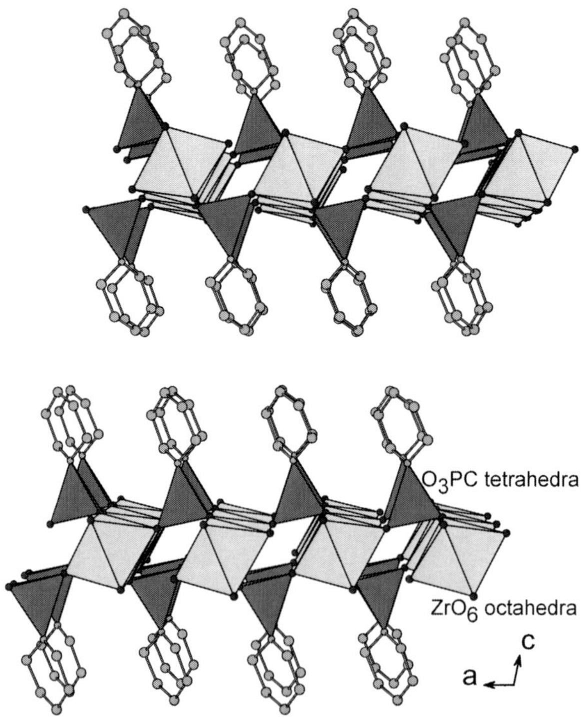

Figure 2.3 View of the crystal structure of α-Zr phenylphosphonate.

Table 2.1 Composition and interlayer distance of some representative layered α-zirconium phosphonates.

Compound	*Interlayer distance/Å*	*Ref.*
$Zr(O_3PCH_3)_2$	7.8	22, 23
$Zr(O_3PC_3H_7)_2$	14.0	24
$Zr(O_3PC_6H_{13})_2$	19.2	36
$Zr(O_3PC_{12}H_{25})_2$	32.4	22
$Zr(O_3PC_{22}H_{45})_2$	56.2	22
$Zr(O_3PCH_2OH)_2$	9.2	25
$Zr(O_3PCH_2COOH)_2$	11.3	25, 32
$Zr(O_3PC_2H_4COOH)_2$	13.4	26, 32
$Zr(O_3PC_2H_4NH_3Cl)_2$	14.3	27
$Zr(O_3PC_2H_4NH_2)_2$	12.3	27
$Zr(O_3PC_6H_5)_2$	14.7	20
$Zr(O_3PCH_2SO_3H)_2$	15.4	22
$Zr(O_3POH)_{0.5}(O_3PC_2H_4COOH)_{1.5} \cdot 0.5H_2O$	13.8	28
$Zr(O_3PCH_2OH)_{0.75}(O_3PC_2H_4COOH)_{1.25}$	13.6	29
$Zr(O_3PCH_2OH)_{1.03}(O_3PC_6H_4SO_3H)_{0.97} \cdot 4.9H_2O$	19.6	31
$Zr(O_3PCH_3)_{1.2}(O_3PC_6H_4SO_3H)_{0.8}$	16.5	30, 31
$Zr(O_3PC_2H_5)_{1.15}(O_3PC_6H_4SO_3H)_{0.85} \cdot 3.7H_2O$	18.5	31

parameters are very similar to those of α-ZrP while the *c* axis is larger because the interlayer region is expanded in order to accommodate phenyl groups.

The phenyl groups are tilted relative to the layers by 30° and Zr atoms are constrained by symmetry to lie in the same plane. The interlayer distance of this compound is 14.82 Å.

During the years a large number of α-Zr phosphonates containing organic groups of different dimension, reactivity and polarity has been presented.[1] Table 2.1 reports the composition and interlayer distance of some representative layered α-Zr phosphonates.

The knowledge of the structure of layers allows an approximate prediction on the arrangement of pendant organic groups and therefore helps the rationalization of different experimental behaviours. For example in Zr carboxyalkylphosphonates, of general formula $\alpha\text{-}Zr(O_3P\text{–}(CH_2)_n\text{–}CO_2H)_2$, the large difference in protonic conductivity between the derivatives with an odd or even n (number of methylene groups in the alkyl chain) has been ascribed to the different arrangement of terminal carboxylic groups in the interlayer region, that was predicted also in absence of accurate structural data, as shown in Figure 2.4.[31] In particular, the conductivity of samples with $n = 1$, 3 and 5 was up to 3 orders of magnitude higher than that of compounds with $n = 2$ and 4.

The use of mixtures of two phosphonic acids with different R groups yields compounds with a mixture of the two pendant R groups, according to the general formula $Zr(O_3P\text{–}R)_x(O_3P\text{–}R')_{2-x} \cdot nS$. The x value can change continuously only if the crystallinity of the compound is poor and the two groups R and R′ are of comparable size and chemically similar. If the crystallinity is high and the two groups are very different in size and chemical behaviour, the range of phase solubility decreases and x assumes discontinuous values. When x approaches to its maximum for a given system, asymmetrical layers may

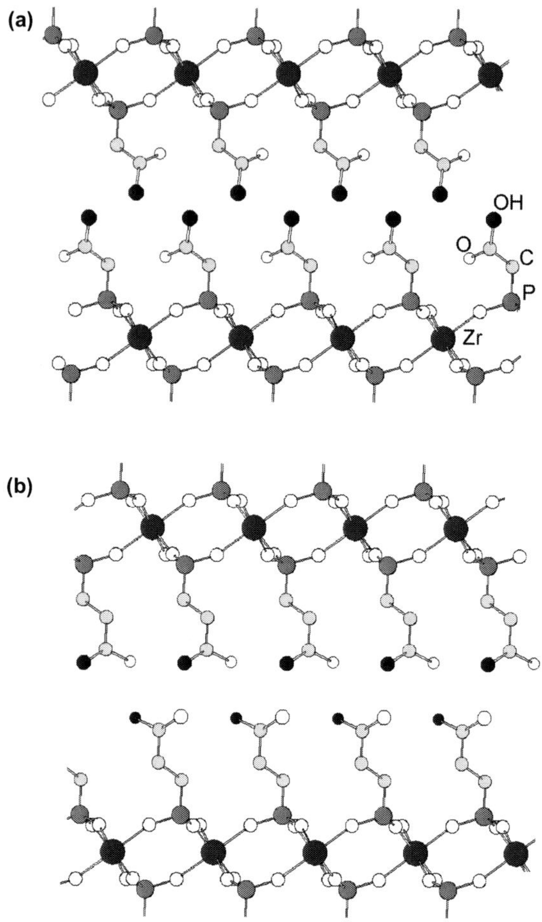

Figure 2.4 Possible arrangement of interlayer organic groups in Zr carboxymethylphosphonate (a) and Zr carboxyethylphosphonate (b).

form, where one side is richer in R groups and the other in R′. The XRPD pattern of these compounds shows a reflection relative to a *d*-spacing that is the sum of the two interlayer distances for the pure R phase and the pure R′ phase.

The mixed Zr derivatives containing sulfophenylphosphonate groups are of special interest. Due to the presence of the strong acidic $-SO_3H$ groups these compounds are good protonic conductors. As the phosphate and sulfophenylphosphonate relative proportion in the solid can be easily changed, a general increasing of the conductivity is observed when the sulfonate group fraction is increased.[32] These materials have also been used with good results as heterogeneous acid catalysts for a large number of organic reactions of various kinds.[10]

Mixed zirconium phosphate phosphonates of general formula $Zr(PO_3OH)_{2-(x+y)}(PO_3R)_x(PO_3R')_y$ (R, R′ = Me, Ph and Pr) with a high surface area (200–380 $m^2 g^{-1}$) have been also used as Brønsted acid catalysts on direct aza-Diels–Alder reactions.[33] The hydrophobic groups on the catalyst surface was found to favour the diffusion of reagents toward the acidic sites, increasing the catalytic activity of these solids compared with that of α-ZrP.

2.2.2 Pillared α-Zirconium Phosphonates

Pillared zirconium diphosphonates with α-type structure have general formula $Zr(O_3P–R–PO_3)$ and they can be obtained in the same synthetic conditions of layered derivatives, by using selected diphosphonic groups in the synthesis.

As described in the previous paragraph, if the cross section of the internal R group is lower than 24 $Å^2$ the formation of an α-type structure is generally observed.

In pillared α-zirconium phosphonates two adjacent inorganic layers are covalently joined to each other with the diphosphonic group that links the zirconium atoms with the terminal $R–PO_3$ heads with the same connectivity previously described for the α-layered phosphonates.

Many zirconium phosphonates containing diphosphonic groups of different length have been prepared (for instance phenyl-, diphenyl-, terphenyl- diphosphonates) leading to a linear increase of the distances between two adjacent layers.[34–36]

However, it should be noted that, since the distance between two neighbour P atoms on one side of the layer is 5.3 Å, the organic pillars remain too close each others and no free space in the interlayer region is available, as shown in the model of Figure 2.5.

In order to increase the interlayer porosity, mixed-component Zr derivatives containing a fraction of small monophosphonic groups such H, OH or Me, randomly interposed between the pillaring diphosphonic groups, were prepared.[34–36] However, due to the intrinsic disorder of the pillaring groups in these compounds it was impossible to have a complete structural control. Furthermore, when increasing the fraction of small groups, such as phosphite, the

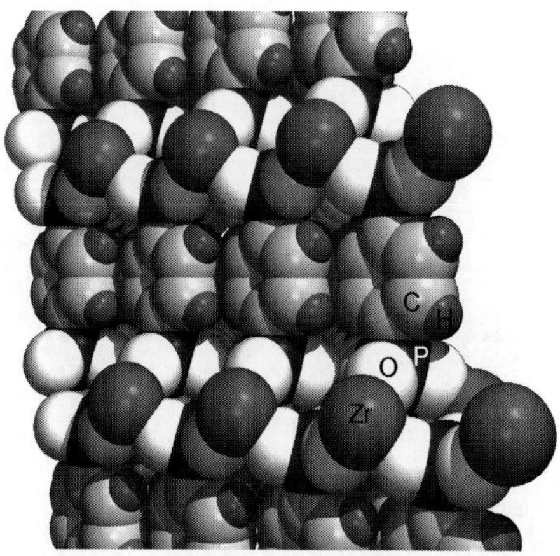

Figure 2.5 Structural model of a pillared Zr phenyldiphosphonate. Due to the crowding of pillars, in the interlayer region there is no porosity.

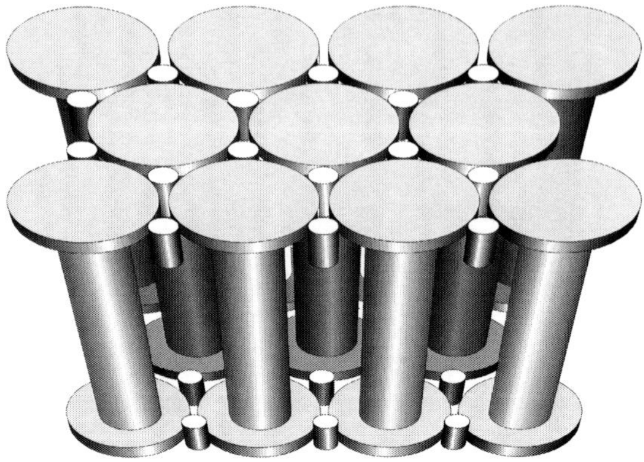

Figure 2.6 Formation of intercolumnar porosity due to the packing of pillars with bases alternating with small groups. (Reproduced with permission from *Microporous Mesoporous Mater.*, 1998, **21**, 297. Copyright 1998 Elsevier.)

percentage of pure phosphite phase was higher than that of the mixed phosphite/diphosphonates one and the measured interlayer microporosity was negligible.[37]

In order to increase the microporosity, a different strategy, based on the use of "pillars with bases", was adopted. With this situation two adjacent positions, in the α-type structure, cannot be occupied by the large pillaring groups, because they have a base cross-section larger than 24 $Å^2$. Therefore six small phosphite groups are expected to surround each pillar. If the pillar cross-section is smaller in the central part, free space, and hence microporosity, can be generated in the intercolumnar region, as shown in Figure 2.6. Using this strategy the first α-pillared Zr derivative with a high surface area (405 $m^2 g^{-1}$), a good micropore volume (0.16 $cm^3 g^{-1}$) and a narrow pore size distribution centred at about 5 Å was prepared.[38] It had formula $Zr(HPO_4)_{1.33}(PO_3–R–O_3P)_{0.33}$ where R was 3,3′,5,5′-tetramethylbiphenyl. In this compound the presence of the methyl groups in *ortho*-position created a steric hindrance at the bases of the pillar that induced the alternation of phosphonic pillars and phosphite groups (Figure 2.7).

Table 2.2 reports the composition and interlayer distance of some representative pillared α-Zr diphosphonates.

2.3 γ-Zirconium Phosphate Phosphonates

2.3.1 Layered γ-Zirconium Phosphate Diphosphonates

Another important archetype structure for zirconium phosphonates is the γ-type structure, and γ-zirconium phosphate, with formula γ-$ZrPO_4(H_2PO_4) \cdot 2H_2O$ (hereafter γ-ZrP), is the progenitor. Its structure[17] consists of a rigid framework of ZrO_6 octahedra placed in two different planes and joined to each other with tetradentate PO_4 inside and H_2PO_4 groups outside these planes (Figure 2.8). Each dihydrogenphosphate group bears two acidic OH groups on the external part of the layer. Crystallization water molecules are located in the

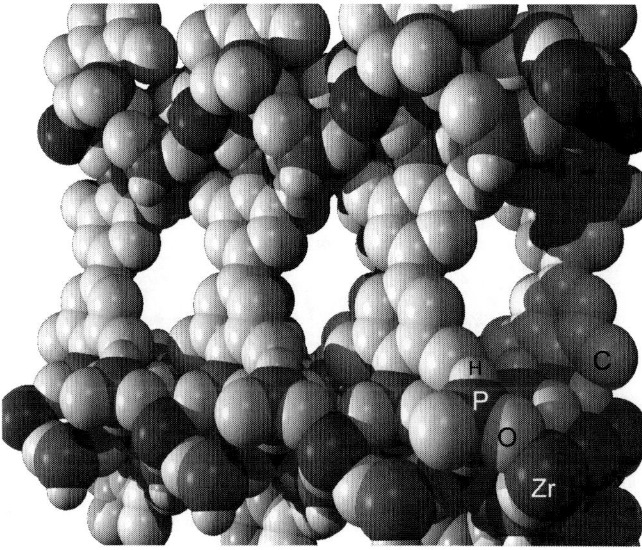

Figure 2.7 Structural model of the microporous α-Zr phosphite diphosphonate of formula $Zr(HPO_3)_{1.2}(O_3P\text{-}R\text{-}PO_3)_{0.4}$ (R = 3,3′,5,5′-tetramethylbiphenyl).

Table 2.2 Composition and interlayer distance of some representative pillared α-zirconium phosphonates.

Compound	*Interlayer distance/Å*	*Ref.*
$Zr(O_3PC_2H_4PO_3)$	7.4	34
$Zr(O_3PC_{10}H_{20}PO_3)$	17.2	34
$Zr(O_3PC_6H_4PO_3)$	9.6	34
$Zr(O_3PCH_2C_6H_4CH_2PO_3)$	10.8	34
$Zr(O_3PC_6H_4C_6H_4PO_3)$	13.9	34
$Zr(O_3PC_6H_4C_6H_4C_6H_4PO_3)$	18.5	34
$Zr(HPO_3)_{1.2}(O_3P\text{-}R\text{-}PO_3)_{0.4}$[1]	14.5	38

[1]R = 3,3′,5,5′-tetramethylbiphenyl

interlayer region and form a network of hydrogen bonds between them and OH groups on both surfaces of layers.

Since the tetradentate phosphate group is an essential building block for this kind of structure, it is evident that pure γ-zirconium phosphonates cannot exist. However, it is possible to replace the interlayer H_2PO_4 groups by monovalent phosphonate or phosphinate anions to obtain layered inorganic–organic hybrids with general formula $ZrPO_4(O_2PRR') \cdot n$S where R and R′ can be H, OH or an organic group, and S is the intercalated solvent (Figure 2.9). These hybrid materials are obtained from a simple topotactic reaction by contacting the original γ-ZrP microcrystals with solutions of phosphonic acids in 1:1 acetone–water mixtures at 80 °C.[39–41] This technique was first used by Yamanaka *et al.*[42,43] since 1975, and was extensively applied by our laboratory in the 1990s, after the

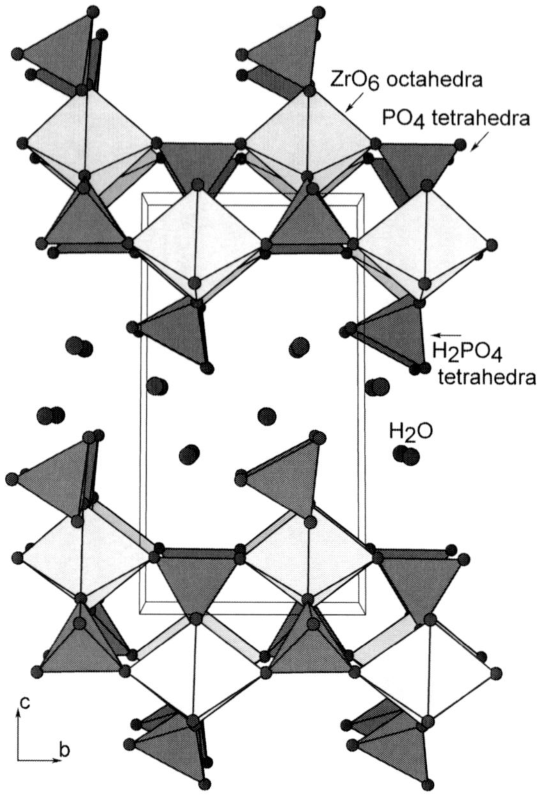

Figure 2.8 Polyhedral representation of the structure of γ-$ZrPO_4(H_2PO_4)\cdot 2H_2O$.

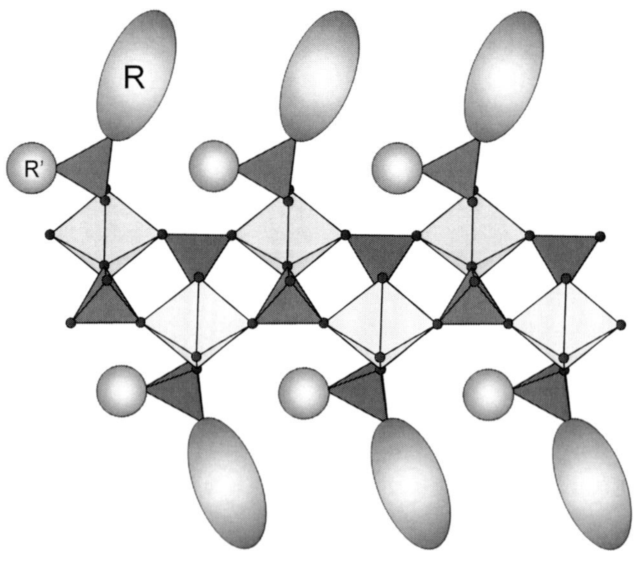

Figure 2.9 Schematic model of a generic γ-Zr phosphate phosphonate.

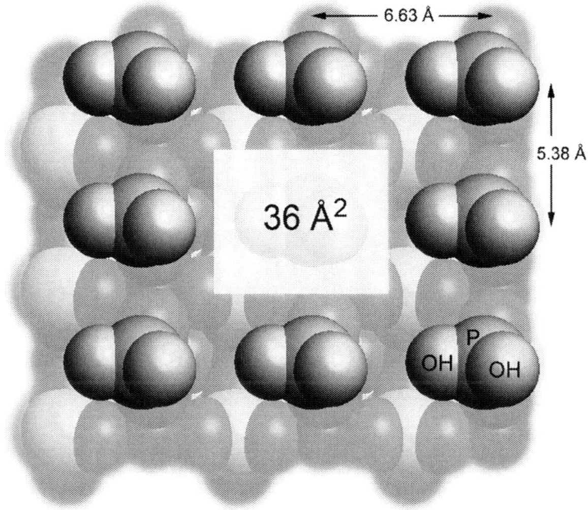

Figure 2.10 Van der Waals representation of the surface of a layer of γ-ZrP. The white region indicates the free area associated to each phosphate or phosphonate site. (Reproduced with permission from *Microporous Mesoporous Mater.*, 2008, **107**, 58. Copyright 2008 Elsevier.)

discovery of the structure of γ-ZrP,[44] and when we realized that exfoliation of γ-ZrP was taking place in the experimental conditions used for the reaction.[45]

Topotactic replacement of original H_2PO_4 groups occurs gradually with the initial formation of a partially exchanged phase, according to the following scheme:

$$ZrPO_4(H_2PO_4) \rightarrow ZrPO_4(H_2PO_4)_{1-x}(O_2PRR')_x \rightarrow ZrPO_4(O_2PRR')$$

The conversion may be complete only when geometrical features of R and R′ organic groups fulfil those of γ-type structure. Figure 2.10 shows the surface of a layer in which H_2PO_4 sites have been highlighted. Each of them has about 36 Å^2 of free area around it. For example, we found that the use of linear chain alkylphosphonates (R = OH, R′ = n-alkyl) can complete the conversion and produce 100% exchanged phases, while dimethylphosphinate derivatives, where R = R′ = Me, have a maximum conversion of 67%.[46] The role of steric hindrance in the determination of the composition of these hybrid γ-Zr phosphate phosphonates was elucidated by the behaviour of phenylphosphonic and phenylphosphinic groups.[41] With the former building block (R = OH, R′ = Ph) we observed a 67% exchange (Figure 2.11), while the latter (R = H, R′ = Ph) gave a 100% exchange. Table 2.3 reports some γ-Zr phosphate phosphonates obtained in this laboratory.

The retention of the γ-type structure after the topotactic reaction was confirmed by a structural study on the propylphosphonate derivative, with formula $ZrPO_4(HO_3PC_3H_7)$, for which samples of good crystallinity were obtained.[47] Depending on the conditions of preparation, two different phases with the same composition and very similar global structure were obtained, although one is

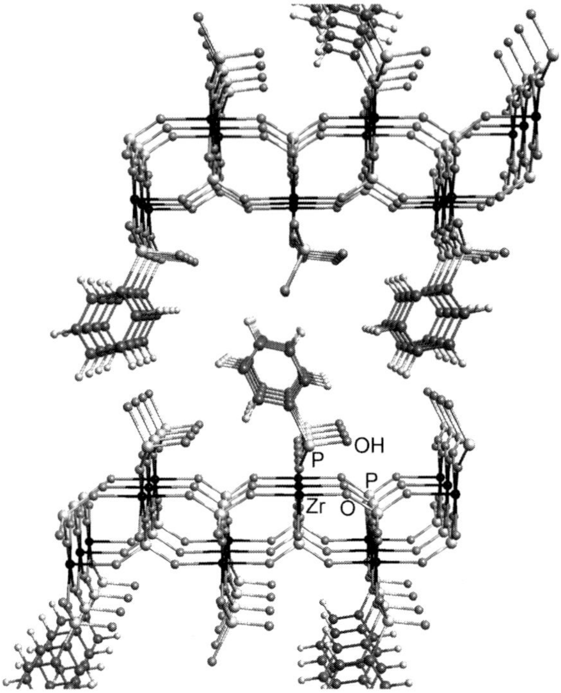

Figure 2.11 Structural model of a γ-Zr phosphate phenylphosphonate of formula $ZrPO_4(H_2PO_4)_{0.5}(HO_3PC_6H_{11})_{0.5}$.

Table 2.3 Composition and interlayer distance of some representative layered γ-zirconium phosphonates.

Compound	*Interlayer distance/Å*	*Ref.*
$ZrPO_4[O_2PH(OH)] \cdot 2H_2O$	12.2	31
$ZrPO_4(O_2PH_2) \cdot H_2O$	8.8	32
$ZrPO_4(HO_3PCH_3) \cdot 2H_2O$	12.8	33
$ZrPO_4(HO_3PC_3H_7) \cdot 1.2H_2O$	15.1	37
$ZrPO_4(H_2PO_4)_{0.33}[O_2P(CH_3)_2]_{0.67} \cdot H_2O$	10.3	32
$ZrPO_4(H_2PO_4)_{0.33}(HO_3PC_6H_5)_{0.67} \cdot 2H_2O$	15.4	33
$ZrPO_4(H_2PO_4)_{0.33}(HO_3PC_6H_{11})_{0.67} \cdot H_2O$	16.9	37
$ZrPO_4[O_2PH(C_6H_5)]$	15.1	1

orthorhombic ($a = 30.212(2)$ Å, $b = 6.6128(4)$ Å and $c = 5.3612(3)$ Å, space group $Pna2_1$) and the other is monoclinic ($a = 5.3593(4)$ Å, $b = 6.6217(4)$ Å and $c = 30.695(2)$ Å, $\beta = 98.810(3)°$, space group $P2_1/c$). Their structure was determined *ab initio* by XRPD data (Figure 2.12). In these structures, propyl groups are not in their extended conformation directed toward the interlayer region, but they tend to remain near the layer and to fill the space present in the γ-type structure between two adjacent phosphonate groups. In γ-ZrP this space is occupied by a water molecule interacting with one of the hydroxyl groups.

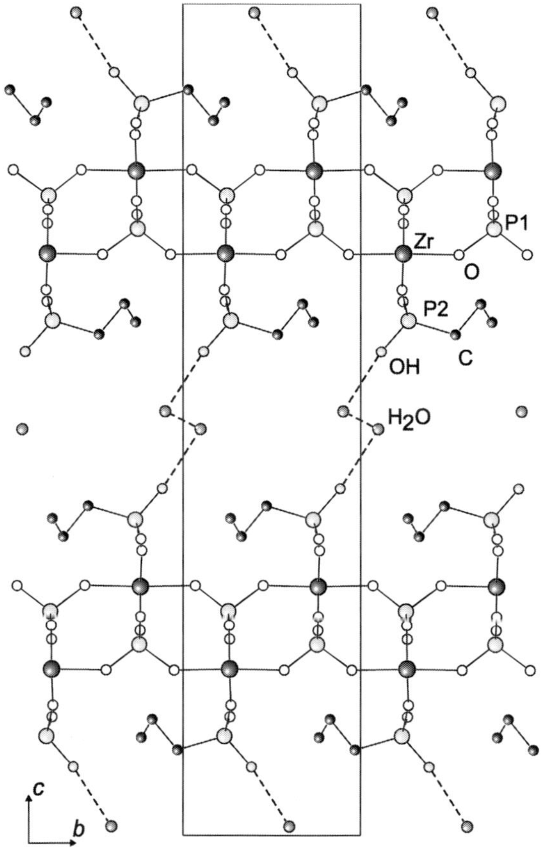

Figure 2.12 Structure of the monoclinic phase of γ-Zr phosphate propylphosphonate, of formula $ZrPO_4(HO_3PC_3H_7)$. (Reproduced with permission from Reference 47. Copyright 1998 Elsevier.)

In the present compound, the hydroxyl group of phosphonate is directed towards the interlayer region and forms a hydrogen bond with one water molecule. These water molecules are located as a corrugated monofilm approximately halfway between two adjacent layers.

γ-derivatives with a high solid state proton conductivity of interest for application in solid state electrochemical devices have been obtained by the replacement of the interlayer H_2PO_4 groups with $HO_3PC_6H_4$–SO_3H. The γ-zirconium phosphate sulfophenylphosphonate obtained has a conductivity of 0.05 S cm^{-1} at 100 °C and 90% relative humidity.[48]

2.3.2 Pillared γ-Zirconium Phosphate Diphosphonates

When γ-ZrP microcrystals are contacted with solutions of diphosphonic acids under the same conditions described above, an analogous topotactic reaction of exchange of H_2PO_4 groups with diphosphonate groups takes place, with the

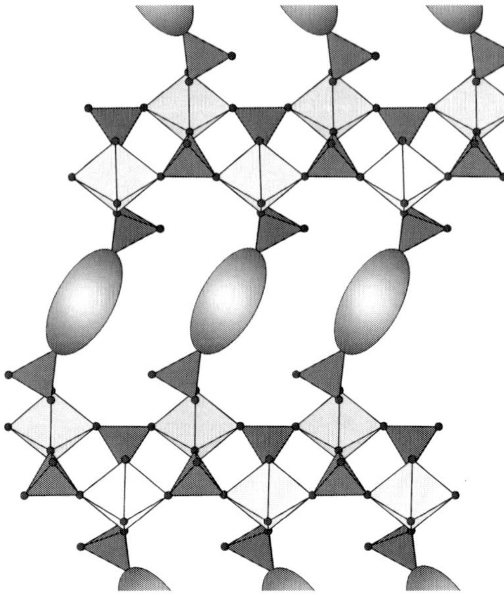

Figure 2.13 Schematic model of a pillared γ-Zr phosphate diphosphonate.

Table 2.4 Composition and interlayer distance of some representative pillared γ-zirconium phosphonates.

Compound	*Interlayer distance/Å*	*Ref.*
$ZrPO_4(H_2PO_4)_{0.56}(HO_3PC_6H_4PO_3H)_{0.22} \cdot 2.2H_2O$	13.1	40
$ZrPO_4(HO_3PC_4H_8PO_3H)_{0.5} \cdot H_2O$	13.0	38
$ZrPO_4(H_2PO_4)_{0.75}[HO_3P(C_6H_4)_2PO_3H]_{0.125}$	16.3	42
$ZrPO_4(H_2PO_4)_{0.8}[HO_3P(CH_2)_{10}PO_3H]_{0.1} \cdot 3H_2O$	17.4	43
$ZrPO_4(H_2PO_4)_{0.76}(HO_3P\text{-DEG-}PO_3H)_{0.12} \cdot H_2O$[1]	15.7	47
$ZrPO_4(H_2PO_4)_{0.76}(HO_3P\text{-PEG-}PO_3H)_{0.12} \cdot H_2O$[2]	17.6	47
$ZrPO_4(H_2PO_4)_{0.6}(HO_3P\text{-BC})_{0.16}(A \cdot H_2O)_{0.24} \cdot 2.6H_2O$[3,4]	27.1	48
$ZrPO_4(H_2PO_4)_{0.28}(HO_3P\text{-DBC-}PO_3H)_{0.24}(A \cdot H_2O)_{0.24} \cdot 2.2H_2O$[3,5]	21	49

[1]DEG = $CH_2(CH_2OCH_2)_3CH_2$
[2]PEG = $CH_2(CH_2OCH_2)_6CH_2$
[3]A = F, OH
[4]BC = benzo 15-crown-5
[5]DBC = dibenzo 18-crown-6

formation of pillared zirconium phosphate diphosphonates, with general formula $ZrPO_4(HO_3P–R–PO_3H)_{0.5} \cdot nS$, in which adjacent γ-structured layers are covalently joined by the organic groups belonging to the diphosphonate moieties (Figure 2.13).[49] Table 2.4 reports some of these derivatives obtained in this laboratory. Also in this case, a complete XRPD structural study, performed on a derivative having good crystallinity, the Zr phosphate butyldiphosphonate, with formula $ZrPO_4$-$(HO_3P–C_4H_8–PO_3H)_{0.5} \cdot H_2O$, confirmed the topotactic nature of the exchange

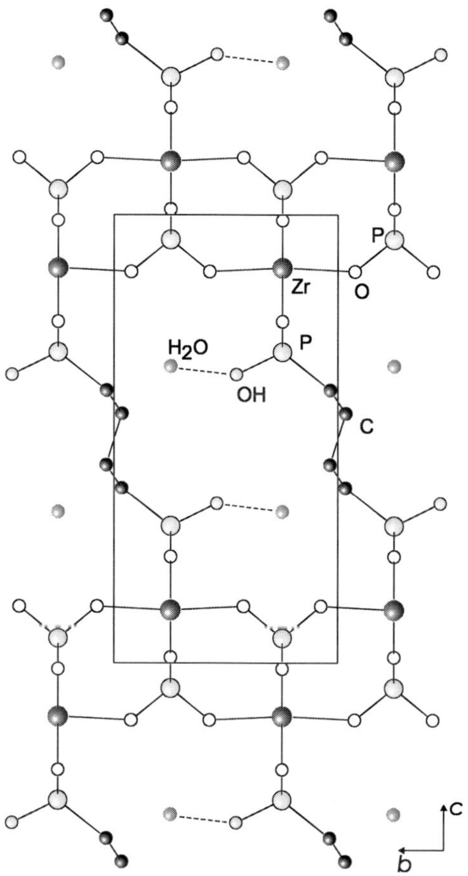

Figure 2.14 Structure of γ-Zr phosphate butyldiphosphonate of formula $ZrPO_4(HO_3P\text{-}C_4H_8\text{-}PO_3H)_{0.5}\cdot H_2O$ viewed down the *a* axis. (Reproduced with permission from Reference 47. Copyright 1998 Elsevier.)

reaction, since the γ-layered structure remained essentially unchanged.[47] This compound crystallizes in the orthorhombic system, $a = 5.3797(2)$ Å, $b = 6.6244(2)$ Å and $c = 13.0111(5)$ Å, space group *Pmmn* (Figure 2.14).

The structure was found to be in good agreement with the more geometrical model that we elaborated some years before on the basis of the structure of γ-ZrP.[49]

We showed that when replacement of the original H_2PO_4 groups with diphosphonate pillars is only partial, the exchanged groups are able to distribute randomly in the interlayer region of the final γ-derivative. This is because exfoliation of γ-ZrP takes place in the experimental conditions used for the reaction, and the topotactic replacement can freely occur at the solid–solvent interface.[50] Therefore, by the right choice of pillaring building blocks, and by controlling the degree of exchange it was possible to finely modulate the geometry and reactivity of the interlayer region. For example we were able to induce microporosity in the interlayer region using rigid pillars with a low degree of pillaring, such as biphenyldiphosphonate groups exchanged at about 25% of pillaring percentage.[51] This derivative showed a permanent porosity, stable as regards its dehydration; a specific surface area of $320\,m^2\,g^{-1}$ and micropore volume of $0.12\,cm^3\,g^{-1}$, with a narrow micropore distribution with an average diameter of 5.8 Å (see Figure 2.15). A model structure of this compound is shown in Figure 2.16.

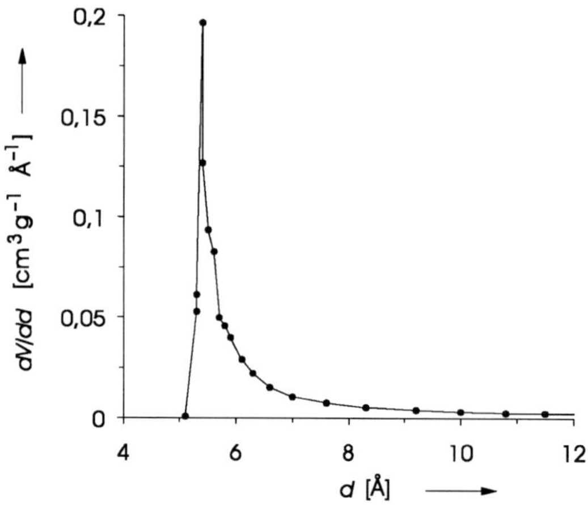

Figure 2.15 Horvath–Kawazoe micropore distribution curve for the γ-pillared compound of formula $ZrPO_4(H_2PO_4)_{0.75}[HO_3P(C_6H_4)_2PO_3H]_{0.125}$. (Reproduced with permission from Reference 51. Copyright 1994 Wiley-VCH.)

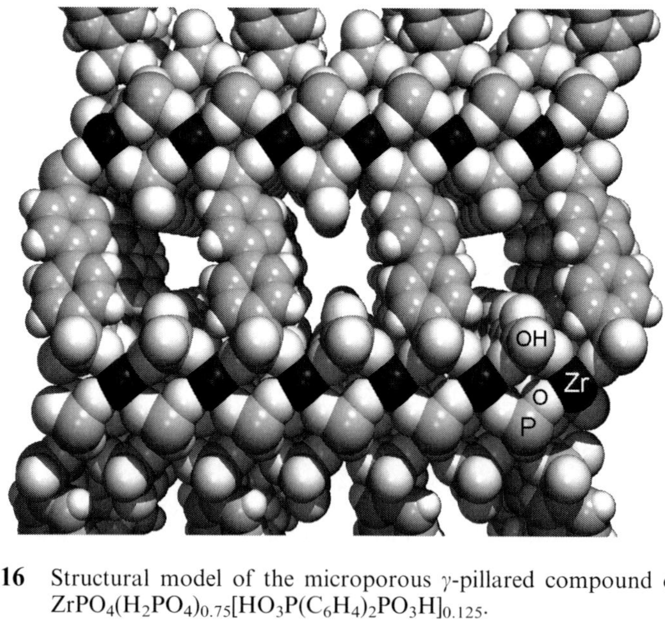

Figure 2.16 Structural model of the microporous γ-pillared compound of formula $ZrPO_4(H_2PO_4)_{0.75}[HO_3P(C_6H_4)_2PO_3H]_{0.125}$.

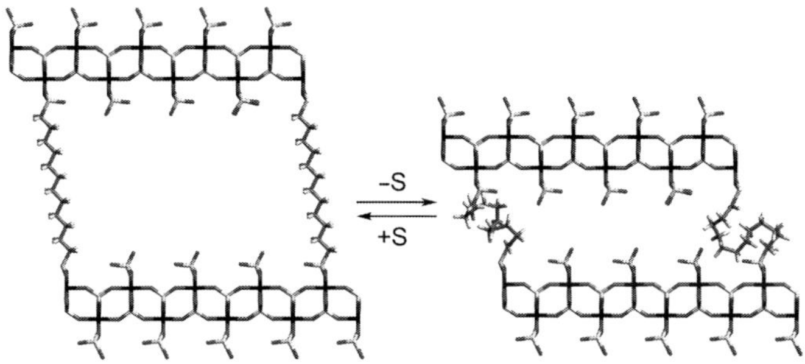

Figure 2.17 Reversible shortening and elongation of γ-pillared compound with nonrigid pillars at low (about 25%) pillaring percentage.

On the other hand, the use of non-rigid pillars, such as long alkyl chains, at a very low degree of pillaring, allowed the preparation of γ-derivatives with chained layers, whose interlayer distance could assume continuous values between two extremes (corresponding to the elongated and the collapsed chain conformations) thanks to the large number of degrees of freedom of the pillars, and depending on the number and nature of intercalated species in the intercolumnar space. As an example, the interlayer distance of the compound with formula $ZrPO_4(H_2PO_4)_{0.8}[HO_3P(CH_2)_{10}PO_3H]_{0.1} \cdot 3H_2O$ can reversibly change from 17.4 to 14 Å as a consequence of the dehydration–rehydration process (Figure 2.17).[52] Therefore, this ability to reversibly change the available space depending on the guest dimensions (jokingly called *accordion effect*) anticipates the "dynamic porosity",[53] or "breathing effect"[54] concepts that were fruitfully introduced many years later in the *Metal Organic Frameworks* field. These interlayer spaces could be seen as *molecular vessels* in which selective reactions could be carried out (molecular reactors) and eventually promote molecular recognition processes.

Other good examples of adaptive structures are the compounds based on γ-Zr phosphate diphosphonate containing non-rigid polyethylenoxide pillars, that were developed in collaboration with the Departamento de Química Organica, of the Universidad Autonoma de Madrid. A series of compounds of general composition $ZrPO_4(H_2PO_4)_{0.76}(HO_3P{-}R{-}PO_3H)_{0.12} \cdot nH_2O$, where the pillar R are diethylene glycol (DEG) or pentaethylene glycol (PEG), with a very low pillaring degree, were prepared.[55] These compounds were able to reversibly intercalate methylamine and water at a specific pH value. Due to this, a dramatic reversible change of the interlayer distance of about 6 Å for DEG (from 15.7 to 21.6 Å) and 9 Å for PEG derivatives (from 17.6 to 26.7 Å) was observed (Figure 2.18). In conclusion, these systems were able to adapt their interlayer space upon a chemical stimulus, that is a pH change of the external solution.

Finally, layered and pillared compounds with γ-structure, containing crown ethers in the interlayer region have been obtained (Figure 2.19).[56,57]

2.4 Zirconium Aminophosphonates

The chemistry of zirconium phosphonates found new stimuli when new families of zirconium derivatives with amino phosphonic acids as building blocks were

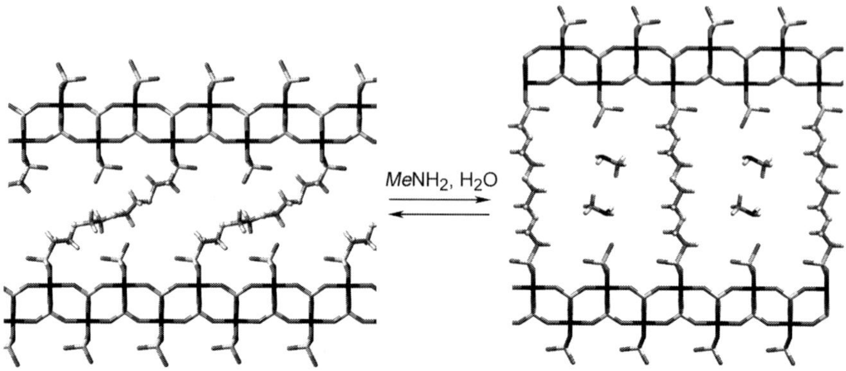

Figure 2.18 Schematic representation of the reversible enlargement – shortening of the interlayer space of the DEG-pillared derivative following intercalation – deintercalation of methylamine and additional water.

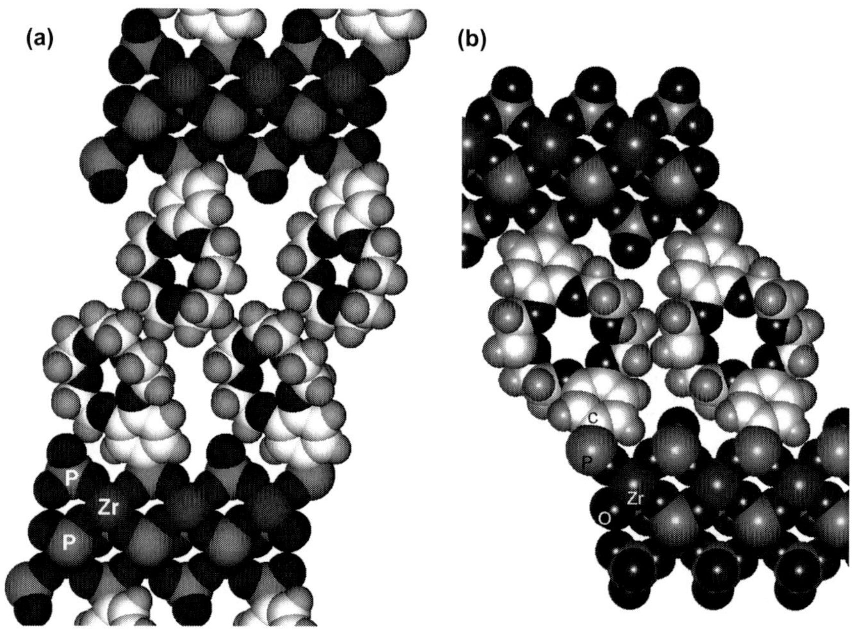

Figure 2.19 Structural model of a γ-layered derivative containing interlayer benzo 15-crown-5 groups (a); and a γ-pillared derivative containing dibenzo 18-crown-6 pillaring groups (b).

prepared. These compounds showed a wider range of structural arrangements, if compared with the lamellar α-type Zr-phosphonates, disclosing new opportunities for their application in many fields. On the other hand, the structural predictability of these materials is very low, starting a tough challenge in crystal engineering.

Most of the phosphonic building blocks used are not commercially available, but they can be synthesized with simple procedures. The best known is the Moedritzer–Irani[58] synthesis: with this reaction it is possible to covalently link the phosphonic moieties to primary or secondary amines, according to the following general schemes:

$$R_2{-}NH + 2CH_2O + H_3PO_3 \rightarrow R_2{-}N(CH_2PO_3H_2) + H_2O \quad (2.1)$$

$$R{-}NH_2 + 4CH_2O + 2H_3PO_3 \rightarrow R{-}N(CH_2PO_3H_2)_2 + 2H_2O \quad (2.2)$$

Reactions (2.1) and (2.2) lead to the formation of amino mono- or bis(methylphosphonic) acids respectively. Obviously each primary or secondary amino group available allows one or two phosphonic moieties to be linked with the formation of poly-phosphonic building blocks.

Since the mechanism involves the formation of an imine intermediate, tertiary amines and heterocyclic aromatic amines, such as pyridine, are not suitable for this reaction. Moreover this reaction does not work with aromatic amines because the nucleophilic character of the nitrogen atom is too low for the reaction with the formaldehyde.

Alternatively, it is possible to prepare phosphonic derivatives from heterocyclic aromatic amines by direct reaction with the diethyl 2-bromoethylphosphonate in aqueous medium.[59] This reaction leaves a positive charge on the nitrogen atom, that is balanced by a counter-ion (normally Cl^-) that could be incorporated into the structure, and introduces an ethylene group as spacer between the nitrogen and the phosphonic group.

The molecular arrangement of some phosphonic building blocks that can be obtained with these reactions are summarized in Figure 2.20. It is worth noting that for sake of simplicity the schemes do not take into account the real number of bonds of the nitrogen atoms: for instance N–R (or N–R–N) could also refer to heterocyclic groups or amino acids.

The building blocks of type (a) and (b) are built from secondary amino groups whereas those of type (c) and (d) are obtained from reaction with primary amines.

(b) and (c) are both diphosphonates but the phosphonic tetrahedra are placed on the opposite sides of the central organic group in type (b), whereas in type (c) they are linked to the same N- atom of the organic residue and thus placed at the same side of the molecule, with the formation of a "candlestick-like" branched structure.

Building block (d) is tetraphosphonic, and may be considered as a "combination" of (b) and (c).

The rational choice of the selected R groups influences the shape, the flexibility and the overall chemical reactivity of the final polyphosphonic building block.

A survey of zirconium derivatives obtained by using aminophosphonic groups and reported in the literature are described here. They can be divided into four main groups:

1) zirconium R-amino-*N*-alkylphosphonates;
2) zirconium R-amino-*N*,*N*-bis(alkylphosphonates);

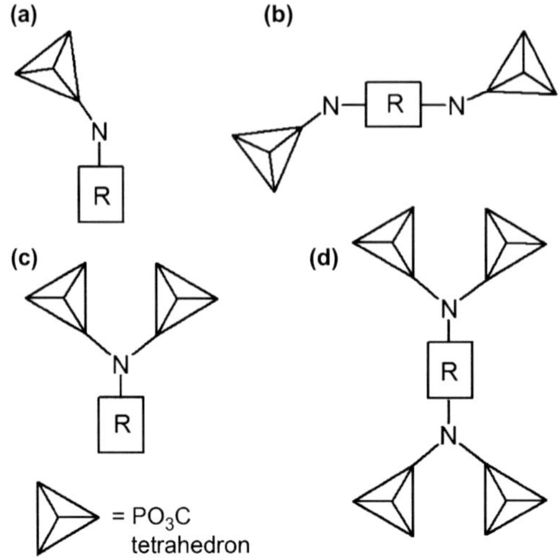

Figure 2.20 Schematic representation of the molecular geometry of the four types of amino phosphonate building blocks used: R-amino-*N*-methylphosphonate (a); R-diamino-*N*,*N*′-bis(methylphosphonate) (b); R-amino-*N*,*N*-bis(methylphosphonate) (c); R-diamino-*N*,*N*,*N*′,*N*′-tetrakis-(methylphosphonate) (d).

3) zirconium R-diamino-*N*,*N*′-bis(alkylphosphonates);
4) zirconium R-diamino-*N*,*N*,*N*′,*N*′-tetrakis(alkylphosphonates).

Table 2.5 reports the composition and crystallographic parameters of the zirconium aminophosphonates discussed in this chapter.

2.4.1 Zirconium R-amino-*N*-alkylphosphonates

The zirconium derivative of *N*-(phosphonomethyl)iminodiacetic acid (Zr-PMIDA) was reported by Zhang *et al.* in 1998.[60]

The compound, with formula $(NH_4)Zr[F_2][H_3\{O_3PCH_2NH(CH_2CO_2)_2\}_2]\cdot 3H_2O\cdot NH_4Cl$, crystallizes in the triclinic *P*-1 space group with the following parameters: $a = 10.743(2)$ Å, $b = 11.483(2)$ Å, $c = 5.330(1)$ Å, $\alpha = 90.63(2)°$, $\beta = 96.65(2)°$, $\gamma = 97.68(2)°$ and $Z = 1$.

The structure of this compound, shown in Figure 2.21, is 1D and is constituted from linear inorganic chains surrounded by the organic pendant groups. These inorganic chains are built from ZrO_4F_2 octahedra bridged by doubly connected PO_3C tetrahedra.

The zirconium atoms are located at the centres of symmetry and are bridged by two phosphonate groups. The carboxylate groups of the phosphonate are not involved in metal bonding and are projected into the inter-chain space. The carboxylate groups of adjacent chains are linked together by hydrogen bonds and by sharing a proton that is located at the centre of symmetry.

Table 2.5 Composition and crystallographic parameters of some zirconium aminophosphonates.

Compound	*a* / Å	*b* / Å	*c* / Å	α / °	β / °	γ / °	*Symmetry/ space group*	*Ref.*
$(NH_4)Zr[F_2][H_3\{O_3PCH_2NH(CH_2CO_2)_2\}_2] \cdot 3H_2O \cdot NH_4Cl$	10.743	11.483	5.330	90.63	96.65	97.68	Triclinic *P*-1	60
$ZrF_3(O_3PCH_2CH_2NC_5H_5)$	12.3634	9.3090	9.8077		112.81		Monoclinic $P2_1/c$	62
$ZrF(O_3PCH_2)_2NHC_5H_{11}$	13.1874	8.8723	10.9793		96.594		Monoclinic $P2_1/c$	63
$ZrHF(O_3PCH_2)_2NHC_3H_6CO_2$	12.9640	8.9900	10.7924		101.85		Monoclinic $P2_1/c$	64
$ZrF(O_3PCH_2)_2NHCH_2C_6H_5$	8.9429	9.1746	31.5654				Orthorhombic *Pbca*	65
$Zr_2H_4[(O_3PCH_2)_2N_2C_4H_8]_3 \cdot 9H_2O$	19.9400		9.5728				Hexagonal *R*-3	67
$ZrF_2(O_3PCH_2)_2(NH)_2C_4H_8$	8.7148	8.1731	9.0134		105.17		Monoclinic $P2_1/c$	67
$(ZrF_3)_2(O_3PCH_2CH_2NC_5H_4)_2 \cdot 2H_2O$	6.6679	12.5871	13.5837		101.73		Monoclinic $P2_1/c$	14, 59
$(ZrF)_2(PO_4)(O_3PCH_2CH_2NC_5H_4)_2F \cdot 3H_2O$	13.589	8.8351	9.2294		100.79		Monoclinic $P2_1/c$	68
$Zr(HPO_3CH_2)_2N\text{-}C_4H_8\text{-}N(CH_2PO_3H)_2 \cdot 4H_2O$	10.7337	12.8604	7.8919		104.91		Monoclinic $P2/c$	69
$Zr(PO_3CH_2)_2N\text{-}C_6H_{10}N(CH_2PO_3)_2Na_2H_2 \cdot 5H_2O$	18.9221	16.5047	7.8440		91.86		Monoclinic $C2/c$	69

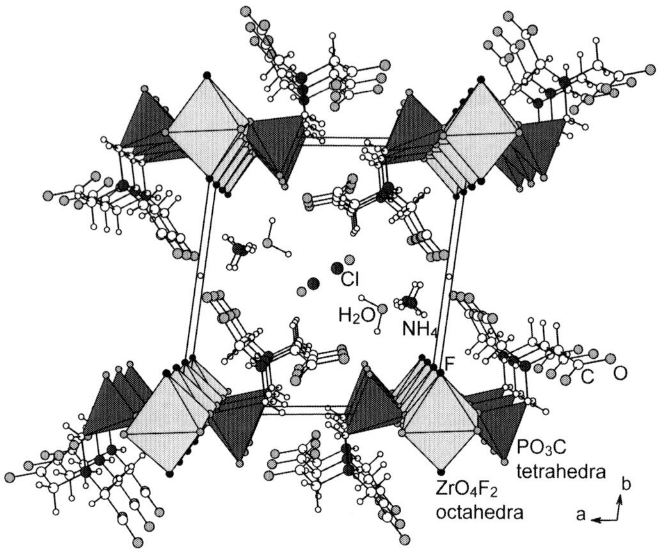

Figure 2.21 Polyhedral representation of the structure of Zr-PMIDA, viewed down the *a* axis. The chains run along the *c* axis.

In the inter-chain space one ionic couple of NH_4^+ and Cl^- is located together with three water molecules that interact via H-bonds with the non-bonded oxygen atoms belonging to the phosphonate and carboxylate moieties.

This compound is one of the first examples of a linear chain Zr phosphonate. The R group belonging to the monophosphonic moiety cannot fit into an α-type structure since it exceeds the 24 Å^2 free area required for the formation of this kind of layered structure. This feature may contribute to induce the formation of the observed low dimensional structure.

The authors also prepared other mixed phosphate/PMIDA and mixed methylphosphonate/PMIDA zirconium derivatives. The former is layered and its structure has been solved from synchrotron XRPD data.[61]

A zirconium fluoride derivative of the *N*-(2-phosphonoethyl)pyridinium (hereafter Zr-PYR) was described by Vermeulen *et al.* in 2002.[62] This compound, which was prepared by the HF method, maintained three fluorine atoms coordinated to Zr; one of them is required to balance the positive extra charge introduced by the pyridinium moiety. Zr-PYR has formula $ZrF_3(O_3PCH_2CH_2NC_5H_5)$ and crystallizes in the monoclinic $P2_1/c$ space group with the following cell parameters: $a = 12.3634(12)$ Å, $b = 9.3090(17)$ Å, $c = 9.8077(13)$ Å, $\beta = 112.819(8)°$, $Z = 4$.

The structure is layered and the layers are corrugated and constituted by unusual ZrO_3F_3 octahedra that corner share their three oxygen atoms, which are arranged in a *fac* (or *cis*) fashion, with the phosphonate tetrahedra (see Figure 2.22).

The organic pyridinium groups are bonded on either side of the inorganic octahedral/tetrahedral layer. Very likely the high number of fluorine atoms on Zr

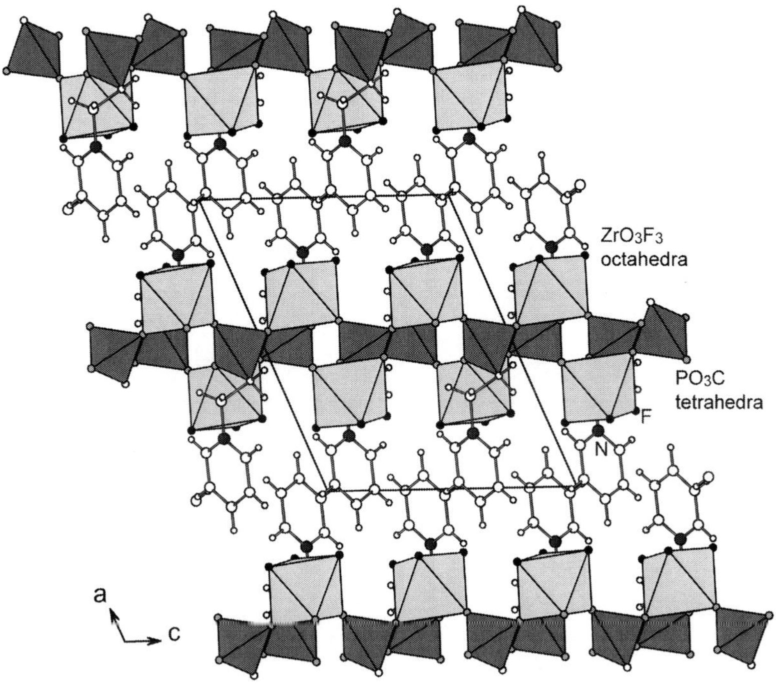

Figure 2.22 View of the structure of Zr-PYR down the *b* axis. The interdigitated arrangement of aromatic rings is shown.

coordination sphere induced by electroneutrality requirements, acts as a structure directing factor, since it influences the value of P/Zr molar ratio, that is unusually 1, and consequently, a connectivity of layers different from that of type α.

2.4.2 Zirconium R-amino-*N*,*N*-bis(alkylphosphonates)

Zr derivatives of *N*,*N*-bis(phosphonomethyl)-R-amines were first reported by Costantino *et al.* in 2002.[63–65] They show a common 2D structural arrangement, based on layers made of ZrO_5F octahedra and phosphonate tetrahedra with the organic functional groups lying in the interlayer space. P/Zr ratio in these compounds is 2, and the two phosphonate groups belonging to the same diphosphonic moiety are not equivalent: one of them is bivalent and tridentate, coordinating three different metal atoms, while the other one is monovalent and bidentate, coordinating two different metal atoms, with a free P–O group, which is involved in a hydrogen bond with the nitrogen atom of an adjacent ligand. It was shown that the nitrogen is protonated, and the P–O group is negatively charged. The presence of such an interaction frees a coordination site on Zr, allowing one fluorine atom of the original $[ZrF_6]^{2-}$ complex, present in the reaction solution, to remain coordinated to the metal atom also in the solid state structure (see Figure 2.23).

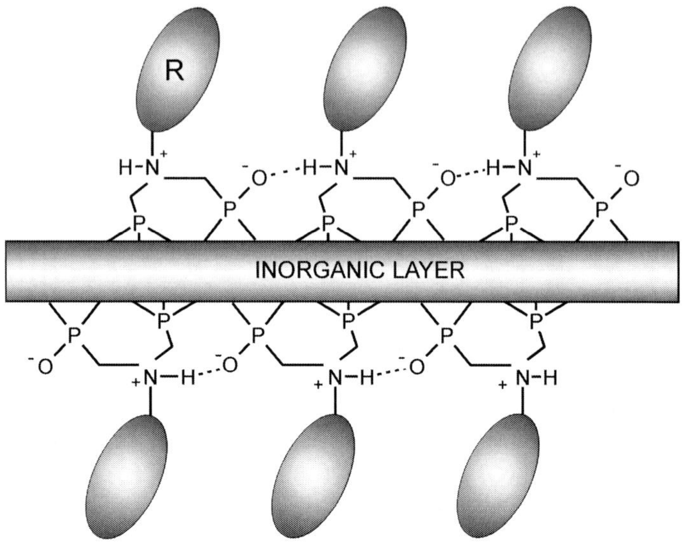

Figure 2.23 Schematic picture representing the layer framework of Zr R-amino-*N*,*N*-bis(alkylphosphonates). The zwitterionic character of the diphosphonate ligand is shown.

These compounds show a XRPD pattern typical of layered materials, characterized by a very intense first peak at low 2θ values, while successive reflections are much less intense. Accordingly, the *d* values of the first peak represent the interlayer distances of the compounds.

Zr derivatives of alkylamino-*N*,*N*-bis(methylenephosphonates) with different number of carbon atoms in the alkyl chain (hereafter Zr-Cn) are very similar. Plotting their *d*-spacings as a function of the number of carbon atoms in the alkyl chain (n_C), a linear correlation that follows the equation d (Å) = $9.28 + 1.04n_C$ can be recognized (Figure 2.24).

The most crystalline sample of the series was the pentylamino derivative[63] with formula $ZrHF(O_3PCH_2)_2NHC_5H_{11}$ (hereafter Zr-C5, Figure 2.25). It crystallizes in the monoclinic space group $P2_1/c$ and its cell parameters are: $a = 13.1874(3)$ Å, $b = 8.8723(3)$ Å, $c = 10.9793(3)$ Å, $\beta = 96.594(3)°$.

In this compound the layers lie on the *bc* plane and alkyl chains bonded to adjacent layers are arranged in an interdigitated fashion, although not elongated to their full extent, since the free area associated to every alkyl chain (48.6 Å^2, $a \times b/2$) is more than twice its estimated van der Waals cross-section (18.6 Å^2). To fill the free space as much as possible, the first three carbon atoms and the nitrogen atom are close to a *trans–trans* conformation, but the two last carbon atoms are bent in a way that allows the best occupation of the free available space.

^{31}P-MAS NMR analysis (Figure 2.26) shows the presence of two signals with approximately the same area related to the two non-equivalent phosphorus atoms present in the compound. The signal at lower chemical shift belongs to

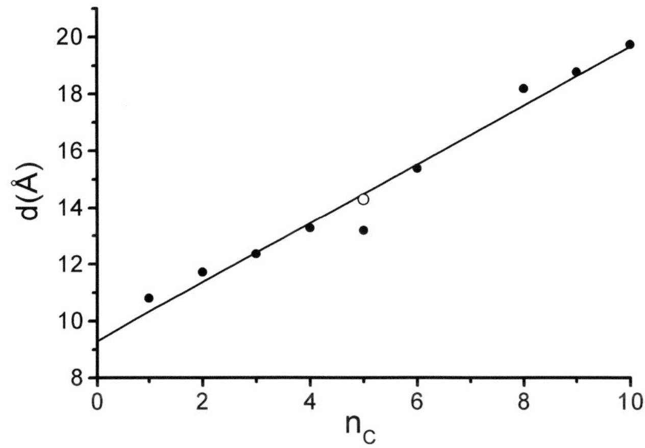

Figure 2.24 Dependence of the interlayer distance (d) versus the number of carbon atoms in the alkyl chain (n_C) in Zr–Cn compounds. Black symbols refer to room temperature data, while the white symbol refers to 150 °C. (Reproduced with permission from Reference 63. Copyright 2002 American Chemical Society.)

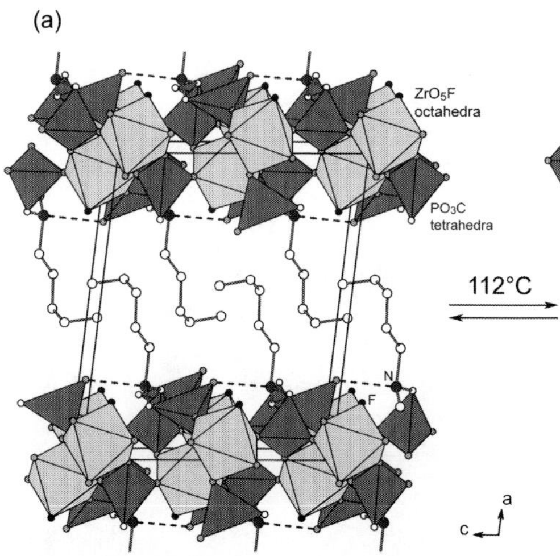

Figure 2.25 Structure of Zr-C5 before (a) and after (b) the phase transition at 112 °C.

the tridentate phosphonate, while the signal at higher chemical shift derives from the bidentate phosphonate.

It can be seen that the interlayer distance of Zr-C5 at room temperature does not fit with the straight line shown in Figure 2.24, being shorter than the predicted one. However, thermal analysis showed that the compound undergoes a phase transition at 112 °C without any weight loss. This transition causes an increase of the interlayer distance larger than 1 Å, as the unit cell parameter

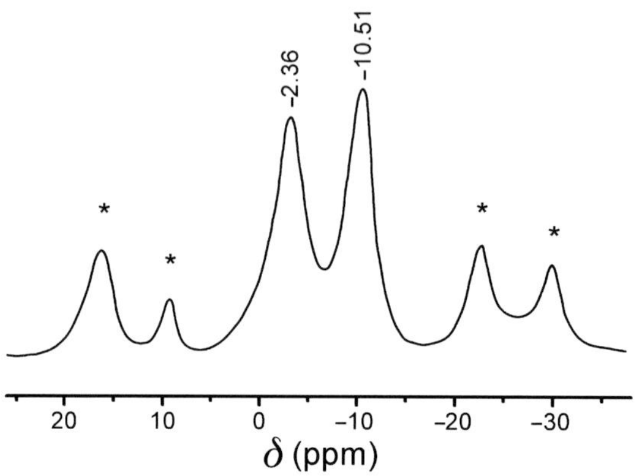

Figure 2.26 ^{31}P-MAS NMR spectrum of Zr-C5. The asterisks mark sidebands of the main signals. (Reproduced with permission from Reference 64. Copyright 2002 American Chemical Society.)

a becomes 14.34 Å, while *b* and *c* are slightly reduced (8.87 and 10.93 Å, respectively). Notwithstanding the partial loss of crystallinity, the structure of the high temperature phase was refined, showing that the alkyl chains are now close to their extended conformation. No chemical reactions occur during this process, that is completely reversible if the material is cooled down to room temperature. The new *d*-spacing is close to the regression straight line in Figure 2.25, suggesting that the structures of analogous compounds, not solved because of their low crystallinity, should be more similar to this phase rather than to the room temperature phase.

The Zr derivative of the *N,N*-bis(phosphonomethyl)-4-aminobutanoic acid has formula $ZrHF(O_3PCH_2)_2NH(CH_2)_3COO$ (hereafter Zr-GABA).[64] Its structure, shown in Figure 2.27, crystallizes in the monoclinic space group $P2_1/c$ with cell parameters $a = 12.9640(3)$ Å, $b = 8.9900(4)$ Å, $c = 10.7924(4)$ Å, $\beta = 101.854(4)°$.

If compared with Zr-C5, it has a very similar structural arrangement, with layers lying in the *bc* plane and interdigitated organic chains, but the carboxylic groups play an important role in the packing of the layers, since they are involved in a hydrogen bond with a fluorine coordinated to a zirconium atom of the adjacent layer. FT-IR analysis reveals that the hydrogen atom is located on the fluorine, as highlighted by the presence of a band at 3197 cm^{-1}, assigned to the H–F stretching vibration.

Thermal analysis of this material reveals that a 5% weight loss occurs between 180 and 280 °C, due to the loss of HF. High temperature XRPD measurements show an irreversible phase transition accompanied by a decrease of the interlayer distance from 12.7 to 11.9 Å (Figure 2.28). This new phase is stable also after cooling to room temperature.

This behaviour is well described by FT-IR analysis: the new phase does not show the H–F stretching band, whereas the C=O stretching frequency (1674 cm^{-1}) is shifted towards lower values than the original compound

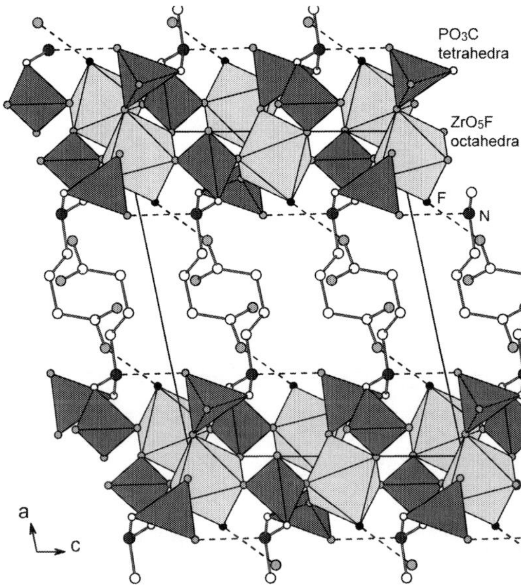

Figure 2.27 Polyhedral representation of the structure of Zr-GABA viewed down the *b* axis.

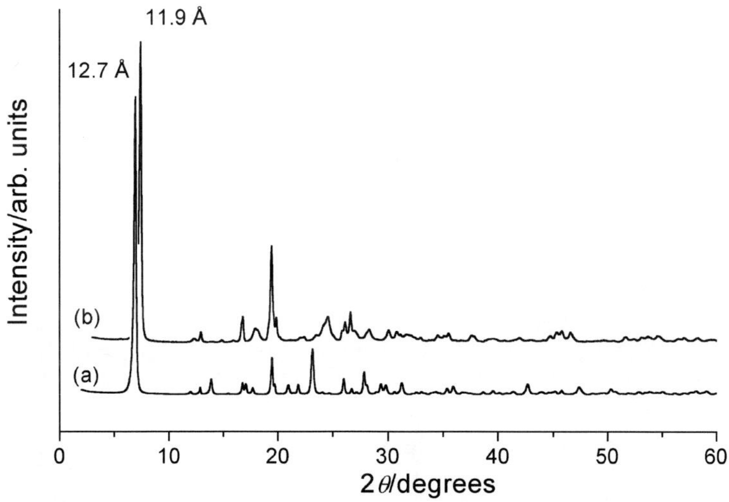

Figure 2.28 XRD patterns before and after the 1 hour thermal treatment at 240 °C.

(1734 cm^{-1}) , indicating that the carboxylate group has filled the coordination vacancy left by the fluorine on the metal atom. A structural scheme of this transformation is shown in Figure 2.29.

The exact method of interaction between the zirconium and the carboxylate has not been clarified: a shift of C=O stretching towards lower frequencies

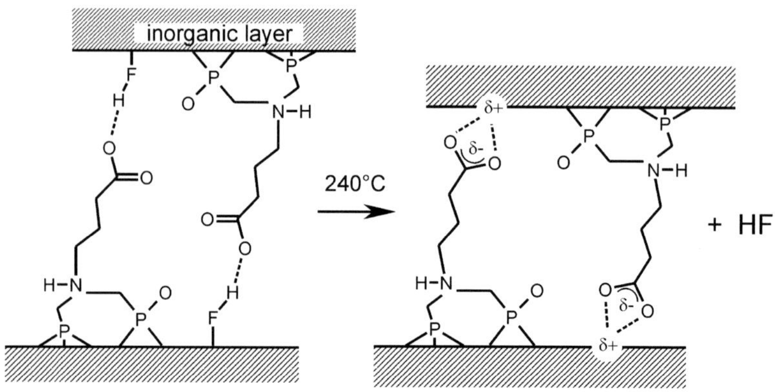

Figure 2.29 Scheme of the phase transition occurring in Zr-GABA at 240 °C. The loss of HF induces the zirconium-carboxylate interaction.

means that the bond order decreases, but this effect can take place either when it acts as a bidentate ligand or when it has ionic interaction with the positively charged metal atom. Coordination as a monodentate ligand has been excluded, since it would cause an increase in stretching frequency.

Another compound belonging to this family is the zirconium derivative of the *N,N*-bis(phosphonomethyl)benzylamine of formula $ZrF(O_3PCH_2)_2$-$NHCH_2C_6H_5$ (hereafter Zr-BENZ).[65] In this compound the pendant groups are aromatic rings and the layer structure remains substantially the same as the aforementioned compounds (see Figure 2.30). It crystallizes in the orthorhombic space group *Pbca*, and its cell parameters are: $a = 8.9429(2)$ Å, $b = 9.1746(2)$ Å, $c = 31.5654(7)$ Å. In this case the layers lie in the *ab* plane and the benzyl moieties point toward the interlayer region from both sides of the layers, forming a herringbone interdigitated arrangement with the rings coming from the adjacent layers. The 3.5 Å distance between two adjacent rings indicates that a π–π stacking interaction takes place. This interaction probably acts as a structure directing factor: to allow the best possible interaction between aromatic rings the structure is compressed in a direction parallel to the layers, while the layer framework is expanded along a direction perpendicular to the layers themselves, if compared with the previously described compounds.

The layered structures of these compounds offer the possibility to use them as hosts for intercalation reaction of various chemical species that can access the interlayer space interacting with the organic pendant groups. The original compounds have no free space permanently available and the organic moieties are arranged in interdigitated fashion, but the presence of an appropriate guest can induce a rearrangement of the structure, thus allowing the intercalation process.

The alkylamino derivatives Zr-C5 and the analogous Zr-C8, based on the *n*-octylamino-*N,N*-bis(methylphosphonate) have been contacted with very dilute aqueous solutions of *n*-alkanols with various chain lengths and have shown a

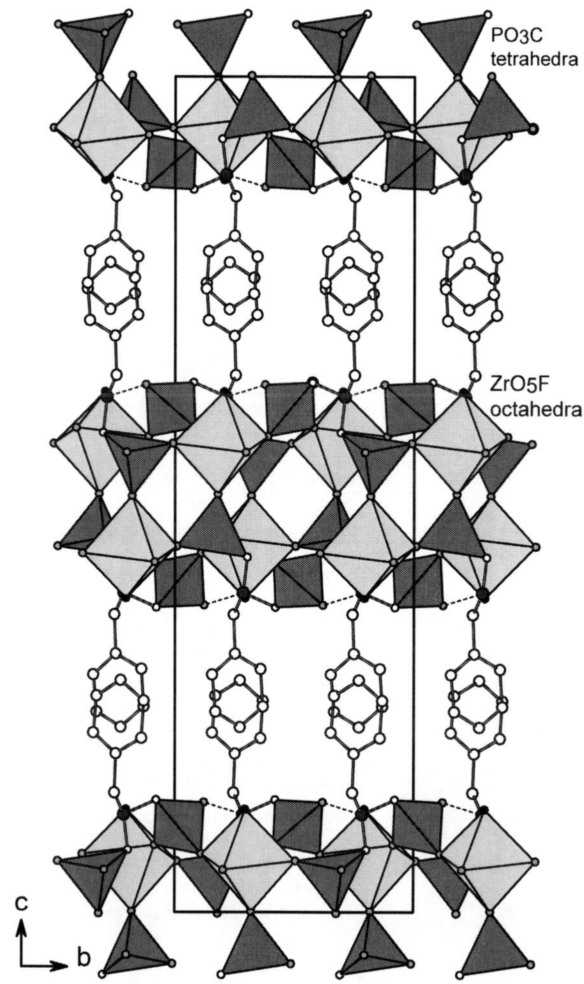

Figure 2.30 Polyhedral representation of the structure of Zr-BENZ, viewed down the *a* axis.

good affinity for these molecules. As a result of the insertion of guest molecules, the interlayer distance increases along with the length of the alkanol, as shown in Figure 2.31.

Zr-C8 has been also used for the intercalation of the photoactive azobenzene molecule, in order to study the behaviour of such a species when included in an inorganic host.[66]

The intercalation of azobenzene causes a 10 Å increase of the interlayer distance, very likely because the guest molecules find space between two adjacent alkyl chains from the same layer, as depicted in Figure 2.32. Photophysical and photochemical studies have been carried out on this intercalation compound, comparing the results with the pure microcrystalline guest and with

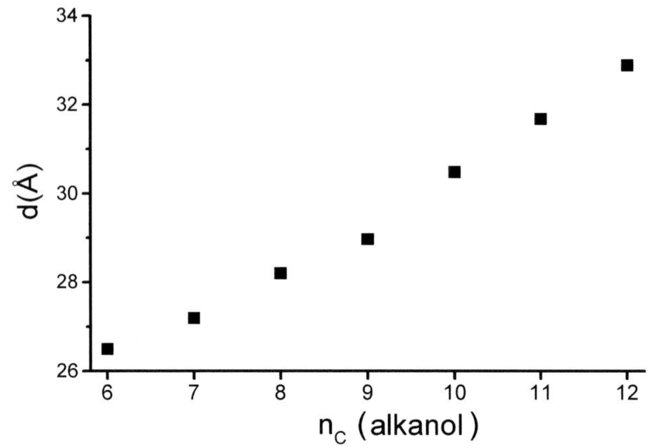

Figure 2.31 Interlayer distance of intercalation compounds of Zr-C8 with *n*-alkanols as a function of the number of carbon atoms in the alkanol chain. (Reproduced with permission from Reference 63. Copyright 2002 American Chemical Society.)

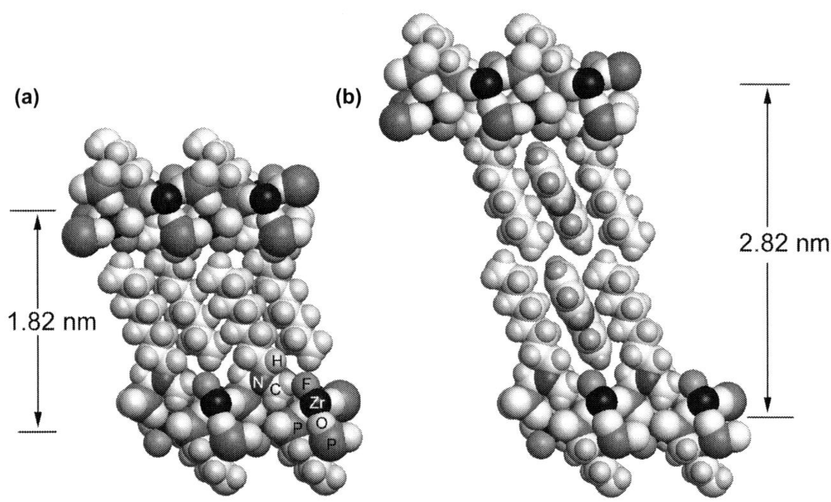

Figure 2.32 Space filling model of Zr-C8 (a) and the intercalation compound with azobenzene (b).

the surface-adsorbed compound. The three materials have similar spectral properties, showing excited states with analogous lifetimes, that are about ten times longer than the azobenzene dissolved in hydrocarbons. Notably, photoisomerization, which does not occur in the pure azobenzene solid, takes place both in the adsorbed and in the intercalated compound, showing a behaviour very similar to that in solution. The *trans–cis* process in the intercalated compound induces the release of the dye, reducing the interlayer distance to the

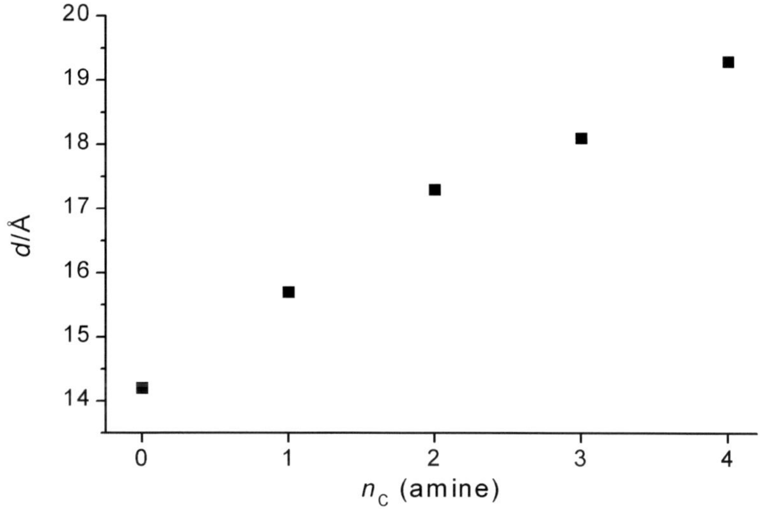

Figure 2.33 Interlayer distance of intercalation compounds of Zr-GABA with *n*-alkylamines as a function of the number of carbon atoms in the amine chain.

original value of Zr-C8, while the thermal *cis–trans* isomerization occurs at rates about 100 times faster than in solution, indicating that the process is favoured by a constrained microenvironment.

Zr-GABA shows good affinity for ammonia and short *n*-alkylamines: intercalation processes were performed from ethanol solutions, obtaining intercalation compounds with interlayer distances increasing linearly with the length of the guest molecules (Figure 2.33). The intercalation should be the result of the interaction between the alkaline guest and the acidic host, thus breaking the COO···F hydrogen bond and allowing the layers to be set apart.

When these intercalation compounds are put in water, they spontaneously exfoliate, producing stable colloidal dispersions. The same result is obtained if the intercalation reaction is carried out directly in aqueous solutions of the amines.

In contrast with these behaviours, Zr-BENZ is practically inert: several attempts to intercalate guest molecules failed, indicating that the cooperative π–π stacking interactions among aromatic rings are so strong that the interlayer region is not accessible to guest species.

2.4.3 Zirconium R-diamino-*N*,*N′*-bis(alkylphosphonates)

Two zirconium derivatives of the piperazino-*N*,*N′*-bis(methylphosphonate) (hereafter Zr-Pip) have been reported by Taddei *et al.* in 2010.[67]

This building block is constituted of a heterocyclic diamino group in which the phosphonate tetrahedra are placed at opposite ends.

Figure 2.34 Polyhedral representation of the structure of Zr-Pip1, viewed down the *c* axis. (Reproduced with permission from Reference 67. Copyright 2010 American Chemical Society.)

This molecule has been also used for the preparation of many transition M^{II}, M^{III} and Ln derivatives.

The zirconium derivatives here described show different structure and composition. One of them has a 3D framework architecture featuring infinite channels able to reversibly adsorb and desorb water, whereas the other has a compact pillared layered structure.

The first compound (Figure 2.34) has formula $Zr_2H_4[(O_3PCH_2)_2N_2C_4H_8]_3 \cdot 9H_2O$ (Zr-Pip1) and a 3D structure, made of infinite inorganic chains of ZrO_6 octahedra and PO_3C tetrahedra, running along the *c* axis direction, connected by piperazine groups in the *ab* plane.

It crystallizes in the hexagonal *R*-3 space group with the following parameters: $a = 19.9400(9)$ Å, $c = 9.5728(6)$ Å, $Z = 3$.

Zr atoms lie in the 3-fold inversion axes. Zr octahedra are bonded to six oxygen atoms belonging to six different phosphonates, and phosphonate tetrahedra act as exobidentate groups, as they share two corners with two different Zr octahedra belonging to the same inorganic chain. The two other phosphonate corners are occupied by the organic group and a terminal P–O group.

The walls of these channels are formed by the piperazine groups arranged in a helicoidal fashion along the channels. Nine water molecules per formula unit are placed in the channels and interact with an intricate network of hydrogen bonds with the terminal P–OH/P–O groups.

Water loss is completely reversible, and the sample can fully rehydrate in some hours. XRPD spectra, taken at different temperatures show that, during

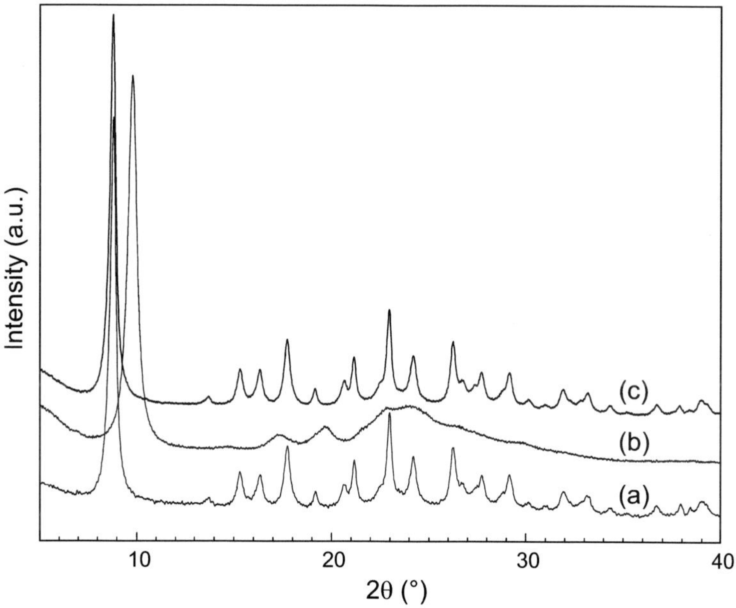

Figure 2.35 Comparison of XRPD patterns of Zr-Pip1 at room temperature (a), at 150 °C (b), and at room temperature after heating (c). (Reproduced with permission from Reference 67. Copyright 2010 American Chemical Society.)

dehydration, the 110 reflection is gradually shifted toward higher 2θ values, the d value changing from 10.02 Å at room temperature to 9.02 Å at 150 °C. In addition, during heating, a gradual degradation of the whole pattern is observed. The pattern taken at 150 °C shows only some large bands, besides the shifted 110 reflection. In agreement with TG data, when cooling, the XRPD changes are fully reversible, and crystallinity is completely recovered, as shown in Figure 2.35.

The second compound (Zr-Pip2), although obtained in similar conditions, has different composition and crystal structure. The chemical formula of Zr-Pip2 is $ZrF_2(O_3PCH_2)_2(NH)_2C_4H_8$ and it crystallizes in the monoclinic $P2_1/c$ space group with the following parameters: $a = 8.7148(2)$ Å, $b = 8.1731(1)$ Å, $c = 9.0134(2)$ Å, $\beta = 105.175(1)°$, $Z = 2$.

Differently from the first compound the latter contains two fluorine atoms in the zirconium coordination sphere. Its structure is layered-pillared in which the inorganic layers are made of Zr octahedra and P tetrahedra, lying on the [100] crystallographic planes, covalently connected by piperazine groups, that act as pillars as depicted in Figure 2.36.

Every Zr atom is placed in the [100] plane. Each of them is trans-coordinated by two fluorine atoms. The octahedron axis defined by the F–Zr–F bond sequence is tilted, alternatively on one side and the other, about 30° to the layer plane, both on b and c directions. The four equatorial positions are occupied by four oxygen atoms belonging to four different phosphonate tetrahedra.

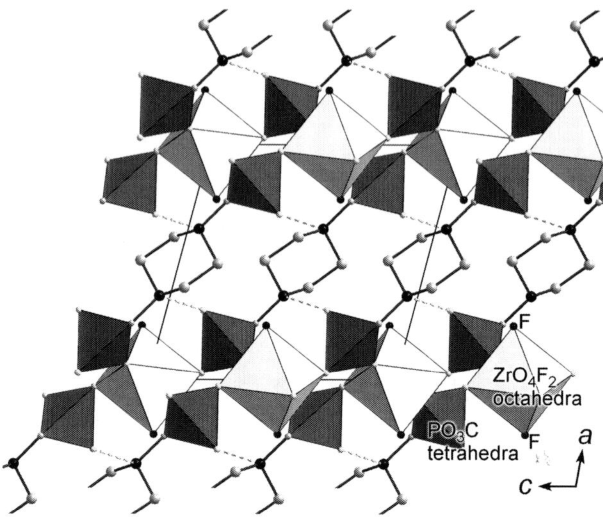

Figure 2.36 Polyhedral representation of the structure of Zr-Pip2, viewed along the *b* axis. (Reproduced with permission from Reference 67. Copyright 2010 American Chemical Society.)

The remaining P–O group, not coordinated to the metal atom, is directed toward the nitrogen atom of the piperazine moiety belonging to the nearest adjacent phosphonate group. The PO···N distance is 2.74 Å, indicating that a hydrogen bond is present between the two atoms. This kind of interaction obviously occurs also on the other nitrogen atom at the opposite side of piperazine moiety, which is crystallographically equivalent. Therefore, all nitrogen atoms are involved in H bonds with P–O groups.

As observed for the above described layered zirconium amino-diphosphonates this non-covalent interaction seems to have an important role in the building of the two compounds.

Zr-Pip1 and Zr-Pip2 were obtained in similar synthetic conditions, the main difference regarding the starting pH value of the reaction solution (about 7 for Zr-Pip1 and about 2 for Zr-Pip2). Therefore this led us to suppose that the protonation level of nitrogen atoms of aminophosphonate groups, which is influenced by pH, can act as a real structure orienting factor: it seems that when the nitrogen atoms of aminophosphonate building blocks are protonated and therefore positively charged, anionic P–O groups tend to closely interact with them, in order to reduce local charge, and thus leaving one or two zirconium coordination sites to fluorine. Moreover, the presence of strong noncovalent interactions inside the structure may induce the formation of a more dense structure.

In 1992, Vermeulen and Thompson[59] prepared a diaminodiphosphonic building block containing an aromatic organic group by reacting 4,4′-bipyridine with two equivalents of diethyl 2-bromoethylphosphonate, and obtained its zirconium derivative, whose structure was then solved two years later thanks to

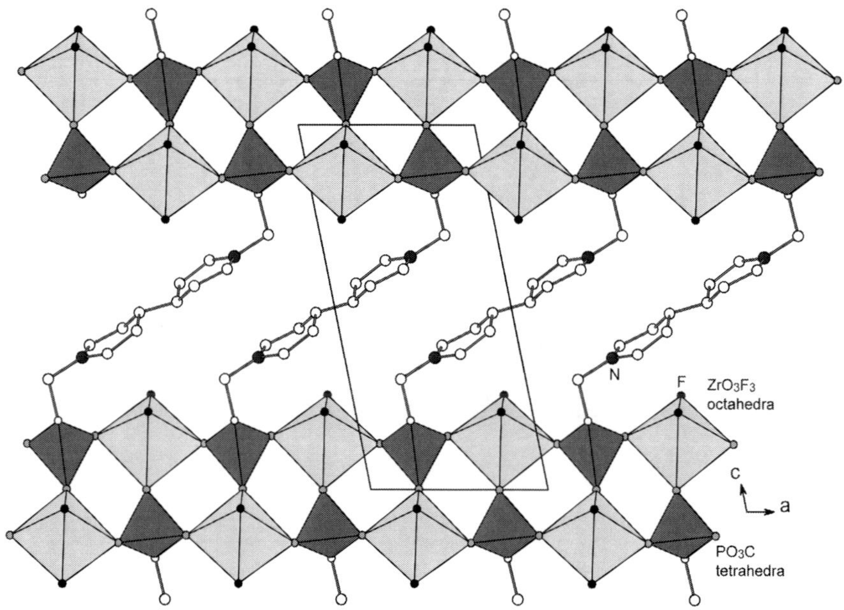

Figure 2.37 Representation of the structure of Zr-PV, viewed down the *b* axis.

the contribution of Clearfield and Poojary.[14] The compound has formula $ZrF_3(O_3PCH_2CH_2NC_5H_4) \cdot H_2O$ (Zr-PV) and it crystallizes in the monoclinic space group $P2_1/c$ with the following lattice parameters: $a = 6.6679(2)$ Å, $b = 12.5871(3)$ Å, $c = 13.5837(4)$ Å, $\beta = 101.739(3)°$ (see Figure 2.37).

Like its analogue, Zr-PYR, the inorganic framework in this compound is built up from ZrO_3F_3 octahedra and, consequently, the P/Zr ratio is 1. However, a difference between the two compounds can be found in the arrangement of ligands around the metal atom: in this case they are in a *mer* (or *trans*) configuration. As a result, the structure of this material is based upon inorganic double chains originated by the association of two symmetry related units, built up from the connection of zirconium octahedra and phosphonate tetrahedra, running along the *a* axis. The viologen moieties act as connectors between two of these double chains on the *ac* plane, thus forming an organo-inorganic hybrid layered structure.

This compound and its analogues with Cl^- and Br^- in place of F^- have been studied for their ability to produce a stable charge separation upon photoinduction.

A mixed phosphate-PV compound was synthesized and structurally characterized by Byrd *et al.* in 1996.[68] It has formula $(ZrF)_2(PO_4)$-$(O_3PCH_2CH_2NC_5H_4$-$C_5H_4NCH_2CH_2PO_3)F \cdot 3H_2O$ (Zr-PO4PV) and crystallizes in the monoclinic space group $P2/c$ with the following lattice parameters: $a = 13.5895(22)$ Å, $b = 8.8351(9)$ Å, $c = 9.2294(10)$ Å, $\beta = 100.79°$. The structure is shown in Figure 2.38. It shows many differences with respect to the previous compound: the layer structure incorporates tetradentate phosphate ligands

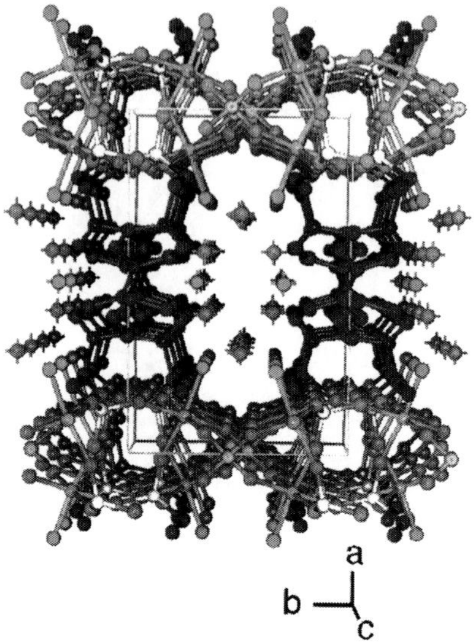

Figure 2.38 Representation of the structure of Zr-PO4PV. (Reproduced with permission from Reference 68. Copyright 1996 American Chemical Society.)

and, despite the connection of phosphonates on zirconium is similar, three oxygen atoms from three different tetrahedra are arranged in *fac* configuration, and only one fluorine is bound to the metal atom, with the other two coordination positions occupied by two oxygens from phosphate groups. As a result, the compound has a porous structure and it has been used for incorporation of platinum ions in order to create an heterogeneous catalyst for hydrogen production.

2.4.4 Zirconium R-diamino-*N*,*N*,*N*′,*N*′-tetrakis-(alkylphosphonates)

Although extensively used for the preparation of many transition and rare-earth lanthanides, the number of zirconium derivatives of the diamino-*N*,*N*,*N*′,*N*′-tetrakis-methylphosphonate groups is very low.

Only two examples of zirconium derivatives of these building blocks have been reported, by Vivani *et al.* in 2006.[69] They are the Zr butyldiamino- and the cyclohexyl-diamino-*N*,*N*,*N*′,*N*′-tetrakis(methylphosphonates). The molecular structure of the building blocks used is shown in Figure 2.39.

The formulas of the two derivatives are $Zr(HPO_3CH_2)_2N–C_4H_8–N(CH_2PO_3H)_2 \cdot 4H_2O$ (hereafter Zr-P4C4) and $Zr(PO_3CH_2)_2N–C_6H_{10}–N(CH_2PO_3)_2Na_2H_2 \cdot 5H_2O$ (hereafter Zr-Cyclo), respectively. The first compound crystallizes in the monoclinic space group $P2/c$ with the following unit cell parameters: $a = 10.7337(3)$ Å, $b = 12.8604(3)$ Å, $c = 7.8919(2)$ Å, $\beta = 104.91(2)°$. The second compound is monoclinic $C2/c$, having the following lattice parameters: $a = 18.923(1)$ Å, $b = 16.5046(7)$ Å, $c = 7.8437(3)$ Å, $\beta = 91.868(3)°$.

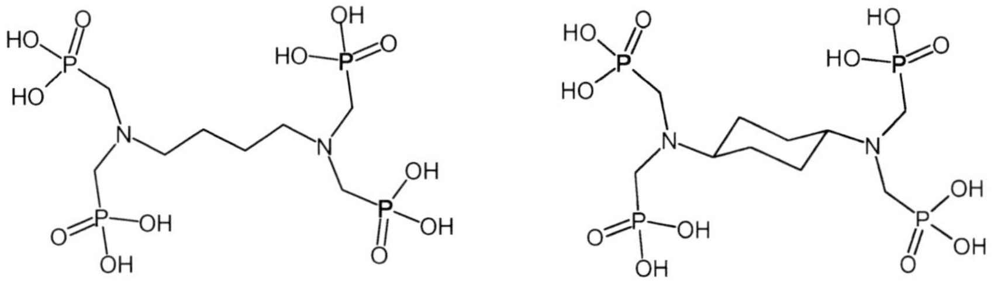

1 2

Figure 2.39 Molecular structure of butyldiamino- (1) and cyclohexyl-diamino-*N*,*N*,*N*′,*N*′-tetrakis(methylphosphonate) (2).

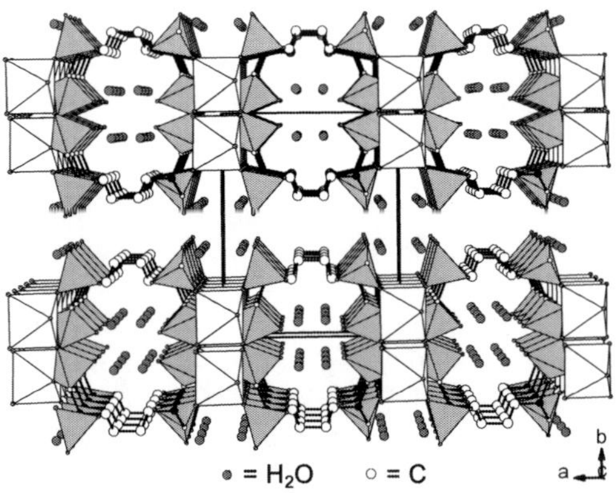

Figure 2.40 Polyhedral representation of the structure of Zr-P4C4, viewed down the *c* axis. (Reproduced with permission from Reference 69. Copyright 2006 American Chemical Society.)

Zr-P4C4 has a layered 2D structure in which the tetraphosphonic group contributes to the formation of hybrid inorganic–organic layers, connecting the ZrO_6 octahedra along the sheet direction, as depicted in Figure 2.40. These layers are about 13 Å thick and are stacked with an A-A sequence. The interlayer region is occupied by two water molecules. Each layer is formed by the connection, through the butyldiamino groups, of inorganic 1D polymeric units running along the *c* axis. These units consist of a double row of ZrO_6 octahedra connected to the phosphonate tetrahedra PO_3C. The butyl chains connect these units along the *a* axis with the formation of intralayer channels with measured size of about 10×3.5 Å. The horizontal size depends on the length of the organic chain that connects two adjacent inorganic building units. Therefore the use of tetraphosphonic groups containing longer chains may generate structure with larger channels. These channels are partially occupied by two water molecules per formula unit.

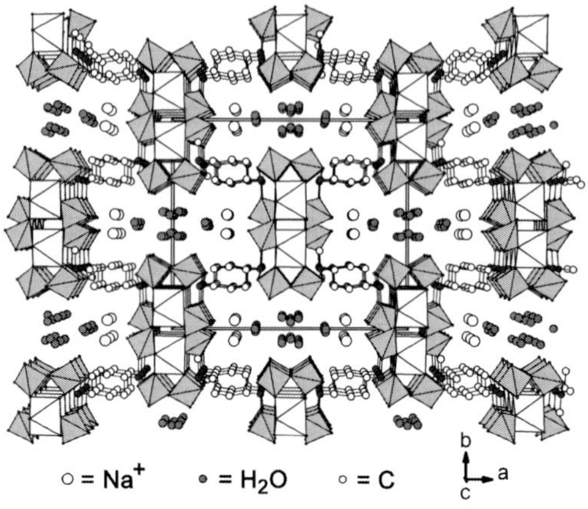

Figure 2.41 Polyhedral representation of the structure of Zr-Cyclo, viewed down the *c* axis. (Reproduced with permission from Reference 69. Copyright 2006 American Chemical Society.)

The second compound, Zr-Cyclo, has a 3D open-framework structure made of inorganic polymeric units bridged by cyclohexyl groups in a "brickwall-like" architecture, as shown in Figure 2.41. This compound has a similar structure with respect to the previous one, despite the higher dimensionality.

The structure, indeed, is built from inorganic building units made of ZrO_6 octahedra and phosphonate tetrahedra PO_3C with the same connectivity observed for Zr-P4C4 connected through the cyclohexyl groups. These inorganic blocks are shifted from each other by $1/2a$ and $1/2b$, and their connection, operated by the organic groups, generates rectangular channels, running along the *c* axis, with dimensions of about 12×5 Å (calculated perpendicularly to the *c* axis and between atomic centres). These channels are occupied by five water molecules and two Na atoms per formula unit. These guest species form a number of strong interactions and H bonds with each other and with the framework P–O groups.

The extra-framework Na^+ ions come from the NaOH used in the synthesis and substitute the acid H^+ belonging to the protonated PO–H groups. They can also be easily exchanged with Li^+, K^+, NH_4^+ or regenerated in acid form by using HCl.

The strong analogies between the structures of Zr-P4C4 and Zr-Cyclo can be presumed as two different ways to covalently connect the same polymeric building unit by organic groups.

In both of these two arrangements, the dimensions of the channels formed inside the structure depend on the size and geometry of the interlinking organic groups. This feature could offer intriguing possibilities to obtain new open framework compounds with a similar structure, in which the cavities can be tuned for specific purposes by selecting an appropriate organic group, as depicted in Figure 2.42.

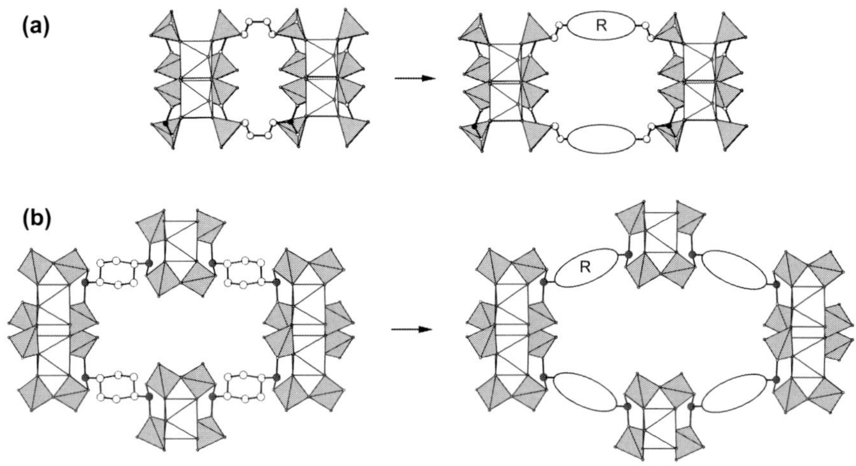

Figure 2.42 Possible tailoring of the dimensions of the channels in the structures of Zr-P4C4 (a) and Zr-Cyclo (b) by using different tetraphosphonate building blocks. The elliptic shapes represent generic organic groups. (Reproduced with permission from Reference 69. Copyright 2006 American Chemical Society.)

2.5 Conclusion

The chemistry of zirconium phosphates and phosphonates is very vast and does not seem to be exhausted. Initially, the research was centred on α- and γ-zirconium derivatives, due to the possibility of easily functionalizing these solids with a large number of organic moieties and the easy predictability of their structural features.

More recently a series of new compounds has been found, and their structures solved. Some of these already showed unusual characteristics and reactivity, although research on most of these new compounds is still in progress. They are characterized by a vast structural variety. This is mainly due to the fact that R–PO_3 phosphonate groups can behave as mono-, bi-, or tridentate, mono- or dianionic ligands. As a result of this, the structure of zirconium phosphonates are still poorly predictable, especially when polyphosphonate building blocks are used, notwithstanding the intense research activity on this topic in the past decade. This peculiarity, on one hand, can be considered a limit for the development of this class of materials; however, on the other hand, it may be seen as a precious source of new structural archetypes, potentially useful for diverse applications. In recent years, aminophosphonate building blocks have been widely employed, probably because of their relatively easy synthesis and their ability to induce the formation of highly ordered structures, owing to the presence of strongly polar amino groups that may act as structure-directing agents.

Future investigations should be devoted to help to understand the connections between the synthetic parameters, the molecular structure of building blocks and the global crystal structure, in the race to discover and design rationally new compounds for materials chemistry.

References

1. G. Alberti, in *Comprehensive Supramolecular Chemistry*, eds. G. Alberti, T. Bein, Pergamon Press, Oxford, 1996, Vol. 7, Chapter 5.
2. A. Clearfield and U. Costantino, in *Comprehensive Supramolecular Chemistry*, eds. G. Alberti, T. Bein, Pergamon Press, Oxford, 1996, Vol. 7, Chapter 4 and references therein.
3. A. Clearfield, *Prog. Inorg. Chem.*, 1998, **47**, 374.
4. G. Alberti, M. Casciola, U. Costantino and R. Vivani, *Adv. Mater.*, 1996, **8**, 291.
5. H. C. Yang, K. Aoki, H. G. Hong, D. D. Sackett, M. F. Arrendt, S. L. Yau, C. M. Bell and T. E. Mallouk, *J. Am. Chem. Soc.*, 1993, **115**, 11855.
6. C. V. Kumar, A. Chaudhari and G. L. Rosenthal, *J. Am. Chem. Soc.*, 1994, **116**, 403.
7. G. Cao, M. E. Garcia, M. Alcalà, L. F. Burges and T. E. Mallouk, *J. Am. Chem. Soc.*, 1992, **114**, 7574.
8. B. Bujoli, S. M. Lane, G. Nonglanton, M. Pipelier, J. Léger, D. R. Talham and C. Tellier, *Chem. Eur. J.*, 2005, **11**, 1980.
9. C. V. Kumar and A. Chaudari, *J. Am. Chem. Soc.*, 2000, **122**, 830.
10. M. Curini, O. Rosati and U. Costantino, *Curr. Org. Chem.*, 2004, **8**, 591.
11. F. Bellezza, A. Cipiciani, U. Costantino and M. E. Negozio, *Langmuir*, 2002, **18**, 8737.
12. B. Zhang, D. M. Poojary, A. Clearfield and G. Z. Peng, *Chem. Mater.*, 1996, **8**, 1333.
13. L. A. Vermeulen and M. E. Thompson, *Chem. Mater.*, 1994, **6**, 77.
14. D. M. Poojary, L. A. Vermeulen, E. Vicenzi, A. Clearfield and M. E. Thompson, *Chem. Mater.*, 1994, **6**, 1845.
15. B. Zhang and A. Clearfield, *J. Am. Chem. Soc.*, 1997, **119**, 2751.
16. D. M. Poojary and A. Clearfield, *Acc. Chem. Res.*, 1997, **30**, 414.
17. D. M. Poojary, D. Zhang, Y. Dong, G. Peng and A. Clearfield, *J. Phys. Chem.*, 1994, **98**, 13616.
18. A. Clearfield and J. A. Stynes, *J. Inorg. Nucl. Chem.*, 1964, **26**, 117.
19. G. Alberti and E. Torracca, *J. Inorg. Nucl. Chem.*, 1968, **30**, 317.
20. G. Alberti, U. Costantino, S. Allulli and N. Tomassini, *J. Inorg. Nucl. Chem.*, 1978, **40**, 1113.
21. D. M. Poojary, H. L. Hu, F. L. Campbell III and A. Clearfield, *Acta Cryst. B*, 1993, **49**, 996.
22. K. Segawa, A. Sugiyama and Y. Kurusu, in *Proceedings of Symposium on Chemistry of Microporous Crystals*, eds. T. Inai, S. Mamba and K. Tatsumi, Kodansha, Tokyo,Elsevier, Amsterdam, 1990.
23. G. L. Rosenthal and J. Caruso, *J. Solid State Chem.*, 1993, **107**, 497.
24. J. Kornyei, L. Szirtes and U. Costantino, *J. Radioanal. Chem.*, 1985, **89**, 331.

25. G. Alberti, U. Costantino and M. L. Luciani Giovagnotti, *J. Chromatogr.*, 1979, **180**, 45.
26. D. A. Burwell and M. E. Thompson, *Chem. Mater.*, 1991, **3**, 14.
27. M. Casciola, U. Costantino, A. Peraio and T. Rega, *Solid State Ionics*, 1995, **77**, 229.
28. D. A. Burwell and M. E. Thompson, *Chem. Mater.*, 1991, **3**, 730.
29. G. Alberti and U. Costantino, *J. Mol. Catal.*, 1984, **27**, 235.
30. U. Costantino, M. Curini, O. Rosati and E. Pisani, *Synlett*, 1996, 333.
31. G. Alberti, M. Casciola, U. Constantino, A. Peraio and V. Vivani, *Solid State Ionics*, 1991, **46**, 61.
32. G. Alberti, M. Casciola, R. Palombari and A. Peraio, *Solid State Ionics*, 1992, **58**, 339.
33. D. Lanari, F. Montanari, F. Marmottini, O. Piermatti, M. Orrù and L. Vaccaro, *J. Catal.*, 2011, **277**, 80.
34. M. B. Dines, P. M. Di Giacomo, K. P. Callahan, P. C. Griffith, R. Lane and R. E. Cooksey, *ACS Symp. Ser.*, 1982, **192**, 223.
35. M. B. Dines and P. C. Griffith, *J. Phys. Chem.*, 1982, **86**, 571.
36. P. M. Di Giacomo and M. B. Dines, *Polyhedron*, 1982, **1**, 62.
37. G. Alberti, U. Costantino, F. Marmottini and G. Perego, *React. Polym.*, 1988, **9**, 267.
38. G. Alberti, U. Costantino, F. Marmottini, R. Vivani and P. Zappelli, *Angew. Chem., Int. Ed. Engl.*, 1993, **32**, 1357.
39. G. Alberti, U. Costantino, R. Vivani and R. K. Biswas, *React. Polym.*, 1992, **17**, 245.
40. G. Alberti, M. Casciola and R. K. Biswas, *Inorg. Chim. Acta*, 1992, **201**, 207.
41. G. Alberti, R. Vivani, R. K. Biswas and S. Murcia Mascaros, *React. Polym.*, 1993, **19**, 1.
42. S. Yamanaka and M. Koizumi, *Clays Clay Miner.*, 1975, **23**, 477.
43. S. Yamanaka, K. Sakamoto and M. Hattori, *J. Phys. Chem.*, 1984, **88**, 2067 .
44. A. Clearfield, R. H. Blessing and J. A. Stynes, *J. Inorg. Nucl. Chem.*, 1968, **30**, 2249.
45. G. Alberti, C. Dionigi, E. Giontella, S. Murcia-Mascaros and R. Vivani, *J. Colloid Interfacial Sci.*, 1997, **188**, 27.
46. G. Alberti, M. Casciola, R. Vivani and R. K. Biswas, *Inorg. Chem.*, 1993, **32**, 4600.
47. G. Alberti, R. Vivani and S. Murcia-Mascaròs, *J. Mol. Struct.*, 1998, **35**, 357.
48. G. Alberti, L. Boccali, M. Casciola, L. Massinelli and E. Montoneri, *Solid State Ionics*, 1996, **84**, 97.
49. G. Alberti, S. Murcia-Mascaros and R. Vivani, *Mater. Chem. Phys.*, 1993, **35**, 187.
50. G. Alberti, E. Giontella, S. Murcia-Mascarós and R. Vivani, *Inorg. Chem.*, 1998, **37**, 4672.
51. G. Alberti, F. Marmottini, S. Murcia-Mascarós and R. Vivani, *Angew. Chem., Int. Ed. Engl.*, 1994, **33**, 1594.

52. G. Alberti, S. Murcia-Mascaros and R. Vivani, *J. Am. Chem. Soc.*, 1998, **120**, 9291.
53. S. Kitagawa and K. Uemura, *Chem. Soc. Rev.*, 2005, **34**, 109.
54. C. Serre, F. Millange, C. Thouvenot, M. Noguès, G. Marsolier, D. Louër and G. Férey, *J. Am. Chem. Soc.*, 2002, **124**, 13519.
55. G. Alberti, E. Brunet, C. Dionigi, O. Juanes, M. J. de la Mata, J. C. Rodríguez-Ubis and R. Vivani, *Angew. Chem., Int. Ed. Engl.*, 1999, **22**, 3351.
56. G. Alberti, L. Boccali, C. Dionigi, R. Vivani, V. I. Kalchenko and L. I. Atamas, *Supramol. Chem.*, 199, **67**, 129.
57. G. Alberti, L. Boccali, C. Dionigi, R. Vivani and V. I. Kalchenko, *Supramol. Chem.*, 1998, **9**, 99.
58. K. Moedritzer and R. R. Irani, *J. Org. Chem.*, 1966, **31**, 1603.
59. L. A. Vermeulen and M. E. Thompson, *Nature*, 1992, **358**, 656.
60. B. Zhang, D. M. Poojary and A. Clearfield, *Inorg. Chem.*, 1998, **37**, 249.
61. D. M. Poojary, B. Zhang and A. Clearfield, *Angew. Chem., Int. Ed. Engl.*, 1994, **33**, 2324.
62. L. A. Vermeulen, R. Z. Fateen and P. D. Robinson, *Inorg. Chem.*, 2002, **41**, 2310.
63. U. Costantino, M. Nocchetti and R. Vivani, *J. Am. Chem. Soc.*, 2002, **124**, 8428.
64. R. Vivani, U. Costantino and M. Nocchetti, *J. Mater. Chem.*, 2002, **12**, 3254.
65. R. Vivani, F. Costantino, M. Nocchetti and G. D. Gatta, *J. Solid State Chem.*, 2004, **177**, 4013.
66. P. L. Gentili, U. Costantino, R. Vivani, L. Latterini, M. Nocchetti and G. G. Aloisi, *J. Mater. Chem.*, 2004, **14**, 1656.
67. M. Taddei, F. Costantino and R. Vivani, *Inorg. Chem.*, 2010, **49**, 9664.
68. H. Byrd, A. Clearfield, D. M. Poojary, K. P. Reis and M. E. Thompson, *Chem. Mater.*, 1996, **8**, 2239.
69. R. Vivani, F. Costantino, U. Costantino and M. Nocchetti, *Inorg. Chem.*, 2006, **45**, 2388.

CHAPTER 3

High-throughput Methods for the Systematic Investigation of Metal Phosphonate Synthesis Fields

PALANIKUMAR MANIAM AND NORBERT STOCK

Christian-Albrechts-Universität zu Kiel, Institut für Anorganische Chemie, Max-Eyth-Straße 2, 24118 Kiel, Germany

3.1 Introduction to the Synthesis of Metal Phosphonates

More than 2000 metal phosphonate structures have been described in the literature[1] and virtually all metal ions have been incorporated. Numerous methods have been employed for their synthesis. At room/ambient temperature conditions, interdiffusion in reaction solutions, sol–gel and solvent evaporation techniques have been employed.[2–4] However, most phosphonates were synthesized by solvothermal methods at higher temperatures using conventional heating (in organic solvents[5] or water[6,7]). Recent work has shown that ionic liquids can function as a solvent (ionothermal) as well as a ligand in the metal phosphonate structure.[8] An alternative method to produce metal phosphonates, which has only been rarely employed, is the use of microwave-assisted heating.[9,10] Another development in the field of metal phosphonate synthesis is the direct employment of phosphonate esters instead of the acids. In contrast to the phosphonic acids, the phosphonate esters can be easily purified by distillation,

Metal Phosphonate Chemistry: From Synthesis to Applications
Edited by Abraham Clearfield and Konstantinos Demadis

Published by the Royal Society of Chemistry, www.rsc.org

recrystallisation or column chromatography. In addition, the use of the phosphonate ester leads often to better crystalline products since the ester undergoes an *in situ* hydrolysis during the reaction.[11,12] The formation of the phosphonic acid by *in situ* oxidation has also been described.[13] In the subsequent step, the resulting organic linker is incorporated into the structure to form a new metal phosphonate. More uncommon synthesis procedures, i.e. sonochemical or mechanochemical syntheses, have not been reported up to now.

The vast number of metal phosphonate structures and the various synthesis methods demonstrate the immense abundance of metal phosphonate chemistry. Although many studies have been performed by various groups, it is hard to compare and correlate the results with each other since different synthesis conditions (i.e. solvent, reaction time, reaction temperature, reactor size and material, etc.) have been employed. Unfortunately, only few systematic studies have been reported that investigate the role of chemical and process parameters. This would be of substantial interest in order to learn more about reaction trends and to "comprehend" the chemistry of metal phosphonates. A methodology has been applied over the last decade, the so-called high-throughput (HT) methods that allow the systematic investigation of complex synthesis fields and thus the elucidation of reaction trends. Here we describe HT methodologies for reactions in liquid media, the equipment employed, the strategy used to investigate complex systems and a summary of their advantages and disadvantages. Studies on the discovery, synthesis optimization and extraction of reaction trends are presented.

3.1.1 High-throughput Technology

A convenient and efficient tool for investigations in the field of materials science is the high-throughput (HT) technology. The pharmaceutical industry has long embraced this method to conduct screening tests for drug discovery and drug design in order to reduce time costs in the research and development phase.[14] The two important and relevant terms used in the literature are "combinatorial" and "high-throughput". In combinatorial chemistry "a combinatorial process is used to prepare sets of compounds from sets of building blocks".[15] This approach is used in medicinal chemistry where a number of components are combined in a single pot reaction process to produce a mixture of compounds. On the other hand, the term "high-throughput synthesis" encompasses reactor arrays with individual miniaturized reaction vessels. For each reaction vessel an individual point in parameter space (defined by temperature, composition, pressure, etc.) can be investigated. Thus a diverse data library can be set up that allows the accelerated discovery as well as the optimization of new materials.[16] By looking closely at high-throughput methods, the words "automation", "parallelization" and "miniaturization" come into play, meaning that tasks will be rapidly and/or simultaneously performed and that small amounts of starting material are needed.[17] The number of experiments to be performed increases dramatically in a multi-parameter system. Each molar ratio of starting materials can lead to a new product. For example, a solvothermal reaction of reactants A and B, whereby the molar fraction A:B is varied in 0.1 steps (0.1 to 0.9) leads to nine individual reactions. However, with the inclusion of a new reactant C and varying the molar fractions A:B:C in 0.1 steps (in the range of 0.1 to 0.9, sum of molar fractions of $A+B+C=1$) already gives 36 reactions and 0.05 steps will then necessitate 171 individual reactions (calculated using equations 3.1 and 3.2).

$$n = \frac{(1-2x)}{x}, \quad x = \text{step increments, i.e. } 0.05,\ 0.1, \text{ etc.} \tag{3.1}$$

$$\sum_{i=1}^{n} i = \frac{n}{2}(n+1) = \text{sum of individual reactions} \tag{3.2}$$

Practically, conventional methods are not applicable for this large number of reactions but utilizing HT reactors this study can be performed in a day. Hence the HT methodology is an efficient tool for finding new compounds, discovering reaction trends and optimizing synthesis conditions. Nonetheless, the goal of a high-throughput study is not to conduct as many reactions as possible, but rather to focus on relevant parameters in order to gain the maximum information about a certain experiment. Therefore, a well thought out design of experiments is needed since "A poorly designed experiment will give bad information with unprecedented speed and in outstanding quantity".[18]

3.1.2 High-throughput Methods for Solvothermal Syntheses

Heterogeneous catalyst impregnation process, sol–gel and solvothermal synthesis are among the most important solvent-based processes where HT methods have been applied.[19,20] Employing HT solvothermal syntheses, i.e. reactions in enclosed containers at temperatures above the boiling point of the solvent and autogenous pressures, many new compounds such as zeolites, metal-phosphates, -carboxylates and -imidazolates have been discovered.[21–24] The harsh temperature and pressure conditions in solvothermal syntheses require the usage of durable and corrosion-resistant materials as well as autoclaves. Most often Teflon®-lined steel autoclaves are used and maximum temperatures of around 200 °C and, thereby, accomplishing autogenous pressure. While single reactors are commercially available and also easy to build, only few research groups have been able to construct multiclaves containing 24, 48 or even 100 miniaturized reaction chambers. One example of such a multiclave is shown in Figure 3.1. The multiclaves are comprised of miniaturized Teflon® liners inserted into a 96-well-plate-based stainless steel block.[21,25–27]

3.1.3 High-throughput Methodology

A typical high-throughput workflow is shown in Figure 3.2.[28] It begins with the design of the experiments (Step 1). This could be done manually or with

Figure 3.1 Stainless steel HT-multiclaves with 48- and 24-Teflon®-lined reaction chambers (120 mm × 80 mm) employed by the group of Stock.

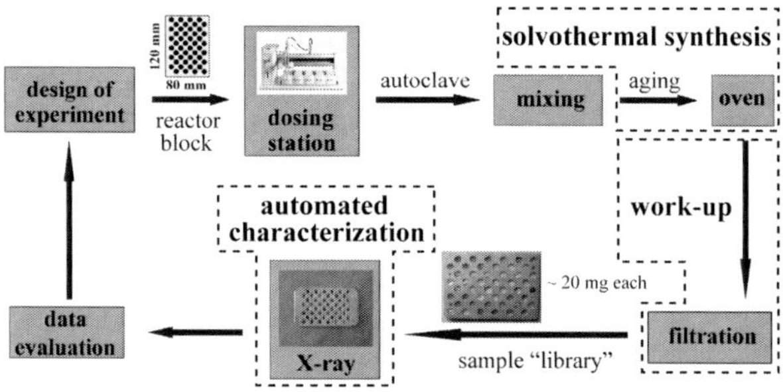

Figure 3.2 A typical HT workflow scheme for the synthesis and characterization of reactions under solvothermal conditions. (Reproduced with permission from ref. 28.)

computer software, depending on the complexity of the experiment. The next step is the preparation of the reaction mixtures (Steps 2 & 3), which will be facilitated by an automated dosing station. It should be able to deliver precise amounts of powder or liquid in the mg and μL range into the designated reaction chambers. One fine example of a fully automated system employing a central robotic arm was described by Corma *et al.* that handles liquid dosing, solid weighing, stirring and solvent evaporation.[29] A more practical, semi-automated approach was described by Bein and Stock. These groups use tools based on the 96-well-plate-format for synthesis, characterization and storage of materials (Figure 3.2).[30] The synthesis in 24 or 48 multiclaves (Step 4) is carried out in an oven with a defined temperature program. Product isolation by filtration is achieved in a custom-made filtration apparatus and the subsequent automated characterization (Steps 5 & 6) of the products is typically done by X-ray powder diffraction (XRPD) in a period of ~3 min per sample.[31] The evaluation of these XRPD data (Step 7) can get complicated due to textural problems, preferred orientation of the crystalline product or the presence of phase mixtures. New methods have been discussed recently to identify phase mixtures from XRPD data.[32]

Most of the HT investigations of metal phosphonates involve the study of three component systems, usually a metal salt, the phosphonic acid ligand and the base ($M:H_xL:B$). Ternary plots, so-called crystallization diagrams, are used to locate the molar composition of the reaction mixtures employed and to summarize the results of the investigation. Thus formation fields of each individual compound can be determined. There are various approaches to the systematic investigation of such chemical systems, which is based on the results from previous investigations and literature data. Very often the molar ratio $M:H_xL$ has only a minor influence while the acid/base content (pH) is the decisive factor in determining the product formation.

3.2 High-throughput Synthesis of Metal Phosphonates

3.2.1 Introduction

High-throughput methods have played a great role in generating numerous new metal phosphonates.[33] These compounds are based on polyphosphonic acids $R–(PO_3H_2)_n$ (n = 2–4, R = aryl or alkyl), phosphonocarboxylic acids $H_2O_3P–R–COOH$ or phosphonosulfonic acids $H_2O_3P–R–SO_3H$. Most often compositional parameters and pH values of the reaction mixtures were investigated since these variables have the most influence on the product formation. The pH values are very important in the synthesis of metal phosphonates because they control the protonation levels and thus vary the coordination modes of the linker molecules. On the other hand, the effects of other important parameters such as concentrations, solvents, reaction temperatures, mode of heating (microwave *vs.* conventional oven), reaction time and the cooling rates have been reported as well (Table 3.1).

The main goals for the application of HT methods are faster and more efficient ways towards the (a) discovery of new compounds, (b) discovery of reaction trends, (c) elucidation of synthesis–structure relationships, and (d) optimization of synthesis recipes or process conditions. Depending on the size/dimensionality of the parameter space under investigation, one can distinguish between discovery and focused arrays. Without any prior knowledge of a reaction system one has to screen many variables, normally compositional parameters. A set-up like this is called a discovery library. Once a new compound is discovered, the so-called focused libraries are set up that screen only the relevant section of the parameter space close to the discovered composition. This approach is shown in Figure 3.3.

Two reactors with 48 distinct molar ratios $Cu^{2+}:H_3L10:NaOH$ were set up (discovery array, Figure 3.3A). The molar compositions are spread out over the whole diagram for the discovery of new products in this system. Three compounds were found to crystallize in a rather small formation field. Therefore, "focused arrays" centering on the area of interest helps to identify the exact formation area of the individual compounds (Figures 3.3B and 3.3C).

Table 3.1 List of metal and organo-ammonium phosphonates synthesized using HT methods.

Type	*Linker*	*Compounds obtained*
Phosphonate	(*L*1) 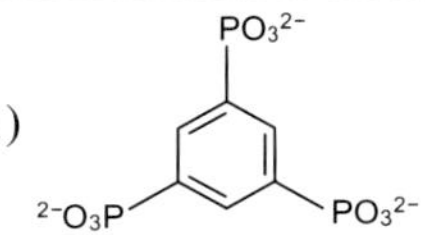	$[Na_2Zn_2(H_3L1)_2]$, $[Zn_2(H_2L1)]$[34]
	(*L*2) 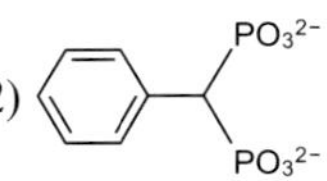	$(NH_3CH_2\text{-}C_6H_4\text{-}CH_2NH_3)_{1.5}(HL2)(H_2O)_4$, $(C_6H_5\text{-}NH_3)_2(H_2L2)$, $(C_5H_{10}NH_2)(H_3L2)$, $(C_5H_5NH)(H_3L2)$, $(NH_3\text{-}C_2H_4\text{-}NH_3)_{0.5}(H_3L2)CH_3OH$, $(C_3N_6H_7)_2(H_2L2)(H_2O)_4$[35]
	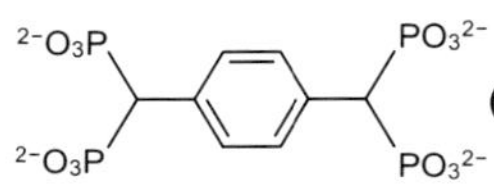 (*L*3)	$(NH_3CH_2\text{-}C_6H_4\text{-}CH_2NH_3)(H_6L3)(H_2O)_{1.5}$[35] $[Ln(H_5L3)]\cdot(H_2O)_4$, $[Ln_2(H_2L3)]\cdot(H_2O)_8$, $[NaLn(H_4L3)]\cdot(H_2O)_4$, Ln: La, Nd, Gd, Dy[36]
	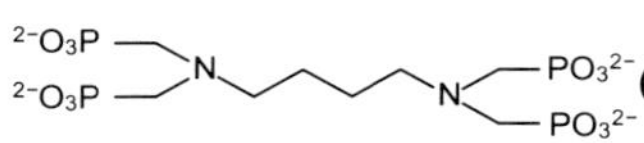 (*L*4)	$[M(H_6L4)]$, M = Mg, Mn, Co, Ni, Cd, Zn[30]
Phosphonocarboxylate	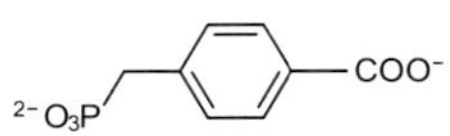 (*L*5)	$[Zn(H_2L5)_2]$, $[Zn(HL5)]$, $[Zn_3(L5)_2(H_2O)_2]\cdot H_2O$, $[Zn(H_2L5)_2(H_2O)_4]$[30] $[Mn(H_2L5)_2(H_2O)_2]$, $[Mn(HL5)(H_2O)]$, $Mn_2(OH)(L5)(H_2O)_2$, $[Cu(H_2L5)_2(H_2O)_2]$, $[Cd(HL5)(H_2O)]$[37]
	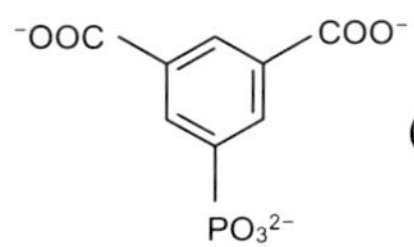 (*L*6)	$[Sm_2(H_2O)_4(HL6)_2]\cdot(H_2O)_2$, $[Cu_3(H_2O)(HL6)_2]\cdot(H_2O)_2$, $[Ca_2(H_2O)(H_2L6)_2]$, $[Ba_2(L6)(H_2O)_3]$[38]

bis(phosphono)aminocarboxylate	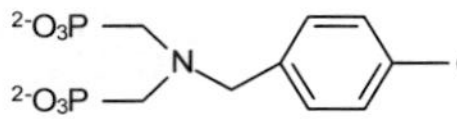 ($L7$)	$[Co_2(HL7)(H_2O)]$, $[Co(HL7^{ox})(H_2O)]$, $Co(H_3L7)$, $Co_2(HL7)(H_2O)_{3.5}$[13] $[Sm(H_2L7)(H_2O)]$, $[Ca(H_3L7)(H_2O)]$, $[Ca(H_4L7)_2(H_2O)_4]$[39] $[Cd(H_4L7)_2(H_2O)_2]\cdot(H_2O)_2$, $[Cd(H_4L7)_2]$, $[Cd(H_3L7)(H_2O)]$, $Cd_3(H_2L7)_2(H_2O)_{14}$, $Cd_2(HL7)(H_2O)_3$, $Cd_2(HL7)H_2O$[40] $Cd_3(OH)(L7)(H_2O)_{4.5}$[41]
Phosphonosulfonate	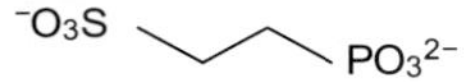 ($L8$)	$[Ln(L8)(H_2O)]$, Ln: La, Ce, Pr, Nd, Sm, Eu, Gd, Tb, Dy[42] $[Ln(L8)]$, Ln: Ho, Er, Tm, Yb, Lu, Y[10] $[Ba(HL8)]$[43] $[Cu_2(L8)(OH)(H_2O)]\cdot H_2O$, $[Cu_{2.5}(L8)(OH)_2]$, $[Cu_{1.5}(L8)(H_2O)]\cdot H_2O$, $[Cu_2[(L8)(OH)(H_2O)_2]\cdot(H_2O)_3$, $NaCu(L8)(H_2O)_3$, $[Cu_{1.5}(L8)(H_2O)]$[44] $[Sr(HL8)]$, $[Sr_3(L8)_2(H_2O)_2]$[45]
	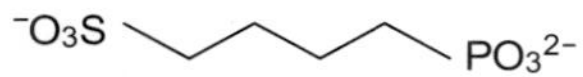 ($L9$)	$[Cu_{1.5}(L9)(H_2O)]\cdot(H_2O)_{1.15}$, $[Cu_{2.5}(L9)(OH)_2]$[46]
	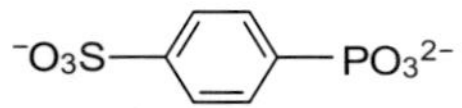 ($L10$)	$[Pb_2(L10)(OH)]$, $[Cu_{1.5}(L10)(H_2O)]$, $NaCu(L10)(H_2O)_3$, $[Cu_2(L10)(OH)(H_2O)]$, $[Cu_3(L10)_2(H_2O)_2]$[47]

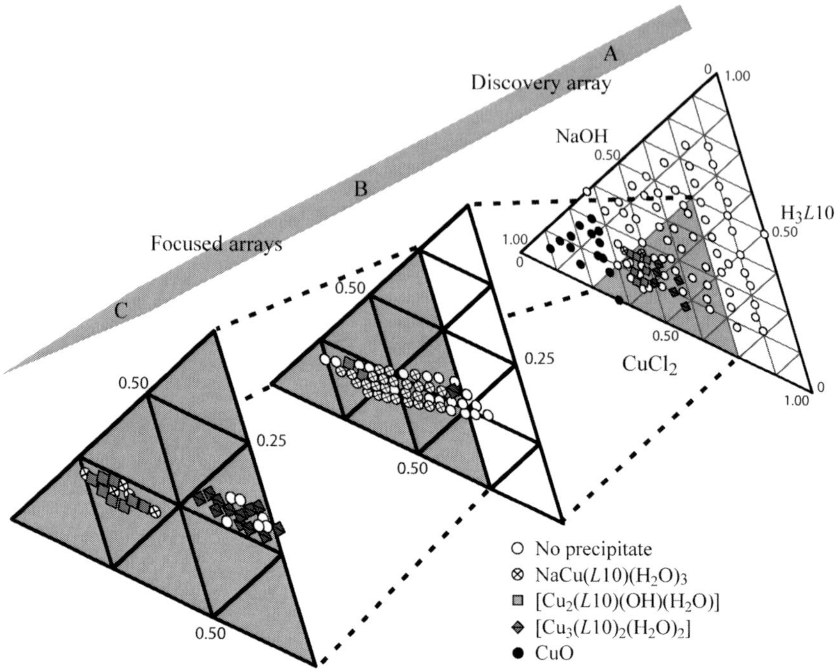

Figure 3.3 HT-investigation of the system $Cu^{2+}/H_3L10/NaOH$ ($H_3L10 = H_2O_3P$-C_6H_4-SO_3H) starting with a discovery array A and followed by focused arrays B and C. Array C reveals the specific formation fields of $[Cu_2(L10)(OH)(H_2O)]$ and $[Cu_3(L10)_2(H_2O)_2]$. Results are based on XRPD data. Relative molar amounts of $CuCl_2$, H_3L10 and NaOH are stated as mole fractions. (Reproduced with permission from ref. 47.)

In the following sections, examples are given that focus on a certain parameter, i.e. concentration, metal salt, pH and temperature. In general these parameters cannot be investigated individually since they mutually affect each other. These chapters highlight examples that are most profoundly influenced by the respective parameter.

3.2.2 Dilution

Metal phosphonates discovered during the early stages of an HT investigation can be easily identified via powder XRD measurement. However, the knowledge of the full crystal structure is fundamental to understand the synthesis–structure relationships. Therefore, single crystals are needed for the structure elucidation via single crystal X-ray diffraction. The formation of single crystals depends on the nucleation and crystal growth rate, which can be influenced by the reaction temperature and time, the cooling rate and the solvent/solvent mixture employed. We have systematically investigated two approaches. The stepwise dilution of the reaction mixtures in the system

Figure 3.4 Crystal size growth of [Zn(H_6L4)] with increasing water content as evidenced by SEM micrographs. Zn^{2+}/H_8L4 molar ratios were fixed at 1:1. (Reproduced with permission from ref. 30.)

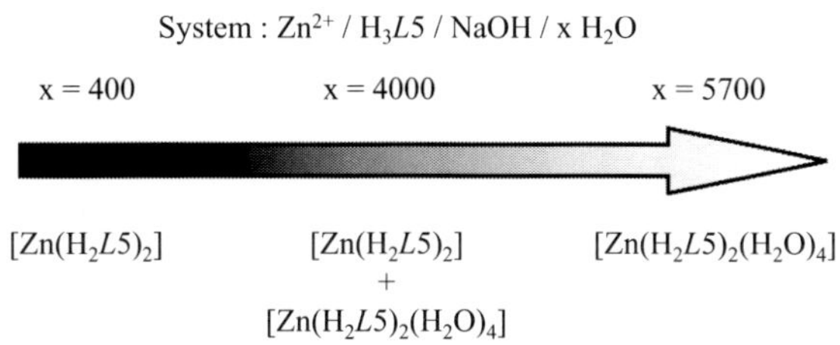

Figure 3.5 Investigation of the impact of water content in the synthesis of [Zn(H_2L5)$_2$] leading to the discovery of the new phase [Zn(H_2L5)$_2$(H_2O)$_4$]. Zn^{2+}:H_3L5:NaOH molar ratios were fixed at 2.5:5:1.5.

Zn^{2+}/H_8L4/H_2O ($H_8L4 = (H_2O_3PCH_2)_2N–(CH_2)_4–N(CH_2PO_3H_2)_2$) generated crystals under high dilution that were suitable for the structure determination (Figure 3.4).[30]

On the other hand, the variation of the amount of water can also lead to the formation of a new compound.[48] In this experiment, the mixtures 2.5 Zn^{2+}:5 H_3L5:1.5 NaOH:$x$$H_2O$ ($H_3L5 = p$-$H_2O_3PCH_2–C_6H_4–COOH$) with $400 \leq x \leq 5700$ were investigated. Two phases were obtained, [Zn(H_2L5)$_2$] at $400 \leq x < 4000$ and a previously unknown phase [Zn(H_2L5)$_2$(H_2O)$_4$] at $x = 5700$; $x = 4000$ resulted in a mixture of both phases (Figure 3.5). Comparing the crystal structures of both compounds, it can be concluded that higher H_2O content tends to produce a new zinc phosphonate with inclusion of water as a ligand. The coordination environment of the zinc ions changed from tetrahedral in [Zn(H_2L5)$_2$] to octahedral in [Zn(H_2L5)$_2$(H_2O)$_4$].

3.2.3 Metal Ion Screening

To study the potential of a phosphonic acid for the formation of new metal phsophonates, different metal ions need to be screened. This is due to the

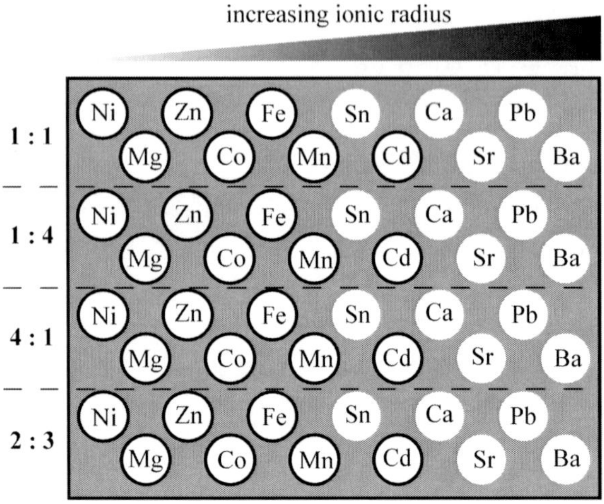

Figure 3.6 Results from the HT metal ion-screening experiment for the system M^{2+}/ H_8L4. Twelve different divalent metal ions (ionic radii range of 69–135 pm[49,50]) were used. Outlined circles indicate crystalline products, normal circles is for X-ray amorphous products.

properties of the metal ions, e.g. the charge, the ionic radius, the preferred coordination number, the presence of sterically active lone pairs of electrons. A metal ion screening experiment was conducted in the system M^{2+}/H_8L4 at four different molar ratios (1:1, 1:4, 4:1, 2:3) using twelve different divalent ions with ionic radii of 69–135 pm and two containing lone pairs of electrons. The set-up and the results of the study are presented in Figure 3.6.[30] Only one crystalline phase is obtained, $M^{II}(H_6L4)$. The XRPD analysis results showed that only metal ions with ionic radii between 69 and 100 pm and without lone pair of electrons were capable of forming crystalline compounds while bigger ions produce only X-ray amorphous products. This screening was followed by a more detailed investigation using focused arrays and varying the water content of the reaction mixtures. Thus, most of the new compounds could be isolated as single-crystalline products.

In another study $H_2O_3P–C_2H_4–SO_3H$ acid (H_3L8) was investigated.[10,42] Trivalent rare earth ions were screened and two new phases are observed. For larger ions (La–Dy) eight-fold coordination and for Y and Ln = Ho–Lu six-fold coordination is observed. On the basis of SEM micrographs a correlation of the crystal size with the pH and the ionic radius could be drawn. Larger rare-earth ions lead under acidic conditions to larger crystals (Figure 3.7).

3.2.4 Counter-ion Screening

The anions of the metal salts used in the synthesis can also have an influence on the formation of the products. They can, for example, influence the pH of the starting solution, be involved in the coordination to the metal ions during the

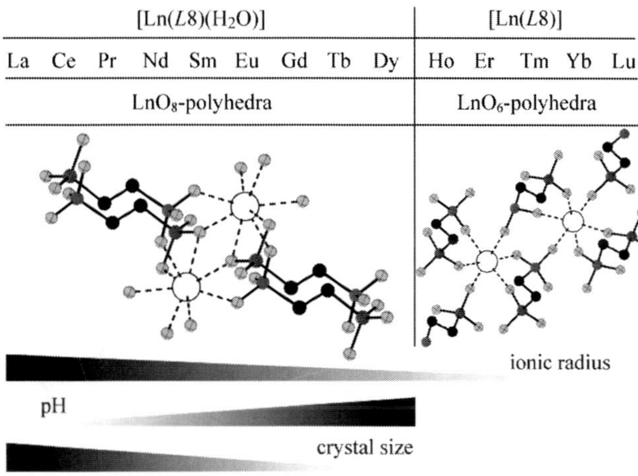

Figure 3.7 Results of HT metal screening experiment of the system $Ln^{3+}/H_3L8/NaOH$. The correlation between ionic radius of lanthanoid ion and pH value with the crystal size were extracted. Ln, white; C, black; light gray, oxygen; dark gray, phosphorus/sulfur.

reaction or in the final product, or they can take part in redox reactions. Thus, depending on the counter-ions, new products or products of varying crystallinity were obtained.

In the system $Co^{2+}/H_5L7/NaOH$, four different cobalt(II) salts, namely the hydrates of $CoCl_2$, $Co(O_2CCH_3)_2$, $Co(NO_3)_2$ and $CoSO_4$, were used employing the same reaction conditions ($H_5L7 = (H_2O_3PCH_2)_2N–CH_2–C_6H_4–COOH$).[13] Four new cobalt carboxyarylphosphonates were discovered. When using $CoCl_2$ or $CoSO_4$ only minor changes in the fields of formation were observed, due to changes in the pH of the starting solutions. Using $Co(O_2CCH_3)_2$ led only to one compound since its solutions are already slightly basic (Figure 3.8). In the case of $Co(NO_3)_2$ as the starting material, an *in situ* oxidation of H_5L7 is observed (Scheme 3.1). The cleavage of the P–C bonds and the oxidation led to the new ligand containing a formamide group which is stabilized by complexation and the formation of the corresponding cobalt phosphonate (Scheme 3.1).

Prior to the investigation of the system $Cu^{2+}/H_3L10/NaOH$ presented in Section 3.2.1 (Figure 3.3) metal and counter-ion screening experiments were performed. Cu^{2+} ions were selected since they led to the formation of different crystalline products. The results of the counter-ion screening (nitrate, chloride, sulfate and acetate) are shown in Figure 3.9. Three parameters were investigated, the increase of (a) metal concentration, (b) ligand concentration and (c) NaOH concentration. $CuSO_4$ and $Cu(O_2CCH_3)_2$ are not the reactants of choice since no metal phosphonate or only Cu_2O is formed over a large part of the parameter space. In comparison to $Cu(NO_3)_2$, $CuCl_2$ was chosen for additional studies (see Figure 3.3) as the ideal reactant since it produces only pure phase products.

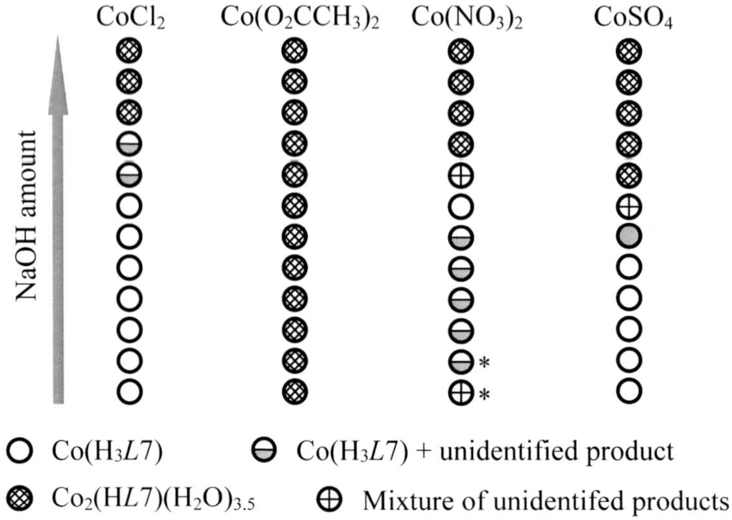

Figure 3.8 Results of HT counter-ion screening of the system Co^{2+}/H_5L7/NaOH employing chloride, acetate, nitrate and sulfate salts. Results are based on XRPD data. * = single crystals of [Co(H$L7^{ox}$)(H_2O) were isolated. (Reproduced with permission from ref. 13.)

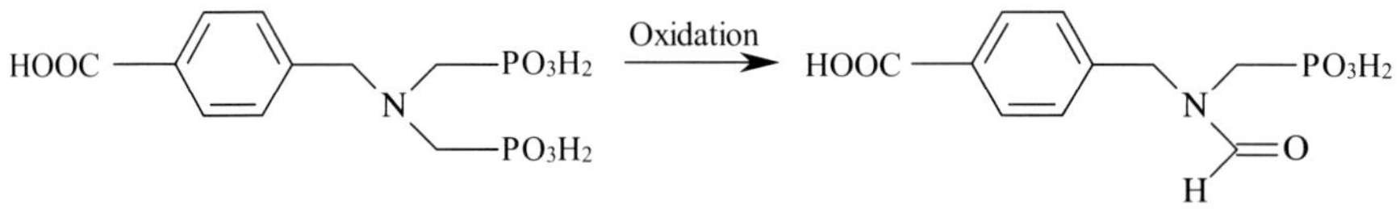

Scheme 3.1 Reaction scheme for the oxidation of H_5L7 producing a new formamide-group containing H_3L7^{ox} ligand which coordinates with Co^{2+} to form [Co(H$L7^{ox}$)(H_2O)].

3.2.5 pH Screening

In the synthesis of metal phosphonates, the pH of the starting solution has in general the most profound influence on the product formation. Therefore, this parameter is screened in most of the HT investigations. It is determined by the type of counter-ions and most importantly, by the amount of base added. The HT methodology is especially powerful in cases where an automated measurement of pH values is possible.[30] This was demonstrated for the system Zn^{2+}/H_3L5/NaOH/H_2O (H_3L5 = *p*-$H_2O_3PCH_2$–C_6H_4–COOH). A discovery library was set up, keeping the water content constant and varying the molar ratios of the three starting materials, according to Figure 3.10. The results are also presented in the same figure. Under basic conditions (pH > 9) the formation of ZnO is observed. In addition, three new phases and their crystal structures were obtained, [Zn(H_2L5)$_2$], [Zn(H$L5$)] and [Zn_3($L5$)$_2$(H_2O)$_2$] · H_2O.

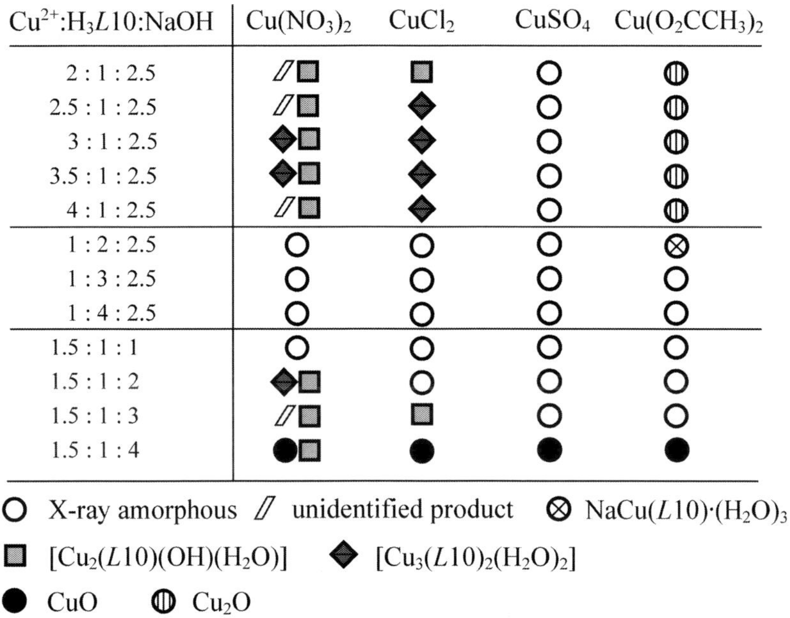

Figure 3.9 Results of HT counter-ion screening of the system $Cu^{2+}/H_3L10/NaOH$ employing nitrate, chloride, sulfate and acetate salts. New copper phosphonates were obtained only with nitrate and chloride salts whereas acetate and sulfate salts gave mostly CuO/Cu_2O. Results are based on XRPD data.[47]

Automated pH measurements showed that a decrease in the H^+ concentration leads to higher condensed structures. While isolated columns are observed in $[Zn(H_2L5)_2]$ (pH ~ 0 in region A), layers that are connected via hydrogen bonds are found in $[Zn(HL5)]$ (pH ~ 1 in region B) and in $[Zn_3(L5)_2(H_2O)_2]\cdot H_2O$ a 3D framework structure is observed (pH ~ 6 in region D).[30]

In most of the studies it has been observed that the pH has a stronger influence on the product formation compared to the molar ratio metal: phosphonic acid. This can clearly been seen in the studies concerned with the synthesis of manganese phosphonates. In the system $Mn^{2+}/H_3L5/NaOH$ the four different molar ratios $Mn^{2+}:H_3L5 = 2:1$, 1:1, 2:3 and 1:2 were used and the amount of NaOH was changed in twelve steps (the diagonal lines in Figure 3.11).[37]

The study allowed the isolation of three new manganese carboxyphosphonates and enabled the extraction of the correlation of ligand deprotonation and structure dimensionality. With increasing pH, $[Mn(H_2L5)_2(H_2O)_2]$ and $[Mn(HL5)(H_2O)]$ exhibiting a 1D and 2D structure, respectively, were obtained. Phase mixtures of both compounds were observed at intermediate levels of NaOH. Further increase of the pH leads to $Mn_2(OH)(L5)(H_2O)_2$. Although the crystal structure could not be determined, spectroscopic results

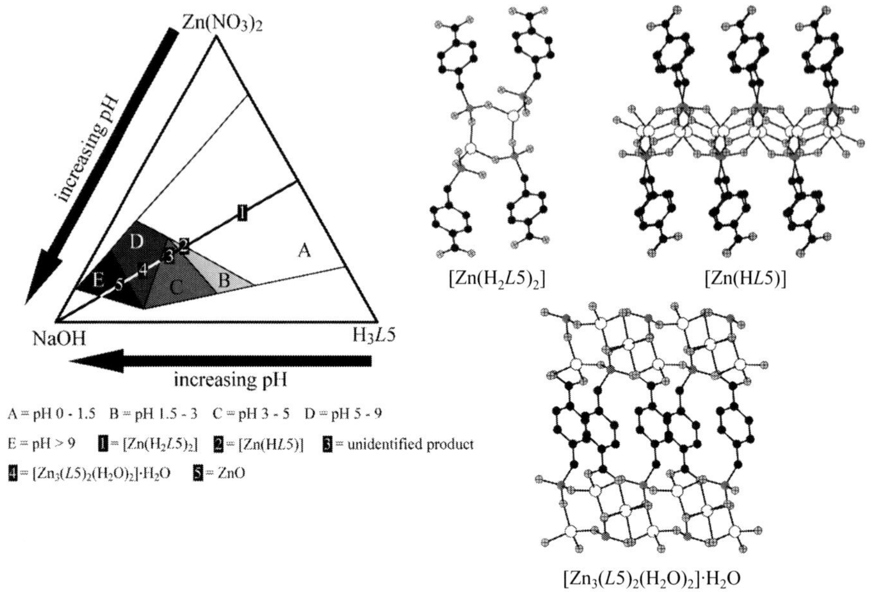

Figure 3.10 Left: crystallization diagram of the phases obtained in the system $Zn(NO_3)_2/H_3L5/NaOH$. Five different pH regions were classified whereby in each region a specific product formation was observed. Right: sections of crystal structures of compound **1**, **2** and **4**. Zn: white, C: black, P: dark gray, O: light gray.

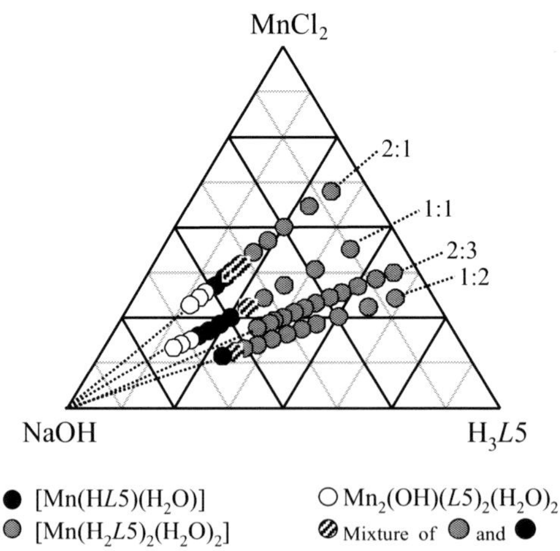

Figure 3.11 Crystallization diagram of the system $MnCl_2/H_3L5/NaOH$. Four different molar ratios Mn^{2+}:H_3L5 were investigated. Results are based on XRPD data.[37]

reveal the deprotonation of the carboxylic acid group and therefore a 3D structure is very likely. Further results on the influence of the pH on the product formation will be described in the following Section 3.2.6 about multiparameter investigations.

3.2.6 Multi-parameter Investigations: Temperature Screening

Solvothermal syntheses are typically considered as "black box" processes since the influence of compositional parameters such as metal salts, ligands, base and the solvent on the product can not be predicted. The complexity is further increased due to process parameters such as reaction time, reactor volume and most significantly, reaction temperature. Generally, it is hard to single out individual parameters since they mutually affect each other. Multi-parameter HT investigations with a focus on the influence of the reaction temperature can be performed in two ways, using HT reactors at multiple isothermal conditions or employing a HT reactor with a temperature gradient. In the investigation of the system $Cu^{2+}/H_3L8/NaOH$ ($H_3L8 = H_2O_3P–C_2H_4–SO_3H$), six different isothermal reactions were conducted using the 48-HT reactor system. The temperature range 90–190 °C in steps of 20 °C was chosen.[44] Five new copper(II) phosphonoethanesulfonates were discovered (Figure 3.12). One compound, marked as light grey circles, forms at all temperatures while certain compounds are only observed in the lower (≤110 °C, marked as dark grey circles and light grey bars) or higher (≥130 °C, marked as dotted circles and fused triangles) temperature range. By correlating the pH values of the solutions (starting and end pH of reactions were identical within error margins) with the reaction temperatures, the following trends can be deduced. At all investigated temperatures, $[Cu_{1.5}(L8)(H_2O)]\cdot H_2O$ crystallized under highly acidic conditions (pH 2–3) and CuO is formed under basic conditions (pH ≥ 8). The formations of the other compounds which favour moderate acidic conditions (pH 3–8) are temperature selective (Figure 3.12, bottom left and right). The following structural trends are observed: (a) the increase in reaction temperature led to a decrease in water content in the crystal structure, and (b) elevated pH values result in more condensed Cu–O–Cu connectivity.

A breakthrough in high-throughput investigation of metal phosphonates is the study of the influence of the reaction temperature using a thermocycler which allows the setting of a temperature gradient. Up to now, this technology enables the simultaneous exploration of reaction temperatures up to 100 °C as well as compositional parameters. The molar ratios $1\,Cd^{2+}:1\,H_5L7:x\,NaOH$ ($x = 0, 0.5, 1, 1.5, 2, 2.5$) were studied at 16 different temperatures (17–99.9 °C, Figure 3.13) ($H_5L7 = (H_2O_3PCH_2)_2N–CH_2–C_6H_4–COOH$). Six new cadmium phosphonocarboxylates (A–F) were synthesized.[40] In addition, reactions were performed using conventional reactors at temperatures up to 200 °C which led to pure phase C as well as the discovery of phase G. In Figure 3.13 it is clearly observable that there are well defined formation fields for most of the phases, i.e. well defined pH and temperature ranges. A detailed structure–reaction

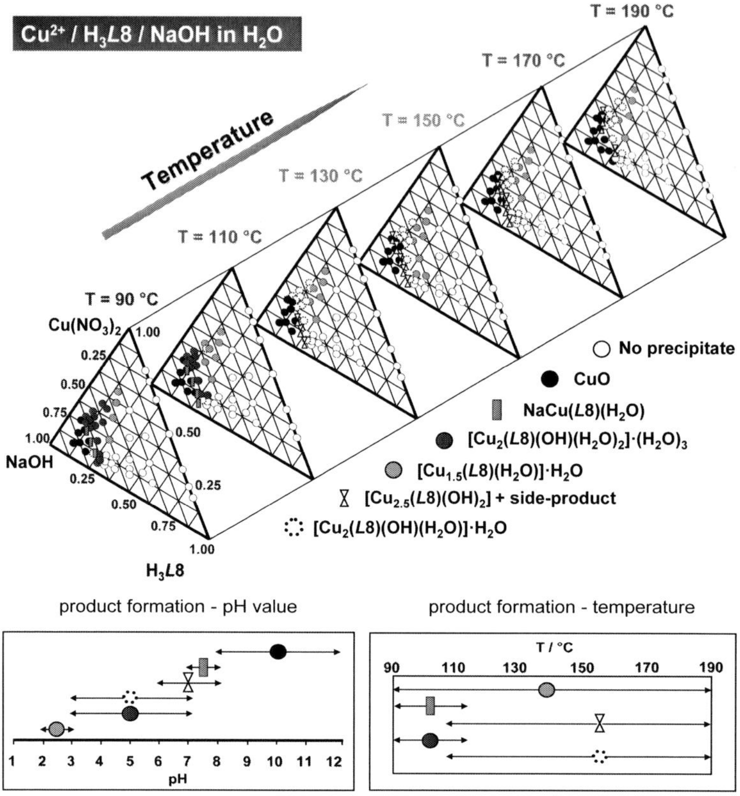

Figure 3.12 Top: results of isothermal HT investigation of the system $Cu^{2+}/H_3L8/$ NaOH are displayed in crystallization diagrams. Six different temperatures were studied and five new copper phosphonates were obtained. Phase mixtures are indicated by the overlapping symbols. Bottom left: overview of pH values with formation fields of compounds. Bottom right: overview of reaction temperatures with formation fields of the compounds. (Reproduced with permission from ref. 44.)

temperature analysis revealed the following correlation: within a fixed molar ratio Cd^{2+}:H_5L7:NaOH, higher reaction temperatures lead to a higher ratio Cd^{2+}:$H_{5-n}L7^{n-}$ (n depending on deprotonation) and a lower water content in the structures. Similar to the results of the previously mentioned copper(II) phosphonatoethanesulfonates, higher reaction temperatures lead to more condensed structures. In compounds A (17 °C < T < 35 °C) and B (45 °C < T < 100 °C) isolated CdO_6 polyhedra are observed that are connected to chains and layers, respectively. Compound C has a similar layered structure but dinuclear Cd_2O_{10} clusters are observed. The correlation between crystal structure and pH of the synthesis mixture reveals that the increase of the molar ratio NaOH:H_5L7 leads to a higher degree of ligand deprotonation and thus to

Code	Cd^{2+}/H_5L7	H_2O/Cd^{2+}	Formula
A	0.5	4	$[Cd(H_4L7)_2(H_2O)_2]\cdot(H_2O)_2$
B	0.5	0	$[Cd(H_4L7)_2]$
C	1	1	$[Cd(H_3L7)(H_2O)]$
D	1.5	4.25	$Cd_3(H_2L7)_2(H_2O)_{14}$
E	2	1.5	$Cd_2(HL7)(H_2O)_3$
F	-	-	unidentified product
G	2	0.5	$Cd_2(HL7)(H_2O)$

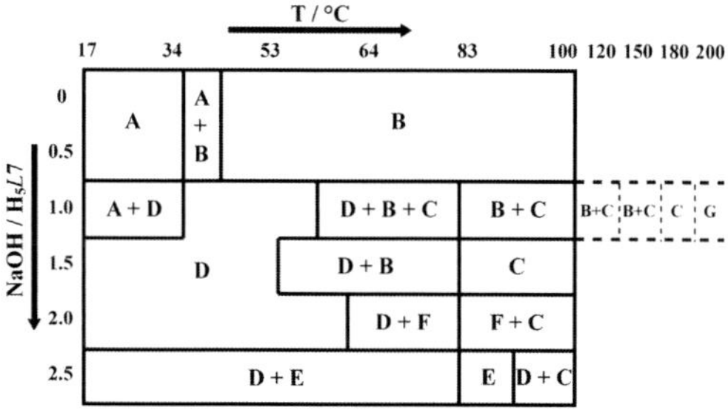

Figure 3.13 Left: table summarizing the cadmium phosphonocarboxylates formed. Compounds A to F were discovered in the HT experiment in a thermocycler with temperature gradient. G was synthesized using a conventional autoclave at 200 °C. Right: schematic diagram illustrating the crystallization fields depending on reaction temperature and NaOH/H_5L7 molar ratio. Phases found above 100 °C were synthesized with conventional autoclaves. (Reproduced with permission from ref. 40.)

compounds with higher molar ratios $Cd^{2+}:H_{5-n}L7^{n-}$, which in general is equivalent to more condensed structures.

3.3 Conclusion

High-throughput methods have proven to be an efficient and valuable tool in the area of discovery, synthesis optimization and determination of synthesis–structure trends. The advantages and disadvantages are summarized in Table 3.2. Due to the small amounts of starting materials employed – equivalent to lower costs per sample – and the possibility to perform a large number of reactions simultaneously, this methodology enables the investigation of complex parameter systems. Thus, parameters are systematically scanned which are

Table 3.2 The advantages and disadvantages of HT methods. (Reproduced with permission from ref. 28.)

Advantages	*Disadvantages*
lower chemical consumption	workflow has to be adjusted to HT system
high reproducibility	high starting investment
time saving	scale-up occasionally problematic
identification of reaction trends	high interdisciplinary skills needed
improved control over reaction conditions	plethora of meaningless data due to poorly designed experiment
enabling systematic investigation	problematic evaluation of X-ray powder diffraction due to preferred orientation
low human error	small amounts of products
lower costs per sample	
less routine work providing more time for creativity	

normally not fully studied or not taken into account at all, for example the influence of dilution or the metal counter-ion, the pH of the starting solution or the reaction temperature. In addition, metal screening allows one to gain an overview of the coordination characteristics of a phosphonic acid ligand and is thus a good starting point for the investigation.

While in solvothermal syntheses compounds are often obtained that are hard to reproduce due to minor impurities or human error, HT methods show quickly the presence of outliers since well-defined fields of formation for the phases can be established. The parallelization guaranties that all samples within a reactor are subjected to the same reaction conditions. Thus, their use leads to an enhanced reproducibility and a better comparison of the results.

There are also a number of disadvantages that need to be mentioned. The most important being the high starting investment and the need to establish a workflow encompassing the steps: synthesis, work-up and characterization and data evaluation in a HT manner. For this, corresponding tools need to be purchased or custom made. The complexity of the methodology demands interdisciplinary cooperation. Thus, the close teamwork of engineers, chemists as well as IT scientists is necessary. Another disadvantage is the formation of only small amounts of samples. Therefore, the synthesis scale-up is often necessary which can sometimes be problematic. Nevertheless the advantages clearly dominate and HT methods minimize the necessity of routine work and hence provide more time for creativity.

Acknowledgement

Financial support from DFG Project STO 643/2-3 is gratefully acknowledged.

References

1. CSD database search (exact entries found in CSD: 2282), 2 August 2010.
2. S. Bauer, H. Müller, T. Bein and N. Stock, *Inorg. Chem.*, 2005, **44**, 9464.
3. R. W. Deemie, M. Rao and D. A. Knight, *J. Organomet. Chem.*, 1999, **585**, 162.
4. T.-Y. Ma, X.-Z. Lin, X.-J. Zhang and Z.-Y. Yuan, *New J. Chem.*, 2010, **34**, 1209.
5. S. Khanra, M. Kloth, H. Mansaray, C. A. Muryn, F. Tuna, E. C. Sañudo, M. Helliwell, E. J. L. McInnes and R. E. P. Winpenny, *Angew. Chem., Int. Ed. Engl.*, 2007, **46**, 5568.
6. G. Cao, V. M. Lynch and L. N. Yacullo, *Chem. Mater.*, 1993, **5**, 1000.
7. N. Stock and T. Bein, *J. Solid State Chem.*, 2002, **167**, 330.
8. P. J. Bryne, D. S. Wragg, J. E. Warren and R. E. Morris, *Dalton Trans.*, 2009, 795.
9. R. Murugavel, P. Davis and M. G. Walawalkar, *Z. Anorg. Allg. Chem.*, 2005, **631**, 2806.
10. A. Sonnauer and N. Stock, *J. Solid State Chem.*, 2008, **181**, 3065.

11. P.-A. Jaffrès, V. Caignaert and D. Villemin, *Chem. Commun.*, 1999, 1997.
12. E. Irran, T. Bein and N. Stock, *J. Solid State Chem.*, 2003, **173**, 293.
13. S. Bauer, T. Bein and N. Stock, *Inorg. Chem.*, 2005, **44**, 5882.
14. D. L. Coffen and J. E. A. Luithle, in *Handbook of Combinatorial Chemistry*, ed. K. C. Nicolaou, R. Hanko and W. Hartwig, Wiley-VCH, Weinheim, 2002, p. 10.
15. D. Maclean, J. J. Baldwin, V. T. Ivanov, Y. Kato, A. Shaw, P. Schneider and E. M. Gordon, *Pure Appl. Chem.*, 1999, **71**, 2349.
16. W. F. Maier, K. Stöwe and S. Sieg, *Angew. Chem., Int. Ed. Engl.*, 2007, **46**, 6016.
17. N. Stock, *Microporous Mesoporous Mat.*, 2010, **129**, 287.
18. J. N. Cawse, in *Experimental Design for Combinatorial and High Throughput Materials Development*, (Hrsg: J. N. Cawse), John Wiley & Sons, New York, 2003.
19. U. Rodemerck, D. Wolf, O. V. Buyevskaya, P. Claus, S. Senkan and M. Baerns, *Chem. Eng. J.*, 2001, **82**, 3.
20. K. Choi, D. Gardner, N. Hildebrandt and T. Bein, *Angew. Chem., Int. Ed. Engl.*, 1999, **38**, 2891.
21. D. E. Akporiaye, I. M. Dahl, A. Karlsson and R. Wendelbo, *Angew. Chem., Int. Ed. Engl.*, 1998, **37**, 609.
22. Y. Song, J. Yu, G. Li, Y. Li, Y. Wang and R. Xu, *Chem. Commun.*, 2002, 1720.
23. S. Bauer, C. Serre, T. Devic, P. Horjacada, J. Marrot, G. Ferey and N. Stock, *Inorg. Chem.*, 2008, **48**, 7568.
24. R. Banerjee, A. Phan, B. Wang, C. Knobler, H. Furukawa, M. O'Keeffe and O. M. Yaghi, *Science*, 2008, **319**, 939.
25. J. Klein, C. W. Lehmann, H.-W. Schmidt and W. F. Maier, *Angew. Chem., Int. Ed. Engl.*, 1998, **37**, 3369.
26. K. Choi, D. Gardner, N. Hildebrandt and T. Bein, *Angew. Chem., Int. Ed. Engl.*, 1999, **38**, 2891.
27. N. Stock and T. Bein, *Solid State Sci.*, 2003, **5**, 1207.
28. S. Bauer and N. Stock, *Chem. Unserer Zeit*, 2007, **41**, 390.
29. M. Moliner, J. M. Serra, A. Corma, E. Argente, S. Valero and V. Botti, *Microporous Mesoporous Mat.*, 2005, **78**, 73.
30. N. Stock and T. Bein, *Angew. Chem., Int. Ed. Engl.*, 2004, **43**, 749.
31. STOE Stadi P Combi X-Ray Powder Diffractometer, STOE & Cie GmbH, Darmstadt, Germany, 2010.
32. L. A. Baumes, M. Moliner, M. Nicoloyannis and A. Corma, *Cryst. Eng. Commun.*, 2008, **10**, 1321.
33. S. Bauer and N. Stock, *Stud. Surf. Sci. Catal.*, 2007, **170**, 2015.
34. A. Sonnauer, Doctoral Dissertation: Systematic Investigation of Inorganic-Organic Hybrid Compounds Based on Polyfunctional Phosphonic Acids Using High-Throughput Methods, available from MACAU database, University of Kiel, Germany, 2008.
35. M. Plabst, N. Stock and T. Bein, *Cryst. Growth Des.*, 2009, **9**, 2049.
36. M. Plabst and T. Bein, *Inorg. Chem.*, 2009, **48**, 4331.

37. N. Stock and T. Bein, *J. Mater. Chem.*, 2005, **15**, 1384.
38. S. Bauer and N. Stock, *J. Solid Sci. Chem.*, 2007, **180**, 3111.
39. S. Bauer, T. Bein and N. Stock, *J. Solid State Chem.*, 2006, **179**, 145.
40. S. Bauer and N. Stock, *Angew. Chem.*, 2007, **119**, 6981.
41. S. Bauer, T. Bein and N. Stock, *Solid State Sci.*, 2008, **10**, 837.
42. A. Sonnauer, C. Näther, H. A. Höppe, J. Senker and N. Stock, *Inorg. Chem.*, 2007, **46**, 9968.
43. A. Sonnauer and N. Stock, *J. Solid State Chem.*, 2008, **181**, 473.
44. A. Sonnauer and N. Stock, *Eur. J. Inorg. Chem.*, 2008, 5038.
45. A. Sonnauer and N. Stock, *Solid State Sci.*, 2009, **11**, 358.
46. A. Sonnauer, M. Feyand and N. Stock, *Cryst. Growth. Des.*, 2009, **9**, 586.
47. P. Maniam, C. Näther and N. Stock, *Eur. J. Inorg. Chem.*, 2010, 3866.
48. N. Stock, private communication.
49. R. D. Shannon, *Acta Crystallogr., Sect. A*, 1976, **32**, 751.
50. L. H. Ahrens, *Geochim. Cosmochim. Acta*, 1952, **2**, 155.

CHAPTER 4

Metal Carboxyphosphonates

AURELIO CABEZA AND MIGUEL A. G. ARANDA

Departamento de Química Inorgánica, Cristalografía y Mineralogía, Universidad de Málaga, 29071 Málaga, Spain

4.1 Introduction

Crystalline organo-inorganic hybrid materials[1] have been widely studied over the last decades due to their valuable properties and potential applications in different areas including gas storage, gas separation, catalysis, optical isomer separation, nonlinear optical materials or drug storage. The term metal-organic frameworks, "MOFs", has been coined for these compounds having 3D porous structures.[2] MOFs are crystalline coordination polymers[3] constructed by coordinate bonds between multidentate organic ligands and metal atoms, or metal-containing clusters also known as secondary building units, "SBUs".[4] By far the most studied MOF hybrids are based on metal-carboxylates but other ligands are emerging which yield hybrids with very promising properties such as zeolite-like imidazol frameworks, "ZIFs".[5] Another type of ligand to construct MOFs is multidentate phosphonic acids.[6] Developing microporosity in phosphonate MOFs is not an easy task as the presence of three oxygens in the phosphonate ligand yield more condensed structures. But it must be highlighted that phosphonate-based MOFs would be very valuable as they will show higher thermal (and chemical) stabilities, due to the stronger $M–O_3PR$ bond, compared to the corresponding carboxylates MOFs and ZIFs.

Carboxyphosphonic ligands, or phosphonocarboxylic acids, are heterofunctional organic building units which are attracting increasing interest. The resulting metal hybrids may have many different properties and they can be considered as an intermediate step between carboxylate MOFs and

Metal Phosphonate Chemistry: From Synthesis to Applications
Edited by Abraham Clearfield and Konstantinos Demadis

Published by the Royal Society of Chemistry, www.rsc.org

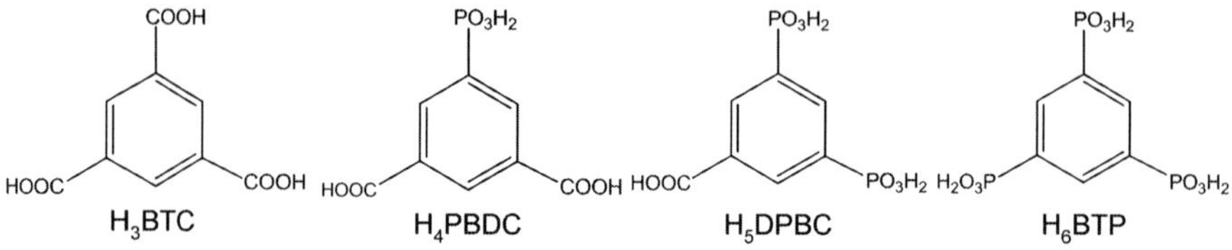

Figure 4.1 Trisubstituted benzene rigid rings with different carboxylic/phosphonic ratios from H_3BTC all the way to H_6BTP. The acronyms for the caboxyphosphonic acids are defined in Table 4.1. Adapted from reference 7.

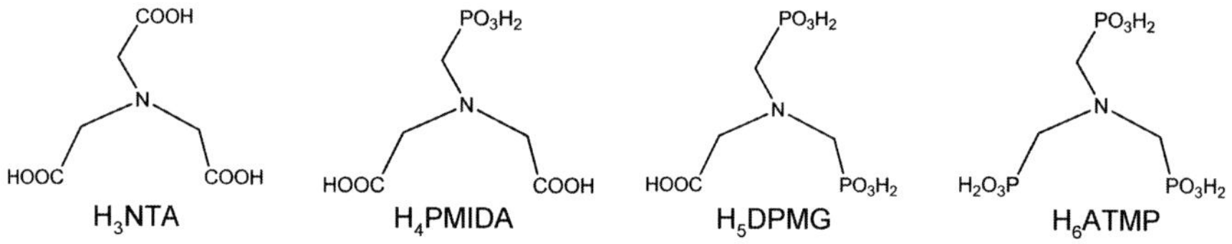

Figure 4.2 Trisubstituted amino groups showing with different carboxylic/phosphonic ratios from H_3NTA all the way to H_6ATMP. The acronyms for the caboxyphosphonic acids are defined in Table 4.1.

phosphonate MOFs. In order to highlight the relationship between the carboxylate MOFs and the phosphonate MOFs, Figure 4.1 displays four ligands of trisubstituted benzene rings from H_3BTC (1,3,5-benzene tricarboxylic acid, also named trimesic acid) to H_6BTP (1,3,5-benzene triphosphonic acid).[7] H_3BTC is a rigid ligand very well known for yielding highly microporous MOFs, and it is expected that these rigid carboxyphosphonic ligands may also result in porous hybrid materials. Metal hybrids derived from H_6BTP are known but they do not exhibit high microporosities. The relationship between carboxylic and phosphonic ligands is not restricted to rigid units but it has also been employed in numerous flexible chains including other functional groups. As an example, Figure 4.2 shows four ligands of trisubstituted amino groups from nitrilotriacetic acid, "H_3NTA", which is a highly stable chelating agent to aminotris(methylenephosphonic acid), "H_6ATMP", which is mainly used to prevent scale formation in water systems due to its excellent chelating ability.[8]

Carboxyphosphonic ligands are heterofunctional organic compounds and therefore they present different acidic groups. The acidities may change due to the presence of other groups within the ligand. Just to give some numbers, the easiest example is carboxyalkylphosphonic acids. For instance, carboxyethylphosphonic acid has two phosphonic and one carboxylic exchangeable protons, with pK_a values of 2.2, 4.6 (corresponding to the –COOH group), and 7.7 (corresponding to the $-PO_3H^-$ group).[9] Much more complex is the case for aminocarboxyphosphonic ligands. Sana *et al.*[8] reported pK_a values for a number of aminocarboxylic, aminophosphonic and aminocarboxyphosphonic ligands. As an example, the pK_a values for *N*-(phosphonomethyl)iminodiacetic acid "H_4PMIDA" were reported being <1, ~ 1.4, 2.3, 5.5 and 10.5. Figure 4.3

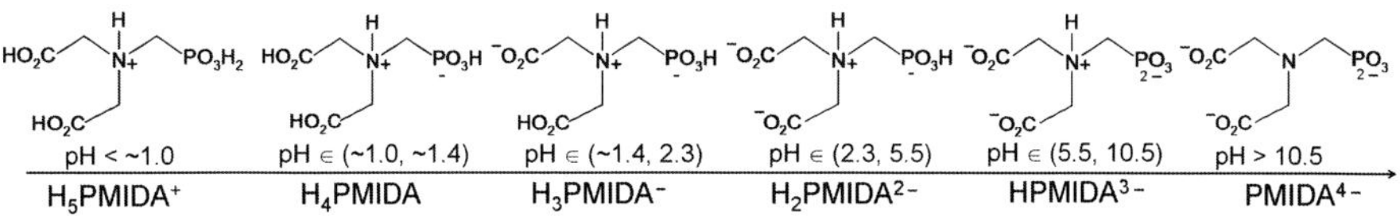

Figure 4.3 Charge evolution of H_4PMIDA, an amino-carboxy-phosphonic ligand, with pH from a cation at very low pH's to the tetranegative anion at high pH values.

shows the evolution of H_4PMIDA with pH from a cation at very low pH values to a tetranegative anion at pH higher than 10.5. The zwitterion nature of PMIDA is displayed in a wide range of pH because the N–H proton is the last to be released out at pH ~ 10.

The acidic nature of the carboxylic and phosphonic groups is just one parameter, but their chelating ability with respect to the metal is another very important factor. In general terms, syntheses at low pH values may yield the carboxylic group protonated meanwhile at higher pH values, both groups may be deprotonated and bonded to the metal. For instance, this is the case for the reaction of 3-phosphonobenzoic acid and a zinc salt. It has been reported that at pH below 4.1 only the phosphonic acid is bonded to the inorganic network yielding a layered hybrid, meanwhile at higher pH both types of functional groups were directly connected to the zinc cations resulting in a 3D framework.[10]

Due to space limitations, it is not possible to discuss here all the work reported on metal carboxyphosphonates. Instead, a personal overview of some developments in the field is provided.

- Initially, a survey of the different type of carboxyphosphonic ligands is presented. Emphasis is placed in the large variety of names and acronyms employed so far, with some ligands being given with four different names, which indicates that some kind of order is needed in the field.
- Afterward, the different synthetic methodologies used are described. The number of crystal structures of metal carboxyphosphonates is sky-rocketing. Here, we will discuss a few examples highlighting different dimensionalities and key features.
- Then, some properties and applications of these hybrids are discussed.
- Finally, a brief personal outlook about the field is reported.

4.2 Carboxyphosphonate Ligands

The number of different carboxyphosphonic (or phosphonocarboxylic) ligands already used to prepare hybrid materials exceeds 30. The coupling of these ligands with all the possible metals yields an endless number of hybrid metal carboxyphosphonates. It must also be noted the trend observed in recent reports of using two different chelating ligands (carboxyphosphonate and carboxylate, for instance, oxalate). However, this will not further discussed here

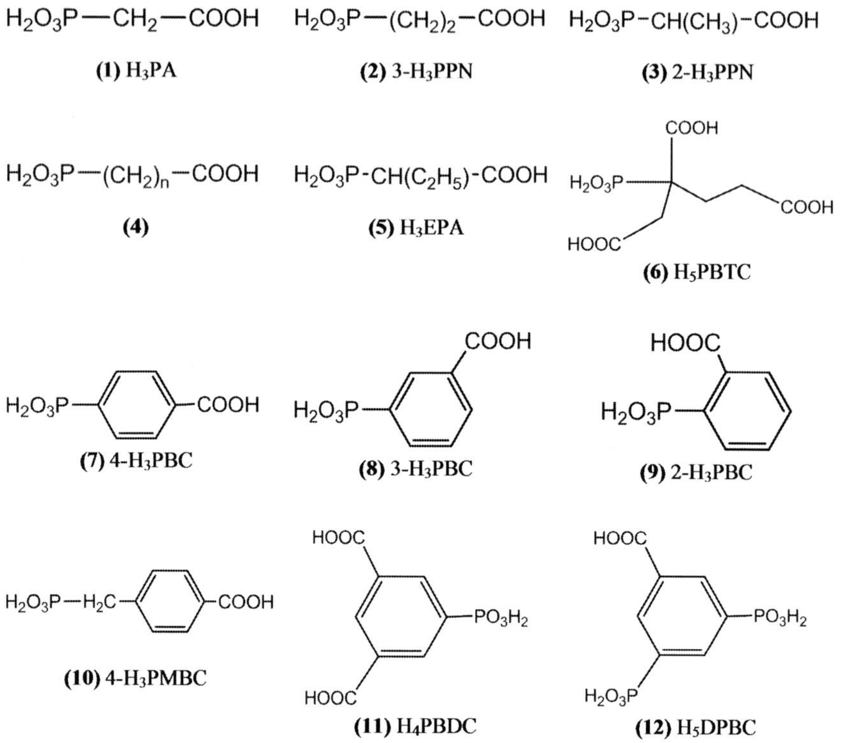

Figure 4.4 Common alkyl- and aryl- carboxyphosphonic acids with a proposed acronym highlighting the number of exchangeable protons. The full names are given in Table 4.1.

as Chapter 5 reviews the use of auxiliary ligands. Further complexity arises from the use of additional compounds: mineralizers and/or organic molecules (to be used as templates, or not). Focusing on the carboxyphosphonate ligands, it must be highlighted that there are acids which have been reported with three, even four, different names and acronyms. This fact adds more complexity to the management of the information within the field as there is not yet a unified criterion to describe the ligands and, therefore, describe the resulting hybrid materials.

The most common ligands are schematically depicted in Figures 4.4 and 4.5 with our proposed unified acronyms. Figure 4.4 displays alkyl- and aryl-carboxyphosphonic ligands. Figure 4.5 displays some multifunctional carboxyphosphonic acids where the amino group is by far the most used additional group. The full name(s) for these alkyl-, aryl- and multifunctional carboxyphosphonic ligands are given in Table 4.1. In that table, we have added the proposed acronyms, most of them already used in bibliography but not uniquely. If the proposed acronyms become generally accepted within the research community, it would allow one to follow the developments within the field much more easily. The acronyms include the number of exchangeable

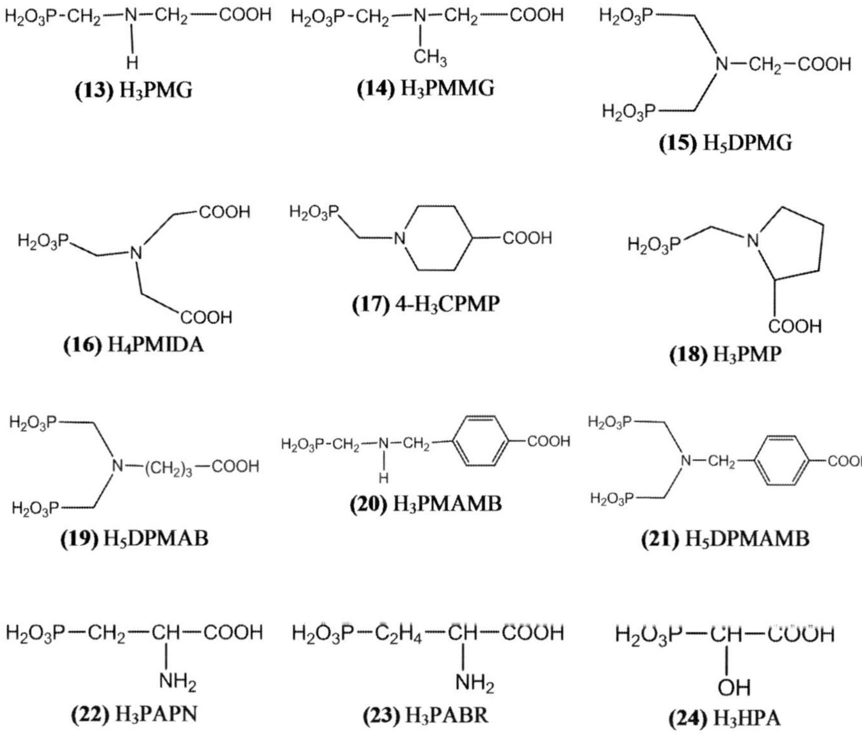

Figure 4.5 Common multifunctional carboxyphosphonic acids with a proposed acronym highlighting the number of exchangeable protons. The full names are given in Table 4.1.

protons, as this is a very important parameter that allows describing both the ligand and the stoichiometry of the resulting hybrid. It has been already mentioned in the introduction that the metal/ligand ratio strongly depends upon the pH value of the synthesis, among several other factors. In any case, it is very helpful to easily describe the number of the remaining protons within the prepared hybrid. Finally, it must be emphasized that the references given in Table 4.1 are just a guide, but by no means is it an exhaustive compilation. We can not include in the table the references dealing with those ligands which exceed a few hundreds.

4.3 Syntheses

4.3.1 Hydrothermal Synthesis

Most of the syntheses of metal carboxyphosphonates are carried out from the direct reaction of a metal salt with the carboxyphosphonic acid under hydrothermal conditions (temperature range 100–200 °C, autogenous pressures) and variable time (from a few hours to several days). Several additives are

Table 4.1 Full name(s), proposed acronyms, including the number of exchangeable protons, and selected references, for the alkyl-, aryl- and multifunctional- carboxyphosphonic ligands shown in Figures 4.4 and 4.5.

Code	*Proposed acronym*	*Full name*	*Other acronyms*	*Ref.*
(1)	**H_3PA**	2-carboxymethylphosphonic acid phosphonoacetic acid phosphonomethylcarboxylic acid	H3ppat PAA	11–15
(2)	**3-H_3PPN**	2-carboxyethylphosphonic acid 3-phosphonopropionic acid	CEPA H3PPN	16–20
(3)	**2-H_3PPN**	2-phosphonopropionic acid		21, 22
(4)	–	n-carboxyalkylphosphonic acid ($n = 3$, 4, up to 15)		23
(5)	**H_3EPA**	ethylphosphonoacetic acid		21, 22
(6)	**H_5PBTC**	2-phosphonobutane-1,2,4-tricarboxylic acid		24, 25
(7)	**4-H_3PBC**	4-carboxyphenylphosphonic acid 4-phosphonobenzoic acid	4-ccpH3	26–29
(8)	**3-H_3PBC**	3-carboxyphenylphosphonic acid 3-phosphonobenzoic acid	3-cppH3	30–33
(9)	**2-H_3PBC**	2-carboxyphenylphosphonic acid 2-phosphonobenzoic acid	2-cppH3	34
(10)	**4-H_3PMBC**	4-(carboxyphenyl)methylphosphonic acid 4-(phosphonomethyl)benzoic acid		35, 36
(11)	**H_4PBDC**	5-phosphonobenzene-1,3-dicarboxylic acid 5-phosphonoisophtalic acid		37, 38
(12)	**H_5DPBC**	3,5-diphosphonobenzoic acid		7
(13)	**H_3PMG**	*N*-(phosphonomethyl)glycine monophosphonic derivative of iminodiacetic Glyphosate	PMG IdaP Npm-Gly	8, 39, 40
(14)	**H_3PMMG**	*N*-(phosphonomethyl)-N-methylglycine	Ndpm-Gly	41, 42
(15)	**H_5DPMG**	*N*,*N*-bis(phosphonomethyl)glycine or glyphosine *N*,*N*-bis(phosphonomethyl)-aminoacetic acid	H5NTA2P GLYP	8, 43–45
(16)	**H_4PMIDA**	*N*-(phosphonomethyl)iminodiacetic acid	H4NTAP	8, 46–49
(17)	**4-H_3CPMP**	(4-carboxypiperidyl)-N-methylenephosphonic acid *N*-(phosphonomethyl)-pipecolinic acid		50–53
(18)	**H_3PMP**	(*S*)-*N*-(phosphonomethyl)proline		54–57
(19)	**H_5DPMAB**	4-[bis(phosphonomethyl)amino]-butanoic acid		58, 59
(20)	**H_3PMAMB**	4-[(phosphonomethylamino)-methyl]benzoic acid		60
(21)	**H_5DPMAMB**	4-[bis(phosphonomethyl)amino-methyl]benzoic acid		61, 62
(22)	**H_3PAPN**	3-phosphono-2-aminopropionic acid		63
(23)	**H_3PABR**	4-phosphono-2-aminobutyric acid		63
(24)	**H_3HPA**	2-hydroxyphosphonoacetic acid	HPAA	64–67

frequently incorporated to the synthesis in order to grow suitable single crystals. Thus, NH_4X ($X = F^-$, Cl^-) are added as a structure-directing or mineralizing agent or NaF, treated with acetic acid to produce HF, which can increase the solubility of the formed compounds allowing the growth of single crystals for X-ray analysis.[68,69]

Rational design of the synthesis of metal carboxyphosphonates is not an easy task because of a large number of experimental factors that must be precisely controlled: temperature, pH of the starting mixture, concentration, metal/acid molar ratio, reaction time, nature and concentration of the organic template, nature and concentration of the inorganic mineralizers, etc. Between these parameters, the pH of the reaction mixture is key in the construction of metal–organic architectures.[70] As was previously mentioned, carboxyphosphonic acids show different functional groups where the protonation degree is pH-dependent, which affects the connectivity of these functional moieties to the metallic centres and, therefore, to the structures (and properties) of the resulting compounds.[18,30,33,67,71] Alkaline solutions, amines or urea are usually added to the reaction mixture in order to adjust the initial pH. However, at least two points have to be taken into account: (a) urea, under hydrothermal conditions, hydrolyses to give ammonium carbonate that gradually increases the pH of the solution and the NH_4^+ cations facilitate the deprotonation of all the donor groups of the carboxyphosphonic acid;[19] and (b) alkaline as well as NH_4^+ cations are frequently incorporated within the frameworks of the obtained products as counterbalance ions showing different structural roles.[13,19,49,67]

The majority of these studies were carried out by defining a set of specific and individual experiments to investigate the role of determined parameters. However, the recent progress in high-throughput (HT) materials science (see Chapter 3) has led to the incorporation of parallel reactor systems to the investigation of hydrothermal syntheses. The methodology is characterized by a complete set of reactors that can be set to work under identical conditions, systematically changing some desired synthesis parameters. This methodology establishes efficiently the phase diagram under the systematically explored conditions.[72] HT methods imply four major steps: design of experiment, parallel synthesis, characterization, and data evaluation which have to be integrated in a workflow in order to reach a maximum of productivity and innovation. These synthesis techniques provide powerful tools to reach the pre-established goals in much shorter times, using lower amounts of chemicals and thus, minimizing the amount of residues. Thus, this methodology has been used to establish the effect of: (a) gradual deprotonation of phosphonocarboxylic acid and the metal/phosphonic acid molar ratio,[36,37,61,72] (b) the concentration of the reagents,[72] even (c) the effect of the temperature[62] on the structure of the resulting compounds. In general, these studies conclude that an increasing amount of base in the syntheses leads to a gradual deprotonation of the ligand and therefore to an increasing metal/ligand molar ratio in the final reaction products, therefore, giving more condensed structures.[36,72] Secondly, the metal/ligand molar ratio has a less profound impact on product formation than the reaction pH value.[72,73] Thirdly, the solids become more dense and less hydrated at higher temperatures,[67,72] which favours the formation of compounds with 3D networks.

On the other hand, counter-ions of metal salts, owing to their own acidic or redox characteristics, can influence the results of the syntheses. For instance, this was observed[61] for the system Co^{2+}/H_5DPMAMB/NaOH in which unexpected results were obtained with the redox-active nitrate as counter-ion. Under acidic conditions, NO_3^- ions induce an *in situ* oxidation of one of the P–C bonds of the ligand and a new phosphonic acid is formed, which it is stabilized by complexation of the Co^{2+} cations.

The direct use of dialkyl- or trialkylphosphonocarboxylate esters is an alternative to employ the corresponding acids as starting compounds in the hydrothermal processes. This route of *in situ* ligand formation usually requires acid- or base-catalyzed hydrolysis of the phosphonocarboxylate esters. However, there are some examples where the acidity of the metal solution, during the course of the hydrothermal treatment, was enough to carry out the hydrolysis yielding the phosphonocarboxylic acid *in situ*.[74,75] This synthetic method can be considered as a new approach to crystal engineering with some of the most notable (potential) benefits including enhanced single-crystal growth (as consequence of the slow *in situ* formation of the phosphonocarboxylic acid), simplified syntheses (removing the need to synthesize the acid form of the ligand), greener routes to materials, and the preparation of compounds not always accessible (directly).[22] There are few examples of metal carboxyphosphonates obtained via dialkyl- or trialkylphosphonates route. For instance, Na[Zn(3-PPN)] · H_2O, a microporous phosphonocarboxylate with the same topology as ABW-type zeolite, was prepared at strongly basic pH values ($pH > 14$).[76] Hou *et al.*[12] reported the synthesis of seven metal phosphonoacetates by *in situ* hydrolysis of triethylphosphonoacetate under acid conditions (pH between 2 and 5) and in presence of different amines as second ligands. Another example is an U^{VI} carboxymethylphosphonate, $(UO_2)_4(H_1PA)(PA)_2$-$(H_2O)_4$ $3H_2O$,[22] obtained by acid-catalyzed hydrolysis of the triethylphosphonoacetate. A last selected example is the preparation of a Cu^{II}-H_1PA series from diethylphosphonic acid, under hydrothermal conditions, but without addition of a mineral acid.[71]

4.3.2 Solvothermal Synthesis

The use of solvothermal conditions for the preparation of metal carboxyphosphonates is much less common than hydrothermal syntheses. For these solvothermal preparations, solvents such as *N,N*-dimethylformamide (DMF), alcohols (ethanol, i-PrOH, ethylene glycol, . . .), polycarboxylates or mixtures DMF/alcohols/water have been employed.[22,28,33,48,77–79] These methods have allowed to crystallize a large variety of compounds with novel structures, including both 2D and 3D open frameworks. For instance, Chen *et al.*[28,77] isolated a series of zeolite-like metal phosphonocarboxylates, based on 4-H_3PBC, using different organic amines as structure-directing agents under mixed-solvothermal condition. It must be underlined, as reported by Liao *et al.*,[38] that the products obtained through this traditional mixed-solvothermal method can differ from those obtained by a layered-solvothermal process carried out under similar conditions. In general, these template molecules may provide non-coordinating species around which an inorganic framework can form. Thus, these molecules may act in a variety of roles: (a) structure-directing agent, (b) space-filling species, (c) second ligand to the metals, (d) sometimes even as charge-balancing counter-ion, and (e) many times with several of these roles.[12,28,52,77–79]

4.3.3 Ionothermal Synthesis

This synthetic method consists of using ionic liquids that may act both as solvent and template.[80] These ionic liquids are commonly defined as salts, or mixtures of salts, whose melting points are lower than 100 °C. The high boiling points and low vapour pressure of these liquids means that even at temperatures typically used in the synthesis of hybrid materials, the pressure within the synthesis is essentially ambient. Other benefits of the ionic liquids are the high thermal stability, and the recyclability of ionic liquids as "green" solvents. Furthermore, the application of an ionic liquid as solvent and template can effectively remove the competition between template-framework and solvent-framework interactions that occurs inevitably in traditional preparation. So, the different chemistry of the ionothermal solvent system, as compared to that of the traditionally used hydrothermal and solvothermal systems, produces conditions under which novel structure types appear to be accessible.[81]

This methodology has already been applied for the synthesis of zeolites, phosphates, phosphites and other inorganic–organic hybrid materials, including metal phosphonates.[81,82] Between the examples of metal carboxyphosphonates obtained via ionothermal synthesis, it is worth highlighting the open-framework solid $Zn(H_1PA)\cdot H_2O$. This solid was obtained using a eutectic mixture of urea and choline chloride, at 160 °C, and starting from zinc acetate and diethylphosphonoacetic acid. However, a standard hydrothermal reaction under the same conditions and reagents resulted in the formation of $Zn_3(PA)_2\cdot nH_2O$ ($n = 3$, 4), a compound with a completely different 3D structure.[83] On the other hand, a third stoichiometry, $[Zn(PA)]\cdot NH_4$, was obtained in a eutectic mixture of urea and choline chloride but under different synthetic conditions: using $Zn(NO_3)_2\cdot 6H_2O$ and H_3PA at 80 °C.[84] This last compound is characterized by adopting a 3D framework with NH_4^+ cations within the tunnels, which were formed *in situ* from the decomposition of urea.

4.3.4 Room Temperature Synthesis

A simple procedure for the preparation of metal carboxyphosphonates is the room temperature crystallization from an aqueous solution containing the metallic salt and the acid. This methodology has been used for the preparation of a series of 1D divalent metal H_3HPA derivatives following a simple "electroneutrality" principle: at the synthesis pH, ~2.5, and in a 1:1 molar ratio, the ligand (H_1HPA^{2-}) and metal (M^{2+}) have the same but opposite charges, thus resulting in neutral frameworks with 1:1 ligand/metal ratio.[85]

Another route, for low temperature preparations, is the slow diffusion of solvent into a filtrate. For instance, this method has been employed for the growth of single crystals of two homochiral metal amino-carboxyphosphonates. Basically, it consists of allowing a very slow diffusion of absolute alcohol, at room temperature, into the filtrate obtained by the reaction of the enantiopure ligand with the metallic salt under acidic condition.[86]

4.3.5 Other Methods

Another synthetic procedure reported in literature, but not very well developed, is the direct preparation of the hybrids by heating a mixture of the metal hydroxide with a stoichiometric quantity of the carboxyphosphonic acid above the ligand melting temperature. For instance, this method was followed for the preparation of $Ni(3\text{-}H_1PPN)\cdot H_2O$[87] and $Ni_{1-x}Zn_x(3\text{-}H_1PPN)\cdot H_2O$.[88] This approach can be used where heating the carboxyphosphonic acid yields its melting instead of its decomposition.

Finally, the use of microwave radiation in hybrid synthesis is beginning to attract a lot of attention, as in some cases the enhancement of the reaction rates is two or three orders of magnitude.[89] Microwave heating has been coupled with high-throughput methodology to screen the phase diagram of the flexible ligand, 2-phosphonoethanesulfonic acid, with several lanthanide cations.[90] Microwave-assisted hydrothermal syntheses have not been frequently used for preparing phosphonocarboxylate hybrids although the synthesis of Zn and Cd derivatives of 2-phosphono-isonicotinic ligand has been already reported.[91]

4.4 Crystal Structures

Phosphonic acids functionalized with carboxylate groups are versatile ligands since both carboxylate and phosphonate groups are capable of binding the metal atom and these ligands can adopt a variety of coordination modes under different experimental conditions giving 0D, 1D, 2D and 3D structures. As an example, Figure 4.6 displays the different connectivity modes that the carboxylate group can show when binding to a metal.[37] There are excellent reviews dealing with the crystal structures of metal phosphonates[6,92–96] including some interesting carboxyphosphonate hybrids. Therefore, here we will only provide a very brief background on this area and we will focus on the discussion of recent work.

4.4.1 1D Frameworks

Metal carboxyphosphonates with a 1D framework are relatively limited compared to those showing 2D or 3D networks. In general, structures of lower dimensionality are obtained by working under more acidic conditions where consecutive protonation, first of the carboxylate and then of the phosphonate group, takes place. The presence of protonated phosphonic acid groups in these

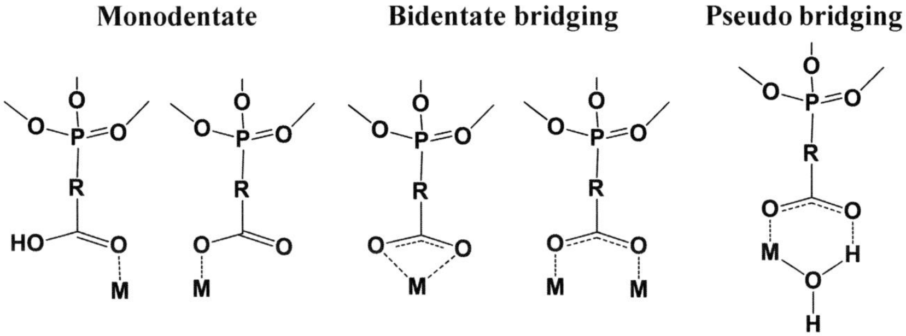

Figure 4.6 Coordination modes of carboxylic groups found in metal carboxyphosphonates. Adapted from reference 37.

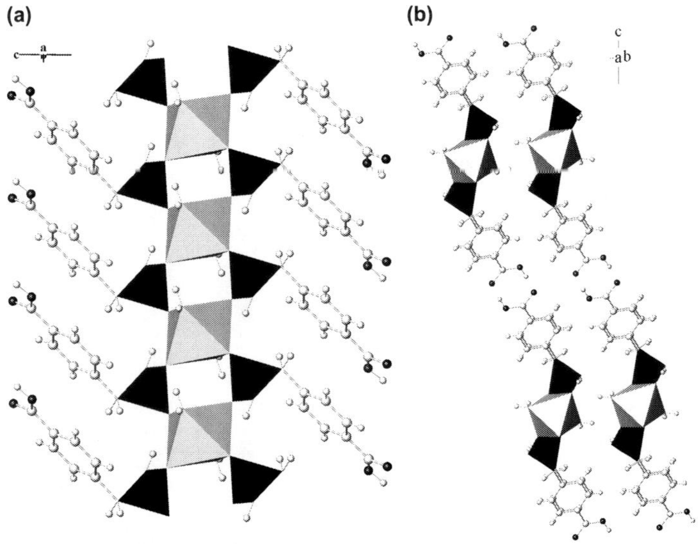

Figure 4.7 (a) Formation of chains of four rings in M[**4-H$_2$PMBC**]$_2\cdot$2H$_2$O (M^{II} = Mn, Cu). (b) Connection of the chains in M[**4-H$_2$PMBC**]$_2\cdot$2H$_2$O (M = Mn, Cu) through O–H · · · O hydrogen bonds. MO$_6$: grey octahedra; CPO$_3$: black tetrahedra; O: black balls; C: white balls; H: small white balls. Adapted from reference 36.

compounds together with the variable modes of connectivity of the phosphonic and carboxylic groups to the metal cation/s (in some heterometallic phosphonocarboxylates[13]) determine the final structural arrangement. For instance, a typical case is that observed in the system M^{II}/4-H$_3$PMBC (M = Cu, Mn)[36] and related hybrids.[26,97] These compounds, prepared under acidic conditions (pH $\approx$ 3), M^{II}[4–H$_2$PMBC]$_2\cdot$2H$_2$O, show chains built of four-membered rings of alternating corner-sharing MO$_6$ and O$_3$PC polyhedra linked to the organic moiety, with the carboxylic group pointing outside (see Figure 4.7). The 3D framework results from the hydrogen bond between the carboxylic and the protonated phosphonic groups of adjacent chains.

A different connectivity is observed for the family of hydroxyphosphonoacetates of divalent metals with general formula $M^{II}(H_1HPA)\cdot nH_2O$ (M^{II} = Mg, Co, Ni, Cu, Zn and $n = 2–4$).[67,71] All of them show a 1D structure formed by infinite chains running parallel to the *a* axis. Each ligand, within the chain, links two M^{II} cations, generating 5- and 6-membered chelate rings, and the carboxylate group bridging the two metal centers in a η^2-μ_2-fashion (see Figure 4.8a). The lattice water molecules, a maximum of two for the compounds with higher hydration degree, are hosted between the chains. This architecture is further stabilized by a number of O–H···O hydrogen bonds involving the different functional groups as well as coordination and lattice water molecules. The loss of the intrachain water molecules yields the topotactic transformation to a 3D structure of composition $M(H_1HPA)$ (M = Mg, Co, Ni, Cu and Zn).[67,98]

A special case is the compound $Ca_3(HPA)_2(H_2O)_{14}$, which is found to be an effective corrosion inhibitor for carbon steel surfaces.[99] This solid was obtained at pH 7.3 and shows a striking molecular linear trimer (see Figure 4.8b). The trimers are formed by a central Ca^{2+} centre in an octahedral environment and two peripheral Ca^{2+} centres in an eight-coordinated environment. The two

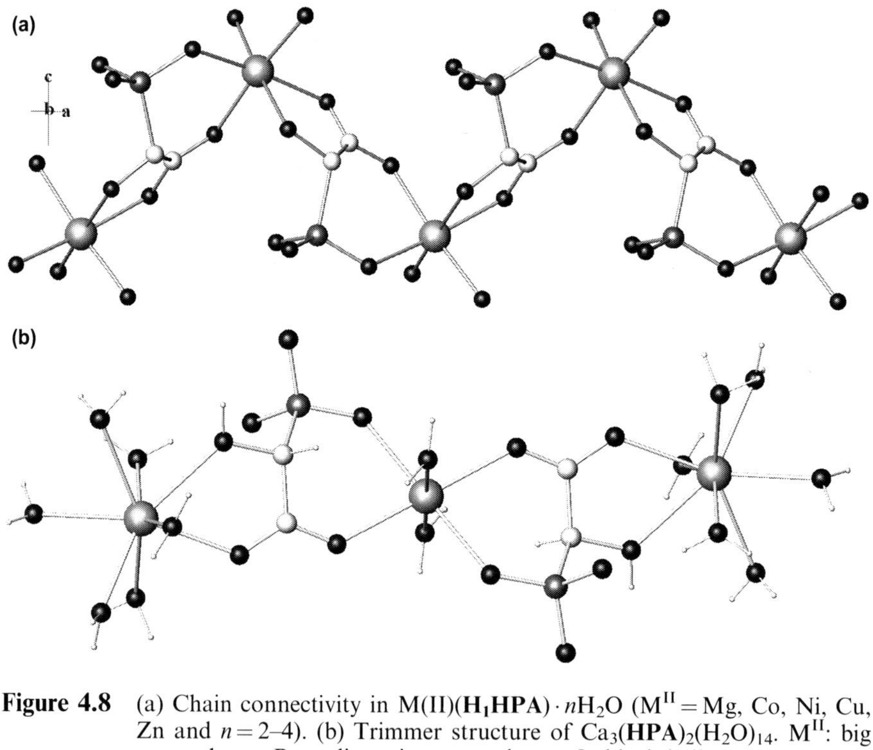

Figure 4.8 (a) Chain connectivity in $M(II)(\mathbf{H_1HPA})\cdot nH_2O$ (M^{II} = Mg, Co, Ni, Cu, Zn and $n = 2–4$). (b) Trimmer structure of $Ca_3(\mathbf{HPA})_2(H_2O)_{14}$. M^{II}: big grey spheres; P: medium-size grey spheres; O: black balls; C: white balls; H: small white balls. Adapted from references 67 and 99.

HPA^{3-} ligands function as bridges between the three Ca^{2+} ions and both phosphonate groups, completely deprotonated, act as a monodentate ligand for the central Ca. These non-coordinating phosphonate oxygens, the water molecules, and the –OH group of H_3HPA participate in strong hydrogen bonds between neighbouring trimers that stabilize the structure.

Finally, Yue *et al.* have reported the synthesis of a series of homochiral microporous lanthanide phosphonates, $Ln(H_2PMP)_3 \cdot 2H_2O$, (Ln = Tb, Dy, Eu, Gd) using an enantiomerically pure derivative of (*S*)-proline, *N*-(phosphonomethyl)proline, as a chiral building unit.[100] The resulting compounds show 1D triple-strand helical chains in which the Ln^{III} centres are interconnected by bridging phosphonate groups of the H_2PMP^- anions giving infinite 1D helical chains (see Figure 4.9). It is noteworthy that the three distinct H_2PMP^- anions in the asymmetric unit adopt the same coordination mode: only one phosphonate oxygen atom is protonated and non-coordinated, while the other two oxygen atoms are coordinated to two adjacent Ln^{III} centres in a monodentate manner. The carboxylate group of the H_2PMP^- anions is deprotonated, but it does not coordinate to the Ln^{III} centre. The organic moieties of the phosphonate groups point outward from the triple-strand helical chain, with the pyrrolidinyl ring approximately perpendicular to the chain. The adjacent triple-strand helical chains are further assembled through O–H···O and N–H···O hydrogen bonds into a 3D supramolecular framework featuring 1D tubular channels with a free diameter of about 4.32 × 3.81 Å. Inside these 1D channels there is helical water chains stabilized by hydrogen bonds.[100]

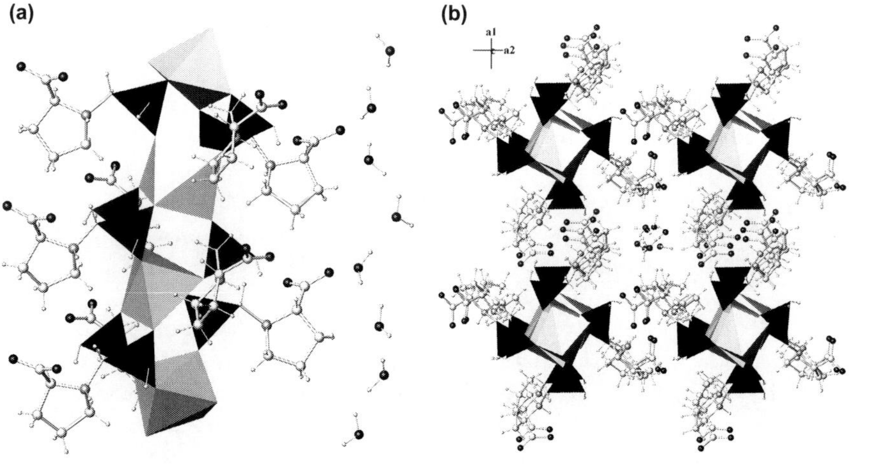

Figure 4.9 Structural views of $Ln(\mathbf{H_2PMP})_3 \cdot 2H_2O$: (a) 1D hydrogen-bonded right-handed helical host chain and the 1D left-handed helical water chain along the *c* axis. (b) 3D supramolecular framework showing the chains packing by H-bonding. LnO_6: grey distorted octahedra; CPO_3: black tetrahedra; O: black balls; C: white balls; N: small grey balls; H: small white balls. Adapted from reference 100.

4.4.2 2D Frameworks

The majority of the metal carboxyphosphonates described in the literature show a layered framework. Basically, different 2D frameworks can be found as a consequence of the arrangement of the phosphonic group and especially by the carboxylic moieties. For instance, for the system M^{n+}/3-H_3PPN ($M^{n+} = Mn^{2+}$, Mn^{3+}, Al^{3+})[101] three structural types of 2D frameworks have been reported (see Figure 4.10) and these can be generalized to many other metal carboxyphosphonates.

(a) 2D layered with the carboxylic groups arranged as a bilayer interacting through weak hydrogen bonds and with the carboxylic groups facing up to each other. This behaviour was found for Mn(3-H_1PPN) · H_2O[101] (Figure 4.10a) and is very common in other metal carboxyphosphonates, especially in alkyl,[35,36,67,102–104] and arylcarboxyphosphonates[10,30,32,35,37] and less common in amino-phosphonates.[105]
(b) 2D layered with the carboxylic groups arranged as a monolayer interacting through weak hydrogen bonds (see Figure 4.10b). This arrangement does not appear so frequently and it has been observed in some compounds such as M(OH)(3-H_1PPN) · H_2O (M = Mn^{2+}, Fe^{2+}),[101,106] Co[H_4DPMAB]$_2$,[59] Sm[H_2DPMAMB] · H_2O,[61] Ca[H_3DPMAMB] · H_2O,[61] Ca[H_3DPMAMB]$_2$ · $4H_2O$,[61] M^{II}(H_3DPMAMB) · $2H_2O$ (M^{II} = Zn and Pb).[107]
(c) 2D pillared with both phosphonic and carboxylic groups bonding the inorganic layers and therefore, this type of compounds can be also considered as 3D structures with 2D inorganic layers. This type of arrangement is found in $Al_3(OH)_3$(3-PPN)$_2$ · $3H_2O$[101] (see Figure 4.10c). This is also a very common type of structure being found in numerous metal carboxyphosphonates.[31,33,60,108–110]

There is a fourth type of layered arrangement which results from the participation of both phosphonic and carboxylic groups in the construction of the inorganic layer. In this case, the layers are held together by hydrogen bonds, i.e. an oxygen of the carboxylic group points out of the sheet,[15,22,42,46,67,71,77,99,111] or through weak van der Waals interactions, i.e. for those cases in which the organic moiety of the ligands are situated toward the interlayer space.[34,57] An example of 2D frameworks with these features is observed for the series [Cu_2M_2(PMIDA)$_2$(H_2O)$_6$] · $3H_2O$ (M = Mn, Co, Cd). These compounds show honeycomb-like 2D layers which, in turn, are made of unusual double-stranded meso-helices. The layers are stacked in ABAB sequence and are held together by numerous interlayer hydrogen bonds that involve coordinated and uncoordinated water as well as uncoordinated carboxylic oxygen atoms.[48]

4.4.3 3D Frameworks

The presence of diverse functional groups on the organo-phosphonate ligand can destabilize the layered structure and favours the formation of 3D (open)

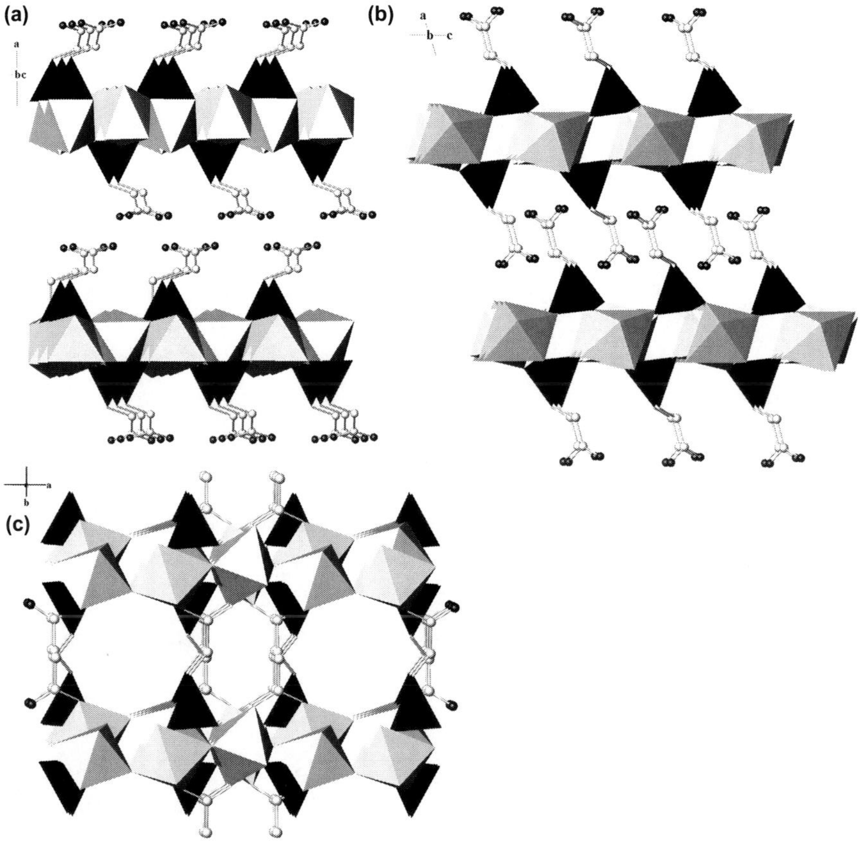

Figure 4.10 Examples of 2D arrangements based on the connectivity of the carboxylic/carboxylate groups: (a) The carboxylic groups point out to the interlayer space in a bilayer fashion, f.i. Mn(**3-HPPN**)$\cdot H_2O$; (b) the organic units are arranged in a zipper-like fashion and point to adjacent layers, f.i. Mn(OH)(**3-HPPN**)$\cdot H_2O$; (c) the carboxylic groups act as linker between adjacent layers giving pillared frameworks, f.i. $Al_3(OH)_3$(**3-PPN**)$_2 \cdot 3H_2O$; MO_6: white octahedra; CPO_3: black tetrahedra; O: black balls; C: white balls; H: not shown. Adapted from reference 101.

frameworks. The number of novel open framework carboxyphosphonates has been widely increasing, although a lot of the 3D frameworks reported correspond in fact to pillared compounds. The first metal carboxyphosphonates reported to show a 3D open framework was Co_3(3-PPN)$_2 \cdot 6H_2O$.[112] More recently, Stucky *et al.* reported the crystal structure of a Zn-carboxyphosphate, Zn_3-$(PA)_2(H_1PA)(BTC)(NH_3CH_2CH_2NH_3)$, with a complex, interrupted tetrahedral zeolite-like structure showing 1D channels composed of 24-membered rings.[78]

More recently, new open frameworks with zeolite-like structure have been described. Thus, Chen *et al.*[28] reported the synthesis of three Zn-phosphonocarboxylates based on the ligand 4-H_3PBC and prepared under mixed

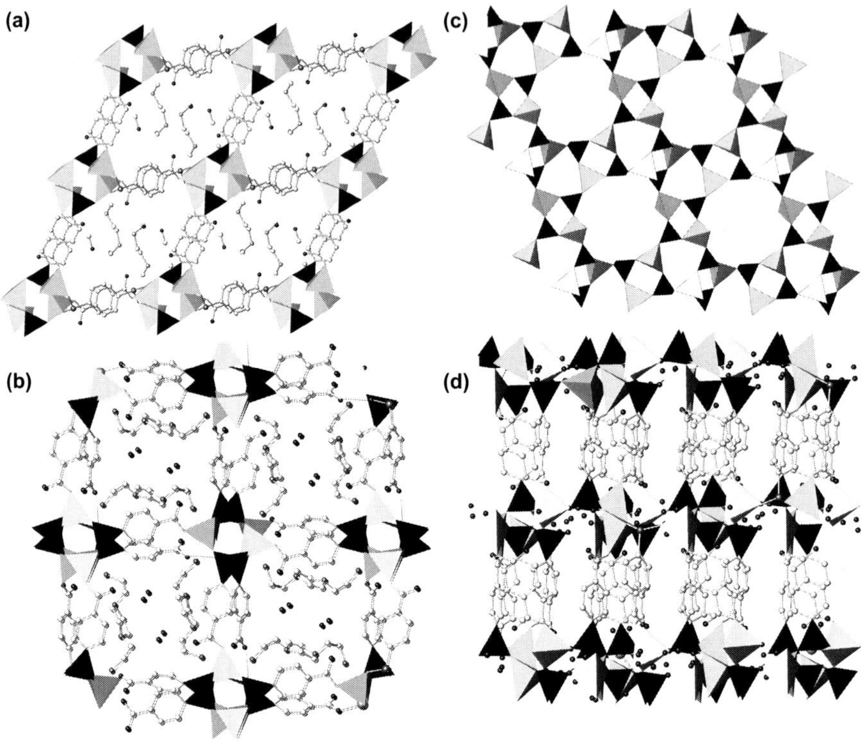

Figure 4.11 Views of four different 3D frameworks: (a) Zn_2(**PBC**)$_2$ $2H_1$dma · DMF · H_2O; (b) Zn_2(**PBC**)$_2$ · H_2teta H_2O, with GIS (gismondine)-zeolite topology; (c) inorganic layer in Zn(**PBC**) · H_3O, with zeolitic sodalite topology (the organic moieties are omitted for clarity); (d) packing of the layers in Zn(**PBC**) · H_3O by the organic moieties giving a pillared structure. ZnO_4: white tetrahedra; CPO_3: black tetrahedra; O: black balls; C: white balls; N, small black balls.

solvothermal synthesis with amines as structure-directing agents. Two of the isolated compounds, $Zn_2(PBC)_2 \cdot 2H_1$dma · DMF · H_2O and $Zn_2(PBC)_2 \cdot H_2$teta H_2O, show GIS (gismondine) and GIS-related zeolite topologies (see Figures 4.11a and 4.11b). In these structures, the sheets are composed of double crankshaft chains constructed by the corner-sharing of ZnO_4 and CPO_3 tetrahedra that are cross-linked by the organic moieties of the ligands. The third compound, Zn(PBC) · H_3O, presents inorganic (4.6.12) sheets as results of the connectivity between ZnO_4 and CPO_3. These sheets are pillared by the benzoic acid group $–C_6H_4–COO^-$ into a 3D open framework structure with zeolitic SOD (sodalite) topology (Figures 4.11c and 4.11d).[28]

An interesting porous phosphonocarboxylate with size-selected properties is $Zn_3(PBDC)_2 \cdot 2H_3O$.[38] This compound shows a 3D rutile-type porous framework. Its structure consists of trimeric zinc phosphonocarboxylate clusters ($Zn_3(CO_2)_4(PO_3)_2$, Zn3-SBU) and PBDC linkers. Considering the Zn3-SBU as

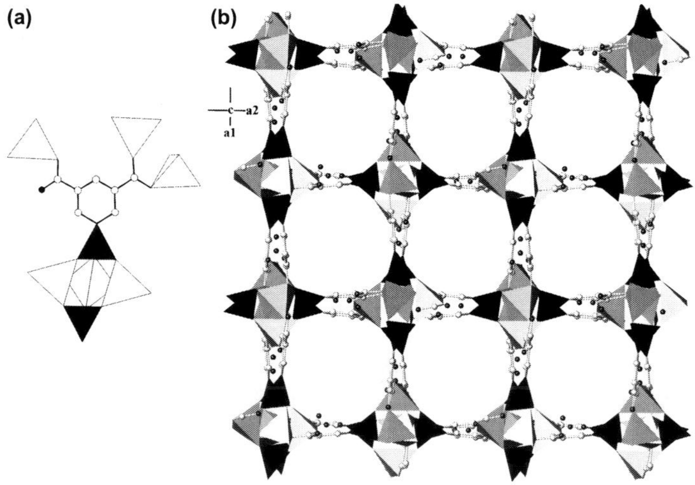

Figure 4.12 Crystal structure of $Zn_3(\mathbf{PBDC})_2 \cdot 2H_3O$: (a) connectivity of the organic ligand with the ZnO_4 tetrahedra; (b) 3D porous framework with regular open channels along the *c* axis. Tetrahedra and atoms as in Figure 4.11. Adapted from reference 38.

a 6-connected node and the PBDC as a 3-connected node (see Figure 4.12a), the 3D porous anionic framework could be described as a (3,6)-connected rutile network, which allows the presence of regular open channels along the *c* axis where the hydrated protons are situated (see Figure 4.12b).[38]

Chiral metal–organic coordination polymers with open frameworks have an especial interest due to their potential applications in enantioselective separation and catalysis. In this aspect, the first chiral metal carboxyphosphonate containing active sites for asymmetric catalysis in achiral channels was reported by Shi *et al.*[54] The homochiral solid, $Zn_2[(S)\text{-PMP}]_2$, was prepared under hydrothermal conditions from enantiopure (*S*)-H_3PMP with $Zn(NO_3)_2 \cdot 6H_2O$ in the presence of triethylamine. Its structure consists of 4,8-nets formed by alternately arranged left- and right-handed helices (composed of alternate ZnO_4 and O_3PC tetrahedral) that are connected through *S*-phosphonomethylproline ligands to generate a 3D open framework. Of particular interest is that the active sites of (*S*)-H_3PMP for asymmetric catalysis – the amino group and the partially uncoordinated carboxylate group – are retained and directed into the channels (see Figure 4.13). The structure contains, along the [100], four- and eight-membered rings (7 × 8 Å, estimated by measuring the distances between the centres of opposite atoms), and two types of microchannels with dimensions of 4 × 6 and 5 × 12 Å along the [001].

Finally, many research activities have recently focused on the synthesis of inorganic–organic hybrid compounds by incorporating a second (auxiliary) organic ligand (bipyridine,[79,113–115] oxalate,[53,116,117] sulfonate,[118] …) in the structures of metal phosphonates to create not only new 3D open frameworks

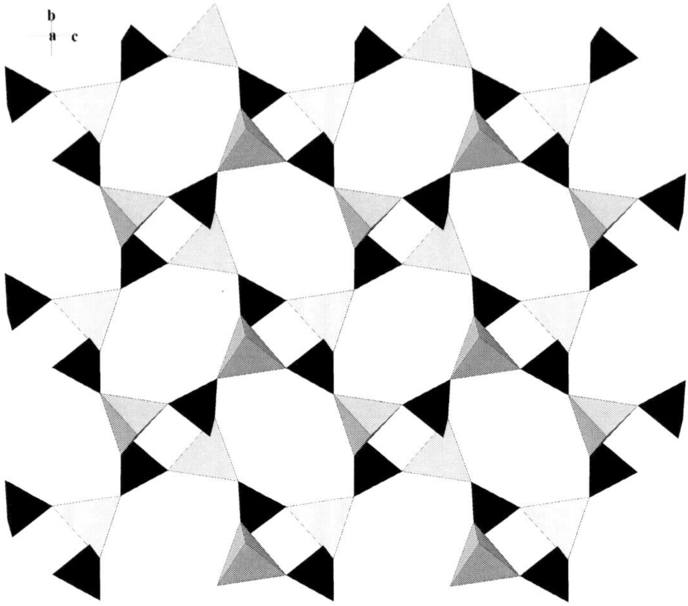

Figure 4.13 Polyhedral view of an inorganic layer in Zn_2[**(S)-PMP**]2 showing the left-handed and right-handed helices, along the *a* axis. S-phosphonomethylproline ligands are situated pointing to the 'pores' but omitted for clarity. ZnO_4: grey tetrahedra; CPO_3: black tetrahedra. Adapted from reference 54.

but also interesting properties. For instance, Zhang *et al.*[53] have reported the hydrothermal synthesis of a lanthanide oxalatophosphonate, $[Nd_2(H_1CPMP)(C_2O_4)_2(H_2O)_3] \cdot H_2O$, with a 3D framework structure (see Figure 4.14). The framework is made up from neodymium oxalate layers bridged by carboxyphosphonate ligands.

4.5 Properties and Applications

It must be noted that here we will only deal with the properties and applications of metal carboxyphosphonate hybrids. However, the reader should be aware that the carboxyphosphonic ligands are already in use for some specific applications. For instance, H_3PMG is an active ingredient of Roundup®[129], a well-known herbicide that which combines high activity with very low toxicity to animals and short residence time in the soil. A related aminocarboxyphosphonic acid, H_5DPMG, is used as a plant growth regulator and also as a herbicide.[119]

4.5.1 Microporosity and Gas Adsorption

There are many reports about open framework carboxyphosphonates.[78,83] However, this statement, derived from the study of the crystal structure, does

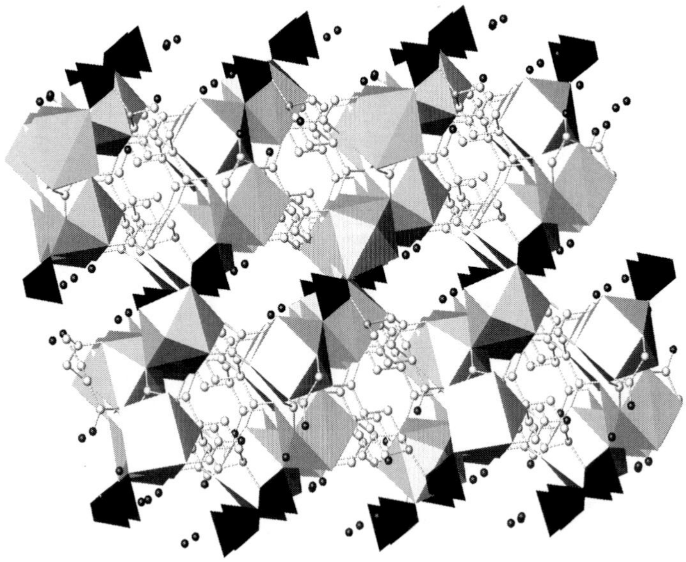

Figure 4.14 3D framework view for $[Nd_2(\mathbf{H_1CPMP})(C_2O_4)_2(H_2O)_3]\cdot H_2O$ along *b* axis showing the 1D channels. NdO_9: grey polyhedra; CPO_3: black tetrahedra; O: black balls; C: white balls; N, small grey balls; H: not shown. Adapted from reference 53.

not mean a microporous material as the "voids" may be filled by different type of species or simply not accessible. Microporosity must be supported by physical measurements such as adsorption isotherms, etc. Furthermore, metal carboxyphosphonates with the topology of zeolites have been published[16,28,76] but no porosity was reported. An initial outstanding example was $Ln(H_2PMP)_3\cdot 2H_2O$ (Ln = Tb, Dy, Eu, Gd), discussed previously.[100] After removal of the water between the chains, the Dy derivative exhibits selective adsorption capacities for N_2, H_2O and CH_3OH, with a BET surface area of $81\,m^2\,g^{-1}$ derived from the N_2 isotherm.

An important step forward has been the preparation of zinc phosphonocarboxylates derived from 4-H_3PBC by mixed solvothermal reaction using DMF and water. Methanol, ethanol and water adsorption isotherms at 298 K were reported for $Zn_2(4\text{-}PBC)_2\cdot H_1dma\cdot H_3O\cdot 2H_2O$ (dma = dimethylamine). The methanol isotherm has a typical type-I shape clearly indicating microporosity, with an estimated Langmuir surface area of $210\,m^2\,g^{-1}$. The ethanol isotherm shows much lower porosity, suggesting that this microporous metal phosphonocarboxylate may show selectivity towards alcohol separation.[77] Hydrogen adsorption has also been reported for a Zn and Cd derivative of enantiomerically pure 1-phosphonomethyl hydroxyproline.[120] This compound, with helical channels, shows a moderate hydrogen adsorption capacity of ~0.9 wt% at 77 K under 0.8 Mpa.

On the other hand, mesoporous carboxyphosphonate hybrids are also known. Aluminium derivatives of H_3PMP were prepared under the presence of phosphoric acid and an appropriate surfactant agent.[121] All surfactant-extracted carboxyphosphonate hybrids presented typical type IV isotherms with H_2 hysteresis loop (according to IUPAC classification) characteristic of the mesoporous material with pores volumes close to $0.5\,cm^3\,g^{-1}$ and BET surface areas close to $400\,m^2\,g^{-1}$.

4.5.2 Catalysis and Proton Conductivity

The use of MOFs (or suitably modified derivatives) in heterogeneous catalysis is attracting a lot of attention. This subject has been very recently reviewed,[122] including some interesting examples from metal phosphonates. Furthermore, the enantioselective catalysis by homochiral MOFs is becoming a popular subfield.[123] Unfortunately, and to the best of our knowledge, there are no reports of metal carboxyphosphonates being tested for this type of application. Some heterogeneous catalytic tests have been reported and we can highlight the recent report by Liao *et al.*[38] This communication reported the layered-solvothermal reaction of Zn^{II} acetate with H_4PBDC and triethylamine to yield a hybrid, $Zn_3(PBDC)_2 \cdot 2H_3O$, which has an anionic open framework containing hydrated protons. Friedel–Crafts benzylation reactions of toluene with various substrates, catalyzed by this hybrid, were tested. The conversion obtained for 4-chlorobenzyl bromide is higher than 99% but it was not active for 3,5-di-*tert*-butylbenzyl bromide. Therefore, this hybrid carboxyphosphonate showed promising catalytic properties.[38]

Proton conduction will be treated in depth in Chapter 15. Here, we will give just some insights about this property for carboxyphosphonate hybrids due to its important implications in new energy technologies, including proton exchange membrane fuel cells. An initial seminal report was the preparation of α-layered zirconium n-alkylcarboxyphosphonates, $Zr(O_3P(CH_2)_nCOOH)_2$ $n = 1–5$, and the measurement of their proton conductivities, which were very low, $\sim 1 \times 10^{-7}\,S\,cm^{-1}$ at 200 °C.[124] The field has advanced a lot, and proton conductivities of $\sim 6.7 \times 10^{-2}\,S\,cm^{-1}$ at 200 °C and fully humid conditions for membranes of Zr-H_5PBTC carboxyphosphonates have been already reported.[125] Even better results can be obtained for composite membranes obtained from Zr-H_5PBTC and sulfonated poly(ether-ether-ketone), SPEEK, which have very low methanol permeabilities and very high proton conductivities, $0.36\,S\,cm^{-1}$ at 160 °C.[126]

4.5.3 Luminescent and Magnetic Properties

The luminescent properties of metal phosphonates are discussed in detail in Chapter 16 and they have also been reviewed.[94] Here, we will discuss just three selected examples for metal carboxyphosphonates, keeping in mind that there are many reports about this subject and, therefore, they can not be extensively discussed. Firstly, Pr and Y derivatives of 4-H_3CPMP have been studied and when the Y hybrid is doped with Eu, a significant red-orange emission under UV radiation was observed.[51] Secondly, Eu, Tb and Dy derivatives of H_3PMAMB showed strong emission bands in red, green and yellow light regions, respectively.[60] Finally, seven lanthanide derivatives of H_4PMIDA were reported and the Tb^{III} hybrid exhibited very strong emissions in the visible region under 380 nm excitation.[127]

There are also numerous reports dealing with the magnetic properties of metal carboxyphosphonate hybrids. However, most compounds are paramagnetic at room temperature and antiferromagnetic at (very) low temperatures. Only canted antiferromagentic behaviour has been observed at very low temperatures (see for instance reference 59) which leads to weak ferromagnetism. Therefore, the magnetic properties of metal carboxyphosphonates are not close to development into applications.

4.5.4 Corrosion Inhibition

Corrosion management can be achieved in several ways, one being based on chemical means (e.g. some chemical additives) that delay or (ideally) stop metallic corrosion. Within these products, metal phosphonates are included as a class of inhibitors that operate by complexation. These materials are composed of a metal ion (usually alkaline earth metals or Zn^{2+}) and polyphosphonate ligands, such as the carboxyphosphonate ligands H_5PBTC and H_3HPA, which form sparingly soluble compounds that eventually precipitate onto the metallic surface to form an ideally 2D protective thin film. A plethora of metal phosphonates, including metal carboxyphosphonates, have been synthesized and evaluated for their anticorrosion properties and they have been recently reviewed, and are also discussed in Chapter 14.[128] We can highlight a very recent work,[99] not included in that review, dealing with the preparation and evaluation of $Ca_3(HPA)_2(H_2O)_{14}$ as a corrosion inhibitor. This compound showed effective corrosion inhibition for carbon steel surfaces.

4.6 Conclusions and Outlook

There are hundreds of metal carboxyphosphonate hybrids reported so far, with much of the work dealing with synthesis and structural characterization. However, useful properties such as microporosity, catalytic behaviour or proton conductivites are rarely reported. This is, at least in part, due to the lack of predictability within the metal phosphonate field, and also within the carboxyphosphonate subfield. A very important step forward would be to define synthetic conditions (including the nature of the chemical reagents) that would yield predictable structures with interesting properties. The quest for light metal carboxyphosphonate hybrids with high porosity is a key example, since combined with good chemical and thermal robustness, they may lead to applications in gas storage and catalysis.

Acknowledgements

We are grateful for the financial support from research grants FQM-113 and MAT2010-15175 from Junta de Andalucía, Spain, and MICINN, Spain, respectively.

References

1. G. Férey, *Chem. Soc. Rev.*, 2008, **37**, 191.
2. O. M. Yaghi, M. O'Keeffe, N. W. Ockwig, H. K. Chae, M. Eddaoudi and J. Kimet, *Nature*, 2003, **423**, 705.
3. S. Kitagawa, R. Kitaura and S. Noro, *Angew. Chem., Int. Ed Engl.*, 2004, **43**, 2334.
4. D. J. Tranchemontagne, J. L. Mendoza-Cortés, M. O'Keeffe and O. M. Yaghi, *Chem. Soc. Rev.*, 2009, **38**, 1257.
5. R. Banerjee, A. Phan, B. Wang, C. Knobler, H. Furukawa, M. O'Keeffe and O. M. Yaghi, *Science*, 2008, **319**, 939.
6. A. Clearfield, *Prog. Inorg. Chem.*, 1998, **47**, 37.
7. J.-M. Rueff, O. Perez, C. Simon, C. Lorilleux, H. Couthon-Gourvès and P.-A. Jaffrès, *Cryst. Growth Des.*, 2009, **9**, 4263.
8. D. Sanna, I. Bódi, S. Bouhsina, G. Micera and T. Kiss, *J. Chem. Soc., Dalton Trans.*, 1999, 3275.
9. F. Serpaggi and G. Férey, *Inorg. Chem.*, 1999, **38**, 4741.
10. J. M. Rueff, V. Caignaert, S. Chausson, A. Leclaire, C. Simon, O. Perez, L. Pluart and P. A. Jaffres, *Eur. J. Inorg. Chem.*, 2008, 4117.
11. N. Stock, S. A. Frey, G. D. Stucky and A. K. Cheetham, *J. Chem. Soc., Dalton Trans.*, 2000, 4292.
12. J.-J. Hou and X.-M. Zhang, *Cryst. Growth Des.*, 2006, **6**, 1445.
13. A. N. Alsobrook, W. Zhan and T. E. Albrecht-Schmitt, *Inorg. Chem.*, 2008, **47**, 5177.
14. K. E. Knope and C. L. Cahill, *Inorg. Chem.*, 2009, **48**, 6845.
15. G. B. Hix, D. S. Wragg, P. A. Wright and R. E. Morris, *J. Chem. Soc., Dalton Trans.*, 1998, 3359.
16. S. Drumel, P. Janvier, P. Barboux, M. Bujoli-Doeuff and B. Bujoli, *Inorg. Chem.*, 1995, **34**, 148.
17. F. Fredoueil, M. Evain, D. Massiot, M. Bujoli-Doeuff, P. Janvier, A. Clearfield and B. Bujoli, *J. Chem. Soc., Dalton Trans.*, 2002, 1508.
18. M. M. Gómez-Alcántara, A. Cabeza, M. A. G. Aranda, A. Guagliardi, J. G. Mao and A. Clearfield, *Solid State Sci.*, 2004, **6**, 479.
19. A. Anillo, A. Altomare, A. G. G. Moliterni, E. M. Bauer, C. Bellitto, M. Colapietro, G. Portalone and G. Righini, *J. Solid State Chem.*, 2005, **178**, 306.
20. K. D. Demadis, Z. Anagnostou and H. Zhao, *Appl. Mater. Interfaces*, 2009, **1**, 35.
21. E. M. Sabbar, M. E. de Roy and J. P. Besse, *Mater. Res. Bull.*, 2000, **35**, 93.

22. K. E. Knope and C. L. Cahill, *Inorg. Chem.*, 2008, **47**, 7660.
23. S. Pawsey, M. McCormick, S. De Paul, R. Graf, Y. S. Lee, L. Reven and Hans W. Spiess, *J. Am. Chem. Soc.*, 2003, **125**, 4174.
24. K. D. Demadis, P. Lykoudis, R. G. Raptis and G. Mezei, *Cryst. Growth Des.*, 2006, **6**, 1065.
25. Y. Yamazaki, M. Y. Jang and T. Taniyama, *Sci. Technol. Adv. Mater.*, 2004, **5**, 455.
26. N. Stock and T. Bein, *Angew. Chem., Int. Ed. Engl.*, 2004, **43**, 749.
27. J. Svoboda, V. Zima, L. Benes, K. Melánová and M. Vlcek, *Inorg. Chem.*, 2005, **44**, 9968.
28. Z. Chen, Y. Zhou, L. Weng and D. Zhao, *Cryst. Growth Des.*, 2008, **8**, 4045.
29. P.-F. Wang, Y. Duan, D.-K. Cao, Y.-Z. Li and L.-M. Zheng, *Dalton Trans.*, 2010, **39**, 4559.
30. J. Rueff, V. Caignaert, A. Leclaire, C. Simon, J. Haelters and P. Jaffrès, *Cryst. Eng. Commun.*, 2009, **11**, 556.
31. J.-M. Rueff, O. Perez, A. Leclaire, H. Couthon-Gourvès and P.-A. Jaffrès, *Eur. J. Inorg. Chem.*, 2009, 4870.
32. J.-M. Rueff, A. Leclaire and P.-A. Jaffrès, *Solid State Sci.*, 2009, **11**, 812.
33. C. Fang, Z. Chen, X. Liu, Y. Yang, M. Deng, L. Weng, Y. Jia and Y. Zhou, *Inorg. Chim. Acta*, 2009, **362**, 2101.
34. J.-T. Li, T. D. Keene, D.-K. Cao, S. Decurtins and L.-M. Zheng, *Cryst. Eng. Commun.*, 2009, **11**, 1255.
35. G. Chaplais, J. Le Bideau, D. Leclercq and A. Vioux, *Chem. Mater.*, 2003, **15**, 1950.
36. N. Stock and T. Bein, *J. Mater. Chem.*, 2005, **15**, 1384.
37. S. Bauer and N. Stock, *J. Solid State Chem.*, 2007, **180**, 3111.
38. T.-B. Liao, Y. Ling, Z.-X. Chen, Y.-M. Zhou and L.-H. Weng, *Chem. Commun.*, 2010, **46**, 1100.
39. E. T. Clarke, P. R. Rudolf, A. E. Martell and A. Clearfield, *Inorg. Chim. Acta*, 1989, **164**, 59.
40. N. Stock, *Solid State Sci.*, 2002, **4**, 1089.
41. C. Lei, J.-G. Mao, Y.-Q. Sun, H.-Y. Zeng and A. Clearfield, *Inorg. Chem.*, 2003, **42**, 6157.
42. C. Lei, J.-G. Mao and Y.-Q. Sun, *J. Solid State Chem.*, 2004, **177**, 2449.
43. J.-G. Mao, Z. Wang and A. Clearfield, *New J. Chem.*, 2002, **26**, 1010.
44. J.-L. Song, J.-G. Mao, Y.-Q. Sun, H.-Y. Zeng, R. K. Kremer and A. Clearfield, *J. Solid State Chem.*, 2004, **177**, 633.
45. A. Mateescu, C. Gabriel, R. G. Raptis, P. Baran and A. Salifoglou, *Inorg. Chim. Acta*, 2007, **360**, 638.
46. J.-G. Mao and A. Clearfield, *Inorg. Chem.*, 2002, **41**, 2319.
47. S.-F. Tang, J.-L. Song, X.-L. Li and J.-G. Mao, *Cryst. Growth Des.*, 2006, **6**, 2323.
48. Z.-G. Gu and S. C. Sevov, *J. Mater. Chem.*, 2009, **19**, 8442.
49. M. Kui-Rong, Z. Yu-Lan and Y. Qi-Fan, *J. Coord. Chem.*, 2009, **62**, 3242.

50. D.-K. Cao, S.-Z. Hou, Y.-Z. Li and L.-M. Zheng, *Cryst. Growth Des.*, 2009, **9**, 4445.
51. C. Serre, N. Stock, T. Bein and G. Férey, *Inorg. Chem.*, 2004, **43**, 3159.
52. Y.-Q. Guo, B.-P. Yang, J.-L. Song and J.-G. Mao, *Cryst. Growth Des.*, 2008, **8**, 600.
53. N. Zhang, C.-Y. Huang, Z.-G. Sun, L. Liu, R.-N. Hua, X. Lu, F. Tong and W.-N. Wang, *Z. Anorg. Allg. Chem.*, 2010, **636**, 449.
54. X. Shi, G. Zhu, S. Qiu, K. Huang, J. Yu and R. Xu, *Angew. Chem., Int. Ed. Engl.*, 2004, **43**, 6482.
55. Q. Yue, J. Yang, G.-H. Li, G.-D. Li and J.-S. Chen, *Inorg. Chem.*, 2006, **45**, 4433.
56. X. Shi, G. Zhu, S. Qiu, K. Huang, J. Yu and R. Xu, *Angew. Chem., Int. Ed. Engl.*, 2004, **43**, 6482.
57. B.-P. Yang, J.-G. Mao, Y.-Q. Sun, H.-H. Zhao and A. Clearfield, *Eur. J. Inorg. Chem.*, 2003, 4211.
58. U. Costantino, M. Nocchetti and R. Vivani, *J. Am. Chem. Soc.*, 2002, **124**, 8428.
59. B.-P. Yang, A. V. Prosvirin, Y.-Q. Guo and J.-G. Mao, *Inorg. Chem.*, 2008, **47**, 1453.
60. T.-H. Zhou, F.-Y. Yi, P.-X. Li and J.-G. Mao, *Inorg. Chem.*, 2010, **49**, 905.
61. S. Bauer, T. Bein and N. Stock, *Inorg. Chem.*, 2005, **44**, 5882.
62. S. Bauer and N. Stock, *Angew. Chem., Int. Ed. Engl.*, 2007, **46**, 6857.
63. S. J. Hartman, E. Todorov, C. Cruz and S. C. Sevov, *Chem. Commun.*, 2000, 1213.
64. R. Fu, S. H. Zhang, J. Zhang and X. Wu, *Cryst. Growth Des.*, 2005, **5**, 1797.
65. L. Liu, J. Li, Z.-G. Sun, D.-P. Dong, N. Zhang, X. Lu, W.-N. Wang and F. Tong, *Z. Anorg. Allg. Chem.*, 2010, **636**, 247.
66. Y. Zhu, Z. Sun, Y. Zhao, J. Zhang, X. Lu, N. Zhang, L. Liu and F. Tong, *New J. Chem.*, 2009, **33**, 119.
67. R. Colodrero, P. Olivera-Pastor, A. Cabeza, M. Papadaki, K. D. Demadis and M. A. G. Aranda, *Inorg. Chem.*, 2010, **49**, 761.
68. Z. Lai, R. Fu, S. Hu and X. Wu, *Eur. J. Inorg. Chem.*, 2007, 5439.
69. J. Zhang, J. Li, Z.-G. Sun, R.-N. Hua, Y.-Y. Zhu, Y. Zhao, N. Zhang, L. Liu, X. Lu, W.-N. Wang and F. Tong, *Inorg. Chem. Commun.*, 2009, **12**, 276.
70. L. Long, *Cryst. Eng. Commun.*, 2010, **12**, 1354.
71. S. Lodhia, A. Turner, M. Papadaki, K. D. Demadis and G. B. Hix, *Cryst. Growth Des.*, 2009, **9**, 1811.
72. P. M. Forster, N. Stock and A. K. Cheetham, *Angew. Chem., Int. Ed. Engl.*, 2005, **44**, 7608.
73. S. Bauer, T. Bein and N. Stock, *Solid State Sci.*, 2008, **10**, 837.
74. P.-A. Jaffres, V. Caignaert and D. Villemin, *Chem. Commun.*, 1999, 1997.
75. G. B. Hix, A. Turner, B. M. Kariuki, M. Tremayne and E. J. MacLean, *J. Mater. Chem.*, 2002, **12**, 3220.

76. X.-M. Zhang, *Eur. J. Inorg. Chem.*, 2004, 544.
77. Z. Chen, Y. Zhou, L. Weng, C. Yuan and D. Zhao, *Chem. Asian J.*, 2007, **2**, 1549.
78. J. Zhu, X. Bu, P. Feng and G. D. Stucky, *J. Am. Chem. Soc.*, 2000, **122**, 11563.
79. J. Li, D. Cao, B. Liu, Y. Li and L. Zheng, *Cryst. Growth Des.*, 2008, **8**, 2950.
80. E. R. Cooper, C. D. Andrews, P. S. Wheatley, P. B. Webb, P. Wormald and R. E. Morris, *Nature*, 2004, **430**, 1012.
81. P. J. Byrne, D. S. Wragg, J. E. Warren and R. E. Morris, *Dalton Trans.*, 2009, 795.
82. J.-D. Feng, K.-Z. Shao, S.-W. Tang, R.-S. Wang and Z.-M. Su, *Cryst. Eng. Commun.*, 2010, **12**, 1401.
83. G. B. Hix, A. Turner, L. Vahter and B. M. Kariuki, *Microporous Mesoporous Mater.*, 2007, **99**, 62.
84. J.-H. Liao, P.-C. Wu and Y.-H. Bai, *Inorg. Chem. Commun.*, 2005, **8**, 390.
85. K. D. Demadis, M. Papadaki, R. G. Raptis and H. Zhao, *Chem. Mater.*, 2008, **20**, 4835.
86. B.-P. Yang and J.-G. Mao, *J. Mol. Struct.*, 2007, **830**, 78.
87. G. B. Hix and K. D. M. Harris, *J. Mater. Chem.*, 1998, **8**, 579.
88. B. Menna and I. J. Shannon, *J. Mater. Chem.*, 2002, **12**, 350.
89. S. H. Jhung, J.-H. Lee, P. M. Forster, G. Ferey, A. K. Cheetham and J.-S. Chang, *Chem. Eur. J.*, 2006, **12**, 7699.
90. A. Sonnauer and N. Stock, *J. Solid State Chem.*, 2008, **181**, 3065.
91. Y.-F. Yang, Y.-S. Ma, S.-S. Bao and L.-M. Zheng, *Dalton Trans.*, 2007, 4222.
92. K. Maeda, *Microporous Mesoporous Mater.*, 2004, **73**, 47.
93. K. Demadis, in *Progress in Solid State Chemistry Research*, ed. Ronald W. Buckley, Nova Science Publishers, Inc., 2007, p. 109.
94. J.-G. Mao, *Coord. Chem. Rev.*, 2007, **251**, 1493.
95. G. K. H. Shimizu, R. Vaidhyanathan and J. M. Taylor, *Chem. Soc. Rev.*, 2009, **38**, 1430.
96. A. Clearfield, *Curr. Opin. Solid State Mater. Sci.*, 2002, **6**, 495.
97. J. Svoboda, V. Zima, L. Benes, K. Melánová and M. Vlcek, *Inorg. Chem.*, 2005, **44**, 9973.
98. K. D. Demadis, M. Papadaki, M. A. G. Aranda, A. Cabeza, P. Olivera-Pastor and Y. Sanakis, *Cryst. Growth Des.*, 2010, **10**, 357.
99. K. D. Demadis, M. Papadaki and I. Císarova, *Appl. Mater. Interfaces*, 2010, **2**, 1814.
100. Q. Yue, J. Yang, G.-H. Li, G.-D. Li and J.-S. Chen, *Inorg. Chem.*, 2006, **45**, 4431.
101. M. M. Gómez-Alcantara, M. A. G. Aranda, P. Olivera-Pastor, P. Beran, J. L. García-Muñoz and A. Cabeza, *Dalton Trans.*, 2006, 577.
102. A. N. Alsobrook and T. E. Albrecht-Schmitt, *Inorg. Chem.*, 2009, **48**, 11079.
103. K. Slepokura and T. Lis, *Acta Cryst.*, 2003, **C59**, m76.

104. M. Sanselme, M. Riou-Cavellec, J.-M. Grenèche and G. Férey, *J. Solid State Chem.*, 2002, **164**, 354.
105. R. Vivani, U. Costantino and M. Nocchetti, *J. Mater. Chem.*, 2002, **12**, 3254.
106. M. Riou-Cavellec, M. Sanselme, J.-M. Grenèche and G. Férey, *Solid State Sci.*, 2000, **2**, 717.
107. J.-L. Song and J.-G. Mao, *J. Mol. Struct.*, 2005, **740**, 181.
108. N. Bestaoui, X. Ouyang, F. Fredoueil B. Bujoli and A. Clearfield, *Acta Cryst.*, 2005, **B61**, 669.
109. N. Stock, K. Karaghiosoff and T. Bein, *Z. Anorg. Allg. Chem.*, 2004, **630**, 2535.
110. A. Cabeza, M. A. G. Aranda and S. Bruque, *J. Mater. Chem.*, 1998, **8**, 2479.
111. R. Fu, S. Hu and X. Wu, *Dalton Trans.*, 2009, 9440.
112. A. Distler and S. C. Sevov, *Chem. Commun.*, 1998, 959.
113. R. Fu, S. Hu and X. Wu, *Dalton Trans.*, 2009, 9843.
114. J.-L. Song, H.-H. Zhao, J.-G. Mao and K. R. Dunbar, *Chem. Mater.*, 2004, **16**, 1884.
115. J.-L. Song, A. V. Prosvirin, H.-H. Zhao and J.-G. Mao, *Eur. J. Inorg. Chem.*, 2004, 3706.
116. S.-M. Ying and J.-G. Mao, *Cryst. Growth Des.*, 2006, **6**, 964.
117. Y. Zhao, J. Li, Z.-G. Sun, J. Zhang, Y.-Y. Zhu, X. Lu, L. Liu and N. Zhang, *Inorg. Chem. Commun.*, 2008, **11**, 1057.
118. J.-L. Song, C. Lei and J.-G. Mao, *Inorg. Chem.*, 2004, **43**, 5630.
119. M. Jezowska-Bojczuk, T. Kiss, H. KozCowski, P. Decock and J. Barycki, *Dalton Trans.*, 1994, 811.
120. H.-Y. Liu, Z.-J. Zhang, W. Shi, B. Zhao, P. Cheng, D.-Z. Liao and S.-P. Yan, *Dalton Trans.*, 2009, 4416.
121. X. Shi, J. Yang and Q. Yang, *Eur. J. Inorg. Chem.*, 2006, 1936.
122. A. Corma, H. García and F. X. Llabres-i-Xamena, *Chem. Rev.*, 2010, **110**, 4606.
123. L. Ma, C. Abney and W. Lin, *Chem. Soc. Rev.*, 2010, **38**, 1248.
124. G. Alberti, U. Constantino, M. Casciola and R. Vivani, *Solid State Ion.*, 1991, **46**, 61.
125. Y. Yamazaki, M. J. Jang and T. Taniyama, *Sci. Tech. Adv. Mater.*, 2004, **5**, 455.
126. Q. Gao, M. Huang, Y. Wang, Y. Cai and L. Xu, *Front. Chem. Eng. China*, 2008, **2**, 95.
127. S.-F. Tang, J.-L. Song, X.-L. Li and J.-G. Mao, *Cryst. Growth Des.*, 2006, **6**, 2322.
128. M. Papadaki and K. D. Demadis, *Comments Inorg. Chem.*, 2009, **30**, 89.
129. see, for example, http://en.wikipedia.org/wiki/Roundup_(herbicide).

CHAPTER 5

Metal Phosphonates and Arsonates Containing an Auxiliary Ligand

JIANG-GAO MAO

State Key Laboratory of Structural Chemistry, Fujian Institute of Research on the Structure of Matter, The Chinese Academy of Sciences, Fuzhou 350002, P. R. China

5.1 Introduction

During the last two decades great research efforts have been devoted to the design and synthesis of metal phosphonates with novel open frameworks or microporous structures due to their potential applications in electro-optical, ion-exchange, catalysis and sensors.[1–7] The strategy of attaching functional groups such as crown ether, amine, hydroxyl, or/and carboxylate groups to the phosphonic acid has been proved to be very effective for the isolation of a variety of metal phosphonates with open framework and microporous structures.[1–7] Furthermore, various types of auxiliary ligands have been employed in the preparations of novel metal phosphonates. These auxiliary ligands may increase the solubility of metal phosphonates, participate in the extended framework formation or lead to various types of cluster compounds, or simply be intercalated between metal phosphonate layers. So far, the role of the auxiliary ligand has not been reviewed. This chapter will be devoted to metal phosphonates and arsonates with auxiliary ligands that have been synthesized

Metal Phosphonate Chemistry: From Synthesis to Applications
Edited by Abraham Clearfield and Konstantinos Demadis

Published by the Royal Society of Chemistry, www.rsc.org

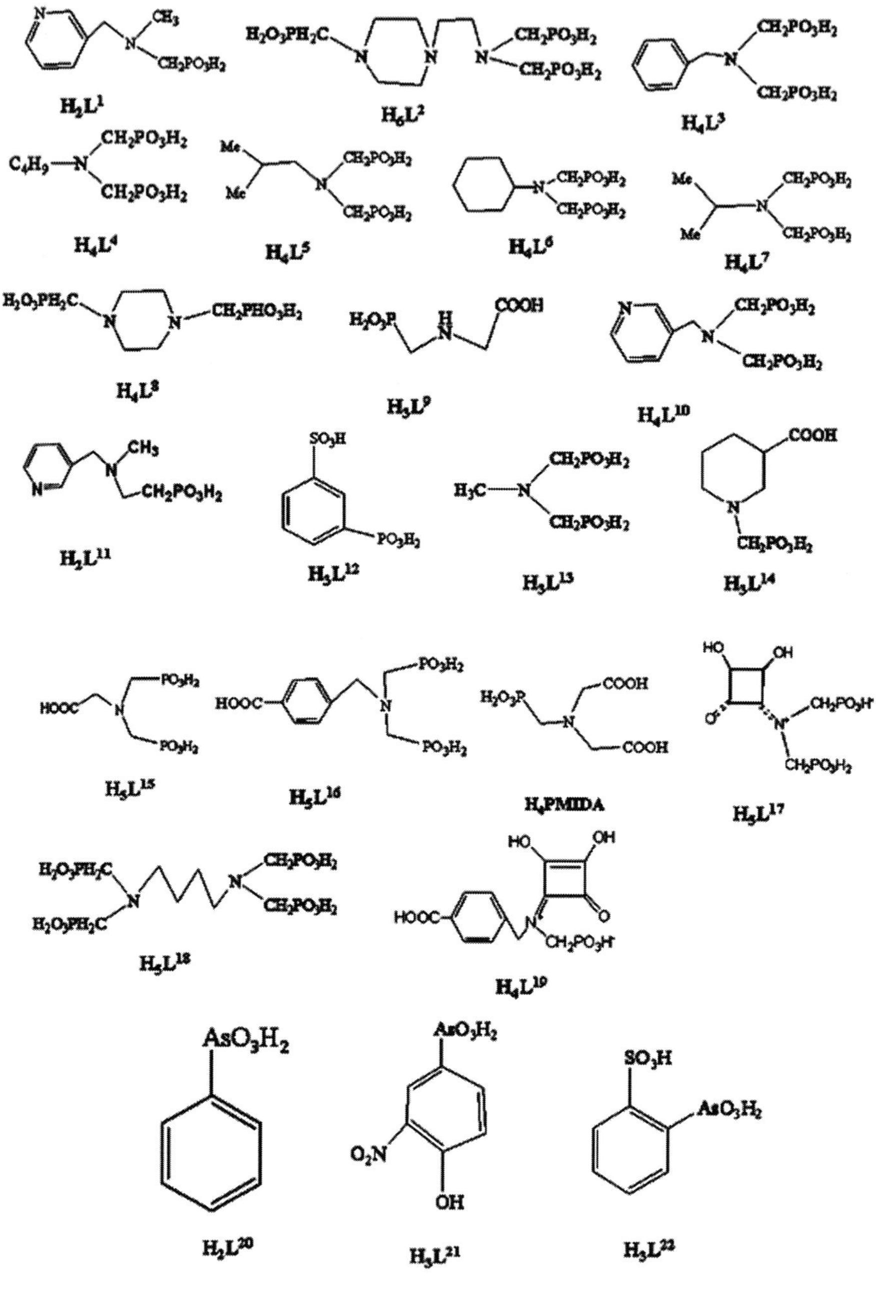

Scheme 5.1 Phosphonic and arsenic acids used in our work.

mainly in our laboratory during the past seven years. The phosphonate, arsonate and auxiliary ligands used are shown in Schemes 5.1 and 5.2 respectively.

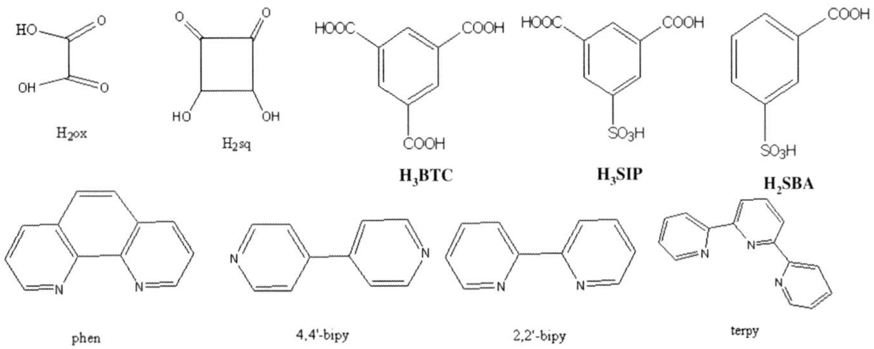

Scheme 5.2 Auxiliary ligands used in our work.

5.2 Main Group Metal Phosphonates with an Auxiliary Ligand

A variety of auxiliary ligands have been used with the main group metal phosphonates, the metal ion is mainly lead(II) ion with a lone pair. The compounds reported are listed in Table 5.1.

Hydrothermal reactions of lead(II) acetate with 1,3,5-benzenetricarboxylic acid (H_3BTC) and a new aminodiphosphonic acid, 3-pyridyl-$CH_2N(CH_2PO_3H_2)_2$ (H_4L^1), or a new aminotriphosphonic acid, 1-[($H_2O_3PCH2)_2$ NCH_2CH_2-]-piperazine-4-$CH_2PO_3H_2$ (H_6L^2), afforded two novel porous carboxylate phosphonate hybrids, namely, $Pb_4(BTC)(HBTC)(HL^1)\cdot 3H_2O$ (**1**) and $Pb_6(BTC)_2(HBTC)(H_2L^2)\cdot 4.5H_2O$ (**2**).[8] The structure of $Pb_4(BTC)(HBTC)(HL^1)\cdot 3H_2O$ (**1**) is built from a 3D porous network of lead(II) tricarboxylate and 1D double chains of lead(II) amino-diphosphonate (Figure 5.1). The BTC^{3-} anion is hexadentate, it chelates with two lead(II) ions and also bridges with two other lead(II) ions, whereas the $HBTC^{2-}$ anion acts as a tridentate ligand and bridges with three Pb^{II} ions using two carboxylate groups. The third carboxylate group remains non-coordinated and protonated. The structure of $Pb_6(BTC)_2(HBTC)(H_2L^2)\cdot 4.5H_2O$ (**2**) contains <002> lead(II) carboxylate-phosphonate hybrid layers which are cross-linked by 1D double chains of lead(II) phosphonate via the organic groups of both types of ligands, resulting in the formation of a complex 3D network (Figure 5.2). The lattice water molecules in both compounds are located at the micro-pores of the structures.

The use of *N*-benzyl-iminobis(methylenephosphonic acid), $C_6H_5CH_2N(CH_2PO_3H_2)_2$ (H_4L^3) gave layered $Pb_3(HL^3)_2\cdot 2(H_3BTC)\cdot 2H_2O$ (**3**).[9] The lead(II) ions are bridged by bisphosphonate ligands into a 2D layer. The neutral H_3BTC ligands are intercalated between two layers, forming hydrogen bonds with lattice water molecules, non-coordinated phosphonate or carboxylate groups (Figure 5.3). The intercalation of H_3BTC greatly increases the interlayer distances from 17.7 Å to 24.4 Å.

Table 5.1 Main group metal phosphonates with an auxiliary ligand.

Compound	*Structural features*	*Ref.*
$Pb_4(BTC)(HBTC)(HL^1)\cdot 3H_2O$ (**1**)	3D porous network composed of lead(II) tricarboxylate and 1D double chains of lead(II) amino-diphosphonate	8
$Pb_6(BTC)_2(HBTC)(H_2L^2)\cdot 4.5H_2O$ (**2**)	3D network composed of lead(II) carboxylate-phosphonate hybrid layers cross-linked by 1D double chains of lead(II) phosphonate	8
$Pb_3(HL^3)_2\cdot 2(H_3BTC)\cdot 2H_2O$ (**3**)	Layered structure with intercalation of neutral H_3BTC ligand	9
$Pb_5(HL^4)_2(L^4)\cdot H_3BTC$ (**4**)	Layered structure with intercalation of neutral H_3BTC ligand	10
$Pb_4(BTC)(L^5)(CH_3CO_2)(H_2O)_2\cdot 2H_2O$ (**5**)	A carboxylate-phosphonate hybrid layer	10
$Pb_3(HL^6)(H_2L^6)(H_2BTC)(H_2O)\cdot H_3BTC\cdot H_2O$ (**6**)	2D hydrogen bonded lead(II) phosphonate layer with H_2BTC anions hanging on the layer and the neutral H_3BTC ligand being intercalated between two layers	11
$Pb_4(L^6)_2\cdot (HO_2CH_2CH_2CO_2H)$ (**7**)	The neutral succinic acid being intercalated between lead(II) phosphonate layers	12
$Pb_7(3\text{-}O_3S\text{-}C_6H_4\text{-}CO_2)(L^7)_3(H_2O)_2\cdot 2H_2O$ (**8**)	Layered structure with the auxiliary hanging on the interlayer space	11
$Pb_4(H_2L^8)(SIP)_2(H_2O)_4\cdot 2H_2O$ (**9**)	A pillared layered architecture composed of $Pb_4(SIP)_2$layers interconnected by H_2L^8 ligands	12
$Pb_3(L^9)(SIP)(H_2O)\cdot H_2O$ (**10**)	3D network built by the interconnection of hexanuclear $Pb_6(L^9)_2$ units and 1D double chains of lead(II) carboxylate-sulfonate	13
$Pb_2(HL^{11})(SIP)\cdot H_2O$ (**11**)	3D network structure built by the cross-linkage of 1D double chains of lead(II) phosphonates and 2D layers of lead(II) carboxylate- sulfonate	13
$[Pb(HL^{12})(phen)]\cdot H_2O$ (**12**)	Layered	14
$[Pb_6(L^{12})_4(phen)_8]\cdot 3H_2O$ (**13**)	Isolated hexanuclear cluster	14
$[Pb_6(L^{12})_4(phen)_{10}]\cdot 2H_2O$ (**14**)	Isolated hexanuclear cluster	14
$[Pb_6(L^{12})_4(4,4'\text{-bipy})(H_2O)_2]\cdot 2H_2O$ (**15**)	3D network of 1D lead(II) phosphonate chains bridged by 4,4′-bipy ligands	14

The use of diphosphonate acid of butylimino-bis(methylenephosphonic acid) ($CH_3CH_2CH_2CH_2N(CH_2PO_3H_2)_2$, H_4L^4), isobutylimino-bis(methylenephosphonic acid) ($Me_2CHCH_2N(CH_2PO_3H_2)_2$, H_4L^5) and *N*-cyclohexylimino-bis(methylene-phosphonic acid, $C_6H_{11}N(CH_2PO_3H_2)_2$ (H_4L^6) afforded three novel other lead(II) carboxylate phosphonate hybrids, namely, $Pb_5(HL^4)_2(L^4)\cdot$

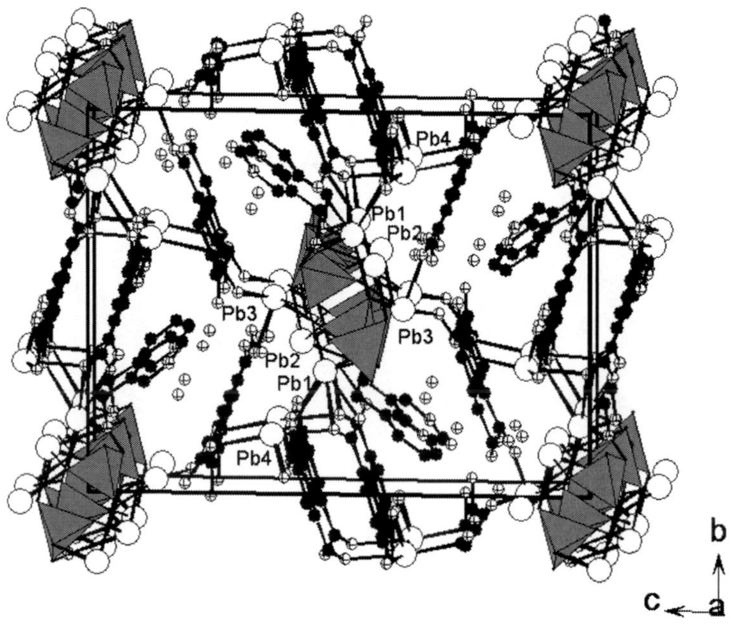

Figure 5.1 View of the structure of $Pb_4(BTC)(HBTC)(HL^1) \cdot 3H_2O$ (**1**) down the *a* axis. The C-PO_3 tetrahedra are shaded in gray, Pb, N, O and C atoms are represented by open, octanded, crossed and black circles, respectively (adapted from Ref. 8).

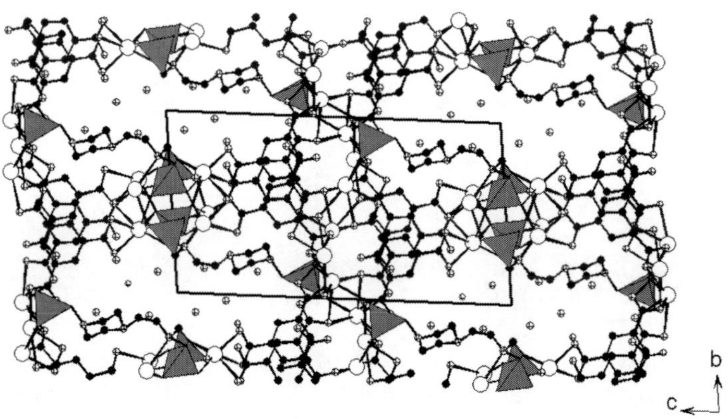

Figure 5.2 View of the porous 3D network of $Pb_6(BTC)_2(HBTC)(H_2L^2) \cdot 4.5H_2O$ (**2**) down the *a* axis (adapted from Ref. 8).

H_3BTC (**4**), $Pb_4(BTC)(L^5)(CH_3CO_2)(H_2O)_2 \cdot 2H_2O$ (**5**) and $Pb_3(HL^6)$-$(H_2L^6)(H_2BTC)(H_2O) \cdot H_3BTC \cdot H_2O$ (**6**).[10,11] The structure of compound $Pb_5(HL^4)_2(L^4) \cdot H_3BTC$ (**4**) contains a lead(II) diphosphonate hybrid layer formed by Pb^{II} ions interconnected by HL^4 and L^4 anions with the neutral

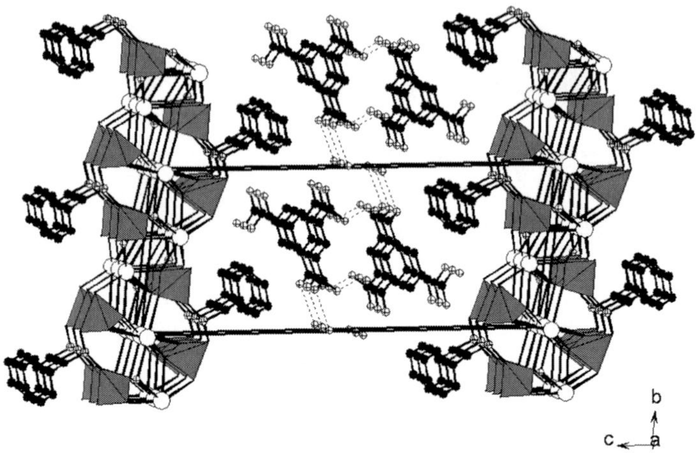

Figure 5.3 View of the structure of $Pb_3(HL^3)_2 \cdot 2(H_3BTC) \cdot 2H_2O$ (**3**) along the *a* axis (adapted from Ref. 9).

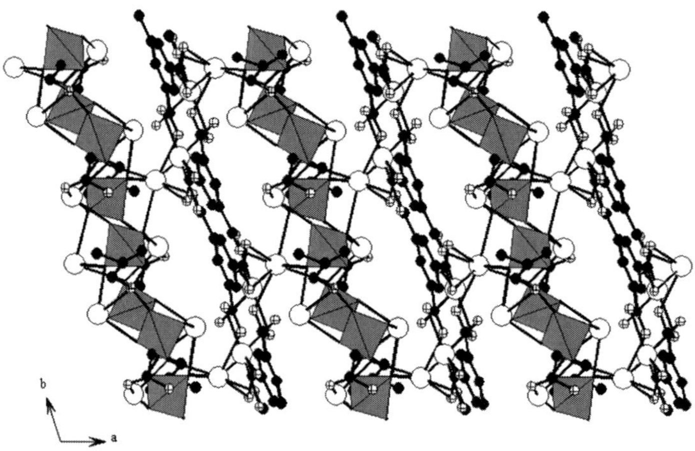

Figure 5.4 A <002> lead carboxylate-phosphonate hybrid layer in $Pb_4(BTC)(L^5)(CH_3CO_2)(H_2O)_2 \cdot 2H_2O$ (**5**) down the *c* axis (adapted from Ref. 10).

H_3BTC as the intercalated species; the interlayer distance is 19.6 Å. $Pb_4(BTC)(L^5)(CH_3CO_2)(H_2O)_2 \cdot 2H_2O$ (**5**) also has a layered structure in which the 1,3,5-benzenetricarboxylate ligand functioned as a multidentate metal linker. The BTC anion is hexadentate, it chelates one Pb atom bidentately and bridges with four other Pb atoms. The interconnection of Pb^{II} ions via bridging L^5 anions led to a 1D chain along the *b* axis, whereas the cross-linkage of Pb^{II} by BTC anions resulted in a 1D double chain along the *b* axis. The interconnection of the above two types of 1D chains leads to the formation of a lead(II) carboxylate-phosphonate layer (Figure 5.4). The interlayer distance is about 16.1 Å.[10] In $Pb_3(HL^6)(H_2L^6)(H_2BTC)(H_2O) \cdot H_3BTC \cdot H_2O$ (**6**), the

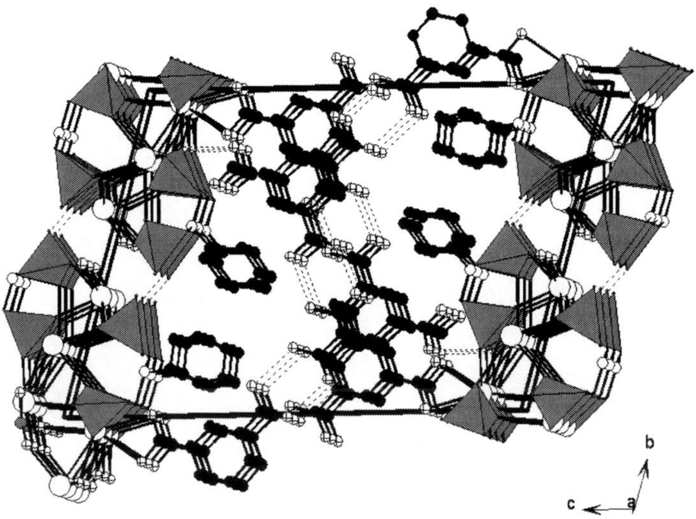

Figure 5.5 View of the structure of $Pb_3(HL^6)(H_2L^6)(H_2BTC)(H_2O) \cdot H_3BTC \cdot H_2O$ (**6**) down the *a* axis (adapted from Ref. 11).

interconnection of the lead(II) ions through bridging diphosphonate ligands resulted in the formation of a lead(II) diphosphonate slab; such slabs are further interlinked via hydrogen bonds between non-coordinated phosphonate oxygen atoms into a <001> layer. The doubly protonated H_2BTC anion is bidentately chelated to a lead(II) ion using a carboxylate group, whereas the neutral H_3BTC ligand is intercalated between two layers, forming hydrogen bonds with the non-coordinated carboxylate groups of the H_2BTC anion (Figure 5.5).[11]

Succinic acid can also be intercalated into metal phosphonates. In $Pb_4(L^6)_2 \cdot (HO_2CH_2CH_2CO_2H)$ (**7**), the interconnection of the lead(II) ions via diphosphonate ligands led to a lead(II) diphosphonate layer. The succinic acid remains protonated and is located in the interlayer space (Figure 5.6).[12]

Four lead(II) phosphonates have been prepared by using 5-sulfoisophalic acid monosodium salt (NaH_2SIP) as auxiliary ligand, namely, $Pb_7(3\text{-}O_3S\text{-}C_6H_4\text{-}CO_2)(L^7)_3(H_2O)_2 \cdot 2H_2O$ (**8**), $Pb_4(H_2L^8)(SIP)_2(H_2O)_4 \cdot 2H_2O$ (**9**), $Pb_3(L^9)(SIP)(H_2O) \cdot H_2O$ (**10**) and $Pb_2(HL^{11})(SIP) \cdot H_2O$ (**11**).[11–13] In $Pb_7(3\text{-}O_3S\text{-}C_6H_4\text{-}CO_2)(L^7)_3(H_2O)_2 \cdot 2H_2O$ (**8**), the interconnection of the PbO_xN_y polyhedra by the diphosphonate L^7 anions as well as carboxylate-sulfonate ligands resulted in a <010> hybrid layer. The interlayer distance is about 14.9 Å. The sulfonate group acts as a pendant group, hanging between two layers (Figure 5.7). The lattice water molecules are also located in the middle of the interlayer space, forming hydrogen bonds with the non-coordinated sulfonate groups as well as the aqua ligands.[11] The structure of $Pb_4(H_2L^8)(SIP)_2(H_2O)_4 \cdot 2H_2O$ (**9**) features a 3D network formed by Pb^{II} ions interconnected by H_2L^8 and the carboxylate-sulfonate SIP anions. The 3D network can be viewed as the neighboring layers of $Pb_4(SIP)_2$ being interconnected by

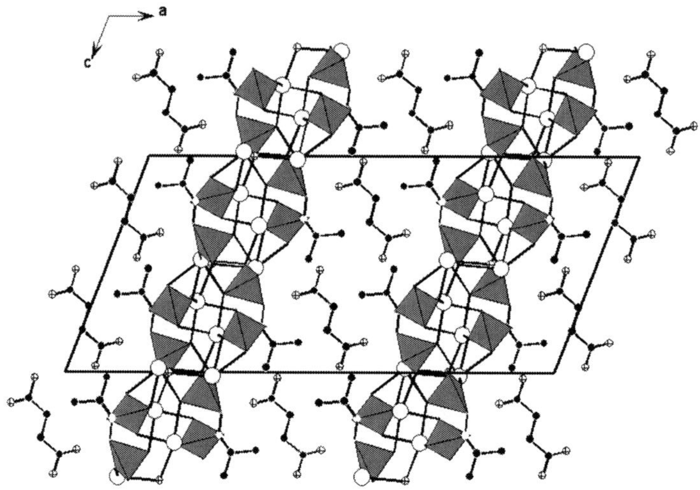

Figure 5.6 View of the structure of $Pb_4(L^6)_2 \cdot (HO_2CH_2CH_2CO_2H)$ (**7**) along the *b* axis (adapted from Ref. 12).

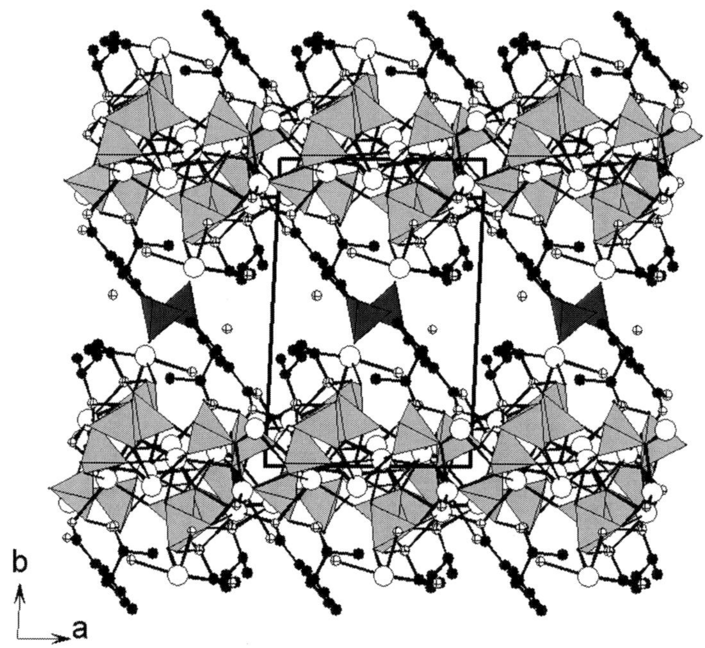

Figure 5.7 View of the structure of $Pb_7(3\text{-}O_3S\text{-}C_6H_4\text{-}CO_2)(L^7)_3(H_2O)_2 \cdot 2H_2O$ (**8**) down the *c* axis (adapted from Ref. 11).

H_2L^8 ligands to form a pillared layered architecture. The phosphonate groups of the diphosphonate ligands bridge the Pb^{II} ions of the $Pb_4(SIP)_2$ layer to form a lead(II) phosphonate-sulfonate hybrid layer and the organic group of phosphonate ligand as pillars.[12] The structure of $Pb_3(L^9)(SIP)(H_2O)\cdot H_2O$ (**10**) features a 3D network built from the interconnection of hexanuclear $Pb_6(L^9)_2$ units and 1D double chains of lead(II) carboxylate-sulfonate. In $Pb_2(HL^{11})(SIP)\cdot H_2O$ (**11**), the H_2L^{11} ligand was formed as a result of the decomposition of one phosphonate group in H_4L^9 during the hydrothermal reaction. $Pb_2(HL^{11})(SIP)\cdot H_2O$ (**11**) displayed a 3D network structure built by the cross-linkage of 1D double chains of lead(II) phosphonates and 2D layers of lead(II) carboxylate- sulfonate.[13] In all three compounds, the auxiliary ligand acts as a multidentate metal linker.

When basic auxiliary ligand is used, the nature of the ligand plays an important role on the structures of the resultant compounds.[14] Using *m*-HO_3S-Ph-PO_3H_2 (H_3L^{12}) as the primary ligand and phen or 4,4′-bipyridine as the auxiliary ligand, four lead(II) compounds have been isolated, namely, $[Pb(HL^{12})(phen)]\cdot H_2O$ (**12**), $[Pb_6(L^{12})_4(phen)_8]\cdot 3H_2O$ (**13**), $[Pb_6(L^{12})_4(phen)_{10}]\cdot 2H_2O$ (**14**) and $[Pb_6(L^{12})_4(4,4'\text{-bipy})(H_2O)_2]\cdot 2H_2O$ (**15**). $[Pb(HL^{12})(phen)]\cdot H_2O$ (**12**) features a layer structure in which 1D chains of Pb_2O_4 clusters are further bridged by sulfonate-phosphonate ligands. $[Pb_6(L^{12})_4(phen)_8]\cdot 3H_2O$ (**13**) and $[Pb_6(L^{12})_4(phen)_{10}]\cdot 2H_2O$ (**14**) represent the first zero-dimensional lead(II) phosphonates and feature novel isolated hexanuclear lead(II) clusters in which six Pb^{II} ions are bridged by two tetradentate and two pentadentate phosphonate groups (Figure 5.8). Each lead(II) ion is bonded to one or two bidentate chelating phen ligands, which reduces the metal coordination sites available to the phosphonate ligand, and hence facilitates the formation of cluster compounds. $[Pb_6(L^{12})_4(4,4'\text{-bipy})(H_2O)_2]\cdot 2H_2O$ (**15**) features a 3D framework in which the 1D lead(II) phosphonate chains are further cross-linked by the 4,4′-bipy ligand. Interestingly, $[Pb_6(L^{12})_4(4,4'\text{-bipy})(H_2O)_2]\cdot 2H_2O$ (**15**) exhibits a strong broad blue fluorescent emission band at 420 nm.[14]

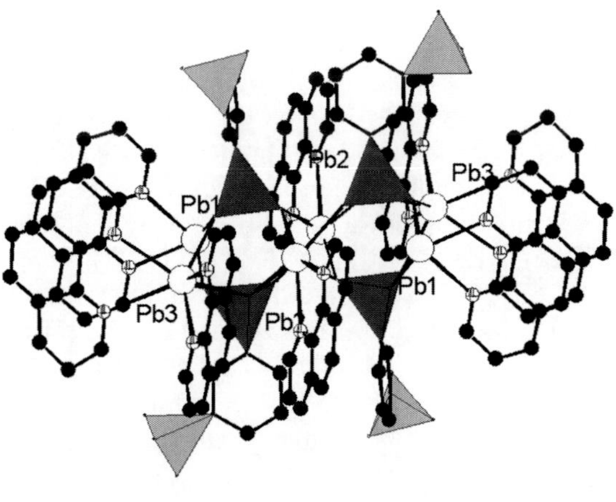

Figure 5.8 A $[Pb_6(L^{12})_4(phen)_8]$ cluster unit in $[Pb_6(L^{12})_4(phen)_8]\cdot 3H_2O$ (**13**) (adapted from Ref. 14).

5.3 Transition Metal Phosphonates with an Auxiliary Ligand

A variety of transition metal phosphonates with an auxiliary ligand have been prepared in our laboratory; they display many types of structures ranging from 0D cluster, 1D chain, 2D layers and 3D networks. These compounds are listed in Table 5.2.

Hydrothermal reactions of *N*-(phosphonomethyl)-*N*-methyl glycine, $MeN(CH_2CO_2H)(CH_2PO_3H_2)$ (H_3L^{13}), with zinc(II) acetate resulted in the formation of a novel zinc carboxylate-phosphonate, $\{Zn_6L_6(Zn)\}\{Zn(H_2O)_6\}_2 \cdot 22H_2O$.[15] Its structure contains a heptanuclear zinc phosphonate cluster anion, $\{Zn_6L_6(Zn)\}^{4-}$ in which seven zinc(II) cations form an unusual $Zn_6(Zn)$ centered octahedron with six of its Zn_3 triangle faces each further capped by a phosphonate group. The Zn^{II} cations of the Zn_6 octahedron are five-coordinated whereas the centered Zn^{II} cation is octahedrally coordinated. Packing of these cluster anions creates micropores occupied by the hydrated zinc(II) cations as well as lattice water molecules. Later, it was found that the hydrated cation, $\{Zn(H_2O)_6\}^{2+}$ can be replaced by other coordination cations such as $\{Ni(4,4'\text{-bipy})(H_2O)_4\}^{2+}$ and $\{Ni(4,4'\text{-bipy})_2(H_2O)_2\}^{2+}$.[16] The structure of $\{Zn_7L^{13}{}_6\}\{Ni(4,4'\text{-bipy})(H_2O)_4\}_2 \cdot 16H_2O$ (**16**) is composed of $\{Zn_6(Zn)L^{13}{}_6\}^{4-}$ cluster anions and 1D cationic chains of $\{Ni(4,4'\text{-bipy})(H_2O)_4\}^{2+}$. In $[H_3O]\{Zn_7L^{13}{}_6\}\{Ni(H_2O)_2(4,4'\text{-bipy})_2\}_{1.5} \cdot 12H_2O$ (**17**), the $\{Zn_6(Zn)L^{13}{}_6\}^{4-}$ cluster anion cages are located at the tunnels formed by the 3D cationic network of $\{Ni(4,4'\text{-bipy})_2(H_2O)_2\}^{2+}$.

When *N*-(phosphonomethyl)pipecolinic acid ($H_2O_3PCH_2\text{-}NC_5H_9\text{-}CO_2H$, H_3L^{14}) was used instead of *N*-(phosphonomethyl)-*N*-methyl glycine, three new zinc(II) phosphonate cluster compounds were isolated, namely $[Zn_7L^{14}{}_6][Zn(phen)_2(H_2O)_2]_2 \cdot 19H_2O$ (**18**), $[Zn_9L^{14}{}_6(H_2O)_8] \cdot 14H_2O$, and $[Zn_7L^{14}{}_6\{Zn(2,2'\text{-bipy})(H_2O)_3\}_2] \cdot 10H_2O$ (**19**).[17] $[Zn_7L^{14}{}_6][Zn(phen)_2(H_2O)_2]_2 \cdot 19H_2O$ (**18**) consists of a centered $\{Zn_6(Zn)L^{14}{}_6\}^{4-}$ core and two $\{Zn(Phen)_2(H_2O)_2\}^{2+}$ counter-cations. The $\{Zn_6(Zn)L^{14}{}_6\}^{4-}$ core in $[Zn_9L^{14}{}_6(H_2O)_8] \cdot 14H_2O$ or **19** is covalently bonded to two $Zn(H_2O)_4{}^{2+}$ or $\{Zn(2,2'\text{-bipy})(H_2O)_3\}^{2+}$ cations through the coordination of two additional carboxylate oxygen atoms to form a nona-nuclear cluster (Figure 5.9).

Hydrothermal reactions of zinc(II) carbonate with *m*-sulfophenylphosphonic acid (H_3L^{12}) and 1,10-phenanthroline (phen) or 4,4′-bipyridine (bipy) lead to three novel zinc(II) sulfonate-phosphonates, namely, $[Zn(phen)_3]_2[Zn_4(L^{12})_4(phen)_4] \cdot 20H_2O$ (**20**), $[Zn_6(L^{12})_4(phen)_8] \cdot 11H_2O$ (**21**) and $[Zn_6(L^{12})_4(4,4'\text{-bipy})_6(H_2O)_4] \cdot 18H_2O$ (**22**).[18] Compound **20** contains a tetranuclear zinc(II) cluster anion in which four Zn^{II} ions are bridged by two tetradentate and two bidentate phosphonate groups, the four negative charges of the cluster are compensated by two $[Zn(phen)_3]^{2+}$ cations. Compound **21** features a hexanuclear zinc(II) cluster in which the same tetranuclear cluster of **20** is bridged with two additional Zn^{II} ions. The structure of **22** features a porous 3D network based on hexanuclear zinc(II) units of $[Zn_6(L^{12})_4]$ interconnected by 4,4′-bipy ligands (Figure 5.10). The hexanuclear cluster in **22** is different from that

Table 5.2 Transition Metal phosphonates with an auxiliary ligand.

Compound	*Structural features*	*Ref.*
$\{Zn_7L^{13}{}_6\}\{Ni(4,4'\text{-bipy})(H_2O)_4\}_2 \cdot 16H_2O$ (**16**)	$\{Zn_6(Zn)L_6\}^{4-}$ cluster anions and 1D cationic chains of $\{Ni(4,4'\text{-bipy})(H_2O)_4\}^{2+}$	16
$[H_3O]\{Zn_7L^{13}{}_6\}\{Ni(H_2O)_2(4,4'\text{-bipy})_2\}_{1.5} \cdot 12H_2O$ (**17**)	the $\{Zn_6(Zn)L_6\}^{4-}$ cluster anion cages being inserted at the tunnels formed by the 3D cationic network of $\{Ni(4,4'\text{-bipy})_2(H_2O)_2\}^{2+}$	16
$[Zn_7L^{14}{}_6][Zn(phen)_2(H_2O)_2]_2 \cdot 19H_2O$ (**18**)	$\{Zn_6(Zn)L^{14}{}_6\}^{4-}$ cluster anions and 0D $[Zn(phen)_2(H_2O)_2]^{2+}$ cations	17
$[Zn_7L^{14}{}_6\{Zn(2,2'\text{-bipy})(H_2O)_3\}_2] \cdot 10H_2O$ (**19**)	a nona-nuclear cluster based on $\{Zn_6(Zn)L^{14}{}_6\}^{4-}$ core covalently bonded with two $\{Zn(2,2'\text{-bipy})(H_2O)_3\}^{2+}$ cations	17
$[Zn(phen)_3]_2[Zn_4(L^{12})_4(phen)_4] \cdot 20H_2O$ (**20**)	tetranuclear zinc(II) cluster	18
$[Zn_6(L^{12})_4(phen)_8] \cdot 11H_2O$ (**21**)	a hexanuclear zinc(II) cluster	18
$[Zn_6(L^{12})_4(4,4'\text{-bipy})_6(H_2O)_4] \cdot 18H_2O$ (**22**)	3D network based on Zn_6 cluster	18
$[Mn_2(HL^{12})_2(phen)_4][Mn_2(HL^{12})_2\text{-}(phen)_4(H_2O)]_2 \cdot 6H_2O$ (**23**)	Two types of dimers	19
$[Mn_4(L^{12})_2(phen)_8(H_2O)_2][ClO_4]_2 \cdot 3H_2O$ (**24**)	Tetranuclear cluster	19
$[Mn(phen)(H_2O)_4]_2[Mn_4(L^{12})_4\text{-}(phen)_4] \cdot 10H_2O$ (**25**)	Tetranuclear cluster and $[Mn(phen)(H_2O)_4]^{2+}$ cation	19
$[Mn_6(L^{12})_4(phen)_8(H_2O)_2] \cdot 4H_2O$ (**26**)	Isolated hexanuclear cluster	19
$[Mn_6(L^{12})_4(phen)_8(H_2O)_2] \cdot 24H_2O$ (**27**)	Isolated hexanuclear cluster	19
$[Mn_6(L^{12})_4(phen)_6(H_2O)_4] \cdot 5H_2O$ (**28**)	1D chain based on hexanuclear cluster	19
$[Cd_4(L^{12})_2(phen)_6(Cl)_2(H_2O)_2] \cdot 14H_2O$ (**29**)	Isolated tetranuclear cluster	20
$[Cd_6(L^{12})_4(phen)_8] \cdot 14H_2O$ (**30**)	Isolated hexanuclear cadmium(II) cage	20
$[Cd_3(L^{12})_2(4,4'\text{-bipy})_3(H_2O)_6] \cdot 4H_2O$ (**31**)	3D network based on 1D chains of $[Cd_3(L)_2]$ cross-linked by 4,4′-bipy ligands	20
$Co_3(H_2L^{15})_2(H_2O)_4(4,4'\text{-bipy})_2 \cdot 11H_2O$ (**32**)	2D layer based on trinuclear clusters	21
$Zn_5L^{15}{}_2(4,4'\text{-bipy})_2 \cdot 9H_2O$ (**33**)	Pillared-layered structure	21
$[Co_3(H_2L^{16})_2(H_2O)_4(4,4'\text{-bipy})]$ (**34**)	a pillar-layered architecture composed of cobalt(II) carboxylate-phosphonate 2D layers that are cross-linked further by 4,4′-bipy ligands	22
$[Co_2(PMIDA)(H_2O)_5(4,4'\text{-bipy})] \cdot 4H_2O$ (**35**)	1D zigzag chains composed of cobalt(II) PMIDA dinuclear units that are further interconnected by 4,4′-bipy ligands	22
$\{Mn[NH(CH_2PO_3H)_2](H_2O)_2\}_2 \{Mn(C_4O_4)(H_2O)_4\} \cdot (C_4H_2O_4)$ (**36**)	3D supramolecular network based on 1D chains of	23

Table 5.2 (*Continued*)

Compound	*Structural features*	*Ref.*
	$Mn[NH(CH_2PO_3H)_2](H_2O)_2$, 1D chains of $\{Mn(C_4O_4)(H_2O)_4\}$ and the neutral squaric acid molecules	
$Cu_2(H_3L^{17})(4,4'\text{-bipy})_2 \cdot 2H_2O$ (**37**)	Layered Cu(I) phosphonates involved in condensation of amino-phosphonate and squaric acid	23

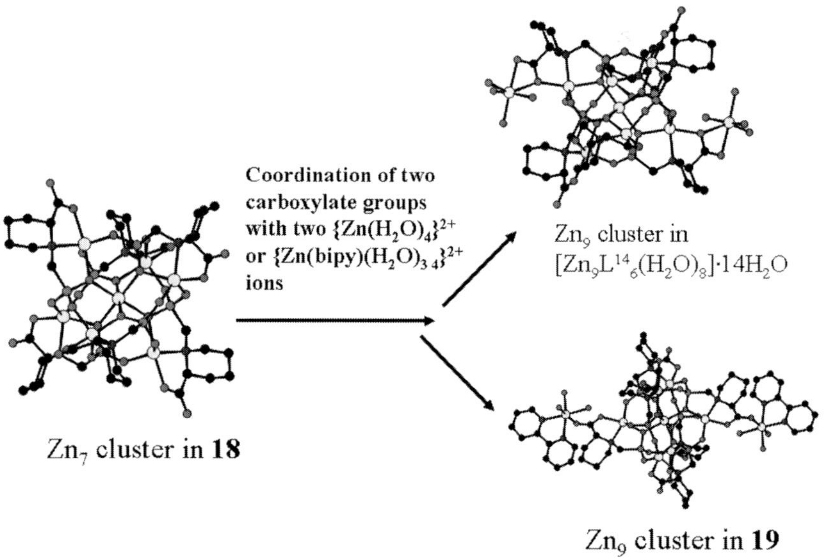

Figure 5.9 The relationship between clusters in compounds **18**, **19** and $[Zn_9L^{14}{}_6(H_2O)_8] \cdot 14H_2O$ (adapted from Ref. 17).

in **21** in that all four phosphonate groups in **3** are tridentate bridging. Compounds **20**, **21** and **22** exhibit broad blue fluorescent emission bands at 378, 409 and 381 nm, respectively.

When manganese(II) salts were used instead of zinc(II) salts six novel manganese(II) sulfonate-phosphonates, namely, $[Mn_2(HL^{12})_2(phen)_4]$-$[Mn_2(HL^{12})_2\ (phen)_4(H_2O)]_2 \cdot 6H_2O$ (**23**), $[Mn_4(L^{12})_2(phen)_8(H_2O)_2][ClO_4]_2 \cdot 3H_2O$ (**24**), $[Mn(phen)(H_2O)_4]_2[Mn_4(L^{12})_4(phen)_4] \cdot 10H_2O$ (**25**), $[Mn_6(L^{12})_4(phen)_8(H_2O)_2] \cdot 4H_2O$ (**26**), $[Mn_6(L^{12})_4(phen)_8(H_2O)_2] \cdot 24H_2O$ (**27**) and $[Mn_6(L^{12})_4(phen)_6(H_2O)_4] \cdot 5H_2O$ (**28**).[19] The structure of compound **23** contains

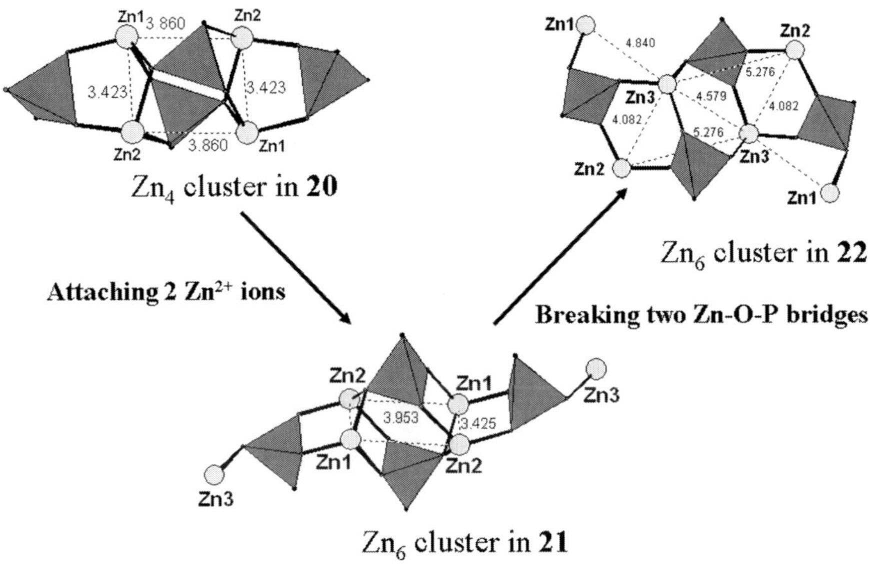

Figure 5.10 The relationship between clusters in compounds **20–22** (adapted from Ref. 18).

two types of dinuclear manganese(II) clusters and compounds **24** and **25** exhibit two types of tetranuclear manganese(II) cluster unit. Compounds **26** and **27** feature two different types of isolated hexanuclear manganese(II) clusters whereas the hexanuclear manganese(II) clusters in compound **28** are bridged by sulfonate-phosphonate ligands into a 1D chain (Figure 5.11). Magnetic property measurements indicate that there exist weak antiferromagnetic interactions between magnetic centers in all six compounds.

Three novel cadmium(II) phosphonatephenylsulfonates were obtained when Cd^{II} was used as the metal ion, namely, $[Cd_4(L^{12})_2(phen)_6(Cl)_2(H_2O)_2]\cdot 14H_2O$ (**29**), $[Cd_6(L^{12})_4(phen)_8]\cdot 14H_2O$ (**30**) and $[Cd_3(L^{12})_2(4,4'\text{-bipy})_3(H_2O)_6]\cdot 4H_2O$ (**31**).[20] Compound **29** contains a tetranuclear cadmium(II) cluster in which four cadmium(II) ions are bridged by two tetradentate phosphonate groups. Compound **30** features a hexanuclear cadmium(II) cage in which six cadmium(II) ions are bridged by one tetradentate and three pentadentate phosphonate groups. The structure of compound **31** features a novel 3D network based on 1D chains of $[Cd_3(L)_2]$ cross-linked by 4,4′-bipy ligands with a topological structure similar to that of $CdSO_4$. Compounds **29–31** exhibit broad blue fluorescent emission bands at 408, 410 and 396 nm, respectively.

The above examples clearly indicate that choosing a suitable phosphonate ligand and an auxiliary bidentate chelating ligand is a promising route to 0D metal phosphonate clusters.

Hydrothermal reactions of transition metal salts with $HO_2CCH_2N(CH_2PO_3H_2)_2$ (H_5L^{15}) and 4,4′-bipy lead to two new metal carboxylate-diphosphonates, namely, $Co_3(H_2L^{15})_2(H_2O)_4(4,4'\text{-bipy})_2\cdot 11H_2O$ (**32**) with a layered

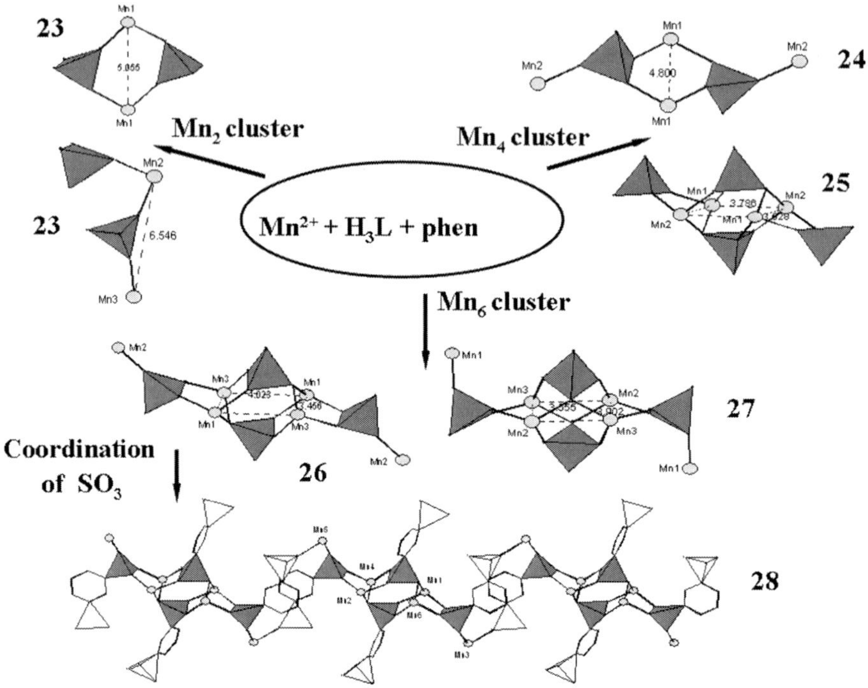

Figure 5.11 The relationship between clusters in compounds **23**–**28** (adapted from Ref. 19).

structure, and a blue luminescent compound, $Zn_5L^{15}{}_2(4,4'\text{-bipy})_2 \cdot 9H_2O$ (**33**) with a pillared layered structure.[21] In compound **1**, three Co^{II} ions are interconnected by two chelating and bridging phosphonate ligands to form a trinuclear unit, and such units are further cross-linked via bridging 4,4′-bipy ligands to form a 2D layer. In compound **2**, the zinc(II) ions in the trigonal bipyramidal or tetrahedral geometries are interconnected via chelating and bridging carboxylate-phosphonate ligands into 2D layers which are further cross-linked by 4,4′-bipy ligands into a pillared layered architecture (Figure 5.12). Luminescent spectrum measurements indicate that $Zn_5L^{15}{}_2(4,4'\text{-bipy})_2 \cdot 9H_2O$ (**33**) is a blue light luminescent material. Under excitation at 370 nm, it displays a very strong blue fluorescent emission band in the solid state at $\lambda_{max} = 457$ nm with a shoulder at 478 nm.

Two novel cobalt carboxylate-phosphonates with 4,4′-bipyridine (4,4′-bipy) as the secondary metal linker, namely, $[Co_3(H_2L^{16})_2(H_2O)_4(4,4'\text{-bipy})]$ (**34**) ($H_5L^{16} = 4\text{-}HO_2C\text{-}C_6H_4\text{-}CH_2N(CH_2PO_3H_2)_2$) with a pillar-layered structure and $[Co_2(PMIDA)(H_2O)_5(4,4'\text{-bipy})] \cdot 4H_2O$ (**35**) ($H_4PMIDA = H_2O_3PCH_2N(CH_2CO_2H)_2$) featuring an infinite 1D chain structure, were also obtained in our laboratory.[22] $[Co_3(H_2L^{16})_2(H_2O)_4(4,4'\text{-bipy})]$ (**34**) features a pillar-layered architecture in which Co^{II} octahedra are linked by carboxylate-phosphonate ligands to form a 2D layer and these layers are cross-linked further by 4,4′-bipy ligands to form a 3D network, the carboxylate groups of the phosphonate

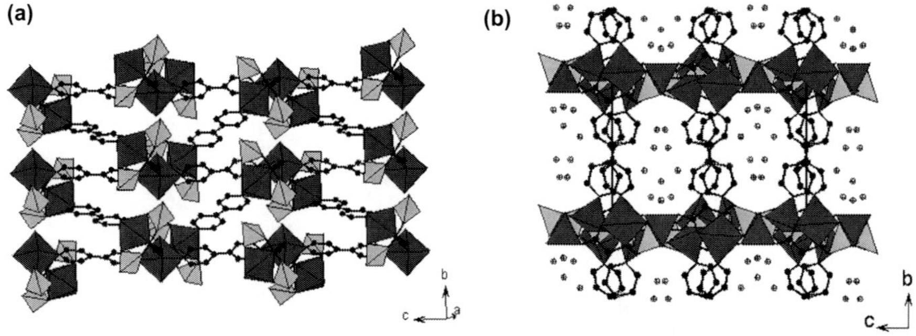

Figure 5.12 The structures of layered $Co_3(H_2L^{15})_2(H_2O)_4(4,4'\text{-bipy})_2 \cdot 11H_2O$ (**32**) (a) and $Zn_5L^{15}{}_2(4,4'\text{-bipy})_2 \cdot 9H_2O$ (**33**) with a pillared-layered structure (b) (adapted from Ref. 21).

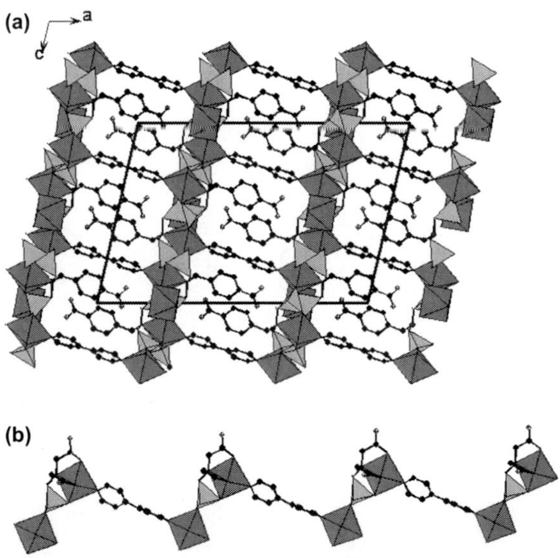

Figure 5.13 View of the 3D network of $[Co_3(H_2L^{16})_2(H_2O)_4(4,4'\text{-bipy})]$ (**34**) along the *b* axis (a) and 1D chain in $[Co_2(PMIDA)(H_2O)_5(4,4'\text{-bipy})] \cdot 4H_2O$ (**35**) (b) (adapted from Ref. 22).

ligands are non-coordinated . The crystal structure of $[Co_2(PMIDA)(H_2O)_5(4,4'\text{-bipy})] \cdot 4H_2O$ (**35**) contains 1D zigzag chains in which two cobalt(II) ions are bridged by PMIDA ligand to form dimeric units, and these dimeric units are further interconnected via bridging 4,4′-bipy ligands (Figure 5.13).

Hydrothermal reaction of a mixture of manganese(II) sulfate, $HN(CH_2PO_3H_2)_2$ (H_5L), squaric acid at 180 °C for 5 days afforded a new mixed-ligand Mn^{II} squarate diphosphonate, $\{Mn[NH(CH_2PO_3H)_2](H_2O)_2\}_2\{Mn(C_4O_4)(H_2O)_4\} \cdot (C_4H_2O_4)$ (**36**).[23] Its structure is composed by 1D chains of

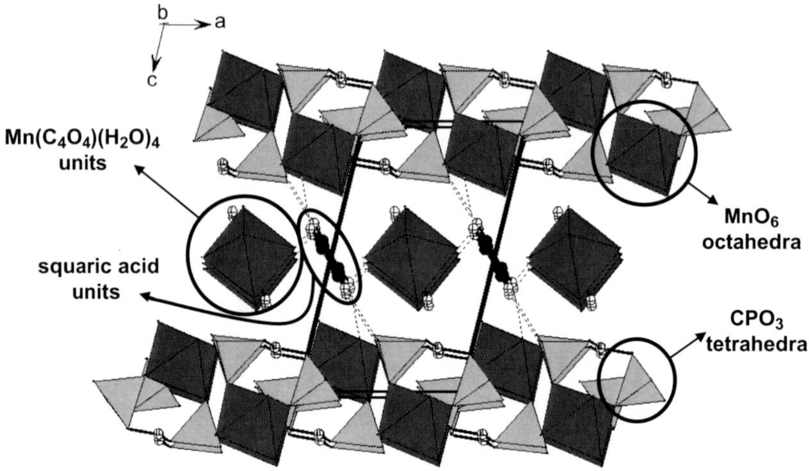

Figure 5.14 View of the structure of $\{Mn[NH(CH_2PO_3H)_2](H_2O)_2\}_2$ $\{Mn(C_4O_4)(H_2O)_4\}\cdot(C_4H_2O_4)$ (**36**) along *b* axis. Hydrogen bonds are drawn as dotted lines (adapted from Ref. 23).

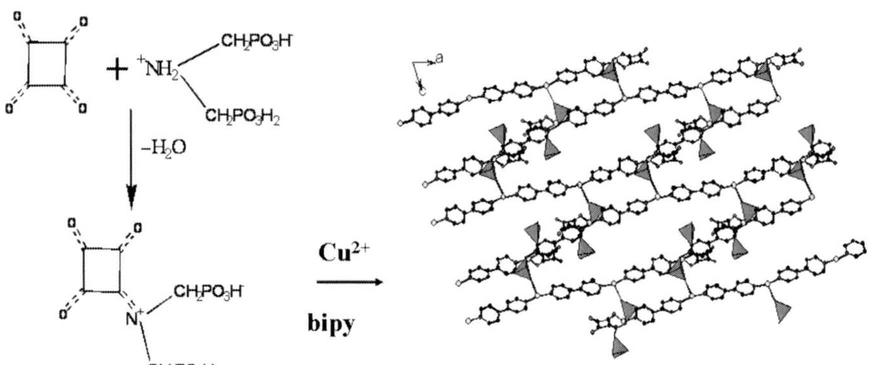

H_5L^{17} **37**

Figure 5.15 The formation and the new phosphonate ligand (H_5L^{17}) and its copper(I) phosphonate (**37**) with a layered structure (adapted from Ref. 23).

$Mn[NH(CH_2PO_3H)_2](H_2O)_2$, 1D chains of $\{Mn(C_4O_4)(H_2O)_4\}$ as well as the neutral squaric acid molecules. These three types of building units are interconnected via hydrogen bonding (Figure 5.14).

Hydrothermal reaction of a mixture of copper(II) acetate, $HN(CH_2PO_3H_2)_2$, squaric acid and 4,4′-bipyridine at 140 °C for 5 days led to a new layered copper(I) phosphonate, namely, $Cu_2(H_3L^{17})(4,4'\text{-bipy})_2\cdot 2H_2O$ (**37**),[23] in which a new multifunctional squarato-imino-diphosphonate ligand (H_5L^{17}) was formed by the condensation reaction between one C=O group of the squaric acid and the amino group of $NH(CH_2PO_3H_2)_2$ (Figure 5.15). The structure of

$Cu_2(H_3L^{17})(4,4'\text{-bipy})_2 \cdot 2H_2O$ (**37**) features a layer architecture in which the Cu^I centers are three coordinated by two N atoms from two 4,4′-bipy ligands and one oxygen atom (phosphonate or carboxylate) from the newly formed ligand. The new ligand acts as a bidentate metal linker, and bridges with two metal centers by one carboxylate and one phosphonate oxygen atoms. Neighboring copper(I) phosphonate layers are further interlinked by strong hydrogen bonds among non-coordination squarate and phosphonate oxygen atoms as well as lattice water molecules.

5.4 Lanthanide Phosphonates Containing a Second Metal Linker

It is found that introducing a second metal linker whose lanthanide compounds have large solubility in water and better crystallinity can improve the solubility and crystallinity of the lanthanide phosphonates. A list of lanthanide phosphonates containing second metal linkers is summarized in Table 5.3.

Hydrothermal reactions of lanthanide metal salts with $MeN(CH_2CO_2H)(CH_2PO_3H_2)$ (H_3L^{13}) and 5-sulfoisophthalic acid monosodium salt (NaH_2SIP) lead to a series of layered Ln^{III} carboxylate-phosphonate-sulfonate hybrids, namely, $Ln(H_2L^{13})(HSIP)(H_2O)_2 \cdot H_2O$ (Ln = La **38**; Pr **39**; Nd **40** and Gd **41**).[24] The asymmetric unit of these isostructural compounds is composed of one Ln^{III} ion, one $(H_2L^{13})^-$ anion, one HSIP anion, two aqua ligands and a lattice water molecule. The lanthanide ion is nine-coordinate with three carboxylate oxygens from two $(H_2L^{13})^-$ anions, two phosphonate oxygens from two other $(H_2L^{13})^-$ anions, one carboxylate oxygen atom from a $HSIP^{2-}$ anion and one sulfonate oxygen atom from another $HSIP^{2-}$ anion as well as two aqua ligands. The $(H_2L^{13})^-$ anion is pentadentate, it chelates with one lanthanide ion bidentately by using its carboxylate group and bridges with four other

Table 5.3 Lanthanide phosphonates containing an auxiliary ligand.

Compound	*Structural features*	*Ref.*
$Ln(H_2L^{13})(HSIP)(H_2O)_2 \cdot H_2O$ (Ln = La **38**; Pr **39**; Nd **40** and Gd **41**)	layered	24
$Ln(C_2O_4)\{H_2L^{13}\} \cdot 0.5H_2O$ (Ln = Nd **42**; Eu **43**; Gd **44**)	3D network	25
$Ln_4(C_2O_4)_5(HL^7)_2(H_2O)_4 \cdot 2H_2O$ (Ln = La **45**; Nd **46**)	3D network	25
$Ln_3(C_2O_4)_4(HL^7)(H_2O)_6 \cdot 6H_2O$ (Ln = Gd **47**; Er **48**)	3D network	25
$La_2(ox)_2(H_6L^{18}) \cdot 2H_2O$ (**49**)	3D network	26
$Ln_2(ox)_2(H_6L^{18})(H_2O)_2 \cdot 4H_2O$ (Ln = Nd **50**; Eu **51**)	3D network	26
$Ln(HL^{19})$ (Ln = La **52**; Pr **53**; Nd **54**; Eu **55**; Gd **56**; Tb **57**; Er **58**)	pillared-layered structure	27
$[La_2(L^{12})_2(phen)_4(H_2O)] \cdot 4.5H_2O$ (**59**)	1D cluster chain	28
$[Ln_2(L^{12})_2(phen)_2(H_2O)_5] \cdot 3H_2O$ (Ln = Nd **60**; Eu **61**; Er **62**)	tetranuclear cluster	28
$[Ln_2(HL^{12})(H_2L^8)_2(H_2O)_4] \cdot 8H_2O$ (Ln = La **63**; Nd **64**)	3D cluster network	28

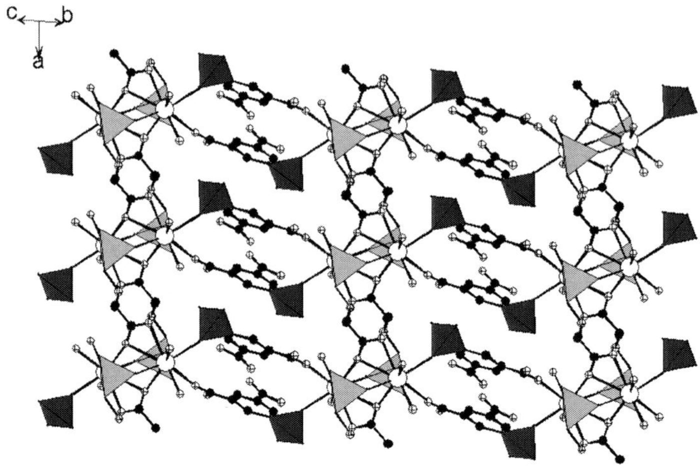

Figure 5.16 The layered architecture of the $Nd(H_2L^{13})(HSIP)(H_2O)_2 \cdot H_2O$ (**40**). The phosphonate and sulfonate tetrahedra are shaded in heavy and light gray, respectively (adapted from Ref. 24).

lanthanide centers. One carboxylate oxygen acts as a μ_2-metal linker. Two phosphonate oxygens are unidentate, whereas the third one is protonated and non-coordinated. The amine group is also protonated and non-coordinated. The $HSIP^{2-}$ anion is bidentate and bridges with two lanthanide ions by using a sulfonate and a carboxylate oxygen atoms. The second carboxylate group of $HSIP^{2-}$ anion is protonated and non-coordinated. The interconnection of Nd^{III} atom by the carboxylate phosphonate $\{H_2L^{13}\}^-$ anions resulted in a $Nd_2(HL^{13})_2$ unit, and such neighboring unit is bridged by the carboxylate groups to form a 1D double chain along the *a* axis. These double chains are cross-linked by bridging HSIP anions to form a lanthanide phosphonate-sulfonate hybrid layer (Figure 5.16). The lattice water molecule is hydrogen bonded to the aqua ligand and the amine group. The non-coordinate phosphonate oxygen also forms hydrogen bonds with non-coordination carboxylate oxygens of the HSIP anion.

Upon complexation of both ligands with the Ln^{III} ions, the luminescent spectra of the compounds formed exhibit emission bands characteristic of the corresponding Ln^{III} ions. Upon excitation at 310 nm, the Nd^{III} compound displays a strong blue fluorescent emission band at 410 nm ($^4D_{3/2} \rightarrow {}^4I_{13/2}$ transition). Under excitation of 356 nm the Nd^{III} compound displays a strong emission band at 902 nm ($^4F_{3/2} \rightarrow {}^4I_{9/2}$), an emission band with much lower intensity at 1067 nm ($^4F_{3/2} \rightarrow {}^4I_{11/2}$) and a very weak band at 1338 nm ($^4F_{3/2} \rightarrow {}^4I_{13/2}$) in the near IR region. The spectrum of the Gd^{III} compound only shows a strong fluorescent emission band at 416 nm ($\lambda_{ex} = 318$ nm). Upon removal of the lattice water and two aqua ligands, the emission band at 410 nm of Nd^{III} compound disappeared whereas other emission bands have been slightly shifted to lower wavelengths and their intensities enhanced greatly. The emission band for the Gd^{III} compound was shifted to 464 nm and its intensity is also significantly enhanced after dehydration.[24]

When the oxalate anion was used as the second metal linker instead of 5-sulfoisophthalic acid, a series of novel lanthanide oxalate phosphonate hybrids with 3D network structures were obtained, namely, $Ln(C_2O_4)\{H_2L^{13}\} \cdot 0.5H_2O$ (Ln = Nd **42**, Eu **43**, Gd **44**).[25] The structure of these isomorphous lanthanide compounds features a 3D network formed by 3D $\{Ln(H_2L^{13})\}^{2+}$ and 1D $\{Ln(C_2O_4)\}^{+}$ cations. The Ln^{III} ion is eight-coordinate with four oxygen atoms from two oxalate anions, two carboxylate and two phosphonate oxygen atoms from four $(H_2L^{13})^{-}$ anions. Each oxalate anion is tetradentate and forms two Ln–O–C–C–O five-member chelating rings. The $(H_2L^{13})^{-}$ anion is tetradentate and bridges with four Ln^{III} ions. The interconnection of Ln^{III} by chelating and bridging oxalate anions results in a 1D helical chain of $\{Ln(C_2O_4)\}^{+}$. The cross-linkage of Ln^{III} ions by bridging $(H_2L^{13})^{-}$ anions lead to a 3D network with tunnels along the *b* axis. The above two building units are interconnected via sharing Ln^{III} ions to form a complex 3D network (Figure 5.17). The structure of this series of compounds can also be viewed as the oxalate anions occupying the tunnels of the 3D network of $\{Ln(H_2L^{13})\}^{2+}$. Luminescence studies indicate that the Nd^{III} compound exhibits several emission bands in the near IR region, the Gd^{III} complex displays a broad blue fluorescent emission band at 451 nm

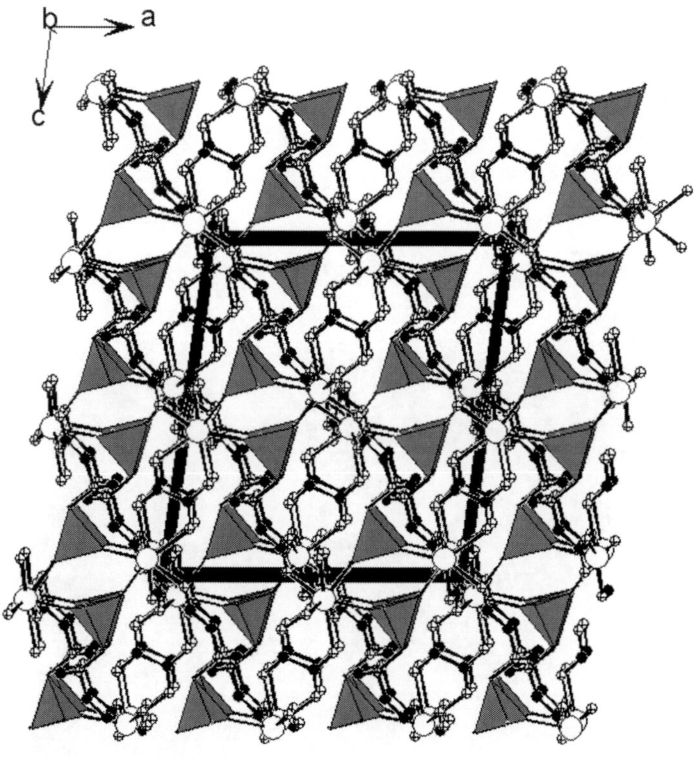

Figure 5.17 View of the structure of $\{Gd(C_2O_4)(H_2L^{13})\} \cdot 0.5H_2O$ (**43**) down the *b* axis (adapted from Ref. 25).

and the Eu^{III} compound shows very strong and sharp emission bands at 592, 616 and 699 nm with a long luminescent lifetime of 1.13 ms.[25]

The use of $Me_2NCH_2PO_3H_2$ (H_2L^7) in the syntheses afforded two different series of new lanthanide oxalate phosphonates, namely, $Ln_4(C_2O_4)_5(HL^7)_2(H_2O)_4 \cdot 2H_2O$ (Ln = La **45**, Nd **46**) and $Ln_3(C_2O_4)_4(HL^7)(H_2O)_6 \cdot 6H_2O$ (Ln = Gd **47**, Er **48**).[25]

Among two unique Ln^{III} ions in the asymmetric unit of $Ln_4(C_2O_4)_5(HL^7)_2(H_2O)_4 \cdot 2H_2O$ (Ln = La **45**, Nd **46**), one is eight-coordinate with six oxygen atoms from four oxalate anions, one phosphonate oxygen from one $\{HL^7\}^-$ anion and an aqua ligand, whereas the other one is nine-coordinate with five oxygen atoms from three oxalate anions, three phosphonate oxygens from two $\{HL^7\}^-$ anions as well as an aqua ligand. The $\{HL^7\}^-$ anion is tetradentate, chelating with one Ln^{III} ion bidentately and also bridging with two other Ln^{III} ions. One phosphonate oxygen is μ_2-bridging, whereas the other two are unidentate. There are two different coordination modes for the five oxalate anions in the formula unit. One oxalate anion is hexadentate. It forms two chelating rings with two Ln^{III} ions and also bridges with two other lanthanide ions. The remaining four oxalate anions are tetradentate.

The interconnection of the Ln^{III} atom by bridging and chelating $\{HL^7\}^-$ anions results in a $Ln_4(HL^7)_2$ unit, in which two Ln^{III} atoms are bridged by a pair of phosphonate groups to form a dimeric unit, and such a dimeric unit further connects with two other Ln^{III} atoms through μ^2-phosphonate oxygens. The interconnection of Ln^{III} ions by chelating and bridging oxalate anions lead to a 3D network of $\{Ln_4(C_2O_4)_5\}^{2+}$ with tunnels running along the *c* axis. The tunnel is formed by 24-member rings composed of six Ln^{III} ions and six oxalate anions. The above two types of building blocks are interconnected into a 3D network (Figure 5.18). The structure of $Ln_4(C_2O_4)_5(HL^7)_2(H_2O)_4 \cdot 2H_2O$

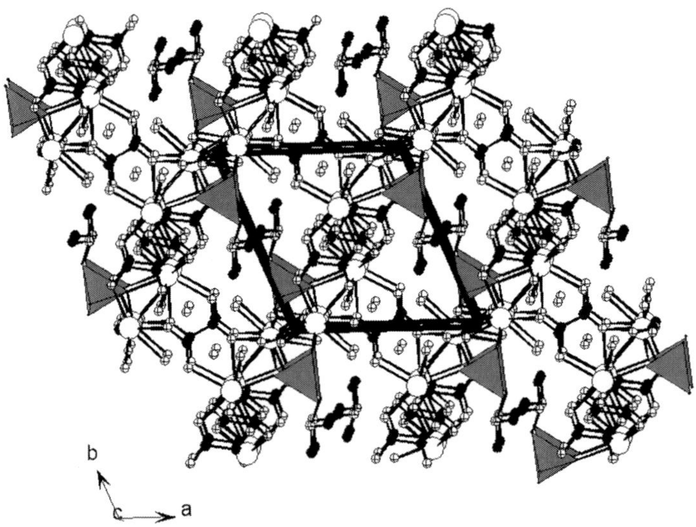

Figure 5.18 View of the structure of $Nd_4(C_2O_4)_5(HL^7)_2(H_2O)_4 \cdot 2H_2O$ (**46**) down the *c* axis (adapted from Ref. 25).

(Ln = La **45**, Nd **46**) can also be viewed as the $\{HL^7\}^-$ anions being inserted in the tunnels of $\{Ln_4(C_2O_4)_5\}^{2+}$. Under excitation of 826 nm, the Nd^{III} compound displays the characteristic emission bands for the Nd^{III} ion: a weak emission band at 896 nm ($^4F_{3/2} \rightarrow {}^4I_{9/2}$), a strong emission band at 1063 nm ($^4F_{3/2} \rightarrow {}^4I_{11/2}$) and a very weak band at 1388 nm ($^4F_{3/2} \rightarrow {}^4I_{13/2}$) in the near IR region. The Nd ($^4F_{3/2}$) lifetimes were measured to be 18.7 μs.[25]

The heavier Ln^{III} complexes form a different type of 3D network, $Ln_3(C_2O_4)_4(HL^7)(H_2O)_6 \cdot 6H_2O$ (Ln = Gd **47**, Er **48**). There are three unique Ln^{III} ions in the asymmetric unit of $Ln_3(C_2O_4)_4(HL^7)(H_2O)_6 \cdot 6H_2O$ (Ln = Gd **47**, Er **48**).[25] One is nine-coordinate with eight oxygen atoms from four oxalate anions and an aqua ligand. The second one is eight-coordinate with six oxygen atoms from three oxalate anions, a phosphonate oxygen from one $\{HL^7\}^-$ anion and an aqua ligand. The third one is eight-coordinate with two oxygen atoms from an oxalate anion, two phosphonate oxygen atoms from two $\{HL^7\}^-$ anions and four aqua ligands. The $\{HL^7\}^-$ anion is tridentate and bridges with three Ln^{III} ions. All phosphonate oxygen atoms are unidentate. All oxalate anions adopt a same coordination mode as that in $Ln_4(C_2O_4)_5(HL^7)_2(H_2O)_4 \cdot 2H_2O$ (Ln = La **45**, Nd **46**).

The interconnection of the first two types of Ln^{III} ions by three bridging and chelating oxalate anions lead to a porous 3D network of $Ln_2(C_2O_4)_3$. The large tunnel is formed by 40 member rings composed of 10 Ln^{III} ions and 10 oxalate anions. Each pair of the third Ln^{III} ions are bridged by two $\{HL^7\}^-$ anions to form a dimeric unit, such units are further interconnected by bridging and chelating oxalate anions into a 1D chain $\{Ln(C_2O_4)(HL^7)\}$ along the *a* axis. These 1D chains of Ln^{III} oxalate phosphonate as well as the lattice water molecules are located at the large tunnels of $Er_2(C_2O_4)_3$ (Figure 5.19).

The emission spectrum of the Gd^{III} exhibits a broad blue fluorescent emission band at $\lambda_{max} = 467$ nm (ex = 375 nm), which corresponds to a ligand-centered (LC) fluorescence. The emission spectrum of the Er^{III} compound shows two comparable strong emission bands at 1545 nm and 1561 nm under 980 nm excitation, both were attributed to the $^4I_{13/2} \rightarrow {}^4I_{15/2}$ transition for the Er^{3+} ion.25

Such a synthetic approach can also be extended to the lanthanide tetraphosphonate system whose crystallinity and solubility problem is much more severe. Using $(H_2O_3PCH_2)_2NCH_2CH_2CH_2CH_2N(CH_2PO_3H_2)_2$ (H_8L^{18}) as the phosphonic acid and oxalate as the second ligand, we were able to prepare three novel lanthanide tetraphosphonates with two types of structures, namely, $La_2(ox)_2(H_6L^{18}) \cdot 2H_2O$ (**49**) and $Ln_2(ox)_2(H_6L^{18})(H_2O)_2 \cdot 4H_2O$ (Ln = Nd **50**, Eu **51**).[26] Their structures can also be viewed as lanthanide phosphonate-based networks being decorated by oxalate anions.

The La^{III} ion in $La_2(ox)_2(H_6L^{18}) \cdot 2H_2O$ (**49**) is eight-coordinate with four oxygen atoms of two chelating oxalate anions, and four phosphonate oxygen atoms of four $\{H_6L^{18}\}^{2-}$ anions. Its coordination geometry can be described as a distorted bi-capped trigonal prism. Each oxalate anion is tetradentate and forms two La–O–C–C–O five-member chelating rings. The $\{H_6L^{18}\}^{2-}$ anion bridges with eight La^{III} ions. All four phosphonate groups of the $\{H_6L^{18}\}^{2-}$ anion are singly protonated, so are the amine groups. The interconnection of La^{III} by chelating and bridging oxalate anions results in a 1D zigzag chain of

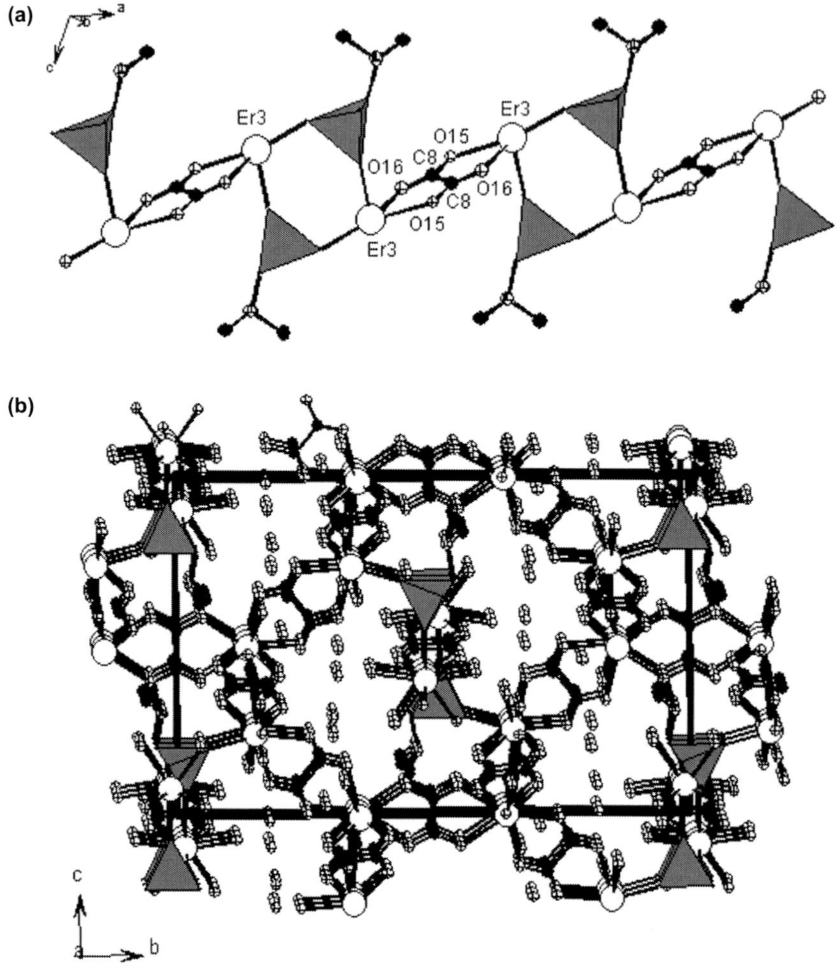

Figure 5.19 A 1D chain of $Er(C_2O_4)(HL^7)$ along *a* axis (a) and view of the crystal structure of $Er_3(C_2O_4)_4(HL^7)(H_2O)_6 \cdot 6H_2O$ (**48**) down the *a* axis (b) (adapted from Ref. 25).

$\{La(C_2O_4)\}^+$ along the *a* axis. The cross-linkage of La^{III} ions by bridging $\{H_6L^{18}\}^{2-}$ anions led to a 3D network with tunnels along the *a* axis. A tunnel is formed by 34-member atomic rings composed of four La^{III} ions, six phosphonate groups and two N–C–C–C–C–N organic groups. The structure of $La_2(ox)_2(H_6L^{18}) \cdot 2H_2O$ can be described as the oxalate anions occupying and bisecting the tunnels of La^{III} tetraphosphonate, with the lattice water molecules located at the thus formed small tunnels along the *a* axis (Figure 5.20).

The structures of $Ln_2(ox)_2(H_6L^{18})(H_2O)_2 \cdot 4H_2O$ (Ln = Nd **50**, Eu **51**) feature a different type of 3D network. The Ln^{III} is eight-coordinate with four oxygen atoms from two oxalate anions, three phosphonate oxygen atoms from three $\{H_6L^{18}\}^{2-}$ anions and an aqua ligand. The coordination geometry around the

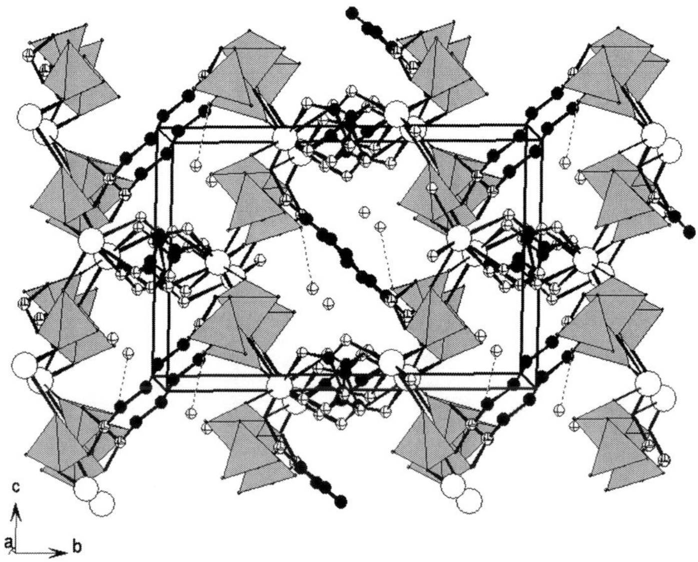

Figure 5.20 View of the structure of $[La_2(ox)_2(H_6L^{18})] \cdot 2H_2O$ (**49**) down the *a* axis (adapted from Ref. 26).

Ln^{III} ion can be also described as a distorted bi-capped trigonal prism. The $\{H_6L^{18}\}^{2-}$ anion is hexadentate and bridges with six Ln^{III} ions, two of its four phosphonate groups are bidentate whereas the remaining two are unidentate. The oxalate anions form a 1D zigzag $\{Ln(C_2O_4)\}^+$ chain along the *b* axis. The interconnection of Ln^{III} ions via bridging $\{H_6L^{18}\}^{2-}$ anions resulted in a 2D lanthanide phosphonate layer (Figure 5.21a). The above two building blocks are interconnected by sharing Ln^{III} ions into a pillared-layered architecture with the oxalate groups acting as the pillars, forming narrow-long shaped tunnels along the *b* axis (Figure 5.21b). Alternatively, the structure of $Ln_2(ox)_2(H_6L^{18})(H_2O)_2 \cdot 4H_2O$ (Ln = Nd **50**, Eu **51**) can also be viewed as the lanthanide-oxalate-phosphonate hybrid layers being interconnected by the organic diamine groups (the pillars).

Under excitation of 360 nm, $La_2(ox)_2(H_6L^{18}) \cdot 2H_2O$ (**49**) displays a broad blue fluorescent emission band at $\lambda_{max} = 435$ nm, which corresponds to a ligand-centered (LC) fluorescence. $Nd_2(ox)_2(H_6L^{18})(H_2O)_2 \cdot 4H_2O$ (**50**) shows the characteristic emission bands for the Nd^{III} ion in the near IR region: two moderate emission bands at 875 and 894 nm ($^4F_{3/2} \rightarrow {}^4I_{9/2}$), a very strong emission band at 1056 nm ($^4F_{3/2} \rightarrow {}^4I_{11/2}$) and a very weak band at 1328 nm ($^4F_{3/2} \rightarrow {}^4I_{13/2}$) in the near IR region. $Eu_2(ox)_2(H_6L^{18})$ $(H_2O)_2 \cdot 4H_2O$ (**51**) exhibits four sets of characteristic emission bands for the Eu^{III} ion in the visible region under excitation at 267 nm: 578.5 nm (very weak, $^5D_0 \rightarrow {}^7F_0$), 587.5, 590.0, 592.5 nm (weak, $^5D_0 \rightarrow {}^7F_1$), 610.5, 617.0, 620.5 nm (very strong, $^5D_0 \rightarrow {}^7F_2$) and 688, 690.5, 694, 698, 701 nm (weak, $^5D_0 \rightarrow {}^7F_4$). The splitting of $^5D_0 \rightarrow {}^7F_1$, $^5D_0 \rightarrow {}^7F_2$, and $^5D_0 \rightarrow {}^7F_4$ transition bands can be attributed to the "crystal field effect" of the Eu^{III} ion in a C_1 symmetry, which led to the

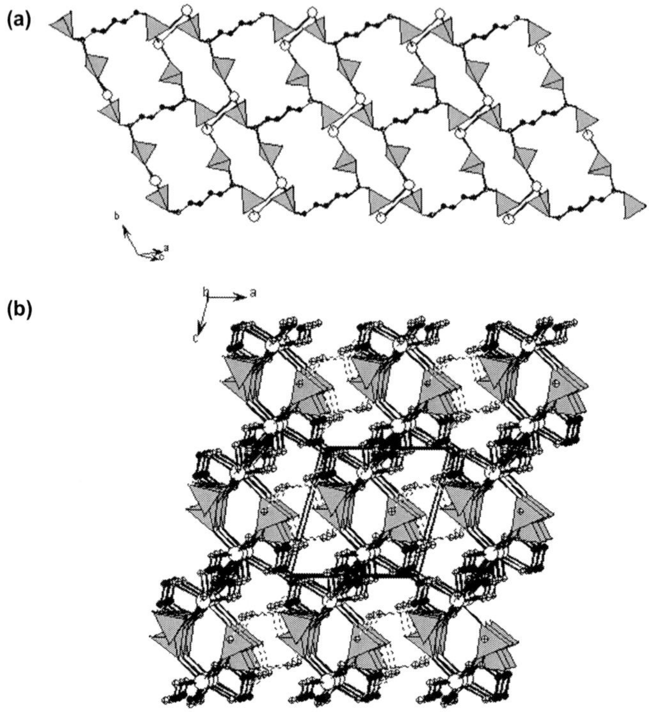

Figure 5.21 A 2D Ln^{III} tetraphosphonate layer (a) and view of the structure of $Nd_2(ox)_2(H_6L^7)(H_2O)_2 \cdot 4H_2O$ (**50**) down the *b* axis (b) (adapted from Ref. 26).

complete lifting of the degeneracy of the $^7F_1$ and the partial lifting of the degeneracy of the ^7F_j (j = 2 and 4) states. The Eu ($^5D_0$) lifetime ($\lambda_{ex,em} = 267$, 617 nm) of $Eu_2(ox)_2(H_6L^7)(H_2O)_2 \cdot 4H_2O$ (**51**) is about 0.38 ms.[26]

Hydrothermal reactions of lanthanide nitrates with aminodiphosphonate ligand ($(H_2O_3PCH_2)_2$-N-CH_2-C_6H_4-COOH, H_5L^{16}) and squaric acid afforded a series of isostructural Ln^{III} squarato-phosphonates, namely, $Ln(HL^{19})$ (Ln = La **52**, Pr **53**, Nd **54**, Eu **55**, Gd **56**, Tb **57**, Er **58**), in which a new multifunctional squarato-phosphonate ligand (H_4L^{19}) was formed by *in situ* condensation reaction between the two types of organic ligands.[27] The seven-coordinated Ln^{III} ions in these compounds are bridged by the squarato-aminophosphonate moieties of the $\{HL^{19}\}^{3-}$ anions into a double layer. Each $\{HL^{19}\}^{3-}$ anion acts as septadentate metal linker, bridging seven lanthanide ions by using one carboxylate oxygen, three squarato and three phosphonate oxygens. These double layers are further cross-linked via the coordination of the carboxylate groups into a pillared layered architecture (Figure 5.22). The Eu, Tb and Nd compounds exhibit strong luminescence in red light, green light and near IR region, respectively.

These results point to a potential improvement of the solubility and crystallinity of lanthanide phosphonates by introducing an acidic carboxylate or

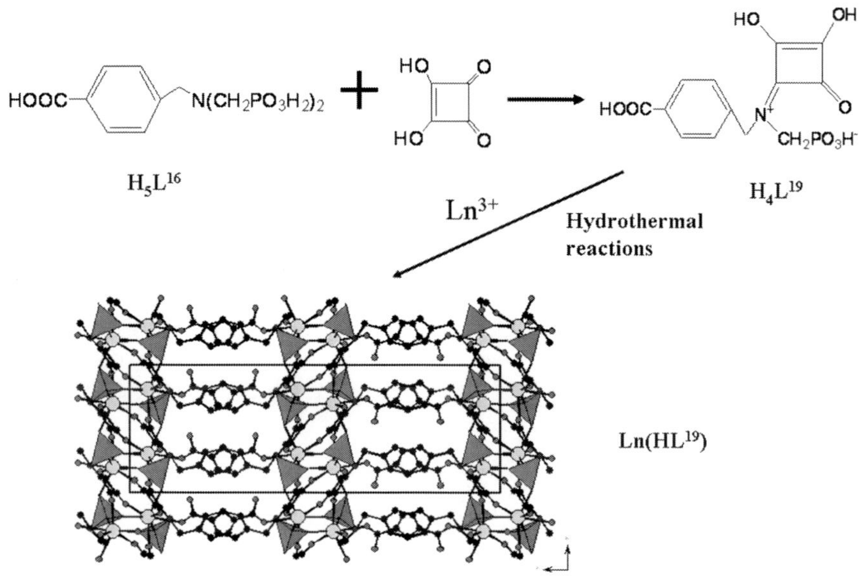

Figure 5.22 Formation of a new squarato-phosphonate ligand and its lanthanide complex (adapted from Ref. 27).

sulfonate ligand as the second metal linker. Such synthetic technique also allows us to design new types of lanthanide complexes with novel crystal structures and luminescent properties.

Basic ligands such as 1,10-phen can also act as the second metal linker, the coordination of such chelating ligand to the Ln^{III} ion reduces the coordination sites available for the phosphonate ligand, thus cluster compounds could be isolated. By using *m*-HO_3S-C_6H_4-PO_3H_2 (H_3L^{12}) as the primary ligand and 1,10-phenanthroline (phen) or *N,N′*-piperazinebis(methylenephosphonic acid) (H_4L^8) as the auxiliary ligand, six novel Ln^{III} sulfonate-phosphonates based on two types of tetranuclear clusters, namely, $[La_2(L^{12})_2(phen)_4(H_2O)]\cdot 4.5H_2O$ (**59**) $[Ln_2(L^{12})_2(phen)_2(H_2O)_5]\cdot 3H_2O$ (Ln = Nd **60**, Eu **61**, Er **62**) and $[Ln_2(HL^{12})(H_2L^8)_2(H_2O)_4]\cdot 8H_2O$ (Ln = La **63**, Nd **64**) were isolated by our group. Their structures feature isolated 1D and 3D tetranuclear lanthanide clusters.[28]

$[La_2(L^{12})_2(phen)_4(H_2O)]\cdot 4.5H_2O$ (**59**) features a 1D chain of tetranuclear clusters (Figure 5.20). There are two unique La^{III} ions in the asymmetric unit, La(1) is nine-coordinate whereas La(2) is eight-coordinate. Among two crystallographically distinct $\{L^{12}\}^{3-}$ anions, one is tridentate with each phosphonate oxygen bridging to a La^{III} ion and the other $\{L^{12}\}^{3-}$ anion is pentadentate by using its three phosphonate oxygens and one sulfonate oxygen. The phosphonate group is tetradentate and bridges with three La^{III} ions, one phosphonate oxygen atom functions as a μ^2-metal linker. The sulfonate group is unidentate. Two La(1) and two La(2) ions are interconnected by four phosphonate groups of four $\{L^{12}\}^{3-}$ anions into a tetranuclear cluster unit.

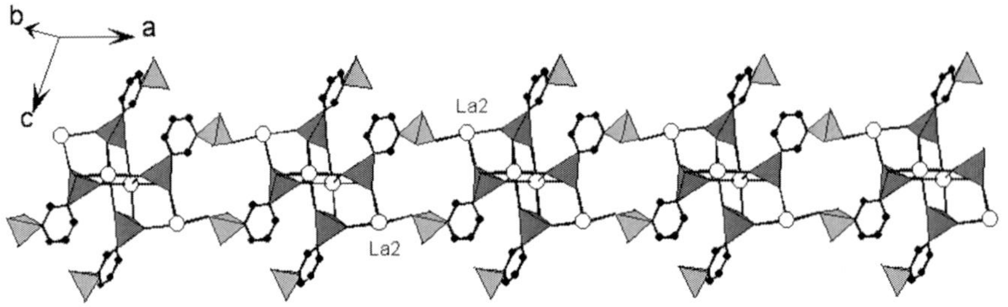

Figure 5.23 A 1D cluster chain along the *a* axis in $[La_2(L^{12})_2(phen)_4(H_2O)] \cdot 4.5H_2O$ (**59**). The CPO_3 and CSO_3 tetrahedra are shaded in medium and light gray, respectively. La and C atoms are drawn as open and black circles, respectively. The phen ligands and aqua ligands have been omitted for clarity (adapted from Ref. 28).

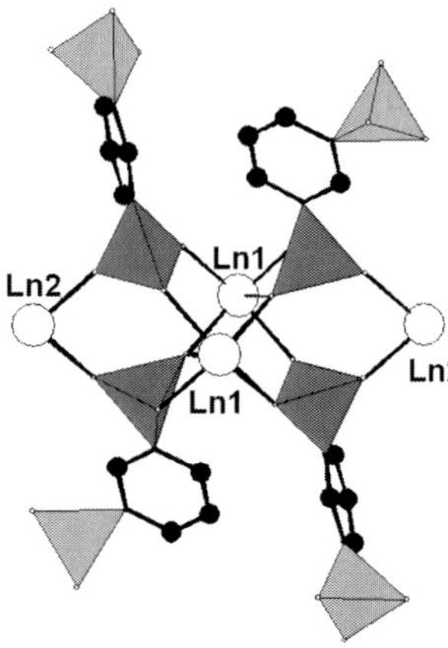

Figure 5.24 A tetranuclear cluster in $[Ln_2(L^{12})_2(phen)_2(H_2O)_5] \cdot 3H_2O$ (Ln = Nd **60**; Eu **61**; Er **62**). The CPO_3 and CSO_3 tetrahedra are shaded in medium and light gray, respectively. The phen ligands and aqua ligands have been omitted for clarity (adapted from Ref. 28).

Each phosphonate ligand caps on a La_3 triangle. Each tetranuclear cluster is connected to two neighboring ones through the coordination of the sulfonate oxygens, forming a 1D chain along the *a* axis (Figure 5.23).

When the La^{III} ions were replaced by other Ln^{III} ions, $[Ln_2(L^{12})_2(phen)_2(H_2O)_5] \cdot 3H_2O$ (Ln = Nd **60**, Eu **61**, Er **62**) were isolated. These compounds feature discrete tetranuclear Ln^{III} clusters (Figure 5.24). There are two unique Ln^{III} ions in the asymmetric unit, both are eight-coordinate. The phosphonate groups of the $\{L^{12}\}^{3-}$ ligands adopt the same coordination fashion as those in the La^{III} compound. However, the sulfonate group remains non-coordinate. Four Ln^{III} ions are bridged by four sulfonate-phosphonate ligands into a similar tetranuclear cluster (Figure 5.24). These clusters are isolated due to the non-coordination nature of the sulfonate groups.

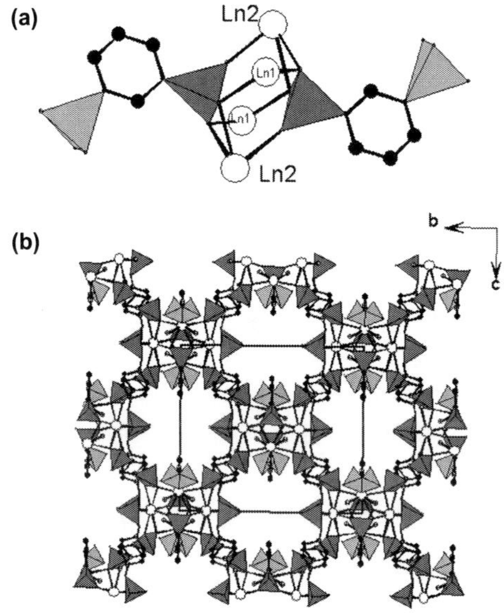

Figure 5.25 A tetranuclear $\{Ln_4(HL^{12})_2\}^{4+}$ cluster (a) and the 3D open framework (b) in $[Ln_2(HL^{12})(H_2L^8)_2(H_2O)_4]\cdot 8H_2O$ (Ln = La **63**; Nd **64**). The CPO_3 and CSO_3 tetrahedra are shaded in medium and light gray, respectively. The lattice water molecules have been omitted for clarity (adapted from Ref. 28).

$[Ln_2(HL^{12})(H_2L^8)_2(H_2O)_4]\cdot 8H_2O$ (Ln = La **63**, Nd **64**) were isolated when *N*,*N'*-piperazinebis(methylenephosphonic acid) (H_4L^8) was applied as the second metal linker instead of phen and the reaction media is more acidic. Their structures feature a 3D architecture based on a different type of lanthanide tetranuclear clusters (Figure 5.25). Among two unique Ln^{III} ions, one is eight-coordinate with six phosphonate oxygen atoms from four $\{H_2L^8\}^{2-}$ anions and two phosphonate oxygens from two $\{HL^{12}\}^{2-}$ anions, whereas the other one is nine-coordinate with three phosphonate oxygen atoms from two $\{HL^{12}\}^{2-}$ anion, two phosphonate oxygen atoms from two $\{H_2L^8\}^{2-}$ anions as well as four aqua ligands. The phosphonate group of the $\{HL^{12}\}^{2-}$ anion is pentadentate, two phosphonate oxygen atoms act as a μ^2-bridging ligand whereas the third oxygen atom is monodentate. The sulfonate group of the $\{HL^{12}\}^{2-}$ anion is severely disordered and remains non-coordinate. Each $\{HL^{12}\}^{2-}$ anion connects with four Ln^{III} ions. The two $\{H_2L^8\}^{2-}$ anions adopt two different coordination modes: one is tetradentate and bridges with four Ln^{III} ions from two cluster units and the other one is bidentate and connects with two Ln^{III} ions from two cluster units. The interconnection of four Ln^{III} ions by two bridging $\{HL^{12}\}^{2-}$ anions resulted in a $\{Ln_4(HL^{12})_2\}^{4+}$ tetranuclear cluster (Figure 5.25a). The four Ln^{III} ions form a slightly distorted square with the two phosphonate groups of $\{HL^{12}\}^{2-}$ ligands cap above and below the square. Each Ln(1)...Ln(2) edge is further bridged by a phosphonate group of a $\{H_2L^8\}^{2-}$ anion. The Ln(1)...Ln(2) edges are significantly shorter than those in first type of tetranuclear cluster. Such clusters are further cross-linked by the bridging $\{H_2L^8\}^{2-}$ anions into a 3D framework with larger tunnels along the *a* axis (Figure 5.25b). Each cluster is bridged to six neighbors. The lattice water molecules are located in the above tunnels. It is found that the open framework of the La compound remains intact after heating at 150 °C for 4 hours based on XRD studies.[28]

Under excitation of 335 nm, $[Nd_2(L^{12})_2(phen)_2(H_2O)_5]\cdot 3H_2O$ (**60**) displays three sets of characteristic emission bands for the Nd^{III} ion: 890 nm and 879 nm ($^4F_{3/2} \rightarrow {}^4I_{9/2}$), 1061 nm and 1072 nm ($^4F_{3/2} \rightarrow {}^4I_{11/2}$) and a weak band at 1335 nm ($^4F_{3/2} \rightarrow {}^4I_{13/2}$) in the near IR region. $[Eu_2(L^{12})_2(phen)_2(H_2O)_5]\cdot 3H_2O$ (**61**) exhibits three strong characteristic emission bands for the Eu^{III} ion in the visible region under excitation at 337 nm: 535 nm ($^5D_0 \rightarrow {}^7F_0$), 591 nm ($^5D_0 \rightarrow {}^7F_1$), 615 nm ($^5D_0 \rightarrow {}^7F_2$) and 685 nm ($^5D_0 \rightarrow {}^7F_4$). The Eu ($^5D_0$) lifetime for $\lambda_{ex,em} = 337$, 614 nm was measured to be 0.29 ms.

5.5 Metal Arsonates with an Auxiliary Ligand

Metal arsonates are expected to display a similar structural chemistry to those of metal phosphonates, but the larger ionic radius of As^V compared to P^V could lead to some different architectures with different physical properties. So far, reports on metal organo arsonates are very limited, and most of them feature isolated clusters in which the arsonate group caps on a triangle formed by three metal ions.[29–38] Our group has been exploring the effects of the auxiliary ligands on the structures and properties of metal arsonates in the last few years (Table 5.4). We hope to find out the similarities and differences between these two classes of compounds.

Hydrothermal reactions of lead(II) acetate with phenylarsonic acid (H_2L^{20}) (or 4-hydroxy-3-nitrophenylarsonic acid, H_3L^{21}) and 5-sulfoisophthalic acid monosodium salt (NaH_2SIP) (or 1,3,5-benzenetricarboxylic acid, H_3BTC) as the second metal linkers afforded three novel mixed-ligand lead(II) carboxylate-arsonates, namely, $Pb_5(SIP)_2(L^{20})_2(H_2O)$ (**65**), $Pb_3(SIP)(L^{21})(H_2O)$ (**66**) and $Pb(H_2L^{21})(H_2BTC)$ (**67**).[39]

The structure of $Pb_5(SIP)_2(L^{20})_2(H_2O)$ (**65**) features a complicated 3D network composed of 2D double layers of lead(II) sulfoisophthalate bridged by 1D chains of lead(II) arsonates along the *b* axis, forming large tunnels along the *b* axis which are occupied by phenyl rings of the arsonate ligands. There are five unique Pb^{II} ions in its asymmetric unit. They display different coordinated geometries. Pb(1) is five-coordinated by one sulfonate oxygen and four carboxylate oxygen atoms from four SIP anions. Pb(2) is five-coordinated by one sulfonate oxygen, three carboxylate oxygen atoms from three SIP^{3-} anions and an aqua ligand. Pb(4) is five-coordinated by one carboxylate oxygen, one sulfonate oxygen atom from two SIP^{3-} anions and three arsonate oxygen atoms from two $\{L^{20}\}^{2-}$ anions. The coordination geometry around Pb(1), Pb(2) and Pb(4) can be described as a severely distorted ψ-PbO_5 octahedron with one site occupied by the lone pair electrons. Pb(3) is four-coordinated by three arsonate oxygen atoms from three $\{L^{20}\}^{2-}$ anions and one oxygen atom from a SIP^{3-} anion. Pb(5) is also four-coordinated but by four arsonate oxygen atoms from four $\{L^{20}\}^{2-}$ anions. The

Table 5.4 Metal arsonates with an auxiliary ligand.

Compound	*Structural features*	*Ref.*
$Pb_5(SIP)_2(L^{20})_2(H_2O)$ (**65**)	a complicated 3D network composed of 2D double layers of lead(II) sulfo-isophthalate and 1D lead(II) arsonate chains	39
$Pb_3(SIP)(L^{21})(H_2O)$ (**66**)	a complicated 3D network composed of lead(II) arsonate double layers and lead(II) sulfonate-carboxylate double layers	39
$Pb(H_2L^{21})(H_2BTC)$ (**67**)	3D supramolecular architecture composed of 1D chains of lead(II) arsonate-carboxylate that are further interconnected via hydrogen bonds	39
$Cd(phen)(HL^{20})_2(H_2O)$ (**68**)	dinuclear clusters	40
$M(phen)(H_2L^{21})_2$ (M = Mn, **69**; Cd, **70**)	1D chains based on dinuclear $M_2(\mu\text{-}O)_2$ cluster units	40
$[M(phen)_2(H_2L^{21})](ClO_4)\cdot(H_2O)$ (M = Mn, **71**; Cd, **72**)	dinuclear clusters	40
$[Mn(phen)_2(HL^{20})](ClO_4)\cdot(H_2O)$ (73)	1D helical chain in which neighboring two metal centers are bridged by one arsonate ligand	40
$[Mn(HL^{22})(phen)_2]_2\cdot 8.5H_2O$ (**74**)	mononuclear	41
$Mn(HL^{22})(phen)_2(H_2O)\cdot 2H_2O$ (**75**)	mononuclear	41
$[Mn(HL^{22})(bipy)_2][Mn(H_2L^{22})(bipy)_2](ClO_4)\cdot 3H_2O$ (**76**)	mononuclear	41
$[Mn(HL^{22})(phen)][Mn(HL^{22})(phen)(H_2O)]$ (**77**)	dinuclear	41
$[Mn_2(HL^{22})(phen)_4(H_2O)](ClO_4)_2\cdot 4H_2O$ (**78**)	dinuclear	41
$[Mn_2(HL^{22})(phen)_4(H_2O)](ClO_4)_2\cdot H_2O$ (**79**)	dinuclear	41
$Mn_2(HL^{22})_2(bipy)_3(H_2O)\cdot H_2O$ (**80**)	dinuclear	41
$[Mn(HL^{22})(terpy)]_2$ (**81**)	dinuclear	41
$Mn_7(OH)_2(L^{22})_4(phen)_8\cdot 10H_2O\cdot phen$ (**82**)	heptanuclear cluster	41
$Mn(HL^{22})(bipy)(H_2O)\cdot 2H_2O$ (**83**)	1D chain	41

coordination geometry around Pb(3) and Pb(5) can be described as a severely distorted ψ-PbO_4 trigonal bipyramid with one site occupied by the lone pair electrons. The two phenylarsonate anions display two different coordination modes. The one containing As(1) atom is tetradentate and bridges with four lead(II) ions whereas the one containing As(2) atom is hexadentate and connects with six lead(II) ions. The two SIP^{3-} anions also exhibit two different types of coordination modes. The SIP^{3-} anion containing S(1) atom is pentadentate and connects with four lead(II) ions and the SIP^{3-} anion containing S(2) atom acts as a heptadentate ligand and bridges with six lead(II) ions.

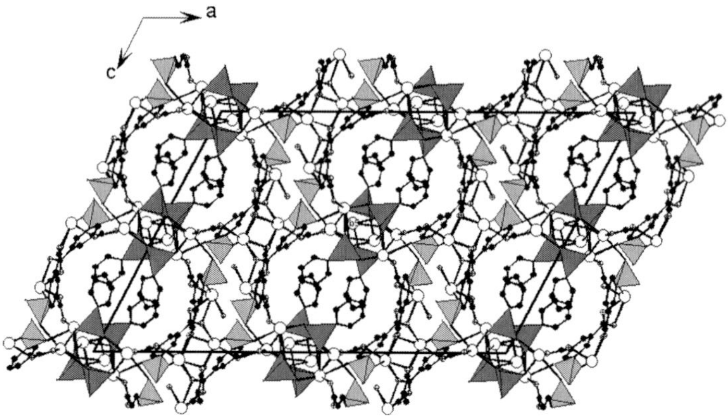

Figure 5.26 View of the structure of compound **65** down the *b* axis. The $CAsO_3$ and CSO_3 tetrahedra are shaded in medium and light gray, respectively. Pb, C and O atoms are represented by open, black and crossed circles, respectively (adapted from Ref. 39).

The interconnection of Pb(3), Pb(4) and Pb(5) atoms by the bridging phenylarsonic ligands groups resulted in a 1D chains whereas the interconnection of the Pb(1), Pb(2), Pb(3) and Pb(4) atoms by bridging and chelating SIP^{3-} anions led to a 2D layer. The cross-linkage of the above two building blocks afforded to a complicated 3D network with large tunnels along the *b* axis, the phenyl rings of the arsonate ligands are located at these tunnels (Figure 5.26).

The formula unit of $Pb_3(SIP)(L^{21})(H_2O)$ (**66**) contains three Pb^{II} ions, one fully deprotonated $\{L^{21}\}^{3-}$ anion, and one fully deprotonated SIP^{3-} anion and one aqua ligand. Pb(1) is four-coordinated by one sulfonate and two carboxylate oxygen atoms from three SIP^{3-} anions as well as an arsonate oxygen atom from a $\{L^{21}\}^{3-}$ anion, its coordination geometry can be described as a ψ-PbO_4 trigonal bipyramid with one corner filled by the lone pair electrons. Pb(3) is four-coordinated by a bidentate chelating carboxylate group of a SIP^{3-} anion, two arsonate oxygens from two $\{L^{21}\}^{3-}$ anions and an aqua ligand, its coordination geometry can be described as a ψ-PbO_4 distorted square pyramid with the pyramidal site filled by the lone pair electrons. Pb(2) is six-coordinated by one carboxylate and one sulfonate oxygen atoms from two SIP^{3-} anions, two arsonate oxygen atoms, one nitro oxygen atom and one hydroxyl group from three $\{L^{21}\}^{3-}$ anions in a severely distorted octahedral geometry. The $\{L^{21}\}^{3-}$ anion is heptadentate and connects with six lead(II) ions: two arsonate oxygens are bidentate whereas the third one is unidentate, the hydroxyl group and one nitro oxygen form a six-member chelating ring with a lead(II) ion, the other nitro oxygen remains non-coordinated. The SIP^{3-} anion is heptadentate and bridges with six lead(II) ions: one carboxylate group chelates bidentately with a Pb^{II} ion and also bridges with another Pb^{II} ion whereas the other carboxylate group bridges with two Pb^{II} ions. The sulfonate group bridges with two Pb^{II} ions by using two oxygen atoms and the third sulfonate oxygen atom

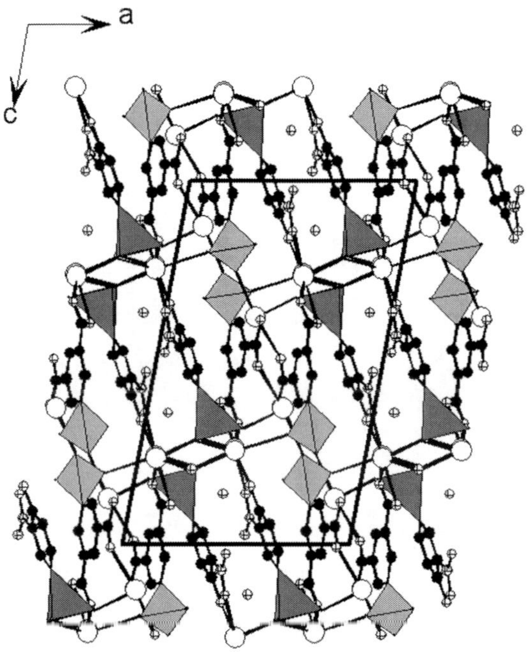

Figure 5.27 View of the structure of compound **66** down the *b* axis. The $CAsO_3$ and CSO_3 tetrahedra are shaded in medium and light gray, respectively. Pb, C and O atoms are represented by open, black and crossed circles, respectively (adapted from Ref. 39).

remains non-coordinated. The interconnection of the lead(II) ions via bridging and chelating $\{L^{21}\}^{3-}$ anions resulted in the formation of a (-202) layer. The interconnection of the Pb^{II} ions via bridging and chelating SIP^{3-} anions led to a (-101) Pb^{II} carboxylate-sulfonate double layer. The cross-linkage of the above two building units gave a complicated 3D network. Such a 3D network can also be viewed as the lead(II) arsonate-sulfonate inorganic layers passing 1/4 and 3/4 of the *c* axis being bridged by organic groups of the above two types of multidentate ligands (Figure 5.27).

The asymmetric unit of $Pb(H_2L^{21})(H_2BTC)$ (**67**) contains one lead(II) ion, one doubly protonated $\{H_2L^{21}\}^-$ anion and one doubly protonated $\{H_2BTC\}^-$ anion. The lead(II) is five-coordinated by a bidentate chelating carboxylate group of a $\{H_2BTC\}^-$ anion and three arsonate oxygens from three $\{H_2L^{21}\}^-$ anions. Its coordination geometry can be described as a severely distorted ψ-PbO_5 octahedron with the lone pair of the lead(II) ion occupying the sixth site. The arsonate ligand is doubly protonated: the hydroxyl group and one oxygen atom of the arsonate group. The $\{H_2L^{21}\}^-$ anion is tridentate: the arsonate group bridges with three Pb^{II} ions by using two oxygen atoms and the third oxygen atom remains non-coordinated. The nitro group and hydroxyl groups are also non-coordinated. The tricarboxylate ligand is also doubly protonated, O(4) and O(8) atoms are most likely protonated based on elongated C–O

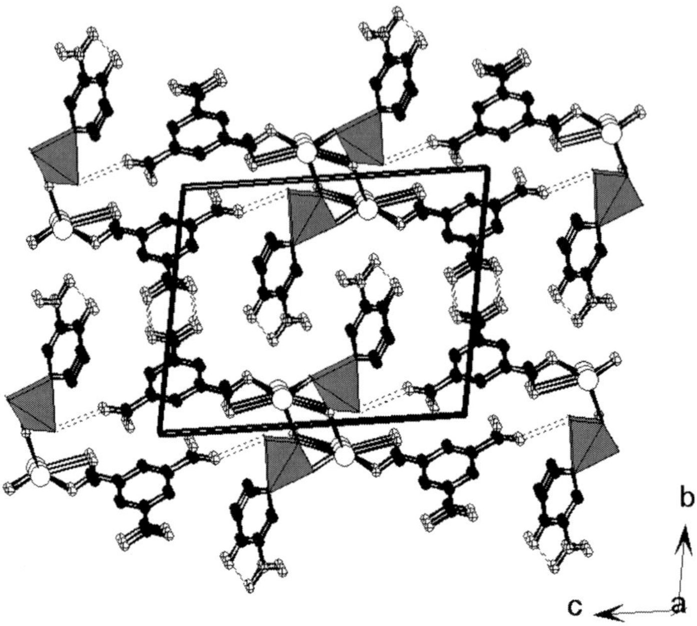

Figure 5.28 View of the structure of compound **3** down the *a* axis. The $CAsO_3$ are shaded in pink. Pb, C and O atoms are represented by open, black and crossed circles, respectively (adapted from Ref. 39).

distances. The $\{H_2BTC\}^-$ anion is bidentate chelating with a Pb^{II} ion by using one carboxylate group, the other two protonated carboxylate groups remain non-coordinated. The interconnection of the Pb^{II} ions via bridging arsonate groups of the $\{H_2L^{21}\}^-$ anions leads to a 1D double chain along the *b* axis. The $\{H_2BTC\}^-$ anions are hanging on the above chains. These 1D chains are further interlinked via hydrogen bonds among non-coordination carboxylate groups and non-coordinate arsonate oxygen atoms into a 3D supramolecular assembly (Figure 5.28). The O...O separations are quite short (2.60(2) and 2.66(1) Å).

Hydrothermal reactions of cadmium(II) or manganese(II) salts with aryl arsenic acids $RAsO_3H_2$ ($R = C_6H_5$-, H_2L^{20}; 3-NO_2-4-OH-C_6H_3-, H_3L^{21}) and 1,10-phenanthroline (phen) led to six new cadmium(II) and manganese(II) organo arsonates, namely, $Cd(phen)(HL^{20})_2(H_2O)$ (**68**), $M(phen)(H_2L^{21})_2$ (M = Mn, **69**, Cd, **70**), $[M(phen)_2(H_2L^{21})](ClO_4)\cdot(H_2O)$ (M = Mn, **71**, Cd, **72**) and $[Mn(phen)_2(HL^{20})]$ $(ClO_4)\cdot(H_2O)$ (**73**).[40] The structure of $Cd(phen)(HL^{20})_2(H_2O)$ (**68**) features a dinuclear cluster in which the two octahedrally coordinated cadmium(II) ions are bridged by two arsonate oxygen atoms from two arsonate anions (Figure 5.29a). $[M(phen)_2(H_2L^{21})]$-$(ClO_4)\cdot(H_2O)$ (M = Mn, **71**, Cd, **72**) feature a different dinuclear $[M(phen)_2(H_2L^2)]^+$ cation in which the two metal centers are bridged by a pair of arsonate ligands through M–O–As–O–M bridges (Figure 5.29b). $M(phen)(H_2L^{21})_2$ (M = Mn, **69**, Cd, **70**) exhibit 1D chains based on dinuclear

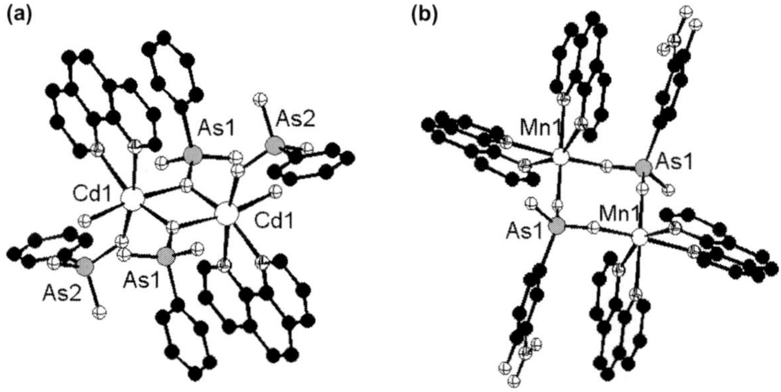

Figure 5.29 The dinuclear cluster unit in compounds **68** (a) and **71** (b) (adapted from Ref. 40).

$M_2(\mu\text{-}O)_2$ cluster units further bridged by arsonate ligands. Hence neighboring two metal centers are interconnected by alternatively a pair of M–O–M and M–O–As–O–M bridges (Figure 5.30a). There are two different types of bidentate arsonate anions, the one containing As(1) is associated with M–O–As–O–M bridge whereas the one containing As(2) is involved in M–O–M bridge (Figure 5.30a). $[Mn(phen)_2(HL^{20})](ClO_4)\cdot(H_2O)$ (**73**) features a 1D helical cationic chain in which neighboring two metal centers are bridged by only one arsonate ligand via M–O–As–O–M connectivity (Figure 5.30b). Each metal ion in **69** or **70** is bonded to one bidentately chelating phen ligand whereas the one in **73** is bonded to two such phen ligands.

Magnetic property measurements on compounds **69**, **71** and **73** indicate that there exist very weak antiferromagnetic interactions between magnetic centers in all three compounds. Compounds **68**–**73** display typical ligand-centered (LC) fluorescence emission bands.

Hydrothermal reactions of manganese(II) salts with *o*-sulfophenylarsonic acid ($o\text{-}HO_3S\text{-}C_6H_4\text{-}AsO_3H_2$, H_3L^{22}) afforded $Mn_3(L)_2(H_2O)_3\cdot H_2O$ with a layered structure. When 1,10-phenanthroline (phen), 2,2′-bipyridine (bipy), and 2,2′:6′,2″-terpyridine (terpy) were used as auxiliary chelating ligands, a series of mixed-ligand manganese(II) sulfonate-arsonates with lower dimensional structures, namely, $[Mn(HL^{22})(phen)_2]_2\cdot 8.5H_2O$ (**74**), $Mn(HL^{22})(phen)_2(H_2O)\cdot 2H_2O$ (**75**), $[Mn(HL^{22})(bipy)_2][Mn(H_2L^{22})(bipy)_2](ClO_4)\cdot 3H_2O$ (**76**), $[Mn(HL^{22})(phen)][Mn(HL^{22})(phen)(H_2O)]$ (**77**), $[Mn_2(HL^{22})(phen)_4(H_2O)](ClO_4)_2\cdot 4H_2O$ (**78**), $[Mn_2(HL^{22})(phen)_4(H_2O)](ClO_4)_2\cdot H_2O$ (**79**), $Mn_2(HL^{22})_2(bipy)_3(H_2O)\cdot H_2O$ (**80**), $[Mn(HL^{22})(terpy)]_2$ (**81**), $Mn_7(OH)_2(L^{22})_4(phen)_8\cdot 10H_2O\cdot phen$ (**82**), and $Mn(HL^{22})(bipy)(H_2O)\cdot 2H_2O$ (**83**) were obtained.[41] **74**–**76** are mononuclear whereas **77**–**81** feature three types of isolated dinuclear cluster units in which the two Mn^{2+} ions are bridged by one or two sulfonate-arsonate ligands. **82** exhibits an interesting heptanuclear cluster in which the Mn^{2+} centers are bridged by arsonate, sulfonate groups and hydroxyl anions. **83** features a 1D chain in which two neighboring Mn^{2+} centers are bridged by

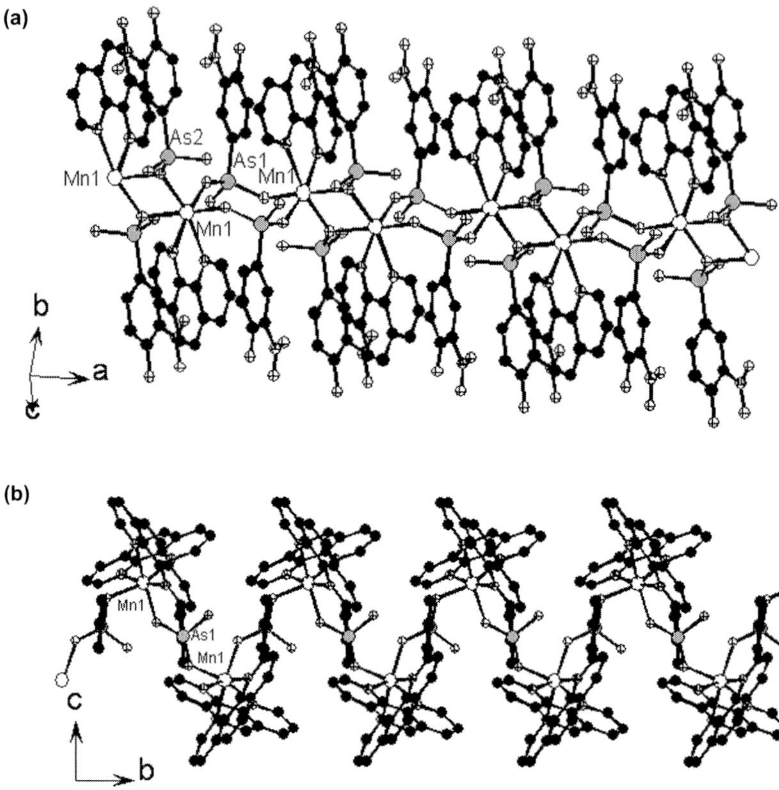

Figure 5.30 The 1D chains in compounds **69** (a) and **73** (b) (adapted from Ref. 40).

an arsonate group of a sulfonate-arsonate ligand (Figure 5.31). Magnetic measurements indicate that $Mn_3(L)_2(H_2O)_3 \cdot H_2O$ exhibits an unprecedented spin topology and behaves as a homospin ferrimagnet whereas **74**–**76** are essentially paramagnetic. **77**–**81** and **83** are weakly antiferromagnetic.[41]

The *o*-sulfophenylarsonic acid is a versatile ligand and is able to adopt many types of coordination modes similar to the corresponding phosphonate ligands. The pK_a values of the phenylphosphonic acid are 1.86 and 7.51 and those of the phenylarsonic acid are 3.39 and 8.25, hence the phenylarsonate ligand is more likely to be partially protonated and binding with fewer metal centers as compared with the corresponding phosphonic acid.

The factors that affect the types of structures formed are rather complicated and difficult to be summarized. It is noticed, however, that the metal sources, reaction temperatures, molar ratios of the starting materials, pH values of the reaction media, as well as the auxiliary ligands all played an important role. It is found that the coordination of an aqua ligand to the metal center will change the coordination mode of the arsonate ligand as in **75**, **78** and **79** compared to other dinuclear compounds. The presence of μ^3-hydroxyl anion facilitates the

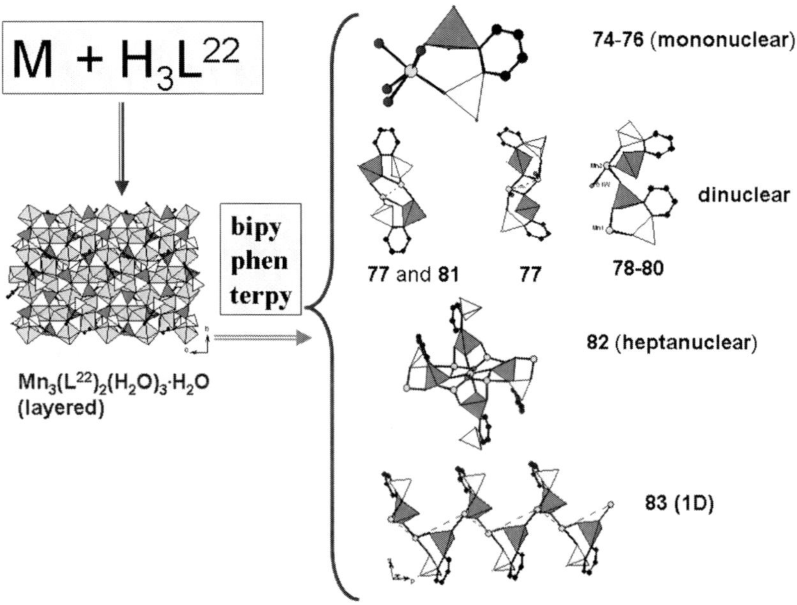

Figure 5.31 Simplified structure diagrams for compounds **74**–**83** (adapted from Ref. 41).

formation of high nuclear cluster such as heptanuclear **82**. However, much more systematic works are needed to further elucidate the factors that control the structures of the resultant complexes. It is also anticipated that more new types of multinuclear clusters of other transition mental or Ln^{III} sulfonate-arsonates can be designed by using such a synthetic route.

5.6 Concluding Remarks

In summary, the introduction of a second metal linker is an effective route to new metal phosphonates and arsonates with interesting structures and novel physical properties. The basic second ligand such as phen can limit the coordination sites of the metal ion available for the phosphonate or arsonate ligand and hence facilitate the formation of metal compounds with isolated cluster units. For an acidic second ligand such as oxalate and benzotricarboxylate the metal compounds obtained usually display complicated 3D network structure, the second ligand can be incorporated into the open framework of metal phosphonates or arsonates (primary ligand dominated); in some cases the phosphonate or arsonate ligand could be incorporated into the open framework formed by the metal ions and the second ligands (second-ligand-dominated). When the reaction media has a low pH value, the auxiliary ligand such as BTC can be partially or completely protonated, it hangs on the interlayer region or merely acts as the intercalated species.

For the future trends in this field, we think that designing 3D open framework with targeted function such as porous materials by careful selection of arsonate or phosphonate ligand and auxiliary ligand would be still a hot topic. On the other hand, low dimensional arrays or aggregations with interesting magnetic properties such as single molecular magnet or single chain magnets would be still an important theme.

Acknowledgements

This work was supported by the National Natural Science Foundation of China (No. 20371047) and NSF of Fujian Province (No. E0420003). I am also indebted to my graduate students and research assistants, Jun-Ling Song, Bing-Ping Yang, Zhong-Ming Sun, Shaoming Ying, Zi-Yi Du, Ya-Qin Guo and Fei-Yan Yi for their productive work.

References

1. A. Clearfield, *Prog. Inorg. Chem.*, 1998, **47**, 371 (and references therein).
2. C. N. R. Rao, S. Natarajan and R. Vaidhyanathan, *Angew. Chem., Int. Ed. Engl.*, 2004, **43**, 1466.
3. K. Maeda, *Micro. Meso. Mater.*, 2004, **73**, 47.
4. E. Matczak-Jon and V. Videnova-Adrabiñska, *Coord. Chem. Rev.*, 2005, **249**, 2458.
5. F. S. Richardson, *Chem. Rev.*, 1982, **82**, 541.
6. J. G. Mao, *Coord. Chem. Rev.*, 2007, **251**, 1493.
7. G. K. H. Shimizu, R. Vaidhyanathan and J. M. Talor, *Chem. Soc. Rev.*, 2009, **38**, 1430.
8. J.-L. Song, J.-G. Mao, Y.-Q. Sun and A. Clearfield, *Eur. J. Inorg. Chem. 2003*, 4218.
9. Z.-M. Sun, J.-G. Mao, B.-P. Yang and S.-M. Ying, *Solid State Sci.*, 2004, **6**, 295.
10. S.-M. Ying and J.-G. Mao, *J. Mol. Struct.*, 2005, **748**, 63.
11. S.-M. Ying and J.-G. Mao, *Eur. J. Inorg. Chem. 2004*, 1270.
12. Z.-Y. Du, S.-M. Ying and J.-G. Mao, *J. Mol. Struct.*, 2006, **788**, 218.
13. J.-L. Song, C. Lei, Y.-Q. Sun and J.-G. Mao, *J. Solid State Chem.*, 2004, **177**, 2557.
14. Z.-Y. Du, H.-B. Xu, X.-L. Li and J.-G. Mao, *Eur. J. Inorg. Chem. 2007*, 4520.
15. C. Lei, J.-G. Mao, Y.-Q. Sun, H.-Y. Zeng and A. Clearfield, *Inorg. Chem.*, 2003, **42**, 6157.
16. C. Lei, J.-G. Mao, Y.-Q. Sun and Z.-C Dong, *Polyhedron*, 2005, **24**, 295.
17. Y.-Q. Guo, B.-P. Yang, J.-L. Song and J.-G. Mao, *Cryst. Growth Des.*, 2008, **8**, 600.
18. Z.-Y. Du, H.-B. Xu and J.-G. Mao, *Inorg. Chem.*, 2006, **45**, 6424.
19. Z.-Y. Du, A. V. Prosvirin and J.-G. Mao, *Inorg. Chem.*, 2007, **46**, 9884.

20. Z.-Y. Du, X.-L. Li, Q.-Y. Liu and J.-G. Mao, *Cryst. Growth Des.*, 2007, **7**, 1501.
21. J.-L. Song, H.-H. Zhao, J.-G. Mao and K. R. Dunbar, *Chem. Mater.*, 2004, **16**, 1884.
22. J.-L. Song, A. V. Prosvirin, H. H. Zhao and J.-G. Mao, *Eur. J. Inorg. Chem.,* 2004, 3706.
23. B.-P. Yang and J.-G. Mao, *Inorg. Chem.*, 2005, **44**, 566.
24. J.-L. Song, C. Lei and J.-G. Mao, *Inorg. Chem.*, 2004, **43**, 5630.
25. J.-L. Song and J.-G. Mao, *Chem. Eur. J.*, 2005, **11**, 1417.
26. S.-M. Ying and J.-G. Mao, *Cryst. Growth Des.*, 2006, **6**, 964.
27. J.-L. Song, F.-Y. Yi and J.-G. Mao, *Cryst. Growth Des.*, 2008, **9**, 3273.
28. Z.-Y. Du, H.-B. Xu and J.-G. Mao, *Inorg. Chem.*, 2006, **45**, 9780.
29. E. Burkholder, E. Wright, V. Golub, C. J. O'Connor and J. Zubieta, *Inorg. Chem.*, 2003, **42**, 7460.
30. B. J. S. Johnson, S. A. Geers, W. W. Brennessel, V. G. Young, Jr. and A. Stein, *Dalton Trans.*, 2003, 4678.
31. M. I. Khan and J. Zubieta, *Angew. Chem., Int. Ed. Engl.*, 1994, **33**, 760.
32. M. I. Khan, Y. Chang, Q. Chen, H. Hope, S. Parking, D. P. Goshorn and J. Zubieta, *Angew. Chem., Int. Ed. Engl.*, 1992, **31**, 1197.
33. B. J. S. Johnson, R. C. Schroden, C. Zhu, V. G. Young, Jr. and A. Stein, *Inorg. Chem.*, 2002, **41**, 2213.
34. B. J. S. Johnson, R. C. Schroden, C. Zhu and A. Stein, *Inorg. Chem.*, 2001, **40**, 5972.
35. J. Salta, Y.-D. Chang and J. Zubieta, *J. Chem. Soc., Chem. Commun.*, 1994, 1039.
36. G. Huan, J. W. Johnson, A. J. Jacobson and J. S. Merola, *Chem. Mater.*, 1990, **2**, 719.
37. M. R. Mason, R. M. Mattews, M. S. Mashuta, J. F. Richardson and A. Vij, *Inorg. Chem.*, 1997, **36**, 6476.
38. Y.-P. Xie, J. Yang, J.-F. Ma, L.-P. Zhang, S.-Y. Song and Z.-M. Su, *Chem. Eur. J.*, 2008, **14**, 4093.
39. F.-Y. Yi, J.-L. Song, N. Zhao and J.-G. Mao, *J. Solid State Chem.*, 2008, **181**, 1393.
40. F.-Y. Yi, N. Zhao, W. Wu and J.-G. Mao, *Inorg. Chem.*, 2009, **48**, 628.
41. F.-Y. Yi, Q.-P. Lin, T.-H. Zhou and J.-G. Mao, *Inorg. Chem.*, 2010, **49**, 3489.

CHAPTER 6

Synthesis of Phosphonic Acids and Their Esters as Possible Substrates for Reticular Chemistry

JERZY ZOŃ, PIOTR GARCZAREK AND MICHAŁ BIAŁEK

Wrocław University of Technology, Faculty of Chemistry, Wybrzeże Wyspiańskiego 27, 50-370 Wrocław, Poland

6.1 Introduction

Phosphonic acids are scientifically and practically an interesting class of organic compounds. Practically important representatives of this class are compounds: herbicide *N*-(phosphonomethyl)glycine (Roundup®), antiviral agent phosphonoformic acid (Foscavir®) and the drug for hypercalcemia treatment 3-amino-1-hydroxypropanediyl-1,1-bisphosphonic acid (Aredia®). Moreover, many phosphonic analogs of biologically active molecules, including metabolities and drugs, have been found as several dozen inhibitors of enzymes or physiological processes, and for that reason focus the attention of many research groups around the world. Some phosphonic acids and their derivatives are also interesting as corrosion inhibitors, flame retardants, heavy metal chelators and organic synthesis reagents. Therefore synthesis of phosphonic acids and their derivatives was discovered as an interesting scientific field. Since Alberti *et al.* obtained the first metal phosphonates[1] and next, metal phosphonates developed, during the last few decades, into a new subclass of functional materials, the interest in phosphonic acids became even bigger. Intensive research led to an increase in the number of methods of carbon–phosphorus bond synthesis and the improvement of the known ones. This chapter provides

Metal Phosphonate Chemistry: From Synthesis to Applications
Edited by Abraham Clearfield and Konstantinos Demadis

Published by the Royal Society of Chemistry, www.rsc.org

background as well as the state of the art information about the synthesis of phosphonic acids. The scope is to describe selected methods of synthesis of phosphonic acids and their esters which have found use or seem to be interesting in synthesis and studies of coordination polymers. One can note that multifunctionality and conformational flexibility seem to be the most important features during design and synthesis of phosphonic acids as ligands.

6.2 Definition and Some Properties of Phosphonic Acids

Phosphonic acids are considered as derivatives of phosphoric acid in which one of the hydroxyl groups is replaced by an organic substituent (R). In consequence, one can give a general formula: $RP(O)(OH)_2$. The phosphorus atom present in the phosphonic group possesses sp^3 hybridization and because of that fact, the geometry of the group is nearly tetrahedral (Figure 6.1).

In terms of acid–base properties of phosphonic acids one can point out one main regularity. Acidity of the first proton expressed by pK_{a1} ranges from 0.5 to about 3, as for the second one pK_{a2} is in the range of 5 to 9. This wide gamut of the values can be explained by strong dependence on the presence of additional electron-withdrawing (EWG) and electron-donating (EDG) groups in the molecule. EDG present in the molecule have more influence on pK_{a2} rather than pK_{a1}.

If one considers phosphonic acids as ligands in coordination chemistry, three coordination states in the group can be outlined: (a) without deprotonation, (b) singly deprotonated, and (c) doubly deprotonated. Due to this fact, the phosphonic acid group has maximally four, five and six binding sites, respectively.

Now a few words giving a comparison with the most common organic functional group in coordination polymers–carboxylic acid. This group is planar, in contrast to the tetrahedral phosphonic acid group, which is due to the presence of sp^2 carbon and has maximally four binding sites. In terms of acidic properties, the pK_a of the carboxylic group is strongly dependent on the presence of EWG or EDG but usually pK_a lies between pK_{a1} and pK_{a2} of the phosphonic group.

6.3 Selected Synthetic Methods

In this section some selected methods of phosphonate and phosphonic acid synthesis are presented. We have chosen eleven reactions to discuss. They

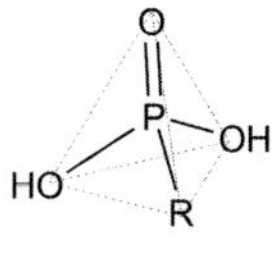

Figure 6.1 General structure of phosphonic acid.

should be useful in the synthesis of phosphonic acids that can be utilized in reticular chemistry. First three methods are highlighted and discussed in more detail and other ones are presented briefly.

6.3.1 The Michaelis–Arbuzov Reaction

The Michaelis–Arbuzov reaction is a method for carbon–phosphorus bond creation.[2] Its product is a dialkyl alkylphosphonate. The reaction was invented by Michaelis and investigated and developed by Arbuzov. It proceeds mainly between primary alkyl halides and trialkyl phosphite (Scheme 6.1) and is usually thermally initiated.

From the mechanistic point of view, the reaction can be divided into two steps – which both are nucleophilic substitutions, as shown on Scheme 6.2.

Initiating nucleophilic attack of trialkyl phosphite, in which the phosphorus atom possesses a lone electron pair, on the α-carbon of the alkyl halide results in the formation of a quasi phosphonium salt. Subsequently, nucleophilic halide anion, which is the leaving group in the first step, attacks the carbon atom of one of the alkoxy groups directly connected to the oxygen atom. Finally after electron pair reorganization, a new C–X bond is formed and

$$RCH_2X + P(OR^1)_3 \longrightarrow RCH_2P(O)(OR^1)_2 + R^1X$$

Scheme 6.1 The Michaelis–Arbuzov reaction.

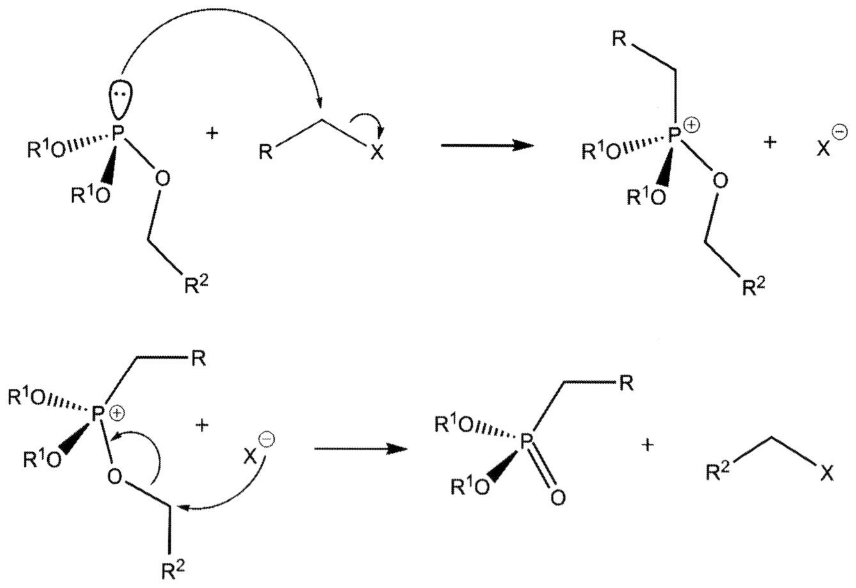

Scheme 6.2 Mechanism of the Michaelis–Arbuzov reaction.

simultaneously cleavage of C–O bond occurs giving the product and as a by-product a new alkyl halide in which the alkyl group comes from trialkyl phosphite. Focusing on substrates of the Michaelis–Arbuzov reaction, from alkyl halides, alkyl bromides and iodides are mainly used due to their higher reactivity. Rarely sulfonates ($ROS(O)_2R'$, R – aliphatic part) are also utilized. Moreover, acyl halides, suitably substituted alkynyl halides[3] and α-bromo esters[4,5] also react with trialkyl phosphite, giving corresponding phosphonates. Aryl and alkenyl halides are not the substrates due to their usually poor or lack of reactivity with trialkyl phosphite. Secondary and tertiary alkyl halides are mainly non-reactive. They usually undergo a dehalogenation reaction after treatment with trialkyl phosphite at elevated temperatures. Anyway, in some cases, it is possible to conduct the Michaelis–Arbuzov reaction using special secondary alkyl halides. This can be illustrated with synthesis of diethyl diphenylmethylphosphonate.[6]

Triaryl phosphites as reagents are also excluded but tris(trimethylsilyl) phosphite, can be used in the reaction.[7] Anyway, the choice of trialkyl phosphite is very important because of the generated by-product – alkyl halide (R^2CH_2X, Scheme 6.2). The reactivity and volatility of this compound is crucial for achieving high yields of desired product. Less reactive and more volatile halides as by-products are preferred. In consequence, triethyl and triisopropyl phosphites are more often used in comparison to trimethyl phosphite. In the last case, a reactive methyl halide is formed and can compete with the substrate (RCH_2X, Scheme 6.1) thus lowering the yield of the desired product.

Recent advances in the Michaelis–Arbuzov rearrangement comprise acceleration of the reaction by microwaves,[8,9] or trials with application of ultrasound,[10] Lewis acid catalysts[11] and trimethylsilyl halides[12] to lower reaction temperatures, or ionic liquids.[13]

Now we will illustrate applications of this method using selected examples of dialkyl phosphonate intermediates, which were used in the synthesis of coordination polymers: 1,2,4,5-tetrakis[(diethoxyphosphoryl)methyl]benzene,[14] tetraisopropyl biphenyldiyl-2,2′-bis(methylphosphonate).[15]

An example procedure for the synthesis of tetraisopropyl biphenyldiyl-2,2′-bis(methylphosphonate),[15] with chapter authors comments in italics and brackets is as follows.

2,2′-Bis(bromomethyl)biphenyl (2.9833 g, 0.008768 mol) and triisopropyl phosphite (4.5 cm^3, 0.01962 mol) *[small excess of phosphite is needed]* was gradually heated. At 126 °C bubbles were first observed *[reaction is initiated in a particular temperature and then exothermic effect occurs]* of isopropyl bromide that indicates the Arbuzov reaction is starting, and then the mixture was kept at 145–150 °C for 1.5 h *[it is best to conduct the Michaelis–Arbuzov using a simple distillation set; sometimes a stream of neutral gas is used to remove the volatile by-product]*. All volatile components were removed under reduced pressure and the residue was subject to column chromatography (flash silica gel, ethyl acetate as eluent) *[sometimes high vacuum distillation is preferred]* to give tetraisopropyl biphenyldiyl-2,2′-bis(methylphosphonate): 2.7307 g (61%), as a colorless oil which solidified under standing, $R_f = 0.20$ (silica gel on PET plates with fluorescent indicator 254 nm; eluent: ethyl acetate; visualization with UV lamp) *[finally the product should be characterized and be subject to hydrolysis; nevertheless hydrolysis of the crude product is sometimes carried out]*.

6.3.2 Catalytic Cross-coupling Reaction

Reaction of aryl or alkenyl halides with trialkyl phosphite (the Michaelis–Arbuzov reaction – see Section 6.3.1) or with dialkyl phosphite (the Michaelis–Becker reaction – see Section 6.3.4) does not occur due to the poor electrophilicity of these organic halides. Nevertheless, there are some exceptions where in the C(sp^2)–X bond the halogen is substituted with phosphorus-centered nucleophiles and the reaction takes place without any catalyst. Examples are activated aromatic and heterocyclic aromatic halides like 2,4-dinitrobromobenzene,[16] 9-chloroacridine,[17] derivatives of nitrothiophene[18] or even non-halogenated *N*-methoxypyridinium salts[19] and polyfluorinated alkenyl halides[20] or ethyl β-substituted vinyl ethers.[21] However, in order to carry out the phosphonylation of common aryl and alkenyl halides a catalyst is needed.

A catalytic cross-coupling reaction is a method of synthesis of dialkyl aryl- and alkenylphosphonates from aryl and alkenyl halides and dialkyl phosphites (recommended name is dialkyl *H*-phosphonates) or trialkyl phosphites in the presence of a transition metal catalyst (Scheme 6.4). However, in a broader sense, apart from dialkyl and trialkyl phosphites, any phosphorus-centered nucleophile, e.g. $RP(OR^1)_2$, also undergoes cross-coupling, like in the non-catalyzed Michaelis–Arbuzov reaction. The cross-coupling reaction, involving monoalkyl esters of phosphonous acids [$R(R^1O)P(O)H$] and phosphinous acids [$RR^1P(O)H$] are, however, out of the scope of this section.

At first, it was discovered that copper bronze, nickel(II) bromide and nickel(II) chloride have the ability to catalyze reaction of aryl halides with triethyl phosphite yielding diethyl arylphosphonates (Scheme 6.3).[22,23]

However, the reaction requires high temperature (e.g. 160 °C) and slow dropwise addition of triethyl phosphite. This method, involving nickel catalyst, is still in use due to its efficiency.[24] In the course of use, this synthetic approach underwent some modification including use of palladium(II) acetate or chloride[25] and solvent.[26–28]

In the search for milder conditions of the coupling reaction of substrates with dialkyl phosphites, the presence of base and new catalyst was introduced.[29,30] Use of the palladium catalyst, namely tetrakis(triphenylphosphine)palladium(0), allowed lowering of reaction temperature to 90 °C (Scheme 6.4). For substrates not liquid at this temperature, addition of a small volume of solvent is necessary. Further modification based on microwave-assisted reaction provides very rapid access to dialkyl vinyl- and arylphosphonates although at 120 °C.[31]

$$ArX + H(O)P(OR)_2 \xrightarrow[\text{base}]{\text{catalyst}} ArP(O)(OR)_2 + HX$$

Scheme 6.3 Nickel-catalyzed cross-coupling reaction.

$$ArX + P(OR)_3 \xrightarrow{\text{catalyst}} ArP(O)(OR)_2 + RX$$

Scheme 6.4 Palladium-catalyzed cross-coupling reaction.

This cross-coupling reaction was also intensively studied in terms of synthesis of various dialkyl arylphosphonates and vinylphosphonates.[32–34] Some dialkyl methyl- and benzylphosphonates can be obtained under very mild reaction conditions using palladium-catalyzed cross-coupling.[35,36] Recently an improvement of reaction conditions was introduced. A change of catalyst to a mixture of palladium(II) acetate and 1,1′-bis(diphenylphosphino)ferrocene (dppf) allowed a decrease in the amount of added palladium.[37]

Additionally to palladium and nickel, copper-catalyzed cross-coupling is also known in the literature, although hardly used. Mainly copper(I) halide is used in conjunction with both dialkyl[38,39] and trialkyl phosphite[40] in both catalytic and stoichiometric amount of copper(I) halide. The reaction was also used in preparation of some hindered arylphosphonates.[41] Copper(I) catalysis works also in reaction of terminal alkynes with dialkyl phosphite giving alkynylphosphonates.[42]

Mechanism of nickel-catalyzed cross-coupling reaction of aryl halides with triethyl phosphite is difficult to investigate. However, a hypothetical mechanism was proposed and involved a few steps as follows (Scheme 6.5).[43–46] At first, reduction of nickel(II) halide with triethyl phosphite results in formation of tetrakis(triethyl phosphite)nickel(0). Aryl halide undergoes oxidative addition to the

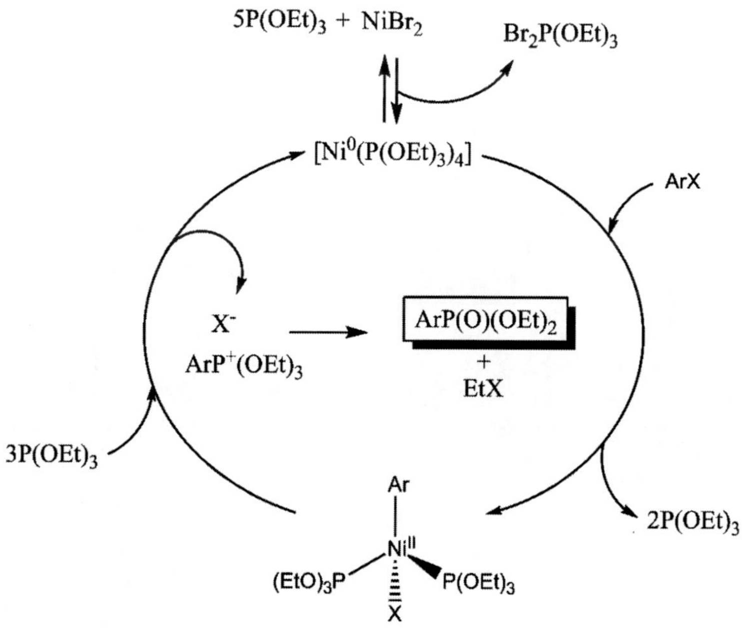

Scheme 6.5 Mechanism of nickel-catalyzed cross-coupling.

nickel(0) complex. Subsequently, in a slow, rate-determining step, the nickel(0) complex is regenerated and arylphosphonium salt is formed. In the final step, which is the same as in the Michaelis–Arbuzov reaction, nucleophilic substitution of the halide anion occurs giving diethyl arylphosphonate and ethyl halide.

Mechanism of palladium-catalyzed reaction is a quite complicated issue and the reaction rate in details depends on several factors, e.g. palladium source, its ligands and anionic additives, electronic and steric factors of organic halides. All these were intensively studied, in general, for cross-coupling reactions[47] and particularly for carbon–phosphorus bond formation.[48–51] Nevertheless, there are few invariable stages which, in the first place, include formation of an active Pd^0 complex from the palladium source (if source contains Pd^{II} then reduction is also needed). After that, oxidative addition of aryl (alkenyl) substrates to palladium complex occurs. In this step Pd loses two electrons and in consequence becomes Pd^{II}. In the next stage ligand exchange is desired so dialkyl phosphite is coordinated by the metal center. After reaction with base, the deprotonated phosphorus atom has a lone electron pair which creates a new Pd–P bond. If this ligand–phosphite–palladium(II)–aryl (or alkenyl) complex is present, the next stage is reductive elimination in which Pd^0 complex is regenerated and a carbon–phosphorus bond is created to form the final phosphonate product. The following mechanism is shown in Scheme 6.6 for a particular palladium(0) catalyst, aryl halide, dialkyl phosphite and base.

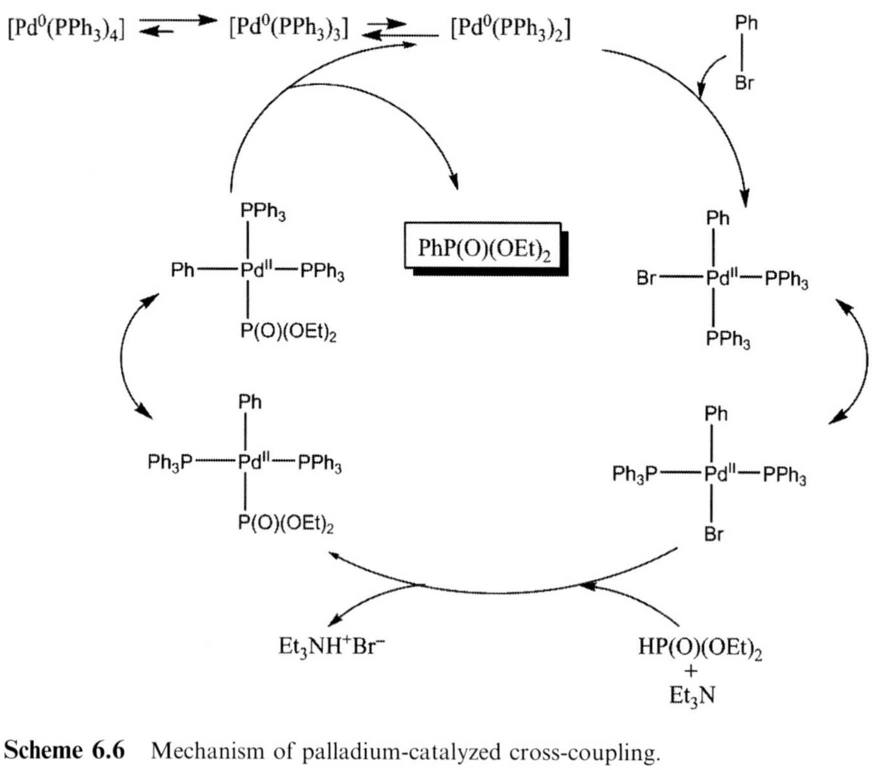

Scheme 6.6 Mechanism of palladium-catalyzed cross-coupling.

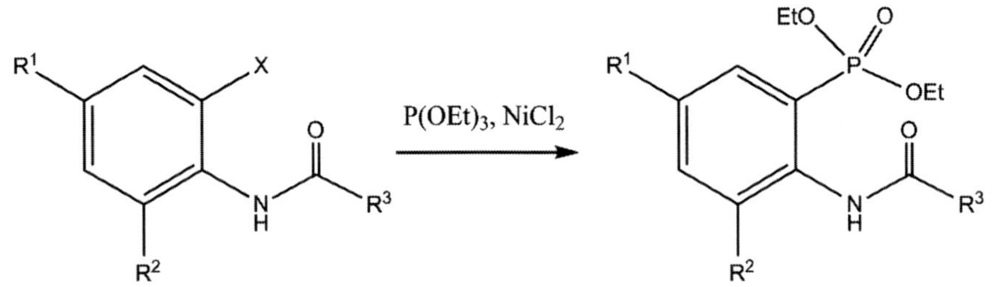

Scheme 6.7 Nickel-catalyzed cross-coupling of hindered *N*-acylated *o*-halogen-anilines.

When a nickel catalyst is used, different trialkyl phosphites can be utilized. Triethyl phosphite is the most common; however, triisopropyl phosphite is sometimes used. It is interesting that using a mixture of triaryl phosphite and triethyl phosphite one can achieve synthesis of diaryl arylphosphonates.[52] In terms of organic substrates, compounds with halogen attached to the sp^2 carbon are used, namely aryl and alkenyl bromides or iodides. Among examples of substrates, sterically hindered *N*-acylated *o*-bromoanilines[53] are noteworthy to mention (Scheme 6.7). Additional research on benzene-substituted *N*-acylated *o*-bromoanilines suggests participation of acylamino group in the reaction mechanism.[24]

Palladium-catalyzed reactions require similar organic halides as substrates. The overall yield of reaction of aryl halides roughly decreases in the following order: iodides, bromides, chlorides. However, recent studies of the reaction mechanism revealed a different role of halogen anion in subsequent steps of the reaction. In addition to the halides, aryl and vinyl sulfonates ($ROS(O)_2R'$, R = aryl and vinyl) are used.[54–56] Diethyl and diisopropyl phosphites are usually utilized as phosphonylating agents. In this case use of base is necessary to arrest the hydrogen halide that is formed and those bases are most often triethylamine or *N*,*N*-diisopropylethylamine. Nonetheless in following review,[32] besides knowledge about cross-coupling P–C bond creation, there are also listed and briefly discussed other hydrogen halides scavengers for palladium-catalyzed reactions, such as propylene oxide or potassium and cesium carbonates in phase-transfer conditions.

In order to accelerate nickel-[9] and palladium-catalyzed[31,57] reactions, microwaves have been successfully utilized.

Many different multifunctional arylphosphonic acids applied as linkers in coordination polymers synthesis were obtained following this method via intermediate esters: some of the largest phosphonates obtained so far are tetrakis-1,3,5,7-(4′-diethoxyphosphorylphenyl)adamantane,[58] tetraethyl 2,2′-bipyridine-based bisphosphonates,[59] amino-substituted diethyl pyridylphosphonates,[60] diethyl and diisopropyl 5-substituted 3-pyridylphosphonates.[61]

An example procedure for synthesis of hexaisopropyl benzenetriyl-1,3,5-triphosphonate[27] using a nickel catalyst with chapter authors comments in italics and brackets follows.

To a suspension of tribromobenzene (50.05 g, 0.159 mol) and anhydrous $NiBr_2$ (8.19 g, 0.037 mol) in mesitylene (10 ml) *[reaction is often conducted under atmosphere of inert gas]* was added dropwise *[reaction mixture at first turns violet and finally yellow]* at 150–155 °C $P(Oi\text{-}Pr)_3$ (133.57 g, 0.714 mol) *[some excess of trialkyl phosphite is desired]*. The i-PrBr formed was collected in a cooling trap. After the addition had been completed the reaction mixture was heated at 155 °C for 1 h. Mesitylene and the excess of $P(Oi\text{-}Pr)_3$ were removed *in vacuo* (15 mbar) at 70 °C. Column chromatography of the residue (silica gel, $CHCl_3$/MeOH 10:1) gave 1,3,5-$[P(O)(i\text{-}PrO)_2]_3C_6H_3$ (74.40 g, 82%) as yellowish solid with a m.p. of 51 °C.

An example procedure for synthesis of diethyl 4-pyridylphosphonate[62] using palladium catalyst with chapter authors comments in italics and brackets follows.

4-Bromopyridine, following toluene extract [obtained after neutralizing 4-bromopyridine hydrochloride (3.11 g, 16 mmol) with sodium carbonate (1.9 g, 18.0 mmol) in water (7 ml) and extracted using toluene (10 and 5 ml) and finally the combined extracts dried with anhydrous potassium carbonate], diethyl phosphite (2.4 ml, 18.0 mmol), and triethylamine (2.5 ml, 18.0 mmol) were coupled using $Pd[(C_6H_5)_3P]_4$ (0.93 g, 0.81 mmol) *[also various combinations of palladium salt and phosphine, including a bisphosphine, can be used as well]*, at 80 °C under argon for 40 h according to the general procedure of Hirao *et al.*[29] A precipitate of triethylamine hydrobromide was formed during the course of the reaction. At the end of the reaction, a mixture of toluene (15 ml) and water (10 ml) was added. The product was separated in toluene from a water–toluene mixture using a separating funnel and dried with anhydrous potassium carbonate. Then toluene was removed and the residue was subjected to column chromatography (silica gel for flash chromatography, ethyl acetate) and afforded an oil (1.83 g, 53%): $R_f = 0.065$ (silica gel on TLC-PET foils with fluorescent indicator 254 nm, ethyl acetate, UV lamp).

6.3.3 The Mannich-type Condensation

The Mannich-type condensation, also called the Moedritzer–Irani reaction,[63] is a common route for synthesizing *N*,*N*-disubstituted aminomethylphosphonic acids or *N*-substituted iminobis(methylphosphonic acids). This synthetic method seems to be the simplest, using easily accessible substrates, and probably the most economic way to obtain the derivatives of aminomethylphosphonic acid usually possessing complexing properties. The Mannich-type condensation is a special case of Kabachnik–Fields reaction.[64,65] In the latter one, substrates are ammonia, primary or secondary amines, carbonyl compounds (aldehydes or seldom ketones) and dialkyl or diaryl phosphites. In the Moedritzer–Irani reaction instead of the phosphites, one uses phosphorous acid (also called phosphonic acid), instead of carbonyl compounds only formaldehyde is used (Scheme 6.8). The reaction is conducted in c.a. 6 M hydrochloric acid.

$$R_{3-n}NH_n + nHCHO + nHP(O)(OH)_2 \xrightarrow[\text{where n = 1, 2, 3}]{HCl} R_{3-n}N[CH_2P(O)(OH)_2]_n + nH_2O$$

Scheme 6.8 The Moedritzer–Irani reaction.

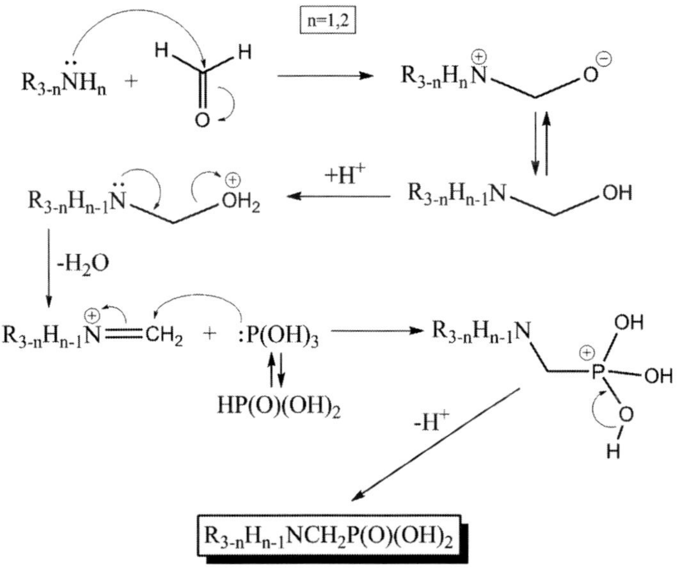

Scheme 6.9 Mechanism of the Moedritzer–Irani reaction.

Because, to the best of our knowledge, there is a lack of mechanistic information in the literature on the reaction, we suggest the following mechanism (Scheme 6.9). The first step of the reaction is a nucleophilic attack of *N*,*N*-dialkylamine nitrogen. Further rearrangement gives *N*-hydoxymethylamine which in strong acidic condition undergoes elimination of water molecule yielding an imine salt.[66] Phosphorous acid in acidic conditions behaves as a nucleophile and therefore attacks the electrophilic imine salt. The charged adduct is stabilized by loss of a proton to give *N*,*N*-disubstituted aminomethylphosphonic acid. *N*-substituted iminobis(methylphosphonic acid) is formed when one starts from primary alkylamine and the multistep reaction is repeated.

Despite it being an inexpensive synthesis, there are also drawbacks. For instance some undesired products are formed. In case of the primary alkylamine as the substrate, a mixture of *N*-mono- and *N*,*N*-bis(methylphosphonic acids) can be produced, as the second methylphosphonylation of amine runs even faster than the first one. Furthermore, hydroxymethylphosphonic acid is a by-product of the reaction. Also *N*-methyl derivatives of aminomethylphosphonic acids can be formed according to the Leuckart–Wallart reaction.[67]

Additionally, part of phosphorous acid can be oxidized to phosphoric acid and this disturbs the stoichiometry. Also phosphorous acid as a component of the crude product can be detected. During the course of the reaction, methanol and methyl formate are observed as by-products[68] but removal of these compounds from the reaction mixture is not a problem due to their volatility. In consequence, control over the course of the reaction as well as purity of the product is desired. It seems that combined ^{1}H, ^{31}P and ^{13}C NMR spectroscopy is a method of choice for achieving this goal.[69] Sometimes an ion-exchange column chromatography enables purification of the crude product but usually crystallization or even recrystallization may provide a reasonably pure product.

Some examples of ligands obtained by this method and used for metal phosphonate formation are the following: *N*,*N*-bis(phosphonomethyl)aminoacetic acid,[70] *N*-(phosphonomethyl)aza-18-crown-6 ether,[71] (*S*)-(–)-*N*-(phosphonomethyl)proline,[72] *N*-(phosphonomethyl)glycine via its *N*-benzyl derivatives,[73] *N*,*N*-bis(phosphonomethyl)-2-amino-2-methylpropionic acid.[74] An example procedure for synthesis of *N*,*N'*-piperazinebis(methylphosphonic acid) hydrochloride[75] with chapter authors comments in italics and brackets follows.

The compound was synthesized using a modified literature procedure.[63] Piperazine (0.1 mol) was mixed with phosphorous acid (0.2 mol), and 50 ml of HCl solution *[molar excess of acid is important]* was mixed with 50 ml of water. Formaldehyde (0.4 mol) was added dropwise over 1 h, and the mixture was refluxed at 120 °C. After 4 h, a white solid had precipitated, which was filtered, washed, and dried. Overall yield 68%.

6.3.4 Other Reactions

Synthetic methods leading to phosphonic acids and their esters presented in previous sections are the most frequently used in the metal phosphonates field; however, they are not the only ones. This section consists of a brief description of other applied reactions.

The Michaelis–Becker reaction proceeds between an anion derived from dialkyl phosphite and mainly alkyl halides leading to dialkyl alkylphosphonates (Scheme 6.10).[76–78]

The reaction involves nucleophilic attack of dialkoxyphosphoryl anion on the electrophilic halide carbon with creation of carbon–phosphorus bond and release of halide anion. The anionic nucleophile is more reactive than a neutral one so the reaction condition is much milder than in the Michaelis–Arbuzov reaction and in consequence fewer side-reactions occur. In some cases the Michaelis–Becker route allows the synthesis of phosphonates while the Michaelis–Arbuzov fails.[79,80] A few different procedures of the reaction exist. The best-known phosphonylating reagent is the *in situ* generated sodium salt of

$$RCH_2X + (R^1O)_2(O)PH \xrightarrow{\text{base}} RCH_2P(O)(OR^1)_2$$

Scheme 6.10 The Michaelis–Becker reaction.

diethyl phosphite. Sometimes, dibutyl phosphite sodium salt, due to better solubility, other anion generators like tertiary amines or potassium and cesium carbonate, with or without phase-transfer catalyst,[81,82] are used. Besides simple alkyl halides, tosylates or mesylates[83] are utilized. Additionally, alkynyl halides undergo the reaction.[77] Microwave-assisted reaction was utilized to improve the reaction rate.[84] Examples of intermediates in ligand synthesis obtained by this method are: tetraethyl methylenediphosphonate[85] and higher homologs of tetraethyl 1,2-ethylenediphosphonate.[86]

Dialkyl phosphonate synthesis can be also achieved by cross-coupling of organometallics with a phosphonylation reagent (Scheme 6.11).[87]

The organometallic route can serve as a supplementary method for the catalytic cross-coupling, and products which are hardly accessible via the latter route can be obtained, e.g. *ortho*-substituted dialkyl arylphosphonates. However, low temperatures, often –78 °C, and the need to use anhydrous solvents are main disadvantages of this method. As a source of dialkylphosphoryl group, dialkyl chlorophosphate, dialkyl chlorophosphite and other chloro derivatives are used. Organometallic substrates are mainly aromatic, sometimes alkyl or alkynyl, Grigniard or lithium reagents. "Reverse addition" procedure, i.e. addition of cooled organometallic substrate to cooled solution of phosphonylating agent, is recommended to avoid side reactions concerning formation of more than one P–C bond. There is no up to date review on synthesis of arylphosphonates through organometallic cross-coupling. For compounds related to them, heterocyclic arylphosphonates, their synthesis via many methods has been recently reviewed.[88]

Examples of ligands whose intermediate esters are obtained by this method are 1,4-phenylenebis(methylidene)tetrakis(phosphonic acid),[89] 2,4,6-(triisopropyl)phenylphosphonic acid,[90] 1-ferrocenylphosphonic acid and ferrocenediyl-1,1′-bisphosphonic acid.[91]

The anionic phospho-Fries rearrangement, called also the phosphate–phosphonate rearrangement, is a convenient method of synthesis of *o*-hydroxyarylphosphonates (Scheme 6.12).[92,93]

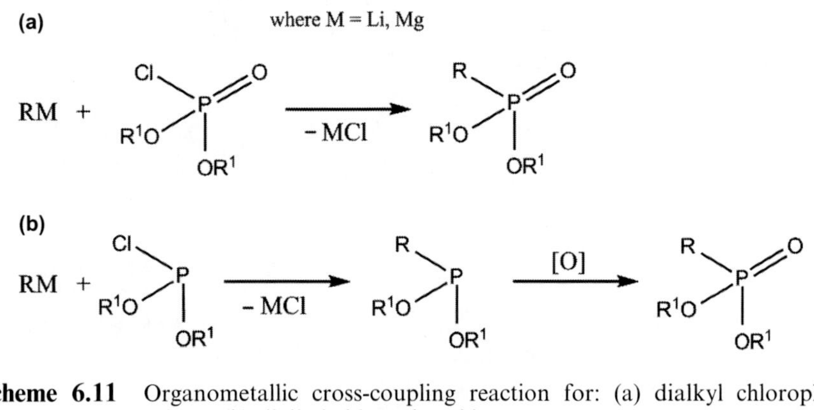

Scheme 6.11 Organometallic cross-coupling reaction for: (a) dialkyl chlorophosphate, (b) dialkyl chlorophosphite.

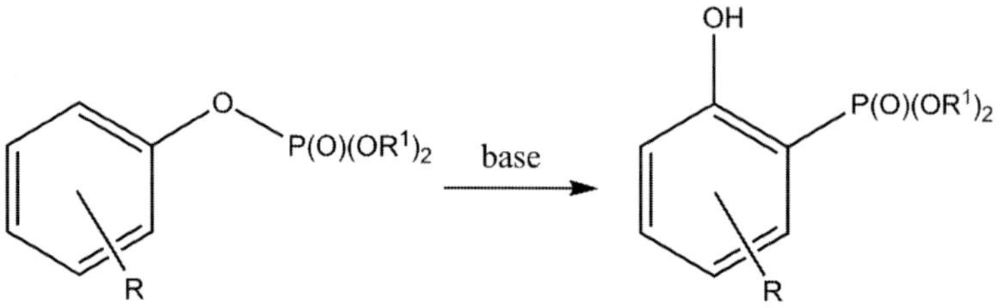

Scheme 6.12 The anionic phospho-Fries rearrangement.

$$\text{(a)}\ RCl + PCl_3 + AlCl_3 \longrightarrow [RPCl_3]^+[AlCl_4]^- \xrightarrow{H_2O} RP(O)Cl_2 \xrightarrow{H_2O} RP(O)(OH)_2$$

$$\text{(b)}\ ArH + PCl_3 + AlCl_3 \xrightarrow[\text{2. [O]}]{\text{1. NaOR, EtOH}} ArP(O)(OEt)_2 \xrightarrow{HCl} ArP(O)(OH)_2$$

Scheme 6.13 Lewis acid assisted phosphonylation of: (a) alkyl halides, (b) aromatic hydrocarbons.

Substrates for this reaction are mainly aryl phosphates, obtained for example from corresponding phenols. An asymmetrical arylphosphate possessing two free *ortho* positions to the phosphate group gives two products. If the *ortho* position is occupied by a halogen, then this position is privileged for intermediate carboanion formation and in consequence one product dominates.[93] Only one product is obtained in the case of an asymmetrical substrate containing a non-reactive substituent in one *ortho* position. Vinyl phosphates undergo the rearrangement although β-oxoalkylphosphonates are obtained.[94] Alkyl phosphates react similarly to aryl derivatives giving α-hydroxyphosphonates.[95] The reaction requires use of very strong bases, typically lithium diisopropylamide (LDA) generated *in situ* from butyllithium and diisopropylamine. In consequence, anhydrous solvents and low temperatures are needed. Methods for synthesis of dialkyl arylphosphates, anion generation, mechanistic aspects and regioselectivity of the rearrangement are comprehensively reviewed.[96] An example of a ligand used for metal phosphonate synthesis is 2,5-dihydroxy-1,4-phenylenebis(phosphonic acid).[97]

Lewis acid-assisted dichlorophosphinoylation and further partial hydrolysis of intermediated complex is an old method for synthesis of alkyl-[98,99] and arylphosphonic acid dichlorides.[100,101] The acid dichlorides can be finally hydrolyzed to the corresponding phosphonic acids (Scheme 6.13).

In case of aliphatic substrates, namely alkyl halides, the reaction proceeds via formation of an ionic complex of usually aluminum chloride, an alkyl halide, most frequently the chloride and phosphorous trichloride.[98] Modification of this reaction involving saturated hydrocarbons in dichloromethane solution has been described.[102] It is important that sterically hindered alkylphosphonic acid dichlorides can be obtained through this method. It is worthy to mention that sometimes mixture of products can be obtained due to isomerization of

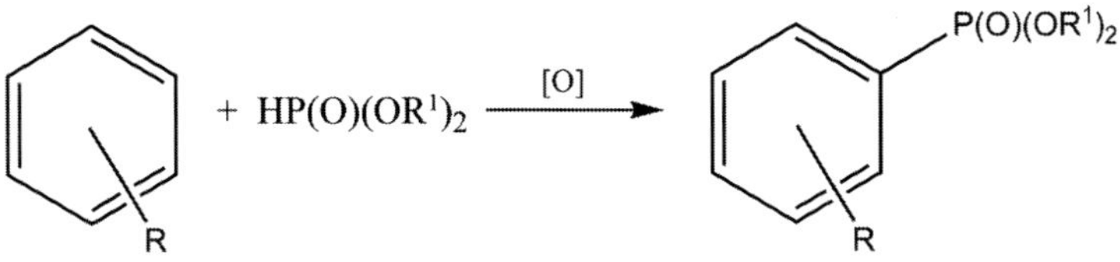

Scheme 6.14 Direct oxidative phosphonylation of arenes.

carbocation intermediate. Direct dichlorophosphinoylation of aromatic hydrocarbons via electrophilic aromatic substitution (the Friedel–Crafts reaction) can also be applied. One can use different Lewis acids catalysts[103,104] and the resulting product may be readily obtained in bigger amounts but regioselectivity is lower than in cross-coupling reactions, for example, and is controlled only by substituent influence.

Now we give some examples of ligands used in metal-phosphonate synthesis that are obtained by this method: *i*-propylphosphonic acid, *t*-butylphosphonic acid[105] and cyclopentylphosphonic acid.[106]

Direct phosphonylation of aromatic hydrocarbons takes place between arenes or some heterocyclic arenes and dialkyl phosphite, in the presence of oxidants, yielding dialkyl arylphosphonates (Scheme 6.14).[107,108]

Some alkenes and alkynes can also undergo the reaction, but in some different conditions addition[109] or substitution[110] products can be obtained.

Various oxidants were used: *t*-butyl peroxide,[111] graphite anode or mixture of sodium peroxydisulfate/silver(I) nitrate(V),[112] ammonium cerium(IV) nitrate(V),[113] manganese(II) acetate/cobalt(II) acetate/O_2-N_2,[107] manganese(III) acetate.[108] It seems to be that dialkylkoxyphosphoryl radical, [·P(O)(OEt)$_2$], is an electrophilic intermediate.[108,110] Recently, using manganese(III) acetate, in acetic acid at 60 °C, arenes bearing electron-donating as well as electron-withdrawing groups were phosphonylated with moderate regioselectivity and high or good yields.[114]

Photo-induced phosphonylation of aryl halides takes place with trialkyl phosphite[115] and dialkyl phosphite alkaline salt in liquid ammonia[116] and other solvents[117] to provide dialkyl arylphosphonates. When aryl iodides were used high yields and high regioselectivity were observed. In case of aryl bromides high yields can be obtained when sodium iodide is used in reaction conditions.[118]

α-Functionalized geminal diphosphonic acids have been obtained on addition of phosphorus(III) species to unsaturated systems.[119] Reaction takes place between nitriles[120,121] or acyl chlorides[122] and phosphorous acid or its tris(-trimethylsilyl) ester, for example (Scheme 6.15).

Other methods for *geminal* diphosphonic acid synthesis are also used.[123–126]

Examples of synthesis of ligands are: 1-aminobenzylidene-1,1-diphosphonic acid[127] and 4-amino-1-hydroxybutylidene-1,1-diphosphonic acid.[128]

Hydrophosphonylation of 1,3,5-trisubstituted hexahydro-1,3,5-triazines with dialkyl phosphites yields dialkyl *N*-substituted aminomethylphosphonates (Scheme 6.16).[129–131]

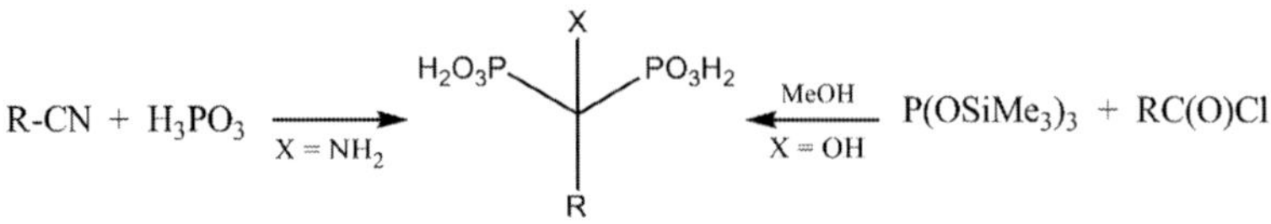

Scheme 6.15 Synthesis of α-functionalized geminal diphosphonic acids.

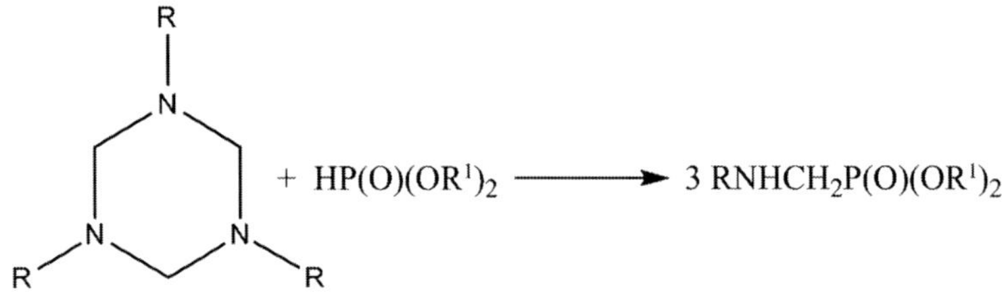

Scheme 6.16 Hydrophosphonylation of 1,3,5-trisubstitutedhexahydro-1,3,5-triazines.

This reaction also takes place with other P–H bond-containing phosphorus compounds but these products are rather out of the scope of this review. *N*-Monosubstituted aminomethylphosphonic acids can be obtained in contrast to the Mannich-type condensation where *N*,*N*-disubstituted aminomethylphosphonic acids are formed. Ring substituted derivatives of *N*-(methoxymethyl)aniline, give also access to *N*-monomethylphosphonate of substituted anilines.[132] When 1,3,5-triacylhexahydro-1,3,5-triazine is used as the substrate and reaction takes place with phosphorus trichloride in glacial acetic acid then aminomethylphosphonic acid is formed.[133]

As one could notice, the set of methods for phosphonate synthesis is very rich and often there is more than one way to obtain the desired product. The main objective for this short review is to gather and discuss selected synthetic methods, that are as general as possible, or at least clever and useful in some special way but, of course, the chemistry of phosphonic acids is incomparably richer, especially for functionalized derivatives of them. For more comprehensive discussion there are a few extensive sources.[77,134,135]

6.4 Transformation of Dialkyl Phosphonates to Phosphonic Acids

Although sometimes dialkyl phosphonates are used for metal phosphonate synthesis,[136–139] usually phosphonic acids are applied for this task. Alkaline hydrolysis usually provides the mono ester,[140] therefore conventional methods for transformation of esters to acids is hydrolysis by refluxing with mixture of concentrated hydrochloric acid and water (Scheme 6.17).

$$RP(O)(OR^1)_2 \xrightarrow[\substack{-R^1Cl \\ \text{or} \\ -R^1OH}]{HCl} RP(O)(OH)_2$$

Scheme 6.17 Acidic hydrolysis of dialkyl phosphonates.

$$RP(O)(OR^1)_2 + 2Me_3SiX \xrightarrow[-2R^1X]{} RP(O)(OSiMe_3)_2 \xrightarrow[-O(SiMe_3)_2]{H_2O} RP(O)(OH)_2$$

Scheme 6.18 Tandem of transsilylation and hydrolysis of dialkyl phosphonates.

The hydrolysis is carried out for a few hours. Isolation of product from reaction mixture is done by filtration of precipitate or evaporation of hydrolyzed mixture. In order to fully remove hydrochloric acid from the residue water or methanol is added and then re-evaporated.

In the case of acids bearing sensitive functional groups a milder procedure at room temperature is used. It considers transformation of alkyl ester into a silyl one by means of trimethylsilyl halides and subsequent water hydrolysis or methanolysis (Scheme 6.18).[141–143] Mainly, bromotrimethylsilane is used as the reagent.

Solid acids are subject to crystallization. Oily acids are precipitated as salts, most often from methanol, and both forms of obtained product are subsequently spectroscopically characterized.

6.5 Spectral Characterization of the Acids and Esters

Newly obtained phosphonic acids and esters are characterized by means of ^{1}H, ^{31}P, ^{13}C nuclear magnetic resonance (NMR) spectroscopy in solution, infrared (IR) spectroscopy, elemental analysis or high resolution mass spectrometry (MS). If it is possible, a crystal structure is desirable. Even for known compounds, for structure and purity check of the synthesized phosphonic acids, verification by spectroscopic methods and comparison of obtained data with the literature data is recommended.

When NMR analysis is considered, a few general remarks can be stated. Chemical shifts (δ in ppm) as well as coupling constants (J in Hz) at given working frequencies of spectrometer, solvent used and chemical shift standards are helpful in characterization of phosphonic acids and their esters.

^{31}P NMR spectra of different types of compounds cover a very broad range of chemical shifts, from –500 to 1400 ppm, and the range of an individual class of compounds tend to be wide and dependent on conditions of the measurement. Additionally, ranges of individual classes may overlap. For the majority of phosphonic acids, chemical shifts lie between 5 and 35 ppm and slightly higher for esters.

Coupling constants that are important in phosphonic acids and their esters are usually as follows: $^{2}J_{P\text{–}C\text{–}H}$ between 10 and 25 Hz, $^{3}J_{P\text{–}C\text{–}C\text{–}H}$ between a few and 30 Hz (dependant on dihedral angle), $^{3}J_{P\text{–}O\text{–}C\text{–}H}$ between 7 and 15 Hz for proton–phosphorus; $^{1}J_{P\text{–}C}$ more than 100 Hz and often about 160–180 Hz, while $^{2}J_{P\text{–}C\text{–}C}$, $^{3}J_{P\text{–}C\text{–}C\text{–}C}$ and $^{4}J_{P\text{–}C\text{–}C\text{–}C\text{–}C}$ from a few to several Hz for carbon–phosphorus.

6.6 Concluding Remarks

As part of our conclusions we present a statistical summary of Current Contents Connect® database searching for the phrase "metal phosphonate*" during the 5 years 2005–2009. We have searched for ligands used in metal phosphonate chemistry and methods of their synthesis.

As one can see on the graphs (Figure 6.2), each year at least 60% of total reactions were carried out according to the first three methods presented in this chapter (Section 6.3.1–6.3.3). Commercial sources, about 35% of all ligands sources, were not included. In most phosphonic acids the P–C(sp^3) linkage is present. Only 16% of ligands were obtained through the cross-coupling method. From an organic chemistry point of view, this fact indicates that non-rigid phosphonic acids are most frequently used in metal phosphonate synthesis, but one should keep in mind that at least one of the described methods actually allows synthesis of ligands possessing an additional functional group protected or free with various conformational flexibilities. Anyway, in the course of the years 2005–2009 there is some progress in using catalyzed cross-coupling in comparison to the Mannich-type condensation. The statistical summary of 2009 indicates that none of the described methods is favored over the others. One can state that phosphonic acid chemistry is getting richer and developing fast to give a chance to obtain new members of this class of compounds, usually more structurally and functionally complex than before, that might be used for metal phosphonate chemistry as well as for biological evaluation or other goals.

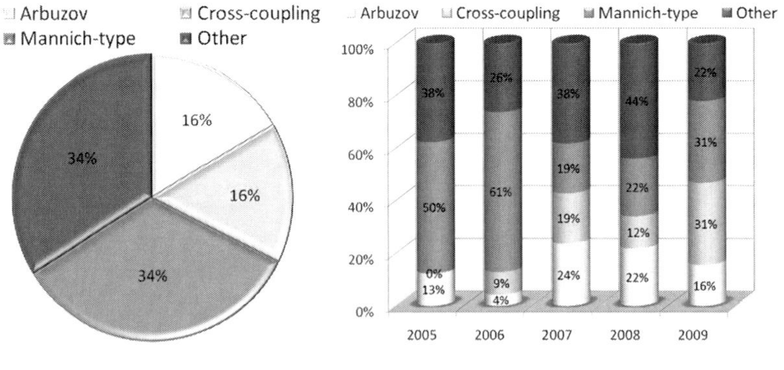

Figure 6.2 Statistical representation of methods of phosphonic acids synthesis: (a) summary for years 2005–2009, (b) annual summaries.

References

1. G. Alberti, U. Costantino, S. Allulli and N. Tomassini, *J. Inorg. Nucl. Chem.*, 1978, **40**, 1113.
2. A. K. Bhattacharya and G. Thyagarajan, *Chem. Rev.*, 1981, **81**, 415.
3. A. A. Petrov, A. V. Dogadina, B. I. Lonin, V. A. Garibina and A. A. Leonov, *Russ. Chem. Rev.*, 1983, **52**, 1030.
4. N. Inguimbert, P. Coric, H. Poras, H. Meudal, F. Teffot, M. C. Fournie-Zaluski and B. P. Roques, *J. Med. Chem.*, 2002, **45**, 1477.
5. A. Ianni and S. R. Waldvogel, *Synthesis*, 2006, 2103.
6. S. Konar and A. Clearfield, *Inorg. Chem.*, 2008, **47**, 5573.
7. L. Woźniak and J. Chojnowski, *Tetrahedron*, 1989, **45**, 2465.
8. D. Villemin, F. Simeon, H. Decreus and P. A. Jaffres, *Phosphorus, Sulfur Silicon Relat. Elem.*, 1998, **133**, 209.
9. J. J. Kiddle and A. F. Gurley, *Phosphorus, Sulfur Silicon Relat. Elem.*, 2000, **160**, 195.
10. J. Y. Winum, M. Kamal, H. Agnaniet, A. Leydet and J. L. Montero, *Phosphorus, Sulfur Silicon Relat. Elem.*, 1997, **129**, 83.
11. P. Y. Renard, P. Vayron, E. Leclerc, A. Valleix and C. Mioskowski, *Angew. Chem., Int. Ed. Engl.*, 2003, **42**, 2389.
12. P. Y. Renard, P. Vayron and C. Mioskowski, *Org. Lett.*, 2003, **5**, 1661.
13. E. V. Matveeva, I. L. Odinets, V. A. Kozlov, A. S. Shaplov and T. A. Mastryukova, *Tetrahedron Lett.*, 2006, **47**, 7645.
14. N. Stock, N. Guillou, J. Senker, G. Ferey and T. Bein, *Z. Anorg. Allg. Chem.*, 2005, **631**, 575.
15. J. Zoń, D. Kong, K. Gagnon, H. Perry, L. Holliness and A. Clearfield, *Dalton Trans.*, 2010, **39**, 11008.
16. G. S. Prasad, M. Manjunath, K. R. K. K. Reddy, O. V. S. Reddy and C. S. Reddy, *ARKIVOC*, 2006, (xvi), 128.
17. G. M. Kosolapoff, *J. Am. Chem. Soc.*, 1947, **69**, 1002.
18. T. Erker and N. Handler, *Synthesis*, 2004, 668.
19. D. Redmore, *J. Org. Chem.*, 1970, **35**, 4114.
20. G. G. Furin, *Russ. Chem. Rev.*, 1993, **62**, 243.
21. N. Kreutzkamp and H. Schindler, *Chem. Ber.*, 1959, **92**, 1695.
22. P. Tavs, *Chem. Ber.*, 1970, **103**, 2428.
23. P. Tavs and F. Korte, *Tetrahedron*, 1967, **23**, 4677.
24. J. Heinicke, N. Gupta, A. Surana, N. Peulecke, B. Witt, K. Steinhauser, R. K. Bansal and P. G. Jones, *Tetrahedron*, 2001, **57**, 9963.
25. M. S. S. Adam, M. K. Kindermann, M. Köckerling and J. W. Heinicke, *Eur. J. Org. Chem.*, 2009, 4655.
26. Z. Wang, J. M. Heising and A. Clearfield, *J. Am. Chem. Soc.*, 2003, **125**, 10375.
27. M. Henn, K. Jurkschat, D. Mansfeld, M. Mehring and M. Schürmann, *J. Mol. Struct.*, 2004, **697**, 213.
28. J. M. Rueff, O. Perez, C. Simon, C. Lorilleux, H. Couthon-Gourvès and P. A. Jaffrès, *Cryst. Growth Des.*, 2009, **9**, 4262.

29. T. Hirao, T. Masunaga, Y. Ohshiro and T. Agawa, *Synthesis*, 1981, 56.
30. T. Hirao, T. Masunaga, N. Yamada, Y. Ohshiro and T. Agawa, *Bull. Chem. Soc. Jpn.*, 1982, **55**, 909.
31. M. Kalek, A. Ziadi and J. Stawinski, *Org. Lett.*, 2008, **10**, 4637.
32. I. P. Beletskaya and M. A. Kazankova, *Russ. J. Org. Chem.*, 2002, **38**, 1391.
33. D. Prim, J. M. Campagne, D. Joseph and B. Andrioletti, *Tetrahedron*, 2002, **58**, 2041.
34. A. L. Schwan, *Chem. Soc. Rev.*, 2004, **33**, 218.
35. G. Lavén and J. Stawinski, *Synlett*, 2009, 225.
36. G. Lavén, M. Kalek, M. Jezowska and J. Stawinski, *New J. Chem.*, 2010, **34**, 967.
37. Y. Belabassi, S. Alzghari and J.-L. Montchamp, *J. Organomet. Chem.*, 2008, **693**, 3171.
38. H. Rao, Y. Jin, H. Fu, Y. Jiang and Y. Zhao, *Chem. Eur. J.*, 2006, **12**, 3636.
39. D. Gelman, L. Jiang and S. L. Buchwald, *Org. Lett.*, 2003, **5**, 2315.
40. G. Axelhad, S. Laosooksathit and R. Engel, *J. Org. Chem.*, 1981, **46**, 5200.
41. A. Osuka, N. Ohmasa, Y. Yoshida and H. Suzuki, *Synthesis*, 1983, 69.
42. Y. Gao, G. Wang, L. Chen, P. Xu, Y. Zhao, Y. Zhou and L. B. Han, *J. Am. Chem. Soc.*, 2009, **131**, 7956.
43. T. M. Balthazor and R. C. Grabiak, *J. Org. Chem.*, 1980, **45**, 5425.
44. E. A. Krasil'nikowa, I. V. Berdnik, V. V. Sentemov, F. Sh. Shagvaleev and T. V. Zykova, *Zh. Obsch. Khim.*, 1986, **56**, 223.
45. V. V. Sentemov, E. A. Krasil'nikowa and I. V. Berdnik, *Zh. Obsch. Khim.*, 1986, **56**, 481.
46. V. V. Sentemov, E. A. Krasil'nikowa and I. V. Berdnik, *Zh. Obsch. Khim.*, 1990, **60**, 1040.
47. C. Amatore, A. Justand, in *Handbook of Organopalladium Chemistry for Organic Synthesis*, Vol. 1, ed. Ei-ichi Negishi, John Wiley & Sons, Inc., New York, 2003, p 943.
48. M. C. Kohler, R. A. Stockland Jr. and N. P. Rath, *Organometallics*, 2006, **25**, 5746.
49. M. Kalek and J. Stawinski, *Organometallics*, 2007, **26**, 5840.
50. M. Kalek and J. Stawinski, *Organometallics*, 2008, **27**, 5876.
51. M. C. Kohler, T. V. Grimes, X. Wang, T. R. Cundari and R. A. Stockland Jr., *Organometallics*, 2009, **28**, 1193.
52. Q. Yao and S. Levchik, *Tetrahedron Lett.*, 2006, **47**, 277.
53. R. K. Bansal, N. Gupta, J. Heinicke, G. N. Nikonov, F. Saguitova and D. C. Sharma, *Synthesis*, 1999, 264.
54. X. Lu and J. Zhu, *Synthesis*, 1987, 726.
55. K. S. Petrakis and T. L. Nagabhushan, *J. Am. Chem. Soc.*, 1987, **109**, 2831.
56. Y. Luo and J. Wu, *Organometallics*, 2009, **28**, 6823.
57. D. Villemin, P. A. Jaffrès and F. Siméon, *Phosphorus, Sulfur Silicon Relat. Elem.*, 1997, **130**, 59.

58. M. V. Vasylyev, E. J. Wachtel, R. Popovitz-Biro and R. Neumann, *Chem. Eur. J.*, 2006, **12**, 3507.
59. V. Penicaud, F. Odobel and B. Bujoli, *Tetrahedron Lett.*, 1998, **39**, 3689.
60. A. Bessmertnykh, C. M. Douaihy, S. Muniappan and R. Guilard, *Synthesis*, 2008, 1575.
61. J. Zon, V. Videnova-Adrabinska, J. Janczak, M. Wilk, A. Samoc, R. Gancarz, M. Samoc, *CystEngComm*, 2011, **13**, 3474.
62. S. Konar, J. Zoń, A. V. Prosvirin, K. R. Dunbar and A. Clearfield, *Inorg. Chem.*, 2007, **46**, 5229.
63. K. Moedritzer and R. R. Irani, *J. Org. Chem.*, 1966, **31**, 1603.
64. R. A. Cherkasov and V. I. Galkin, *Russ. Chem. Rev.*, 1998, **67**, 857.
65. N. S. Zefirov and E. D. Matveeva, *ARKIVOC*, 2008, (i), 1.
66. J. Clayden, N. Greeves, S. Warren and P. Wothers, *Organic Chemistry*, Oxford University Press, New York, 2001.
67. F. Lebouc, I. Dez and P.-J. Madec, *Polymer*, 2005, **46**, 319.
68. M. Soroka, *personal communication.*
69. P. B. Iveson, M. P. Lowe and J. C. Lockhart, *Polyhedron*, 1993, **12**, 2313.
70. J.-G. Mao, Z. Wang and A. Clearfield, *New J. Chem.*, 2002, **26**, 1010.
71. M. Tazaki, K. Nita, M. Takagi and K. Ueno, *Chem. Lett.*, 1982, 571.
72. X. Shi, G. Zhu, S. Qiu, K. Huang, J. Yu and R. Xu, *Angew. Chem., Int. Ed. Engl.*, 2004, **43**, 6482.
73. L. Maier, *Phosphorus Sulfur Silicon*, 1991, **62**, 29.
74. J. Kasser, A. A. Nazarov, C. G. Hartinger, B. Wdzienkonski, C. Dani, M. L. Kuznetsov, V. B. Arion and B. K. Keppler, *Bioorg. Med. Chem.*, 2009, **17**, 3388.
75. J. A. Groves, P. A. Wright and P. Lightfoot, *Inorg. Chem.*, 2005, **44**, 1736.
76. A. Michaelis and T. Becker, *Chem. Ber.*, 1897, **30**, 1003.
77. R. S. Edmundson, in *The Chemistry of Organophosphorus Compounds*, Vol. 4, Ter- and quinque-valent phosphorus acids and their derivatives, ed. F. R. Hartley, John Wiley & Sons, Chichester, 1996, p. 47.
78. M. D. Fletcher, in *Organophosphorus Reagents. A Practical Approach in Chemistry*, ed. P. J. Murphy, Oxford University Press, New York, 2004, p. 185.
79. R. K. Boeckman Jr. and A. J. Thomas, *J. Org. Chem.*, 1982, **47**, 2823.
80. Z. Hasnik, R. Pohl and N. Hocek, *Tetrahedron Lett.*, 2010, **51**, 2464.
81. R. J. Cohen, D. L. Fox, J. F. Eubank and R. N. Salvatore, *Tetrahedron Lett.*, 2003, **44**, 8617.
82. H. Fakhraian and A. Mirzaei, *Phosphorus, Sulfur Silicon Relat. Elem.*, 2006, **181**, 511.
83. C. Benezra and J. L. Bravet, *Can. J. Chem.*, 1975, **53**, 474.
84. D. Meziane, J. Hardouin, A. Elias, E. Guénin and M. Lecouvey, *Heteroatom Chem.*, 2009, **20**, 369.
85. O. E. O. Hormi, E. O. Pajunen, A. K. C. Avall, P. Pennanen, J. H. Näsman and M. Sundell, *Synth. Commun.*, 1990, **20**, 1865.
86. K. Moedritzer and R. R. Irani, *J. Inorg. Nucl. Chem.*, 1961, **22**, 297.
87. A. Burger and N. D. Dawson, *J. Org. Chem.*, 1951, **16**, 1250.

88. S. Van der Jeught and C. V. Stevens, *Chem. Rev.*, 2009, **109**, 2672.
89. M. Plabst and T. Bein, *Inorg. Chem.*, 2009, **48**, 4331.
90. V. Chandrasekhar, P. Sasikumar, R. Boomishankar and G. Anantharaman, *Inorg. Chem.*, 2006, **45**, 3344.
91. S. R. Alley and W. Henderson, *J. Organomet. Chem.*, 2001, **637–639**, 216.
92. L. S. Melvin, *Tetrahedron Lett.*, 1981, **22**, 3375.
93. R. C. Cambie and B. D. Palmer, *Aust. J. Chem.*, 1982, **35**, 827.
94. T. Calogeropoulou, G. B. Hammond and D. F. Wiemer, *J. Org. Chem.*, 1987, **52**, 4185.
95. F. Hammerschmidt and S. Schmidt, *Monatsh. Chem.*, 1997, **128**, 1173.
96. C. M. Taylor and A. J. Watson, *Curr. Org. Chem.*, 2004, **8**, 623.
97. B. Dhawan and D. Redmore, *J. Org. Chem.*, 1984, **49**, 4018.
98. J. P. Clay, *J. Org. Chem.*, 1951, **16**, 892.
99. A. M. Kinnear and E. A. Perren, *J. Chem. Soc.*, 1952, 3437.
100. G. N. Kosolapoff and W. F. Huber, *J. Am. Chem. Soc.*, 1947, **69**, 2020.
101. G. N. Kosolapoff, *J. Am. Chem. Soc.*, 1952, **74**, 4119.
102. G. A. Olah, O. Farooq, Q. Wang and A. Wu, *J. Org. Chem.*, 1990, **55**, 1224.
103. J. A. Miles, M. T. Beeny and K. W. Ratts, *J. Org. Chem.*, 1975, **40**, 343.
104. F. Siméon, P. A. Jaffrès and D. Villemin, *Tetrahedron*, 1998, **54**, 10111.
105. P. C. Crofts and G. M. Kosolapoff, *J. Am. Chem. Soc.*, 1953, **75**, 3379.
106. V. Chandrasekhar, T. Senapati and E. C. Sañudo, *Inorg. Chem.*, 2008, **47**, 9553.
107. T. Kagayama, A. Nakano, S. Sakaguchi and Y. Ishii, *Org. Lett.*, 2006, **8**, 407.
108. X. J. Mu, J. P. Zou, Q. F. Qian and W. Zhang, *Org. Lett.*, 2006, **8**, 5291.
109. O. Tayama, A. Nakano, T. Iwahama, S. Sakaguchi and Y. Ishii, *J. Org. Chem.*, 2004, **69**, 5494.
110. X. Q. Pan, J. P. Zou, G. L. Zhang and W. Zhang, *Chem. Commun.*, 2010, **46**, 1721.
111. E. F. Jason and E. K. Fields, *J. Org. Chem.*, 1962, **27**, 1402.
112. F. Effenberger and H. Kottmann, *Tetrahedron*, 1985, **41**, 4171.
113. H. Kottmann, J. Skarzewski and F. Effenberger, *Synthesis*, 1987, 797.
114. W. Xu, J. P. Zou and W. Zhang, *Tetrahedron Lett.*, 2010, **51**, 2639.
115. R. Obrycki and C. E. Griffin, *J. Org. Chem.*, 1968, **33**, 632.
116. J. F. Bunnett and X. Creary, *J. Org. Chem.*, 1974, **39**, 3612.
117. J. F. Bunnett, R. G. Scamehorn and R. P. Traber, *J. Org. Chem.*, 1976, **41**, 3677.
118. A. Boumekouez, E. About-Jaudet, N. Collignon and P. Savignac, *J. Organomet. Chem.*, 1992, **440**, 297.
119. A. N. Pudovik and I. V. Konovalova, *Synthesis*, 1979, 81.
120. S. H. Szajnman, E. L. Ravaschino, R. Docampo and J. B. Rodriguez, *Bioorg. Med. Chem. Lett.*, 2005, **15**, 4685.
121. L. V. Kaabak, N. E. Kuz'mina, A. V. Khudenko and A. P. Tomilov, *Russ. J. Gen. Chem.*, 2006, **76**, 1673.

122. M. Lecouvey, I. Mallard, T. Bailly, R. Burgada and Y. Leroux, *Tetrahedron Lett.*, 2001, **42**, 8475.
123. M. S. Wu, R. Y. Chen and Y. Huang, *Synth. Commun.*, 2004, **34**, 1393.
124. M. B. Martin, J. S. Grimley, J. C. Lewis, H. T. Heath III, B. N. Bailey, H. Kendrick, V. Jardley, A. Caldera, R. Lira, J. A. Urbina, S. N. J. Moreno, R. Docampo, S. L. Croft and E. Oldfield, *J. Med. Chem.*, 2001, **44**, 909.
125. D. Lecerclé, S. Gabillet, J. M. Gomis and F. Taran, *Tetrahedron Lett.*, 2008, **49**, 2083.
126. V. Chaleix and M. Lecouvey, *Tetrahedron Lett.*, 2007, **48**, 703.
127. B. J. Chai, W. Covina and F. D. Muggee, *1980*, U.S. Patent 4,239,695.
128. G. R. Kieczykowski, R. B. Jobson, D. G. Melillo, D. F. Reinhold, V. J. Grenda and I. Shinkai, *J. Org. Chem.*, 1995, **60**, 8310.
129. R. W. Ratcliffe and B. G. Christensen, *Tetrahedron Lett.*, 1973, **14**, 4645.
130. E. W. Petrillo Jr. and E. R. Spitzmiller, *Tetrahedron Lett.*, 1979, **20**, 4929.
131. M. K. Mao and J. E. Franz, *Synthesis*, 1991, 920.
132. J. Barluenga, P. J. Campos, G. Canal and G. Arsensio, *Synlett*, 1990, 261.
133. M. Soroka, *Synthesis*, 1989, 547.
134. P. Savignac and B. Iorga, *Modern Phosphonate Chemistry*, CRC Press, Boca Raton, FL, 2003.
135. R. Engel and J. L. I. Cohen, *Synthesis of Carbon–Phosphorus Bonds*, CRC Press, Boca Raton, FL, 2004.
136. S. Bauer and N. Stock, *J. Solid. State Chem.*, 2007, **180**, 3111.
137. S. Bauer and N. Stock, *Z. Anorg. Allg. Chem.*, 2008, **634**, 131.
138. F. N. Shi, T. Trindade, J. Rocha and F. A. A. Paz, *Cryst. Growth Des.*, 2008, **8**, 3917.
139. S. Lodhia, A. Turner, M. Papadaki, K. D. Demadis and G. B. Hix, *Cryst. Growth Des.*, 2009, **9**, 1811.
140. D. W. Allen, B. G. Hutley and M. T. J. Mellor, *J. Chem. Soc. Perkin Trans. 2*, 1977, 789.
141. R. Rabinowitz, *J. Org. Chem.*, 1963, **28**, 2975.
142. C. E. McKenna, M. T. Higa, N. H. Cheung and M. C. McKenna, *Tetrahedron Lett.*, 1977, **18**, 155.
143. T. Morita, Y. Okamoto and H. Sakurai, *Bull. Chem. Soc. Jpn.*, 1978, **51**, 2169.

CHAPTER 7

Structural Chemistry of Bimetallic Oxides Constructed from Molybdodiphosphonate Building Blocks

STEPHANIE JONES AND JON ZUBIETA

Department of Chemistry, Syracuse University, Syracuse, NY, 13244, USA

7.1 Introduction

The considerable contemporary interest in solid state inorganic oxide materials reflects their diverse properties which result in significant applications to areas of chemistry and materials science ranging from heavy construction to molecular electronics.[1–15] The range of properties is related to the vast compositional range, allowing modifications in covalency, geometry, and oxidation states, and to a diverse crystal chemistry, which can provide a range of pore structures and dimensions, coordination sites, and juxtapositions of functional groups. However, while some naturally occurring oxides and minerals possess complex crystal structures, the majority are of simple composition and have highly symmetrical structures with rather small unit cells. Most silicates, important ores, gems, rocks, and soils are examples of such materials. Although such simple oxides can possess unique and specific properties, such as piezoelectricity, ferromagnetism, or catalytic activity, as a general rule there is a correlation between the complexity of the structure of a material and its functionality.

While synthesis, as the source of new compounds with potentially novel properties, is the essence of chemistry, the rational design of inorganic materials remains a challenge in solid state chemistry. In contrast to organic materials

Metal Phosphonate Chemistry: From Synthesis to Applications
Edited by Abraham Clearfield and Konstantinos Demadis

Published by the Royal Society of Chemistry, www.rsc.org

which can self-assemble into ordered arrays at low temperature, corresponding advances in synthesis of inorganic solids have been hampered by the lack of suitable molecular precursors and well-defined mechanisms that allow their low-temperature assembly into crystalline materials. However, the ubiquity of inorganic phases in the geosphere and the biosphere[16–22] suggests that synthetic strategies mimicking Nature's design of complex materials may provide guidelines for the preparation of synthetic phases and the modification of inorganic microstructures. One approach to the design of new solids exploits the use of organic molecules as templates or structure-directing components to imprint or transmit structural information.[23–26] Such materials are mixtures of inorganic subunits coexisting with organic molecules which dramatically influence the inorganic microstructure and allow higher order organization of hierarchical structures.[27–29]

7.1.1 Hybrid Organic–Inorganic Materials[30–44]

A classical approach to complex materials combines organic molecules with inorganic entities to provide crystalline hybrid materials. The inorganic component can provide magnetic, dielectric, or optical properties, mechanical hardness, and thermal stability while the organic molecules offer processability, structural diversity, a range of polarizabilities, and luminescent properties.[45] Furthermore, the use of metal atoms in the hybrid materials provides advantages over purely organic materials. Metal–ligand coordination bonds are stronger than hydrogen bonds; metal–ligand bonds are more directional than other weak interactions; metal sites offer variable coordination polyhedra, as well as differences in size, ligand preferences, and crystal field energies, providing versatility of construction modes.

The most common type of hybrid organic–inorganic materials consists of extended arrays of isolated metal atoms or clusters bridged by polyfunctional organic ligands. The materials have witnessed dramatic development in the past decade through the efforts of Yaghi,[46,47] Ferey[48], Kitagawa[49] and others.[50–53] An important subclass of this family of materials are the hybrid metal oxides, which contain metal–oxygen–metal (M–O–M) arrays as part of their structures. In such materials, the inorganic oxide contributes to the increased complexity, and hence functionality, through incorporation as one component in a multilevel structural material where there is a synergistic interaction between organic and inorganic components.

7.1.2 Hybrid Metal Oxides

Our strategy for the synthesis of new metal oxides has exploited the "bottom up" strategy of linking molecular oxide building blocks through appropriate organic molecules or secondary metal complex cations (SMCC).[54,55] The vast family of molecular oxides commonly referred to as polyoxometalates (POMs)[56–60] provides robust clusters of different sizes, topologies, and oxidation states for the construction of solids with more or less predictable connectivity in the crystalline state.[61–66] We have demonstrated this approach using both a two-component system based on a POM cluster and a SMCC (Figure 7.1) and a three-component system consisting of a POM cluster, a tethering organic component, and a SMCC (Figure 7.2). In the course of these investigations, it became apparent that a number of structural determinants were significant, including the identity of the ligand of the SMCC.

7.1.3 Properties[47–53,67–72]

The interest in hybrid materials reflects their properties and applications. Microporosity offers a variety of important applications, including catalysis, separation, and gas storage. The chemistry of hybrid organic–inorganic materials provides rational design of solid materials with the appropriate pore topology, environment, and molecular juxtaposition for reactivity. Several potential advantages may be identified: the materials combine the size and

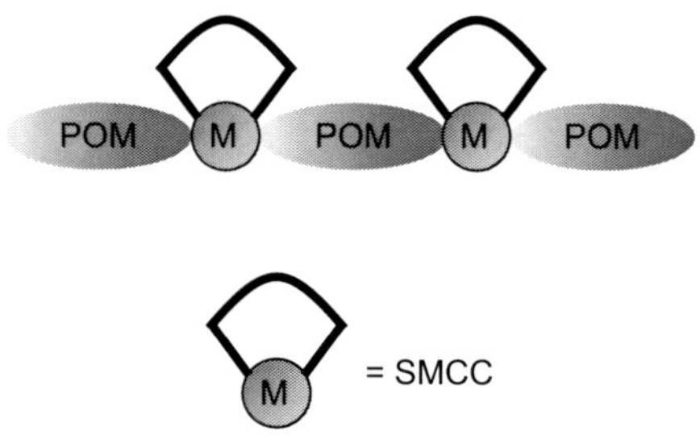

Figure 7.1 Schematic representation of the construction of a 1D material constructed from polyoxometalate anions (POM) and secondary metal complex cations (SMCC).

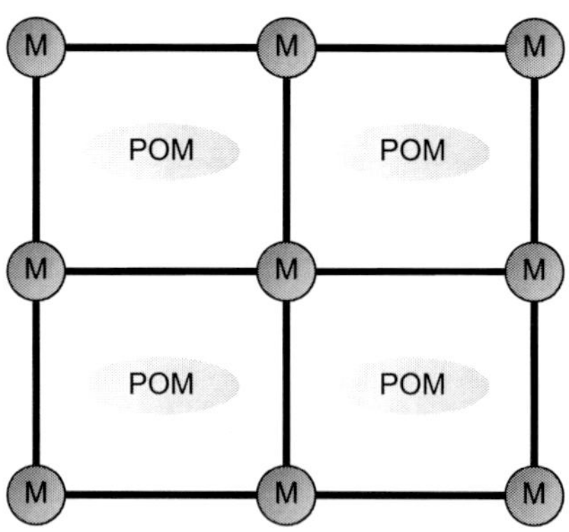

Figure 7.2 Encapsulation of POM clusters within a SMCC framework. In this view, the heavy black bars represent the tethering ligand.

shape selectivity of zeolites[67] with the reactivity and properties of transition metal cations; the inorganic substructure confers chemical and thermal stability; transition metal sites provide nodes for ligand coordination with directional preferences and may display coordinative unsaturation. On the other hand, the organic substructure allows the construction of interior cavities and channels that may be functionalized with organic groups, allowing shape, size, and function of the cavity to be modified. Furthermore, since transition metal sites are necessary components of the hybrid material, properties such as magnetism[70] and luminescence[83] can also be developed, suggesting the possibility of multifunctional materials. For example, the synergistic interplay of the organic ligands, the transition metal identity, and the nuclearity of the structural building blocks can provide magnetic correlations that are expressed in zero, one, two, or three dimensions.

7.1.4 Metal Organophosphonates

The extensive contemporary interest in metal organophosphonate chemistry[73–79] reflects their role as prototypical organic–inorganic hybrid materials that exhibit a range of structures, including molecular clusters, chains, networks, and 3D frameworks. The structural diversity is reflected in a range of physical properties giving rise to applications in catalysis, proton conductivity, ion exchange, intercalation chemistry, and photochemistry.[80–87]

A significant subclass of these materials is the oxometal organophosphonates, represented by the oxovanadium and oxomolybdenum organophosphonates.[88–115] Of particular interest is the design of such organic–inorganic oxide materials with diphosphonate ligands, H_2O_3P-tether-PO_3H_2, a family of dipodal ligands that characteristically provide "pillared layer" structures with a variety of metals, including oxovanadates (Figure 7.3). These are 3D frameworks with alternating inorganic and organic domains. The separations between inorganic layers may be varied by modification of the

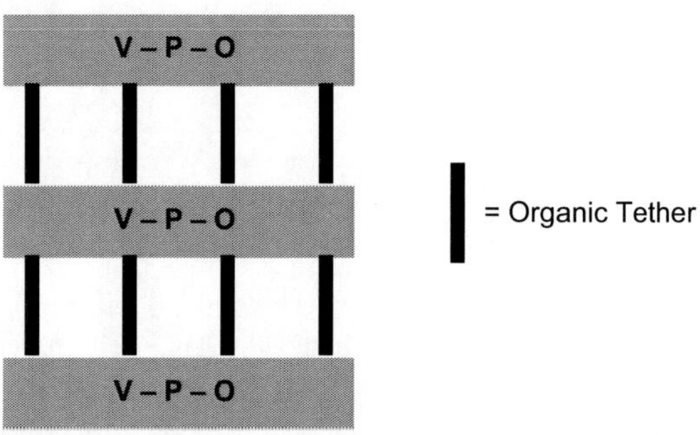

Figure 7.3 Schematic representation of a "pillared" layered structure consisting of oxovanadium-phosphonate layers linked through organic tethers.

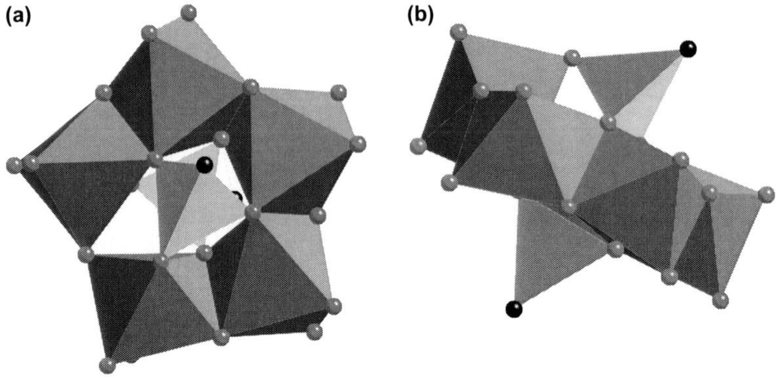

Figure 7.4 Top and side polyhedral views of the $\{Mo_5O_{15}(O_3PR)_2\}^{4-}$ cluster.

tether lengths, and the interlamellar domain may be sculpted by appropriate functionalization of the organic linkers.[93,116]

In contrast, the oxomolybdate chemistry with diphosphonate ligands is characterized by the incorporation of discrete polyoxomolybdate organophosphonate clusters as building blocks, most notably the $\{Mo_5O_{15}(O_3PR)_2\}^{4-}$ cluster (Figure 7.4). The 1D $\{Mo_5O_{15}(O_3PR)_2\}_n^{4n-}$ chains can undergo spatial expansion through appropriate secondary metal–ligand complexes, such as the dipodal $\{Cu_2(tpyprz)\}^{4+}$, to provide 2D and 3D materials (tpyprz = tetrapyridylpyrazine).

This strategy represents a common approach for the preparation of organic–inorganic hybrid materials, one that exploits dipodal ligands, such as 4,4′-bipyridine, dicarboxylic acids, or diphosphonates as the organic tethers. However, a related type of dipodal ligand is one that exhibits different donor group termini, such as 1,4-carboxy-phenylphosphonic acid, $H_2O_3PC_6H_4CO_2H$. The incorporation of different donor groups offers the potential for designing materials incorporating two different metals with different donor group preferences. While two-component systems, such as the well-documented metal organic frameworks,[117] have revealed some design principles, the chemistry of three-component systems remains relatively unexplored. The materials of the M^{II}-ligand/Mo_xO_y/organophosphonate family are examples of this latter class of compounds.

7.2 Hydrothermal Synthesis

Conventional solid state syntheses are carried out at high temperatures, which preclude the retention of the structural characteristics of organic starting materials and which produce thermodynamic products. However, numerous reactions relevant to solid state synthesis such as cation exchange, dehydration, intercalation, redox, sol–gel, and hydrothermal, occur at ambient or relatively mild temperatures.[118] These routes to the preparation of new materials are collectively referred to as "Chimie douce", a concept first introduced by Rouxel and Livage.[119] Although well established for the preparation of aluminosilicates, hydrothermal techniques have only recently been adopted for the preparation of a wide variety of metastable materials, including transition metal phosphates, metal organophosphonates, and complex polyoxoalkoxometalates.[120] The term "hydrothermal synthesis" is conventionally applied to heterogeneous reactions in aqueous media at temperatures above 100 °C and under autogenous pressures. The method mimics natural conditions for the formation of structurally complex mineral species under relatively mild conditions.[121] Consequently, the technique has been recognized for some time as offering convenient approaches to metastable compounds, low-temperature phases, and materials with elements in unusual oxidation states. Hydrothermal methods have been applied to the preparation of a variety of materials, including microporous phases, chemical sensors, conducting solids, ceramics, magnetic solids, and phosphors.[122] The applications of hydrothermal methods to crystallization processes,[123] zeolite synthesis,[124] the preparation of metastable oxide materials,[125] and general preparative chemistry have been reviewed.[121–127]

7.2.1 Hydrothermal Synthesis of Organic–Inorganic Hybrid Materials

One approach to the preparation of novel solid state materials exploits the use of organic molecules which can profoundly alter the inorganic microstructure. Such organic–inorganic hybrid materials combine the unique characteristics of the organic and inorganic components in a complementary fashion to provide novel solid state structures and composite materials with unexpected properties. However, traditional solid state syntheses of thermodynamic products conventionally rely on solid–solid interactions at temperatures of 500–1000 °C, conditions which will not retain the structural elements of the organic component. On the other hand, hydrothermal synthesis has been found to provide a powerful technique for the preparation of organic–inorganic hybrid materials with retention of the structural elements of the reactants in the final product phase.[54a] Hydrothermal reactions, typically carried out in the temperature range 120–260 °C under autogenous pressure, exploit the self-assembly of the product from soluble precursors. The reduced viscosity of water under these conditions enhances diffusion processes so that solvent extraction of solids and crystal growth from solution are favored. Since problems associated with different solubility properties of reactants are minimized, a variety of simple precursors may be introduced, as well as a number of organic and/or inorganic structure-directing agents from which those of appropriate size and shape may be selected for efficient crystal packing during the crystallization process. Under such non-equilibrium crystallization conditions, metastable kinetic phases rather than the thermodynamic phase are often isolated. While several pathways, including that resulting in the most stable phase, are available in such non-equilibrium mixtures, the kinetically favored structural evolution results from the smallest perturbations of atomic positions. Consequently, nucleation of a metastable phase may be favored. Thus, our favored synthetic approach is to employ an organic component at low temperature to modify or control the surface of growing composite crystals in a hydrothermal medium.

7.2.2 Reaction Conditions

While hydrothermal reactions in the 700 °C and 3 kbar range are not uncommon for some applications, these conditions require specially designed autoclaves.[128] Most syntheses involving organic components are performed in the 100–250 °C temperature range under autogenous pressure and, consequently, may be carried out in more conventional reaction vessels such as acid digestion bombs, thick-walled borosilicate tubes, or even Teflon pouches.[129,130] While the materials described in this chapter were prepared in water, a variety of organic solvents (acetonitrile, ethylene glycol, chloroform) and mixed aqueous–organic solvents have also been exploited in the more generally designated "solvatothermal" method.

It is noteworthy that, in our experience, the crystals of the hybrid product form at the reaction temperature and pressure. Crystallization of the hybrid material does not occur in most cases during the cooling phase to room temperature or upon standing for various time periods. It is also apparent that the hydrothermal parameter space is extensive, including time, temperature, fill volume, relative stoichiometries of reactants, nucleation and crystal growth rates, pH, and identity of the starting materials. While pH and reaction temperature are critical variables in the syntheses of this study, other parameters have not been systematically investigated. A more complete understanding of the factors influencing the syntheses and properties of such hybrid materials will require a continuing expansion of the structural database. In fact, the solids discussed in this chapter, and the many examples from other investigators, suggest that these materials possess an extensive compositional range and exceptional structural versatility.

7.3 The Building Block Approach: the $\{Mo_5O_{15}(O_3PR)_2\}^{4-}$ Cluster as the Oxide Component

We had demonstrated in 1997 that polyoxomolybdate clusters, specifically the octamolybdates $\{Mo_8O_{26}\}^{4-}$, can be readily exploited as building blocks in the construction of mixed metal oxides incorporating organic components as ligands, structure-directing moieties, and/or charge-compensating units.[54a,131] Since then, we and others have demonstrated the considerable scope of this approach.[132,133]

This strategy is somewhat limited as a consequence of the unpredictability of the coordinate covalent bonding between the oxide clusters and the secondary metal–ligand components that serve to link the clusters into 1D, 2D, or 3D assemblies. The venerable phosphomolybdate cluster $\{Mo_5O_{15}(O_3PR)_2\}^{4-}$ (R = alkyl, aryl, or H) suggested itself as a more versatile metal–organic cluster.[134] The most obvious approach to design is to link the parent clusters $\{Mo_5O_{15}(O_3PR)_2\}^{4-}$ with R = C_6H_6 or H, through appropriate dipodal linkers, such as {M-(ligand bridge)-M}$^{n+}$ as shown in Figure 7.5. A typical binuclear secondary metal–ligand tether is $\{Cu_2(tpyprz)\}^{4+}$ (tpyprz = tetrapyridylpyrazine) that assumes the anticipated role in the 1D structure of $[\{Cu_2(tpyprz)(H_2O)\}Mo_5O_{15}(O_3PC_6H_5)_2]$ (**2**), shown in Figure 7.6.[112] As anticipated, the structure consists of $\{Mo_5O_{15}(O_3PC_6H_5)_2\}^{4-}$ clusters linked through $\{Cu_2(tpyprz)\}^{4+}$ rods.

However, when the R group of the phosphonate is –OH rather than an aryl or alkyl group, 2D and 3D structures are observed (Figure 7.7). The chain structure of **2** exhibits two attachment points to the secondary metal–ligand tethers per molybdenum oxide cluster. In contrast, α- and β-$[\{Cu_2(tpyprz)(H_2O)_2\}Mo_5O_{15}(O_3POH)_2]\cdot 2H_2O$ (**12** and **12a**) exhibit four and three attachment points per cluster, respectively. Each $\{Cu_2(tpyprz)(H_2O)_2\}^{4+}$ rod links three phosphomolybdate clusters to generate the overall 2D network. However, the trihydrate $[\{Cu_2(tpyprz)(H_2O)_2\}Mo_5O_{15}(O_3POH)_2]\cdot 3H_2O$ (**19**) is 3D, with four attachment points per cluster and each cationic rod linking four cluster sites. As we shall see, this variability in connectivity sites on the oxide clusters and by the secondary metal–ligand components provides a means for structural versatility but also renders predictability problematical.

Figure 7.5 Schematic representation of the construction of a 1D material from $\{Mo_5O_{15}(O_3PR)_2\}^{4-}$ clusters (ellipse labeled phosphomolybdate) and secondary metal cations (M) linked through dipodal tethering ligands (solid lines).

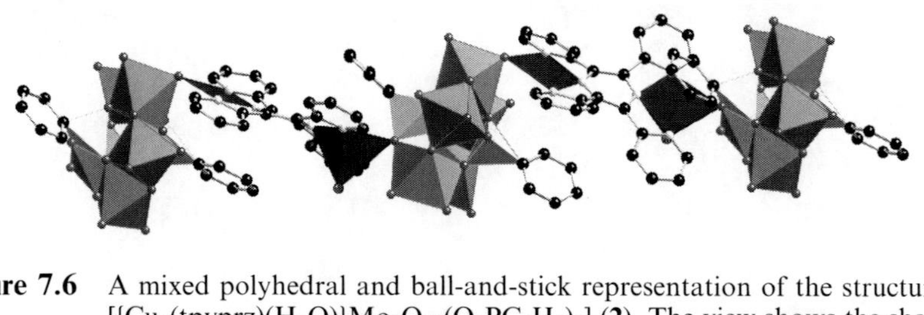

Figure 7.6 A mixed polyhedral and ball-and-stick representation of the structure of $[\{Cu_2(tpyprz)(H_2O)\}Mo_5O_{15}(O_3PC_6H_5)_2]$ (**2**). The view shows the shading pattern common to the figures: molybdenum, medium gray polyhedra; copper or other M^{II} cation, dark gray polyhedra; phosphorus, light gray polyhedra; oxygen, medium gray spheres; nitrogen, light gray/white spheres; carbon, black spheres.

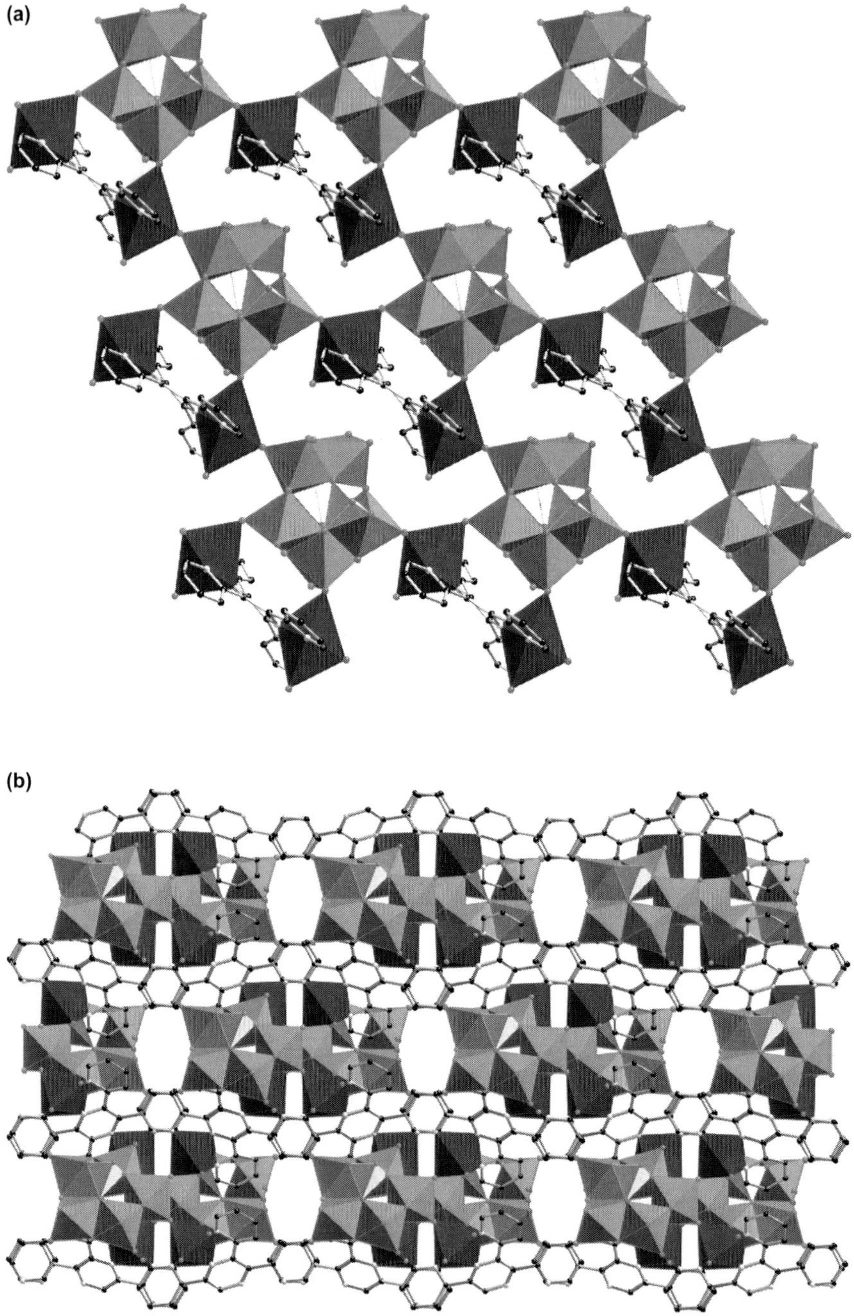

Figure 7.7 (a) The 2D structure of α-[{Cu_2(tpyprz)(H_2O)$_2$}Mo_5O_{15}(O_3POH)$_2$]·2H_2O (**12**) and (b) the 3D structure of [{Cu_2(tpyprz)(H_2O)$_2$}Mo_5O_{15}(O_3POH)$_2$]·3H_2O (**19**).

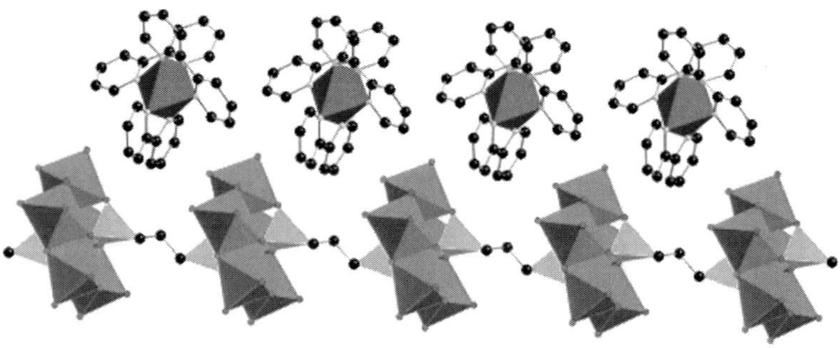

Figure 7.8 A view of the 1D chain of $[Co(bpy)_3][Mo_5O_{14}(OH)\{HO_3P(CH_2)_3PO_3\}]$ (**4**) and of the $\{Co(bpy)_3\}^{2+}$ cations.

7.3.1 Diphosphonate Ligands as Tethers: One-dimensional Structures

An alternative synthetic strategy is to link the phosphomolybdate clusters through organic tethering groups, using organodiphosphonates of the type $\{O_3P(CH_2)_nPO_3\}^{4-}$ ($n \geq 2$). The prototypical structure of this class is the 1D $[Co(bpy)_3][Mo_5O_{14}(OH)\{HO_3P(CH_2)_3PO_3\}]$ (**4**). As shown in Figure 7.8, the phosphomolybdate substructure is a chain constructed from $\{HMo_5O_{15}(HO_3PR)_2\}^-$ clusters linked through the $(CH_2)_3$ tethers of the organodiphosphonate ligand.

While the "naked" chain of **4** reveals the prototypical 1D substructure, phosphomolybdate chains decorated by secondary metal–ligand subunits are the most readily isolated. This observation reflects the need to passivate the surface of the oxide building block to prevent oligomerization or polymerization and possibly more effective charge density matching by the secondary metal–ligand complex cation than by organonitrogen or alkali metal cations.[135] Typical examples of this class of 1D materials are provided by $[\{Cu_2(phen)_3(H_2O)_2\}Mo_5O_{15}\{O_3P(CH_2)_3PO_3\}]$ (**3**) and $[\{Co_2(phen)_3(H_2O)\}Mo_5O_{15}\{O_3P(CH_2)_3PO_3\}]$ (**5**), shown in Figure 7.9. Once again, the different number of attachment points per cluster, 2 in **3** and 5 in **5**, are reflected in the folding of the 1D chain.

7.3.2 Two- and Three-dimensional Structures

The construction of 1D phosphomolybdate chains by the simple expedient of using organodiphosphonate ligands with appropriate tethering groups suggested a straight forward strategy for the design of 2D and 3D materials. The noteworthy feature of the 1D structures was the facile incorporation of secondary metal–ligand subunits as charge-compensating groups, decorating the periphery of the oxide clusters, an observation that demonstrated that the oxo groups of the clusters were sufficiently basic to form coordinate covalent bonds with oxophilic metal cations of metal–ligand subunits. Consequently,

(a)

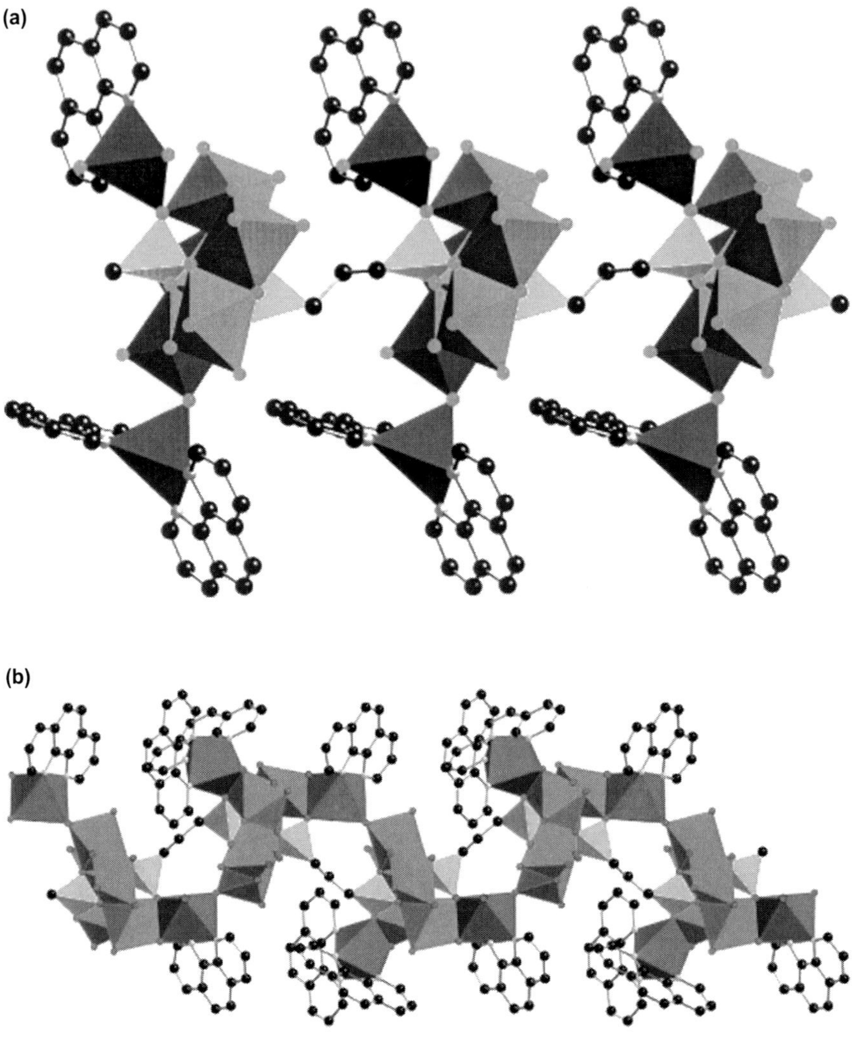

(b)

Figure 7.9 The 1D structures of $[\{Cu_2(phen)_3(H_2O)_2\}Mo_5O_{15}\{O_3P(CH_2)_3PO_3\}]$ (**3**) (a) and $[\{Co_2(phen)_3(H_2O)\}Mo_5O_{15}\{O_3P(CH_2)_3PO_3\}]$ (**5**) (b).

introduction of a binucleating organonitrogen ligand offers the potential for tethering the 1D phosphomolybdate chains $\{Mo_5O_{15}(O_3P\text{-linker-}PO_3)\}_n^{4n-}$ into 2D or 3D structures, as shown in Figure 7.10. In such structures, the molybdate cluster acts as a node and structural expansion depends on the number of surface accessible binding sites and their relative dispositions, in a manner reminiscent of metal organic frameworks.

An example of a typical 2D structure is provided by $[\{Cu_2(tpyprz)(H_2O)_2\}$-$Mo_5O_{15}(O_3PCH_2CH_2PO_3)]$ (**14**), shown in Figure 7.11. The 1D $\{Mo_5O_{15}(O_3PCH_2CH_2PO_3)\}_n^{4n-}$ chains are linked through the $\{Cu_2(tpyprz)(H_2O)_2\}^{4+}$ rods into the network structure. It is noteworthy that the connectivity of rods and

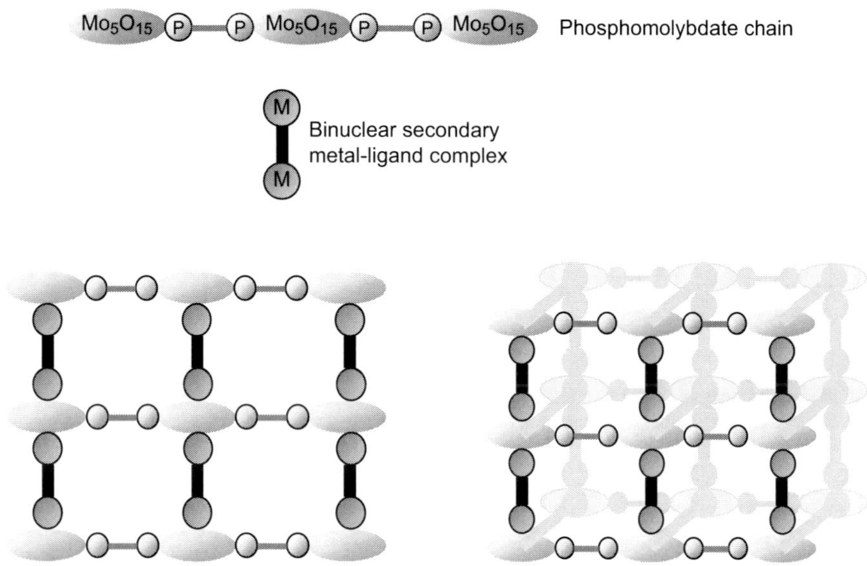

Figure 7.10 Schematic representation of the expansion into 2D and 3D structures, exploiting organodiphosphonate ligands and rod-like binuclear SMCC.

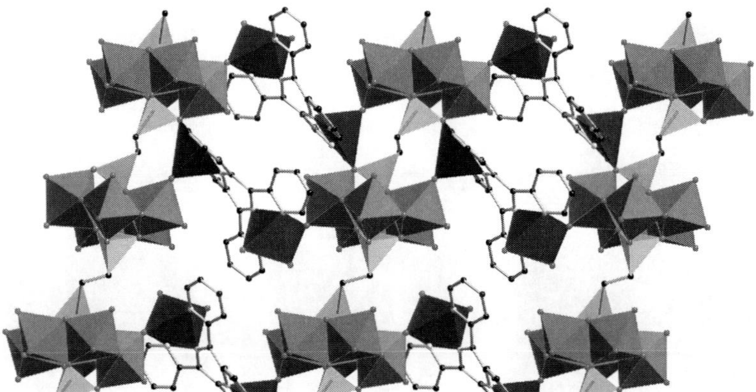

Figure 7.11 The 2D structure of $[\{Cu_2(tpyprz)(H_2O)_2\}Mo_5O_{15}(O_3PCH_2CH_2PO_3)]$ (**14**).

nodes is more complex than the naive construction of Figure 7.10. Thus, each $\{Cu_2(tpyprz)(H_2O)_2\}^{4+}$ rod bonds to two clusters of one chain at one Cu terminus and to a single cluster of an adjacent chain at the other terminus. The resultant grid pattern is folded rather than the simple rectangular grid of the schematic.

Figure 7.12 illustrates the expansion into three dimensions in the structure of $[\{Cu_2(tpyprz)(H_2O)_2\}Mo_5O_{15}\{O_3P(CH_2)_3PO_3\}]$ (**20**). In this case, each phosphomolybdate chain is linked to six adjacent chains through the $\{Cu_2(tpyprz)(H_2O)_2\}^{4+}$ rods. In the 2D structure of **14**, each cluster contributed three attachment points, while in the 3D structure of **20**, the expansion is accomplished as a result of using four attachment points.

In fact, the connectivities between nodes and rods for the 2D and 3D structures exhibit considerable variability in terms of attachment points and

Figure 7.12 The 3D structure of $[\{Cu_2(tpyprz)(H_2O)_2\}Mo_5O_{15}\{O_3P(CH_2)_3PO_3\}]$ (**20**). The $[Mo_5O_{15}\{O_3P(CH_2)_3PO_3\}]_n^{4n-}$ chains propagate normal to the plane of the page.

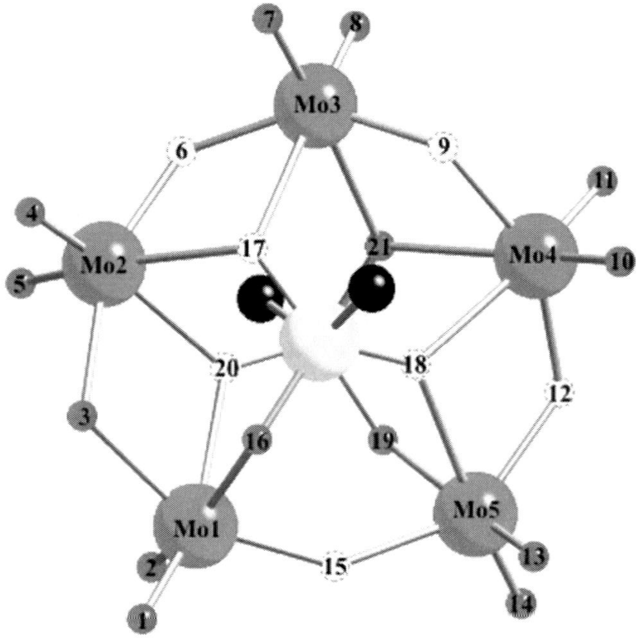

Figure 7.13 The $\{Mo_5O_{15}(O_3PR)_2\}^{4-}$ cluster, showing the oxygen atom labeling scheme.

linkage modes. The linkage modes between the molybdate clusters and the cationic components are quite diverse, ranging in number from zero to five and employing a variety of cluster oxygen sites. As indicated by Figure 7.13, oxygen atoms at sites 1–5, 7, 8, 10, 13, 14, 16, and 21 participate in bonding in at least one structure. While the majority of structures of this general class exhibit M^{II} interactions with terminal {Mo=O} oxo groups exclusively, five also exhibit {M–O(P)–Mo} linkage to phosphonate oxygen doors. There is a single example $[\{Cu(terpy)\}_2Mo_5O_{15}\{O_3P(CH_2)_4PO_3\}]$ of participation by bridging oxo groups {Mo–O–Mo} in bonding to the secondary metal.

7.4 Observations on Factors that Influence Structure of Materials of the $\{Mo_5O_{15}(O_3PR)_2\}_n^{4n-}/M^{II}$ Ligand Family

The range of structural types containing the $\{Mo_5O_{15}(O_3PR)_2\}^{4-}$ building block is quite extensive, reflecting factors such as hydrothermal reaction conditions, the identity of the secondary metal, the nature of the binucleating ligand and the diphosphonate tether length. A discussion of structural trends and comparisons may prove instructive.

7.4.1 Hydrothermal Conditions

The hydrothermal parameter space allows considerable scope in reaction conditions. Small changes in one or a few of the reaction variables of time, temperature, stoichiometries, concentrations, template, identity of the starting materials, addition of mineralizers, and pH can have dramatic effects on the outcome of the reactions. This number and degree of coupling between the variables renders prediction of reaction products quite problematic. However, this complexity is often reflected in the rich structural chemistry associated with the hydrothermal domain.

As an example, let us consider the phases of the $\{Co_2(tpyprz)\}^{4+}$/$Mo_5O_{15}\{O_3P(CH_2)_3PO_3\}^{4-}$ family, illustrated in Figure 7.14, that are isolated through minor changes in reaction conditions. The 1D structure of $[\{Co_2(tpyprz)_2\}Mo_5O_{15}\{O_3P(CH_2)_3PO_3\}]$ (**9**) exhibits an unusual $\{M_2(tpyprz)_2\}^{4+}$ unit decorating a distinctly puckered phosphomolybdate chain.

The remaining examples of propylenediphosphonic acid-based materials are the 3D $[\{Co_2(tpyprz)(H_2O)_3\}Mo_5O_{15}\{O_3P(CH_2)_3PO_3\}]\cdot 7H_2O$ (**24**), $[\{Co_2(tpyprz)(H_2O)_3\}Mo_5O_{15}\{O_3P(CH_2)_3PO_3\}]\cdot H_2O$ (**23**) and $[H_3O]_2[\{Co_3(tpyprz)_2(H_2O)_3\}(Mo_5O_{15})_2\{O_3P(CH_2)_3PO_3\}_2]$ (**25**). The structures are shown schematically in Figure 7.15. It should be emphasized that for **25** the ratio of rods to clusters is 1:2 resulting in a low density of cross-linking attachments between the anionic chains and the cationic rods. The connectivity pattern generates large open galleries of dimensions 21.6 Å × 14.6 Å × 10.6 Å, accommodating ca. 21.4 H_2O molecules, consistent with 28.6% accessible void volume per cell.

7.4.2 The Secondary Metal

While the oxomolybdodiphosphonate building block $\{Mo_5O_{15}(O_3PR)_2\}^{4-}$ is nearly ubiquitous in the structural chemistry of the Co^{II}, Ni^{II}, and Cu^{II} families, the details of the connectivities between building blocks are often quite distinct, as summarized in Tables 7.1–7.3. This observation is most readily apparent in the overall dimensionality of the material. As illustrated in Figure 7.16, for the structure of the general type M^{II}-organonitrogen/oxomolybdate-organophosphonate, 1D, 2D, and 3D structures have been identified, although the naïve expectation for the linking of anionic oxomolybdodiphosphonate chains and cationic rods of the $\{M_2(tpyprz)\}^{4+}$ subgroup is the formation of 2D materials. In fact, 3D phases are more common than 2D for this limited series of materials. However, the relatively small number of structures renders any conclusions on the dimensional preferences tentative at best. The identities of the organic ligands and the secondary metal are also likely to prejudice the results.

This variability is related to a number of structural determinants, including the identity of the secondary metal component, the tether length of the diphosphonate ligand, the number of attachment points on the molybdate cluster employed in bonding to the secondary metal–ligand subunit, and the

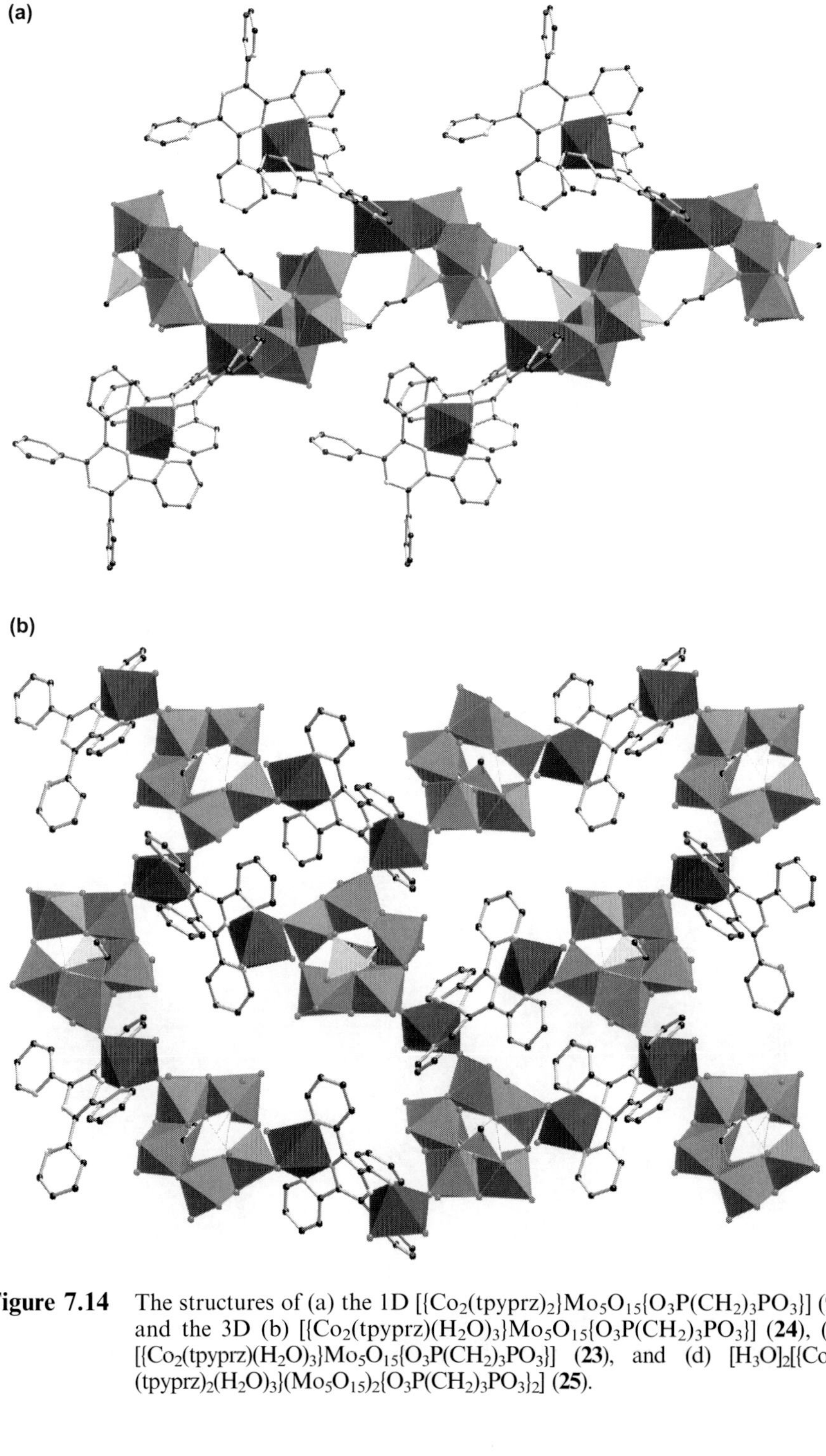

Figure 7.14 The structures of (a) the 1D $[\{Co_2(tpyprz)_2\}Mo_5O_{15}\{O_3P(CH_2)_3PO_3\}]$ (**9**) and the 3D (b) $[\{Co_2(tpyprz)(H_2O)_3\}Mo_5O_{15}\{O_3P(CH_2)_3PO_3\}]$ (**24**), (c) $[\{Co_2(tpyprz)(H_2O)_3\}Mo_5O_{15}\{O_3P(CH_2)_3PO_3\}]$ (**23**), and (d) $[H_3O]_2[\{Co_3\text{-}(tpyprz)_2(H_2O)_3\}(Mo_5O_{15})_2\{O_3P(CH_2)_3PO_3\}_2]$ (**25**).

(c)

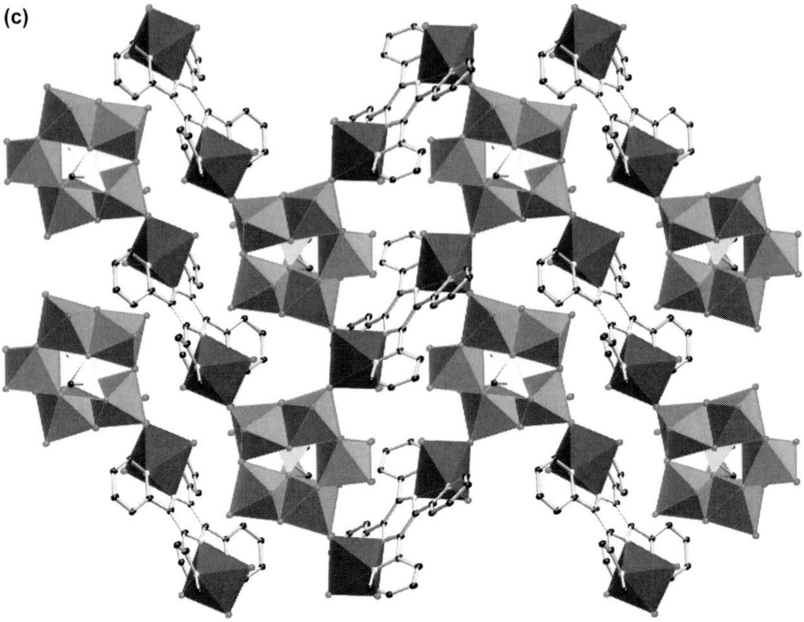

(d)

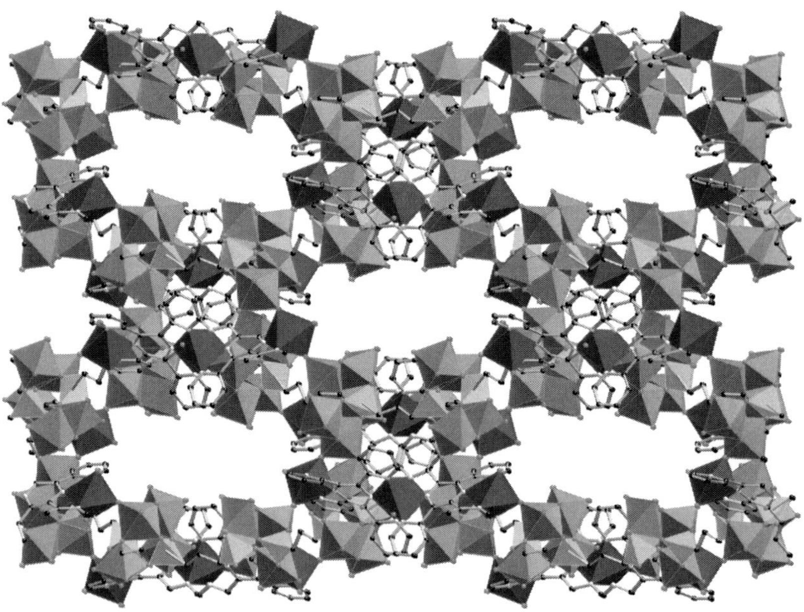

Figure 7.14 Continued

coordination mode adopted by the secondary metal component. The influences of the coordination requirements of the secondary metal are most apparent in the contrasting characteristics of Cu^{II} with respect to Co^{II} and Ni^{II}.

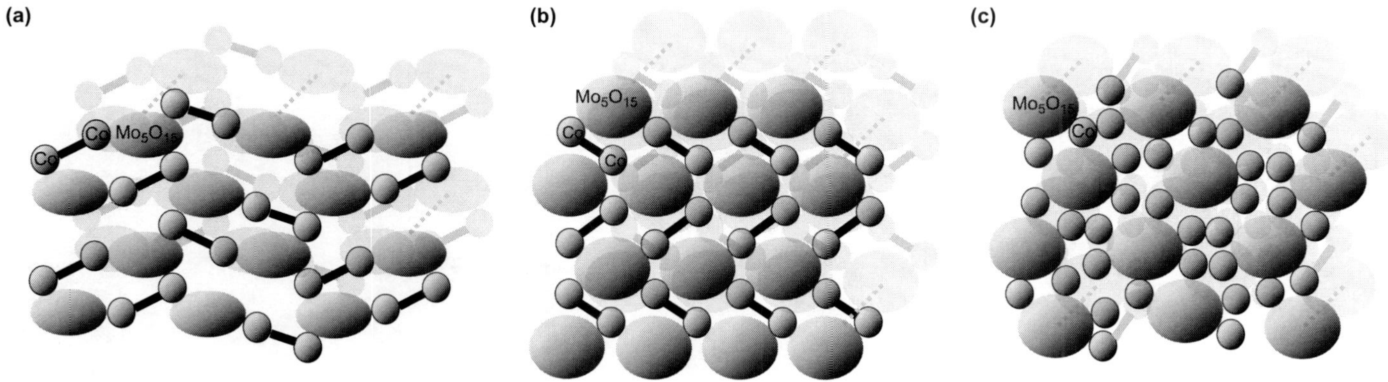

Figure 7.15 Schematic representations of the structures of (a) **24**, (b) **23**, and (c) **25**. The phosphomolybdate clusters are the large ovals; the diphosphonate tethers are shown as dashed lines; the cobalt cations are small spheres; and the tpyprz ligand is illustrated as thick black lines.

Table 7.1 Summary of selected structural characteristics of one-dimensional materials containing the $\{Mo_5O_{15}(O_3PR)_2\}^{4-}$ subunit.

Compound	*M^{II} attachments per cluster*	*Attachment type*	*Clusters linked per cationic unit*
$[\{Cu_2(tpyprz)(H_2O)_3\}Mo_5O_{15}(HOPO_3)(O_3PCH_2CO_2H)]$ (**1**)	2	Mo=O	2
$[\{Cu_2(tpyprz)(H_2O)\}$-$Mo_5O_{15}(O_3PC_6H_5)_2]\cdot 2H_2O$ (**2**)	2	Mo=O P-O-Mo	2
$[\{Cu_2(phen)_3(H_2O)_2\}$-$Mo_5O_{15}\{O_3P(CH)_3PO_3\}]$ (**3**)	2	2 × {Mo=O}	
$[Co(bpy)_3][Mo_5O_{14}(OH)\{HO_3P(CH_2)_3PO_3\}]$ (**4**)	0	–	
$[\{Co_2(phen)_3(H_2O)\}$-$Mo_5O_{15}\{O_3P(CH_2)_3PO_3\}]$ (**5**)	5	3 × {Mo=O} 2 × {P-O-Mo}	2
$[\{Cu_2(bpy)_3(H_2O\}$-$Mo_5O_{15}\{O_3P(CH_2)_4PO_3\}]$ (**6**)	3	2 × {Mo=O} 1 × {P-O-Mo}	
$[\{Cu(terpy)(H_2O)\}_2$-$Mo_5O_{15}\{O_3P(CH_2)_2PO_3\}]$ (**7**)	2	2 × {Mo=O}	
$[\{Co(H_2(tpyprz)\}$-$Mo_5O_{15}(O_3PCH_2CH_2PO_3)]$ (**8**)	3	2 × {Mo=O} 1 × {P-O-Mo}	2
$[\{Co_2(tpyprz)_2\}Mo_5O_{15}\{O_3P(CH_2)_3$-$PO_3\}]\cdot 5H_2O$ (**9**)	3	2 × {Mo=O} 1 × {P-O-Mo}	2
$[\{Ni_2(tpyprz)_2\}Mo_5O_{15}\{O_3P(CH_2)_n$-$PO_3\}]$, $n=4, 5$ (**10, 11**)	3	2 × {Mo=O} 1 × {P-O-Mo}	2

Table 7.2 Summary of selected structural characteristics of two-dimensional materials containing the $\{Mo_5O_{15}(O_3PR)_2\}^{4-}$ core.

Compound	*M^{II} attachments per cluster*	*Attachment type*	*Clusters linked per cationic unit*
α-$[\{Cu_2(tpyprz)(H_2O)_2\}Mo_5O_{15}(HO$-$PO_3)_2]\cdot 2H_2O$ (**12**)	4	Mo=O	3
β-$[\{Cu_2(tpyprz)(H_2O)_2\}Mo_5O_{15}(HO$-$PO_3)_2]\cdot 2H_2O$ (**12a**)	3	Mo=O	3
$[\{Cu(terpy)\}_2Mo_5O_{15}\{O_3P(CH_2)_3PO_3\}]$ (**13**)	4	2 × {Mo=O} 2 × {Mo-O-Mo}	2
$[\{Cu_2(tpyprz)(H_2O)_2\}$-$Mo_5O_{15}\{O_3P(CH_2)_2PO_3\}]$ (**14**)	3	2 × {Mo=O} 1 × {P-O-Mo}	3
$[\{Ni_4(tpyprz)_3\}Mo_5O_{15}\{O_3P(CH_2)_2PO_3\}]$ (**15**)	3	2 × {Mo=O} 1 × {P-O-Mo}	4
$[\{Co_4(tpypry)_3\}$-$Mo_5O_{15}\{O_3P(CH_2)_2PO_3\}].18H_2O$ (**16**)	3	2 × {Mo=O} 1 × {P-O-Mo}	4
$[\{Co_2(tpyprz)(H_2O)_3\}$-$Mo_5O_{15}\{O_3P(CH_2)_9PO_3\}]\cdot 2.5H_2O$ (**17**)	3	3 × {Mo=O}	3
$[\{Cu(bpyrm)\}_2Mo_5O_{15}(O_3PCH_3)_2]$ (**18**)	3	3 × {Mo=O}	4

Table 7.3 Summary of selected structural characteristics of three-dimensional materials containing the $\{Mo_5O_{15}(O_3PR)_2\}^{4+}$ core.

Compound	*M^{II} attachments per cluster*	*Attachment type*	*Clusters linked per cationic unit*
$[\{Cu_2(tpyprz)(H_2O)_2\}Mo_5O_{15}(HO\text{-}PO_3)_2]\cdot 3H_2O$ (**19**)	4	4 × {Mo=O}	4
$[\{Cu_2(tpyprz)(H_2O)_2\}\text{-}Mo_5O_{15}\{O_3P(CH_2)_3PO_3\}]$ (**20**)	4	4 × {Mo=O}	3
$[\{Ni_2(tpyprz)(H_2O)_3\}\text{-}Mo_5O_{15}\{O_3P(CH_2)_3PO_3\}]\cdot H_2O$ (**21**)	4 and 2	4 × {Mo=O} 2 × {Mo=O}	4 2
$[\{Ni_3(tpyprz)_2(H_2O)_2\}(Mo_5O_{15})(Mo_2F_2O_4)\{O_3P(CH_2)_3PO_3\}]$ (**22**)	4	2 × {Mo=O} 2 × {P=O}	4
$[\{Co_2(tpyprz)(H_2O)_3\}\text{-}Mo_5O_{15}\{O_3P(CH_2)_3PO_3\}]\cdot H_2O$ (**23**)	4 and 2	4 × {Mo=O} 2 × {Mo=O}	4 2
$[\{Co_2(tpyprz)(H_2O)_3\}\text{-}Mo_5O_{15}\{O_3P(CH_2)_3PO_3\}].7H_2O$ (**24**)	3	3 × {Mo=O}	3
$(H_3O)_2[\{Co_3(tpyprz)_2(H_2O)_3\}(Mo_5O_{15})_2\{O_3P(CH_2)_3PO_3\}_2].21.4H_2O$ (**25**)	2	2 × {Mo=O}	2
$[\{Cu_2(tpyprz)(H_2O)_2\}\text{-}Mo_5O_{15}\{O_3P(CH_2)_4PO_3\}]\cdot H_2O$ (**26**)	4	4 × {Mo=O}	3
$[\{Ni_2(tpyprz)(H_2O)_2\}\text{-}Mo_5O_{15}\{O_3P(CH_2)_4PO_3\}]$ (**27**)	4	4 × {Mo=O}	4
$[\{Co_2(tpyprz)(H_2O)_2\}\text{-}Mo_5O_{15}\{O_3P(CH_2)_4PO_3\}]2\cdot H_2O$ (**28**)	4	4 × {Mo=O}	4
$[\{Co_2(tpyprz)(H_2O)_3\}\text{-}Mo_5O_{15}\{O_3P(CH_2)_5PO_3\}]$ (**29**)	3	3 × {Mo=O}	3
$[\{Cu_4(bpyrm)_3(H_2O)_6\}\text{-}Mo_5O_{15}(O_3PCH_2CH_2CH_2PO_3)_2]$ (**30**)	3	3 × {Mo=O}	4
$[\{Cu_2(bpyrm)(H_2O)_2\}\text{-}Mo_5O_{15}(O_3PCH_2CH_3)_2]$ (**31**)	4	2 × {Mo=O} 2 × {P=O}	8
$[\{Cu_2(bpyrm)(H_2O)_4\}\text{-}Mo_5O_{15}\{O_3P(CH_2)_3PO_3\}]$ (**32**)	4	4 × {Mo=O}	2 and 4
$[\{Ni_2(tpyrm)(H_2O)_4\}\text{-}Mo_5O_{15}\{O_3P(CH_2)_3PO_3\}]$ (**33**)	4	3 × {Mo=O}	3
$[\{Cu_2(bpyrm)(H_2O)_4\}\text{-}Mo_5O_{15}\{O_3P(CH_2)_4PO_3\}]$ (**34**)	2	2 × {Mo=O}	2
$[\{Cu_2(bpyrm)(H_2O)_4\}\text{-}Mo_5O_{15}\{O_3P(CH_2)_5PO_3\}]$ (**35**)	4	4 × {Mo=O}	2 and 4
$[\{Cu_2(bpyrm)(H_2O)_2\}\text{-}Mo_5O_{15}\{O_3P(CH_2)_6PO_3\}]$ (**36**)	4	2 × {Mo=O} 2 × {P=O}	3
$[\{Cu_2(bterpy)(H_2O)\}\text{-}Mo_5O_{15}\{O_3P(CH_2)_4PO_3\}]$ (**62**)	3	3 × {Mo=O}	2
$[\{Cu_4(4\text{-}Hpt)_3(H_2O)\}\text{-}Mo_5O_{15}\{O_3P(CH_2)_2PO_3\}]$ (**63**)	4	4 × {Mo=O}	$\{Cu_4(4\text{-}Hpt)_3(H_2O)\}_n^{4n+}$ framework

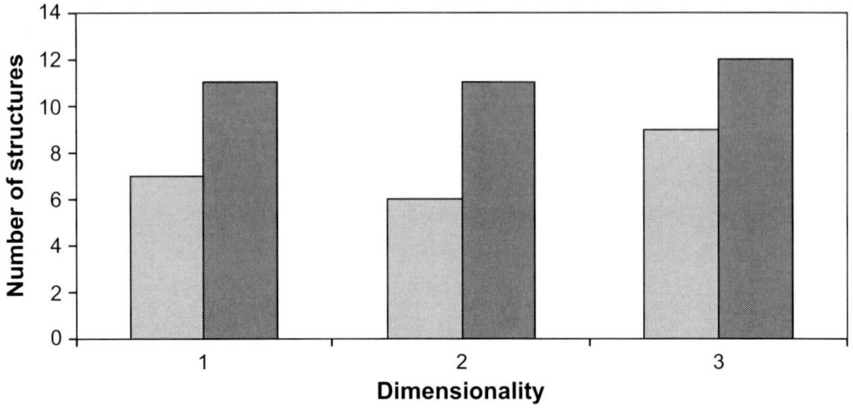

Figure 7.16 A bar graph illustrating the number of structures for a given dimensionality of members of the M^{II}-organonitrogen/oxomolybdate-organophosphonate family. The lighter shading indicates structures constructed from $[Mo_5O_{15}(O_3PR)_2]^{4-}$ building blocks, while the darker shading is for all structures.

While Co^{II} and Ni^{II} display a preference for more or less regular octahedral coordination, Cu^{II} displays Jahn–Teller distorted geometries (square planar, "4 + 1" square pyramidal, and "4 + 2" distorted six coordinate). The structural consequences are most apparent when tpyprz is the ligand to the secondary metal. A Cu^{II} cation will bind to a terminus of a single tpyprz ligand, thus favoring the $\{Cu_2(tpyprz)\}^{4+}$ cationic rod with $\{CuN_3O\}$, $\{CuN_3O_2\}$, or $\{CuN_3O_3\}$ coordination geometries adopted upon linking to the oxide chain through one, two, or three attachment points, respectively. In contrast, Co^{II} and Ni^{II} can adopt regular octahedral geometry by bonding to two tpyprz ligands to produce $\{MN_6\}$ coordination. The extreme examples of this tendency occur in $[Ni_2(tpyprz)_2][Mo_6O_{18}(H_2O)(O_3PC_6H_3)_2]$ and $[Ni_2(tpyprz)_2][Mo_7O_{21}(O_3PCH_2PO_3)]$, where the cationic substructure is present as an isolated $\{Ni(tpyprz)\}_n{}^{2n+}$ chain. More commonly, there is a balance between the idealized octahedral geometry and the oxophilicity of the M^{II} centers which can result in oligomerization which provides $\{M_2(tpyprz)(H_2O)_x\}^{4+}$, $\{M_2(tpyprz)_2\}^{4+}$, $\{M_3(tpyprz)_2\}^{6+}$, and $\{M_4(tpyprz)_3\}^{8+}$ cationic rods. One consequence of this potential oligomerization is the expansion of structural chemistry of the Co^{II} and Ni^{II} families in comparison to the Cu^{II} materials. This catenation phenomenon is observed in the structures of compounds **15**, **16**, **22**, and **25**.

As might be expected, there are few examples of structural isomorphism between members of the Cu^{II} family and those of the Co^{II} and Ni^{II} types. Thus, while the compounds $[\{M_2(tpyprz)(H_2O)_2\}(Mo_3O_8)_2(O_3PCH_2PO_3)_2]$, M = Ni, Cu, are isomorphous, no other isostructural pairs of the Cu/Ni or Cu/Co series are encountered. On the other hand, there are several parallels in the structures of the Co^{II} and Ni^{II} series. Thus, compounds **16** and **28** are isostructural with their Ni^{II} analogs, while **9** is the structural analog of the Ni^{II} counterpart with $n = 4$ for the diphosphonate tether length.

The structural consequences of the identity of the secondary metal cation are illustrated by comparing the structures of $[\{Cu_2(tpyprz)(H_2O)_2\}Mo_5O_{15}\{O_3P(CH_2)_3PO_3\}]$ (**20**), $[\{Ni_2(tpyprz)(H_2O)_3\}Mo_5O_{15}\{O_3P(CH_2)_3PO_3\}]$ (**21**), and $[\{Co_2(tpyprz)(H_2O)_3\}Mo_5O_{15}\{O_3P(CH_2)_3PO_3\}]$ (**24**). While all are 3D and constructed from the same secondary building blocks, $[Mo_5O_{15}\{O_3P(CH_2)_3PO_3\}]_n^{4n-}$ chains, and $M_2(tpyprz)(H_2O)_n\}^{4+}$ rods, the detailed connectivity patterns between the building blocks are quite distinct. It is most noteworthy that the $\{M_2(tpyprz)(H_2O)_n\}^{4+}$ secondary building units adopt different linkage modes with respect to the anionic clusters of the chains. This observation reflects in part the variable numbers of aqua ligands n which may be incorporated into the rod. Furthermore, the number of available coordination sites at the M^{II} centers for linking to the oxygen sites of the phosphomolybdate clusters will be limited by the aqua ligation. There can be considerable variability in the number of aqua ligands per cationic unit of the type $\{M_2(tpyprz)(H_2O)_n\}^{4+}$ with n taking on values 0–3. Since the aqua ligation cannot be directly controlled in the synthesis, this characteristic of the chemistry renders structural predictability rather problematic.

It is also apparent that the secondary metal subunit does not act simply as a rod connecting two cluster sites on adjacent anionic chains. In fact a survey of the structural chemistry of the M^{II}-tpyprz/oxomolybdate diphosphonate materials reveals that $\{M_2(tpyprz)(H_2O)_n\}^{4+}$ subunits may be associated with one, two, three, or four phosphomolybdate clusters. Furthermore, the cationic subunit may bridge clusters of the same chain or clusters on adjacent chains.

7.4.3 The Binucleating Organonitrogen Ligand

In view of the structural role of the $\{M_2(\text{binucleating ligand})\}^{n+}$ subunit of these structures, it is natural to focus on the dimensions of the binucleating ligand, most specifically the distance between metal sites. This constraint can be studied by expanding the distance between the ligand donor groups in the progression bipyrimidine (bpyrm), tetrapyridylpyrazine (tpyprz), bis-terpyridine (bterpy), shown in Figure 7.17.

The structural influences of the binucleating ligand can be relatively subtle, as in the series $[\{Cu_2(ligand)(H_2O)_x\}Mo_5O_{15}\{O_3P(CH_2)_4PO_3\}]$ (ligand = bpyrm, $x = 4$ (**34**); ligand = tpyprz, $x = 2$ (**26**); ligand = bterpy, $x = 1$ (**62**), shown in Figure 7.18. While the structure of **34** projects a typical grid structure with large pores, the structure contains three interpenetrating frameworks to produce a dense framework material. In **34**, each phosphomolybdate chain is linked to four adjacent chains, while in **26** and **62** each chain is linked to six adjacent chains. Since the cationic rods do not project to generate rectangular grids in profile but rather "accordion" to form acute rhombi (Figure 7.19), the large spatial extension of the bterpy relative to bpyrm and tpyprz is not exploited to generate more open frameworks.

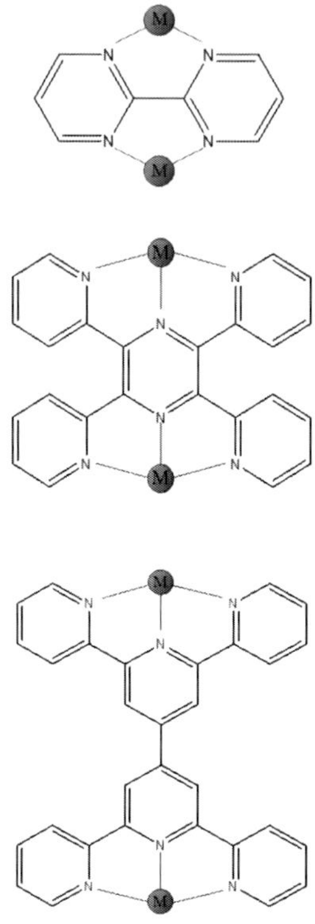

Figure 7.17 The binucleating ligands of this study.

The structural effects are more pronounced in the pentyldiphosphonate series [{Cu_2(ligand)(H_2O)$_x$}Mo_5O_{15}{$O_3P(CH)_5PO_3$}] where the bpyrm structure (**35**) is 3D, as is the tpyprz structure which is isomorphous with **62**, while the bterpy structure is 1D, adopting a most unusual double chain structure (Figure 7.20).

If more complex ligands are used for the secondary metal–ligand complex subunit, more unusual structures are encountered. For example, when 4-pyridyltetrazole (4-pt) is the ligand to the secondary metal site, the 3D [{Cu_4(4-Hpt)$_3$(H_2O)}Mo_5O_{15}{$O_3P(CH_2)_2PO_3$}] (**63**) is isolated. The structure, shown in Figure 7.21, consists of the common $\{Mo_5O_{15}(O_3PR)_2\}_n^{4n-}$ chains embedded in a $\{Cu_4(4\text{-Hpt})_3(H_2O)\}_n^{4n+}$ framework. The pyridyl group is protonated and the copper has been reduced to Cu(I) in this unusual material.

7.5 Other Phosphomolybdate Core Structures

While the pentamolybdate cluster $\{Mo_5O_{15}(O_3PR)_2\}^{4-}$ is the prototypical building unit for structures of the Mo_xO_y/organodiphosphonate/M^{II} ligand family, factors such as reaction conditions, the identity of the diphosphonate ligand, and the introduction of fluoride anions can result in a variety of phosphomolybdate substructures. These will be surveyed briefly.

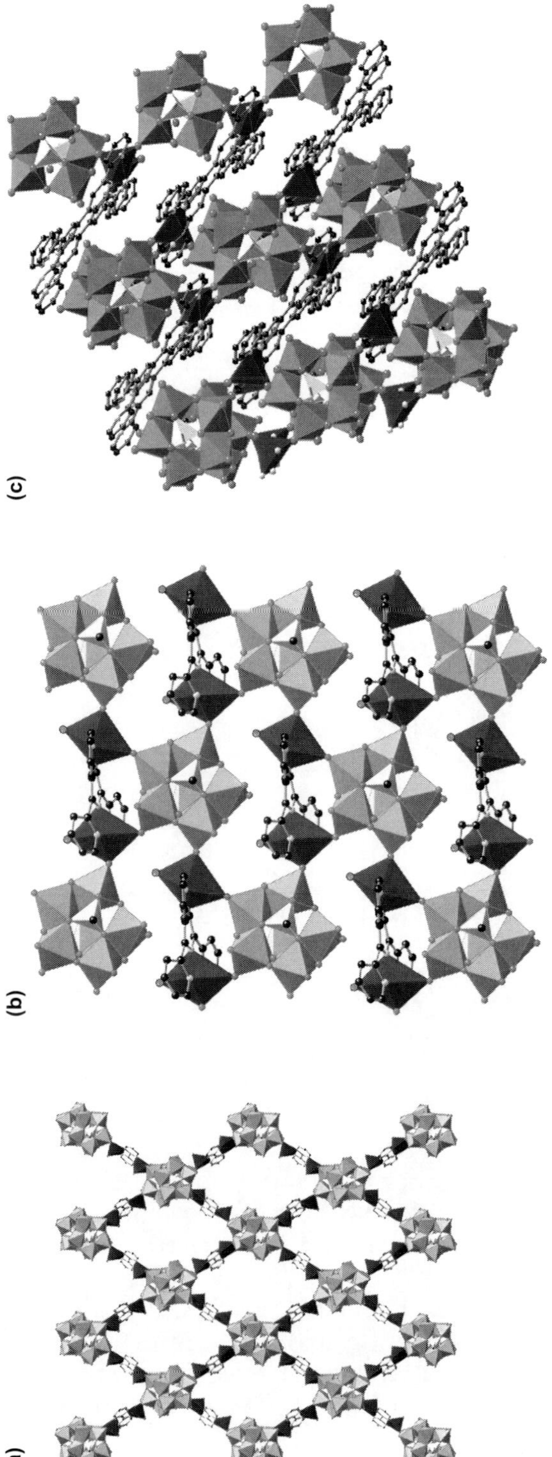

Figure 7.18 The structure of (a) [{Cu_2(bpyrm)$(H_2O)_4$}Mo_5O_{15}{$O_3P(CH_2)_4PO_3$}] (**34**), (b) [{Cu_2(tpyprz)$(H_2O)_2$}Mo_5O_{15}{$O_3P(CH_2)_4PO_3$}] (**26**), and (c) [{Cu_2(bterpy)(H_2O)} Mo_5O_{15}{$O_3P(CH_2)_4PO_3$}] (**62**).

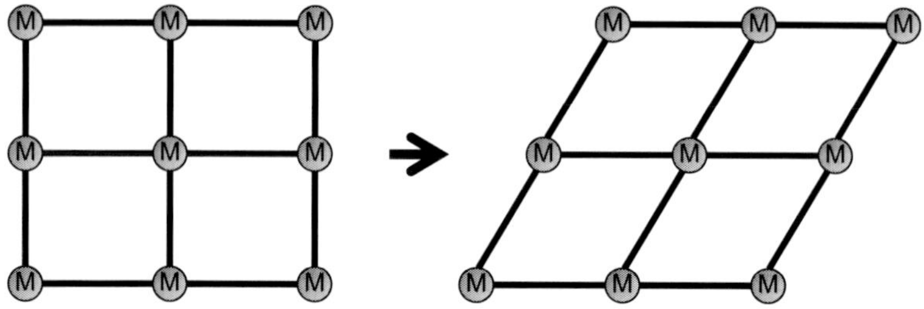

Figure 7.19 A schematic representation of the "accordion" effect.

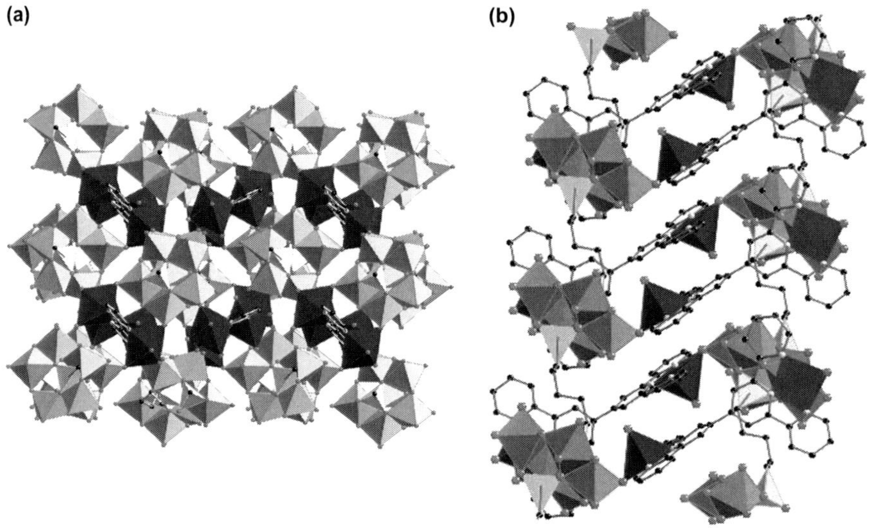

Figure 7.20 The structures of (a) $[\{Cu_2(bpyrm)(H_2O)_4\}Mo_5O_{15}\{O_3P(CH_2)_5PO_3\}]$ (**35**) and (b) $[\{Cu_2(bterpy)(H_2O)\}Mo_5O_{15}\{O_3P(CH_2)_5PO_3\}]$.

7.5.1 The $\{Mo_3O_8\}^{2+}$ Core

The influence of the diphosphonate tether length is revealed most clearly in the structures incorporating methylenediphosphonate. The donor groups of this ligand are juxtaposed to favor the formation of six-membered {M–O–P–C–P–O–M} chelate rings rather than to provide spatial expansion in one dimension through cluster bridging. The structural consequences are apparent in the 1D phase $[\{Cu_2(tpyprz)(H_2O)\}Mo_3O_8(HO_3PCH_2PO_3)_2]$ (**42**) of Figure 7.22. The folded chain is constructed from $\{(Mo_3O_8)(HO_3PCH_2PO_3)_2\}^{4-}$ clusters linked through $\{Cu_2(tpyprz)(H_2O)\}^{4+}$ rods. This is a persistent subunit which also appears in the structures of the copper phases $[\{Cu_2(tpyprz)(H_2O)\}_2(Mo_3O_8)_2(O_3PCH_2PO_3)_3]$ (**43**) and $[\{Cu_2(tpyprz)(H_2O)_2\}(Mo_3O_8)_2(O_3PCH_2PO_3)_2]$ (**45**) which are 1D and 3D, respectively (Figure 7.23). The Ni^{II} species $[\{Ni_2(tpyprz)(H_2O)_2\}(Mo_3O_8)_2(O_3PCH_2PO_3)_2]$ (**46**) is isomorphous with the Cu^{II} analog.

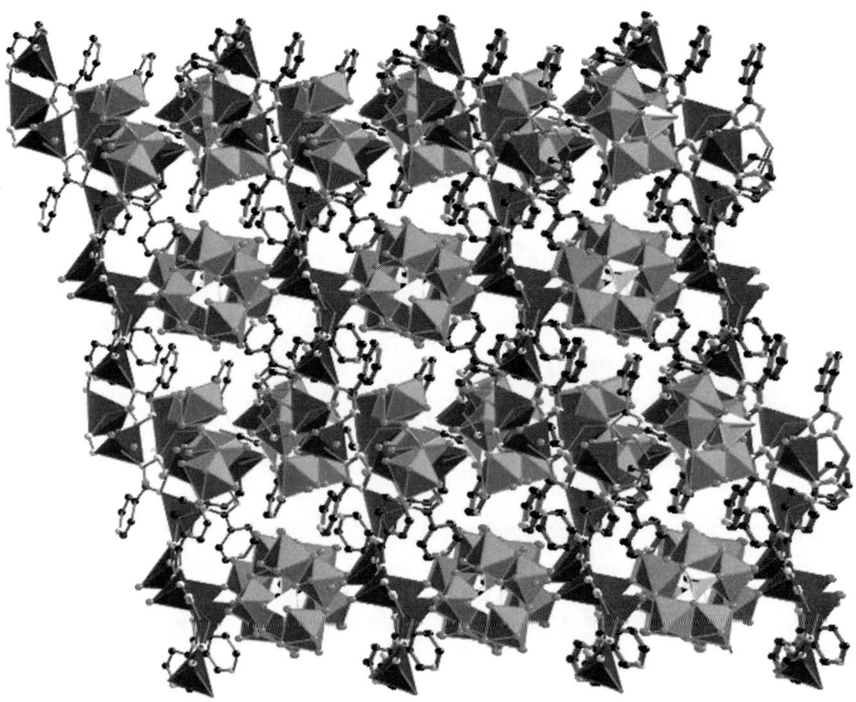

Figure 7.21 A view of the 3D structure of $[\{Cu_4(4\text{-}Hpt)_3(H_2O)\}\text{-}Mo_5O_{15}\{O_3P(CH_2)_2PO_3\}]$ (**63**). The $[Mo_5O_{15}\{O_3P(CH_2)_2PO_3\}]_n^{4n-}$ chains propagate normal to the plane of the page. The $\{Cu_4(4\text{-}Hpt)_3\}_n^{4n+}$ substructure is 3D.

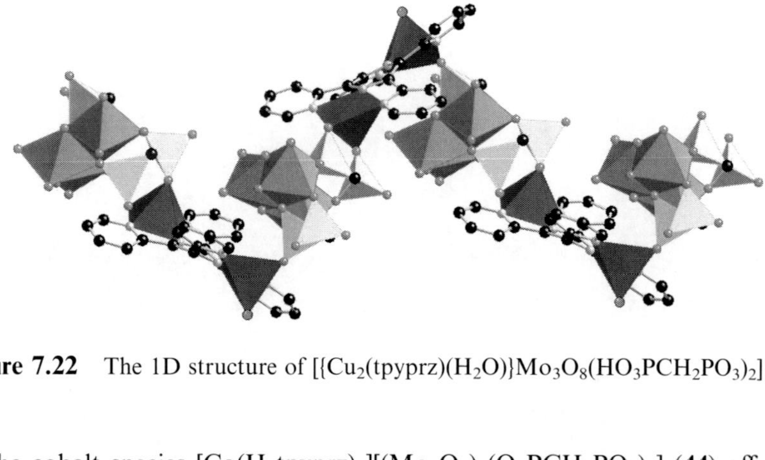

Figure 7.22 The 1D structure of $[\{Cu_2(tpyprz)(H_2O)\}Mo_3O_8(HO_3PCH_2PO_3)_2]$ (**42**).

The cobalt species $[Co(H_2tpyprz)_2][(Mo_3O_8)_3(O_3PCH_2PO_3)_3]$ (**44**) offers an unusual example of "naked" $[(Mo_3O_8)_3(O_3PCH_2PO_3)_3]_n^{6n-}$ chains constructed from $\{Mo_3O_8\}^{2+}$ triads linked through the diphosphonate ligands (Figure 7.24).

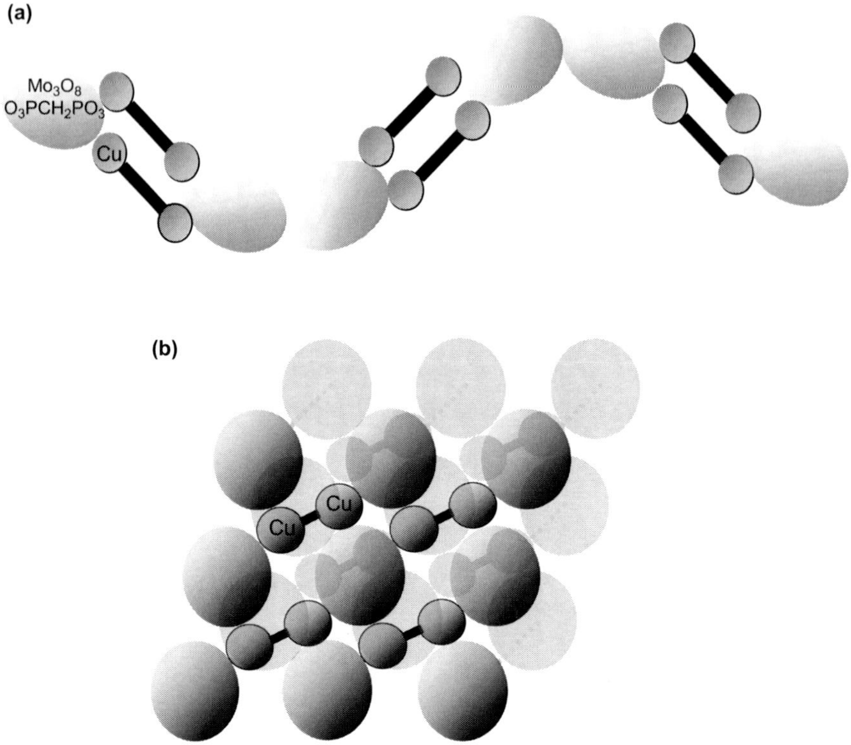

Figure 7.23 Schematic representations of the structures of (a) $[\{Cu_2(tpyprz)(H_2O)\}_2(Mo_3O_8)_2(O_3PCH_2PO_3)_3]$ (**43**) and (b) $[\{Cu_2(tpyprz)(H_2O)_2\}(Mo_3O_8)_2(O_3PCH_2PO_3)_2]$ (**45**). Each oval represents an $\{Mo_3O_8(O_3PR)_2\}^{2-}$ unit. In (a) the $\{(Mo_3O_8)_2(O_3PCH_2PO_3)_2\}^{4-}$ unit is in the plane of the page and shown as two touching ovals. In (b) the $\{(Mo_3O_8)_2(O_3PCH_2PO_3)_2\}^{4-}$ clusters propagate normal to the page and the $\{Mo_3O_8(O_3PR)_2\}^{2-}$ unit is shown linked through a dashed line representing the $-CH_2$ tether of the diphosphonate ligand.

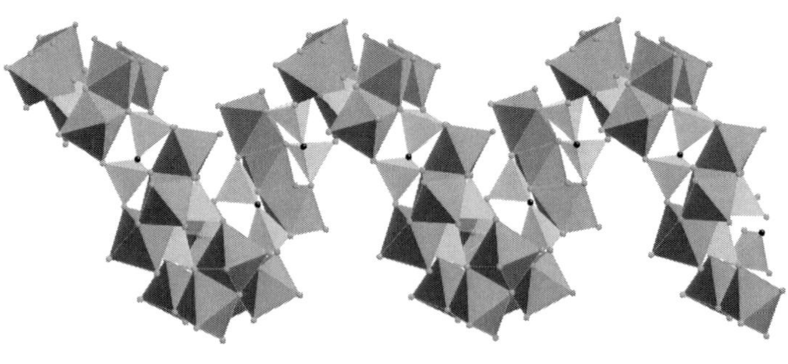

Figure 7.24 Polyhedral representation of the $\{(Mo_3O_8)_3(O_3PCH_2PO_3)_3\}_n^{6n-}$ chain of $[Co(H_2tpyprz)_2][(Mo_3O_8)_3(O_3PCH_2PO_3)_3]$ (**44**).

Table 7.4 Summary of selected structural features of materials with $\{MoO_2\}$, $\{Mo_2O_5\}$, $\{Mo_3O_8\}$, $\{Mo_4O_{10-12}\}$ and $\{H_xMo_6O_{18}\}$ building blocks

Compound	*Overall dimensionality*	*M^{II} attachment points per building block*	*Attachment type*	*Substructures*
$[Cu(bpy)MoO_2(H_2O)(O_3PCH_2PO_3)]$ (**37**)	1-D	3	$2\times P{=}O$ $Mo{=}O$	$\{MoO_2(H_2O)(O_3PCH_2PO_3)\}_n{}^{2n-}$ chain
$[Cu(phen)(Mo_2O_5)(H_2O)(O_3PCH_2PO_3)]$ (**38**)	1-D	3	$2\times P{=}O$ $Mo{=}O$	$\{Mo_2O_5(O_3PCH_2PO_3)\}_n{}^{2n-}$ chain
$[Cu(bpy)(Mo_2O_5)(H_2O)(O_3PCH_2PO_3)]$ (**39**)	2-D	3	$2\times P{=}O$ $Mo{=}O$	$\{Mo_2O_5(H_2O)(O_3PCH_2PO_3)\}_n{}^{2n-}$ network
$[Cu(bpy)(Mo_2O_5)\{O_3P(CH_2)_3PO_3\}]$ (**40**)	2-D	4	$2\times P{=}O$ $2\times Mo{=}O$	$\{Mo_2O_5(O_3PCH_2CH_2CH_2PO_3)\}_n{}^{2n-}$ network
$[\{Cu(terpy)(Mo_2O_5)(O_3PCH_2PO_3)]$ (**41**)	2-D	1	$P{=}O$	$\{Mo_2O_5(O_3PCH_2PO_3)\}_n{}^{4n-}$ chain
$[\{Cu_2(tpyprz)(H_2O)\}Mo_3O_8(HO_3PCH_2PO_3)_2]$ (**42**)	1-D	3	$3\times P{=}O$	$\{Mo_3O_8(HO_3PCH_2PO_3)_2\}^{4-}$ clusters
$[\{Cu_2(tpyprz)(H_2O)\}_2(Mo_3O_8)_2(O_3PCH_2PO_3)_3]$ (**43**)	1-D	4	$4\times P{=}O$	$\{(Mo_3O_8)_2(O_3PCH_2PO_3)_2\}^{8-}$ clusters
$[Co(H_2tpyprz)_2][(Mo_3O_8)_3(O_3PCH_2PO_3)_3]$ (**44**)	1-D	–	–	$\{Mo_3O_8(O_3PCH_2PO_3)\}_n{}^{2n-}$ spiral chains
$[\{M_2(tpyprz)(H_2O)_2\}(Mo_3O_8)_2(O_3PCH_2PO_3)_2]$ M=Cu, Ni (**45**, **46**)	3-D	2	$2\times Mo{=}O$	$\{Mo_3O_8(O_3PCH_2PO_3)\}_n{}^{2n-}$ chains
$[\{Cu(bpy)\}_2(Mo_4O_{12})(H_2O)_2(O_3PCH_2PO_3)]$ (**47**)	2-D	4	$2\times P{=}O$ $2\times Mo{=}O$	$\{Mo_4O_{12}(H_2O)_2(O_3PCH_2CH_2PO_3)\}_n{}^{4n-}$ network
$[\{Cu(phen)\}_2(Mo_4O_{12})(H_2O)_2(O_3PCH_2CH_2PO_3)]$ (**48**)	2-D	3	$2\times P{=}O$ $Mo{=}O$	$\{Mo_4O_{12}(O_3PCH_2CH_2PO_3)\}_n{}^{4n-}$ chain
$[\{Cu_2(bpyrm)\}(MoO_2)(HO_3PCH_2PO_3)_2]$ (**49**)	3-D	4	$P{=}O$	$\{Cu_2(bpyrm)(HO_3PCH_2CH_2PO_3)_2\}_n{}^{2n-}$ chain
$[\{Cu(bpy)\}_2(Mo_4O_{10})(O_3PCH_2C_6H_4CH_2PO_3)_2]$ (**50**) $\{Mo_4O_{10}(O_3PR)_4\}_n{}^{4n-}$ 3D framework	3-D	6	$4\times P{=}O$	$2\times Mo{=}O$
$[\{Cu(terpy)\}_2(Mo_6O_{17})(H_2O)(O_3PCH_2NH_2CH_2PO_3)_2]$ (**51**)	3-D	3	$3\times P{=}O$ $Mo{=}O$	$\{Mo_6O_{17}(H_2O)(O_3PCH_2NHCH_2PO_3)_2\}^{4-}$ cluster
$[\{Co_2(tpyprz)(H_2O)_2\}(Mo_6O_{18})\{O_3P(CH_2)_5PO_3\}]$ (**52**)	3-D	2	$Mo{=}O$	$\{Mo_6O_{18}(O_3P(CH_2)_5PO_3)\}_n{}^{4n-}$ chain

7.5.2 Infrequently Encountered Molybdate Cores (Table 7.4)

The structural variability of the chemistry is revealed by a number of molybdate building blocks which appear unexpectedly from hydrothermal reaction mixtures. While these structures generally include diphosphonate ligands with

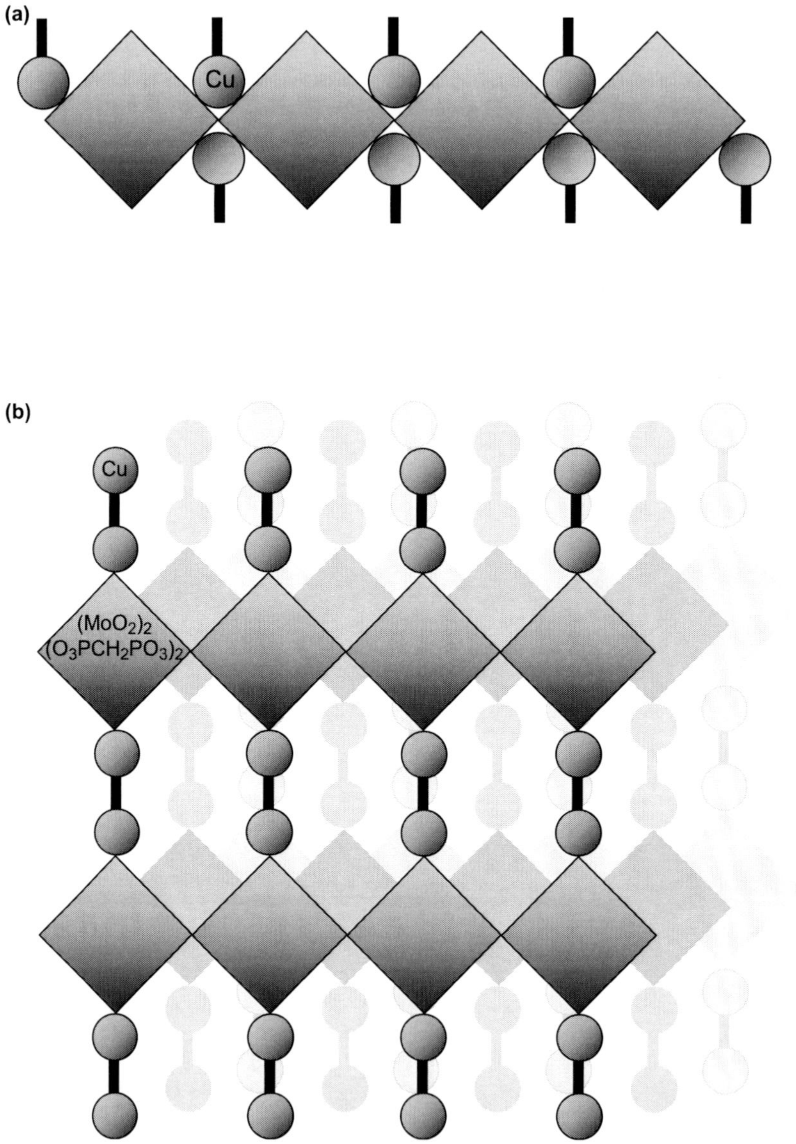

Figure 7.25 Schematic representations of the structures of (a) [Cu(bpy)MoO_2(H_2O)($O_3PCH_2PO_3$)] (**37**) and (b) [Cu_2(bpyrm)MoO_2($HO_3PCH_2PO_3$)$_2$] (**49**). The corner-sharing squares represent the molybdophosphonate chains.

unusual constraints, for example, methylenediphosphonate or bis-*N*,*N*′-(methylphosphonate)amine, several examples exhibit simple alkyldiphosphonates $\{O_3P(CH_2)_nPO_3\}^{4-}$ with $n \geq 2$.

The simplest molybdate building block is the mononuclear $\{MoO_2\}^{2+}$ unit found in the 1D $[Cu(bpy)MoO_2(H_2O)(O_3PCH_2PO_3)]$ (**37**) and the 3D $[Cu_2(bpyrm)MoO_2(HO_3PCH_2PO_3)_2]$ (**49**), shown schematically in Figure 7.25.

The second recurring motif is the binuclear $\{Mo_2O_5\}^{2+}$ unit which occurs in the 1D $[Cu(phen)(Mo_2O_5)(H_2O)(O_3PCH_2PO_3)]$ (**38**) and the 2D $[Cu(ligand)(Mo_2O_5)(O_3PCH_2PO_3)_2]$ (**39, 41**) (ligand = bpy and terpy, respectively) and $[Cu(bpy)(Mo_2O_5)\{O_3P(CH_2)_3PO_3\}]$ (**40**), shown schematically in Figure 7.26.

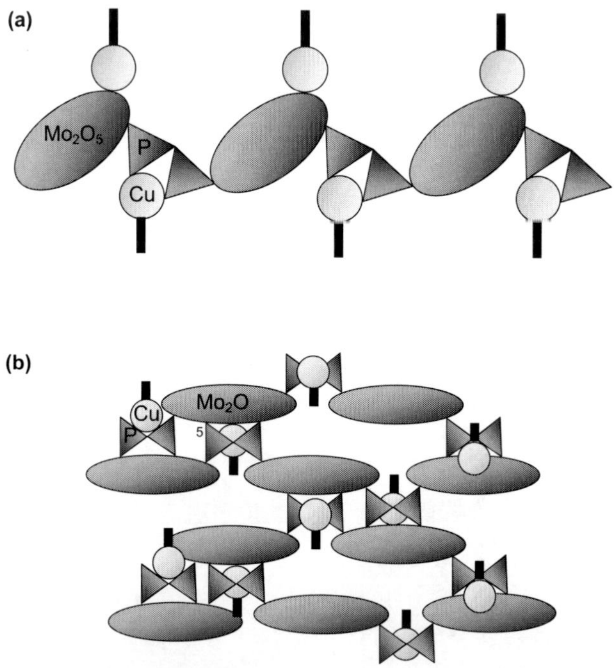

Figure 7.26 Schematic representations of the structures of (a) $[Cu(phen)(Mo_2O_5)(H_2O)(O_3PCH_2PO_3)]$ (**38**), (b) $[Cu(bpy)(Mo_2O_5)(H_2O)(O_3PCH_2PO_3)]$ (**39**), (c) $[Cu(terpy)(Mo_2O_5)(O_3PCH_2PO_3)]$ (**41**) and (d) $[Cu(bpy)(Mo_2O_5)\{O_3P(CH_2)_3PO_3\}]$ (**40**). For (a), the $\{Mo_2O_5\}^{2+}$ units are illustrated as ovals, the $\{O_3PCH_2PO_3\}^{4-}$ ligands as bow ties, and the $\{Cu(bpy)\}^{2+}$ fragments as circles with a protruding dark line; in (b) the same schematic representations are used to show the very different connectivity pattern adopted; in (c) the fused circles suggest the bridging of Cu^{II} sites through phosphonate oxygens; and in (d), the ovals are again the $\{Mo_2O_5\}^{2+}$ clusters, while one terminus of the $\{O_3PCH_2PO_3\}^{4-}$ ligand appears as dark circles on the surface layer connected by a dashed line representing the $-CH_2$ bridge linked to a light triangle representing the second $-PO_3$ terminus and the circles with the dark connectors represent the $\{Cu(bpy)\}^{2+}$ units.

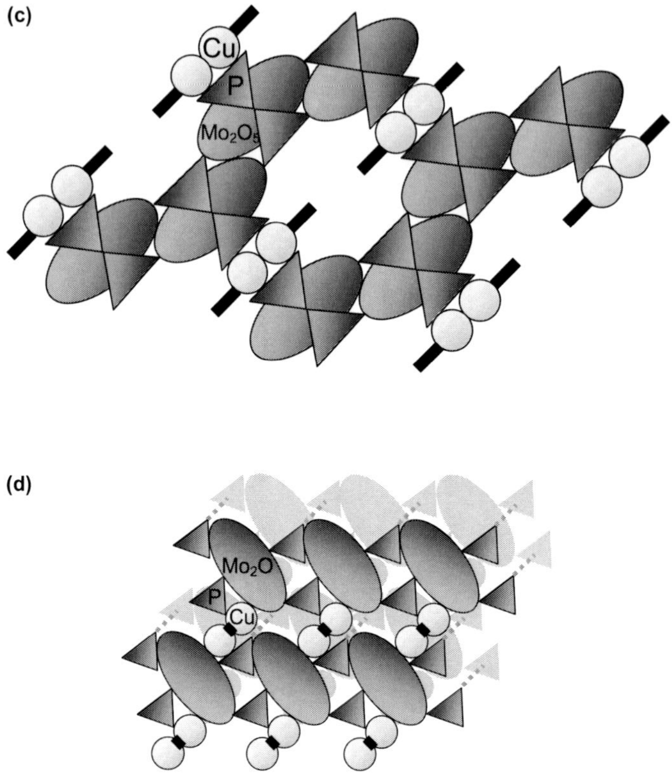

Figure 7.26 Continued

The $\{Mo_4O_x(H_2O)_y\}^n$ core ($x = 10$, $y = 0$, $n = +4$; $x = 12$ $y = 2$, $n = 0$) is encountered in the 2D $[\{Cu(bpy)\}_2(Mo_4O_{12})(H_2O)_2(O_3PCH_2PO_3)]$ (**47**) and $[\{Cu(phen)\}_2(Mo_4O_{12})(H_2O)_2(O_3PCH_2CH_2PO_3)]$ (**48**) and the 3D $[\{Cu(bpy)\}_2(Mo_4O_{10})(O_3PCH_2C_6H_4CH_2PO_3)_2]$ (**50**) (Figure 7.27).

Finally, the hexanuclear cores $\{Mo_6O_{18}\}$ and $\{Mo_6O_{17}(H_2O)\}^{2+}$ are found in the 3D structures of $[\{Co_2(tpyprz)(H_2O)_2\}(Mo_6O_{18})\{O_3P(CH_2)_5PO_3\}]$ (**52**) and $[\{Cu(terpy)\}_2(Mo_6O_{17})(H_2O)(O_3PCH_2NH_2CH_2PO_3)_2]$ (**51**). While the cobalt structure is constructed from the common phosphomolybdate chains, the coordination requirements of the amino diphosphonate ligand in **51** result in phosphomolybdate clusters linked through the $\{Cu(terpy)\}^{2+}$ units to form the overall framework structure (Figure 7.28).

7.5.3 Molybdenum Oxy Fluoride Cores, $\{Mo_xO_yF_z\}^{n-}$

The hydrothermal reaction mixtures used to prepare the materials of this study often contain HF as a solubilizing agent and to promote suitable crystal growth. In general, fluoride is not incorporated into the final product.

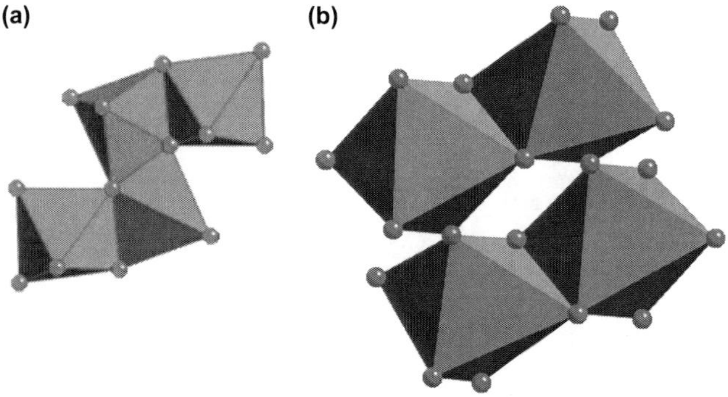

Figure 7.27 (a) The $\{Mo_4O_{12}(H_2O)_2\}$ cluster of $[\{Cu(bpy)\}_2(Mo_4O_{12})(H_2O)_2(O_3PCH_2PO_3)]$ (**47**) and (b) the $\{Mo_4O_{10}\}^{4+}$ cluster of $[\{Cu(bpy)\}_2(Mo_4O_{10})(O_3PCH_2C_6H_4CH_2PO_3)_2]$ (**50**).

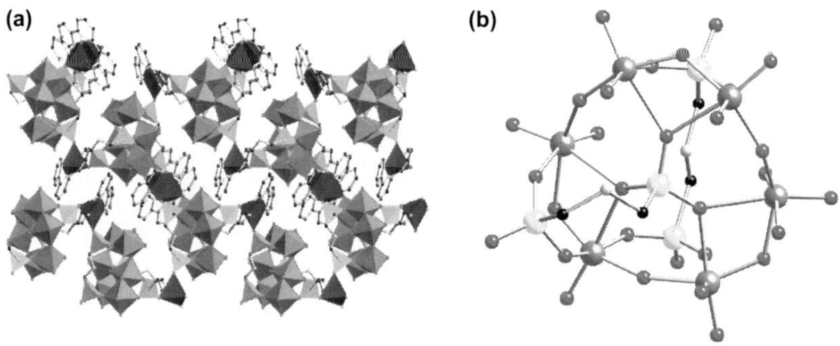

Figure 7.28 (a) The structure of $[\{Cu(terpy)\}_2(Mo_6O_{17})(H_2O)\{O_3PCH_2NH_2CH_2PO_3\}_2]$ (**51**). (b) The $\{Mo_6O_{17}(H_2O)(O_3PCH_2NH_2CH_2PO_3)_2\}^{4-}$ cluster of **51**.

However, if the HF concentration is increased such that the F:Mo stoichiometry is greater than 5:1, fluoride incorporation is often effected. The structural chemistry of these phases appears to be quite complex and a structural systematic has yet to be manifested (see Table 7.5). 1D, 2D, and 3D structures are observed (Figure 7.29), with binuclear and tetranuclear oxofluoromolybdate cores. However, the compositions of these cores span a considerable range, for example, $\{Mo_2O_4F_2\}^{2+}$, $\{Mo_2O_4F_4\}^0$, and $\{Mo_2O_4F_6\}^{2-}$ for the bimolybdate cores and $\{Mo_4O_xF_y\}^n$ with x in the 8–12 range and y in the 4–8 range for the tetranuclear cores (Figure 7.30). It is noteworthy that the structural data set for these oxofluoromolybdates is quite limited, suggesting that this is an area ripe for development. Onc of the primary conclusions to be drawn is that huge numbers of representatives of this class with unusual, unprecedented and inconceivable structures remain to be discovered.

Table 7.5 Summary of selected structural characteristics of materials of the $(Mo_xO_yF_z)$/organodiphosphonate/M^{II}-ligand type.

Compound	*Overall dimensionality*	*M^{II} attachments per cluster*	*Attachment type*	*Clusters or chains linked per cationic unit*	*Substructures*
$[\{Cu_2(bterpy)\}(Mo_2F_4O_4)\{O_3P(CH_2)_5PO_3\}]$ (**53**)	1D	4	Mo– F P=O	2	$\{Mo_2F_4O_4(O_3P(CH_2)_5PO_3)\}^{4-}$ cluster
$[\{Cu_2(bterpy)\}(Mo_2F_6O_4)\{HO_3P(CH_2)_4PO_3H\}]$ (**54**)	1D	4	Mo– F P=O	2	$\{Mo_2F_6O_4(HO_3P(CH_2)_4PO_3H)\}^{4-}$ cluster
$[\{Cu_2(bterpy)\}(Mo_4F_4O_{10})\{O_3P(CH_2)_6PO_3\}]$ (**55**)	1D	4	Mo=O P=O	2	$\{Mo_4F_4O_{10}(O_3P(CH_2)_6PO_3\}^{4-}$ cluster
$[\{Cu_2(bterpy)\}(Mo_4F_8O_8)\{O_3P(CH_2)_9PO_3\}]$ (**56**)	1D	2	P=O	2	$\{Mo_4F_8O_8(O_3P(CH_2)_9PO_3\}^{4-}$ cluster
$[\{Ni_3(tpyprz)_2(H_2O)_2\}(Mo_5O_{15})(Mo_2F_2O_4)\{O_3P(CH_2)_3PO_3\}_2]$ (**57**)	2D	2	Mo=O P=O	2	Chains of alternating $\{Mo_2F_2O_4(O_3PR)_2\}^{2-}$ and $\{Mo_5O_{15}(O_3PR)_2\}^{4-}$ clusters
$[\{Cu_2(bterpy)(OH)\}(Mo_2F_3O_4)(O_3PCH_2CH_2PO_3)]$ (**58**)	2D	3	Mo=O 2 × P=O	4	$\{Mo_2F_3O_4(O_3PCH_2PO_3)\}^{3-}$ clusters
$[\{Cu_4(phenbterpy)_2(H_2O)_2\{HO_3P(CH_2)_4PO_3H\}\}(Mo_4FO_{12})_2\{O_3P(CH_2)_4PO_3\}]$ (**59**)	2D	4	P=O	3	$\{(Mo_4FO_{12})_2(O_3P(CH_2)_4PO_3\}^{6-}$ clusters
$[\{Cu_2(bterpy)(H_2O)_2\}Mo_4F_4O_{10}(O_3PCH_2CH_2PO_3)]$ (**60**)	2D	–	2 × P=O	2	$\{Mo_4F_4O_{10}(O_3P(CH_2)_2PO_3)\}_n{}^{4n-}$ chains
$[\{Cu_2(bterpy)\}(Mo_4F_6O_9)(O_3P(CH_2)_3PO_3)]$ (**61**)	3D	4	4 × P=O	4	$\{Mo_4F_6O_9(O_3P(CH_2)_3PO_3)\}_n{}^{4n-}$ chains $\{Cu_2(bterpy)\}_n{}^{4n+}$ chains

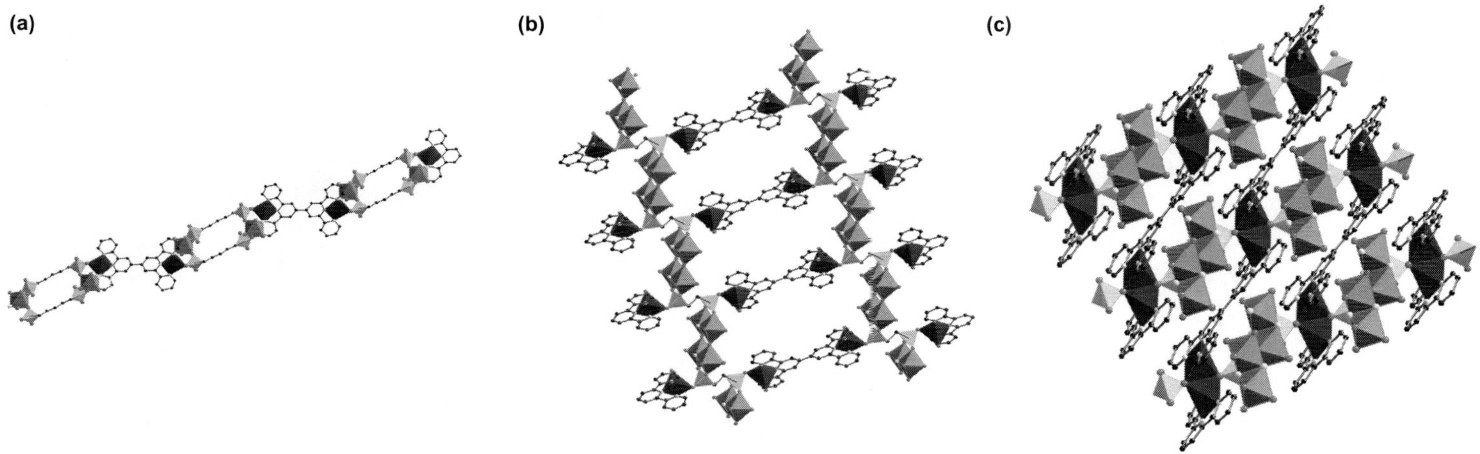

Figure 7.29 The structures of (a) [{Cu_2(bterpy)}($Mo_2F_4O_4$){$O_3P(CH_2)_5PO_3$}] (**53**), (b) [{Cu_2(bterpy)($H_2O)_2$}($Mo_4F_4O_{10}$)($O_3PCH_2CH_2PO_3$)] (**60**), and (c) [{Cu_2(bterpy)}($Mo_4F_6O_9$){$O_3P(CH_2)_3PO_3$}] (**61**).

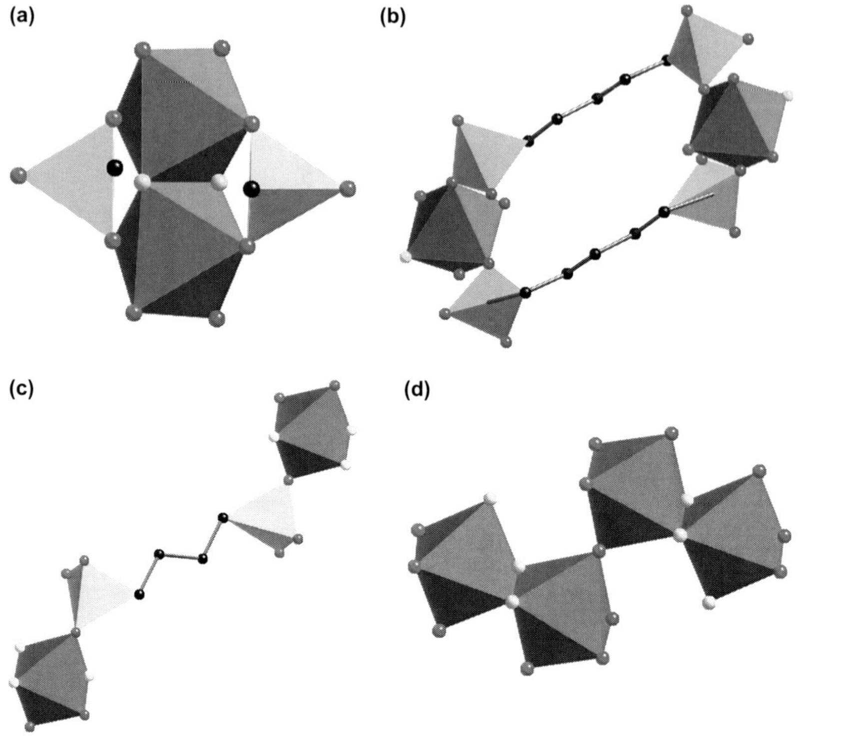

Figure 7.30 (a) The $\{Mo_2O_4F_2\}^{2+}$ unit of **57**; (b) the $[(Mo_2F_4O_4)\{O_3P(CH_2)_5PO_3\}_2]^{8-}$ unit of **53**; (c) the $[(Mo_2F_6O_4)\{HO_3P(CH_2)_4PO_3H\}]^{4-}$ unit of **54**; and (d) the $\{Mo_4F_6O_9\}^0$ cluster of **61**.

7.6 Conclusions

The structural chemistries of three series of materials of the Mo_xO_y/diphosphonate/M^{II}-tpyprz type have now been extensively investigated, namely, those with $M^{II} = Co^{II}$, Ni^{II}, and Cu^{II}. The identity of the secondary metal M^{II} is clearly a significant structural determinant. This is most evident in comparing the structures of the Cu^{II} series to those of the Co^{II} and Ni^{II} families. In the case of the former materials, the Jahn–Teller distorted geometries associated with the Cu^{II} site provide a range of coordination geometries in association with the tpyprz ligand and the phosphomolybdate clusters: $\{CuN_3O\}$, $\{CuN_3O_2\}$, and $\{CuN_3O_3\}$. In contrast, for the Co^{II} and Ni^{II} series, more regular octahedral geometries prevail, and the $\{MN_3O_3\}$ coordination geometry is ubiquitous.

As expected from these arguments, structural isomorphism between the Cu^{II} series and the two series with M = Co, Ni is rarely encountered. In fact, there is a single case: $[\{M_2(tpyprz)(H_2O)\}(Mo_3O_8)(HO_3PCH_2PO_3)_2]$ (M = Ni, Cu). However, since Co^{II} and Ni^{II} both favor six-coordinate geometries, we might naively expect that materials with the same diphosphonate component would adopt identical structures. In fact, there is a single instance of isomorphism for the Co^{II} and Ni^{II} series, $[\{M_4(tpyprz)_3\}(Mo_5O_{15})_2(O_3PCH_2CH_2PO_3)_2]\cdot xH_2O$.

The relevant observation is the variability of the numbers and locations on the phosphomolybdate clusters of the secondary metal-tpyprz rods that link clusters. The variability of attachment modes of $\{(M^{II})_2(tpyprz)\}^{4+}$ units to peripheral cluster oxygen donors reflects the relatively uniform charge distribution of the cluster surfaces. The attachment of M^{II}-tpyprz groups presumably reflects charge density matching,[86] steric, and packing considerations.

It is noteworthy that aqua coordination to the $\{MN_3O_6\}$ sites also has profound structural consequences. Thus, the M^{II} site for Co and Ni upon coordinating to a single tpyprz ligand may complete its coordination geometry by bonding to a second tpyprz group to give catenation through $\{MN_6\}$ units or to provide $\{M(tpyprz)_2\}^{2+}$ cations, by attaching to three oxygen donors from phosphomolybdate clusters, or by bonding to some combination of phosphomolybdate donors and aqua ligands. Focusing on structures containing the pentamolybdate core, $[\{M_2(tpyprz)(H_2O)_x\}\text{-}Mo_5O_{15}\{O_3P(CH_2)_nPO_3\}]$, three of six Ni^{II} structures exhibit aqua coordination and five of nine Co^{II} species incorporate aqua ligands. Such variable coordination modes render predictability of structure an elusive goal. While some structural systematics are beginning to emerge, the exquisite synthetic control of organic chemistry has yet to be realized.

Finally, it is noteworthy that while hydrothermal syntheses of aluminosilicates proceed under kinetic control, thermodynamics appear to play a dominant role in the formation of metal organic hybrid scaffolds.[87] For example, there is no apparent change in compositions as a function of reaction times. Similarly, studies of cobalt succinates have shown the importance of thermodynamic control.[88,89] On the other hand, highly porous materials prepared at low temperature, such as $[Zn_4O(O_2CC_6H_4CO_2)_3]$,[90] form under kinetic control, suggesting that this question remains open and in need of further study.

Acknowledgements

This work has been supported by grants from the National Science Foundation (Grants CHE-0242153, CHE-0604527, and CHE-0907787).

References

1. M. Jansen, *Nachrichten Chem.*, 2007, **55**, 622.
2. L. A. D. Carlos, R. A. Sá Ferreira, V. de Zea Bermudez, *Hybrid Materials*, ed. G. Kickelbick, Wiley-VCH Verlag GmbH & Co. KGaA, Weinheim, Germany, 2007, p. 337.

3. P. Rabu and M. Drillon, *Adv. Eng. Mater.*, 2003, **5**, 189.
4. G. H. Salley, O. S. Wenger, K. W. Krämer and H. U. Gudel, *Curr. Opin. Solid State & Mater. Sci.*, 2002, **6**, 487.
5. C. T. Kresge, S. S. Dhingra, Kirk-Othmer Separation Technology, 2nd edn, 2008, 2, 732.
6. S. M. Auerbach, K. A. Carrado, P. K. Dutta, *Handbook of Zeolite Science and Technology*, Marcel Dekker, Inc., New York, 2003.
7. *Metal Oxides: Chemistry and Applications*, ed. J. L. G. Fierro, CRC Press, Boca Raton, FL, 2005.
8. D. W. Bruce and D. O'Hare (eds.), *Inorganic Materials*, Wiley, Chichester, 1992.
9. A. K. Cheetham, *Science*, 1994, **264**, 794.
10. W. Büchner, R. Schliebs, G. Winter and K. H. Büchel, *Industrial Inorganic Chemistry*, VCH, New York, 1989.
11. W. H. McCarroll, *Encyclopedia of Inorganic Chemistry*, ed. R. B. King, John Wiley and Sons, New York, 1994, Vol. 6, p. 2903.
12. G. H. Haertling, *Ceramic Materials for Electronics*, 2nd edn., ed. R. C. Buchanan, Marcel Dekker, New York, 1991, p. 129.
13. H. W. Leverenz, *Luminescence of Solids*, Wiley, New York, 1980.
14. A. Clearfield, *Chem. Rev.*, 1988, **88**, 125.
15. J. M. Newsam, *Zeolites, Solid State Chemistry: Compounds*, ed. A. K. Cheetham and P. Day, Clarendon Press, Oxford, 1992, p. 234.
16. A. F. Wells, *Structural Inorganic Chemistry*, 6th edn, Oxford University Press, New York, 1987.
17. N. N. Greenwood and A. Earnshaw, *Chemistry of the Elements*, 2nd edn, Butterworth-Heinemann, Oxford, England, 1997.
18. J. R. Smyth, S. D. Jacobsen and R. M. Hazen, *Rev. Mineral. Geochem.*, 2001, **41**, 157.
19. B. Mason, *Principles of Geochemistry*, 3rd edn, Wiley, New York, 1966.
20. G. Sen, *Earth's Materials: Minerals and Rocks*, Prentice Hall, Upper Saddle River, NJ, 2001.
21. L. L. Hench, *Materials Chemistry, an Emerging Discipline*, ed. L. V. Interrante, L. A. Casper, A. B. Ellis, ACS Series 245, Chapter 21, 1995, p. 523.
22. *Handbook of Biomineralization: Biomimetic and Bioinspired Chemistry*, ed. S. Mann, Wiley-VCH Verlag GmbH & Co. KGaA, Weinheim, Germany, 2007.
23. H. Maekawa, *Materia*, 2006, **45**, 359.
24. S. L. Suib, *Ann. Rev. Mater. Sci.*, 1996, **26**, 135.
25. M. E. Davis, *CATTECH*, 1997, **1**, 19.
26. R. F. Lobo, S. I. Zones and M. E. Davis, *J. Inclusion Phenom. Mol. Recog. Chem.*, 1995, **21**, 47.
27. H. Imai, *Topics Curr. Chem.*, 2007, 270 (Biomineralization I), 43.
28. H. Cölfen and S. Mann, *Angew. Chem., Int. Ed. Engl.*, 2003, **42**, 2350.
29. G. J. Soler-Illia, A. A. de, C. Sanchez, B. Lebeau and J. Patarin, *Chem. Rev.*, 2002, **102**, 4093.

30. L. R. MacGillivray, G. S. Papaefstathiou, T. Friscic, T. D. Hamilton, D.-K. Bucar, Q. Chu, D. B. Varshney and I. G. Georgiev, *Acc. Chem. Res.*, 2008, **41**, 280.
31. S. Natarajan and S. Mandal, *Angew. Chem., Int. Ed. Engl.*, 2008, **47**, 4798.
32. C. N. R. Rao, A. K. Cheetham and A. Thirumurugan, *J. Phys.: Condens. Matter*, 2008, **20**, 1.
33. D. J. Collins and H.-C. Zhou, *J. Mater. Chem.*, 2007, **17**, 3154.
34. E. R. Parnham and R. E. Morris, *Acc. Chem. Res.*, 2007, **40**, 1005.
35. B. M. Foxman and M. D. Ward, *Mater. Res. Bull.*, 2007, **32**, 7.
36. A. K. Cheetham and C. N. R. Rao, *Science*, 2007, **318**, 58.
37. C. N. R. Rao, A. Müller and A. K. Cheetham (eds.), *Nanomaterials Chemistry: Recent Developments and New Directions*, Wiley-VCH Verlag GmbH & Co. KGaA, Weinheim, Germany, 2007.
38. A. K. Cheetham, C. N. R. Rao and R. K. Feller, *Chem. Commun.*, 2006, 4780.
39. U. Mueller, M. Schubert, F. Teich, H. Puetter, K. Schierle-Arndt and J. Pastré, *J. Mater. Chem.*, 2006, **16**, 626.
40. W. Lin, *J. Solid State Chem.*, 2005, **178**, 2486.
41. O. Yaghi, M. O'Keeffe, *J. Solid State Chem.*, Special Edition, 2008, 178.
42. S. L. James, *Chem. Soc. Rev.*, 2003, **32**, 276.
43. *Accounts of Chemical Research*, Special Edition, 2001, 34.
44. S. Kitagawa and S. Noro, *Comprehensive Coordination Chemistry II*, 2004, Chap. 7.5, p. 231.
45. D. B. Mitzi, *Dalton Trans.*, 2001, 1.
46. (a) W. Morris, C. J. Doonan, H. Furukawa, R. Banerjee and O. M. Yaghi, *J. Am. Chem. Soc.*, 2008, **130**, 12626, and references therein; (b) H. Furukawa, J. Kim, N. W. Ockwig, M. O'Keeffe and O. M. Yaghi, *J. Am. Chem. Soc.*, 2008, **130**, 11650; (c) O. M. Yaghi, *Nature Materials*, 2007, **6**, 92.
47. (a) B. Wang, A. P. Côté, H. Furukawa, M. O'Keeffe and O. M. Yaghi, *Nature*, 2008, **453**, 207; (b) R. Banerjee, A. Phan, B. Wang, C. Knobler, H. Furukawa, M. O'Keeffe and O. M. Yaghi, *Science*, 2008, **319**, 939.
48. (a) G. Férey, *Stud. Surf. Sci. Catal.*, 2007, **170A**, 66, and references therein; (b) G. Férey, *Stud. Surf. Sci. Catal.*, 2007, **168**, 327; (c) G. Férey, *Chem. Soc. Rev.*, 2008, **37**, 191.
49. Y. Kubota, M. Takata, T. C. Kobayashi and S. Kitagawa, *Coord. Chem. Rev.*, 2007, **251**, 2510.
50. (a) B. Moulton and M. J. Zaworotko, *Curr. Opinion Solid State & Mater. Sci.*, 2002, **6**, 117; (b) M. J. Zaworotko, *Crystal Engineering: The Design and Application of Functional Solids* ed. K. R. Seddon, M. Zaworotko, NATO Series C, 1999, **539** 383.
51. (a) L. Ma, J. Y. Lee, J. Li and W. Lin, *Inorg. Chem.*, 2008, **47**, 3955; (b) W. J. Rieter, K. M. L. Taylor and W. Lin, *J. Am. Chem. Soc.*, 2007, **129**, 9852.

52. (a) H. J. Choi, M. Dinca and J. R. Long, *J. Am. Chem. Soc.*, 2008, **130**, 7848; (b) M. Dinca, A. Dailly, C. Tsay and J. R. Long, *Inorg. Chem.*, 2008, **47**, 11.
53. (a) M. Eddaoudi and J. F. Eubank, *Org. Nanostruct.*, 2008, 251; (b) D. F. Sava, V. Ch. Kravtsov, F. Nouar, L. Wojtas, J. F. Eubank and M. Eddaoudi, *J. Am. Chem. Soc.*, 2008, **130**, 3768.
54. (a) P. J. Hagrman, D. Hagrman and J. Zubieta, *Angew. Chem., Int. Ed. Engl.*, 1999, **38**, 2638; (b) D. Hagrman and J. Zubieta, *ACA Transactions*, 1999, **33**, 105; (c) D. J. Chesnut, D. Hagrman, P. J. Zapf, R. P. Hammond, R. LaDuca, Jr., R. C. Haushalter and J. Zubieta, *Coord. Chem. Rev.*, 1999, **190–192**, 737; (d) D. Hagrman, P. J. Hagrman and J. Zubieta, *Commun. Inorg. Chem.*, 1999, **21**, 225.
55. (a) E. Burkholder and J. Zubieta, *Inorg. Chim. Acta*, 2004, **357**, 279; (b) E. Burkholder and J. Zubieta, *Inorg. Chim. Acta*, 2004, **357**, 301; (c) E. Burkholder, V. Golub, C. J. O'Connor and J. Zubieta, *Inorg. Chem.*, 2003, **42**, 6729; (d) D. G. Allis, R. S. Rarig, E. Burkholder and J. Zubieta, *J. Mol. Struct.*, 2004, **688**, 11; (e) E. Burkholder, S. Wright, V. Golub, C. J. O'Connor and J. Zubieta, *Inorg. Chem.*, 2003, **42**, 7460; (f) E. Burkholder and J. Zubieta, *Inorg. Chim. Acta.*, 2004, **357**, 1229; (g) E. Burkholder, V. Golub, C. J. O'Connor and J. Zubieta, *Inorg. Chem. Commun.*, 2004, **7**, 363; (h) T. Soumahoro, E. Burkholder, W. Ouellette and J. Zubieta, *Inorg. Chim. Acta*, 2005, **358**, 606; (i) E. Burkholder and J. Zubieta, *Solid State Sci.*, 2004, **6**, 1421; (j) E. Burkholder and J. Zubieta, *Inorg. Chim. Acta*, 2005, **358**, 116; (k) E. Burkholder, V. Golub, C. J. O'Connor and J. Zubieta, *Inorg. Chem.*, 2004, **43**, 7014; (l) N. G. Armatas, E. Burkholder and J. Zubieta, *J. Solid State Chem.*, 2005, **178**, 2430; (m) R. S. Rarig, Jr. and J. Zubieta, *J. Solid State Chem.*, 2002, **167**, 370; (n) R. S. Rarig and J. Zubieta, *Polyhedron*, 2003, **22**, 177; (o) R. S. Rarig Jr, L. Bewley, V. Golub, C. J. O'Connor and J. Zubieta, *Inorg. Chem. Commun.*, 2003, **6**, 539; (p) R. C. Finn, E. Burkholder and J. Zubieta, *Persp. Supramol. Chem.*, 2003, **7**, 241; (q) R. S. Rarig, L. Bewley, E. Burkholder and J. Zubieta, *Ind. J. Chem.*, 2003, **42A**, 2235; (r) D. G. Allis, E. Burkholder and J. Zubieta, *Polyhedron*, 2004, **23**, 1145.
56. A. Proust, R. Thouvenot and P. Gouzerh, *Chem. Commun.*, 2008, 1837.
57. M. T. Pope, *Heteropoly and Isopoly Oxometalates*, Springer-Verlag, Berlin, 1983.
58. *Polyoxometalate Chemistry From Topology via Self-Assembly to Applications*, ed. M. T. Pope, A. Muller, Kluwer Academic Publishers, Dordrecht, Netherlands, 2001.
59. *Polyoxometalate Chemistry for Nano-Composite Design*, ed. T. Yamase, M. T. Pope, Kluwer Academic/Plenum Publishers, New York, 2002.
60. (a) L. Cronin, *Tomorrow's Chemistry Today*, 2008, 31; (b) D.-L. Long, E. Burkholder and L. Cronin, *Chem. Soc. Rev.*, 2007, **36**, 105; (c) L. Cronin, *Annual Reports on the Progress of Chemistry, Section A: Inorganic Chemistry*, 2006, **102**, 353.

61. (a) A.-X. Tian, J. Ying, J. Peng, J. –Q. Sha, Z.-G. Han, J.-F. Ma, Z.-M. Su, N.-H. Hu and H.-Q. Jia, *Inorg. Chem.*, 2008, **47**, 3274; (b) B. Botar, P. Köegerler and C. L. Hill, *Inorg. Chem.*, 2007, **46**, 5398.
62. C. Sun, E. Wang, D. Xiao, H. An and L. Xu, *J. Mol. Struct.*, 2005, **741**, 149.
63. Y.-Q. Lan, S.-L. Li, K.-Z. Shao, X.-L. Wang and Z.-M. Su, *Dalton Trans.*, 2008, 3824.
64. X. Zhao, D. Liang, S. Liu, C. Sun, R. Cao, C. Gao, Y. Ren and Z. Su, *Inorg. Chem.*, 2008, **47**, 7133.
65. V. Shivaiah and S. K. Das, *Inorg. Chem.*, 2005, **44**, 8846.
66. S. Bareyt, S. Piligkos, B. Hasenknopf, P. Gouzerh, E. Lacote, S. Thorimbert and M. Malacria, *Angew. Chem., Int. Ed. Engl.*, 2003, **42**, 3404.
67. (a) D. W. Beck, *Zeolite Molecular Sieves*, John Wiley, New York, 1974; (b) *Verified Syntheses of Zeolitic Materials*, ed. H. Robson, 2nd revised edition, Elsevier, New York, 2001; (c) H. S. Sherry, *Handbook of Zeolite Science and Technology*, ed. S. M. Auerbach, K. A. Carrado, P. K. Dutta, M. Dekker, New York, p. 1007; (d) *International Symposium on Zeolite Microporous Crystals 2006, Micropor. Mesopor. Mater.*, 2007, **101**, 1112.
68. K. S. Min and M. P. Suh, *J. Am. Chem. Soc.*, 2000, **122**, 6834.
69. (a) C.-D. Wu and W. Lin, Angew. Chem., *Int. Ed. Engl.*, 2007, **46**, 1075; (b) C.-D. Wu, A. Hu, L. Zhang and W. Lin, *J. Am. Chem. Soc.*, 2005, **127**, 8940; (c) S.-H. Cho, B. Ma, S. B. T. Nguyen, J. T. Hupp and T. E. Albrecht-Schmitt, *Chem. Commun.*, 2006, 2563; (d) L. Pan, H. Liu, X. Lei, X. Huang, D. H. Olson, N. J. Turro and J. Li, *Angew. Chem., Int. Ed. Engl.*, 2003, **42**, 542.
70. (a) O. Kahn, *Acc. Chem. Res.*, 2000, **33**, 647; (b) O. Fabelo, J. Pasán, F. Lloret, M. Julve and C. Ruiz-Pérez, *Inorg. Chem.*, 2008, **47**, 3568; (c) G. Beobide, W.-G. Wang, O. Castillo, A. Luque, P. Román, G. Tagliabue, S. Galli and J. A. R Navarro, *Inorg. Chem.*, 2008, **47**, 5267; (d) G. Férey, *Nature Materials*, 2003, **2**, 136; (e) N. Yanai, W. Kaneko, K. Yoneda, M. Ohba and S. Kitagawa, *J. Am. Chem. Soc.*, 2007, **129**, 3496.
71. (a) S. Bordiga, C. Lamberti, G. Ricchiardi, L. Regli, F. Bonino, A. Damin, K.-P. Lillerud, M. Bjorgen and A. Zecchina, *Chem. Commun.*, 2004, 2300; (b) W.-W. Zhou, J.-T. Chen, G. Xu, M.-S. Wang, J.-P. Zou, X.-F. Long, G.-J. Wang, G.-C. Guo and J.-S. Huang, *Chem. Commun.*, 2008, 2762; (c) B. Chen, L. Wang, F. Zapata, G. Qian and E. B. Lobkovsky, *J. Am. Chem. Soc.*, 2008, **130**, 6718; (d) Z. Pan, Y. Song, Y. Jiao, Z. Fang, Y. Li and H. Zheng, *Inorg. Chem.*, 2008, **47**, 5162.
72. (a) X.-S. Wang, S. Ma, P. M. Forster, D. Yuan, J. Eckert, J. J. López, B. J. Murphy, J. B. Parise and H. C. Zhou, Angew. Chem., *Int. Ed. Engl.*, 2008, **47**, 7263; (b) M. Xue, S. Ma, Z. Jin, R. M. Schaffino, G.-S. Zhu, E. B. Lobkovsky, S.-L. Qiu and B. Chen, *Inorg. Chem.*, 2008, **47**, 6825; (c) J. Luo, H. Xu, Y. Liu, Y. Zhao, L. L. Daemen, C. Brown, T. V. Timofeeva, S. Ma and H.-C. Zhou, *J. Am. Chem. Soc.*, 2008, **130**, 9626.

73. A. Vioux, J. LeBideau, P. H. Mutin and D. Leclercq, *Top. Curr. Chem.*, 2004, **232**, 145.
74. A. Clearfield, *Curr. Opin. Solid State Mater. Sci.*, 2002, **6**, 495.
75. A. Clearfield, *Prog. Inorg. Chem.*, 1998, **47**, 371.
76. G. Alberti, *Comprehensive Supramolecular Chemistry*, ed. J. L. Atwood, J. E. D. Davis, F. Vogel, Pergamon Press, New York, 1996, Vol. 9, p. 152.
77. A. Clearfield, *Comprehensive Supramolecular Chemistry*, ed. J. L. Atwood, J. E. D. Davis, F. Vogel, Pergamon Press, New York, 1996, Vol. 9, p. 107.
78. A. Clearfield, *Chem. Mater.*, 1998, **10**, 2801.
79. L. A. Vermeulen, *Prog. Inorg. Chem.*, 1997, **44**, 143.
80. A. Clearfield, *Design of New Materials*, D. L. Cocke, ed. A. Clearfield, Plenum Press, New York, 1986, p. 121.
81. A. Clearfield, *Solvent Extr. Ion Exch.*, 2000, **18**, 655.
82. H. Byrd, A. Clearfield, D. M. Poojary, K. P. Reis and M. E. Thompson, *Chem. Mater.*, 1996, **8**, 2239.
83. S.-M. Ying and J.-G. Mao, *Cryst. Growth Des.*, 2006, **6**, 964.
84. K. Maeda, *Microporous Mesoporous Mater.*, 2004, **73**, 47 and references therein.
85. A. K. Cheethan, G. Férey and T. Loiseau, *Angew. Chem., Int. Ed. Engl.*, 1999, **38**, 3268.
86. E. W. Stein, A. Clearfield and M. A. Subramanian, *Solid State Ionics*, 1996, **83**, 113.
87. S. Konar, J. Zon, A. V. Prosvirin, K. R. Dunbar and A. Clearfield, *Inorg. Chem.*, 2007, **46**, 5229 and references therein.
88. J. W. Johnson, A. J. Jacobson, W. M. Butler, S. E. Rosenthal, J. F. Brody and J. T. Lewandowski, *J. Am. Chem. Soc.*, 1989, **111**, 381.
89. G. Huan, A. J. Jacobson, J. W. Johnson and E. W. Corcoran, *Chem. Mater.*, 1990, **2**, 91.
90. W. T. A. Harrison, L. L. Dussack and A. J. Jacobson, *Inorg. Chem.*, 1996, **35**, 1461.
91. D. Riou, C. Serre and G. Férey, *J. Solid State Chem.*, 1998, **141**, 89.
92. D. Riou, O. Roubeau and G. Férey, *Microporous Mesoporous Mater.*, 1998, **23**, 23.
93. D. Riou and G. Férey, *J. Mater. Chem.*, 1998, **8**, 2733.
94. C. Ninclaus, C. Serre, D. Riou and G. Férey, *C.R. Acad. Sci., Ser. IIc: Chim.*, 1998, **1**, 551.
95. D. Riou, C. Serre, J. Provost and G. Férey, *J. Solid State Chem.*, 2000, **155**, 238.
96. D. Riou, P. Baltazar and G. Férey, *Solid State Sci.*, 2000, **2**, 127.
97. M. Riou-Cavellec, M. Sanselme and G. Férey, *J. Mater. Chem.*, 2000, **10**, 745.
98. R. C. Finn, J. Zubieta and R. C. Haushalter, *Prog. Inorg. Chem.*, 2003, **51**, 451.
99. M. I. Khan and J. Zubieta, *Prog. Inorg. Chem.*, 1995, **43**, 1.
100. M. I. Khan, Y.-S. Lee, C. J. O'Connor, R. C. Haushalter and J. Zubieta, *Inorg. Chem.*, 1994, **33**, 3855.

101. V. Soghomonian, Q. Chen, R. C. Haushalter and J. Zubieta, *Angew. Chem. Int. Ed. Engl.*, 1995, **34**, 223.
102. V. Soghomonian, R. Diaz, R. C. Haushalter, C. J. O'Connor and J. Zubieta, *Inorg. Chem.*, 1995, **34**, 4460.
103. W. Ouellette, B.-K. Koo, E. Burkholder, V. Golub, C. J. O'Connor and J. Zubieta, *Dalton Trans.*, 2004, 1527.
104. G. Yucesan, M.-H. Yu, W. Ouellette, C. J. O'Connor and J. Zubieta, *Cryst. Eng. Commun.*, 2005, **7**, 480.
105. G. Yucesan, M.-H. Yu, C. J. O'Connor and J. Zubieta, *Cryst. Eng. Commun.*, 2005, **7**, 711.
106. W. Ouellette, V. Golub, C. J. O'Connor and J. Zubieta, *Dalton Trans.*, 2005, 291.
107. P. DeBurgomaster, A. Aldous, H. Liu, C. J. O'Connor and J. Zubieta, *Cryst. Growth & Des.*, 2010, **10**, 2209.
108. W. Ouellette, G. Wang, H. Liu, G. T. Yee, C. J. O'Connor and J. Zubieta, *Inorg. Chem.*, 2009, **48**, 953 and references therein.
109. N. G. Armatas, W. Ouellette, K. Whitenack, J. Pelcher, H. Liu, E. Romaine, C. J. O'Connor and J. Zubieta, *Inorg. Chem.*, 2009, **48**, 8897.
110. N. G. Armatas, D. G. Allis, A. Prosvirin, G. Caruntu, C. J. O'Connor, K. Dunbar and J. Zubieta, *Inorg. Chem.*, 2008, **47**, 832.
111. E. Burkholder, V. Golub, C. J. O'Connor and J. Zubieta, *Chem. Commun.*, 2003, 2128.
112. E. Burkholder, V. Golub, C. J. O'Connor and J. Zubieta, *Inorg. Chem.*, 2003, **42**, 6729.
113. E. Burkholder and J. Zubieta, *Inorg. Chim. Acta*, 2004, **357**, 1229.
114. N. G. Armatas, E. Burkholder and J. Zubieta, *Solid State Chem.*, 2005, **178**, 2430.
115. E. Burkholder, V. Golub, C. J. O'Connor and J. Zubieta, *Inorg. Chem.*, 2004, **43**, 7014.
116. J. G. Nguyen and S. M. Cohen, *J. Am. Chem. Soc.*, 2010, **132**, 4560.
117. G. Férey, *Dalton Trans.*, 2009, 4400.
118. J. Livage, *Chemica Scripta*, 1988, **28**, 9.
119. J. Pouxel, *Chemica Scripta*, 1988, 33.
120. M. I. Khan and J. Zubieta, *Prog. Inorg. Chem.*, 1995, **43**, 1.
121. A. Rabenau, *Angew. Chem., Int. Ed. Engl.*, 1985, **24**, 1026.
122. S. Feng and R. Xu, *Acc. Chem. Res.*, 2001, **34**, 239 and references therein.
123. A. N. Lobachev and G. D. Archard, *Crystallization Processes Under Hydrothermal Conditions*, 1973, Consultants Bureau, New York.
124. R. M. Barrer, *Hydrothermal Chemistry of Zeolites*, 1982, Academic Press, New York.
125. J. Gopalakrishnan, *Chem. Mater.*, 1995, **7**, 1265.
126. S. Somiya, *Hydrothermal Reactions for Materials Science and Engineering*, 1986, Elsevier Applied Science, New York.
127. J. Zubieta, in, *Comprehensive Coordination Chemistry II*, 2004, **1**, 697.
128. R. A. Laudise, *Chem. Eng. News*, 1987, Sept. 28, 30.

129. P. Halasyamani, M. J. Willis, C. L. Stern, P. M. Lundquist, G. K. Wong and K. R. Poeppelmeier, *Inorg. Chem.*, 1996, **35**, 1367.
130. T. E. Gier and G. D. Stucky, *Nature*, 1991, **349**, 508.
131. D. Hagrman, C. Zubieta, D. J. Rose and J. Zubieta, *Angew. Chem., Int. Ed. Engl.*, 1997, **36**, 873.
132. J. Niu, D. Guo, J. Zhao and J. Wang, *New J. Chem.*, 2004, **28**, 980, and references therein.
133. S. Jones, H. Liu, W. Ouellette, K. Schmidtke, C. J. O'Connor and J. Zubieta, *Inorg. Chem. Commun.*, 2010, **13**, 491.
134. W. Kwak, M. T. Pope and T. F. Scully, *J. Am. Chem. Soc.*, 1975, **97**, 5735.
135. P. A. Maggard and P. D. Boyle, *Inorg. Chem.*, 2003, **42**, 4250.

CHAPTER 8

Structural and Magnetic Studies of Cobalt Phosphonates

LI-MIN ZHENG AND YAN DUAN

State Key Laboratory of Coordination Chemistry, School of Chemistry and Chemical Engineering, Nanjing University, Nanjing 210093, P. R. China

8.1 Introduction

Metal phosphonates have gained increasing attention owing to their potential applications in such fields as catalysis,[1] absorption and separation,[2] proton conductivity,[3] photochemistry,[4] magnetism,[5] and biotechnology,[6] etc. Among the variety of metal phosphonates, cobalt phosphonates are rather interesting because they can display versatile architectures with a wide range of colors arising from different geometries of cobalt ions such as octahedral, tetrahedral, square pyramidal, and trigonal-bipyramid, etc. Such compounds are also magnetically important because the octahedrally coordinated Co^{II} with $^4T_{1g}$ ground state is magnetically highly anisotropic due to spin–orbit coupling, which leads to very interesting magnetic behavior such as single molecule magnet, single chain magnet, canted antiferromagnetism, ferrimagnetism, and ferromagnetism, etc.[7–9] A search of CCDC database reveals that there are over 139 cobalt phosphonates reported in the literature. In this review, the structures and magnetic properties of these cobalt phosphonates are presented. A collection of the crystallographic data for cobalt phosphonates with 1D chain, 2D layer, and 3D framework structures is provided in Tables 8.1–8.3. The abbreviations of some ligands used in this paper are shown in Schemes 8.1–8.3.

8.2 Zero-dimensional Cobalt Phosphonates

Metal phosphonates are well known for their ability to form polymeric structures.[10] Nevertheless, there are a few examples in the literature that

Metal Phosphonate Chemistry: From Synthesis to Applications
Edited by Abraham Clearfield and Konstantinos Demadis

Published by the Royal Society of Chemistry, www.rsc.org

Table 8.1 Unit cell parameters and structural features of cobalt phosphonates with chain structures.

Compound	*Unit cell parameters*	*Structure features*	*Ref.*
$(NH_4)_2Co_2(hedpH)_2$	triclinic, P-1, $a = 5.464(1)$, $b = 7.721(2)$, $c = 9.976(2)$ Å, $\alpha = 71.198(4)$, $\beta = 82.385(4)$, $\gamma = 87.920(5)°$, $V = 394.82(16)$ Å^3	ladder-like chain bridged by both O(P) and O-P-O	23
$[NH_3(CH_2)_4NH_3]Co_2(hedpH)_2 \cdot 2H_2O$	monoclinic, $P2_1/c$, $a = 5.4868(8)$, $b = 12.9116(18)$, $c = 15.251(2)$ Å, $\beta = 98.843(2)°$, $V = 1067.6(3)$ Å^3, $Z = 2$	ladder-like chain bridged by both O(P) and O-P-O	24
$[NH_3(CH_2)_5NH_3]Co_2(hedpH)_2 \cdot 2H_2O$	monoclinic, $P2_1/c$, $a = 5.4757(7)$, $b = 12.7740(16)$, $c = 15.794(2)$ Å, $\beta = 98.797(2)°$, $V = 1091.7(2)$ Å^3, $Z = 2$	ladder-like chain bridged by both O(P) and O-P-O	24
$(NH_3C_6H_4NH_3)Co_2(hedpH)_2 \cdot H_2O$	triclinic, P-1, $a = 5.4651(14)$, $b = 12.406(3)$, $c = 15.362(4)$ Å, $\alpha = 87.452(4)$, $\beta = 88.909(4)$, $\gamma = 78.415(4)°$, $V = 1019.3(5)$ Å^3, $Z = 2$	ladder-like chain bridged by both O(P) and O-P-O	25
$Co(ahpdp) \cdot H_2O$	triclinic, P-1, $a = 5.4483(19)$, $b = 8.828(3)$, $c = 9.809$ (4) Å, $\alpha = 91.748(5)$, $\beta = 103.976(5)$, $\gamma = 97.928(5)°$, $V = 452.5(3)$ Å^3, $Z = 2$.	ladder-like chain bridged by both O(P) and O-P-O	26
$Co(hpedpH_2)(H_2O)$	monoclinic, $P2_1/n$, $a = 6.5514(15)$, $b = 12.477(3)$, $c = 13.595(3)$ Å, $\beta = 98.029(4)°$, $V = 1100.4(4)$ Å^3, $Z = 4$	ladder-like chain bridged by both O(P) and O-P-O	27
$Co(pomp)(H_2O)$	triclinc, P-1, $a = 6.525(3)$, $b = 7.487(3)$, $c = 10.005(4)$ Å, $\alpha = 88.645(7)$, $\beta = 78.272(7)$, $\gamma = 70.516(8)°$, $V = 450.7(3)$ Å^3, $Z = 2$	ladder-like chain bridged by both O(P) and O-P-O	28
$Co(H_2L^2)(H_2O)$ ($H_4L^2 = 4$-$MeC_6H_4CH_2N(CH_2PO_3H_2)_2$)	monoclinic, $P2_1/c$, $a = 8.354(1)$, $b = 29.211(4)$, $c = 6.217(1)$ Å, $\beta = 110.621(3)°$, $V = 1420.0(4)$ Å^3, $Z = 4$	2_1 helical chain bridged by both O(P) and O-P-O	29
$Co(pnaH)(H_2O)_3 \cdot H_2O$	monoclinic, $P2_1/n$, $a = 13.191(2)$, $b = 7.316(2)$, $c = 13.954(2)$ Å, $\beta = 119.774(10)°$, $V = 1168.2(3)$ Å^3, $Z = 4$	2_1 helical chain bridged by O-P-O	30
$Co\{(C_7H_5N_2)CH_2N(CH_2PO_3H)_2\}$	orthorhombic, Pbca, $a = 15.207(2)$, $b = 10.4626(16)$, $c = 16.794(3)$ Å, $V = 2672.0(7)$ Å^3, $Z = 8$	zigzag chain bridged by O-P-O	31
$Co(H_2L^7)_2(H_3L^7)_2$ ($H_3L^7 = HOOCCH_2NH_2CH_2PO_3H_2$)	triclinc, P-1, $a = 5.501(1)$, $b = 10.983(2)$, $c = 11.381(2)$ Å, $\alpha = 69.848(2)$, $\beta = 86.645(2)$, $\gamma = 79.488(2)°$, $V = 634.6(2)$ Å^3, $Z = 1$	linear chain bridged by O-P-O	32
$[Co_2(L^8)(H_2O)_5(4,4'\text{-bipy})] \cdot 4H_2O$ ($H_4L^8 = H_2O_3PCH_2N(CH_2CO_2H)_2$)	triclinic, P-1, $a = 7.801(1)$, $b = 13.456(1)$, $c = 13.545(1)$ Å, $\alpha = 65.207(2)$, $\beta = 88.035(2)$, $\gamma = 77.135(2)°$, $V = 1255.6(2)$ Å^3, $Z = 2$	chain containing dimeric units bridged by 4,4′-bipy	33
Co(phen)(4,4′-bpy)$(PhPO_3H)_2 \cdot 0.5H_2O$	tetragonal, $P4_32_12$, $a = 22.989(1)$, $c = 12.077(1)$ Å, $V = 6380.2(5)$ Å^3, $Z = 8$	linear chain bridged by 4,4′-bipy	34
$[Co(H_2L)(H_2O)_4(H_4L)] \cdot 2H_2O$ (H_4L = 1,8-octylenediphosphonic acid)	triclinic, P-1, $a = 5.5415(8)$, $b = 8.6382(8)$, $c = 16.794$ (2) Å, $\alpha = 87.694(2)$, $\beta = 80.859(2)$, $\gamma = 76.005(2)°$, $V = 770.11(19)$ Å^3, $Z = 1$	linear chain bridged by H_2L	35

Table 8.2 Unit cell parameters and structural features of cobalt phosphonates with layer structures.

Compound	*Unit cell parameters*	*Structure features*	*Ref.*
$Co(CH_3PO_3)(H_2O)$	orthorhombic, $Pna2_1$, $a = 17.413(6)$, $b = 4.7856(3)$, $c = 5.6638(10)$ Å, $V = 471.99(18)$ Å^3	inorganic layer containing 3-member rings	37
$Co(C_2H_5PO_3)(H_2O)$	monoclinic, *Pn*, $a = 4.806(2)$, $b = 10.243(3)$, $c = 5.674\ (1)$ Å, $\beta = 90.56(3)°$, $V = 279.28(14)$ Å^3	inorganic layer containing 3-member rings	37
$Co(PhPO_3)(H_2O)$	orthorhombic, $Pnn2_1$, $a = 5.624(1)$, $b = 14.391(3)$, $c = 4.862(1)$ Å, $V = 393.44(12)$ Å^3, $Z = 2$	inorganic layer containing 3-member rings	39
$Co_3(aedpH_2)_2(aedpH_3)_2(H_2O)_4 \cdot 2H_2O$	monoclinic, $C2/c$, $a = 27.727(4)$, $b = 7.109(1)$, $c = 19.827(3)$ Å, $\beta = 118.488(3)°$, $V = 3434.9(9)$ Å^3, $Z = 2$	inorganic layer containing 12-member rings	44
$Co_3(ImhedpH)_2(H_2O)_4 \cdot 2H_2O$	triclinic, P-1, $a = 6.945(1)$, $b = 9.573(2)$, $c = 9.788(2)$ Å, $\alpha = 77.284(3)$, $\beta = 78.623(4)$, $\gamma = 87.507(4)°$, $V = 622.4(2)$ Å^3, $Z = 1$	inorganic layer containing 4-, 8- and 16-membered rings	12
$Co_3(hpedpH)_2(H_2O)_4 \cdot H_2O$	triclinic, P-1, $a = 6.968(2)$, $b = 9.880(2)$, $c = 10.313(2)$ Å, $\alpha = 78.820(5)$, $\beta = 86.037(4)$, $\gamma = 78.203(5)°$, $V = 681.5(3)$ Å^3, $Z = 1$	inorganic layer containing 4-, 8- and 16-membered rings	45
$Co_3(cmdpH)_2(H_2O)_2$	not determined	double-layer structure	
$Co(H_3L^1)_2(H_2O)_2$ $H_4L^1 = NH(CH_2PO_3H_2)_2$	monoclinic, $P2_1/n$, $a = 7.361(3)$ Å, $b = 8.133(3)$ Å, $c = 14.078(5)$ Å, $\beta = 104.40(1)°$, $V = 816.3(5)$ Å^3, $Z = 2$	hybrid layer containing tetranuclear 16-member rings	47
$Co(H_2L^1)(H_2O)_2 \cdot H_2O$	monoclinic, $P2_1/c$, $a = 8.7876(6)$, $b = 7.2981(4)$, $c = 15.633(1)$ Å, $\beta = 90.266(2)°$, $V = 1002.5(1)$ Å^3, $Z = 4$	hybrid layer containing 4-, 6- and 8-member rings	48
$Co(HL^7)(H_2O)_2$ $H_3L^7 = HOOCCH_2NHCH_2PO_3H_2$	monoclinic, $P2_1/c$, $a = 8.967(1)$, $b = 7.367(1)$, $c = 12.758(1)$ Å, $\beta = 101.433(2)°$, $V = 826.0(1)$ Å^3, $Z = 4$	inorganic layer containing 4- and 8-member rings	32
$Co_3(H_2L^5)_2(H_2O)_4(4,4'\text{-bipy})_2 \cdot 11H_2O$ $H_5L^5 = HO_2CCH_2N(CH_2PO_3H_2)_2$	$P2/c$, $a = 14.8651(14)$Å, $b = 7.5694(7)$Å, $c = 22.796(2)$ Å, $\alpha = 90.0°$, $\beta = 106.833(1)°$, $\gamma = 90.0°$, $V = 2455.1(4)$ Å^3, $Z = 2$	trimers linked by 4,4′-bipy	49
			50

Table 8.2 *(Continued)*

Compound	*Unit cell parameters*	*Structure features*	*Ref.*
$Co(H_4L^6)_2$ $H_5L^6 = HO_2C(CH_2)_3N(CH_2PO_3H_2)_2$	monoclinic, $P2_1/c$, $a = 12.586(3)$, $b = 8.891(1)$, $c = 9.727(3)$ Å, $\beta = 97.349(4)°$, $V = 1079.5(1)$ Å^3, $Z = 2$	inorganic layer with a 4,4-net topology	
$Co_2(L^8)(H_2O)_5 \cdot H_2O$ $H_4L^8 = H_2O_3PCH_2N(CH_2CO_2H)_2$	$P2_1/c$, $a = 14.172(3)$, $b = 10.412(2)$, $c = 9.814(2)$ Å, $\beta = 106.037(4)°$, $V = 1391.9(5)$ Å^3, $Z = 4$	hybrid layer	53
$(enH_2)_{0.5}[Co(hpa)(H_2O)] \cdot H_2O$ $hpaH_3 = HO_2CCH(OH)PO_3H_2$	monoclinic, $P2_1/n$, $a = 8.422(1)$, $b = 9.785(1)$, $c = 10.797(1)$ Å, $\beta = 106.23(1)°$, $V = 854.26(12)$ Å^3, $Z = 4$	inorganic layer containing 4- and 8-member rings	54
$Co_3(O_3PCH_2NC_4H_7CO_2)_2 \cdot 5H_2O$	triclinic, P-1, $a = 9.985(2)$, $b = 10.378(3)$, $c = 11.942(3)$ Å, $\alpha = 90.238(5)$, $\beta = 106.810(5)$, $\gamma = 111.346(4)°$, $V = 1095.0(3)$ Å^3, $Z = 2$	inorganic layer containing 10-member rings	55
$(H_3O)_2[Co_3(OH)_2(OOCCH_2PO_3)_2]$	triclinic, P-1, $a = 4.711(2)$, $b = 8.051(3)$, $c = 9.276(3)$ Å, $\alpha = 107.164$, $\beta = 94.284$, $\gamma = 96.846°$, $V = 331.5(2)$ Å^3, $Z = 1$	chains of edge-sharing CoO_6 linked by PO_3C	56
$Co(3\text{-}HOOCC_6H_4PO_3)(H_2O)$	monoclinic, $P2_1/n$, $a = 4.912(1)$, $b = 32.316(5)$, $c = 5.748(1)$ Å, $\beta = 93.86(1)°$, $V = 910.4(2)$ Å^3, $Z = 4$	inorganic layer containing 3-member rings	57
$Co(2\text{-}OOCC_6H_4PO_3H)(H_2O)$	orthorhombic, *Pbca*, $a = 8.999(5)$, $b = 8.614(5)$, $c = 22.506(13)$ Å, $V = 1744.6(17)$ Å^3, $V = 8$	hybrid layer containing Co_2O_2 dimers inter-linked by O-C-O and O-P-O	58
$Co(4\text{-}hpmp)(H_2O)$ $4\text{-}hpmp^{2-} = C_5H_4NCH(OH)PO_3$	monoclinic, $C2/c$, $a = 22.737(4)$, $b = 5.1613(8)$, $c = 16.318(2)$ Å, $\beta = 121.818(2)°$, $V = 1627.2(4)$ Å^3, $V = 8$	inorganic "staircase" chain linked by organic moieties	59
$Co(2\text{-}pmp)(H_2O)_2$ $2\text{-}pmpH_2 = C_5H_4NCH_2PO_3H_2$	monoclinic, $P2_1/c$, $a = 10.697(2)$, $b = 7.346(1)$, $c = 12.439(2)$ Å, $\beta = 110.358(3)°$, $V = 916.5(3)$ Å^3, $Z = 4$	inorganic layer containing 4- and 8-member rings	60
$Co(2\text{-}pmp)(H_2O)$	monoclinic, $P2_1/n$, $a = 8.8520(2)$, $b = 7.8961(1)$, $c = 12.7913(3)$ Å, $\beta = 112.206(1)°$, $V = 827.75(3)$ Å^3, $Z = 4$	inorganic layer containing 4- and 8-member rings	60
$Co(2\text{-}pmp)$	monoclinic, $P2_1/n$, $a = 9.8450(3)$, $b = 8.8954(4)$, $c = 9.6732(4)$ Å, $\beta = 112.331(2)°$, $V = 783.61(2)$ Å^3, $Z = 4$	inorganic layer containing 4- and 8-member rings	60

Table 8.3 Unit cell parameters and structural features of cobalt phosphonates with three-dimensional structures.

Compound	*Cell parameters*	*Structure features*	*Ref.*
$Co_2(O_3PC_6H_4PO_3)(H_2O)_2$	orthorhombic, Pnnm, $a = 19.306(5)$, $b = 4.8293(12)$, $c = 5.6390(14)$ Å, $V = 525.7(2)$ Å^3, $Z = 2$	pillared layered	61
$Co_2(H_2O)_2[O_3PCH_2(C_6H_4)CH_2PO_3]$	orthorhombic, $Pca2_1$, $a = 9.6214(19)$, $b = 5.6855(11)$, $c = 21.725(4)$ Å, $V = 1188.4(4)$ Å^3	pillared layered	62
$Co_2(H_2O)_2(O_3PC_2H_4PO_3)$	monoclinic, $P2_1/n$, $a = 5.6500(11)$, $b = 4.7800(10)$, $c = 15.330(3)$ Å, $\beta = 98.50(3)$, $V = 409.47(14)$ Å^3, $Z = 2$	pillared layered	63a
$Co(O_3PCH_2CH_2NH_2)$	monoclinic, $P2_1/c$, $a = 6.4172(3)$, $b = 8.4131(4)$, $c = 9.8360(4)$ Å, $\beta = 90.997(1)°$, $V = 523.00(4)$ Å^3, $Z = 4$	pillared layered	64
$Co_3(O_3PCH_2\text{-}NHC_5H_9\text{-}COO)_2(O_3PCH_2\text{-}NC_5H_{10})(H_2O)$	monoclinic, $P2_1/c$, $a = 31.15(2)$, $b = 9.011(7)$, $c = 10.166(8)$ Å, $\beta = 91.15(3)°$, $V = 2853(4)$ Å^3, $Z = 4$	pillared layered	66
$Co_3(pna)_2(H_2O)_2$	triclinic, P-1, $a = 6.072(2)$, $b = 8.089(2)$, $c = 9.1[illegible]2(3)$ Å, $\alpha = 88.314(5)$, $\beta = 73.692(5)$, $\gamma = 77.059(5)°$, $V = 418.4(2)$ Å^3, $Z = 1$	pillared layered	67
$Co_3(H_2L^3)_2(H_2O)_4(4,4'\text{-bipy})$ $H_5L^3 =$ 4-$HO_2CC_6H_4CH_2N(CH_2PO_3H_2)_2$	monoclinic, $C2/c$, $a = 25.9281(5)$, $b = 9.76310(10)$, $c = 17.3671(5)$, $\beta = 103.290(2)°$, $V = 4278.6(2)$ Å^3, $Z = 4$.	pillared layered	33
$Co_2(cmdpH_2)_2(4,4'\text{-bipy})_{0.5}(H_2O)$	monoclinic, $P2_1/c$, $a = 5.629$ (18), $b = 18.673(7)$, $c = 27.932(10)$, $\beta = 100.98(9)°$, $V = 2882.6(18)$ Å^3, $Z = 4$	pillared layered	46
$Co_2(O_3PCH_2PO_3) \cdot H_2O$	monoclinic, $C2/c$, $a = 18.820(4)$, $b = 8.246(2)$, $c = 8.916(2)$ Å, $\beta = 106.68(3)°$, $V = 1325.4(5)$ Å^3	open framework	68
$Na_6Co_7(hedp)_2(hedpH)_4(H_2O)_4 \cdot 8H_2O$	triclinic, P-1, $a = 10.712(2)$, $b = 11.203(2)$, $c = 12.311(3)$ Å, $\alpha = 73.62(3)$, $\beta = 86.29(3)$, $\gamma = 86.74(3)°$, $V = 1413.4(5)$ Å^3, $Z = 1$	open framework	69
$NaCo_2(amdpH_{1.5})_2(H_2O)_2 \cdot xH_2O$	monoclinic, $C2/c$, $a = 18.772(4)$, $b = 9.417(2)$, $c = 9.612(2)$ Å, $\beta = 102.593(4)°$, $V = 1658.4(6)$ Å^3, $Z = 4$.	open framework	70
$Co_3(O_3PCH_2NH_2CH_2PO_3)_2$	orthorhombic, $P2_12_12$, $a = 10.733(3)$, $b = 13.387(4)$, $c = 5.3953(13)$ Å, $V = 775.31(4)$ Å^3, $Z = 4$	open framework	55
$Co_3(O_2CC_2H_4PO_3)_2(H_2O)_6$	monoclinic, $P2_1/c$, $a = 4.6086(9)$, $b = 10.278(3)$, $c = 17.504(3)$ Å, $\beta = 95.74(1)°$, $V = 825.0(3)$ Å^3, $Z = 2$	open framework	72
$NaCo[O_3PCH(OH)CO_2]$	orthorhombic, Pbca, $a = 10.1827(8)$, $b = 9.6673(8)$, $c = 10.7853(9)$ Å, $V = 1061.7(2)$ Å^3, $Z = 8$	open framework	73
$Co(O_3PCH_2OH)$	trigonal, R-3, $a = 16.0312(1)$, $c = 7.7499(1)$ Å, $V = 1724.88(5)$ Å^3, $Z = 18$	open framework	74
$[Co_2Cl(S\text{-}HL^9)(H_2O)_5]Cl \cdot H_2O$	orthorhombic, $P2_12_12_1$, $a = 8.7142(11)$, $b = 10.8075(14)$, $c = 18.074(2)$ Å, $V = 1702.2(4)$ Å^3, $Z = 4$	hybrid 3D framework	75
$Tb_2Co_3(C_5H_4NPO_3)_6 \cdot 4H_2O$	cubic, $I2_13$, $a = 16.108(2)$, $V = 4179.4(6)$, $Z = 4$	open framework	76
$DyCo_3(C_5H_4NPO_3)_6 \cdot 4H_2O$	cubic, $I2_13$, $a = 16.084(2)$, $V = 4160.5(4)$, $Z = 4$	open framework	76

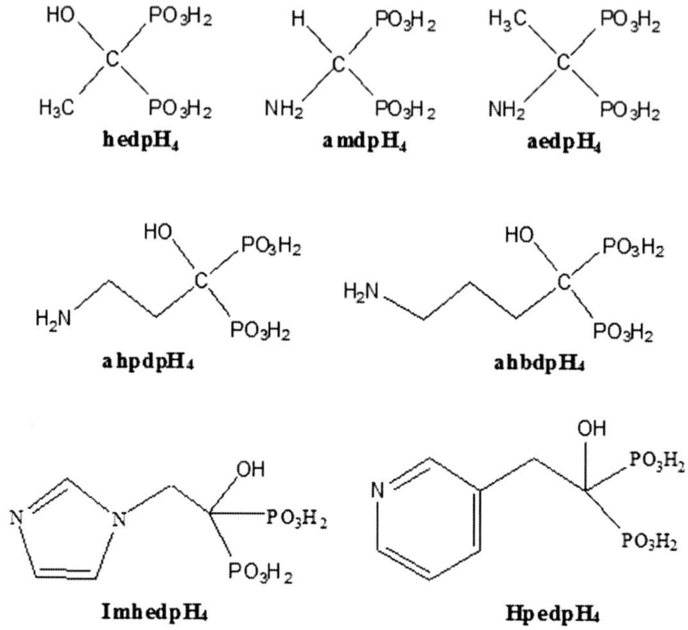

Scheme 8.1

mononuclear metal complexes can be formed under some experimental conditions using multifunctional phosphonate ligands.

Based on *N*,*N*-bis(phosphonomethyl)glycine (H_5L^5), complex $Co(H_3L^5)(H_2O)_2 \cdot 2H_2O$ is isolated in aqueous solution at pH 1.5.[11] The complex consists of a mononuclear, octahedral Co^{II} center coordinated by oxygen atoms belonging to the terminal phosphonate and carboxylate groups and the nitrogen atom of the ligand, and two bound water molecules. The magnetic properties have been investigated. The room temperature value of $\chi_M T$ (2.98 $cm^3\ K\ mol^{-1}$) is higher than 1.875 $cm^3\ K\ mol^{-1}$ expected for a Co^{II} system with $S = 3/2$, attributed to the presence of a significant orbital contribution to the anisotropic nature of the Co^{II} system. The susceptibility data are fitted according to the following general Hamiltonian (Equation 8.1):

$$H = D\left[S_z^2 - \frac{1}{3}S(S+1)\right] + E\left(S_x^2 - S_y^2\right) + g\beta H \cdot S \tag{8.1}$$

where all the parameters have their usual definitions, and $S = 3/2$. The best fit yields parameters: $D = 41.1\ cm^{-1}$ and $g = 2.53$.

The other mononuclear cobalt phosphonates include $Co(ImhedpH_3)_2(H_2O)_2$ based on 2-(1-imidazole)-1-hydroxyl-1,1′-ethylidenediphosphonic acid ($ImhedpH_4$),[12] and $(H_3O)[Co(H_2L^5)(H_2O)_2] \cdot 2H_2O$ based on the same H_5L^5 ligand.[13] The magnetic susceptibility data of the latter was fitted to the Curie–Weiss law: $\chi_M = C/(T - \theta)$, leading to a Weiss constant of –5.4 K.

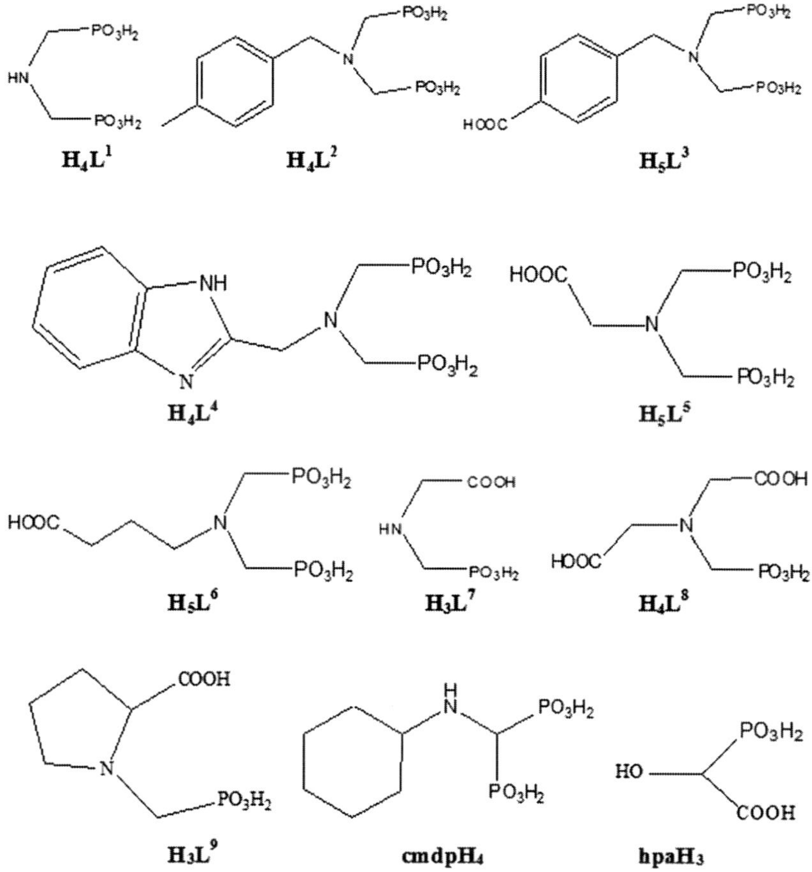

Scheme 8.2

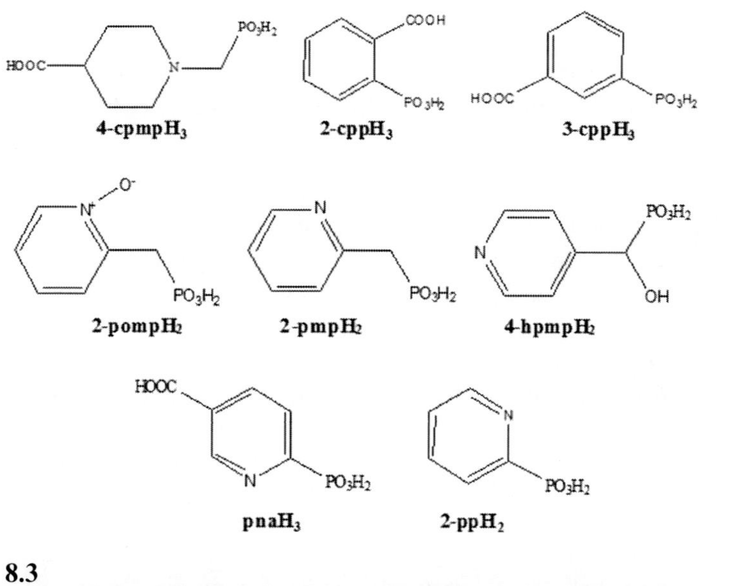

Scheme 8.3

Compound $(NH_4)_4[Co(H_2O)_6][(H_2O)_2Co(HL^5)Co(NH_3)_2(H_2O)_3]_2[Co(L^5)(H_2O)_2]_2 \cdot 10H_2O \cdot 1.36CH_3CH_2OH$[14a] contains discrete mononuclear species of $[Co(L^5)(H_2O)_2]^{3-}$ and neutral dinuclear species of $[(H_2O)_2Co(HL^5)Co(NH_3)_2(H_2O)_3]$ in which the two Co^{II} ions are bridged by O–P–O unit of the ligand. Reactivities of five phosphonate esters coordinated to a dinuclear Co^{III} complex, *e.g.* $[Co_2(tacn)_2(OH)_2\{O_2P(Me)OAr\}]^{3+}$ (tacn = 1,4,7-triazacyclononane; substituent = *m*-F, *p*-NO_2, *p*-NO_2, *m*-NO_2, *p*-Cl, H) were also investigated as model systems of protein phosphatase-1.[14b]

Higher nuclearity cage complexes of cobalt phosphonates can be obtained, mostly from solutions of organic solvent and in the presence of co-ligands. In a few cases, cobalt phosphonate clusters are synthesized via solvothermal reactions. Such cluster compounds include Co_4,[15] Co_6,[16,18,19] Co_8,[17] Co_{10},[18] Co_{12},[18,19] Co_{13},[18,19] Co_{14},[19] Co_{15}[19] and Co_{20}.[18] Since a special chapter is devoted by Winpenny *et al.* in this book (Chapter 12), no details will be discussed herein.

8.3 One-dimensional Cobalt Phosphonates

8.3.1 Ladder-like Chain

The design and syntheses of paramagnetic molecular solids with ladder-like chain structures are of great importance because of the current interest in quantum antiferromagnetic spin ladders.[20] In an ideal spin ladder, the exchange coupling along the rungs ($J_\perp$) is very similar to that along the chains ($J_{||}$), but each ladder is well isolated from its equivalent neighbors. In the presence of an external magnetic field, spin ladders can experience a transition from a nonmagnetic ground state to a spin-polarized state. When doped with charge carriers, an even-leg ladder is predicted to become a superconductor. Therefore, many efforts are being made to search for new spin ladder systems, mostly with spin $S = 1/2$. Examples of spin-1/2 two leg ladders include copper oxides such as $SrCu_2O_3$.[20,21] Due to the strong magnetic coupling between the Cu^{II} ions over the oxygen bridge, the spin gap becomes very large (*ca.* 420 K) such that the field-induced magnetic transition can not be observed. By using halides as the bridging unit, compounds $(5IAP)_2CuBr_4 \cdot 2H_2O$ and $Cu_2(C_5H_{12}N_2)_2Cl_4$ with ladder structures are obtained which show significantly reduced spin gaps.[22] The critical fields to observe the field induced magnetic transitions are over 7 T, which are not accessible in normal laboratories.

It is well known that the phosphonate ligands can link the metal ions through single O(P) or three atom bridge O–P–O. Both can mediate only weak magnetic interactions. The construction of a spin ladder in which the metal ions are bridged through phosphonate linkages could largely reduce the spin gap and hence the critical fields to observe the transition.

Based on 1-hydroxyethylidenediphosphonate [hedpH_4, $CH_3C(OH)(PO_3H_2)_2$], Zheng and co-workers reported a series of metal-hedp compounds with 1D ladder-like chain structures. The cobalt compounds among these are $(NH_4)_2Co_2(hedpH)_2$,[23] $[NH_3(CH_2)_nNH_3]Co_2(hedpH)_2 \cdot 2H_2O$ ($n = 4, 5$)[24] and

$(NH_3C_6H_4NH_3)Co_2(hedpH)_2 \cdot H_2O$.[25] Their magnetic properties are investigated in detail.

Figure 8.1 shows the chain structure and packing diagram of compound $(NH_4)_2Co_2(hedpH)_2$.[23] The compound crystallizes in space group P-1, and contains $\{Co_2(hedpH)_2\}_n^{2n-}$ ladder-like double chains and NH_4^+ counter-ions. The double chain can be described as O–P–O bridged single chains of cobalt, interconnected through edge-sharing of the CoO_6 octahedra. The resulted $\{Co_2(\mu\text{-}O)_2\}$ dimers can be viewed as the rung. The Co1–O4–Co1B angle is 100.16(9)°. The Co · · · Co distance over the μ-O4 bridge is 3.236 Å, and those across the O–P–O bridges are 5.463 and 4.754 Å for Co1 · · · Co1A and Co1 · · · Co1D, respectively. Between the adjacent double chains, there exist very strong hydrogen bonds between the protonated oxygen O3 and the pendant O6 (O3 · · · O6 2.544 Å), thus forming a layer in the *ab* plane. The NH_4^+ cations occupy the intra-layer cavities, and are stabilized by hydrogen bonds toward the phosphonate oxygens of the $\{Co_2(hedpH)_2\}_n$ chains. The shortest interchain Co · · · Co distance is 7.010 Å within the layer. The layers are connected by the hydrogen bond contacts through the hydroxyl O7 atom from one layer and the phosphonate O3 atom from the other (O7 · · · O3 2.882 Å), hence generating a 3D network.

Temperature-dependent molar magnetic susceptibility (χ_M) measurements reveal a typical low-dimensional antiferromagnetic behavior with a Weiss constant of –34.1 K, determined by the data in the temperature range 100–300 K. The *ac* susceptibility measurements performed in the range 5–20 K, at $H_{ac} = 5$ Oe and frequencies of 111, 199, 355, 633, and 1111 Hz, show neither an out-of-phase signal nor frequency dependent behavior. The in-phase signal shows a broad maximum around 10 K, consistent with a low-dimensional antiferromagnetic material. Considering that the dominant antiferromagnetic interaction should be propagated within the ladder-like chain through μ-O and/or O–P–O bridges, the system can be viewed as a quasi 1D antiferromagnet. A sharp drop of χ_M toward zero at very low temperature suggests that the ground state of the system is $S = 0$. The field-dependent magnetizations at 1.9 and 5.0 K show a clear sigmoid shape of the curve, indicating the occurrence of a magnetic transition from the antiferromagnetic ground state to a spin polarized state. The transition field, determined by dM/dH derivative curve, is *ca.* 35–40 kOe

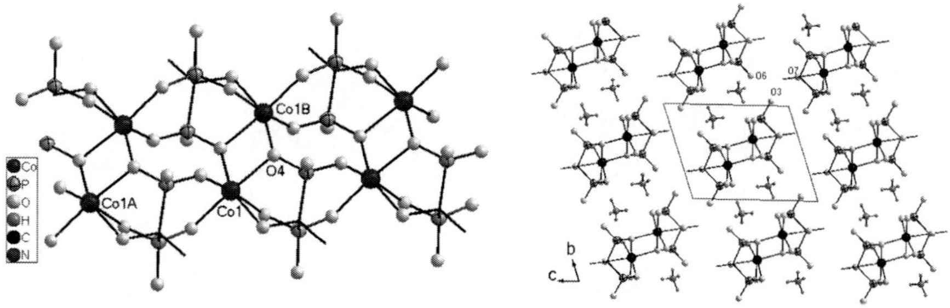

Figure 8.1 Chain fragment and packing diagram of $(NH_4)_2Co_2(hedpH)_2$ (adapted from ref. 23).

depending on the temperature. The magnetization of $(NH_4)_2Co_2(hedpH)_2$ at 70 kOe is 4.0 Nβ per Co_2 unit, close to the saturation value of 4.1 Nβ anticipated for a pair of $S = 1/2$ spins with $g = 4.1$ (Figure 8.2).

By using bigger template amines, isostructural compounds $[NH_3(CH_2)_nNH_3]Co_2(hedpH)_2 \cdot 2H_2O$ ($n = 4$, 5) are obtained.[24] They contain similar $\{Co_2(hedpH)_2\}_n$ species to that in $(NH_4)_2Co_2(hedpH)_2$. In this case, however, each double chain is connected to its four equivalent neighbors through very strong hydrogen bonds [O3···O6: 2.428(3), 2.433(4) Å], thus forming a 3D open network with channels generated along [100] directions. The $NH_3(CH_2)_nNH_3^{2+}$ cations and lattice water reside in the channels (Figure 8.3).

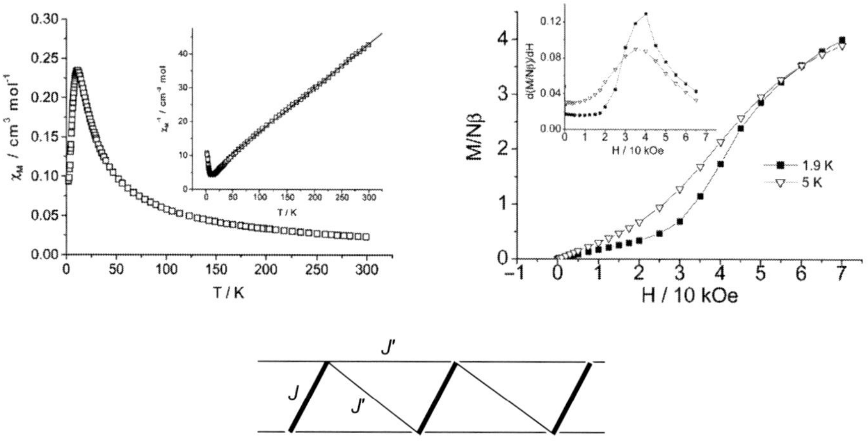

Figure 8.2 Plots of χ_M and $1/\chi_M$ versus T and M versus H for $(NH_4)_2Co_2(hedpH)_2$. The magnetic exchange pathways within the ladder-like chain is given in the bottom. (Reproduced with permission from ref. 23. Copyright 2003 American Chemical Society.)

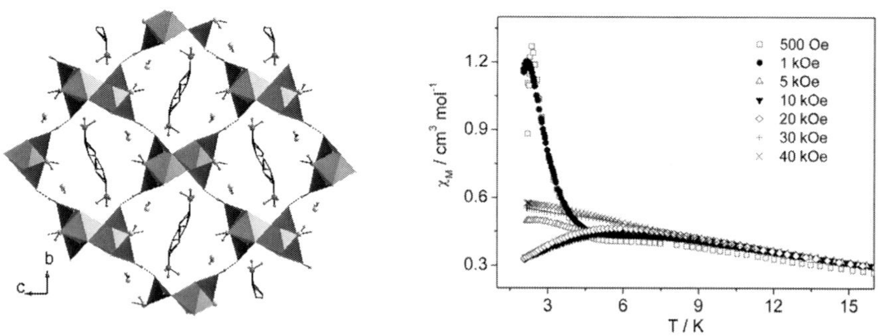

Figure 8.3 Packing diagram viewed along the a axis (left), and χ_M versus T plots at different external fields (right) for $[NH_3(CH_2)_5NH_3]Co_2(hedpH)_2 \cdot 2H_2O$. (Reproduced with permission from ref. 24. Copyright 2004 American Chemical Society.)

The magnetic behavior of $[NH_3(CH_2)_4NH_3]Co_2(hedpH)_2 \cdot 2H_2O$ is quite similar to that of $(NH_4)_2Co_2(hedpH)_2$ except that the critical field of magnetic transition at 1.8 K becomes 25 kOe. For compound $[NH_3(CH_2)_5NH_3]Co_2$-$(hedpH)_2 \cdot 2H_2O$, a sharp peak appears in the χ_M versus T curve at *ca.* 4 K, followed by a broad maximum around 7 K, suggesting the occurrence of a long-range antiferromagnetic ordering at very low temperature (Figure 8.3). The upturn of $\chi_M T$ below 4.1 K suggests an uncompensated magnetic moments of the system arising from spin canting of the antiferromagnetically coupled Co^{II} ions. The zero-field *ac* susceptibility measurements show both in-phase $[\chi_M'(T)]$ and out-of-phase $[\chi_M''(T)]$ signals at *ca.* 2.5 K, in agreement with a canted antiferromagnet. The field-dependent magnetization, measured at 1.8 K, shows a two-step field-induced transition. The first step is related to the canted antiferromagnetism, while the second step corresponds to the transition to a polarized state. The critical field of the second step transition, indicated by dM/dH curve, is 27.5 kOe at 1.8 K. The occurrence of a long-range canted antiferromagnetism at very low temperature could be explained by its structural phase transition at low temperature and/or the very strong hydrogen bond interactions between the chains.

To achieve more "isolated" chains of $\{Co_2(hedpH)_2\}_n^{2n-}$ and thus to remarkably reduce the inter-chain couplings, a rigid *p*-phenylenediamine is introduced as template which leads to the formation of compound $(NH_3C_6H_4NH_3)Co_2(hedpH)_2 \cdot H_2O$.[25] The overall structure of this compound is similar to those of $[NH_3(CH_2)_nNH_3]Co_2(hedpH)_2 \cdot 2H_2O$ except that two crystallographically distinguished double chains of $\{Co_2(hedpH)_2\}_n$ are found in $(NH_3C_6H_4NH_3)Co_2(hedpH)_2 \cdot H_2O$. The inter-chain hydrogen bond interactions are also weakened in one direction with the O···O distances becoming 2.451 and 2.535 Å. The magnetic behavior of $(NH_3C_6H_4NH_3)Co_2$-$(hedpH)_2 \cdot H_2O$ is similar to that of $[NH_3(CH_2)_4NH_3]Co_2(hedpH)_2 \cdot 2H_2O$ with the critical field of 27.5 kOe at 1.8 K.

The observation of field-induced magnetic transitions in these Co-hedp compounds is related to the double chain nature of $\{Co_2(hedpH)_2\}_n$. The dominant antiferromagnetic interaction could be propagated within the ladder-like chain through μ-O and/or O–P–O bridges. As the Co···Co distance over the μ_3-O bridge (~3.3 Å) is much shorter than those over the O–P–O bridges along the leg (~5.4 Å) and the diagonal (~4.8 Å), the exchange coupling through the μ_3-O bridge ($J_\perp$) should be dominant. When the external magnetic field is increased from 0 to 70 kOe, magnetic transitions arising from a change of the electronic ground state occur. This phenomenon is similar to those for spin ladder systems. The critical fields, which reflect the spin gap between the singlet ground state and the first excited state, may be related to the Co–O–Co bond angles. Table 8.4 gives the Co–O–Co angles and the critical fields obtained for these compounds. Although no clear clue can be followed, a trend may be found for compounds $[NH_3(CH_2)_nNH_3]Co_2(hedpH)_2 \cdot 2H_2O$ ($n = 4, 5$) and $(NH_3C_6H_4NH_3)Co_2(hedpH)_2 \cdot H_2O$ with similar open-network structures. That is, with the Co–O–Co bond angle increasing, the critical field (H_c) becomes larger. Additionally, although low-dimensional antiferromagnetic behaviors are clearly observed for these compounds down to 1.8 K, long range magnetic ordering below 1.8 K cannot be entirely excluded.

Similar ladder-like chains are also observed in compounds Co(ahpdp)·H_2O and Co(ahbdpH_2)·H_2O where 3-ammonium-1-hydroxypropylidene-1,1-bisphosphonate (ahpdp^{4-}) or 4-amino-1-hydroxybutane-1,1-diphosphonate (ahbdp^{4-}) are used instead of hedp^{4-}.[26] The magnetic properties of the latter have been investigated. Field-induced magnetic transition is again observed. The critical field is *ca.* 35 kOe at 1.8 K.

A related compound, Co(hpedpH_2)(H_2O),[27] where hpedpH_4 is 1-hydroxy-2-(3-pyridyl)ethylidene-1,1-diphosphonic acid, shows a new type of chain structure (Figure 8.4). A significant feature of the structure is the unique coordination mode of hpedpH_2^{2-} within the chain. One of its two phosphonate groups caps on top of the {Co_2(μ-O)$_2$} dimer while the other connects the adjacent {Co_2(μ-O)$_2$} dimers through O–P–O units, hence leading to a novel type of ladder-like structure. The shortest Co···Co distances across the μ-O(P) and O–P–O unit are 3.050 and 5.140 Å, respectively. The Co–O–Co angle within the {Co_2(μ-O)$_2$} dimer is 90.8(1)°. The equivalent double chains are connected into a 3D network through π–π contacts as well as the hydrogen bond interactions (Figure 8.4). The hydrogen bond between the chains is very strong [O···O: 2.485(3) Å], similar to those of the {Co_2(hedpH)$_2$}$_n^{2n-}$ systems.

Table 8.4 The Co-O-Co angles (°), Co...Co distances (Å) and critical fields at 1.8 K (kOe) for some related compounds with double chain structures.

Compound	*Co-O-Co*	*Co···Co*	H_c	*Ref.*
$(NH_4)_2Co_2(hedpH)_2$	100.2(1)	3.236, 5.463, 4.754	35	23
$[NH_3(CH_2)_4NH_3]Co_2(hedpH)_2 \cdot 2H_2O$	98.8(1)	3.276, 5.487, 4.867	25	24
$[NH_3(CH_2)_5NH_3]Co_2(hedpH)_2 \cdot 2H_2O$	100.5(1)	3.236, 5.464, 4.865	27.5	24
$(NH_3C_6H_4NH_3)Co_2(hedpH)_2 \cdot H_2O$	100.4(1), 101.9(1)	3.307, 3.336, 5.465, 4.77	27.5	25
Co(ahbdpH_2)·H_2O	—	—	35	26a
Co(H_2O)(hpedpH_2)	90.8(1)	3.050, 5.140, 6.551	15	27
Co(2-pomp)$(H_2O)_2$	98.2(2)°	3.239, 5.109, 6.525	27.6	28

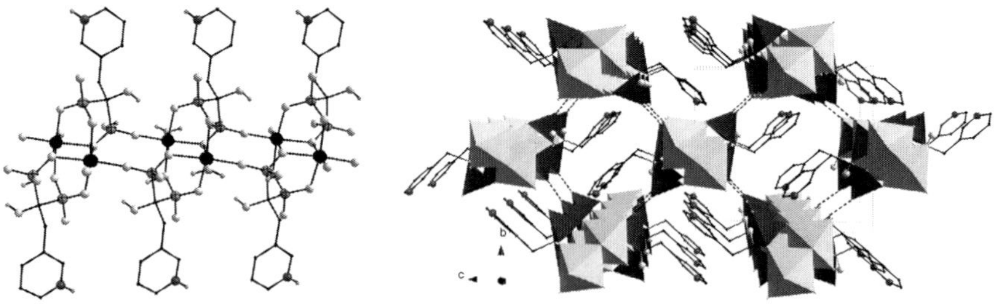

Figure 8.4 Chain fragment and packing diagram in Co(hpedpH_2)(H_2O). The inter-ladder strong hydrogen bonds are represented by dashed line (adapted from ref. 27).

Although structurally different, the magnetic pathways in Co(hpedpH_2)(H_2O) are identical to those in the Co-hedp double chain compounds. Again, the field-dependent magnetization of Co(hpedpH_2)(H_2O) at 1.8 K reveals a pronounced S-shaped curve, indicating the occurrence of a magnetic transition from the nonmagnetic ground state to a spin-polarized state. The transition field is *ca.* 15 kOe, which is much lower than those of $\{Co_2(hedpH)_2\}_n^{2n-}$ systems. Considering the smaller Co–O–Co bond angle noted above, the small transition field observed in Co(hpedpH_2)(H_2O) is in accordance with the trend found in $\{Co_2(hedpH)_2\}_n^{2n-}$ ladder-like compounds, e.g. the critical field (H_c) decreases with decreasing Co–O–Co bond angle.

It is interesting that the reaction of a monophosphonic acid, e.g. (2-pyridyl *N*-oxide) methylphosphonic acid [2-pompH_2, (2-C_5H_4NO)$CH_2PO_3H_2$] with metal salts results in compound Co(2-pomp)$(H_2O)_2$ with a similar chain structure.[28] Within the structure, the 2-pomp^{2-} ligand chelates and bridges the Co atoms to form a double chain along the *a* axis which contains $\{Co_2(\mu\text{-}O)_2\}$ dimers and O–P–O linkages (Figure 8.5). Within the dimer, the Co···Co distance is 3.239 Å, and Co–O–Co angle 98.2(2)°. The Co···Co distances across the O–P–O bridges are 5.109 and 6.525 Å, respectively, which are significantly larger than those found in the related Co-hedp compounds. Besides, the packing modes of the double chains are completely different from those of (NH_3–R–NH_3)$M_2(hedpH)_2 \cdot xH_2O$. In Co(2-pomp)$(H_2O)_2$, moderately strong hydrogen bonds and π–π stacking of pyridyl groups are found between the chains. The elongation of Co···Co distances over the O–P–O bridges would significantly reduce the magnetic exchanges through this pathway.

The magnetic responses of Co(2-pomp)$(H_2O)_2$ were investigated over a wide range of magnetic fields (up to 10 T) and temperatures (down to 50 mK). The dominant magnetic interactions are antiferromagnetic. Isothermal magnetic field sweeps at 50 mK provide signatures in the magnetic responses that are associated with antiferromagnetic to field-induced fully polarized (magnetically saturated) transitions. Analysis of the magnetic data indicates that the compound forms magnetic dimer-like clusters with weak dimer–dimer interactions present.

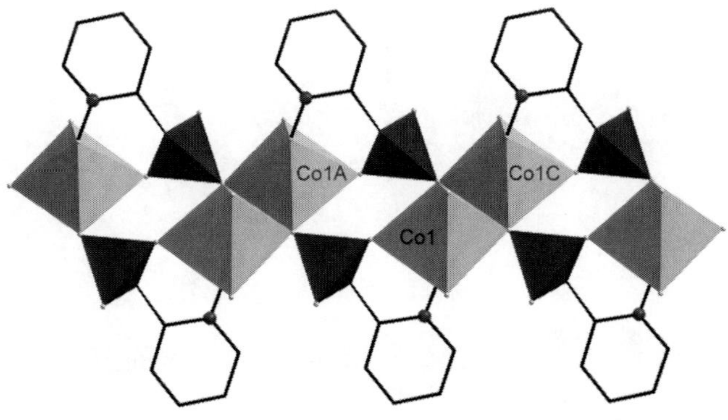

Figure 8.5 Chain fragment in Co(pomp)$(H_2O)_2$. (Adapted from ref. 28.)

8.3.2 Other Chains

The hydrothermal reaction of $4\text{-MeC}_6H_4CH_2N(CH_2PO_3H_2)_2$ (H_4L^2) with cobalt(II) acetate in a tetramethylammonium chloride aqueous solution leads to the formation of $Co(H_2L^2)(H_2O)$ with a 1D chain structure.[29] This compound crystallizes in monoclinic space group $P2_1/c$. The cobalt(II) ions are in a pseudo-octahedral environment consisting of a tridentate chelating diphosphonate ligand, two phosphonate oxygen atoms from adjacent $Co(H_2L^2)$ chelating units, and a water molecule. The cobalt(II) octahedra are interconnected into a 1D zigzag chain motif via corner-sharing phosphonate oxygen atom (Figure 8.6). The intrachain Co···Co separations are 3.775(1) Å over the O(P) atom and 6.217(1) Å over the bridging phosphonate group. Neighboring 1D chains in $Co(H_2L^2)(H_2O)$ are involved in hydrogen bonds between phosphonate oxygen atoms to form an overall layered architecture in the *ac* plane. The $MeC_6H_4CH_2$– substituents on the phosphonate groups extend into the spaces between layers. The nearest inter-chain Co···Co distance is 8.108(2) Å.

Magnetic studies reveal that the Co^{II} ions in the chain compound are antiferromagnetically coupled, but there is a net moment that is Ising in behavior due to spin canting. The spin canting arises from the fact that the crystallographic positions of neighboring Co^{II} ions are inequivalent because of the presence of 2_1 rotating axis. The magnetic exchange parameter is calculated to be $J = -10.5\ \text{cm}^{-1}$. The *ac* susceptibilities measurements show that both the real and imaginary components are strongly frequency dependent below 2.4 K (Figure 8.6). The nearly semicircle shape of plots χ'' versus χ' is evident in the Cole–Cole plots, indicating that there is only one relaxation process with an α of 0.15–0.35. The frequency dependent fit to the Arrhenius expression [$\tau(T) = \tau_0 \exp(\Delta/k_BT)$] results in parameters $\Delta = 18.6\ \text{cm}^{-1}$; $\tau_0 = 3.4 \times 10^{-9}\ \text{s}^{-1}$ for the in-phase signal and $\Delta = 20.2\ \text{cm}^{-1}$; $\tau_0 = 8.4 \times 10^{-10}\ \text{s}^{-1}$ for the out-of-phase signal (Δ: energy barrier, τ_0: the pre-exponential factor). The τ_0 values fall in the

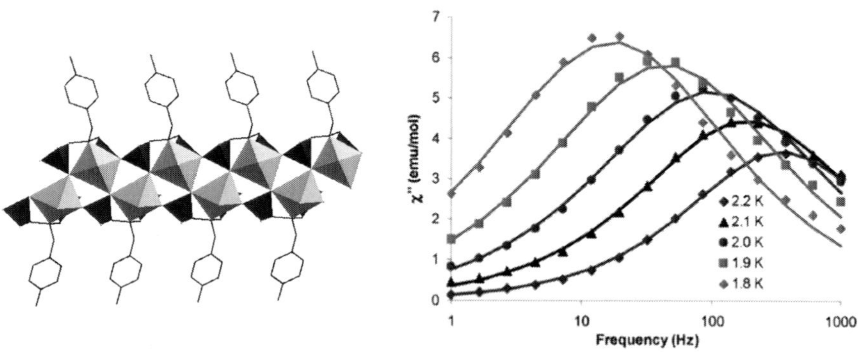

Figure 8.6 Chain fragment in $Co(H_2L^1)(H_2O)$ (left); The frequency dependent out-of-phase *ac* signals for $Co(H_2L^1)(H_2O)$ at different temperatures (right). (Reproduced with permission from ref. 29c. Copyright 2008 American Chemical Society.)

range for single-molecule and single-chain magnets ($10^{-8} - 10^{-10}$). The energy gap Δ obtained is in accord with the predictions of Glauber's theory for a 1D Ising system: $\Delta_{Glauber} = 2J_{||} = 21\ cm^{-1}$ considering the Hamiltonian $H = -2J\Sigma_i S_{zi} S_{z(i+1)}$ for an Ising chain. Thus, compound $Co(H_2L^2)(H_2O)$ provides the first example of a "single chain magnet" based on antiferromagnetically coupled homospin centers.

Chain compound $Co(pnaH)(H_2O)_3 \cdot H_2O$[30] ($pnaH_3$ = 6-phosphonic nicotinic acid) crystallizes in monoclinic space group $P2_1/n$. The Co^{II} ions again have a distorted octahedral environment. The CoO_5N octahedra are linked by PO_3C tetrahedra through vertex-sharing, forming 2_1 helical chains running along the *b* axis. The Co···Co distance over the O–P–O bridge is 4.889(1) Å. The chains are packed and stabilized by extensive hydrogen bonds among the phosphonate oxygen, carboxylate oxygen, and water molecules (Figure 8.7). The shortest inter-chain Co···Co distance is 4.802(1) Å. Considering that structure of Co(pnaH)$(H_2O)_3 \cdot H_2O$ contains 2_1 helical chains, spin canting could occur because of the lack of an inversion center between the adjacent Co^{II} spins. However, this possibility is excluded above 1.8 K by the magnetization measurements as well as the zero-field *ac* magnetic susceptibility measurements. Both in-phase and out-of-phase *ac* signals are frequency independent and no peaks appear above 1.8 K. The very weak magnetic interaction mediated through the O–P–O bridges could be responsible for the absence of the canted antiferromagnetic behavior down to 1.8 K.

Based on (benzimidazol-2-ylmethyl)iminobis(methylenephosphonic acid) [$(C_7H_5N_2)CH_2N(CH_2PO_3H_2)_2$, H_4L^4], compound $Co(H_2L^4)$[31] can be synthesized under hydrothermal conditions. It crystallizes in orthorhombic space group *Pbca*, and shows a 1D chain structure where CoO_3N_2 trigonal bipyramids are connected by PO_3C tetrahedra through corner-sharing. The chains are further connected to each other through aromatic stacking of benzimidazol rings and extensive hydrogen bonds, forming a supramolecular 3D structure (Figure 8.8). Magnetic susceptibility studies reveal that weak antiferromagnetic interactions are mediated between the metal centers. A theoretical analysis

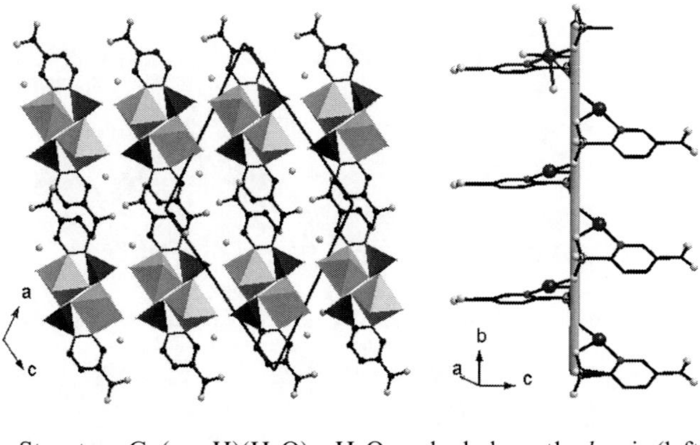

Figure 8.7 Structure $Co(pnaH)(H_2O)_3 \cdot H_2O$ packed along the *b* axis (left), and the 2_1 helical chain in the structure (right) (adapted from ref. 30).

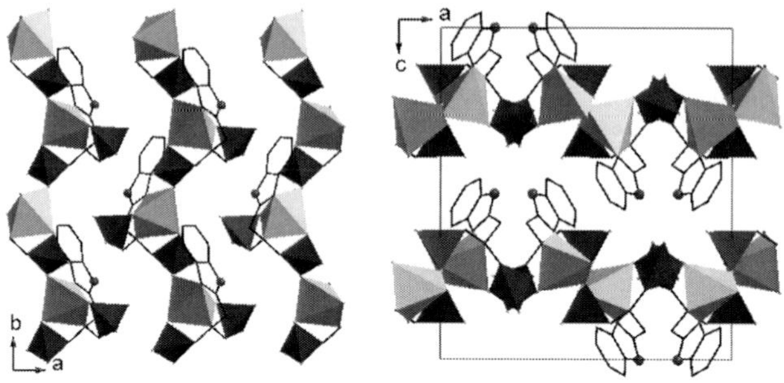

Figure 8.8 Structure Co{$(C_7H_5N_2)CH_2N(CH_2PO_3H)_2$} packed along *c* and *b* directions (adapted from ref. 31).

considering both zero field splitting (ZFS) and a small antiferromagentic interaction has been performed by means of the full-matrix diagonalization program MAGPACK, resulting in parameters $J = -0.15\ cm^{-1}$, $D = 7.6\ cm^{-1}$, $g_z = 2.33$, $g_{xy} = 2.08$ and TIP = 1.1×10^{-3}. The small negative *J* value is in accordance with the weak antiferromagnetic interactions over the O–P–O bridges. A chain compound in which the Co^{II} ions are bridged by O–P–O units is also reported for compound $Co(H_2L^7)_2(H_3L^7)_2$ based on $HOOCCH_2NH_2CH_2PO_3H_2$ (H_3L^7).[32]

Another type of chain structure is found in compound $[Co_2(L^8)(H_2O)_5(4,4'\text{-bipy})] \cdot 4H_2O$ $[H_4L^8 = H_2O_3PCH_2N(CH_2CO_2H)_2]$.[33] In this case, the 1D zigzag chains contain dimeric units in which two cobalt(II) ions are bridged by L^8 ligand, and these dimeric units are connected further by bridging 4,4′-bipy ligands. Compound $Co(phen)(4,4'\text{-bpy})(PhPO_3H)_2 \cdot 0.5H_2O$ (phen = 1,10-phenanthroline, 4,4′-bpy = 4,4′-bipyridine)[34] crystallizes in the tetragonal system with a $P4_32_12$ space group, and exhibits a 1D cobalt chain connected by 4,4′-bpy bridges. Each $PhPO_3H^-$ coordinates to the cobalt atom as a terminal ligand. Since the 4,4′-bpy is not an efficient magnetic mediator, the magnetic property of this compound is not interesting.

Chain compound $[Co(H_2L)(H_2O)_4(H_4L)] \cdot 2H_2O$ (H_4L = 1,8-octylenediphosphonic acid, $H_2O_3PCH_2(CH_2)_4CHCHCH_2PO_3H_2$) was synthesized in an acidic aqueous solution.[35] The octahedrally coordinated Co atoms are bridged by H_2L to form an infinite chain along the *c* axis. The remaining four coordination sites are filled with water molecules. The undissociated ligand H_4L is involved in the lattice (Figure 8.9).

8.4 Two-dimensional Cobalt Phosphonates

8.4.1 Monophosphonates without Additional Functional Groups

In the late 1970s, Cunningham *et al.*[36] reported the synthesis of divalent phosphonates $M(RPO_3)(H_2O)$ (M = Mg, Mn, Co, Cu, Ni). They correctly surmised that the metal was octahedrally coordinated but did not provide further structural detail. The X-ray single-crystal structures of Co(CH_3

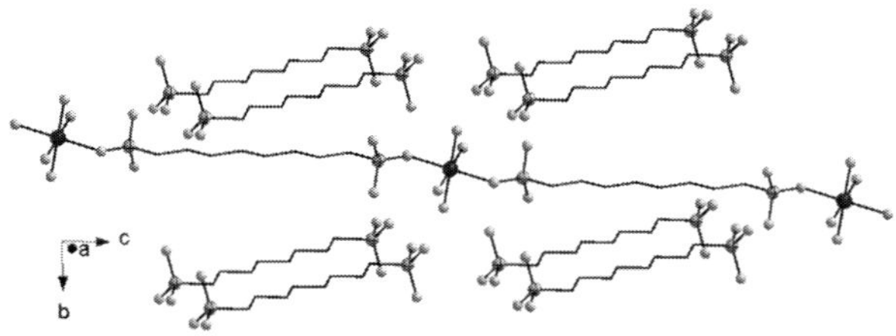

Figure 8.9 View of the crystal structure of compound $Co(H_2L^3)H_2O)_4 \cdot (H_4L^3) \cdot 2H_2O$ along the *a* axis (adapted from ref. 35).

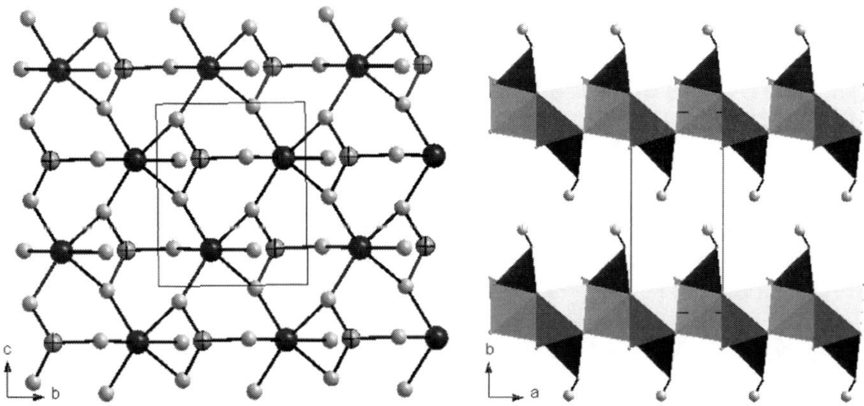

Figure 8.10 The layer structure and packing diagram of $Co(C_2H_5PO_3)(H_2O)$ (adapted from ref. 37).

$PO_3)(H_2O)$ and $Co(C_2H_5PO_3)(H_2O)$ were not determined until 2006 by Bauer *et al.*[37]

The structures of $Co(CH_3PO_3)(H_2O)$ and $Co(C_2H_5PO_3)(H_2O)$ are rather similar. Figure 8.10 shows the structure of $Co(C_2H_5PO_3)(H_2O)$ which is typical for layered metal monophosphonates with a general formulae $M(RPO_3)(H_2O)$ (R = alkyl, aromatic). The Co^{II} ion is six-coordinated by five phosphonate oxygen atoms and one from the water molecule. Each PO_3C tetrahedron is corner- and edge-shared with four CoO_6 octahedra, forming an inorganic layer containing three-membered rings. These layers are separated by bi-layers of alkyl groups and van der Waals contacts are established between them.

The compounds obey the Curie–Weiss law at temperatures above 100 K. The Curie (C) and Weiss (θ) constants are $C = 3.36\ cm^3\ K\ mol^{-1}$ and $\theta = -53$ K for $Co(CH_3PO_3)(H_2O)$ and $C = -3.62\ cm^3\ K\ mol^{-1}$ and $\theta = -75$ K for $Co(C_2H_5$-$PO_3)(H_2O)$, respectively. The effective magnetic moment of 5.18 μ_B is in the range of values observed for several high-spin octahedral Co^{II} ions. The negative values of θ are indicative of the presence of antiferromagnetic exchange couplings between the near-neighbor Co^{II} ions within the layers. The quantitative analysis of the magnetic properties for systems involving non-isolated metal ions with a significant orbital contribution is a difficult task.[38] Attempts to fit the susceptibility data, especially over the entire range of measured temperatures, by using the classical Heisenberg 2D square model and 2D Ising model were unsuccessful.

The single crystal structure of $Co(PhPO_3)(H_2O)$ was reported by Salami *et al.*[39] which is identical to those of $Co(C_nH_{2n+1}PO_3)(H_2O)$ ($n = 1$, 2) except that the phenyl groups occupy the interlayer spaces. The material undergoes an antiferromagnetic transition at a relatively low Néel temperature of 4.0 K. The Weiss constant is –76.5 K, indicating the dominant antiferromagnetic interactions between the Co^{II} ions. Yao *et al.*[40] studied the magnetic properties of $Co(PhPO_3)(H_2O)$ by first principles within the density-functional theory (DFT) and the full potential linearized augmented plane wave (FP-LAPW) method. The calculations reveal that this compound is a metallic antiferromagnet.

Cao *et al.*[41] found that compound $Co(CH_3PO_3)(H_2O)$ can be dehydrated and that the resulting dehydrated phosphonates remain crystalline. The topochemically dehydrated metal phosphonate shows shape selectivity in forming intercalation compounds; *i.e.* only primary amines are absorbed, while amines with branching at the α-position are excluded. The new intercalation compounds have the general formula $Co(CH_3PO_3)(RNH_2)$. A pure compound $Co(CH_3PO_3)(NH_3)$ obtained by reacting $Co(CH_3PO_3)$ with pure gaseous ammonia can be indexed with a set of unit cell constants $a = 5.89$ Å, $b = 8.92$ Å, and $c = 4.85$ Å in space group *Pmn*2, which is isostructural to $Co(CH_3PO_3)(H_2O)$. In a parallel study, Frink *et al.*[42] found that size-selective intercalation of the closely related anhydrous phenylphosphonate $Co(PhPO_3)$ occurs. In this case the size of the bonding pocket is restricted by the phenyl rings, so that only ammonia and water, and not larger amines, can intercalate.

Langmuir–Blodgett (LB) films of cobalt octadecylphosphonate were prepared by Talham and co-workers[43] and shown to form mixed organic/inorganic layered films of the formula $Co(O_3PC_{18}H_{37}) \cdot H_2O$ possessing 2D continuous solid metal phosphonate layers. The films were shown to be LB analogs of the known solid-state layered cobalt phosphonates. Like the solid-state analogs, the cobalt phosphonate LB films can be dehydrated and reacted to uptake ammonia from the vapor phase.

8.4.2 Phosphonates with Additional Functional Groups

By introducing one or two additional functional groups such as amino, carboxylate, pyridyl and a second phosphonate group into the monophosphonate ligand, metal phosphonates with versatile structures can be obtained, many of which show layer structures. Methylenediphosphonate and its derivatives are good candidates for the construction of layered metal phosphonates.

Based on 1-aminoethylidenediphosphonate [aedpH_4, $CH_3C(NH_2)(PO_3H_2)_2$], compound $Co_3(aedpH_2)_2(aedpH_3)_2(H_2O)_4 \cdot 2H_2O$ was obtained which crystallizes in monoclinic space group C2/c.[44] The structure contains a trimer unit of $Co_3(aedpH_2)_2$ in which two equivalent phosphonate ligands chelate and bridge

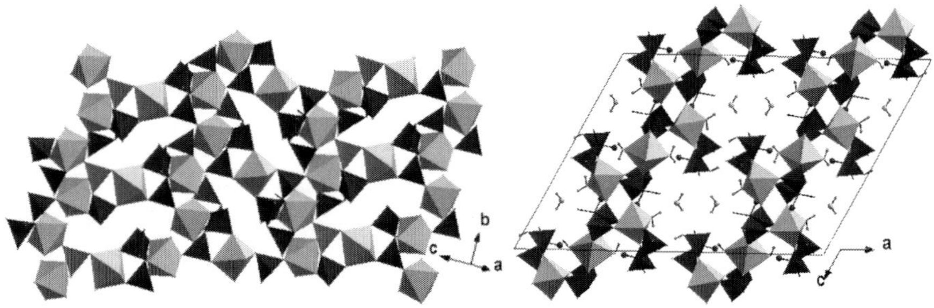

Figure 8.11 The layer structure and packing diagram of $Co_3(aedpH_2)_2(aedpH_3)_2(H_2O)_4 \cdot 2H_2O$ (adapted from ref. 44).

the three cobalt ions. Each trimer unit is further linked to its four equivalent neighbors through corner-sharing of CoO_6 octahedra and PO_3C tetrahedra, forming a 2D inorganic layer in the *bc* plane which contains 12-membered rings. These layers are connected to each other by extensive hydrogen bonds (Figure 8.11). Magnetic studies show that weak antiferromagnetic interactions are mediated between the cobalt ions, in accordance with the fact that the Co^{II} ions are connected to each other purely through the O–P–O bridges within the layer. The Weiss constant determined by the magnetic susceptibility data above 100 K is –13 K.

Using 2-(1-imidazole)-1-hydroxyl-1,1′-ethylidenediphosphonic acid [$ImhedpH_4$, $(1\text{-}C_3H_3N_2)CH_2C(OH)(PO_3H_2)_2$], compound $Co_3(ImhedpH)_2(H_2O)_4 \cdot 2H_2O$[12] was isolated which crystallizes in triclinic space group P-1 and shows a layered structure made up of $Co_3(ImhedpH)_2(H_2O)_4$ trimer units. These trimers are connected by edge-sharing of the CoO_6 octahedra, forming an undulating chain along the *c* axis which contains $\{Co_2O_2\}$ dimer unit. The Co···Co distance and Co–O–Co angle within this dimer are 3.023 Å and 92.8(1)°, respectively. The adjacent chains are fused through the coordination of the phosphonate oxygen atom from one chain to the Co2 atom from the other, thus generating a 2D layer containing 4-, 8- and 16-membered rings. The Co···Co distances across the O–P–O units are 5.372, 4.561, 4.859, and 6.336 Å. The imidazole groups are grafted on the two sides of the layer (Figure 8.12).

Magnetic measurements reveal a ferrimagnetic behavior with dominant antiferromagnetic interactions between the Co^{II} ions. The Weiss constant is –12.3 K, determined by the susceptibility data between 90 and 300 K. Below 2.6 K, compound $Co_3(ImhedpH)_2(H_2O)_4 \cdot 2H_2O$ experiences a long-range magnetic ordering due to the spin canting with the canting angle α of 1.1°, calculated from the expression $\sin(\alpha) = M_R/M_S$. No magnetic hysteresis is present even at 1.8 K which classifies this material as a soft weak ferromagnet. When the external field is further increased, the magnetization experiences an S-shaped increase, corresponding to a metamagnetic transition from a weak ferromagnetic state to a ferrimagnetic state. The critical field of this metamagnetic transition is *ca.* 38 kOe (Figure 8.13). It is interesting that slow

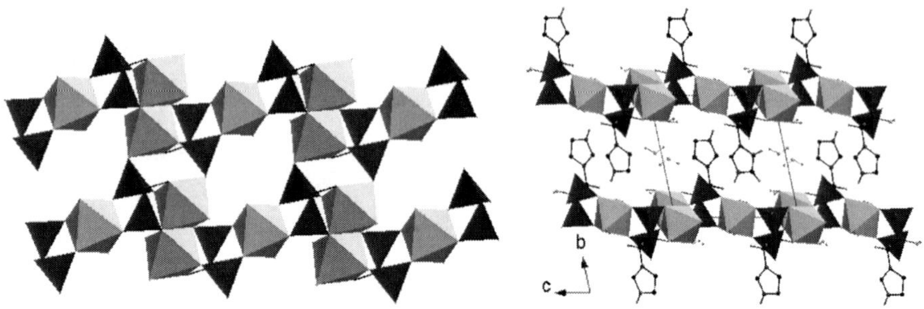

Figure 8.12 The layer structure and packing diagram of $Co_3(ImhedpH)_2(H_2O)_4 \cdot 2H_2O$ (adapted from ref. 12).

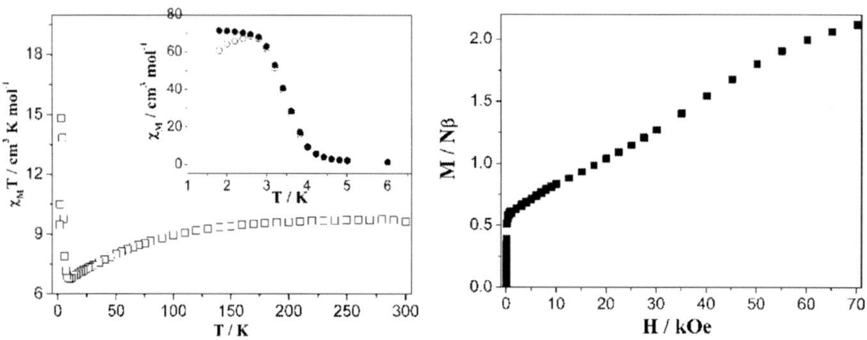

Figure 8.13 The $\chi_M T$ versus T (inset: ZFCM/FCM) and M versus H (at 1.8 K) plots for $Co_3(ImhedpH)_2(H_2O)_4 \cdot 2H_2O$. (Reproduced with permission from ref. 12. Copyright 2007 American Chemical Society.)

relaxation is observed from the *ac* susceptibility measurements below 2.6 K. The shift of the peak temperature (T_p) of χ_M' is measured by a parameter $\phi = (\Delta T_p/T_p)/\Delta(\log f) = 0.03$ (f is the frequency of H_{ac}), which is in the range of a normal spin-glass. The frequency dependence of T_p on χ_M'' can be fitted well to the Arrhenius law: $\tau(T) = \tau_0 \exp(\Delta/k_B T)$. The best fit results in parameters $\tau_0 = 1.2 \times 10^{-10}$ s, $\Delta = 46$ K. The activation energy may be compared with those typical for the domain wall movements.

Compound $Co_3(hpedpH)_2(H_2O)_4 \cdot H_2O$ ($hpedpH_4$ = 1-hydroxy-2-(3-pyridyl) ethylidene-1,1-diphosphonic acid)[45] possesses an identical layer structure to that of $Co_3(ImhedpH)_2(H_2O)_4 \cdot 2H_2O$.[12] In this case, the Co···Co distances are 3.031(1) Å over the μ_3-O bridge and 4.561(1), 4.910(1), and 5.412(1) Å across the O–P–O units. The Co–O2–Co angle is 93.4(2) Å. These layers stack through extensive hydrogen bonds and π–π contact of pyridyl groups to complete a 3D network with lattice water molecules sitting in the inter-layer cavities (Figure 8.14). The interlayer separation, calculated by c sinα, is 10.117 Å which is much larger than that in $Co_3(ImhedpH)_2(H_2O)_4 \cdot 2H_2O$ (8.131 Å, calculated by b sinα). Unlike $Co_3(ImhedpH)_2(H_2O)_4 \cdot 2H_2O$ which shows canted antiferromagnetism below 2.6 K, compound $Co_3(hpedpH)_2(H_2O)_4 \cdot H_2O$ turns out to be a soft ferrimagnet with $T_c = 3.8$ K. This is attributed to the larger interlayer space in the latter case which favors the ferromagnetic interactions between the layers. Slow magnetic relaxation is also observed below 3.8 K. The relaxation time (τ_0) and the energy gap (Δ) are 1.8×10^{-9} s and 35 K, respectively, suggesting a thermally activated mechanism typical for the domain wall movements.

Based on *N*-cyclohexylaminomethanediphosphonic acid ($cmdpH_4$), $Co_3(cmdpH)_2(H_2O)_2$ is obtained.[46] Although its single crystal structure is not solved, it is isostructural to $Zn_3(cmdpH)_2(H_2O)_2$ and shows a double-layer structure. The double layer is composed of two single layers, interconnected by MO_4 tetrahedra. Within the single layer, the MO_6 and PO_3C polyhedra are corner-shared with each other to form a layer containing 3- and 7-membered rings. The organic moieties of $cmdpH^{3-}$ are grafted on the two sides of the double-layer. Dominant antiferromagnetic interactions are found between the Co^{II} centers, confirmed by the negative Weiss constant ($\theta = -31.6$ K) determined by Curie–Weiss fit of the magnetic susceptibility data above 50 K. The isothermal field dependence of magnetization at 1.8 K reveals that the magnetization increases gradually with increasing external field and displays an inflexion at about 30 kOe, characteristic of a spin-flop behavior. The magnetization at 70 kOe is 3.89 $N\beta$ which is far from the saturation value for three Co^{II}.

Jankovics *et al.*[47] isolated compound $Co(H_3L^1)_2(H_2O)_2$ in aqueous solution at pH 2 using an amino-dimethylphosphonic acid (H_4L^1, $H_2O_3PCH_2NH_2^+CH_2PO_3H^-$) as the starting material. The compound crystallizes in monoclinic space group $P2_1/n$. The octahedrally coordinated cobalt(II) ions are each bridged by four phosphonate ligands forming an inorganic–organic hybrid layer containing tetranuclear 16-membered rings, with cavities extending in two dimensions and holes in the third dimension throughout the lattice.

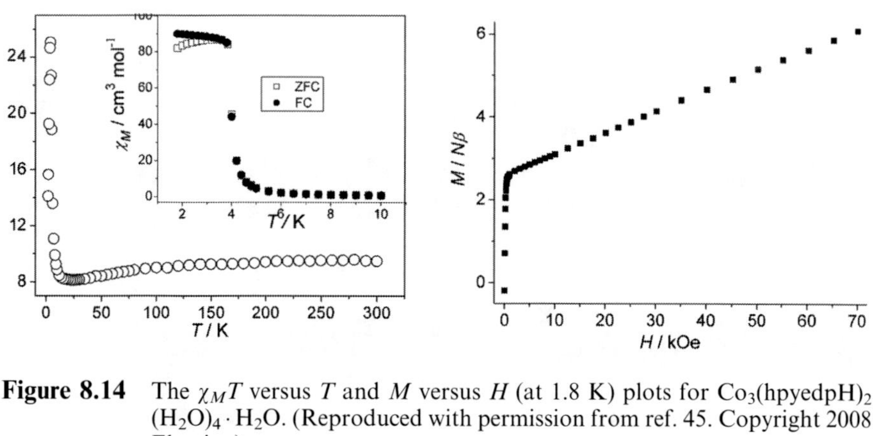

Figure 8.14 The $\chi_M T$ versus T and M versus H (at 1.8 K) plots for $Co_3(hpyedpH)_2(H_2O)_4 \cdot H_2O$. (Reproduced with permission from ref. 45. Copyright 2008 Elsevier.)

The remaining two sites of the Co^{II} ion are completed by two water molecules (Figure 8.15).

The magnetic behavior of $Co(H_3L^1)_2(H_2O)_2$ is in principle a single-ion property. Thus the susceptibility data were fitted by the model derived for isolated $S=3/2$ spin based on Hamiltonian (Equation 8.1). The fitting taking into account only the axial part of distortion of the crystal field D, and an isotropic g-value, provide the following parameters: $D=40\ cm^{-1}$ and $g=2.33$. The large value of the D parameter is in accordance with an octahedral Co^{II}, where the ground state doublet is well-isolated from the excited ones. X-Band EPR measurements were carried out in powder samples of $Co(H_3L^1)_2(H_2O)_2$ as well as in solutions of it in water. As a consequence of the fast spin-lattice relaxation time of high-spin Co^{II}, signals were observed only below 40 K. The g-values of the powder sample are $g_{||}=6.4$, $g_{\perp}=2.93$, which give an effective g-value of 4.1. In solution, the g_{eff} value of Co^{II} is isotropic and equal to the value of 4.1.

Based on the same ligand (H_4L^1), Mao and co-workers obtained compound $Co(H_2L^1)(H_2O)_2 \cdot H_2O$.[48] The compound crystallizes in monoclinic space group P2$_1$/c. Each CoO_6 octahedron is corner-shared with four PO_3C tetrahedra, while each PO_3C is linked to two CoO_6 octahedra and the nitrogen atom. Thus a 2D hybrid layer structure is constructed which contains 4-, 6- and 8-membered rings. The Co···Co distances over the O–P–O bridges are 5.298(1) and 5.483(1) Å. The layers are stacked along the [100] direction with the interlayer distance of about 8.7 Å. The lattice water fills in the interlayer space with extensive hydrogen bond interactions (Figure 8.16). The temperature-dependent magnetic susceptibility data obey the Curie–Weiss law in 50–300 K with a Weiss constant of –23.3 K.

Compound $Co(HL^7)(H_2O)_2$ based on $HOOCCH_2NHCH_2PO_3H_2$ displays a layer structure where the CoO_6 and PO_3C polyhedra are corner-shared to form a layer containing 4- and 8-membered rings.[32] The Weiss constant determined by the magnetic susceptibility data above 9 K is –29.4(3) K.

Based on $HO_2CCH_2N(CH_2PO_3H_2)_2$ (H_5L^5), Song *et al.*[49] synthesized compound $Co_3(H_2L^5)_2(H_2O)_4(4,4'\text{-bipy})_2 \cdot 11H_2O$ in the presence of 4,4′-bipy. It crystallizes in monoclinic space group P2/c. Three Co^{II} ions are chelated and

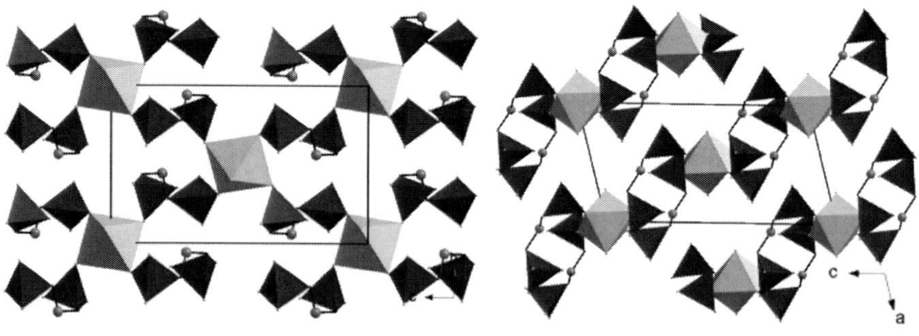

Figure 8.15 The layer structure and packing diagram of $Co(H_3L^1)_2(H_2O)_2$ (adapted from ref. 47).

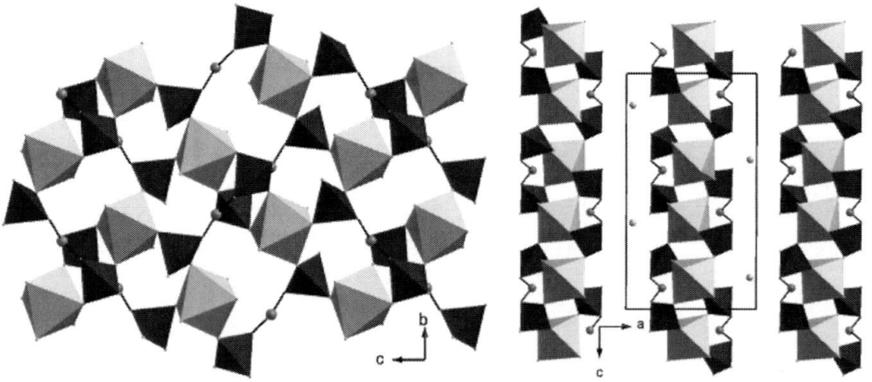

Figure 8.16 The layer structure and packing diagram of $Co(H_2L^1)(H_2O)_2 \cdot H_2O$ (adapted from ref. 48).

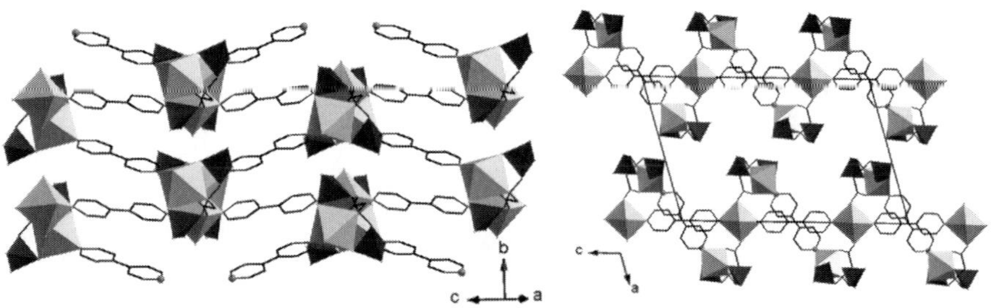

Figure 8.17 The layer structure and packing diagram of $Co_3(H_2L^5)_2(H_2O)_4(4,4'\text{-bipy})_2 \cdot 11H_2O$ (adapted from ref. 49).

bridged by the phosphonate ligands to form a trinuclear unit. Each trimer unit is linked to four equivalents via bridging 4,4′-bipy ligands to form a 2D layer. The non-coordinated phosphonate oxygen atoms are involved in hydrogen bonding with aqua ligands and interstitial water molecules (Figure 8.17). Magnetic measurements in the temperature range of 2–300 K reveal that antiferromagnetic interactions are propagated between Co^{II} ions within the trinuclear $Co_3(H_2L^5)_2$ unit (Co · · · Co 5.317(1) Å) together with the zero-field splitting effects of the single Co^{II} ion.

The reaction of cobalt(II) salt with amino-dimethylphosphonic acid containing a longer carboxylate tail, e.g. $HO_2C(CH_2)_3N(CH_2PO_3H_2)_2$ (H_5L^6), results in compound $Co(H_4L^6)_2$ with a layered architecture shown in Figure 8.18.[50] The Co^{II} ion is octahedrally coordinated by six phosphonate oxygen atoms from four ligands. Neighboring CoO_6 octahedra are interconnected by phosphonate groups into a 2D layer with a 4,4-net topology. Adjacent layers are further cross-linked via hydrogen bonds between the non-coordinate carboxylate groups and non-coordinate phosphonate oxygens.

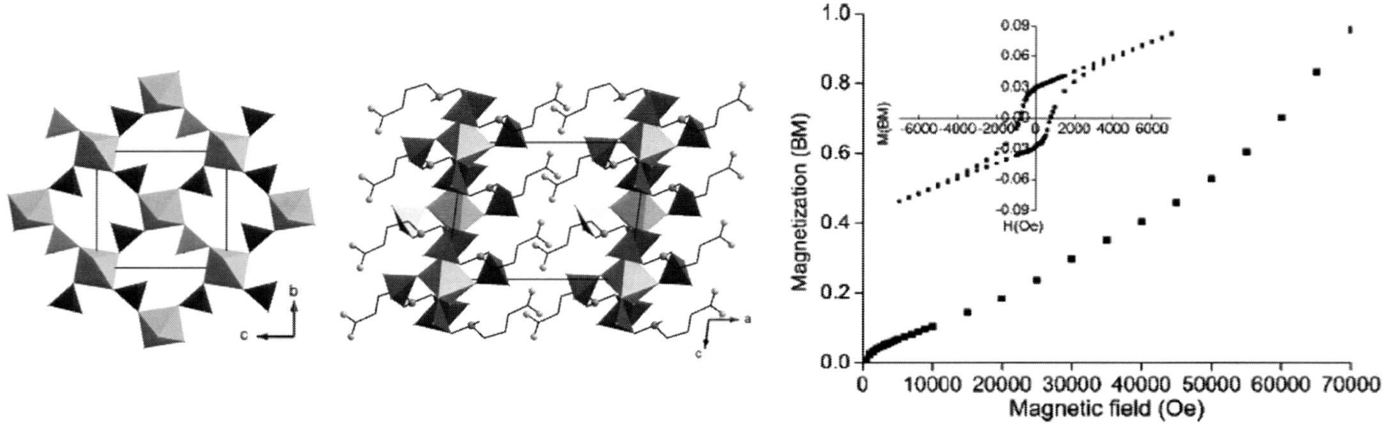

Figure 8.18 The layer structure and packing diagram of $Co(H_4L^6)_2$, and the field-dependent magnetization and hysteresis loops measured at 1.8 K (adapted from ref. 50).

The magnetic properties of this compound have been investigated in detail. The susceptibility data obey the Curie–Weiss law above 10 K, with a Curie constant of 3.86 cm^3 K mol^{-1} and a Weiss constant of –42 K, indicating dominant antiferromagnetic interactions and the spin-orbit coupling of the cobalt(II) ion in a slightly distorted octahedral geometry. The data can be fit by an expression for a mononuclear Co^{II} complex in an octahedral environment, based on Hamiltonian $H = -\lambda LS$.[51] The exchange parameter (J) is estimated to be –4.2 cm^{-1}. The low-temperature part of magnetic susceptibility was alternatively interpreted through a model based on a 2D Heisenberg quadratic-layer antiferromagnet, giving the exchange parameter of –4.1 cm^{-1}.[52] The FC/ZFC measurements exhibit bifurcation at $T_c = 8.75$ K indicating the occurrence of long-range ferromagnetic ordering, which is supported by the *ac* measurements. The magnetization curves below T_c exhibit a hysteresis effect with a coercive field and a small net magnetization M_r. The coercive field is 800 Oe, and $M_r = 0.03$ μ_B at 1.8 K. The spontaneous magnetization is due to the spin canting with the canting angle estimated to be 0.75°. The field-dependent behavior is indicative of a metamagnetic or spin-flop phase transition exhibited for weakly coupled antiferromagnets.

Based on *N*-(phosphonomethyl)iminodiacetic acid ($H_2O_3PCH_2N(CH_2CO_2H)_2$, H_4L^8), compound $Co_2(L^8)(H_2O)_5 \cdot H_2O$ is obtained.[53] It crystallizes in monoclinic space group $P2_1/c$ and displays a layered structure. The cobalt atoms in the structure are all octahedrally coordinated. The Co2 atoms are chelated and bridged by the L^8 ligand, forming a Co/carboxylate chain running along the *c* axis; such chains are held together by hydrogen bonds formed between carboxylic oxygen atoms and lattice water molecules to form a layer in the *bc* plane. The chains are also linked to the $Co1O_6$ octahedra through corner-sharing with PO_3C tetrahedra, forming a covalently connected layer. These layers are packed along the *a* direction, resulting in a 3D network (Figure 8.19). Magnetic properties were not investigated.

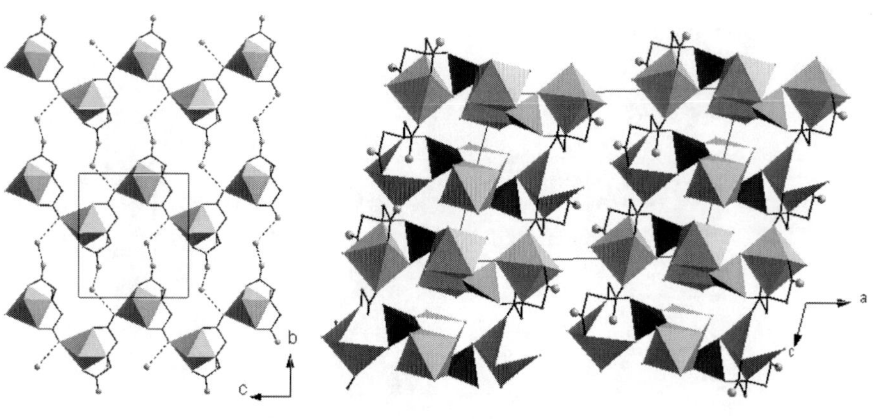

Figure 8.19 The layer structure and packing diagram of $Co_2(L^8)(H_2O)_5 \cdot H_2O$ along the *b* axis (adapted from ref. 53).

In cobalt(II) phosphonate $(enH_2)_{0.5}[Co(hpa)(H_2O)]\cdot H_2O$ ($hpaH_3$ = hydroxyphosphonoacetic acid, $HO_2CCH(OH)PO_3H_2$; en = ethylenediamine),[54] the inorganic layer is made up of corner-sharing CoO_6 and PO_3C polyhedra, which contains 4- and 8-membered rings. The protonated ethylenediamine and lattice water occupy the interlayer space with extensive hydrogen bond interactions (Figure 8.20). Magnetic studies indicate that the magnetic behavior follows the Curie–Weiss law above 25 K with $C = 3.63\ cm^3\ K\ mol^{-1}$ and $\theta = -26.2$ K. Long-range magnetic ordering arising from the spin canting is observed below 3.3 K. The canting angle is estimated to be 0.5°.

Compound $Co_3(O_3PCH_2NC_4H_7CO_2)_2\cdot 5H_2O$ based on a chiral (*S*)-$HO_3PCH_2NHC_4H_7CO_2H$ (H_3L^9) displays a different layer structure.[55] The inorganic layers are constructed from pairs of edge-sharing Co-centered octahedra which are linked to the PO_3C tetrahedra through corner-sharing. The layers stack in the [001] direction, generating channels in the same direction in which the water molecules reside. The organic part of the phosphonate anion is directed into the interlayer region (Figure 8.21). It is interesting to note that the

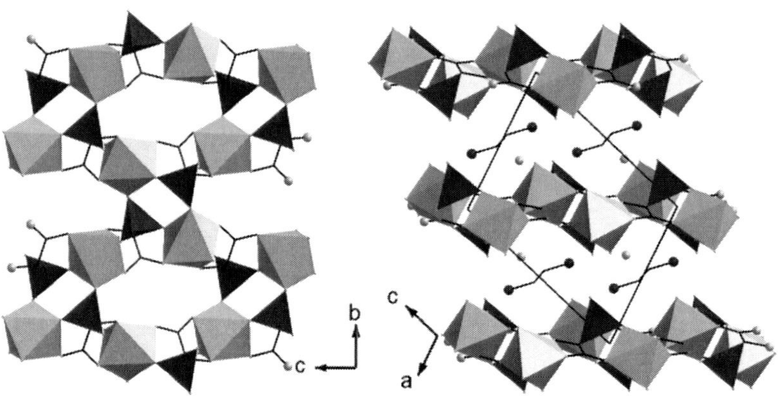

Figure 8.20 The layer structure and packing diagram of complex $(enH_2)_{0.5}[Co(hpa)(H_2O)]\cdot H_2O$ (adapted from ref. 54).

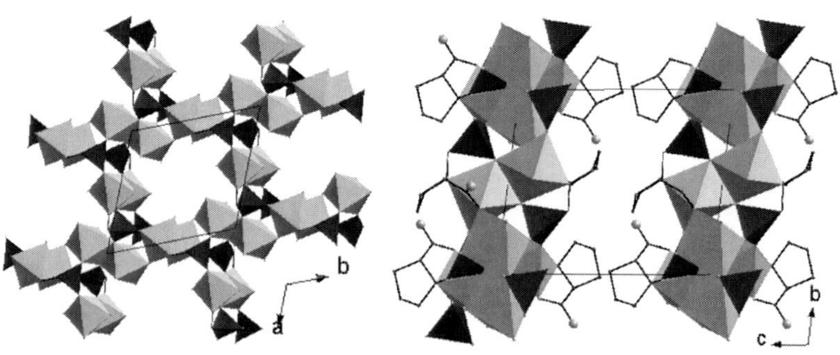

Figure 8.21 The layer structure and packing diagram of $Co_3(O_3PCH_2NC_4H_7CO_2)_2\cdot 5H_2O$ (adapted from ref. 55).

chirality of the proline-derived phosphonic acid is not kept in the product. Half of the phosphonate anions have been inverted and are now present as the *R* enantiomer, *i.e.* the acid has been racemized during the synthetic procedure. The racemization could be caused by the experimental conditions employed in the preparation of this compound.

Layer structure is also found in $(H_3O)_2[Co_3(OH)_2(OOCCH_2PO_3)_2]$, which crystallizes in the triclinic *P*-1 space group.[56] The topology of the inorganic layer in the structure is different from those found in the other compounds, and can be best described as the cobalt octahedra ribbons cross-linked by PO_3C tetrahedra through corner-sharing. Within the ribbon, the Co1 and Co2 octahedra are edge-shared by one μ-OH and one phosphonate oxygen, while the two Co2 octahedra are edge-shared by two phosphonate oxygens (Figure 8.22). The Co1 · · · Co2 and Co2 · · · Co2 distances are 3.187(1) and 2.184(1) Å, respectively. The Co1–O–Co2 angles are 94.0(1)° and 99.4(1)° over the O(H) and O(P) bridges. The Co2–O–Co2 angle is 99.1°. The temperature dependence of the magnetic susceptibility, measured in the temperature range of 2–300 K under 1 kOe, reveals that the $\chi_M T$ value decreases continuously upon cooling, indicating antiferromagnetic coupling. The data above 20 K were fitted to the Curie–Weiss law, resulting in parameters $C = 9.559$ cm^3 K mol^{-1} and $\theta = -13.2$ K.

Compound $Co(3\text{-}HOOCC_6H_4PO_3)(H_2O)$[57] shows a layer structure very similar to those found in $Co(RPO_3)(H_2O)$. Between the layers, the organic molecules are bonded together through hydrogen bonds formed by the carboxylate groups, resulting in a long interlayer distance (~ 16.2 Å). Temperature dependent magnetic susceptibilities have been measured. The Weiss constant is $\theta = -13$ K.

The involvement of rigid aromatic or heteroaromatic rings into the phosphonate ligands could result in layered cobalt phosphonates with new architectures. Figure 8.23 shows the structure of compound $Co(2\text{-}OOCC_6H_4PO_3H)(H_2O)$[58] where

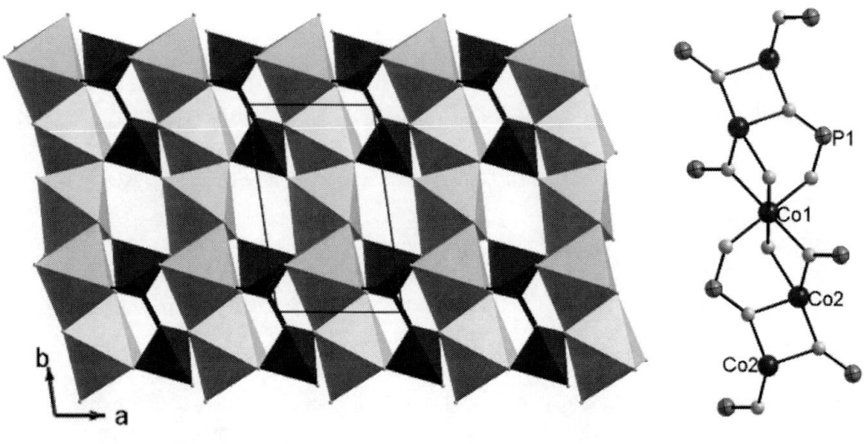

Figure 8.22 The layer structure and the single chain structure of cobalt in $(H_3O)_2[Co_3(OH)_2(OOCCH_2PO_3)_2]$ (adapted from ref. 56).

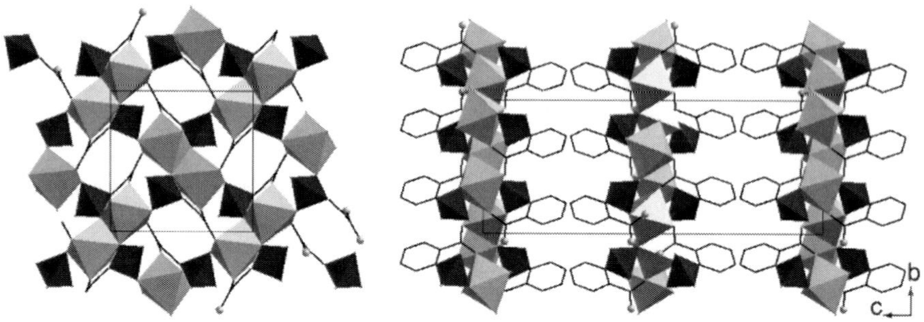

Figure 8.23 The layer structure and packing diagram of Co(2-$OOCC_6H_4$ PO_3H)(H_2O) (adapted from ref. 58).

the 2-carboxylphenylphosphonate (2-cpp^{3-}) ligand is incorporated. It crystallizes in the orthorhombic space group *Pbca*. The Co atom has a distorted octahedral environment, with the six coordination sites provided by three phosphonate oxygen, two carboxylate oxygen atoms, and one water molecule. The two equivalent CoO_6 octahedra are edge-shared, forming a {Co_2O_2} dimer. The Co···Co distance within this dimer is 3.181(2) Å. The Co–O–Co angle is 98.7(1)°. These dimers are connected by both carboxylate and phosphonate groups, thus leading to a 2D layer. Neighboring layers are connected through van der Waals interactions. The inter-layer separation is 11.253 Å. A dominant antiferromagnetic interaction is found in this compound. A net magnetic moment exists at low temperature (<26 K), possibly due to the weak ferromagnetic interaction between the antiferromagnetic chains.

Based on [hydroxy(4-pyridyl)methyl]phosphonate (4-$hpmp^{2-}$), a layer compound Co(4-hpmp)(H_2O) is obtained.[59] The compound crystallizes in the monoclinic space group C2/c. The Co atom is five-coordinated with a distorted trigonal dipyramidal geometry, coordinated by one pyridyl nitrogen, three phosphonate oxygens, and one water molecule. Each CoO_4N trigonal dipyramid is corner-shared by three PO_3C tetrahedra to generate an inorganic "staircase" chain which is linked together through the organic moieties of the 4-$hpmp^{2-}$ ligand, forming a 2D layer (Figure 8.24). Moderately strong hydrogen bonds are found between the layers. Magnetic study shows very weak antiferromagnetic exchange between the magnetic centers with a Weiss constant of −12.7 K.

Compound Co(2-pmp)$(H_2O)_2$ (2-$pmpH_2$ = 2-pyridylmethylphosphonic acid) shows a buckled layer structure in which the $CoNO_5$ octahedron is corner-shared with three PO_3C tetrahedra and vice versa.[60] Within the layer, the Co···Co distances over the O–P–O units are 5.240(1), 5.262(1), and 6.221(1) Å. The organic part of 2-pmp^{2-} is located between the layers with an interlayer distance of 10.697 Å (Figure 8.25). Two water molecules are coordinated to Co^{II} which can be completely removed in two-steps upon heating in the temperature ranges 50–130 °C and 165–216 °C, respectively. The dehydration process is accompanied by a distinct color change from red to purple to deep

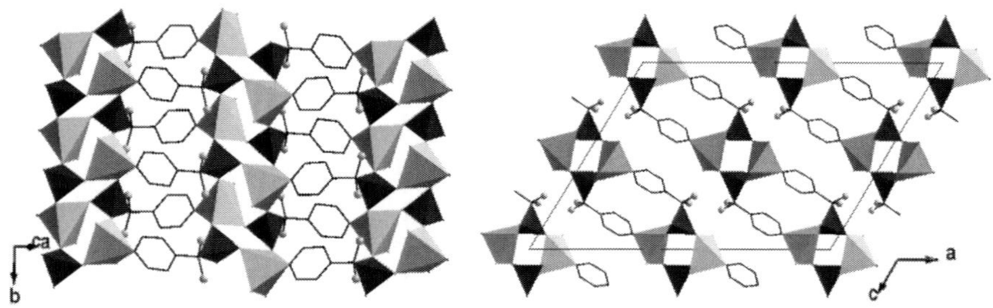

Figure 8.24 The layer structure and packing diagram of Co(4-hpmp)(H_2O) (adapted from ref. 59).

blue, suggesting the formation of monohydrate Co(2-pmp)(H_2O) and anhydrous Co(2-pmp), respectively. After immersing in water, Co(2-pmp) changes back to the initial hydrated form of Co(2-pmp)(H_2O)$_2$ with the color returning back to red within a few minutes.

Ab initio crystal structural determinations with consecutive Rietveld refinements of powder samples Co(2-pmp)(H_2O) and anhydrous Co(2-pmp) at ambient conditions reveal that both compounds have monoclinic crystal structures in $P2_1/n$ space group. They maintain similar layer structures where corner-sharing of $CoNO_x$ and PO_3C polyhedra leads to an inorganic layer containing 4- and 8-membered rings. But the coordination geometries around the Co atoms change remarkably from octahedral to distorted trigonal bipyramidal in Co(2-pmp)(H_2O) and a distorted tetrahedral environment in Co(2-pmp). Consequently, the Co$\cdots$Co distances over the O–P–O units are 5.101(1), 5.563(1), and 6.248(1) Å in Co(2-pmp)(H_2O), and 4.702(5), 5.566(5), and 5.816(4) Å in Co(2-pmp). Furthermore, the interlayer spaces are significantly shortened due to the release of coordination water (6.525 Å for Co(2-pmp)(H_2O) and 5.614 Å for Co(2-pmp)).

Magnetic measurements reveal a dominant antiferromagnetic interaction between the Co^{II} centers together with strong spin-orbit couplings with an antiferromagnetic ground state in all three cases. Since the Co atoms are tetrahedrally coordinated in Co(2-pmp), the susceptibility data can be fitted using an isotropic dimer model based on the Heisenberg Hamiltonian $H = -2J\mathbf{S}_A \cdot \mathbf{S}_B$, with zJ' accounting for the inter-dimer couplings, leading to parameters $g = 2.41$, $2J = -4.85$ cm^{-1}, $zJ' = -2.48$ cm^{-1}. Metamagnetism is observed in Co(2-pmp)(H_2O)$_2$ with critical fields at 1.8 K being 30 kOe. It is interesting that the metamagnetic behavior of Co(2-pmp)(H_2O)$_2$ can be tuned by a reversible dehydration–hydration process. The critical fields of the field-induced transition become 18.1 kOe for Co(2-pmp)(H_2O) and 66.2 kOe for Co(2-pmp), respectively. After Co(2-pmp) is rehydrated, the magnetic properties are recovered reversibly. Thus, compound Co(2-pmp)(H_2O)$_2$ provides the first example of metal phosphonates that show reversible metamagnetic behaviors upon dehydration and rehydration process.

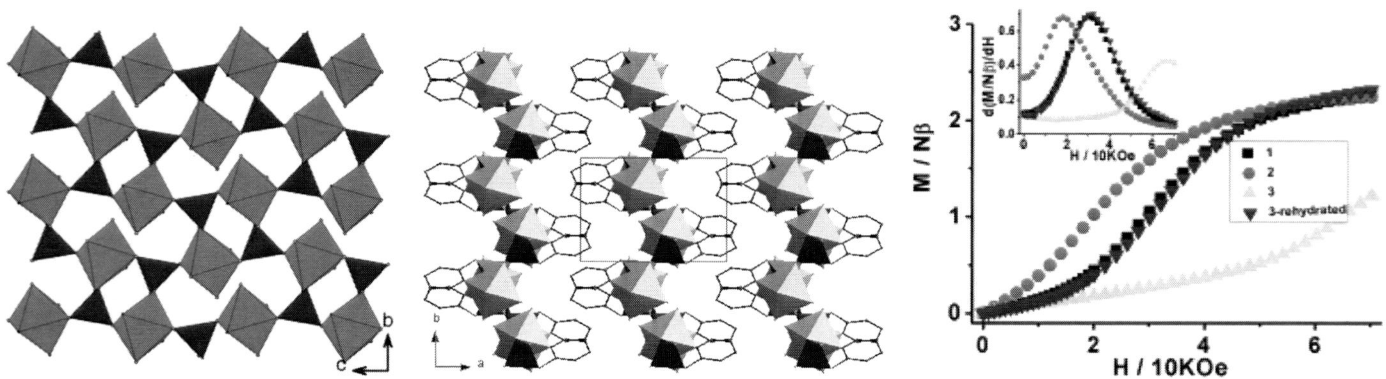

Figure 8.25 The layer structure and packing diagram of Co(2-pmp)(H_2O)$_2$, and the isothermal field dependence of magnetization at 1.8 K for Co(2-pmp)(H_2O)$_2$ (squares), Co(2-pmp)(H_2O) (circles), Co(2-pmp) (triangles) and the rehydrated sample (inverted triangles). Inset: the dM/dH versus H curves (adapted from ref. 60).

8.5 Three-dimensional Cobalt Phosphonates

8.5.1 Pillared Layered Structures

Compound $Co_2(O_3PC_6H_4PO_3)(H_2O)_2$ exhibits a pillared layered structure.[61] The inorganic layer containing 3-membered rings is identical to those in $Co(RPO_3)(H_2O)$ (R = alkyl) in which each PO_3C tetrahedron is connected to four CoO_6 octahedra through corner- or edge-sharing. The inorganic layers are linked by the phenyl groups of the diphosphonate ligand, resulting in a pillared layered structure (Figure 8.26). The phenyl group is disordered over two orientations. The interlayer distance is about 6.3 Å. Magnetic studies reveal dominant antiferromagnetic interactions between the magnetic centers. The susceptibility data obey the Curie–Weiss law in the temperature range 50–300 K with a Weiss constant of −60.5 K. A pillared layered structure is also found in compound $Co_2(H_2O)_2[O_3PCH_2(C_6H_4)CH_2PO_3]$[62] which crystallizes in the orthorhombic space group $Pca2_1$. Antiferromagnetic interactions are again observed. The Weiss constant is estimated to be −50.97 K. Other closely related compounds include $Co_2(H_2O)_2(O_3PC_2H_4PO_3)$,[63a] $Co_2(H_2O)_2(O_3P\text{-}C_6H_4OC_6H_4PO_3)$[63b] and $Co_2(H_2O)_2(O_3PC_8H_{16}PO_3)$;[35] the crystal structures of the latter two have not been determined.

Monophosphonate compound $Co(O_3PCH_2CH_2NH_2)$ also shows a pillared layered structure,[64] but the layer topology is completely different from those in $Co_2(H_2O)_2(O_3PRPO_3)$. In this case, the tetrahedrally coordinated cobalt(II) are corner-shared with the PO_3C tetrahedra, forming an inorganic layer containing four and eight members. The $-NH_2$ groups are involved in the coordination with the Co^{II} ions and link the adjacent layers into a 3D structure. The layers are stacked in AAA mode resulting in 1D channels along the *a* axis (Figure 8.27). Magnetic susceptibility data are consistent with weak antiferromagnetic interactions mediated between the tetrahedral Co^{II} centers. The Weiss constant obtained in the temperature range 100–300 K is −1 K.

Rabu *et al.* performed a detailed magnetic study of $Co_3(O_3PC_2H_4CO_2)_2$.[65] Although the crystal structure of this compound is not determined, it is found to be isostructural to the pillared layered metal phosphonates $M_3(O_3PC_2H_4CO_2)_2$ (M = Zn, Mn) according to the powder X-ray patterns.

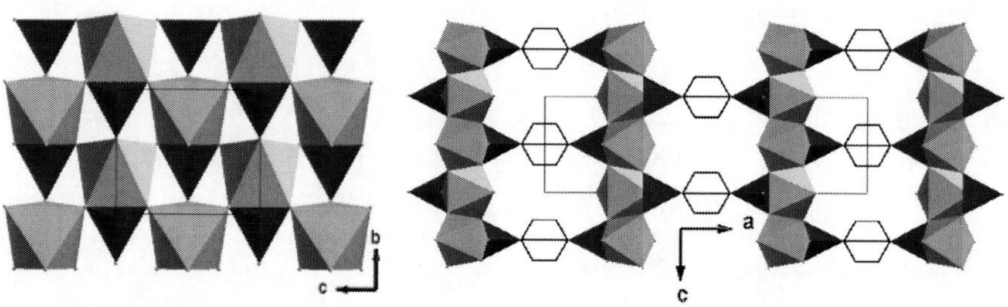

Figure 8.26 The layer structure and packing diagram of $Co_2(O_3PC_6H_4PO_3)(H_2O)_2$ (adapted from ref. 61).

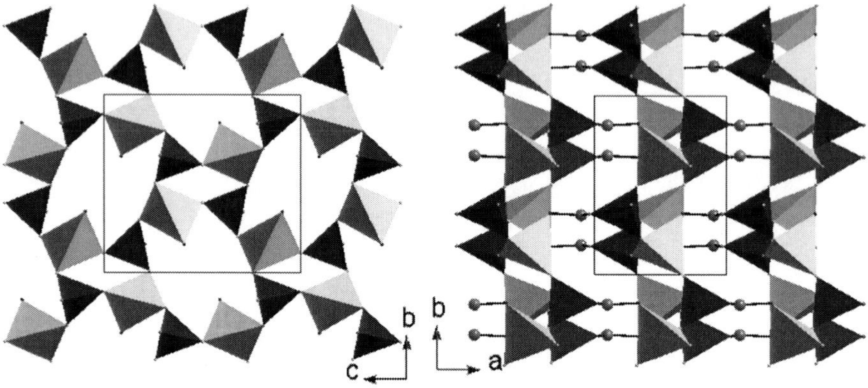

Figure 8.27 The layer structure and packing diagram of $Co(O_3PCH_2CH_2NH_2)$ (adapted from ref. 64).

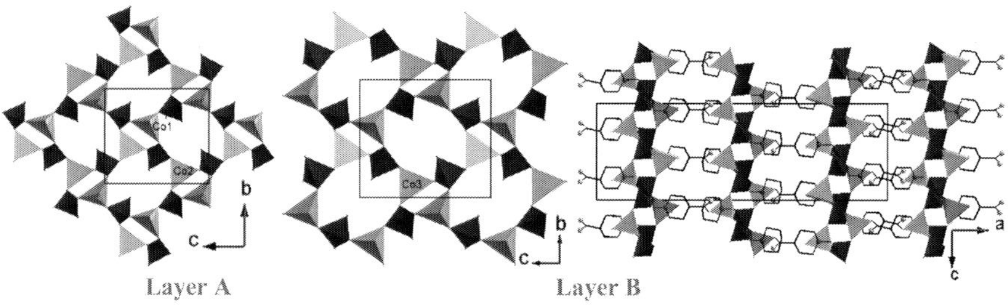

Figure 8.28 Layer A, layer B and the packing diagram in $Co_3(O_3PCH_2NHC_5H_9COO)_2$ $(O_3PCH_2NC_5H_{10})(H_2O)$ (adapted from ref. 66).

Hence the edge-sharing CoO_4 tetrahedra and CoO_6 octahedra are connected by PO_3C tetrahedra to form the inorganic layer. The *ac* and *dc* magnetic susceptibilities and magnetization measurements show that $Co_3(O_3PC_2H_4CO_2)_2$ is a canted antiferromagnet at $T_N = 15.5$ K. The canting angle is 4°. Below T_N, the hysteresis cycles exhibit an S-shape related to a spin-flop effect. The specific heat measurements indicate that large short-range correlations are present above T_N in relation with the layered character of the structure.

Hydrothermal reaction of (4-carboxypiperidyl)-*N*-methylenephosphonic acid (4-cpmpH_3) and cobalt sulfate leads to the *in situ* partial ligand decarboxylation and the formation of a novel mixed ligated compound $Co_3(O_3PCH_2\text{-}NHC_5H_9\text{-}COO)_2(O_3PCH_2\text{-}NC_5H_{10})(H_2O)$.[66] It crystallizes in the monoclinic space group $P2_1/c$. All Co atoms are tetrahedrally coordinated. The overall structure of this compound is pillared layered containing two types of inorganic layers (Figure 8.28). In layer A, the CoO_4 tetrahedra are connected by the PO_3C groups from the 4-cpmpH^{2-} ligands through corner-sharing, forming an inorganic layer containing 4- and 8-membered rings. Layer B has a similar topology to layer A, but the CoO_4 tetrahedra are corner-shared by

PO_3C tetrahedra from the piperidyl-*N*-methylenephosphonate (*N*-pmp^{2-}) ligands. Neighboring A layers are connected by the coordination of carboxylate oxygen O5 with Co2, forming a double layer AA. The equivalent double layers are further linked by layer B through the coordination of carboxylate oxygen O9 and Co3 atoms, leading to a 3D pillared layered structure with the layers stacked in a···AABAAB···sequence (Figure 8.28). The CH_2–NC_5H_{10} moieties of *N*-pmp^{2-} are grafted on the two sides of layer B.

Considering that Co^{II} ions are in tetrahedral environments with a ground state of $^4A_2$, the magnetic susceptibility data of $Co_3(O_3PCH_2\text{-}NHC_5H_9\text{-}COO)_2(O_3PCH_2\text{-}NC_5H_{10})(H_2O)$ can be analyzed by an expression for a noninteracting spin of $S=3/2$ using the Hamiltonian given in equation (8.1). The weak magnetic interactions between magnetic centers through O–P–O bridges are accounted (zJ'). Assuming $g_x=g_z=g$, the best fit is obtained in the temperature range 300–8 K, leading to parameters $g=2.22$, $D=27.36$ cm^{-1}, $zJ'=-2.52$ cm^{-1}. At very low temperature, this compound behaves as a metamagnet, and experiences a field-induced magnetic transition from an antiferromagnetic ground state to a ferrimagnetic state with the critical field of 40.1 kOe at 1.8 K.

Based on 6-phosphonic nicotinic acid (pnaH_3), compound $Co_3(pna)_2(H_2O)_2$ is obtained.[67] It possesses a pillared layered structure. Within the inorganic layer, chains made up of dimers of edge-sharing CoO_6 octahedra through O1W, O–P–O and O–C–O units are inter-connected by PO_3C tetrahedra. The pyridyl groups of pna^{3-} serve as the pillars (Figure 8.29). The Co···Co distance and Co–O–Co angle over the μ_3-O bridge are 3.148 Å and 97.93(1)°, respectively. The Co···Co distances over the O1W, O–P–O and O–C–O bridges are 4.127 Å and 4.543 Å, respectively. The interlayer distance is 8.740 Å. Magnetic studies reveal that compound $Co_3(pna)_2(H_2O)_2$ has an antiferromagnetic ground state, and experiences a field-induced magnetic transition at low temperature. The critical field is *ca.* 3 kOe at 1.8 K.

By incorporating a second bridging ligand, $Co_3(H_2L^3)_2(H_2O)_4$(4,4′-bipy) was obtained where H_5L^3 is $4\text{-}HO_2CC_6H_4CH_2N(CH_2PO_3H_2)_2$. It crystallizes in the monoclinic space group C2/c.[33] The CoO_6 octahedra are corner-shared with the PO_3C tetrahedra to form a 2D inorganic layer containing 4- and 8-membered

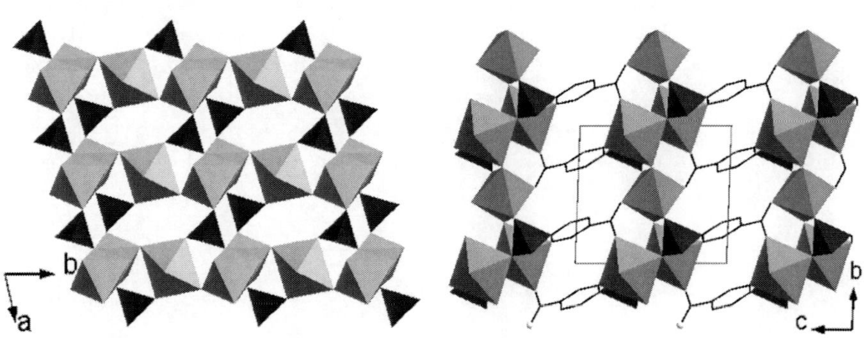

Figure 8.29 The packing diagram of $Co_3(pna)_2(H_2O)_2$ (adapted from ref. 67).

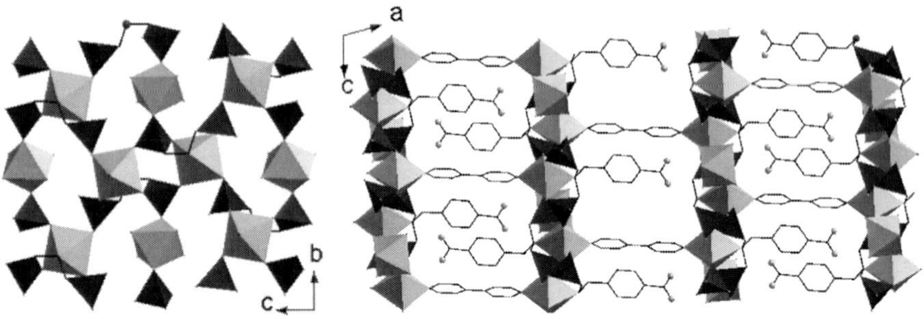

Figure 8.30 The layer structure and packing diagram of $Co_3(H_2L)_2(H_2O)_4$(4,4′-bipy) (adapted from ref. 33).

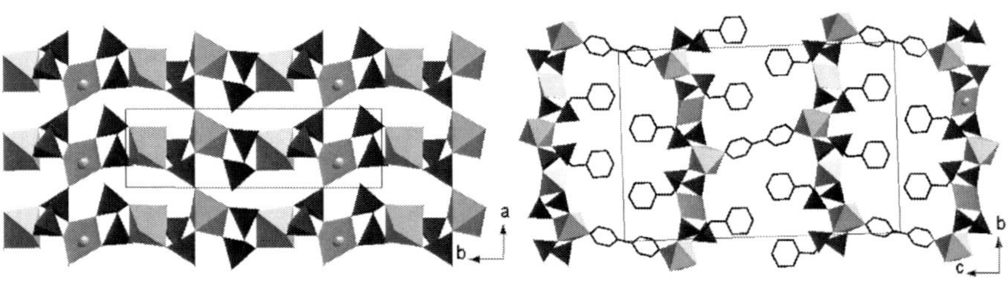

Figure 8.31 The layer structure and packing diagram of $Co_2(cmdpH_2)_2$(4,4′-bipy)$_{0.5}$ (H_2O) (adapted from ref. 46).

rings. These layers are cross-linked further by 4,4′-bipy ligands to form a 3D framework structure (Figure 8.30). The magnetic properties of this compound were not studied.

The 4,4′-bipy pillared structure is also found in $Co_2(cmdpH_2)_2$(4,4′-bipy)$_{0.5}$(H_2O) ($cmdpH_4$ = *N*-cyclohexylaminomethanediphosphonic acid), which crystallizes in monoclinic space group $P2_1/c$.[46] Within the inorganic layer, the Co1O_5 tetragonal pyramids and Co2NO_5 octahedra are each linked by PO_3C tetrahedra through corner-sharing, forming a layer in the *ab* plane which contains 3- and 10-membered rings. These layers are pillared by 4,4′-bipyridine via coordination with the Co2 atoms from the adjacent layers, leading to a 3D open framework structure with channels generated along the *a* axis. The organic groups of $cmdpH_2^{2-}$ locate within the channels (Figure 8.31). Dominant antiferromagnetic interactions are found between the Co^{II} ions. The Curie–Weiss fit of the susceptibility data in the temperature range of 50–300 K gives a Weiss constant (θ) of –26.5 K.

8.5.2 Open Framework Structures

Metal phosphonates with open framework or microporous structures are of great importance as molecular sieves, catalysts, ion-exchangers, and sensor materials. However, cobalt phosphonates with open framework structures are still rather few in number.

Sevov *et al.* reported the first microporous cobalt phosphonate $Co_2(O_3PCH_2PO_3)(H_2O)$ with an inorganic backbone and hydrophobic channels.[68] Within the structure, there are three crystallographically distinguished cobalt atoms. The octahedrally coordinated Co2 and Co3 centers are edge-shared and form zigzag chains running along the *b* axis. The Co2 octahedra also share a corner with the Col tetrahedra, linking the zigzag chains into sheets of cobalt polyhedra. One site of the Co2 octahedron is occupied by a water molecule, while all the other sites of the cobalt polyhedra are filled with the phosphonate oxygen atoms. The sheets of Co polyhedra are fused to form a 3D open framework with channels generated along the *c* axis. The channels have hydrophobic walls built from methylene groups with the short dimension of the channel being about 4.3 Å ($CH_3 \cdots CH_3$ distance) (Figure 8.32). The framework structure is stable up to about 500 °C.

Compound $Na_6Co_7(hedp)_2(hedpH)_4(H_2O)_4 \cdot 8H_2O$[69] was obtained in the presence of pyrazine. It crystallizes in triclinic space group P-1, and shows a framework structure composed of cobalt phosphonate layers and the NaO_x linkages. Each Co atom has a distorted octahedral coordination environment. The cobalt phosphonate layer consists of two types of chains which are interconnected in an alternative way by corner-sharing of CoO_6 and PO_3C polyhedra. Chain I has a composition of $\{Co_3(hedp)_2\}_n$ in which the cobalt ions are bridged purely through O–P–O units of $hedp^{4-}$. While chain II with a composition of $\{Co_2(hedpH)_2\}_n$ is analogous to those found in the ladder-like chain compounds $(cat)Co_2(hedpH)_2 \cdot xH_2O$, where the Co atoms are bridged by both μ-O(P) and O–P–O units. The shortest $Co \cdots Co$ distances are 4.655 Å within chain I, 3.243 Å within chain II, and 5.280 Å between chains I and II, respectively (Figure 8.33).

The magnetic property of this compound is quite interesting. It shows an antiferromagnetic ground state derived from antiparallel alignment of

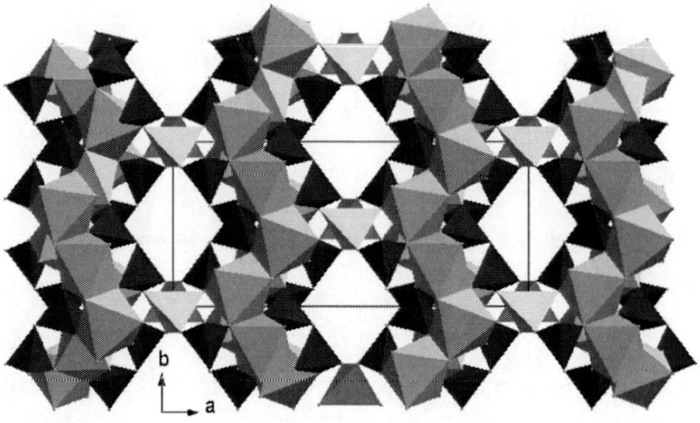

Figure 8.32 The packing diagram of $Co_2(O_3PCH_2PO_3)(H_2O)$ (adapted from ref. 68).

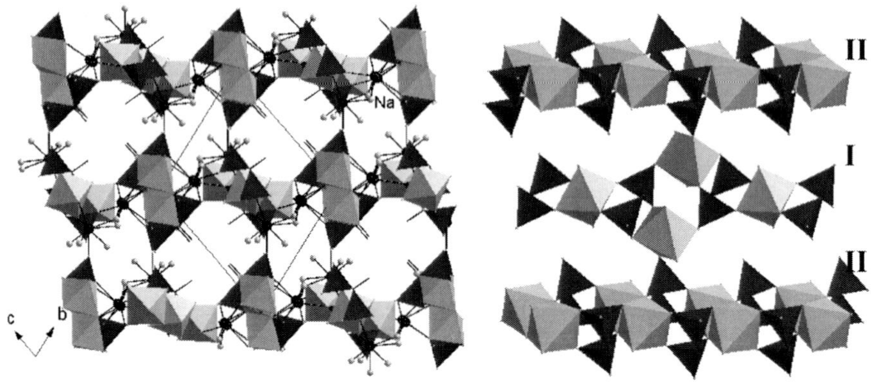

Figure 8.33 The packing diagram and layer structure of $Na_6Co_7(hedp)_2(hedpH)_4(H_2O)_4 \cdot 8H_2O$ (adapted from ref. 69).

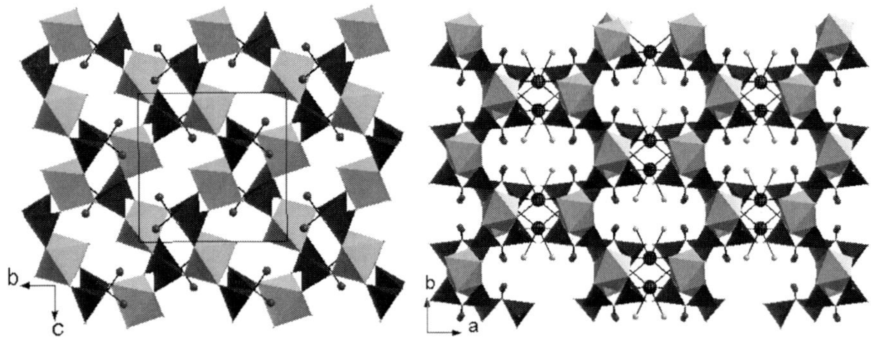

Figure 8.34 The layer structure and packing diagram of $NaCo_2(amdpH_{1.5})_2(H_2O)_2 \cdot xH_2O$ (adapted from ref. 70).

ferrimagnetic layers. The ferrimagnetism within the layer arises from the non-zero net magnetic moment due to the odd number of the antiferromagnetically interacted Co^{II} ions within the layer. The field dependent magnetization at 1.8 K reveals a two-step magnetic phase transition in the field range 0–70 kOe, corresponding to an antiferromagnetic to a ferrimagnetic state and further to a ferromagnetic state. The two critical fields are *ca.* 6.5 kOe and 30 kOe, respectively.

Compound $NaCo_2(amdpH_{1.5})_2(H_2O)_2 \cdot xH_2O$, incorporating an aminomethylenediphosphonate ($amdp^{4-}$) ligand, shows another type of open-framework structure, composed of anionic layers with composition $Co(amdpH_{1.5})(H_2O)$ and the NaO_6 linkages (Figure 8.34).[70] Each Co^{II} has a distorted octahedral environment, five positions occupied by phosphonate oxygens and one filled with a water molecule. The aminomethylenediphosphonate bridges the Co^{II} ions into an infinite zigzag chain along the *c* axis, which are connected by vertex-sharing of the CoO_6 octahedra from one chain and the PO_3C tetrahedra from the other, forming a 2D undulating layer in the *bc* plane containing 4- and 8-membered rings. The anionic layers of cobalt diphosphonate are connected through NaO_6 units, forming a 3D open framework with channels generated along the *c* direction (Figure 8.34). The lattice water resides in the channels. The protonated amino group and the coordinated water molecules point toward the channel. Dominant antiferromagnetic interactions are mediated between the magnetic centers. The magnetic susceptibility data above 50.0 K obey the Curie–Weiss law with a Weiss constant $\theta = -37$ K.

Compound $Co_3(O_3PCH_2NH_2CH_2PO_3)_2$ crystallizes in the orthorhombic space group $P2_12_12$ and displays a complicated 3D framework structure (Figure 8.35).[55] The material has three channel systems in the [001] direction formed by the cross-linking of chains of tetrahedrally coordinated Co atoms. The channel dimensions, defined by the longest and shortest distance atoms are *ca.* 5.9 Å × 3.2 Å, 3.4 Å × 4.6 Å and 2.7 Å × 3.4 Å, respectively. The same structure was also described by Kong *et al.*[71] They studied the magnetic properties of this compound. The susceptibility data follows the Curie–Weiss behavior with $C = 7.5$ cm^3 K mol^{-1} and $\theta = -12$ K, indicating antiferromagnetic and/or zero field splitting effects. Since the Co^{II} ions are tetrahedrally coordinated, possessing a $^4A_2$ ground state, the susceptibility data can be treated in a similar way to that for $Co_3(O_3PCH_2\text{-}NHC_5H_9\text{-}COO)_2(O_3PCH_2\text{-}NC_5H_{10})(H_2O)$.[66] A least-squares fitting leads to $D = 17$ cm^{-1}, $g_{||} = 2.8$, $g_{\perp} = 2.0$ and $zJ' = -0.5$ cm^{-1}.

By introducing a carboxylate group into the monophosphonate ligand, compound $Co_3(OOCC_2H_4PO_3)_2(H_2O)_6$ with an open framework structure is obtained, crystallizing in monoclinic space group $P2_1/c$.[72] Within the structure, trimers of edge-sharing CoO_6 octahedra are found which are connected by the PO_3C tetrahedra via corner-sharing to form an infinite chain along the *a* axis. Each chain is connected to the four other equivalents through the organic

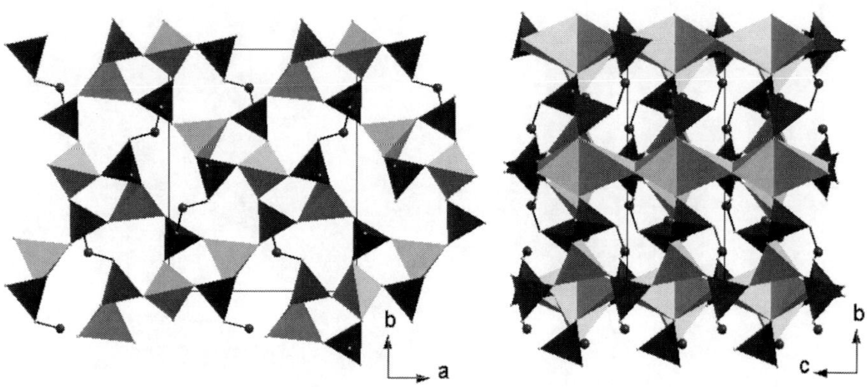

Figure 8.35 The packing diagram of $Co_3(O_3PCH_2NH_2CH_2PO_3)_2$ viewed along the *c* and *a* axis (adapted from ref. 55).

groups of the phosphonate ligands, leading to a 3D open framework structure (Figure 8.36). There are two types of cobalt atoms, Co2 at an inversion center at the center of the trimer, and Co1 inside the other two octahedral. The three water molecules are also of two types, two of them complete the octahedral coordination sphere around the Co1 atoms while the other bridges Co1 and Co2 atoms. Magnetic susceptibility measurements in the range 10–300 K show a Curie-type paramagnetic behavior, consistent with the structure of "isolated" islands of cobalt trimers ($\mu_{eff} = 9\mu_B$).

The reaction of 2-hydroxyl(phosphono)carboxylic acid (hpaH_3) with cobalt (II) salt leads to compound $NaCo[O_3PCH(OH)CO_2]$, crystallizing in orthorhombic space group Pbca.[73] Each cobalt(II) ion in the structure is octahedrally coordinated by three phosphonate oxygen, two carboxylate oxygen and one hydroxyl oxygen atoms. All the oxygen atoms from the $(L^{10})^{3-}$ ligand are involved in coordination with cobalt ions. Thus the CoO_6 octahedra are corner-shared with the PO_3C tetrahedra, forming a 3D open framework with channels generated along the *a* axis (Figure 8.37). The charge-compensating Na^+ cations are located at the intersections of these channels.

The structure of compound $Co(O_3PCH_2OH)$ was solved by the Rietveld refinement of powder XRD data.[74] The octahedrally coordinated Co^{II} ions are connected by the phosphonate groups through edge- or corner-sharing, forming infinite zigzag chains of Co atoms along the c axis. The chains are then cross-linked by O–P–O and P–C–O–Co bridges, forming hexagonal channels parallel to the *c* axis, which have a diameter of *ca.* 5.7 Å between two opposite methylene C atoms. The magnetic properties of this compound are not studied.

Compound $[Co_2Cl(S\text{-}HL^9)(H_2O)_5]Cl \cdot H_2O$ ($S\text{-}H_3L^9 = S\text{-}HO_2CC_4H_7NCH_2PO_3H_2$) crystallizes in the chiral space group $P2_12_12_1$.[75] The asymmetric unit contains two unique Co^{II} ions. The Co1 and Co2 octahedra are edge-shared to

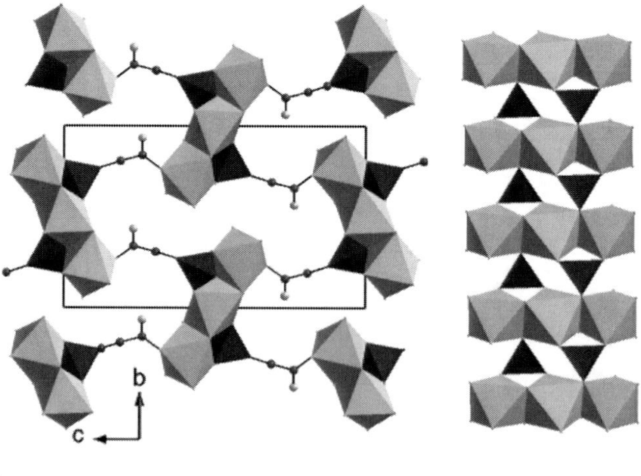

Figure 8.36 The packing diagram of $Co_3(O_2CCH_2CH_2PO_3)_2 \cdot 6H_2O$ (adapted from ref. 72).

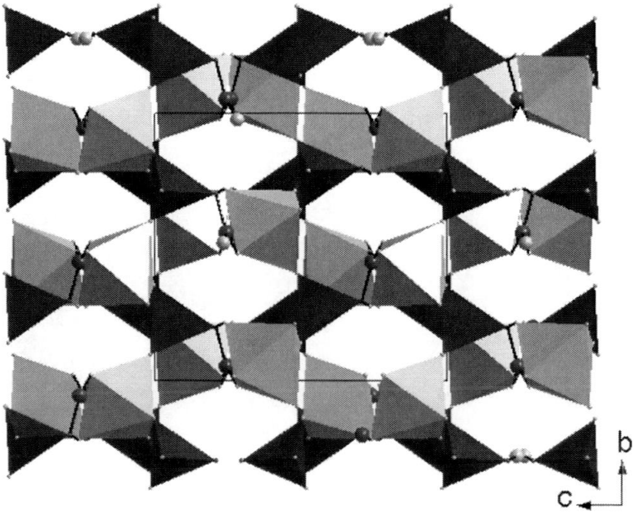

Figure 8.37 The packing diagram of $NaCo[O_3PCH(OH)CO_2]$ (adapted from ref. 73).

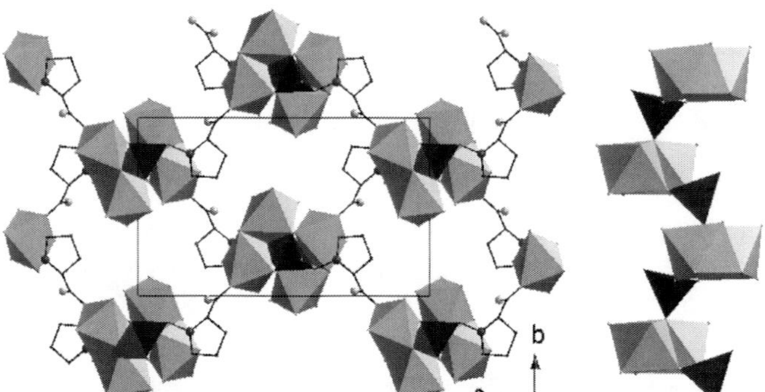

Figure 8.38 Packing diagram of $[Co_2Cl(S\text{-}HL)(H_2O)_5]Cl \cdot H_2O$ along *a* axis, and 1D right-handed cobalt(II) phosphonate chain along *a* axis (adapted from ref. 75).

form a dimeric unit. Such dimers are bridged by the phosphonate groups into a left-handed helical chain along the *a* axis. Neighboring chains are interconnected by the organic groups of the phosphonate ligands into a hybrid 3D framework with helical channels running along *a*- and *b*-axes. The non-coordinated water molecules and chlorine anions are located at these channels. The proline rings of the phosphonate ligands also orient to the above channels (Figure 8.38).

The reactions of $Ln(NO_3)_3 \cdot xH_2O$, $CoSO_4 \cdot 7H_2O$, and 2-pyridylphosphonic acid (2-ppH_2) under hydrothermal conditions result in a series of heterometallic

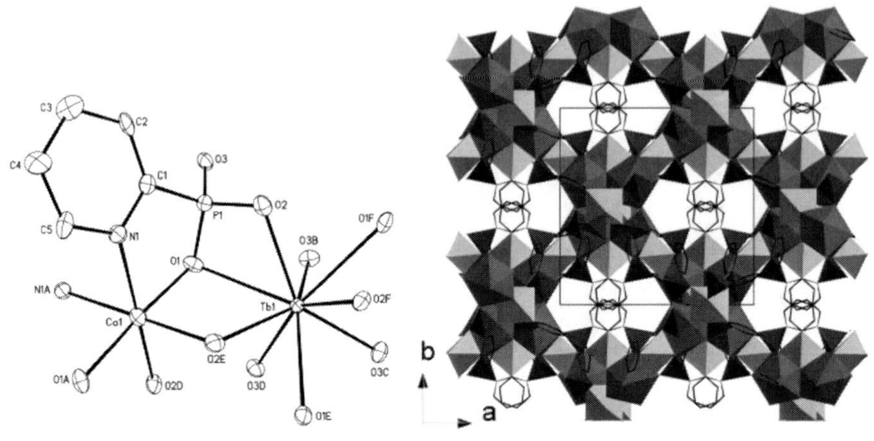

Figure 8.39 Building unit and packing diagram of $Tb_2Co_3(C_5H_4NPO_3)_6 \cdot 4H_2O$ (adapted from ref. 76).

phosphonate compounds with formula $Ln_2Co_3(C_5H_4NPO_3)_6 \cdot 4H_2O$ [Ln^{III} = La, Ce, Pr, Nd, Sm, Eu, Gd, Tb, Dy).[76] These compounds are isostructural and crystallize in a chiral cubic space group $I2_13$. Each structure contains the LnO_9 polyhedra and CoN_2O_4 octahedra which are connected by edge-sharing to form an inorganic open-framework structure. Figure 8.39 shows the structure of $Tb_2Co_3(C_5H_4NPO_3)_6 \cdot 4H_2O$ as an example. The structure can be simply viewed as a cubic 3-connected 10-gon (10,3) net containing uniform 2_1 helical channels, where the $Co(C_5H_4NPO_3)_2$ units are considered as linkers and the La atoms as nodes. The Co1–O1–Tb1 and Co1–O2E–Tb1 angles within the TbO_2Co four-member ring are 101.1(2)° and 109.4(3)°, respectively. The dihedral angle (δ) between the Co1O1O2E and Tb1O1O2E planes is 14.8°. The Co···Tb distance across the μ_3-O bridge (3.701 Å) is much shorter than the M···M separations across the O–P–O units (Co···Tb 6.534 Å, Tb···Tb 5.698 Å, Co···Co 5.904 Å).

The nature of Ln^{III}-Co^{II} magnetic interactions in Ln_2Co_3 is investigated by a comparison with their Ln^{III}-Zn^{II} analogs. It was found that the Ln^{III}-Co^{II} interaction is weak antiferromagnetic for Ln = Ce and ferromagnetic for Ln = Sm, Gd, Tb, and Dy. In the cases of Ln = Pr, Nd, and Eu, no significant magnetic interaction is observed.

8.6 Conclusions

By introducing additional functional groups and/or co-ligands, cobalt phosphonates with a variety of structures ranging from clusters, 1D chains, 2D layers, and 3D frameworks may be obtained through solution or hydrothermal methods. These materials could have shown very interesting magnetic behaviors such as single molecule magnet, single chain magnet, quasi-spin ladder, metamagnet, weak ferromagnet, and ferrimagnet behaviors. Besides, cobalt phosphonates with layered or open framework structures could show absorption and exchange properties. The results demonstrate that the cobalt phosphonates are promising not only as low-dimensional magnetic materials, but also as multifunctional materials.

Acknowledgments

We acknowledge financial support by the National Natural Science Foundation of China, NSF of Jiangsu Province (No. BK2009009) and the National Basic Research Program of China (2007CB925102), and the support of the students and colleagues who have made great contributions to these work.

References

1. (a) A. Clearfield, *Prog. Inorg. Chem.*, 1998, **47**, 371; (b) L.-Q. Ma, C. Abney and W.-B. Lin, *Chem. Soc. Rev.*, 2009, **38**, 1248; (c) S. Chessa, N. J. Clayden, M. Bochmann and J. A. Wright, *Chem. Commun.*, 2009, 797.
2. (a) K. Maeda, *Microporous Mesoporous Mater.*, 2004, **73**, 47; (b) M. Plabst, L. B. McCusker and T. Bein, *J. Am. Chem. Soc.*, 2009, **131**, 18112.
3. (a) J. Umeda, M. Suzuki, M. Kato, M. Moriya, W. Sakamoto and T. Yogo, *J. Power Sources*, 2010, **195**, 5882; (b) J. W. Traer and G. R. Goward, *Phys. Chem. Chem. Phys.*, 2010, **12**, 263; (c) G. Alberti and M. Casciola, *Solid State Ionics*, 2001, **145**, 3; (d) E. Montoneri, V. Boffa, S. o. Bottigliengo, M. Casciola, M. Sganappa, A. Marigo, G. Speranza, L. Minati, S. Torrengo, G. Alberti and L. Bertinetti, *Solid State Ionics*, 2010, **181**, 578.
4. (a) J.-G. Mao, *Coord. Chem. Rev.*, 2007, **251**, 1493; (b) E. Brunet, H. M. H. Alhendawi, O. Juanes, L. Jiménez and J. C. Rodríguez-Ubis, *J. Mater. Chem.*, 2009, **19**, 2494.
5. (a) S. Maheswaran, G. Chastanet, S. J. Teat, T. Mallah, R. Sessoli, W. Wernsdorfer and R. E. P. Winpenny, *Angew., Chem. Int. Ed. Engl.*, 2005, **44**, 5044; (b) D. Cave, F. C. Coomer, E. Molinos, H.-H. Klauss and P. T. Wood, *Angew. Chem., Int. Ed. Engl.*, 2006, **45**, 803; (c) H.-C. Yao, J.-J. Wang, Y.-S. Ma, O. Waldmann, W.-X. Du, Y. Song, Y.-Z. Li, L.-M. Zheng, S. Decurtins and X.-Q. Xin, *Chem. Commun.*, 2006, 1745.
6. (a) J. Monot, M. Petit, S. M. Lane, I. Guisle, J. Léger, C. Tellier, D. R. Talham and B. Bujoli, *J. Am. Chem. Soc.*, 2008, **130**, 6243; (b) J. Amalric, P. H. Mutin, G. Guerrero, A. Ponche, A. Sotto and J. P. Lavigne, *J. Mater. Chem.*, 2009, **19**, 141.
7. M. Kurmoo, *Chem. Soc. Rev.*, 2009, **38**, 1353.
8. M. Murrie, *Chem. Soc. Rev.*, 2010, **39**, 1986.
9. (a) T. F. Liu, D. Fu, S. Gao, Y. Z. Zhang, H. L. Sun, G. Su and Y. J. Liu, *J. Am. Chem. Soc.*, 2003, **125**, 13976; (b) X.-N. Cheng, W.-X. Zhang and X.-M. Chen, *J. Am. Chem. Soc.*, 2007, **129**, 15738; (c) S.-i. Ohkoshi,

S. Ikeda, T. Hozumi, T. Kashiwagi and K. Hashimoto, *J. Am. Chem. Soc.*, 2006, **128**, 5320; (d) S. Hayami, Y. Shigeyoshi, M. Akita, K. Inoue, K. Kato, K. Osaka, M. Takata, R. Kawajiri, T. Mitani and Y. Maeda, *Angew. Chem., Int. Ed. Engl.*, 2005, **44**, 4899.
10. G. K. H. Shimizu, R. Vaidhyanathan and J. M. Taylor, *Chem. Soc. Rev.*, 2009, **38**, 1430.
11. A. Mateescu, C. P. Raptopoulou, A. Terzis, V. Tangoulis and A. Salifoglou, *Eur. J. Inorg. Chem.*, 2006, 1945.
12. D. K. Cao, Y. Z. Li and L. M. Zheng, *Inorg. Chem.*, 2007, **46**, 7571.
13. K. Ma, J. Xu, L. Zhang, J. Shi, D. Zhang, Y. Zhu, Y. Fan and T. Song, *New J. Chem.*, 2009, **33**, 886.
14. (a) A. Mateescu, C. Gabriel, R. G. Raptis, P. Baran and A. Salifoglou, *Inorg. Chim. Acta*, 2007, **360**, 638; (b) G. Feng, E. A. Tanifum, H. Adams, A. C. Hengge and N. H. Williams, *J. Am. Chem. Soc.*, 2009, **131**, 12771.
15. (a) R. Murugavel and S. Shanmugan, *Dalton Trans.*, 2008, 5358; (b) V. Baskar, M. Shanmugam, E. C. Sanudo, M. Shanmugam, D. Collison, E. J. L. McInnes, Q. Wei and R. E. P. Winpenny, *Chem. Commun.*, 2007, 37.
16. S. Langley, M. Helliwell, J. Raftery, E. I. Tolis and R. E. P. Winpenny, *Chem. Commun.*, 2004, 142.
17. S. J. Langley, M. Helliwell, R. Sessoli, P. Rosa, W. Wernsdorfer and R. E. P. Winpenny, *Chem. Commun.*, 2005, 5029.
18. S. Langley, M. Helliwell, R. Sessoli, S. J. Teat and R. E. P. Winpenny, *Inorg. Chem.*, 2008, **47**, 497.
19. S. Langley, M. Helliwell, R. Sessoli, S. J. Teat and R. E. P. Winpenny, *Dalton Trans.*, 2009, 3102.
20. E. Dagotto, *Rep. Prog. Phys.*, 1999, **62**, 1525.
21. A. N. Vasil'ev, M. M. Markina and E. A. Popova, *Low Temperature Phys.*, 2005, **31**, 203.
22. (a) W. Zheng, R. R. P. Singh and J. Oitmaa, *Phys. Rev. B*, 1997, **55**, 8052; (b) R. D. Willett, C. Galeriu, C. P. Landee, M. M. Turnbull and B. Twamley, *Inorg. Chem.*, 2004, **43**, 3804; (c) A. Shapiro, C. P. Landee, M. M. Turnbull, J. Jornet, M. Deumal, J. J. Novoa, M. A. Robb and W. Lewis, *J. Am. Chem. Soc.*, 2007, **129**, 952.
23. P. Yin, S. Gao, L.-M. Zheng and X.-Q. Xin, *Chem. Mater.*, 2003, **15**, 3233.
24. L.-M. Zheng, S. Gao, P. Yin and X.-Q. Xin, *Inorg. Chem.*, 2004, **43**, 2151.
25. P. Yin, S. Gao, Z.-M. Wang, C.-H. Yan, L.-M. Zheng and X.-Q. Xin, *Inorg. Chem.*, 2005, **44**, 2761.
26. (a) Z.-C. Zhang, S.-S. Bao and L.-M. Zheng, *Inorg. Chem. Commun.*, 2007, **10**, 1063; (b) Y. Gong, W. Tang, W.-B. Hou, Z.-Y. Zha and C.-W. Hu, *Inorg. Chem.*, 2006, **45**, 4987.
27. Z.-C. Zhang, S. Gao and L.-M. Zheng, *Dalton Trans.*, 2007, 4681.
28. T.-H. Yang, E. S. Knowles, D. M. Pajerowski, J.-S. Xia, L. Yin, S. Gao, M. W. Meisel and L.-M. Zheng, *Inorg. Chem.*, 2010, **49**, 8474.
29. (a) Z. M. Sun, A. V. Prosvirin, H. H. Zhao, J. G. Mao and K. R. Dunbar, *J. Appl. Phys.*, 2005, **97**, 10B305/1; (b) A. V. Palii, S. M. Ostrovsky, S. I. Klokishner, O. S. Reu, Z. M. Sun, A. V. Prosvirin, H. H. Zhao, J. G. Mao

and K. R. Dunbar, *J. Phys. Chem. A.*, 2006, **110**, 14003; (c) A. V. Palii, O. S. Reu, S. M. Ostrovsky, S. I. Klokishner, B. S. Tsukerblat, Z. M. Sun, J. G. Mao, A. V. Prosvirin, H. H. Zhao and K. R. Dunbar, *J. Am. Chem. Soc.*, 2008, **130**, 14729.
30. P.-F. Wang, Y. Duan and L.-M. Zheng, *Science in China B*, 2010, **53**, 2112.
31. D.-K. Cao, J. Xiao, Y.-Z. Li, J.-M. Juan, E. Coronado and L.-M. Zheng, *Eur. J. Inorg. Chem.*, 2006, 1830.
32. G.-F. Han, H.-Z. Luo, Q. Ye and R.-G. Xiong, *Z. Anorg. Allg. Chem.*, 2008, **634**, 1991.
33. J.-L. Song, A. V. Prosvirin, H.-H. Zhao and J.-G. Mao, *Eur. J. Inorg. Chem.*, 2004, 3706.
34. M. Wang, C.-B. Ma, C.-N. Chen and Q.-T. Liu, *J. Mol. Struct.*, 2008, **891**, 292.
35. E. V. Bakhmutova, X. Ouyang, D. G. Medvedev and A. Clearfield, *Inorg. Chem.*, 2003, **42**, 7046.
36. D. Cunningham, P. J. D. Hennelly and T. Deeney, *Inorg. Chim. Acta*, 1979, **37**, 95.
37. E. M. Bauer, C. Bellitto, M. Colapietro, S. A. Ibrahim, M. R. Mahmoud, G. Portalone and G. Righini, *J. Solid State Chem.*, 2006, **179**, 389.
38. J. P. Garcia-Teran, O. Castello, A. Luque, U. Garcia-Courceiro, P. Roman and F. Loret, *Inorg. Chem.*, 2004, **43**, 5761.
39. T. O. Salami, X. Fan, P. Y. Zavalij and S. R. J. Oliver, *Dalton Trans.*, 2006, 1574.
40. K. L. Yao, X. L. Li, Z. L. Liu, Y. L. Li and G. Y. Gao, *J. Magn. Magn. Mater.*, 2007, **312**, 16.
41. G. Cao and T. E. Mallouk, *Inorg. Chem.*, 1991, **30**, 1434.
42. K. J. Frink, R.-C. Wang, J. L. Colon and A. Clearfield, *Inorg. Chem.*, 1991, **30**, 1438.
43. C. T. Seip and D. R. Talham, *Mater. Res. Bull.*, 1999, **34**, 437.
44. P. Yin, X.-C. Wang, S. Gao and L.-M. Zheng, *J. Solid State Chem.*, 2005, **178**, 1049.
45. Z.-C. Zhang and L.-M. Zheng, *Inorg. Chem. Commun.*, 2008, **11**, 1243.
46. Y.-H. Su, D.-K. Cao, Y. Duan, Y.-Z. Li and L.-M. Zheng, *J. Solid State Chem.*, 2010, **183**, 1588.
47. H. Jankovics, M. Daskalakis, C. P. Raptopoulou, A. Terzis, V. Tangoulis, J. Giapintzakis, T. Kiss and A. Salifoglou, *Inorg. Chem.*, 2002, **41**, 3366.
48. B.-P. Yang, A. V. Prosvirin, H.-H. Zhao and J.-G. Mao, *J. Solid State Chem.*, 2006, **179**, 175.
49. J.-L. Song, H.-H. Zhao, J.-G. Mao and K. R. Dunbar, *Chem. Mater.*, 2004, **16**, 1884.
50. B.-P. Yang, A. V. Prosvirin, Y.-Q. Guo and J.-G. Mao, *Inorg. Chem.*, 2008, **47**, 1453.
51. F. E. Mabbs and D. J. Machin, *Magnetism and Transition Metal Complexes*, Chapman and Hall, London, 1973, p. 99.
52. (a) M. E. Lines, *J. Phys. Chem. Solids*, 1970, **31**, 101; (b) J. Darriet, M. S. Haddad, E. N. Duesler and D. N. Hendrickson, *Inorg. Chem.*, 1979, **18**, 2679.

53. J.-G. Mao and A. Clearfield, *Inorg. Chem.*, 2002, **41**, 2319.
54. Y.-Y. Zhang, M.-H. Zeng, Y. Qi, S.-Y. Sang and Z.-M. Liu, *Inorg. Chem. Commun.*, 2007, **10**, 33.
55. A. Turner, P.-A. Jaffrès, E. J. MacLean, D. Villemin, V. McKee and G. B. Hix, *Dalton Trans.*, 2003, 1314.
56. J.-J. Hou and X.-M. Zhang, *Cryst. Growth Des.*, 2006, **6**, 1445.
57. J.-M. Rueff, V. Caignaert, S. Chausson, A. Leclaire, C. Simon, O. Perez, L. Le Pluart and P.-A. Jaffrès, *Eur. J. Inorg. Chem.*, 2008, 4117.
58. J.-T. Li, T. D. Keene, D.-K. Cao, S. Decurtins and L.-M. Zheng, *Cryst EngChem.*, 2009, **11**, 1255.
59. X.-Y. Yi, L.-M. Zheng, W. Xu and J.-S. Chen, *Dalton Trans.*, 2003, 953.
60. T.-H. Yang, Y. Liao, L.-M. Zheng, R. E. Dinnebier, Y.-H. Su and J. Ma, *Chem. Commun.*, 2009, 3023.
61. D.-K. Cao, S. Gao and L.-M. Zheng, *J. Solid State Chem.*, 2004, **177**, 2311.
62. H. Li, G.-S. Zhu, X.-D. Guo, F.-X. Sun, H. Ren, Y. Chen and S.-L. Qiu, *Eur. J. Inorg. Chem.*, 2006, 4123.
63. (a) C. A. Merrill and A. K. Cheetham, *Inorg. Chem.*, 2007, **46**, 278; (b) M. M. Gomez-Alcantara, A. Cabeza, M. Martinez-Lara, M. A. G. Aranda, R. Suau, N. Bhuvanesh and A. Clearfield, *Inorg. Chem.*, 2004, **43**, 5283.
64. W. R. Gemmill, M. D. Smith and B. A. Reisner, *J. Solid State Chem.*, 2005, **178**, 2658.
65. P. Rabu, P. Janvier and B. Bujoli, *J. Mater. Chem.*, 1999, **9**, 1323.
66. S.-Z. Hou, D.-K. Cao, X.-G. Liu, Y.-Z. Li and L.-M. Zheng, *Dalton Trans.*, 2009, 2746.
67. P.-F. Wang, S.-S. Bao, S.-M. Zhang, D.-K. Cao, X.-G. Liu and L.-M. Zheng, *Eur. J. Inorg. Chem.*, 2010, 895.
68. D. L. Lohse and S. C. Sevov, *Angew. Chem., Int. Ed. Engl.*, 1997, **36**, 1619.
69. P. Yin, S. Gao, L.-M. Zheng, Z.-M. Wang and X.-Q. Xin, *Chem. Commun.*, 2003, 1076.
70. S.-S. Bao, L.-M. Zheng, Y.-J. Liu, W. Xu and S.-H. Feng, *Inorg. Chem.*, 2003, **42**, 5037.
71. D. Kong, Y. Li, X. Ouyang, A. V. Prosvirin, H. Zhao, J. H. Ross, Jr., K. R. Dunbar and A. Clearfield, *Chem. Mater.*, 2004, **16**, 3020.
72. A. Distler and S. C. Sevov, *Chem. Commun.*, 1998, 959.
73. L.-Y. Cui, Z.-G. Sun, H. Chen, L. Meng, D.-P. Dong, C.-H. Tian, Z.-M. Zhu and W.-S. You, *J. Coord. Chem.*, 2007, **60**, 1247.
74. G. B. Hix, A. Turner, B. M. Kariuki, M. Tremayne and E. J. MacLean, *J. Mater. Chem.*, 2002, **12**, 3220.
75. B.-P. Yang and J.-G. Mao, *J. Mol. Struct.*, 2007, **830**, 78.
76. Y.-S. Ma, Y. Song and L.-M. Zheng, *Inorg. Chim. Acta*, 2008, **361**, 1363.

CHAPTER 9

How Laminar Zirconium Phosphates Can Make Organic Molecules Display New Behavior and Properties at the Supramolecular Level in the Solid State

ERNESTO BRUNET

Departamento de Química Orgánica, Facultad de Ciencias 01.01.602, Universidad Autónoma de Madrid, 28049-Madrid, Spain

9.1 Introduction

The design of porous solids of controlled molecular geometry for various applications is a challenge of enormous technological and scientific importance. Many years ago Feynman made the query "What would the properties of materials be if we could really arrange the atoms (molecules) the way we want them?"[1] Chemistry, Physics, Engineering and even Biology are nowadays highly intertwined in the defying of Feynman's challenge. Yet, half a century has passed since the challenge was placed and "fortunately" we are still relatively far from accomplishing a once-and-for-all solution to this problem. The fields of Supramolecular Chemistry ("chemistry beyond the molecule") and Crystal Engineering are achieving chemical processes, especially based on self assembly, giving rise to interesting materials with a large variety of properties. In general, the design of supramolecular materials is aimed along the lines of Feynman's challenge and the literature contains many smart routes to

Metal Phosphonate Chemistry: From Synthesis to Applications
Edited by Abraham Clearfield and Konstantinos Demadis

Published by the Royal Society of Chemistry, www.rsc.org

approach it, such as sol–gel chemistry,[2] self-assembly,[3] film making (Langmuir–Blodgett, *etc.*),[4] and the use of classical porous materials like zeolites, among with a long list of other methods.[5] However, despite all these huge efforts and intelligent developments, the finding of new materials is as yet, for the time being, quite open a field that allows for many serendipity driven discoveries, the alma mater of basic research.

Besides, Feynman conjectured "What could we do with layered structures with just the right layers?" The placing of organic molecules between the layers of certain inorganic salts leads to sturdy, solid materials where the *act of confinement* makes the organic molecules change their properties or even display new ones at the supramolecular level. Along the years we and others have found that zirconium phosphate is a versatile layered inorganic salt that acts as a "carving board" where organic compounds can be attached, either covalently or by ionic forces, by mild hydrothermal reactions, to form 3D structures with properties and potential applications only limited by the imagination of the researcher. A representative coverage of the literature in this field will be presented in this chapter, embroidering past and ongoing results and ideas of our research group concerning the chemistry of metal phosphates/phosphonates essentially in relation with the following topics: (a) molecular recognition; (b) chemically driven porosity changes; (c) chiral memory and supramolecular chirality; (d) luminescence signaling; (e) photo-induced electron transfer; (f) gas storage; and (g) drug confinement.

9.2 Tools for the Construction

The notion of confinement is ubiquitous. As always, Nature masters the concept and expresses it in a myriad of manners. For Mankind this is an important concept too in many aspects of our everyday living. For instance, Figure 9.1 shows two typical manifestations of popular art. The Chinese *chi* puzzle balls

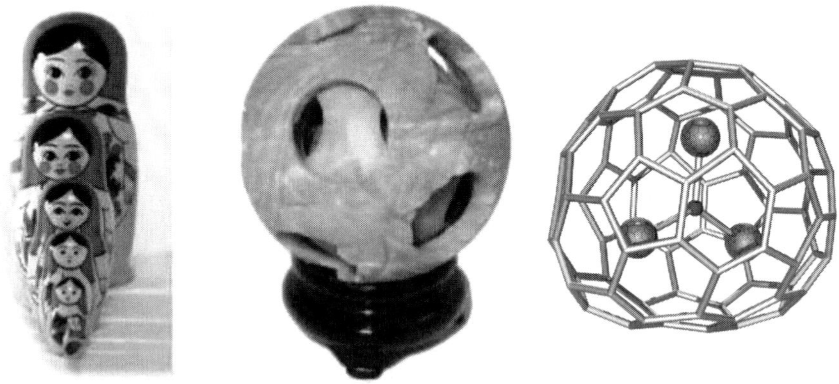

Figure 9.1 Examples of the concept of confinement in popular art and chemistry: Russian matriuskas (left) and Chinese *chi* puzzle balls (center) and endohedral fullerene (right).

are awesome because the concentric spheres are carved from within, from a single piece of jade, through the various decorative holes left. This is an outstanding example which puts us back on track to chemistry because it reminds us of the recent achievements in fullerene chemistry.[6]

Chemical confinement can easily be achieved using layered inorganic salts. The formation of layers in the mineral world is quite common. Natural clays are an excellent example[7] but there are many other instances, either natural or artificial, like layered double hydroxides[8] (LDHs), layered perovskites, niobates,[9] *etc*. A remarkable feature of these different layered materials is their general ability to form intercalates by various chemical means, namely ion-exchange processes. This outstanding property allows their structure to be modified in many ways to produce advanced multifunctional composites with unusual combinations of properties that are difficult to achieve otherwise. Among all these natural and man-made layered structures, tetravalent metal phosphates stand out, especially those of Zr due to their high stability and ease of use.

9.2.1 Zirconium Phosphate (ZrP)

Figure 9.2 shows the three most common allotropic forms of this layered inorganic salt:[10] α- (a), γ- (b) and λ-ZrP (c). The structures found in the literature can be considered derived from these three arrangements with

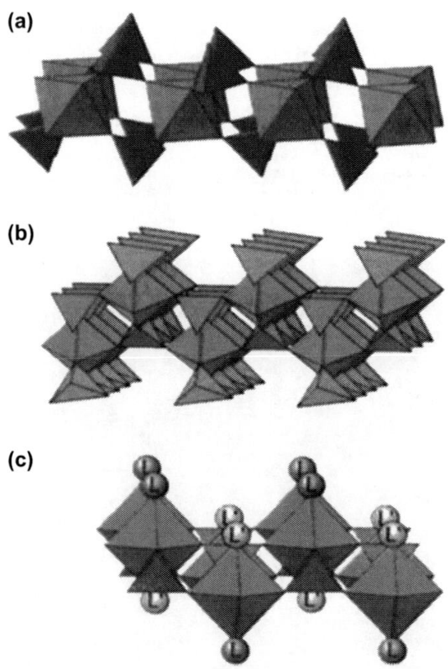

Figure 9.2 Most common structures of zirconium phosphates: α- (a), γ- (b) and λ-ZrP (c).

slight variations, if at all. They all contain octahedral Zr atoms. In the α-ZrP (a) there is only one kind of phosphate whereas in γ-ZrP (b) there are two, one internal and another external. The λ-ZrP (c) bears an internal phosphate, similar to that in γ-ZrP, but in this case every Zr atom is bonded to negatively charged (L) and neutral ligands (L′) to maintain electro neutrality. All structural features of these architectures will be carefully described in the following paragraphs.

9.2.2 Alpha Zirconium Phosphate (α-ZrP)

α-ZrP was first synthesized in 1964 by Clearfield and Stynes,[11] yet its structure had to wait 13 years[12] to be accurately determined, in one of the rare instances where single crystals of enough quality could be collected from these lamellar materials. In the sketch of Figure 9.3 it may be seen that the structure of α-ZrP comprises a set of octahedral Zr atoms bridged by a series of equivalent tetrahedral phosphates bonded by three oxygen atoms to three different metals. In every phosphate there is left an acidic OH group pointing to the interlayer region that forms a hydrogen-bond network with water molecules, the latter acting as H donors. Figure 9.4 shows a space-filling model which visualizes the relative high density of pristine α-ZrP, *i.e.* when no additional species are held between the layers.

Pertinent crystal data and main inter-atomic distances are gathered in Figure 9.5.

The calculated density using the Hyperchem program[13] from the model indicated in Figure 9.4 is 2.3 g cm^{-3}, reasonably close to the experimental value determined by X-ray data (Figure 9.5). The two parallelepipeds enclosing two consecutive models of molecular formula $[Zr(HPO_4)_2]_{60}$, taken as a representative fragment (MW = 33 981.42 g; see Figure 9.4), have a volume of 25.2×10^{-21} cm^3 or 1.51×10^4 cm^3 considering Avogadro's number.

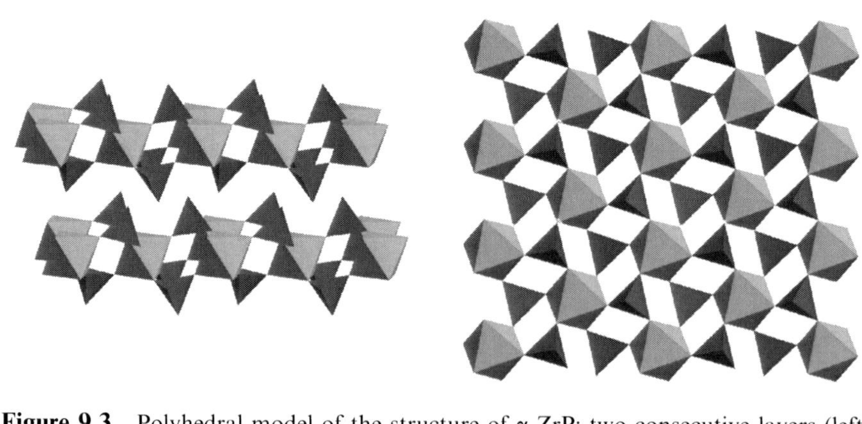

Figure 9.3 Polyhedral model of the structure of α-ZrP: two consecutive layers (left) and top view of a single layer.

Figure 9.4 Space-filling spheres model of α-ZrP showing the tight packing of atoms (Zr; P; O; H) and interlayer distance.

Space group	$P2_1/n$
a, Å	9.060 (2)
b, Å	5.297 (1)
c, Å	15.414 (3)
β, deg	101.71 (2)
V, Å^3	724.3 (5)
d_{calcd}, g cm^{-3}	2.762
d_{obsd}, g cm^{-3}	2.72
Z	4

Zr–O4	2.048 (2)
Zr–O5	2.074 (2)
Zr–O6	2.071 (2)
Zr–O8	2.054 (2)
Zr–O9	2.065 (2)
Zr–O11	2.075 (2)
P2–O4	1.510 (2)
P2–O5	1.512 (2)
P2–O6	1.524 (2)
P2–O7	1.564 (2)
P3–O8	1.517 (2)
P3–O9	1.518 (2)
P3–O10	1.551 (2)
P3–O11	1.519 (2)

O4–Zr–O5	91.14 (8)
O4–Zr–O6	90.45 (7)
O4–Zr–O8	89.76 (8)
O4–Zr–O9	89.35 (8)
O4–Zr–O11	178.85 (8)
O5–Zr–O6	88.96 (7)
O5–Zr–O8	88.88 (7)
O5–Zr–O9	179.51 (7)
O5–Zr–O11	89.96 (7)
O6–Zr–O8	177.83 (8)
O6–Zr–O9	91.04 (7)
O6–Zr–O11	89.91 (7)
O8–Zr–O9	91.12 (7)
O8–Zr–O11	89.92 (8)
O9–Zr–O11	89.55 (7)

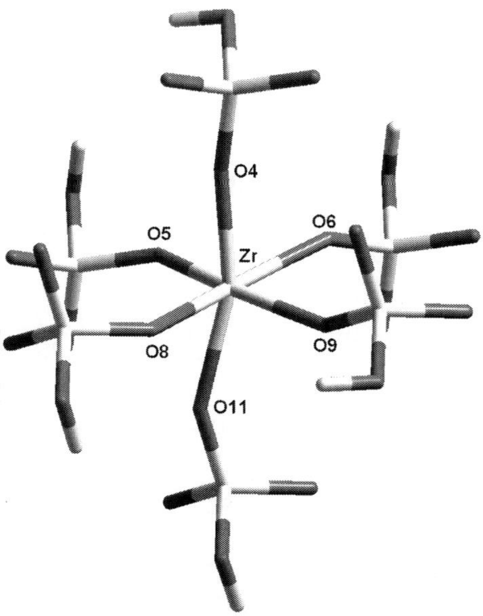

Figure 9.5 Main bond distances and angles in the Zr octahedron of α-ZrP.

Crystallinity of α-ZrP, low (LC) or high (HC), may vary depending on the synthesis procedure[14] which in essence comprises a simple reflux (LC) or hydrothermal treatment at 200 °C (HC) of zirconium oxychloride in 3–6 M phosphoric acid. Figure 9.6 shows powder X-ray diffraction (XRD) patterns and the corresponding SEM images of two samples of α-ZrP with very different degrees of crystallinity. Despite the quality of the latter, interlayer distance was measured to be 7.6 Å (XRD) as is also shown in the model of Figure 9.4.

If the reaction medium contains phosphorus acid ($H\text{-}PO_3H_2$) or organic phosphonates ($R\text{-}PO_3H_2$), they can be incorporated into the α-ZrP structure where the phosphates are partially or totally replaced by phosphite or phosphonate tetrahedrons, respectively, with their correspondingly H or R groups

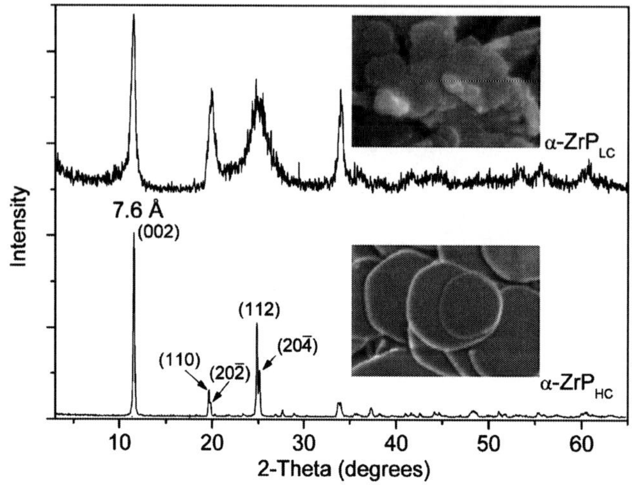

Figure 9.6 XRD diffraction patterns of α-ZrP of low (LC) and high crystallinity (HC) and their corresponding SEM images.

pointing to the interlayer region. There are excellent reviews,[15] profusely cited along the years,[16] covering this aspect that allows the direct synthesis of umpteen different organic–inorganic structures.

9.2.3 Gamma Zirconium Phosphate (γ-ZrP)

As happened with its α counterpart, γ-ZrP was discovered by Clearfield in 1968.[17] At that time, γ-ZrP was believed to bear the structure of α-ZrP with a different hydration degree. In fact, α- and γ-ZrP share the same empirical formula. Yamanaka's paper in 1979 was also important in this field,[18] yet the riddle was solved by Clayden,[19] whose finding in 1987 of two signals in the solid-state ^{31}P-NMR (α-ZrP gave only one) was instrumental in the realization that γ-ZrP was truly different from its α counterpart and not a mere hydrate variation.[20] Clearfield's Rietveld refinement of the powder X-ray structure put a definite end to previous speculations.[21] Figure 9.7 shows the structure of γ-ZrP. In the polyhedral model it may be seen that γ-ZrP comprises two sets of octahedral Zr atoms bridged by two series of mirror-image planes of tetrahedral phosphates bonded by four oxygen atoms to four different metals. In the surfaces of the layer, one may find a different set of phosphates bonded through two oxygen atoms to two different metals. This is the reason why $Zr(PO_4)(H_2PO_4)$ stands for the molecular formula of γ-ZrP while $Zr(HPO_4)_2$ corresponds to α-ZrP, the number of atoms being in fact identical for the two phases. Electro neutrality requires the external phosphates be bonded to the metals using neutral P=O and anionic P–O bonds. Therefore, every surface phosphate points two acidic P–OH bonds towards the interlayer region. The space-filling model of Figure 9.7 visualizes pristine γ-ZrP whose density is smaller than that of its α equivalent.

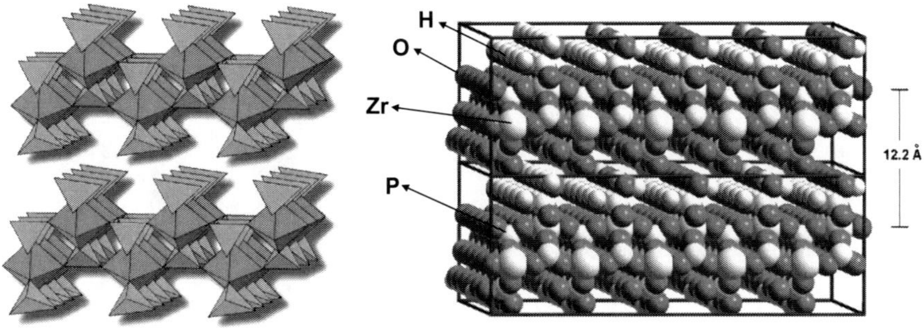

Figure 9.7 Polyhedral and space-filling models of the structure of γ-ZrP.

Space group	$P2_1$ (no. 4)
a/Å	5.3825(2)
b/Å	6.6337(1)
c/Å	12.4102(4)
β/°	98.687(2)
Z	2
D_c/Mg m^{-3}	2.417

Zr–O(1)	2.028(7)
Zr–O(2)	2.059(8)
Zr–O(3)	2.026(8)
Zr–O(4)	2.032(7)
Zr–O(5)	2.086(9)
Zr–O(6)	2.014(7)
P(1)–O(1)	1.571(7)
P(1)–O(2)	1.527(8)
P(1)–O(3)	1.531(8)
P(1)–O(4)	1.568(7)
P(2)–O(5)	1.545(7)
P(2)–O(6)	1.545(7)
P(2)–O(7)	1.586(7)
P(2)–O(8)	1.545(8)
O(1)–Zr–O(2)	89.7(5)
O(1)–Zr–O(3)	85.6(6)
O(1)–Zr–O(4)	95.8(5)
O(1)–Zr–O(5)	84.3(5)
O(1)–Zr–O(6)	174.8(5)
O(2)–Zr–O(3)	173.5(4)
O(2)–Zr–O(4)	89.8(5)
O(2)–Zr–O(5)	87.9(4)
O(2)–Zr–O(6)	91.2(4)
O(3)–Zr–O(4)	95.0(7)
O(3)–Zr–O(5)	87.3(5)
O(3)–Zr–O(6)	93.2(4)
O(4)–Zr–O(5)	177.7(6)
O(4)–Zr–O(6)	89.3(4)
O(5)–Zr–O(6)	90.6(3)

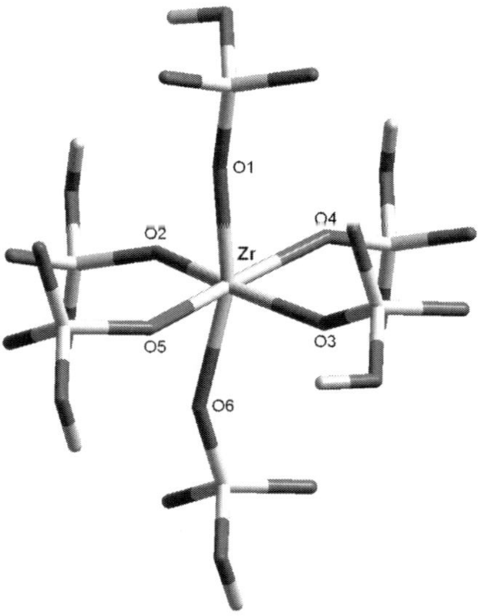

Figure 9.8 Main bond distances and angles in the Zr octahedron of γ-ZrP.

The smaller calculated density using the Hyperchem program (2.2 g cm^{-3}) as compared to that estimated for its α companion agrees with the experimental tendency (Figure 9.8). The two parallelepipeds enclosing two consecutive models of molecular formula $[Zr(PO_4)(H_2PO_4)]_{60}$, taken as a representative fragment (MW = 33 981.42 g; see Figure 9.7), have a volume of 24.5×10^{-21} cm^3 or 1.48×10^4 cm^3 considering Avogadro's number. Bond angles and distances are very similar in the α and γ phases (compare Figures 9.5 and 9.8).

9.2.4 Lambda Zirconium Phosphate (λ-ZrP)

This phase, sketched in Figure 9.9, was discovered, as it could not be otherwise, by Clearfield in 1994 – who determined its structure as well.[22] Its high resemblance to γ-ZrP can be appreciated in that it contains an internal layer of

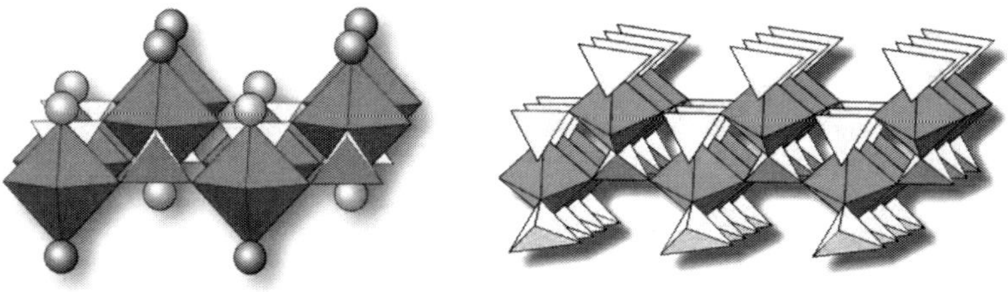

Figure 9.9 Polyhedral models of λ- (left) and γ-ZrP.

sustaining phosphates, yet the surface phosphates have been replaced by anionic and neutral species, namely F^- and DMSO, respectively, in Clearfield's original structure. The internal phosphates had to rearrange a little with respect to the γ phase to allow the metals to coordinate to these new species. Its synthesis was accomplished by heating at 120 °C for several days a solution of 0.05 mol of $ZrOCl_2$ in 44% HF (Zr:F ratio 1:25), 0.1 mol of 85% H_3PO_4 and 0.8 mol of DMSO. Three years later Alberti described the same structure, but with Cl^- anions instead of F^-.[23]

The trend of diminishing density is again qualitatively well reproduced by the Hyperchem program. The value calculated from the space-filling model indicated in Figure 9.10 is 1.9 g cm^{-3}. The two parallelepipeds enclosing two consecutive models of molecular formula $[Zr(PO_4)F(CH_3)_2SO)]_{60}$, taken as a representative fragment (MW = 33 824.45 g; see Figure 9.11), have a volume of 30.0×10^{-21} cm^3 or 1.80×10^4 cm^3 considering Avogadro's number. Again, bond distances and angles are alike with the previously described phases (compare Figures 9.5, 9.8, and 9.11).

9.2.5 Miscellanea

Other structures different from the α-, γ-, and λ-phases have been described in the literature, although the majority of them are slight variations of the three. An excellent review about the different archetypes of zirconium phosphates has been recently published.[24]

9.3 Reactions

9.3.1 Intercalation

The intercalation chemistry of ZrP has been studied for more than 40 years. In the case of the acidic α- and γ-phases, it is possible to intercalate basic compounds that react with the Brønsted acid OH groups of the surface phosphates pointing towards the interlayer region. Hydrogen bonding donor-acceptor species like alkanols, polyols, ketones, sulfoxides, *etc.* of far lower basicity than amines can be intercalated too.

Formula	$C_2H_6FO_5PSZr$
Pattern range, 2θ/°	15–80
Step scan increment, 2θ/°	0.02
Step scan time/s	20
Space group	*P*4/*n* (no. 85)
a/Å	6.6093(2)
c/Å	9.2371(3)
Z	2
D_c/Mg m^{-3}	2.33

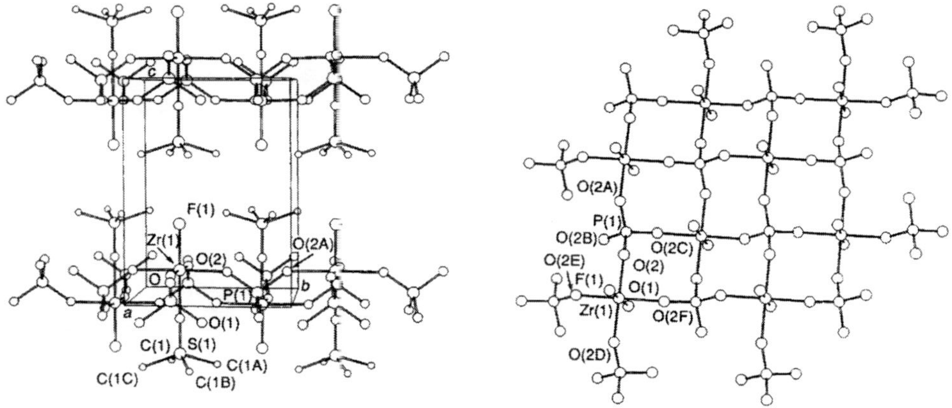

Zr–O(1)	2.107(8)		Zr–F(1)	1.991(7)	
Zr–O(2)	2.046(3)	(4 ×)			
O(1)–Zr–O(2)	89.1(2)	(4 ×)	O(2)–Zr–F(1)	90.87(15)	(4 ×)
O(1)–Zr–F(1)	180		O(2)–Zr–O(2)[a]	178.3(3)	(2 ×)
O(2)–Zr–O(2)[a]	89.987(4)	(4 ×)			
P(1)–O(2)	1.563(3)	(4 ×)			
O(2)–P(1)–O(2)[b]	109.1(3)	(2 ×)	O(2)–P(1)–O(2)[b]	109.7(2)	(4 ×)
S(1)–O(1)	1.546(8)		S(1)–C(1)	1.708(8)	(2 ×)
O(1)–S(1)–C(1)	107.2(5)	(2 ×)	C(1)–S(1)–C(1)[a]	85.0(3)	(2 ×)
P–O(2)–Zr	144.0(4)		S–O(1)–Zr	180	

Figure 9.10 Main bond distances and angles in λ-ZrP.

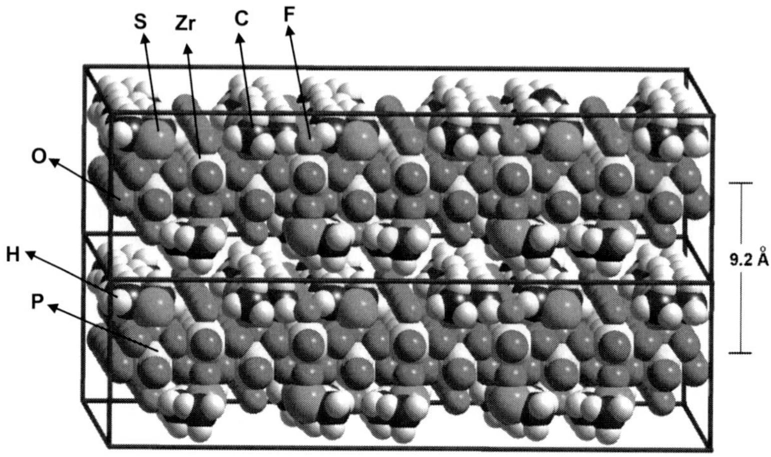

Figure 9.11 Space-filling spheres model of λ-ZrP showing the tight packing of atoms and interlayer distance.

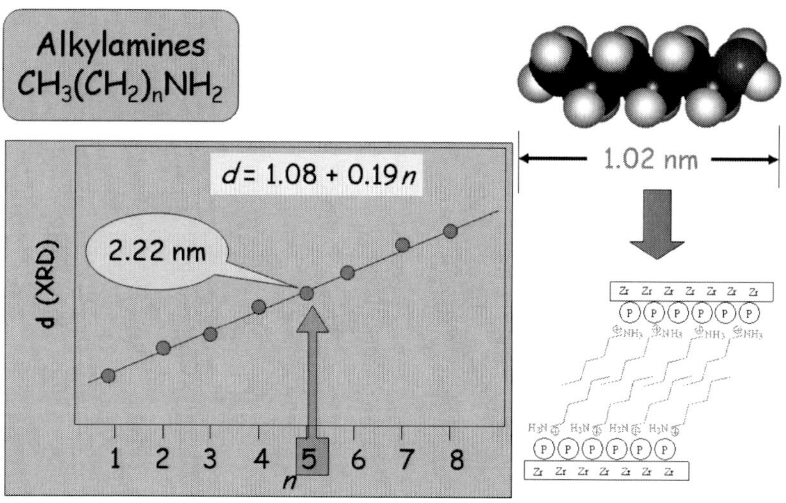

Figure 9.12 Double-layer formation in the intercalation of alkylamines in γ-ZrP.

The intercalation of alkylamines is a classic.[25] Figure 9.12 shows for example how the basal spacing of γ-ZrP increases with the length of the alkyl chain, as it has been experimentally measured by powder X-ray diffraction.

It may be seen in Figure 9.12 that the interlayer distance (XRD measurement) steadily increases with the length of the amine, the plot showing an excellent correlation coefficient, an intercept that coincides with the basal spacing of anhydrous γ-ZrP (10.8 Å) and a slope of 1.9 Å that is larger than the C–C distance (1.54 Å). Elemental analyses of the intercalated compounds show that there is a 1:1 Zr/amine ratio, thus suggesting that the amines may form a double layer, in a very similar fashion to how the phospholipids arrange in a cellular membrane. Therefore, the slope of the plot of Figure 9.12, lower than double the C–C bond distance, is compatible with the *ca.* 60° tilting, relative to average plane of the layers, of the alkyl chains as Figure 9.12 sketches.

These intercalation processes turn out to be essential in the chemistry of ZrP because most of its transformations cannot be performed until the inorganic layers are separated enough either by these intercalation reactions or appropriate exfoliation (see Section 9.3.2). The literature on ZrP intercalation was covered by an excellent revision up to 2004.[26] Table 9.1 summarizes the intercalated species into ZrP found in the more recent literature.

Table 9.1 Recent examples of intercalation into ZrP phases.

Intercalating species	*Uses and references*
$[Ru(phend)_2bpy]^{2+}$	biosensing,[28] electrochemical studies[29]
Co^{II}- and Ni^{II}-1,10-phenanthroline complexes	thermal and kinetic studies,[30] structural studies[31]
diethylenetriamine	absorption of carboxylic acids[32]
melamine	formaldehyde absorption[33]
$Ru(bpy)_3Cl_2$	photo physical measurements[34]
tris(2,2′-bipyridine)ruthenium(2+)	structural studies,[35, 36] luminescence studies[37]
metalloporphyrin	catalysis[38]
butylamine	sorption of silver(I)[39]
biogenic amines	structural studies[40]
proteins	structural studies[41]
hemoglobin	electrochemical properties[42]
enzymes	structural studies[43]
sulfanilic acid	proton conductivity[44]
Pt^{II} complexes	luminescence studies[55]
ionic liquids	structural studies[46]
diaminonaphthalenes	structural studies[47]
ferrocene	redox studies[48]
cetyltrimethylammonium bromide	structural studies[49]
1,2-alkanediols	structural studies[50]
rhenium complexes	structural studies[51]
methylviologens	photochemical studies[52]
2,9-dimethyl-1,10-phenanthroline	complexation studies[53]
Zn and Cd complexes	structural studies[54]
amines	adsorption of aldehyde fragrances[55]
DNA and proteins	protein activity[56]
glucosamine and chitosan	structural studies[57]
polymers	Structural studies[58]
jeffamines	structural studies[59]
hexadecylamine	structural studies[60]
1-pyremethylamine	photolysis studies[61]
propylamine	gel formation[62]
1,10-phenanthroline	complex formation[63]
porphyrins	structural studies[64]
glucose oxidase	glucose sensing[65]
proteins	structural studies[66]
surfactants	structural studies[67]
acrylamide	polymerization studies[68]

9.3.2 Topotactic Replacement of Phosphates

This is a unique feature of the γ phase of ZrP. Its external phosphates are amenable of replacement by other phosphorus functions without altering the overall structure of the layers. This is what the word "topotactic" means. The internal phosphates are thus untouched and play a key role in maintaining the integrity of the γ-ZrP structure. Figure 9.13 compares the process as to the laying of octopus's eggs on the ceiling of a small cave, the eggs representing the organic part of a phosphonate.

This reaction cannot be carried out on α-ZrP because this phase lacks the sustaining internal phosphates (see above). It is nevertheless true that α-ZrP may be made to contain additional phosphorus functions, *i.e.* phosphonates, but the materials must be obtained by direct reaction of all components. The same appears to be true for the λ phases. The versatility of the topotactic exchange carried out upon preformed γ-ZrP is way much higher, as will be shown later on.[27]

The exchange of the surface phosphates in γ-ZrP is a fascinating reaction. Let us carefully analyze what it takes. First of all, we have to consider that γ-ZrP is not soluble in any solvent, including water, acetone or alcohols. Therefore, the reaction has to take place in the liquid–solid interface. As Alberti put it in 1997,[69] "It is noteworthy that the topotactic preparation of the γ-organic derivatives from exfoliated γ-ZrP is a rare example (maybe the first) in which a crystalline compound is first transformed into a colloidal dispersion and then, after chemical modifications obtained by fast topotactic reactions, again recomposed". Therefore, Alberti's studies[70] strongly suggested that the lamellae of γ-ZrP reacted with phosphonates because the microcrystals exfoliated first, forming a colloidal dispersion in appropriate conditions. In this way, the

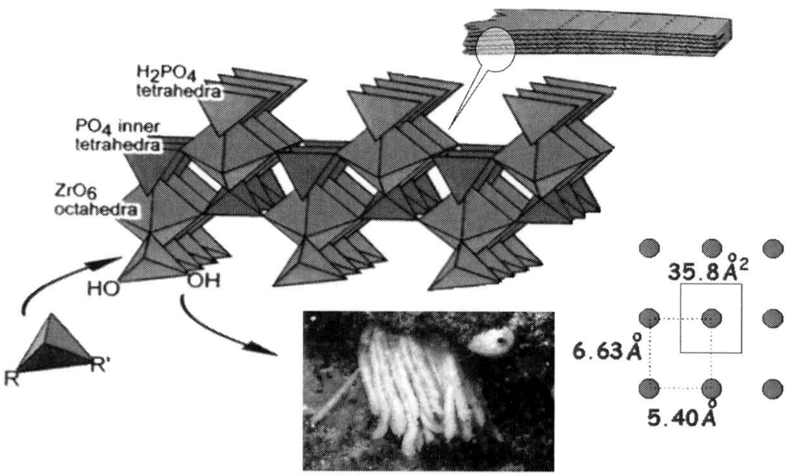

Figure 9.13 Sketch of the topotactic replacement on γ-ZrP of the surface phosphates by other phosphorus functions as organic phosphonates.

attacking phosphorus functions are able to reach the layer surface and replace the phosphates. The best conditions to achieve exfoliation were found to be in mixtures of water–acetone (from 0.7:0.3 to 0.2:0.8 in volume) at 60–80 °C.

Once the lamellae are sufficiently apart from each other, some phosphates have to be detached first. How are the phosphates expelled from the lamellae surface? Figure 9.14 outlines the most plausible process. As yet, no direct evidence of this mechanism or any other has been published. Taking into account that the octahedrons $Zr^{IV}O_6$ have to be formed by the contribution of four P–O$^-$ and two P=O groups to keep electroneutrality, the most probable possibility is that one of the two Zr–O bonds coming from the P=O groups is broken by the action of water (A). The surface phosphate still remains attached to another metal (B). The attacking phosphorus function gets near the reactive Zr bonded to water (C). The attack has been exemplified in Figure 9.14 with methylphosphonic acid that then displaces water and becomes bonded to Zr through the P=O bond (D). Finally, the surface phosphorus is completely detached by the action of one of the P–OH groups of the phosphonic acid (E). This process has to be repeated many times, until a certain amount of surface phosphates is replaced, and stops when the material flocculates. There are no studies in the literature concerning the thermodynamic parameters of this process. Flocculation takes place faster or slower depending primarily on the chemical structure of the attacking phosphorus functions. The extent of the topotactic exchange thus depends on the flocculation rate. Nevertheless, the strong stirring of the flocculated dispersion may allow further exchange although the process is much slower because it now depends on the hampered diffusion of the attacking phosphorus derivative along the already heavily filled interlayer space.

A particle may easily have areas of several μm^2. The distance and available area between phosphates was indicated in Figure 9.13. This means that a particle of 1 μm^2 bears *ca.* 28 000 phosphates on each of its two faces. Therefore, for the exchange reaction to be as efficient as it is, the flocculation rate has to be at least *ca.* five orders of magnitude slower than the individual phosphate/phosphonate replacement.

In principle, the organic phosphonates used for the topotactic exchange may be of any kind: rigid or flexible, polar or non-polar, chiral or non-chiral or with any conceivable chemical function. The topotactic exchange may be multiple, either simultaneously performed by mixing different phosphorus derivatives or in sequence, by adding one after another. We thus have to our disposal (Figure 9.15) a set of conceptual and practical tools to designing organic–inorganic materials of any imaginable nano-scaled structure whose arrangement can be predicted with quite reasonable accuracy.

In the case of using diphosphonates, the final structure ends up as a pillared 3D arrangement of inorganic lamellae covalently bonded to each other by appropriate organic moieties. In this case, the mechanism of the topotactic exchange is even more intriguing. In order to shine a light on it we can compare the results of the topotactic exchange of four different diphosphonates, namely, alkyl, ethylenpolyoxa, polyphenyl and polyphenylethynyl (Figure 9.16). The first two are flexible but the polarity of their chains is very different. The last

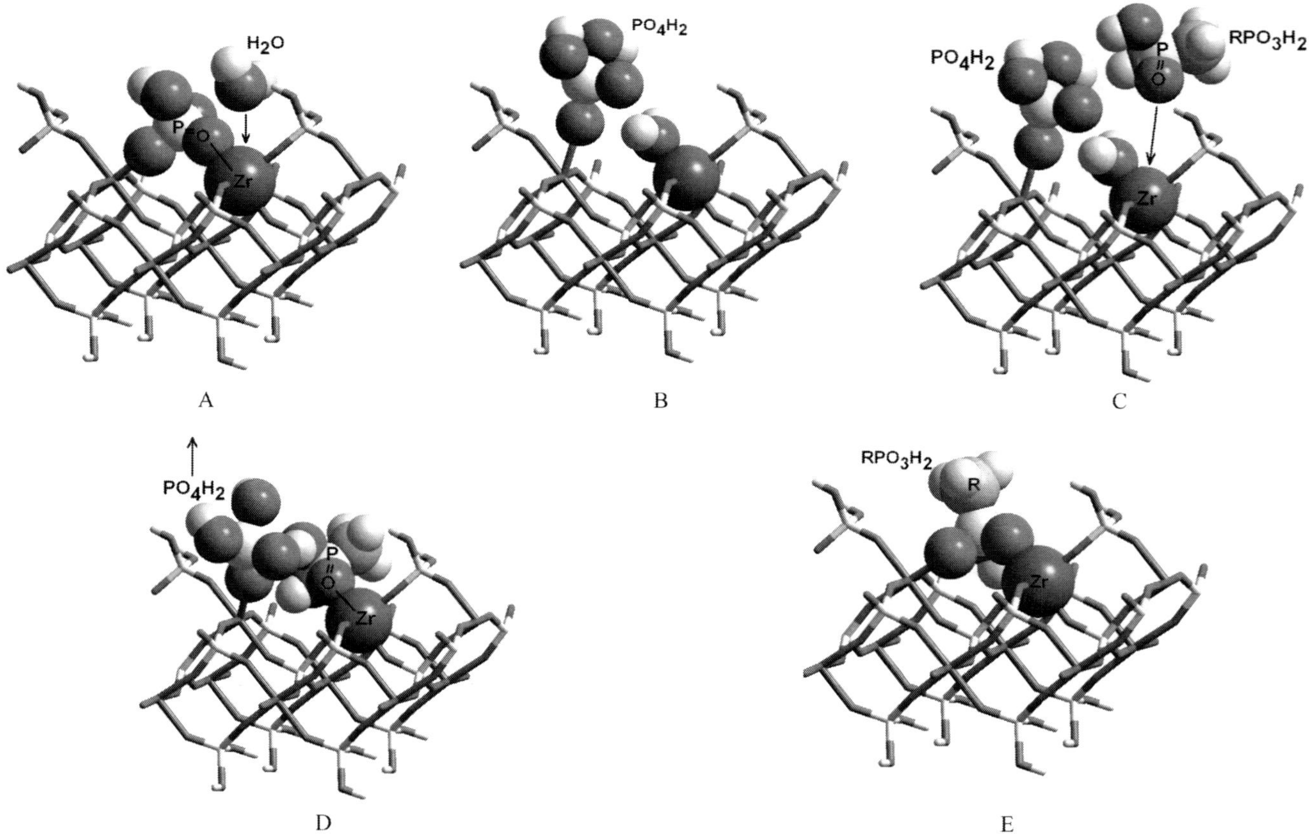

Figure 9.14 Cartoon of the most plausible mechanism for the topotactic exchange of phosphate by methylphosphonate in γ-ZrP.

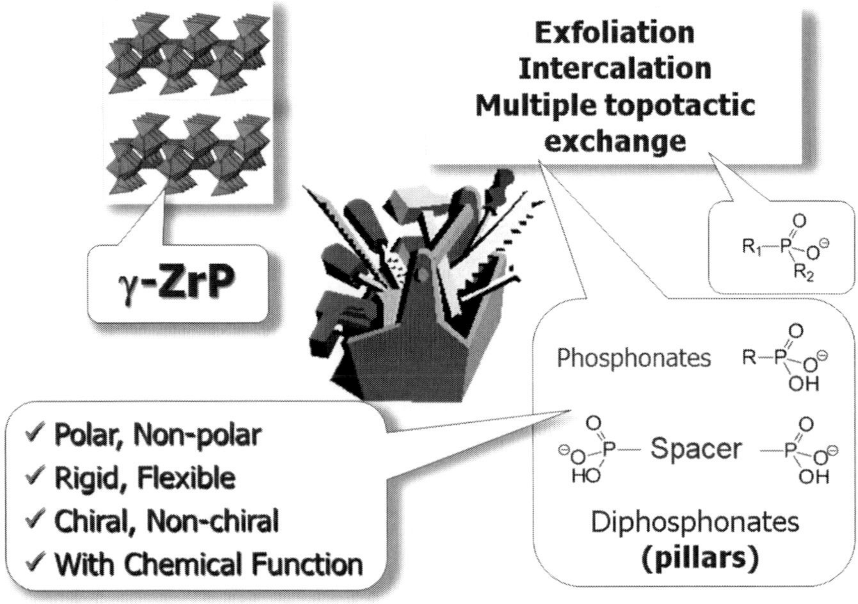

Figure 9.15 Schematic view of our conceptual and practical tools for the construction of organic–inorganic materials.

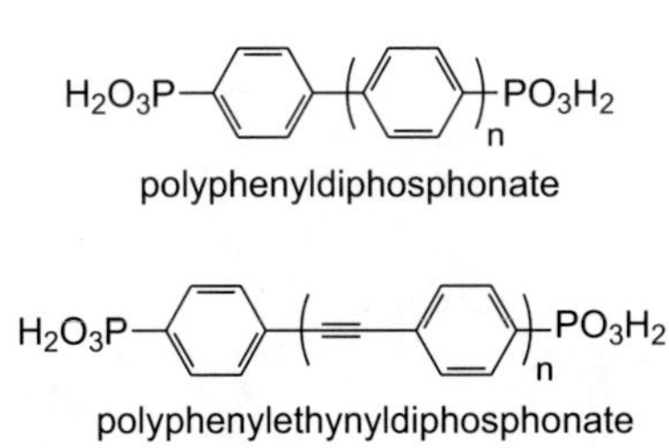

Figure 9.16 Diphosphonates considered in the mechanism of topotactic exchange.

two are rigid but the presence or absence of the ethynyl bond turns out to induce very important differences in the final outcome.

Figure 9.17 shows the level of exchange reached at various initial molar ZrP:diphosphonate ratios for alkyl- and ethylenpolyoxa diphosphonates. An important difference may immediately be seen. Almost all alkyl diphosphonate

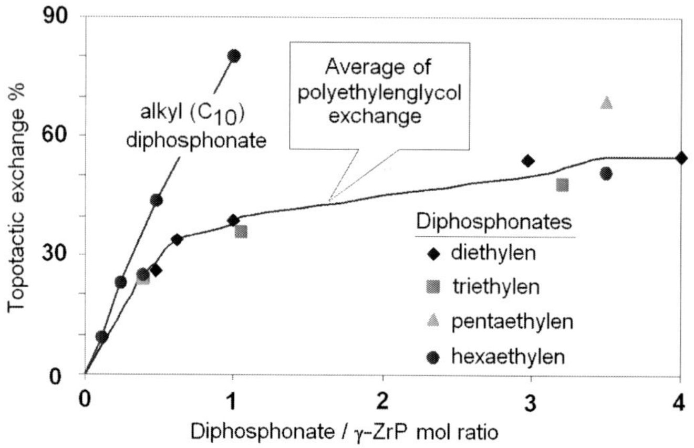

Figure 9.17 Plot of the final level of exchange reached at various initial molar ZrP:diphosphonate ratios for the indicated alkyl- and ethylenpolyoxa diphosphonates.

is incorporated to the γ-ZrP phase whereas only a maximum of 50–60% of the phosphates can be exchanged by ethylenpolyoxa diphosphonates, regardless of either their excess put into play in the reaction or their chain length. The cross-sectional area of alkyl- and ethylenpolyoxa diphosphonates is very similar and does not exceed the available area around the surface phosphates (Figure 9.13). The different behavior should then arise from the different polarity of the chains.

Two slightly different mechanisms may explain this different behavior, as shown in Figures 9.18 and 9.19.

The exfoliated lamellae are attacked by one end of the diphosphonates (A), the alkyl chain of the latter, not having any affinity for the polar γ-ZrP surface, remain standing up (B). When a certain amount of phosphates have been replaced, the neighboring lamellae can start piling up, favored by the interaction among the protruding phosphonate groups (C). But to achieve pillaring the alkyl chains must interdigitate or imbricate (D), which probably is a relatively slow process. Once this barrier has been overcome, the formation of pillars takes place (E) and the material flocculates (Figure 9.18).

The process for the ethylenpolyoxa counterparts can be simpler due to the polarity of the chains and their affinity for the polar γ-ZrP surface. After the initial exchange reaction (F), the chains lay down close to the surface because their oxygen atoms may form a network of hydrogen bonds with the acidic phosphates (G) which are then protected from further reaction. The neighboring lamellae then have no hitch to get close and flocculate quickly (H), thus precluding high levels of phosphate/phosphonate exchange (Figure 9.19).

The case of the rigid diphosphonates may in principle be similar to that of the alkyl diphosphonates because the stiffness of the molecules forces them to definitely stand up. However the ^{31}P-NMR spectra of the final materials show striking differences (Figure 9.20). Materials γT resulting from the exchange of the

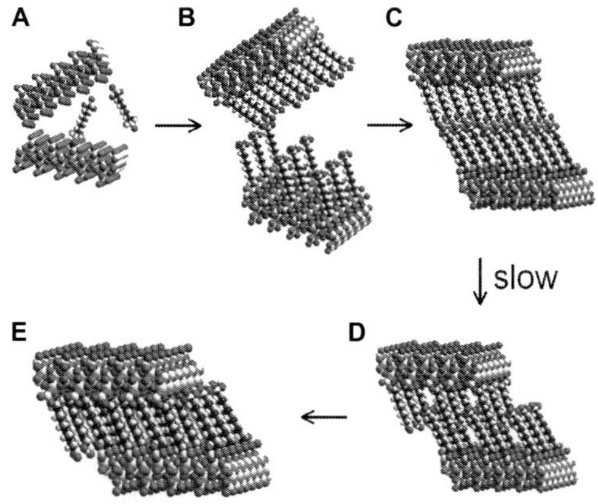

Figure 9.18 Mechanism for the exchange of alkyl diphosphonates.

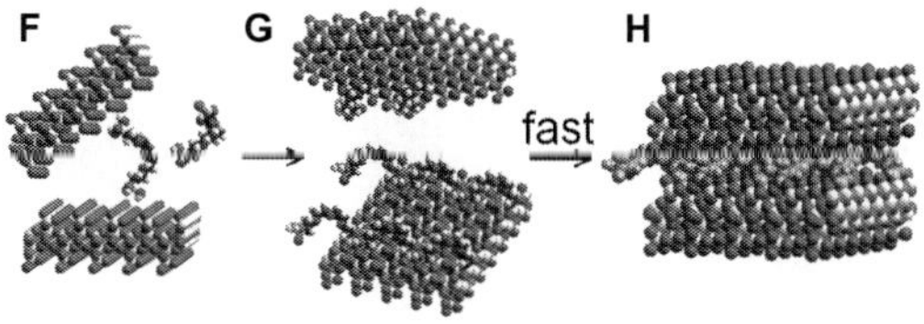

Figure 9.19 Mechanism for the exchange of ethylenpolyoxa diphosphonates.

terphenyl diphosphonate ($n = 2$ in Figure 9.16) at various stoichiometries (25%, 35%, 40%, and 60%) shows the expected signals for the internal phosphate (–27 ppm), the remaining surface phosphate (–14 ppm), and the attached phosphonate (4 ppm). In contrast, materials γTt_2 coming from the reaction of γ-ZrP with the corresponding polyphenylethynyl diphosphonate ($n = 2$ in Figure 9.16), again at various stoichiometries (25%, 35%, 45%, and 50%), shows a new signal at 18.5 ppm, especially at high stoichiometric ratios, attributable to unreacted phosphonate groups. Figure 9.21 sketches how this can be understood.

The presence or absence of the triple bond makes the exchange mechanism different, as is outlined in Figure 9.22.

It seems reasonable that these rigid diphosphonates aggregate in solution but our hypothesis is that polyphenyl derivatives do it in a head–tail fashion (A) whereas polyphenylethynyl diphosphonates interact laterally (B). Head–tail aggregation prevents one end of the polyphenyl diphosphonate from reacting with the γ-ZrP surface before all available phosphate places have been taken. The pillaring is thus severely hampered and the diphosphonates finally imbricate to display the experimental interlayer spacing which is higher than that expected for true pillaring. However, "lateral" aggregation (B) enables polyphenylethynyl diphosphonates to preserve room for one another leading to a stage of previous imbrication that favors the formation of factual pillars.

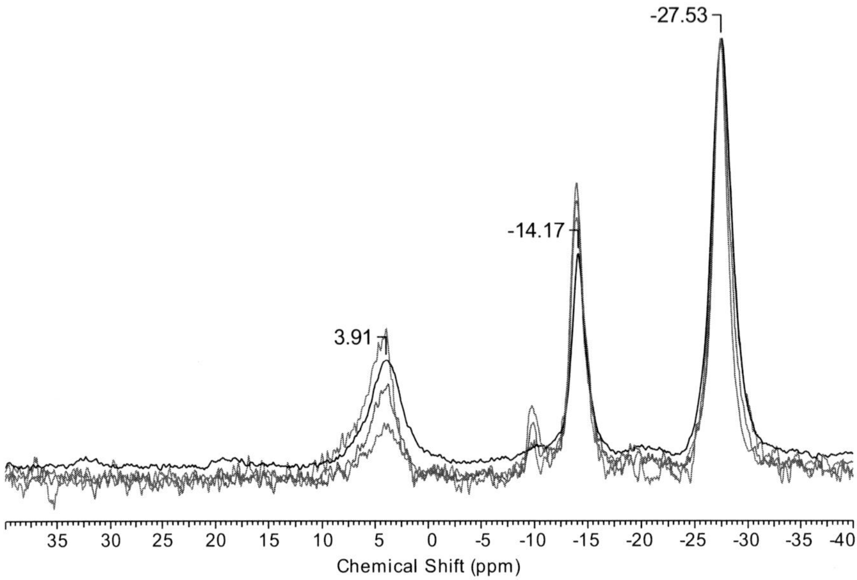

A. γT25 (red), **35** (blue), **40** (green), **60** (black).

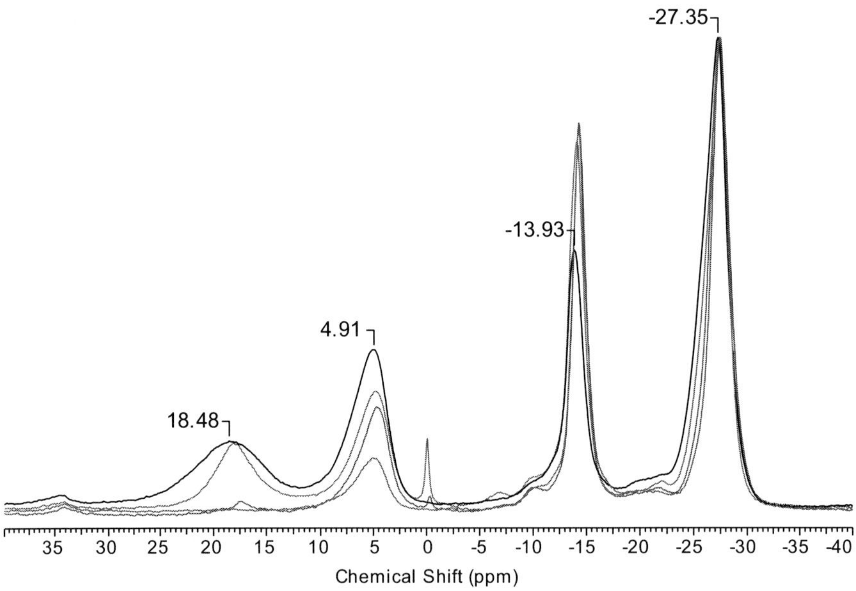

B. γTt$_2$25 (red), **35** (blue), **45** (green), **50** (black).

Figure 9.20 Solid-state "non-CP" MAS (10 kHz) ^{31}P-NMR (relaxation delay 20 s) of the indicated materials.

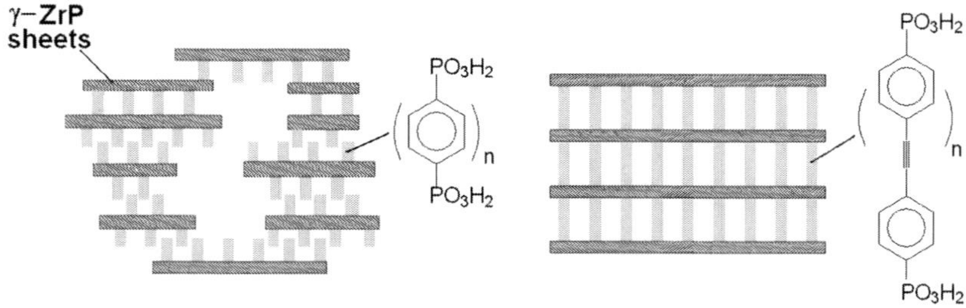

Figure 9.21 Schematic representation of a defectively pillared structure (left) where some of the polyphenyldiphosphonate groups have failed to cross-link consecutive layers of the inorganic salt. This was not observed with the polyphenylethynyldiphosphonate counterparts (right).

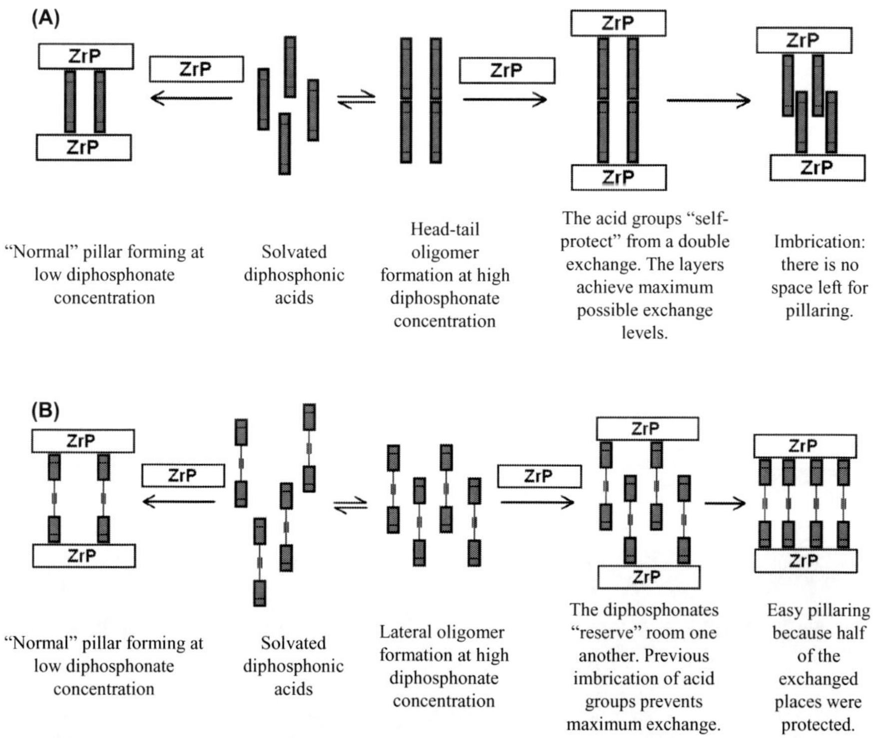

Figure 9.22 Possible mechanisms of phosphate/diphosphonate topotactic exchange in γ-ZrP.

The high complexity of the mechanism of the topotactic exchange reaction has been clearly illustrated with these four examples and the outcome of the process is determined by the following factors:

1. initial stoichiometry of reactants
2. cross-sectional area of the phosphonates

3. conformational flexibility of phosphonates
4. affinity of the organic chain for the lamellae surface
5. mode of self-aggregation of phosphonates in the reaction media.

9.4 Applications

9.4.1 Confinement of Crown Ethers

The first report considering the idea of forming supramolecular crown ethers inside the layers of ZrP was published in 1984.[71] γ-ZrP was reacted with $CH_3(OCH_2CH_2)_nOPO_3^{2-}$ ($n = 1$–3) and half of the phosphates were replaced by the phosphonates. The topotactically exchanged material displayed recognition capabilities towards alkali salts of soft anions not shown by the phosphonates in solution.

Real crown ethers were inserted into the layers of γ-ZrP either by intercalation or topotactic exchange for the first time in 1996.[72] Within the intercalated materials one may find a series of cyclams, such as those shown in Figure 9.23.[73] The lactams (left) led to larger interlayer distances (2 nm) than those expected considering the average diameter of the rings (0.8 nm). Elemental analyses indicated a high level of incorporation (66–94%), well above that likely (<50%) taking into account the cross-section of the lactams. The experimental results were thus explained by inserting a double layer of hydrogen-bonded dimers, slightly tilted with respect to the perfect perpendicular

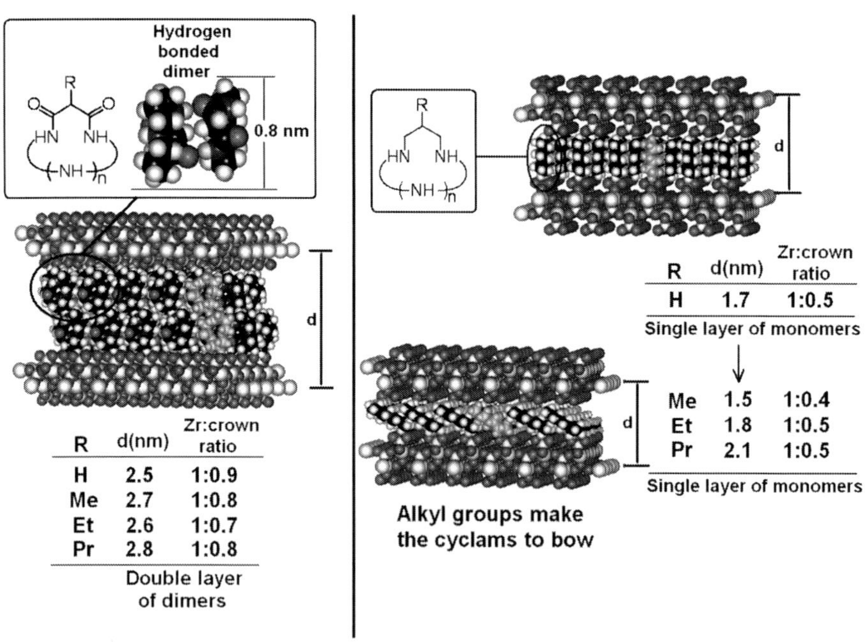

Figure 9.23 Intercalation of cyclams into γ-ZrP.

arrangement. The presence of alkyl groups vaguely affected the tilting and the level of incorporation as well. When the lactams were reduced to the corresponding amines (Figure 9.23; right), the intercalation behavior was very different. The parent compound (R = H) led to a material with an interlayer distance (1.7 nm) and incorporation level (46%) compatible with a monolayer of cyclam between the γ-ZrP lamellae. The replacement of H by Me made the cyclam to bow severely, thus showing how very small structural changes in the organic intercalate may strongly affect the whole scaffold.

The seminal paper on the topotactic exchange with various crown ethers[74] of different sizes, containing one or two phosphonate groups, showed that the phosphate/phosphonate replacement occurred with an intriguing odd/even effect depending on the size of the crowns. Figure 9.24 summarizes the observations. In exfoliating conditions, even-membered crowns incorporated as an imbricated single layer (short interlayer distance of 1.6–1.8 nm) whereas the odd-membered counterpart did as a non-imbricated double layer (larger interlayer distance of 2.0 nm). The previous intercalation of butylamine produced slightly different results. This work also describes for the first time ever the result of multiple exchange reactions, namely the further replacement with methylphosphonate of the remaining phosphates from the initial crown ether/phosphonate reaction. In this way, the polarity and acidity of the inorganic surface was drastically altered. The effect on the recognition abilities of the material towards Na^+ and K^+ ions was quite visible (Figure 9.25). Competitive experiments showed that the recognition abilities towards the alkaline ions, exerted in the solid–liquid interface by the confined crowns, were very different to those commonly exhibited in solution. Furthermore, the second phosphate/methylphosphonate exchange completely altered the crowns' response to the ions, thus showing that the nature of the whole matrix structure is crucial in the overall recognition process.

9.4.2 Chemically Driven Porosity Changes

Diphosphonates derived from polyethylenglycol chains of different lengths (from di- to hexa-ethylenglycol) were prepared and exchanged at 25% level into γ-ZrP.[75] The corresponding pillared materials (Figure 9.26) were thus formed, bearing the general molecular formula of $Zr(PO_4)(H_2PO_4)_{0.75}[HO_3P(CH_2)_2(OCH_2CH_2O)_n(CH_2)_2PO_3H]_{0.125}$ ($n = 2$–6). The remaining surface phosphates $[(H_2PO_4)_{0.75}]$ were quantitatively replaced by hypophosphite $[(H_2PO_2)_{0.75}]$ leading to two sets of materials, polar/polar (polar columns and polar surface phosphates) and polar/non-polar (polar columns and non-polar surface hypophosphites). Figure 9.26 summarizes the observed variation of interlayer distance (as measured by XRD) when the materials containing pentaethylenglycol diphosphonate ($n = 5$), as a representative example, were treated with methylamine in aqueous dispersion. It may be seen that the polar/polar material steadily augmented its interlayer distance with the increasing amount of intercalating amine. When no amine is present, the polar columns interact by hydrogen bonding with the surface phosphates and the layers are compressed to

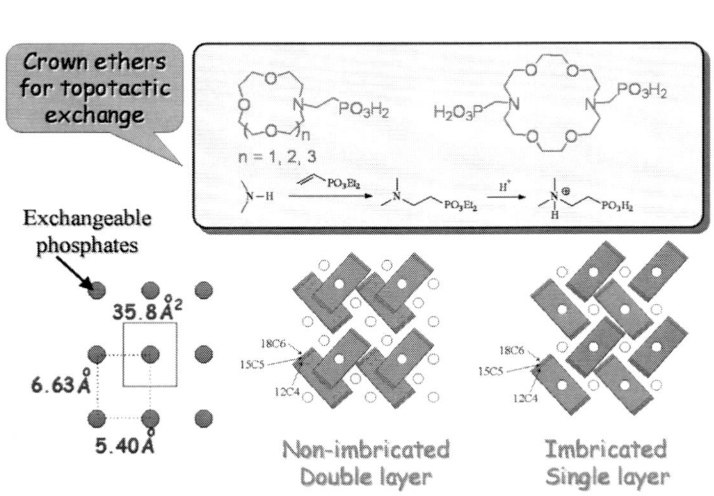

$ZrPO_4(H_2PO_4)_{1-x}(crown)_x(H_2O)_y$
$ZrPO_4(H_2PO_4)_{1-x}(diaza)_{x/2}(H_2O)_y$

Crown	Direct exchange	Previous intercalation
12-*C*-4	x = 0.29 (1.6 nm) imbricated	x = 0.47 (1.9 nm) non-imbricated
15-*C*-5	x = 0.13 (2.0 nm) non-imbricated	x = 0.45 (2.0 nm) non-imbricated
18-*C*-6	x = 0.25 (1.8 nm) imbricated	x = 0.28 (1.9 nm) imbricated
diaza 18-*C*-6	x = 0	x = 0.44 (1.6 nm) pillared

Figure 9.24 Results of the topotactic exchange on γ-ZrP of the indicated crown-ethers by direct reaction of the exfoliated salt or with previous intercalation of butylamine.

Material	Singly exchanged	Doubly exchanged
ZrP	0.14/0.16	0.06/0.02
12C4	0.66/0.10	0.50/0.23
15C5	0.94/1.27	-
18C6	0.27/0.99	0.84/0.68
diaza-18C6	0.42/0.86	0.72/0.22

Figure 9.25 Amount of Na^+ and K^+ ions relative to Zr found by atomic absorption measurements in the indicated materials (see Figure 9.23); in the doubly exchanged materials, the remaining phosphates of the first crown-ether phosphonate reaction were replaced by methylphosphonate.

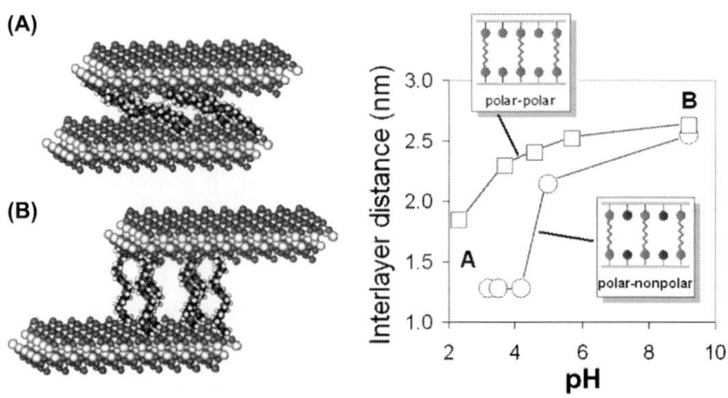

Figure 9.26 Molecular models of γ-ZrP exchanged with pentaethylenglycol diphosphonate with variable interlayer distance; plot of interlayer distance variation with pH of the indicated materials.

one another (Figure 9.26A). The amine reacts with the surface phosphates and progressively disrupts the web of hydrogen bonding interactions, thus making the columns to stand up (Figure 9.26B). The overall porosity of the material is thus heavily increased by the mild acid–base reaction. In the case of the polar/non-polar material the increase of interlayer distance occurs all of a sudden, within a very narrow pH range (see the plot in Figure 9.26). The replacement of the surface phosphates by hypophosphite greatly diminishes the overall acidity of the material, the only acidic OH groups being those of the phosphonates. In the absence of amine, the polar columns still can establish O · · · H–P hydrogen bonds with the surface hypophosphite groups. When sufficient amount of amine is present, the few remaining acidic OH groups are quickly neutralized and the methylammonium ions act as wedges that pull up the columns to suddenly rise. Therefore, the interlayer distance is doubled and the porosity is on the whole profoundly changed. To the best of our knowledge, this odd supramolecular behavior in the solid state, responding to a simple acid–base reaction in the solid–liquid interface, has never been observed before.[76]

9.4.3 Supramolecular Chirality and Chiral Memory

When we performed molecular models of the materials described in the previous section, we realized that to fit the largest achieved interlayer distance, the polyethylenoxa columns had to be helicoidally arranged (Figure 9.26B). This is not surprising since polyethylenglycols are known to attain that conformation in aqueous solution. Yet, what makes a notable difference is that in our materials, the polyethylenoxa columns are covalently attached by both ends to the inorganic layers, thus conferring them a supramolecular order. Therefore, assuming the necessary conditions are established, the P/M helicity of the chains could be somewhat directed to render a certain degree of homochirality which could be detected by simple optical rotation measurements. Figure 9.27 outlines the experiments aimed to that observation. Dispersions of 5 mg in 10 mL of 1:1 water/acetone of either native γ-ZrP or exchanged at 25% with hexaethylenglycol diphosphonate ($n=6$; see the molecular formula in previous section) displayed no sizeable optical rotation. However, when optically pure (+)-phenethylamine was intercalated in the material containing hexaethylenglycol, a relatively large value of optical rotation was measured. We then smoothly replaced the enantiomerically pure amine, by treating the material in water dispersion with an excess of hexylamine, and the optical rotation kept showing through. In absence of any (+)-phenethylamine, as shown by pertinent NMR experiments, the only source of optical activity should be the homochirality of the polyethylenoxa columns that was arranged by the concourse of the chiral amine and was maintained when the latter was replaced by the achiral one. This is an expression of supramolecular chiral memory; no similar cases can be found in the literature.[77]

We wonder what would happen if the chiral influence were intrinsic to the polyethylenoxa chains. To this end, we prepared enantiomerically pure diphosphonates derived from the addition of *R*-glycidol to diethyl vinylphosphonate and subsequent epoxide aperture with hexaethylenglycol (the reaction is shown in Figure 9.28).[78] The topotactic exchange of this diphosphonate into γ-ZrP was performed at 25% level and the material was positively characterized using the usual techniques. It occurred to us that the conformation of the polyethylenoxa chains attached to the inorganic layers could be changed if amines of appropriate length were intercalated. Figure 9.28 summarizes the results. As expected, the interlayer distance steadily increased with the length of the amine. Molecular modeling showed that the organic chains are continuously changing their conformation from compressed random coiling at the shortest layer separation to stretched zigzag arrangement at the largest basal

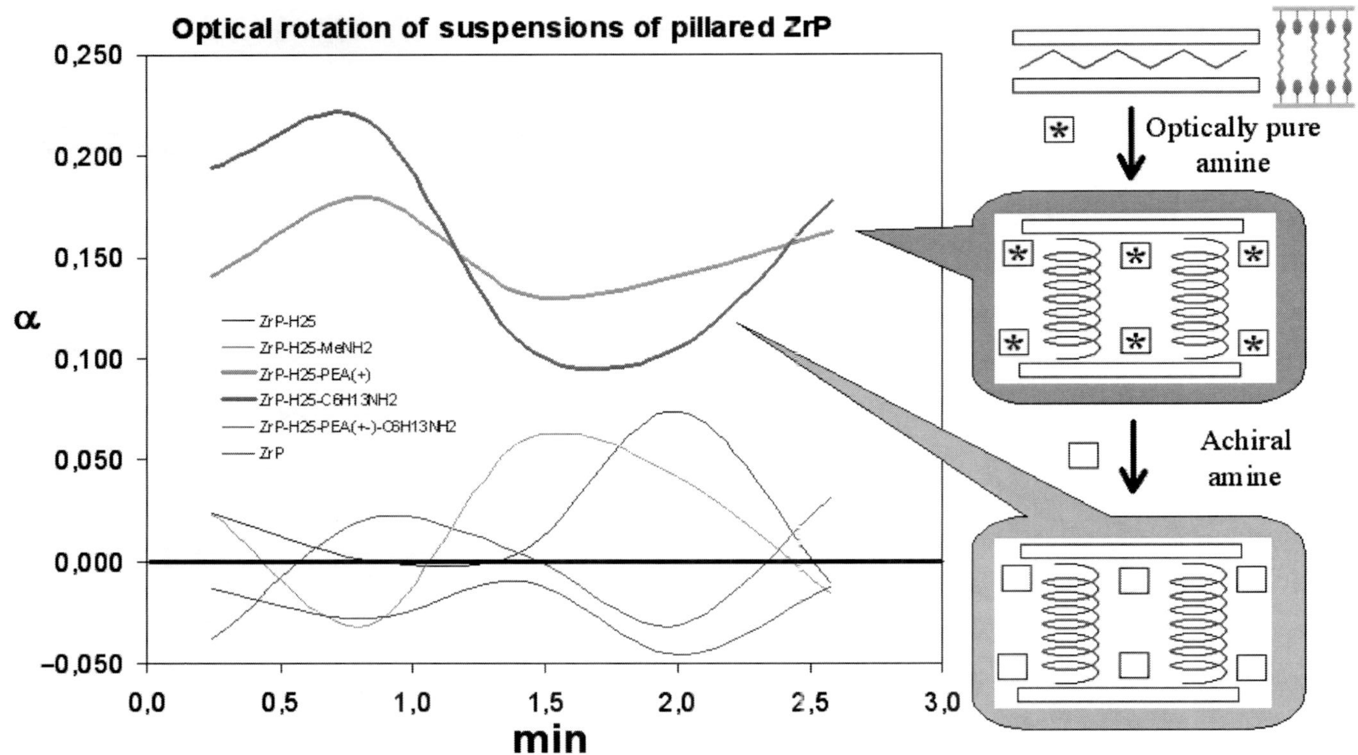

Figure 9.27 Optical rotation measurement of the indicated samples (abbr.: H25, hexaethylenglycol exchanged at 25%; PEA, 1-phenethylamine).

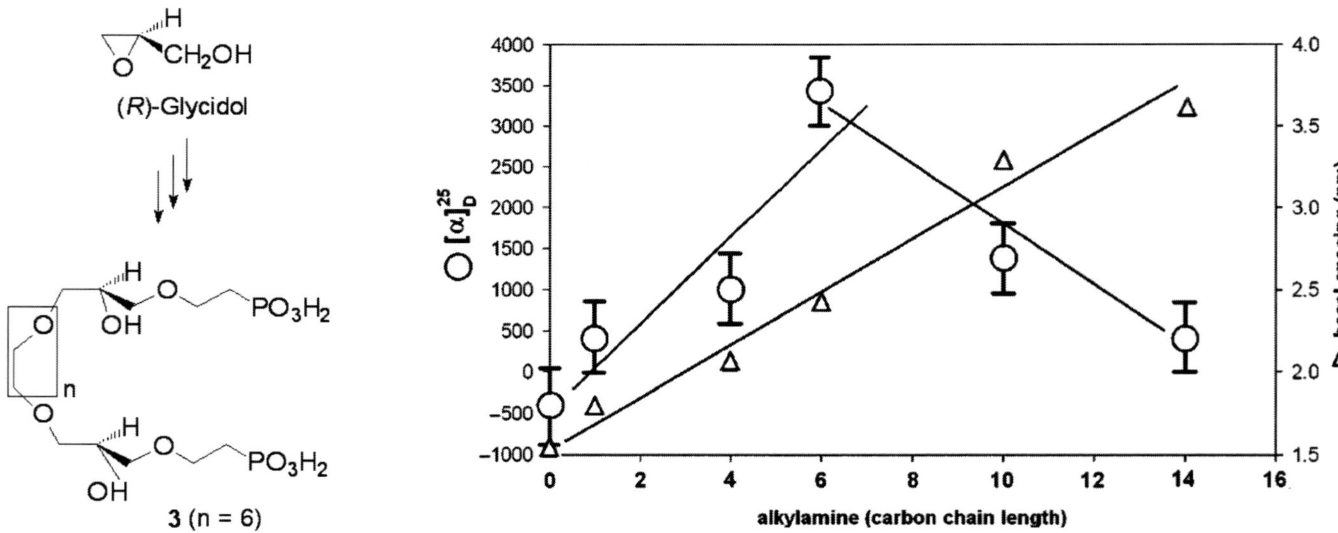

Figure 9.28 Variation of interlayer distance and optical rotation of γ-ZrP exchanged with the compound of the scheme with the intercalation of alkylamines of increasing length.

spacing. Optical rotation measured in 1:1 water/acetone dispersion passed through a maximum value when hexylamine was intercalated and the resulting interlayer distance allowed for the columns to be in their most ordered helicity, as the molecular models strongly suggested.

In our opinion this experiment clearly proved that chirality can be created at the supramolecular level in the solid state and finely tuned by mild reactions in the solid–liquid interface. There are no examples in the literature achieving this effect in such a clear-cut manner.[79]

9.4.4 Luminescence Signaling

The building of molecular systems able to display the strong luminescence of lanthanide metals is very active a field in chemistry.[80] To accomplish it two stringent conditions have to be met: (a) a suitable organic chromophore should absorb light and efficiently transfer energy to the metal (the *antenna effect*); (b) the coordination sphere of the metal should be free of water molecules because the OH oscillators easily quench metal emission. The white powder of γ-ZrP does not absorb visible light and remains reasonably transparent in the UV region. On the other hand, the pillared materials described in the previous section with polyethylenoxa columns may constitute an excellent spider web to enshroud the oxygenphylic lanthanide metals and isolate them within the solid matrix. Acid–base reactions with the surface phosphates may be taken as the driving force to intercalate suitable chromophores in the organic–inorganic lattice. Figure 9.29 shows part of the achieved results.

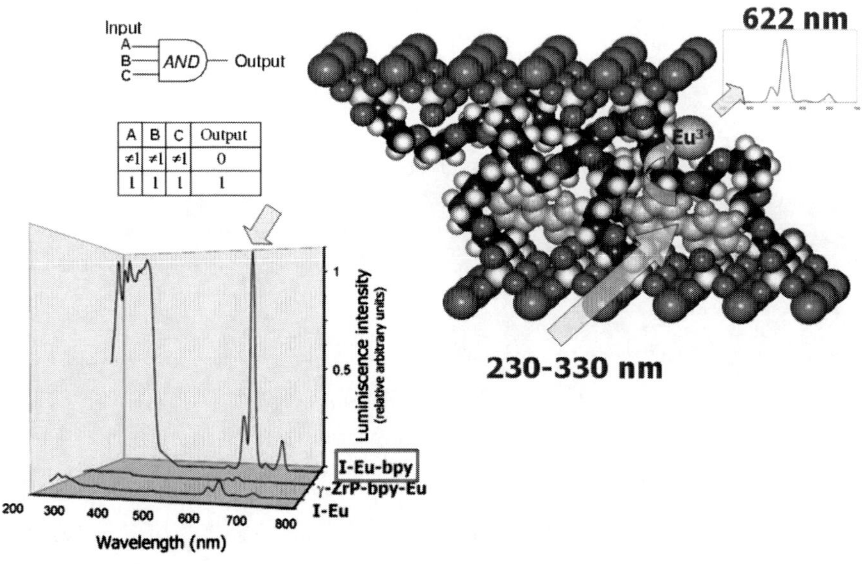

Figure 9.29 The lanthanide emission was only observed when polyethylenoxa columns, the 2,2′-bipiridyl and the metal were confined within γ-ZrP.

In electronics' terms the composite behaves as a triple AND gate because, for the metal luminescence to take place, it is necessary to gather the three components: lanthanide metal, polyethylenoxa columns, and 2,2′-bipyridyl as sensitizer. These materials may find a number of different applications either in solid or in the solid–liquid interface. More sophisticated chromophores are being tried (Figure 9.30) and preliminary results showed the pursued strong luminescence of the metals.[81] The bis-triazolpyridine diphosphonates have been provided with chirality (Figure 9.31) in order to check whether the resulting materials from the topotactic exchange are able to show circularly polarized luminescence (CPL) of lanthanides.[82] This property, seldom measured in the solid state, might be very important in the technology of optical handling of information. Work is under way.

9.4.5 Photo-induced Electron Transfer

There is no fundamental principle running against the possibility of attaining the proper arrangement of chemical components on a surface to accomplish long-lived photo-induced electron transfer,[83] the first step in either the achievement of artificial photosynthesis or the construction of efficient solar cells. Inorganic γ-ZrP may constitute an excellent carving board to realize that goal. We have attached to it phosphonate derivatives of the $Ru^{II}(bpy)_3$ complex, well known by its excellent light-absorption and electron-donor properties, and placed side-by-side to it different electron donor acceptors ranging from relatively simple viologens to more elaborated fullerene derivatives. Figure 9.32 contains some molecular models of these complex structures which were positively characterized by the usual techniques.[84]

The first important observation was that the phosphorescence emission of the $Ru^{II}(bpy)_3$ complex was heavily quenched by the presence of the electron-acceptor species, suggesting that the pursued electron transfer took place. Very recently, flash photolysis revealed some of these materials to have outstanding properties in that the separation of charges is long lived and the inorganic layer that resulted, initially thought to be inert, was semiconducting thus driving the separated electrons and/or holes along the material.[85] Additional results showed that these powders can be arranged as solar cells with promising efficiency.[86]

9.4.6 Gas Storage

Hydrogen is a very appealing energy vector – the release of its energy does not involve the noxious carbon dioxide. Yet, environmentally friendly production and safe storage/transportation methods must be devised if hydrogen is to be efficiently used as the clean energy carrier of the future. Although there are already many reasonably useful technical approaches, neither of the two problems is nowadays at a level of resolution which would make the use of hydrogen routinely possible. Concerning hydrogen storage, it may be attained among other procedures by *physisorption* in porous matrices.[87] We conjectured

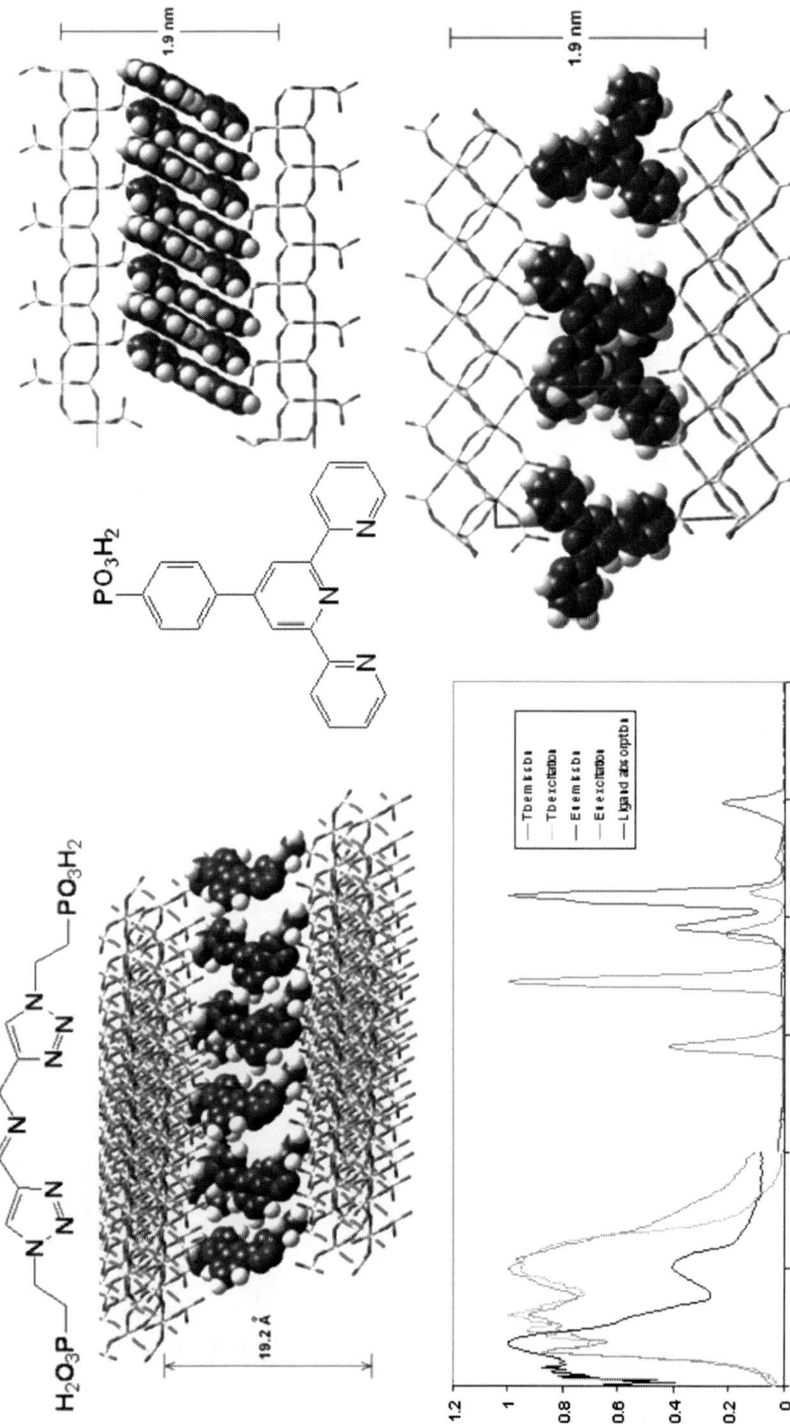

Figure 9.30 Exchanged γ-ZrP with chromophores derived from bis-triazolylpyridine and phenylterpyridine, showing the characteristic emission of Eu^{3+} and Tb^{3+}.

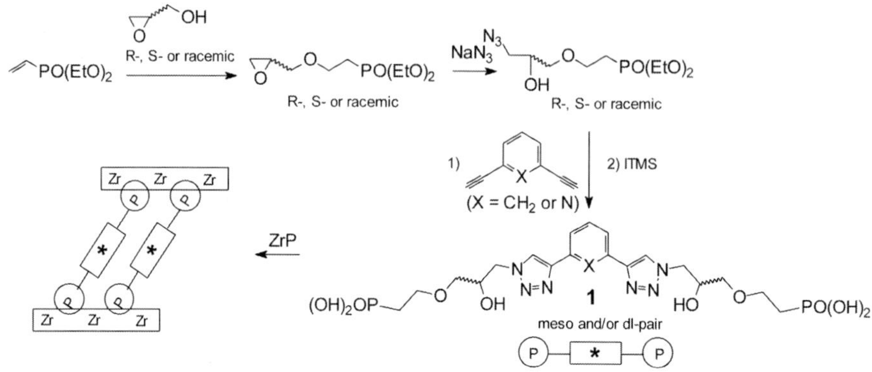

Figure 9.31 Pillaring of γ-ZrP with chiral chromophores derived from bis-triazolylpyridine to achieve circularly polarized luminescence (CPL) of Eu^{3+} and Tb^{3+}.

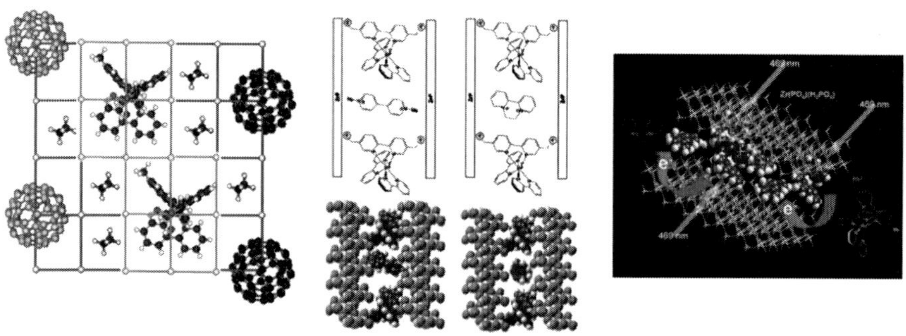

Figure 9.32 Idealized arrangement of electron donor $Ru^{II}(bpy)_3$ and electron acceptors (fullerene derivatives or viologens) on the simplified surface of γ-ZrP.

whether appropriate matrices based on ZrP could serve the latter purpose. A thorough study of the literature allows one to find numerous organic, inorganic, and hybrid systems in which the storage of hydrogen has been tested, from the delusive carbon nanotubes to the very cleverly designed systems with almost unbelievable specific surface areas close to 5000 m^2 g^{-1}.[88] The analysis of the different available structures points to the conclusion that a large accessible volume is desirable, but by means of micropores or ultramicropores, and therefore presenting a large internal contact surface to the elusive hydrogen molecules. Some interesting theoretical studies and experimental ones on zeolites also point to the benefit of having polarizing centers, namely small cations as for instance Li^+. Also the presence of transition metals with open coordination sites appears to be important. We believe that our set of tools (Figure 9.16) might allow us to design appropriate materials for hydrogen storage. Figure 9.33 summarizes our rationale and the best achieved results.[89]

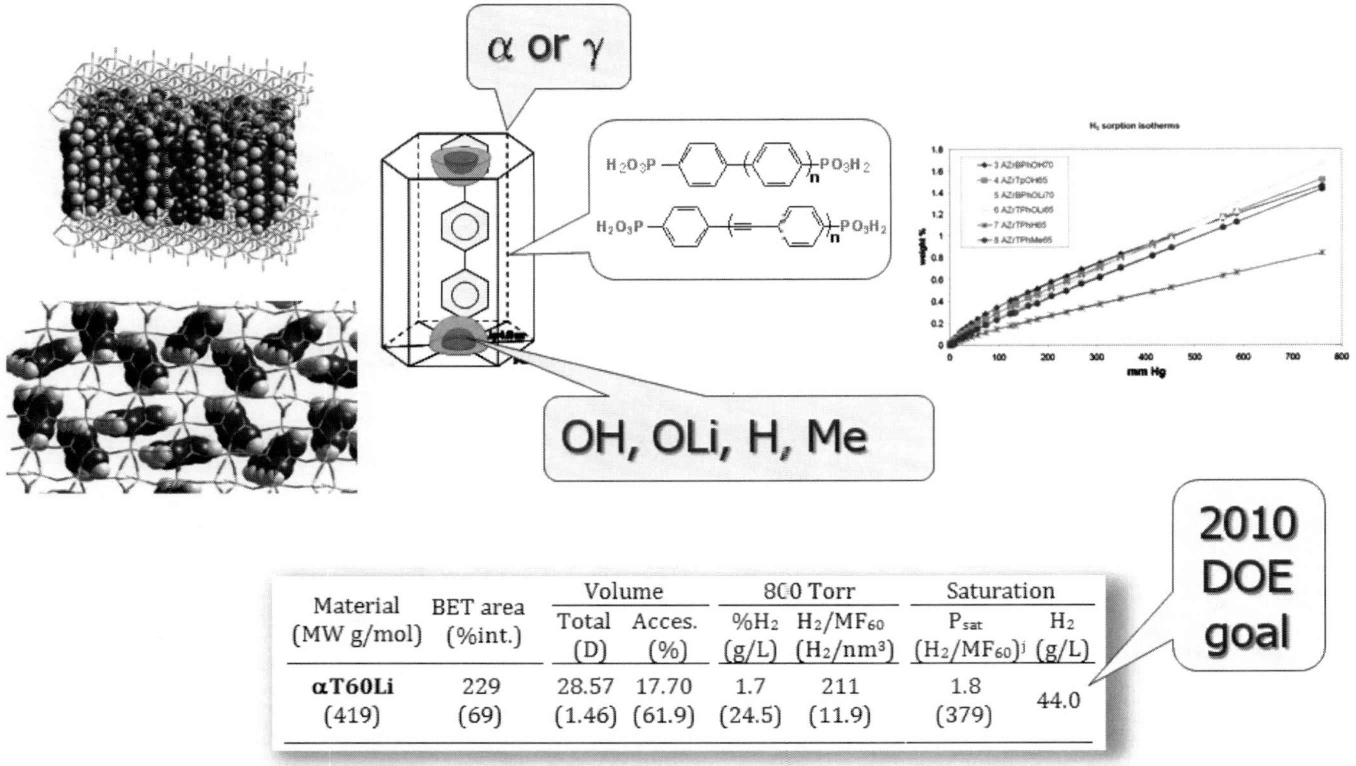

Material (MW g/mol)	BET area (%int.)	Volume		800 Torr		Saturation	
		Total (D)	Acces. (%)	$\%H_2$ (g/L)	H_2/MF_{60} (H_2/nm^3)	P_{sat} (H_2/MF_{60})[j]	H_2 (g/L)
αT60Li (419)	229 (69)	28.57 (1.46)	17.70 (61.9)	1.7 (24.5)	211 (11.9)	1.8 (379)	44.0

Figure 9.33 Phosphonate-exchanged ZrP materials for hydrogen storage.

The attachment of polyphenyl or polyphenylethynyl diphosphonates to either α- or γ-ZrP led to materials with slit-like ultramicropores of different length. Further exchange reactions led to polar or non-polar groups at both ends of the pores. This set of materials allowed us to check for a large number of different arrangements. The best results were obtained with the material named αT60Li, α-ZrP with 60% of terphenyldiphosphonates,[90] the remainder of the phosphates bearing Li^+ as counter-ions. At 800 Torr and 77 K, 1.7% w/w of hydrogen could be stored and the Department of Energy goal for 2010 (45 g of hydrogen per L of material) can be attained at 1.8 atm at 77 K (Figure 9.33).

Another important aspect of gas or vapor storage and/or detection is the finding of a "chemical nose" [91] for the *in situ* sampling of certain volatile components of, for instance, explosives. The design of simple, non-intrusive devices to detect dangerous substances within different objects like air cargo, luggage, *etc* is an important activity of many security companies. Such a device could be based in the piezo-resistive detector indicated in Figure 9.34, where the micro cantilever resonates at a frequency that depends on its mass.

Should a porous material be deposited on it, its total mass would change when a given vapor is absorbed in the pores. The extent of the resonance frequency thus changes and, together with a simple calibration protocol, it enables the quantification of the absorbed component. To this effect, many porous matrices have been used, for example porous and mesoporous silica, zeolites,

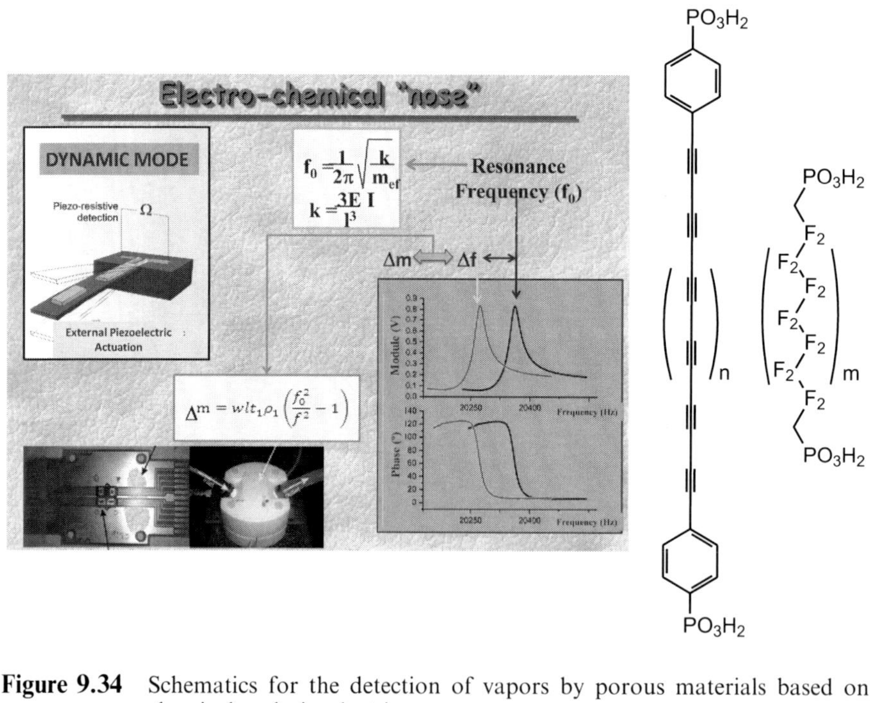

Figure 9.34 Schematics for the detection of vapors by porous materials based on chemical and piezolectric means.

and ordered mesoporous materials. Modified γ-ZrP could be considered within the latter, yet it has never been tested for this purpose. Our material, named αT60, *i.e.* α-ZrP pillared with 60% of terphenyldiphosphonates, had the ability of reversibly detecting *ca.* 400 ppm of toluene vapor (used as a model for the detection of the aromatic compounds contained in certain explosives) in N_2 with a response time lower of 30 s. This is a very promising result upon which we are devoting further research in order to improve sensitivity and selectivity. What would the capabilities of ZrP phases pillared with the molecules of Figure 9.34, or with any other molecules?

9.4.7 Drug Confinement

Confinement of drugs into molecular matrices might be a good solution to their slow release at particular locations on the body.[92] We have preliminary studies where it can be shown that the erythromycin family antibiotics can intercalate into γ-ZrP, the driving force being the acid–base interaction between the amine group(s) of the organic compound and the acidic surface phosphates. The measurement of the interlayer distance versus the stoichiometry of the reactants (Figure 9.35) shows that mono- and bilayers of

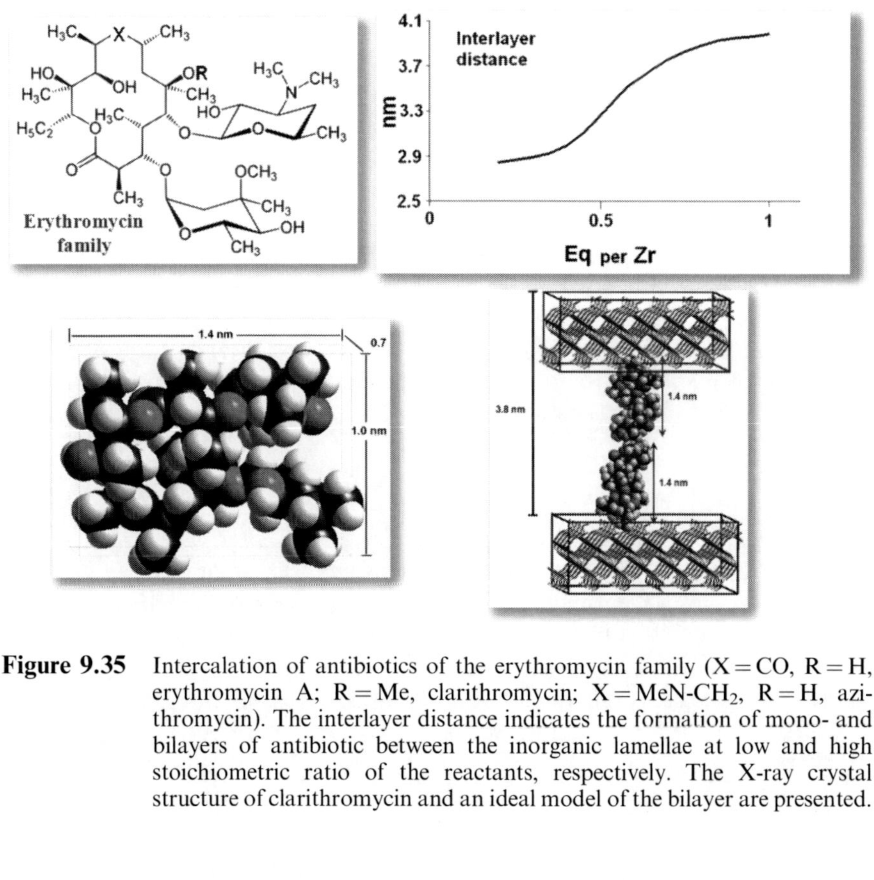

Figure 9.35 Intercalation of antibiotics of the erythromycin family (X = CO, R = H, erythromycin A; R = Me, clarithromycin; X = MeN-CH_2, R = H, azithromycin). The interlayer distance indicates the formation of mono- and bilayers of antibiotic between the inorganic lamellae at low and high stoichiometric ratio of the reactants, respectively. The X-ray crystal structure of clarithromycin and an ideal model of the bilayer are presented.

erythomycin derivatives may be enclosed between the inorganic lamellae. More detailed studies will follow.

9.5 Conclusion

It has been our aim to show, within the limited length of this chapter, the enormous possibilities that the building of organic–inorganic 3D structures based on ZrP does have in numerous fields of either basic or applied research. It allows the construction of rather complex composites where the confinement of the organic molecules confers them new and unexpected properties at the supramolecular level.

Acknowledgements

This work would not have been possible during the last 15 years without the smart contribution at different levels of many people to whom I am greatly indebted. Among all of them, my old friend Prof. Juan Carlos Rodríguez-Ubis does stand out.

Throughout these years financial support has relatively been difficult to obtain from governmental sources as the Ministries of Science and Education and Foreign Affairs, the Autonomous Community of Madrid, and the European Union. As a matter of fact, it unfortunately ceased in 2010 for reasons difficult to understand save for the impact of the global economic crisis. On the contrary, private indirect funding from ERCROS Farmacia S.A. (Aranjuez, Spain) has been and is being generous. I wish to expressly thank Dr. Carmen Cruzado, its R&D manager, for her generosity in letting us develop this chemistry as an important bonus to our consulting work with the company.

Another important source of scientific income has been related to our activity in the preparation of chemical standards for the diagnosis of inborn errors of metabolism. We are highly indebted to many laboratories around the world for their confidence in our synthetic work.

Last, but not least, I do wish to thank my good friend Prof. Abraham Clearfield for giving me the opportunity to humbly contribute to this book.

References

1. R. Feynman, *Eng. Sci.*, 1960, 22.
2. A. Walcarius and M. M. Collinson, *Annu. Rev. Anal. Chem.*, 2009, **2**, 121.
3. (a) C. A. Hunter and H. L. Anderson, Angew. Chem., *Int. Ed. Engl.*, 2009, **48**, 7488; (b) E. Kharlampieva, V. Kozlovskaya and S. A. Sukhishvili, *Adv. Mater.*, 2009, **21**, 3053; (c) Y. Liu and H. Yan, *Science*, 2009, **325**, 685.
4. J. Yang and D. Yan, *Chem. Soc. Rev.*, 2009, **38**, 2634.
5. D. Farrusseng, S. Aguado and C. Pinel, *Angew. Chem., Int. Ed. Engl.*, 2009, **48**, 7502.
6. N. Chen, M. N. Chaur, C. Moore, J. R. Pinzon, R. Valencia, A. Rodriguez-Fortea, J. M. Poblet and L. Echegoyen, *Chem. Commun.*, 2010, **46**, 4818.

7. S. W. Bailey, *Am. Mineral.*, 1980, **65**, 1.
8. P. Nalawade, B. Aware, V. J. Kadam and R. S. Hirlekar, *J. Sci. Ind. Res.*, 2009, **68**, 267.
9. M. A. Bizeto, A. L. Shiguihara and V. R. L. Constantino, *J. Mater. Chem.*, 2009, **19**, 2512.
10. (a) G. Alberti, S. Murcia-Mascaros and R. Vivani, *Mater. Sci. Forum*, 1994, 152; (b) R. Vivani, G. Alberti, F. Costantino and M. Nocchetti, *Microp. Mesop. Mater.*, 2007, **107**, 58.
11. A. Clearfield and J. A. Stynes, *J. Inorg. Nucl. Chem.*, 1964, **26**, 117.
12. J. M. Troup and A. Clearfield, *Inorg. Chem.*, 1977, **16**, 3311.
13. Hyperchem program, http://www.hyper.com.
14. L. Sun, J. Y. O'Reilly, D. Konga, J. Y. Sub, W. J. Boob, H.-J. Sueb and A. Clearfield, *Coll. Interf. Sci.*, 2009, **333**, 503.
15. A. Clearfield, *Prog. Inorg. Chem.*, 1998, **47**, 371.
16. There is a number of recent review papers concerning the use of phosphonates in the formation of open frameworks (MOFs) which are out of the scope of this chapter. See for example G. K. H. Shimizu, R. Vaidhyanathan, J. M. Taylor, *Chem. Soc. Rev.*, 2009, 38, 1430; T. Devic, in *Ordered Porous Solids*, ed. C. Serre, V. Valtchev, S. Mintova, M. Tsapatsis, 2009, p. 77; A. Clearfield, *Dalton Trans.*, 2008, **44**, 6089.
17. A. Clearfield, R. H. Blessing and J. A. Sytnes, *J. Inorg. Nucl. Chem.*, 1968, **30**, 2249.
18. S. Yamanaka and M. Tanaka, *J. Inorg. Nucl. Chem.*, 1979, **41**, 45.
19. N. Clayden, *J. Chem. Soc., Dalton Trans.*, 1987, 1877.
20. It is rumored that the key sample of γ-ZrP analyzed by N. Clayden was provided to him by A. Clearfield, but it seems that the latter was never credited with this.
21. D. M. Poojary, B. Shpeizer and A. Clearfield, *J. Chem. Soc., Dalton Trans. Inorg. Chem.*, 1995, **1**, 111.
22. M. P. Poojary, B. Zhang and A. Clearfield, *J. Chem. Soc., Dalton Trans.*, 1994, 2453.
23. G. Alberti, M. Bartocci, M. Santarelli and R. Vivani, *Inorg. Chem.*, 1997, **36**, 3574.
24. R. Vivani, G. Alberti, F. Costantino and M. Nocchetti, *Microp. Mesop. Mater.*, 2007, **107**, 58.
25. (a) L. Sun, W. J. Boo, R. L. Browning, H.-J. Sue and A. Clearfield, *Chem. Mater.*, 2005, **17**, 5606; (b) L. Sun, J. Y. O'Reilly, D. Kong, J. Y. Su, W. J. Boo, H.-J. Sue and A. Clearfield, *J. Coll. Interf. Sci.*, 2009, **333**, 503.
26. C. V. Kumar, A. Bhambhani, N. Hnatiuk, in *Handbook of Layered Materials*, ed. S. M. Auerbach, K. A. Carrado, P. K. Dutta, 2004, p. 313.
27. (a) G. Alberti, U. Costantino, C. Dionigi, S. Murcia-Mascaros and R. Vivani, *Supramol. Chem.*, 1995, **6**, 29; (b) A. Clearfield, U. Costantino, *Comprehensive Supramolecular Chemistry*, 1996, **7**, p. 107; (c) P. Olivera-Pastor, P. Maireles-Torres, E. Rodriguez-Castellon, A. Jimenez-Lopez, T. Cassagneau, D. J. Jones and J. Roziere, *J. Chem. Mater.*, 1996, **8**, 1758.

28. M. B. Santiago, G. A. Daniel, A. David, B. Casanas, G. Hernandez, A. R. Guadalupe and J. L. Colon, *Electroanal.*, 2010, **22**, 1097.
29. M. B. Santiago, M. M. Velez, S. Borrero, A. Diaz, C. A. Casillas, C. Hofmann, A. R. Guadalupe, J. L. Colon and L. Jorge, *Electroanal.*, 2006, **18**, 559.
30. S. Vecchio, R. Rocco and C. Ferragina, *J. Thermal Anal. Calorim.*, 2009, **97**, 805.
31. Y. Du, Q. Pan, J. Li, J. Yu and R. Xu, *Inorg. Chem.*, 2007, **46**, 5847.
32. H. Nakayama, *Phosph. Res. Bull.*, 2009, **23**, 1.
33. A. Hayashi, H. Nakayama and M. Tsuhako, *Solid State Sci.*, 2009, **11**, 1007.
34. S. Shi, Y. Peng and J. Zhou, *J. Nanosci. Nanotech.*, 2009, **9**, 2746.
35. C. Ferragina, R. Di Rocco, P. Giannoccaro, P. Patrono and L. Petrilli, Lucantonio, *J. Incl. Phenom. Macroc. Chem.*, 2009, **63**, 1.
36. S. Shi, J. Zhou, R. Zong and J. Ye, Jianping, *J. Lumin.*, 2007, **122–123**, 218.
37. M. H. Xiang, X. B. Shi, N. Li and K. A. Li, *Chinese Chem. Lett.*, 2007, **18**, 89.
38. H. Y. Wang, W. D. Ji and D. X. Han, *Chinese Chem. Lett.*, 2008, **19**, 1330.
39. B. Wozniak, W. Apostoluk and J. Wodka, *Solv. Extrac. Ion Exch.*, 2008, **26**, 699.
40. A. Hayashi, Y. Yoshikawa, N. Ryu, H. Nakayama, M. Tsuhako and T. Eguchi, *Phosph. Res. Bull.*, 2008, **22**, 48.
41. A. Bhambhani and C. V. Kumar, *Microp. Mesop. Mater.*, 2008, **110**, 517.
42. Y. Liu, C. Lu, W. Hou and J.-J. Zhu, *Anal. Biochem.*, 2008, **375**, 27.
43. A. Bhambhani and C. V. Kumar, *Microp. Mesop. Mater.*, 2008, **109**, 223.
44. Z. P. Xu, Y. Jin, J. C. Diniz da Costa and G. Q. Lu, *Solid State Ionics*, 2008, **178**, 1654.
45. E. J. Rivera, C. Figueroa, J. L. Colon, L. Grove and W. B. Connick, *Inorg. Chem.*, 2007, **46**, 8569.
46. H. Y. Wang and D. X. Han, *Chinese Chem. Lett.*, 2007, **18**, 764.
47. L. Benes, K. Melanova, J. Svoboda, V. Zima and M. Kincl, *J. Phys. Chem. Solids*, 2007, **68**, 803.
48. M. B. Santiago, C. Declet-Flores, A. Diaz, M. M. Velez, M. Z. Bosques, Y. Sanakis and J. L. Colon, *Langmuir*, 2007, **23**, 7810.
49. S. Shi, R. Zong, Y. Liu, Y. Wang and J. Zhou, *Key Eng. Mater.*, 2007, **336–338**, 2589.
50. K. Melanova, L. Benes, V. Zima, J. Svoboda, M. Trchova and J. Dybal, *J. Incl. Phenom. Macroc. Chem.*, 2007, **58**, 95.
51. A. A. Marti, N. Rivera, K. Soto, L. Maldonado and J. L. Colon, *Dalton Trans.*, 2007, **17**, 1713.
52. A. A. Marti, G. Paralitici, L. Maldonado and J. L. Colon, *Inorg. Chim. Acta*, 2007, **360**, 1535.
53. S. Vecchio, R. Di Rocco and C. Ferragina, *Therm. Acta*, 2007, **453**, 105.
54. C. Ferragina, R. Di Rocco, S. Foglia and L. Petrilli, *Coll. Surf., A: Phys. Eng. Aspects*, 2007, **293**, 114.

55. A. Hayashi, H. Katsuta, A. Tsuruta, H. Nakayama and M. Tsuhako, *Phosp. Res. Bull.*, 2004, **17**, 107.
56. A. Bhambhani and C. V. Kumar, *Adv. Mater.*, 2006, **18**, 939.
57. A. Hayashi, Y. Nakabayashi, J. Yasutomi, H. Nakayama and M. Tsuhako, *Bull. Chem. Soc. Jpn.*, 2006, **79**, 262.
58. J. Liu, W.-J. Boo, A. Clearfield and H.-J. Sue, *Mater. Manuf. Proc.*, 2006, **21**, 143.
59. N. Bestaoui, N. A. Spurr and A. Clearfield, *J. Mater. Chem.*, 2006, **16**, 759.
60. B. Ha, K. Char and H. S. Jeon, *J. Phys. Chem. B*, 2005, **109**, 24434.
61. R. A. Bermudez, R. Arce and J. L. Colon, *J. Photochem. Photobiol. A: Chem.*, 2005, **175**, 201.
62. M. Casciola, G. Alberti, A. Donnadio, M. Pica, F. Marmottini, A. Bottino and P. Piaggio, *J. Mater. Chem.*, 2005, **15**, 4262.
63. S. Vecchio, R. Di Rocco, C. Ferragina and S. Materazzi, *Therm. Acta*, 2005, **435**, 181.
64. H. Wang, D. Han, L. Na and K. Li, *J. Incl. Phen. Macroc. Chem.*, 2005, **52**, 247.
65. S. Park, T. D. Chung, S. K. Kang, R.-A. Jeong, H. Boo and H. C. Kim, *Anal. Sci.*, 2004, **20**, 1635.
66. A. Chaudhari, J. Thota and C. V. Kumar, *Microp. Mesop. Mater.*, 2004, **75**, 281.
67. C. Ferragina, R. Di Rocco, A. Fanizzi, P. Giannoccaro and L. Petrilli, *J. Therm. Anal. Calorim.*, 2004, **76**, 871.
68. J. Wang, Y. Hu, B. Li, Z. Gui and Z. Chen, *Ultrason. Sonochem.*, 2004, **11**, 301.
69. G. Alberti, E. Giontella and S. Murcia-Mascaros, *Inorg. Chem.*, 1997, **36**, 2844.
70. G. Alberti, E. Giontella, S. Murcia-Mascarós and R. Vivani, *Inorg. Chem.*, 1998, **37**, 4672.
71. S. Yamanaka, K. Yamasaka and M. Hattori, *J. Incl. Phen.*, 1984, **2**, 297.
72. E. Brunet, M. Huelva, R. Vazquez, O. Juanes and J. C. Rodriguez-Ubis, *Chem. Eur. J.*, 1996, **2**, 1578.
73. E. Brunet, *unpublished results*.
74. E. Brunet, M. Huelva and J. C. Rodriguez-Ubis, *Tetrahedron. Lett.*, 1994, **35**, 8697.
75. G. Alberti, E. Brunet, C. Dionigi, O. Juanes, M. J. Mata, J. C. Rodriguez-Ubis and R. Vivani, *Angew. Chem., Int. Ed. Engl.*, 1999, **38**, 3351.
76. (a) E. Brunet, M. J. Mata, O. Juanes and J. C. Rodriguez-Ubis, *Angew. Chem. Int. Ed. Engl.*, 2004, **43**, 619; (b) E. Brunet, M. J. Mata, H. M. H. Alhendawi, C. Cerro, M. Alonso, O. Juanes and J. C. Rodriguez-Ubis, *Chem. Mater.*, 2005, **17**, 1424.
77. E. Brunet, *Chirality*, 2002, **14**, 135.
78. E. Brunet, M. J. Mata, O. Juanes, H. M. H. Alhendawi, C. Cerro and J. C. Rodriguez-Ubis, *Tetrahedron: Asym.*, 2006, **17**, 347.
79. E. Brunet, O. Juanes and J. C. Rodriguez-Ubis, *J. Mex. Chem. Soc.*, 2009, **53**, 154.

80. E. Brunet, O. Juanes and J. C. Rodriguez-Ubis, *Curr. Chem. Biol.*, 2007, **1**, 11.
81. (a) E. Brunet, M. J. Mata, O. Juanes and J. C. Rodriguez-Ubis, *Chem. Mater.*, 2004, **16**, 1517; (b) E. Brunet, H. M. H. Alhendawi, O. Juanes, L. Jimenez and J. C. Rodriguez-Ubis, *J. Mater. Chem.*, 2009, **19**, 2494.
82. G. Muller, *Dalton Trans.*, 2009, **44**, 9692.
83. P. G. Hoertz and T. E. Mallouk, *Inorg. Chem.*, 2005, **44**, 6828.
84. (a) E. Brunet, M. Alonso, M. J. Mata, S. Fernandez, O. Juanes, O. Chavanes and J. C. Rodriguez-Ubis, *Chem. Mater.*, 2003, **15**, 1232; (b) E. Brunet, M. Alonso, C. Cerro, O. Juanes, J. C. Rodriguez-Ubis and A. E. Kaifer, *Adv. Funct. Mater.*, 2007, **17**, 1603.
85. E. Brunet, M. Alonso, M. C. Quintana, P. Atienzar, O. Juanes, J. C. Rodriguez-Ubis and H. Garcia, *J. Phys. Chem. C*, 2008, **112**, 4029.
86. L. Teruel, M. Alonso, M. C. Quintana, A. Salvador, O. Juanes, J. C. Rodriguez-Ubis, E. Brunet and H. Garcia, *Phys. Chem. Chem. Phys.*, 2009, **11**, 2922.
87. E. Brunet, C. Cerro, O. Juanes, J. C. Rodriguez-Ubis and A. Clearfield, *J. Mater. Sci.*, 2008, **43**, 1155.
88. See for example J. G. Vitillo, L. Regli, S. Chavan, G. Ricchiardi, G. Spoto, P. D. C. Dietzel, S. Bordiga, A. Zecchina, *J. Am. Chem. Soc.*, 2008, **130**, 8386 and references cited therein.
89. E. Brunet, H. M. H. Alhendawi, C. Cerro, M. J. Mata, O. Juanes and J. C. Rodriguez-Ubis, *Angew. Chem., Int. Ed. Engl.*, 2006, **45**, 6918.
90. This material was produced by an unexpected isomerization of γ-ZrP exchanged with polyphenylphosphonates; see E. Brunet, H. M. H. Alhendawi, C. Cerro, M. J. Mata, O. Juanes, J. C. Rodriguez-Ubis, *Chem. Eng. J.*, 2010, **158**, 333.
91. M. Urbiztondo, P. Pina, J. Santamaria, in *Ordered Porous Solids* ed. V. Valtchev, S. Mintova, M. Tsapatsis, 2009, p. 387.
92. J. Salonen, A. M. Kaukonen, J. Hirvonen and V.-P. Lehto, *J. Pharm. Sci.*, 2008, **97**, 632.

CHAPTER 10

Open Framework and Microporous Metal Phosphonate MOFs with Piperazine-based Bisphosphonate Linkers

MICHAEL T. WHARMBY AND PAUL A. WRIGHT

School of Chemistry, University of St Andrews, Purdie Building, North Haugh, St Andrews, Fife KY16 9ST, UK

10.1 Introduction

The discovery by Maeda *et al.* of the aluminium methylphosphonate polymorphs AlMePO-α and -β (composition $Al_2(CH_3PO_3)_3$)[1,2] opened up the field of crystalline, three dimensionally connected microporous metal phosphonates, adding to those known porous materials derived from the modification of layered solids by substitution with bisphosphonate ligands, such as zirconium phosphites.[3,4] The frameworks of these AlMePO polymorphs contain Al^{3+} cations in both tetrahedral and octahedral geometry, with bridging phosphonate groups fully coordinated through all three O atoms. This leaves hexagonal arrays of channels lined with methyl groups, with free diameters of ca. 5.5 Å. Solid state ^2H NMR indicates that these structural methyl groups rotate rapidly, even at low temperatures, clicking between low energy positions.[5] This results in unique adsorption properties for both small gas molecules and for hydrocarbons.[6] As well as being hydrophobic (and showing Type III isotherms for H_2O adsorption), the AlMePOs readily adsorb monoaromatic hydrocarbon molecules, for example, which remain mobile and able to re-orient at much lower temperatures than in inorganic zeolites or aluminophosphates, so that their entropy loss upon adsorption is much lower.[7] Furthermore, the AlMePO-α

Metal Phosphonate Chemistry: From Synthesis to Applications
Edited by Abraham Clearfield and Konstantinos Demadis

Published by the Royal Society of Chemistry, www.rsc.org

polymorph has been shown to be a highly selective adsorbent of ethyl chloride in the presence of vinyl chloride, which is of interest in the purification of streams of vinyl chloride, the monomer of the commercially important polyvinyl chloride.[8] However, subsequent synthetic studies have not yielded any further porous AlMePO polymorphs, and indeed only a few other porous metal methylphosphonates have been discovered. The sodium scandium methylphosphonate, $NaSc(MePO_3)_2 \cdot 0.5H_2O$ is one example,[9] and possesses a scandium methylphosphonate framework containing charge-balancing sodium cations, from which 'zeolitic' water molecules can be reversibly removed, but its pores are too small to admit larger molecules.

Meanwhile, the discovery of permanently porous metal carboxylate metal–organic frameworks (MOFs),[10] such as the zinc terephthalate MOF-5,[11] the copper trimesate HKUST-1[12] and the trivalent and tetravalent metal terephthalates MIL-47,[13] MIL-53,[14] MIL-88[15] and MIL-101[16] revolutionized the field of permanently porous organic–inorganic hybrid solids, and there has been an exponential increase in the number of known porous carboxylates (and carboxylate amines[17]). Attention has extended to tricarboxylate and tetracarboxylate linkers, the use of functionalized linkers and reactions with larger linkers with similar geometry and the same functional groups. This latter approach has in some cases demonstrated examples of *isoreticular synthesis*[18] where frameworks with the same topology but with different dimensions can be prepared. The synthesis of this ever expanding family of porous solids, with its diverse structural chemistry and properties, is currently being attended by the search for applications in gas storage, separation, drug delivery catalysis and materials properties such as magnetism, ferroelectricity or multiferroic behaviour (for examples, see *Chemical Society Reviews* special issue on Metal Organic Frameworks[19]). Over the same period, progress in porous phosphonates has been slower. The use of bis-, tris- and tetraphosphonic acids to prepare metal phosphonate MOFs is one area that has received attention, and has recently yielded solids porous to larger molecules up to 2 nm in dimensions. Members of this family also demonstrate open metal sites, flexibility, structural 'breathing', selective cation exchange and luminescence.

Many bis-, tris- and tetraphosphonic acids have been synthesized and used to prepare phosphonate MOFs. Maeda[20] and Shimizu *et al.*[21] have reviewed some of the structures prepared with them. Here we discuss the our work and the work of others on metal bisphosphonates formed with the ligand *N,N′*-piperazinebis(methylenephosphonic acid) (H_4**L**), which includes the synthesis of the first large pore phosphonate MOFs to display full crystalline order.[22] In some ways the H_4**L** linker occupies a similar place in metal phosphonate chemistry to that taken by the terephthalic acid linker in metal carboxylates, but it also exhibits important differences, including the ability to bind metals via one, two or three phosphonate O atoms, with or without coordination to the piperazinyl N atom. It is also able to exist in different configurations, with methylenephosphonate groups either equatorial or axial to the ring, which itself can take either chair or boat configurations. Coupled with variation of geometry about different metal cations, this yields solids of considerable structural variety, many of which display a motif of parallel inorganic 'metal phosphate' chains linked by the organic rings. Here we discuss the structures observed to crystallize with this ligand using a variety of different cations, and the physical properties of these materials. Further, we consider how modified ligands affect the structure types that crystallize, and discuss general structural principles for this type of phosphonate MOF.

10.2 Synthesis of *N,N′*-piperazinebis(methylenephosphonic acid) and Related Bisphosphonic Acids

There are many relatively facile routes to the synthesis of polyfunctional phosphonic acids for the preparation of open framework materials, the majority relying on the Arbuzov reaction or the Mannich reaction to form phosphorus–carbon bonds. As these have been reviewed more fully elsewhere in this book, this section will focus on the synthesis of ligands based on piperazine and its derivatives using the Mannich reaction.

The synthesis of α-aminomethylphosphonic acids was first reported by Moedritzer and Irani, via a modified Mannich reaction.[23] The synthetic approach is readily extended to the synthesis of a bis(α-aminomethylphosphonic acid) by using a diamine, such as piperazine, as the substrate. The modified Mannich reaction proceeds by the formation of a stabilized imine intermediate followed by attack by phosphorous acid (Scheme 10.1). Products are then recrystallized from acidic reaction solutions, using ethanol/water solvent mixtures to induce crystallization. Both hydrochloric acid and

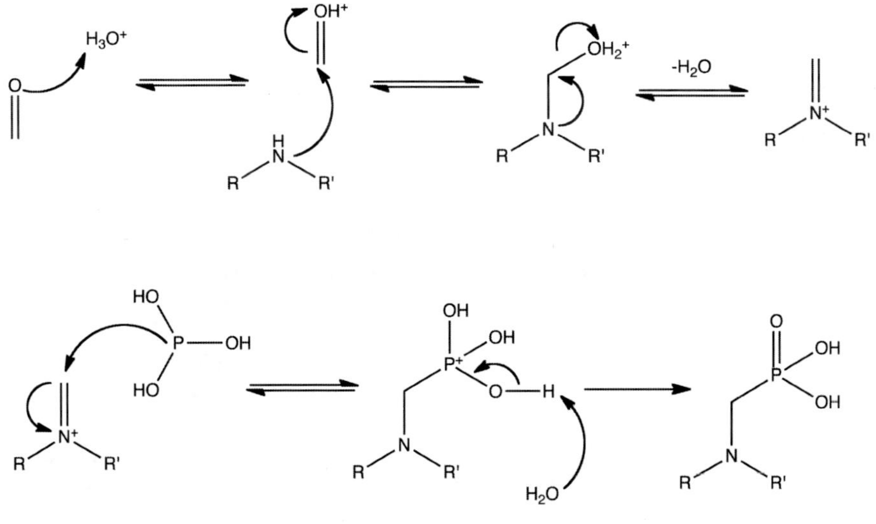

Scheme 10.1

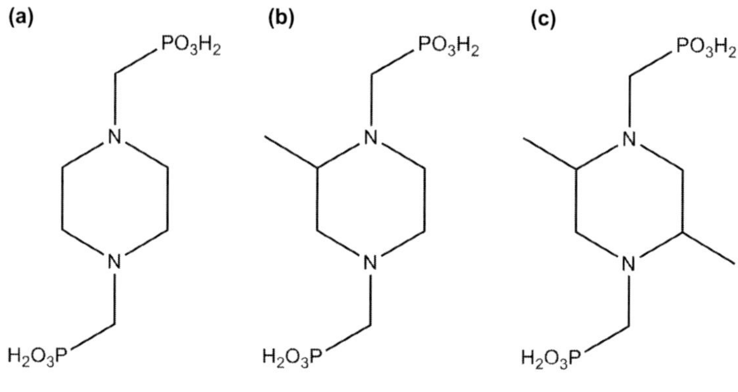

Scheme 10.2

hydrobromic acid have been found to catalyze this reaction, though a small amount of acid residue is incorporated into the final product. As Cl^- ions have been observed to have templating effects in the synthesis of framework materials with piperazine-based linkers,[24] it can be preferable to use hydrobromic acid as the catalyst.

The crystal structure of the piperazine-derived *N*,*N*′-piperazinebis (methylenephosphonic acid), H_4**L**, was reported by Choi *et al.*,[25] prepared by the method of Markhaeva *et al.*[26] In the crystal structure of the free acid form of H_4**L**, the ligand adopts a chair conformation with both methylenephosphonic acid groups equatorial. This places the relatively bulky methylenephosphonic acid groups away from other substituents on the ring, reducing steric interactions to give a lower energy conformation than with the groups axial. Subsequent work by Mowat *et al.*[27] details the synthesis of two derivatives of this ligand, *N*,*N*′-2-methylpiperazinebis(methylenephosphonic acid) (H_4**L**′) and *N*,*N*′-2,5-dimethylpiperazinebis(methylenephosphonic acid) (H_4**L**″), with structures confirmed for each by ^{1}H, ^{13}C and ^{31}P solution-state NMR spectroscopy. All three ligands have been used in the synthesis of open framework materials (Scheme 10.2).

10.3 Early Examples of Phosphonate MOFs with *N*,*N*′-piperazinebis(methylenephosphonic acid)

The first phosphonate MOFs prepared with ligand H_4**L** were $VO(H_2O)(LH_2)$[28] (**1**) and $M(LH_2)\cdot xH_2O$ (M = Mn, Co),[29] (**2**). In the vanadyl phosphonate structure, there are layers of corner-sharing VO_6 octahedra and PO_3C tetrahedra, which are linked by protonated piperazinyl units in the third dimension through covalent bonds (Figure 10.1). Two of the three phosphonate O atoms of each group coordinate to the metal cations, leaving the third free and able to engage in H-bonding to a protonated ring N atom, which is found to be a common feature of many structures described in this chapter. In structure **2**,

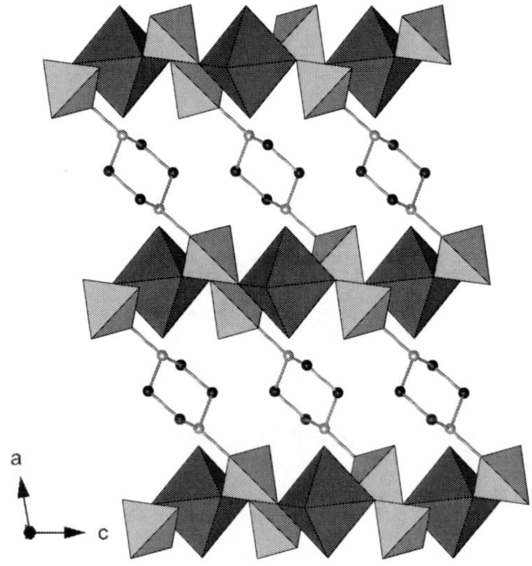

Figure 10.1 View of the pillared layer structure, **1**, $VO(H_2O)L$ (**L** = *N*,*N*′-piperazinebis-(methylenephosphonate) parallel to the inorganic 'vanadyl phosphate' sheets, showing the connection via the piperazine rings (VO_6 octahedra in dark grey, PO_3C tetrahedra in light grey, C atoms in black, N in grey, H atoms omitted).

inorganic chains containing tetrahedrally coordinated metal cations and bridging phosphonate groups are linked by the ligand in chair configuration to give a channel-based structure: unbound O atoms from P=O groups again form H-bonds with protonated piperazinyl N atoms (Figure 10.2). The narrow channels permit reversible uptake of water molecules, but the dehydrated form shows no permanent porosity for larger molecules. We have since prepared the Fe^{2+} and Zn^{2+} analogues of this material.[22,30]

10.4 Phosphonate MOFs with Divalent Metals and *N*,*N*′-piperazinebis(methylenephosphonic acid)

During exploratory syntheses using divalent metal cations and H_4**L**, raised synthesis pH was found to prepare Zn_2**L** (**3**) in which the tetrahedral Zn^{2+} cation is coordinated by three O atoms and one ring N atom.[30] The undulating inorganic layer structure contains alternating ZnO_3N and PO_3C groups within rings defined by 8 and 4 tetrahedra. These layers are pillared by the bisphosphonate linkers, which coordinate via all their phosphonate O atoms and piperazinyl N atoms. Contemporary with this work, Wang *et al.* reported another pillared layered structure, $Cu_2(H_2O)_2$**L** (**4**) where the inorganic sheets are made up of alternating CuO_4N and PO_3C polyhedra, within six-membered rings, and the ligand O and N atoms are fully bound to metal cations.[31] Neither structure is porous.

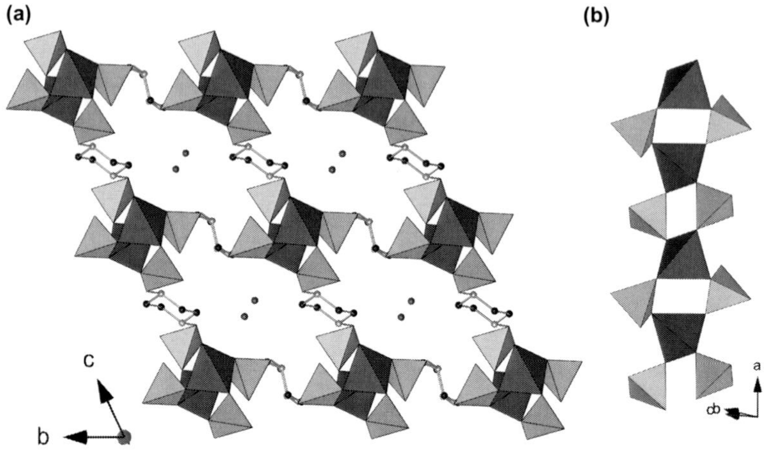

Figure 10.2 (a) The bisphosphonate framework $M(\mathbf{L}H_2)\cdot H_2O$ (M = Mn, Fe, Co, Zn), **2**, viewed down the channel axis, showing the position of extra framework water molecule O atoms. (b) The inorganic zinc phosphonate chain that runs parallel to the channel axis in **2**. (ZnO_4 tetrahedra in dark grey, water molecule O atoms as grey spheres, representation otherwise as Figure 10.1.)

At similar elevated synthesis pH values and using Mn^{2+}, Fe^{2+} and Co^{2+} it is possible to prepare the novel solid STA-12 (**5**, $M_2(H_2O)_2\mathbf{L}\cdot xH_2O$), in which the metal cations exhibit octahedral, rather than tetrahedral geometry (Figure 10.3).[22,32] The same structure is prepared with Ni^{2+} over a wider synthesis pH range since Ni^{2+} preferentially adopts octahedral rather than tetrahedral coordination, so that competitive crystallization of a Ni^{2+} analogue of either structure **2** or **3** is not possible. The structure of as-prepared STA-12(Ni) was solved by powder X-ray diffraction (PXRD), combined with molecular modelling, in a rhombohedral space group (Figure 10.3). This approach is often necessary since metal phosphonates commonly form as microcrystalline powders that are not amenable to laboratory single crystal diffraction. The framework is built from helical chains of edge-sharing MO_5N octahedra which are linked to three others via the piperazinyl groups of the ligand. This gives a hexagonal 'honeycomb' arrangement of large channels. All phosphonate groups are identical, with two of three O atoms coordinating the metal and a P=O group projecting into the pore. The N atom from the ligand also coordinates to the metal, the octahedral coordination of which is completed by a water molecule, which engages in H-bonding to the pendant P=O group. In the fully hydrated solid at low temperature these surface groups are H-bonded to an array of water molecules within the cages that bears similarity to the structure of ice.[32]

Upon complete dehydration, the structure distorts from rhombohedral to triclinic and a remarkable structural transformation occurs, as determined by structure solution from high resolution synchrotron PXRD (Figure 10.4).[32]

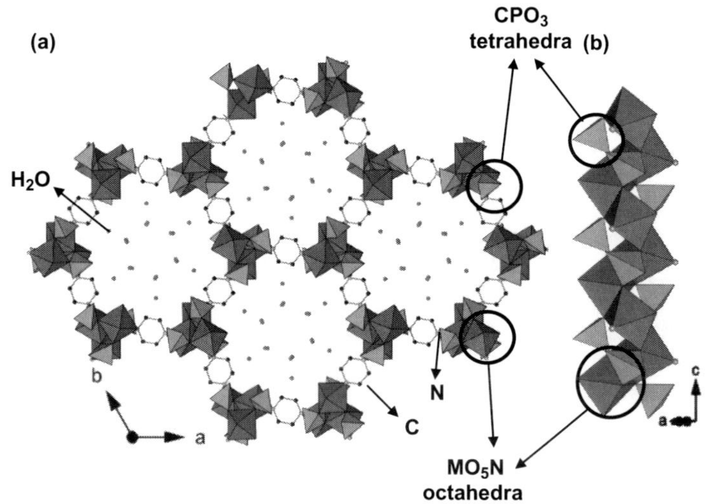

Figure 10.3 (a) The fully hydrated large pore metal bisphosphonate STA-12, $M_2(H_2O)_2\mathbf{L}\cdot xH_2O$ (M = Mn, Fe, Ni, Co), viewed down its channels. The helical chains of edge-sharing MO_5N octahedra (b), linked by phosphonate tetrahedra, run parallel to the channels.

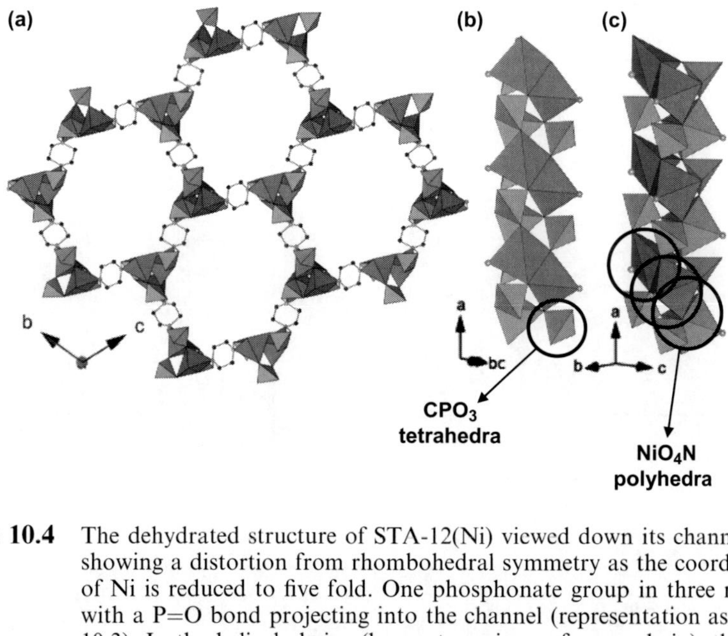

Figure 10.4 The dehydrated structure of STA-12(Ni) viewed down its channel axis, showing a distortion from rhombohedral symmetry as the coordination of Ni is reduced to five fold. One phosphonate group in three remains with a P=O bond projecting into the channel (representation as Figure 10.3). In the helical chains (b, c – two views of same chain), the three crystallographically distinct NiO_4N polyhedra are indicated by arrows.

After initial loss of the physisorbed water, the coordinated water is lost from the metal cations and there is sufficient flexibility in the phosphonate ligand for rearrangement. In the dehydrated structure, all the Ni^{2+} centres exhibit 5-fold coordination and two out of the three phosphonate groups become fully coordinated to Ni^{2+}, leaving the third phosphonate with a P=O group projecting into the channel. The channel also distorts to an elliptical cross-section with a free diameter of around 0.8 nm. The details of the transformation have been followed by TGA, XRD and IR, which show that STA-12(Ni) responds in a rapid and reversible way to loss or re-adsorption of water at the metal site. Furthermore, the large pore structure that results presents several different sites to adsorbed molecules, including Lewis acidic coordinatively unsaturated sites (CUS), of widespread interest in studies of carboxylate MOFs, and also unprecedented P=O adsorption sites. These are confirmed by adsorption, calorimetry and IR studies of probe molecules such as CO and CD_3CN adsorbed on STA-12(Ni) (Figure 10.5).[32] Adsorbed CO molecules show a peak at 2170 cm^{-1} (Figure 10.5b), shifted $+30$ cm^{-1} from physisorbed CO, which is indicative of adsorption on a Lewis acid site, whereas a second peak at 2133 cm^{-1} is thought to arise from adsorption of CO on pendant P=O sites. Dehydrated STA-12 also shows a ten-fold higher uptake of CO_2 ($\Delta H = 30$–35 kJ mol^{-1}) over methane (14–15 kJ mol^{-1}) at 1 bar and 300 K, which is thought to result partly from the interaction of CO_2 with the open metal sites.[32]

The STA-12 structure crystallizes with a range of divalent metals capable of exhibiting octahedral coordination. The dehydration behaviour, and in

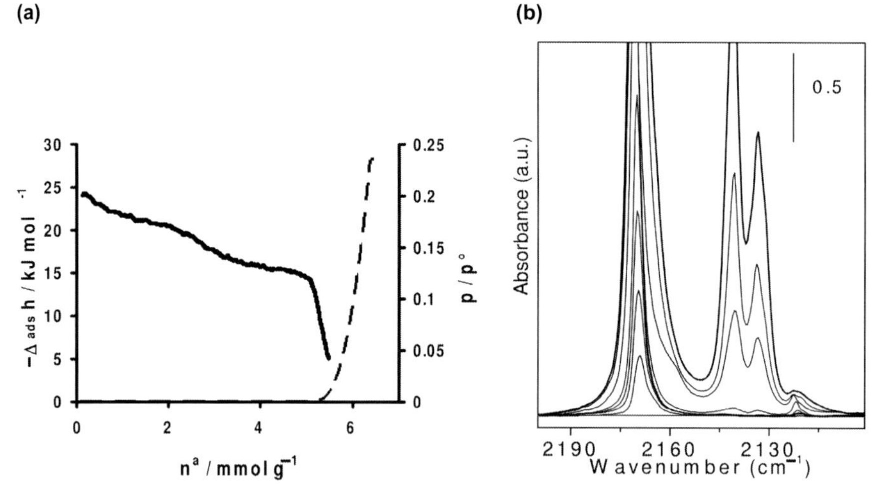

Figure 10.5 (a) Calorimetric data for the chemisorption of CO on dehydrated STA-12(Ni) shows strong interaction for ca. 5 mmol g^{-1}, corresponding to bonding at the open metal sites. Low temperature IR spectroscopy of adsorbed CO indicates (b) chemisorption is dominated by the Lewis acidic Ni^{2+} interaction, seen as the band at 2170 cm^{-1}. (Reproduced with permission from ref. 32.)

particular the porosity of the dehydrated forms, varies strongly between analogues with different metal cations: whereas STA-12(Co) shows a maximum pore volume for N_2 of 0.10 $cm^3\ g^{-1}$, STA-12(Mn) adsorbs almost no N_2, despite remaining fully crystalline.[33] This type of variability is also observed between different cationic forms of other MOFs – variation in the behaviour of the different cationic forms of CPO-27, which has a very similar helical arrangement of edge-sharing metal octahedra as its inorganic backbone, is closely analogous.[34] We speculate that surface modifications resulting in pore mouth blockage may be a possible cause.

10.5 Phosphonate MOFs with Trivalent Metals and *N*,*N*′-piperazinebis(methylenephosphonic acid)

A series of reactions with trivalent metals (Al, Sc, Fe^{III}, Ga, In, Y, rare earths La–Yb) has given a wealth of different structure types (Table 10.1).[35] Whereas the divalent metals discussed so far adopt 4-, 5- or 6-fold coordination, the trivalent metals examined exhibit 6-, 7- and 8-fold coordination, with some cations in the intermediate size range (e.g. Y – see Section 10.5.2) able to exhibit a range of these geometries, depending on the structure. In addition to the frameworks reported here, a number of other new phases, identified from their diffraction patterns, remain unsolved. The known structure types are discussed below according to cation coordination.

10.5.1 Structure Types with Metals in Octahedral Coordination

Two structure types with linker H_4L have trivalent metals in octahedral coordination, MIL-91(Al, V, Fe, In)[36,37] (**6**, $M(OH)(LH_2)\cdot xH_2O$) and $M_2(LH_2)_3\cdot xH_2O$ (**7**, M = Y, Gd).[38,39] The MIL-91 structure crystallizes for those trivalent metal cations at the lower end of the range of ionic radii of those adopting an octahedral coordination.

Table 10.1

Coordination		Al^{3+}	Fe^{3+}	Sc^{3+}	In^{3+}	Yb^{3+}	Y^{3+}	Gd^{3+}	Nd^{3+}	Ce^{3+}	La^{3+}
VI	Radius*/Å	0.54	0.65**	0.75	0.80	0.87	0.90	0.94	0.99	1.01	1.03
	H_4L	**6**	**6**		**6**	**7**	**7**	**7**			
	H_4L'	**6**		**17**	**6**	**17**	**17**	**17**			
VII	Radius/Å					0.93	0.96	1.00		1.07	1.10
	H_4L										**11, 12**
	H_4L'										**16**
VIII	Radius/Å		0.78	0.87	0.92	0.99	1.02	1.05	1.11	1.14	1.16
	H_4L						**8**		**8, 10**	**8, 9**	**8, 9**
	H_4L'								**14**	**14**	

*Reference 35
**High spin value as Fe^{3+} thought to be high spin in MIL-91(Fe)

The MIL-91 structure (Figure 10.6)[36] is based on corner-sharing 'tancoite' chains[40] of MO_6 octahedra, with protonated, corner-sharing O atoms. These each link to four other chains via two crystallographically distinct sets of ligands. Each phosphonate group binds to two metal cations via two O atoms, leaving a P=O group 'free' to H-bond to an adjacent protonated N atom of the piperazinyl ring to establish the framework. H-bonding provides an enthalpic stabilization of the framework, driving structure selection. However, above 200 °C H-bonding interactions are no longer able to stabilize the structure, leading to framework disintegration. Structure **6** is also adopted by tetravalent Ti^{4+} cations, where $[TiO]^{2+}$ units replace $[M^{III}OH]^{2+}$ units. Physisorbed water molecules within the pores are readily and reversibly removed without significant framework rearrangement. MIL-91(Al,Ti) was the first porous metal bisphosphonate MOF to be reported, with 4 Å free diameter channels and pore volumes (from N_2 adsorption measurements) of 0.195 $cm^3\ g^{-1}$ (Al) and 0.19 $cm^3\ g^{-1}$ (Ti). Iron, indium and vanadium analogues have also been prepared.[37]

Slightly larger trivalent cations such as Gd^{3+} and Y^{3+} crystallize with structure **7** (Figure 10.7).[38,39] In these materials, inorganic metal phosphonate chains contain isolated MO_6 octahedra linked to neighbouring octahedra above and below by three phosphonate units, each binding to two different metal cations in a bridging mode. The third O atom forms a pendant P=O

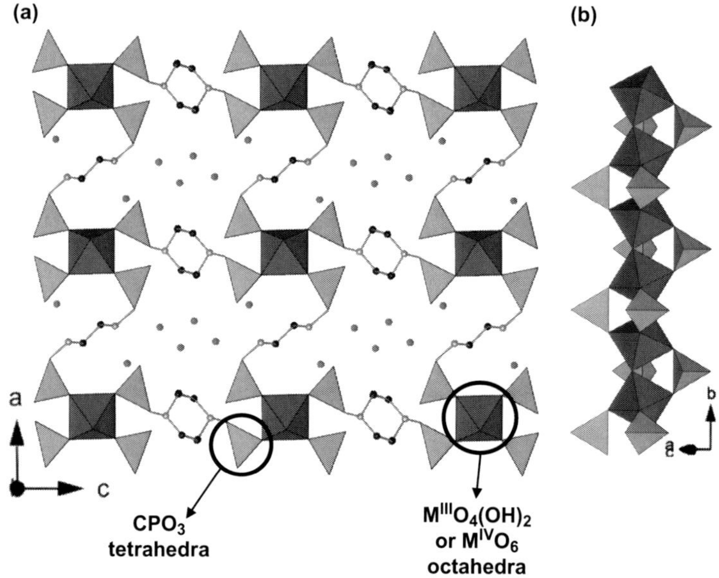

Figure 10.6 The hydrated small pore MIL-91 ($M^{III}(OH)LH_2 \cdot xH_2O$ (M^{III} = Al, Fe, In) or $M^{IV}OLH_2 \cdot xH_2O$ (M^{IV} = Ti)), **6**, viewed down its channel axis (a). The chains of corner-sharing $M^{III}O_4(OH)_2$ or $M^{IV}O_6$ octahedra, linked by bridging phosphonate tetrahedra, run along the channel axis (b). (Representation otherwise as Figure 10.3.)

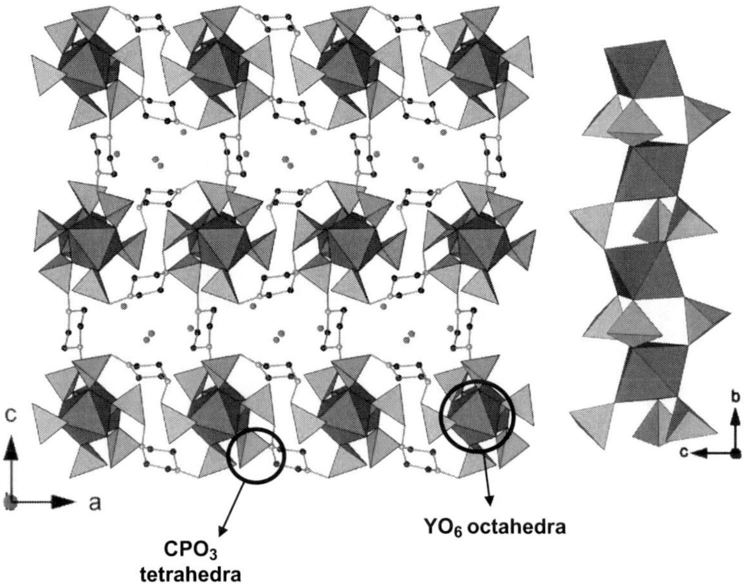

Figure 10.7 (a) The framework structure of $Y_2(LH_2)_3 \cdot 5H_2O$, **7**, viewed down the *b* axis and (b) the inorganic chains in **7**.

bond. The inorganic chains link to four others to make up the framework. Again H-bonding between protonated N atoms and pendant P=O groups is present in the structure. Upon thermal removal of the physisorbed water molecules from the pores, the unit cell volume greatly decreases (due to a 12% reduction along the *c* axis) and the resulting solid is not porous, though dehydration is reversible. The preference for structure **7** over MIL-91 (**6**) for Y^{3+} and Gd^{3+} is thought to arise because of the isolated nature of the octahedra in **7**. This allows the structure to support cations of larger radius without imposing constraints on the rest of the framework. For MIL-91, by contrast, increasing cation size would result in a potentially unfavourable decrease in M–(OH)–M angles along the chains of corner-sharing octahedra.

10.5.2 Structure Types with Metal Cations in Higher Coordination

For the larger trivalent rare earth cations, two main structure types are observed to crystallize upon hydrothermal reaction with H_4**L** at M:**L** ratios in the range 1.0:1.5 to 1.0:3.0 depending on the pH of the synthesis gel.[27] If the pH is raised by the addition of alkali metal hydroxides the ligand is fully deprotonated (including the ring N atoms) and the phase MLn**L**$\cdot 4H_2O$ (**8**, M = Na, K, Cs; Ln = La–Nd) results (Figure 10.8). In the absence of added base the ligand remains partially protonated and the phase Ln**L**(**L**H_2)$\cdot 4.5H_2O$ (**9**, Ln = La, Ce) forms (Figure 10.9).

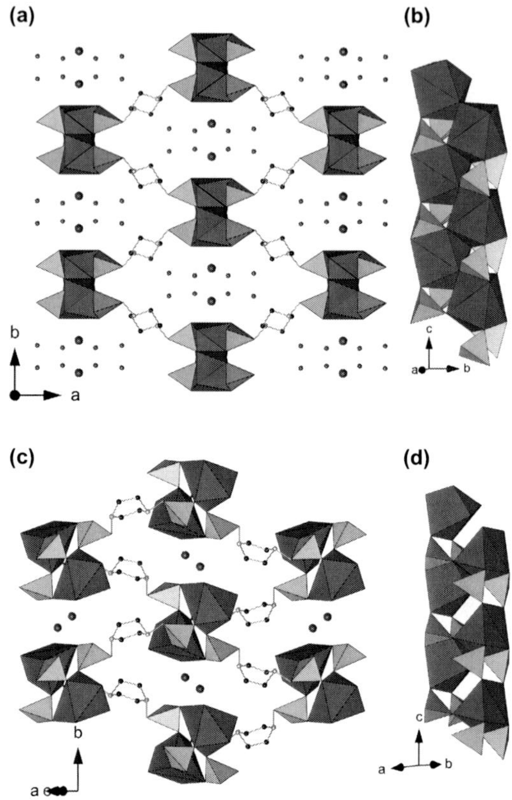

Figure 10.8 Framework structure of hydrated MLnL · xH_2O (M = Na, K, Cs; Ln = La–Nd), **8**, viewed down the channel axis (a), and a single inorganic chain (b). In the dehydrated structure (c-viewed off the channel axis; d-single inorganic chain) the piperazinyl rings adopt a boat, rather than a chair, configuration. (LnO_8 polyhedra in dark grey, charge balancing cations, M, large dark grey spheres, otherwise representation as Figure 10.2.)

Structure **8** consists of chains of edge-sharing LnO_8 polyhedra linked by the bisphosphonate ligand. Each phosphonate group is fully coordinated to metal cations via three O atoms of the methylenephosphonate group and each inorganic chain is linked to four others to give a structure with a trellis-like cross-section. Hydrated alkali metal cations within the channels balance the framework's negative charge. These extra-framework cations show ion exchange properties, a feature also observed in recently reported lanthanide tetraphosphonates.[41] Dehydration of **8** results in a structural transition in which the piperazinyl ring adopts the boat conformer, the channels close and there is no permanent porosity (Figure 10.8c); the dehydration is reversible. An yttrium form of this material has also been prepared, with the metal adopting 8-fold coordination rather than the 6-fold coordination observed in structure **7**.[37]

Structure **9** (Figure 10.9; Ln = La in the current discussion), which forms in the absence of alkali metal cations, has the same motif of metal phosphonate

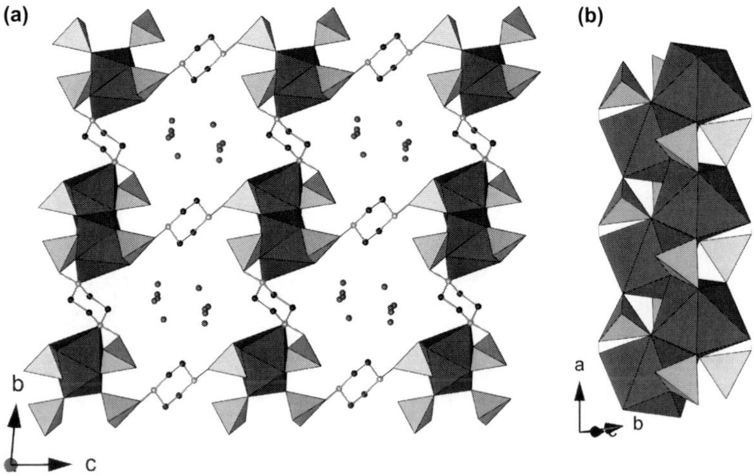

Figure 10.9 Channel (a) and chain (b) views of $Ln_2\mathbf{L}(\mathbf{L}H_2)\cdot xH_2O$ (Ln = La, Ce), **9**, which possesses LnO_7N polyhedra if the piperazinyl N atom is considered (at 2.87 Å) to be coordinated (LnO_7N polyhedra in dark grey, otherwise representation as Figure 10.2).

chains linked by organic groups as in structure **8**, but the inorganic chains comprise edge-sharing LaO_7N polyhedra (with N at 2.87 Å from the La^{3+} cation). There are two crystallographically and coordinatively distinct linkers, both with the piperazinyl ring in the chair conformation and with the methylene-phosphonate groups in equatorial positions. For one of the linkers, all six of the phosphonate O atoms are coordinated to La^{3+} cations and both N atoms are protonated. The second linker has both N atoms and two out of three phosphonate O atoms coordinated to La. The other phosphonate O atom engages in H-bonding to NH groups on the first ligand type. As for structure **8** the material shows reversible dehydration behaviour but no permanent porosity when dehydrated.

For lanthanide cations of ionic radius intermediate between Ce and Gd, such as Nd (Ce^{3+}: 1.01 Å; Nd^{3+}: 0.98 Å Gd^{3+}: 0.94 Å – in octahedral coordination),[35] a further framework has been reported with $H_4\mathbf{L}$, $Nd_2(\mathbf{L}H_2)_3\cdot 9H_2O$ (**10**).[38] Although this framework has the same metal:ligand stoichiometry as structure **7**, in this case the slightly larger lanthanide Nd is 8-fold rather than 6-fold coordinated (Nd: 0.98 Å, 6-fold coordinate; 1.11 Å 8-fold coordinate). The inorganic chain is made up of dimers of NdO_8 edge-sharing polyhedra: Nd cations edge-share and are coordinated by two phosphonates in a bridging mode and two in an edge-sharing mode; dimers are linked in the *a*-direction via two phosphonate groups in a bridging mode (Figure 10.10). All of the phosphonate groups in each of the three crystallographically distinct linkers coordinate through two O atoms, and each inorganic chain is connected to six others. In this material the eighth coordination site is filled by a water molecule.

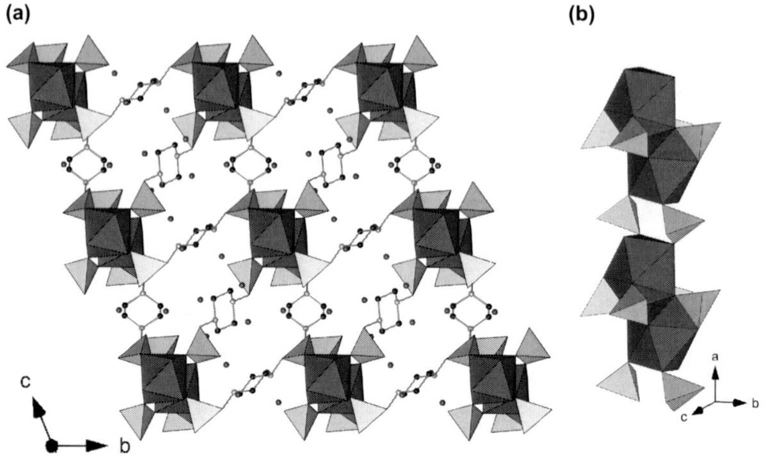

Figure 10.10 Channel (a) and chain (b) views of $Nd_2(\mathbf{L}H_2)_3 \cdot 9H_2O$, **10**, showing edge-sharing dimers of NdO_8 polyhedra. (NdO_8 polyhedra in dark grey, otherwise representation as Figure 10.2.)

Slight variations in synthetic conditions may also give other lanthanide structures. In the presence of Cl^- ions,[24] for example, two related phases, $La_2(\mathbf{L}H_2)_2(\mathbf{L}H_4)Cl_2$ (**11**) and $La_2(\mathbf{L}H_2)_2(\mathbf{L}H_3)Cl$ (**12**) form, based on similar 'lanthanum phosphonate' chain motifs, constructed from isolated LaO_7 polyhedra linked by fully coordinated phosphonate tetrahedra (Structure **11** is shown in Figure 10.11). In both structures the linker demonstrates two different coordination modes: monodentate coordination through both phosphonates and fully coordinating (through all three phosphonate O atoms) through both phosphonates, with twice as many linkers demonstrating the latter mode. The structures differ in the configuration of the monodentate linker, which is present in structure **11** with both methylenephosphonate groups axial. This axial configuration allows a more extended H bonding network to form between the linker and both the Cl^- ion and framework phosphonate O atoms. In structure **12** by contrast one methylenephosphonate is axial and one equatorial. Structure **11** also shows non-centric occupancy of the channels by Cl^- ions, indicating polar ordering.

10.6 Phosphonate MOFs with Tetravalent Metals and *N*,*N*′-piperazinebis(methylenephosphonic acid)

As discussed in Section 10.5, tetravalent metal phosphonate MOFs have also been prepared with ligand $H_4\mathbf{L}$, for example the microporous MIL-91(Ti). Very recently, two Zr phosphonates have been prepared with this ligand: $Zr_2\mathbf{L}_3H_4.9H_2O$ (**13**) and $ZrF_2(\mathbf{L}H_2)$ (**14**)[42] In both cases the presence of fluoride ions as mineralizing agents keeps Zr in solution long enough to allow

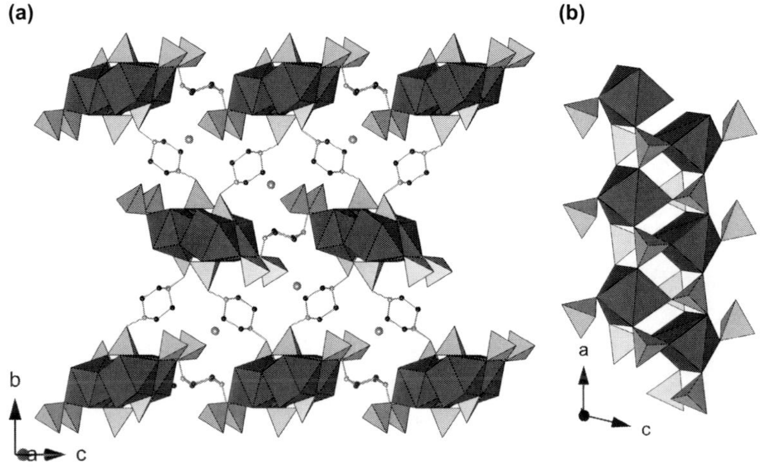

Figure 10.11 Off channel axis (a) and chain (b) views of $La_2(\mathbf{L}H_2)_2(\mathbf{L}H_4)Cl_2$, **11**. Cl^- ions shown as large light grey spheres, otherwise representation as Figure 10.2. Note that some phosphonate tetrahedra only coordinated through one O atom.

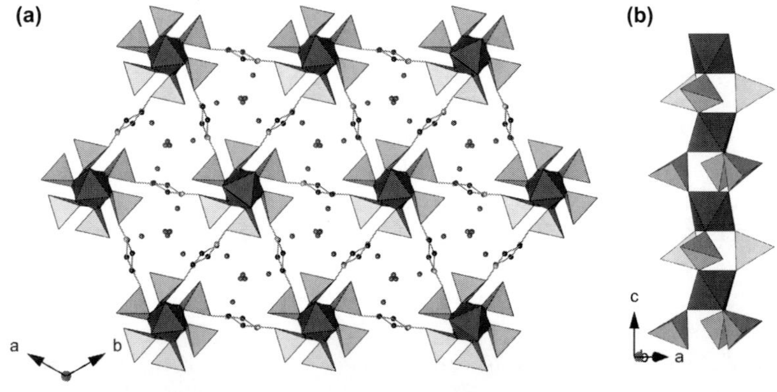

Figure 10.12 Channel (a) and chain (b) views of $Zr_2H_4\mathbf{L}_3 \cdot 9H_2O$, **13**, showing similar chain units to structures **7** and **17**. (ZrO_6 octahedra in dark grey, otherwise representation as Figure 10.2.)

phosphonate crystallization. Structure **13** shows inorganic chains similar to those observed in structure **7**, with isolated ZrO_6 octahedra bridged above and below by three phosphonate groups. In structure **13** these chains possess three-fold symmetry and are connected to six other chains via the piperazine-based linker (Figure 10.12). Charge balance is maintained in the structure by four disordered protons distributed across three ligands. The as-prepared material contains channels with a free diameter of around 3 Å, which are occupied by solvent water molecules. The structure can reversibly lose and re-adsorb these water molecules, though the channels contract on dehydration, resulting in the dehydrated material showing no permanent porosity to N_2. The second Zr phosphonate, **14**, which forms at lower pH, is a layered structure pillared by the piperazinyl groups, which has the same topology as structure **1**. Layers in structure **14** are composed of ZrO_4F_2 octahedra, with the F^- cations replacing the divalent O and H_2O of structure **1**, connected in a square array with phosphonate tetrahedra bridging between two Zr cations. These layers are connected by the organic groups, so that there is no porosity.

10.7 Hydrothermal Synthesis with *N,N′*-2-methylpiperazinebis(methylenephosphonic acid) or *N,N′*-2,5-dimethylpiperazinebis (methylenephosphonic acid)

The modified Mannich reaction described in Section 10.2 can also be used to give the 2-methyl- (H_4**L′**) and 2,5-dimethylpiperazine (H_4**L″**) derivatives of the *N,N′*-piperazinebis(methylenephosphonic acid) ligand.[27] 2-methylpiperazine is chiral at the 2-position on the ring, and both racemic and enantiopure *R* and *S* forms have been successfully used as substrates in ligand synthesis. In hydrothermal crystallizations, the replacement of the original piperazine-derived ligand by the racemic H_4**L′** or H_4**L″** can result in either the same phase with slightly modified cell parameters or in some cases where, for example, the presence of the methyl group would otherwise result in unfavourable steric interactions, can lead to novel structures.

Instances of the former type of behaviour with the H_4**L′** are observed for MIL-91 (**6**), STA-12 (**5**) and MLn**L**·$x$$H_2O$ (**8**), where there is sufficient space for the methyl group to be included in the as-prepared materials with only minor changes in the unit cell size. In structure **8** for example, inclusion of methyl groups causes a scissoring motion of the framework to make space for it. For STA-12(Ni), dehydration of the methyl form does not result in the triclinic distortion observed for the non-methyl version, and the pores are 'propped' open giving a larger pore volume (rather than a smaller pore volume, which might have been expected from methyl groups occupying space in the channels). Syntheses using the 2,5-dimethylpiperazine-derived ligand are successful in preparing structure type **8**, where the ligand is accommodated by further opening up of the framework, but attempts to prepare the STA-12 analogue resulted in ligand breakdown and crystallization of the nickel phosphate VSB-5.[43]

In some cases, use of H_4**L′** instead of H_4**L** in syntheses gives rise to new structures. A structure type has been reported that contains octahedrally coordinated divalent Co^{2+}, $[Co_4L'_{1.5}(AcO)_{1.5}(OH,H_2O)_3]\cdot(NH_4)_{0.5}H_2O_{5.5}$ (**15**, Figure 10.13),[22] which like STA-12 contains a hexagonal arrangement of channels. This is prepared during reactions using acetates as the metal source and with higher M:L ratios. In this structure the metal polyhedra form

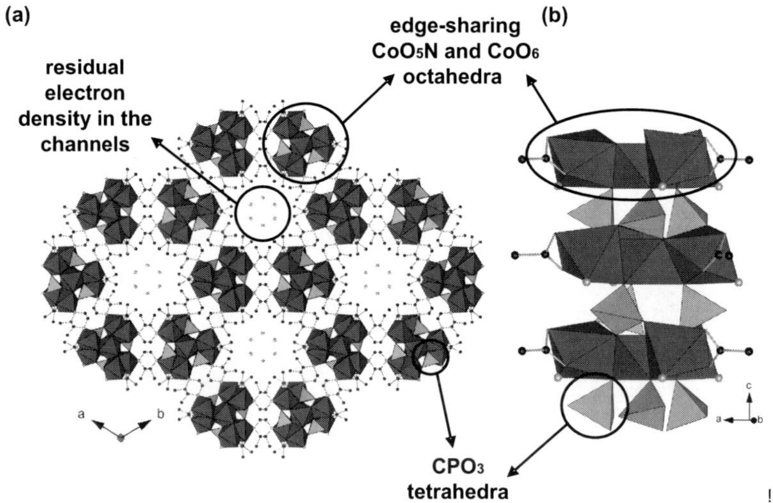

Figure 10.13 Views (a) down the channel axis of $Co_4L'_{1.5}(OAc)_{1.5}(OH,H_2O)$, **15**, and (b) of its inorganic chains. Three CoO_5N and one CoO_6 octahedra share edges to give tetramers which are bridged by phosphonate groups

edge-sharing tetramers linked by phosphonate groups to other tetramers above and below, and via the linker to three adjacent chains to form an array of unidirectional channels. Additional metal coordination is provided by bidentate acetate groups. Although this should be a large pore solid, only low permanent porosity has so far been measured.

For the lighter lanthanides La–Nd, the structure formed without addition of base to the synthesis, structure **9**, is replaced as a product by $Ln(\mathbf{L}'H)\cdot 4H_2O$ (**16**, $\mathbf{L}' = \mathbf{L}'$ or $\mathbf{L}''$).[27] This phase (Figure 10.14) is based on metal phosphonate chains with isolated LnO_7 polyhedra, similar to those observed previously for the chloride-templated frameworks **11** and **12**. Again the structure shows reversible dehydration but no permanent porosity.

Finally, with smaller lanthanides and related trivalent cations (Gd–Yb, Y, Sc) the addition of $H_4\mathbf{L}'$ in place of $H_4\mathbf{L}$ results in crystallization of a novel structure, $M_2(\mathbf{L}'H_2)_3\cdot xH_2O$, named STA-13 (**17**, Figure 10.15).[39] This material contains metal cations in octahedral coordination within inorganic chains with the same topology as those of structure **7**, but linked to three rather than four others to give a hexagonal array of unidirectional channels. In each phosphonate group, two O atoms engage in coordinative bonding, each to one metal cation, with the third P=O forming an H-bond to the protonated amine group of a neighbouring ring. The methyl groups project into the channels. Dehydration leaves a permanently porous structure ($0.12\ cm^3\ g^{-1}$) with narrow (ca. 3 Å) channels able to take up small molecules such as CO_2, N_2 and CH_4. The structure is a true racemate, requiring equal numbers of the two enantiomers to construct the framework. The importance of the methyl group in determining that this structure forms is through its role in controlling the ring configuration

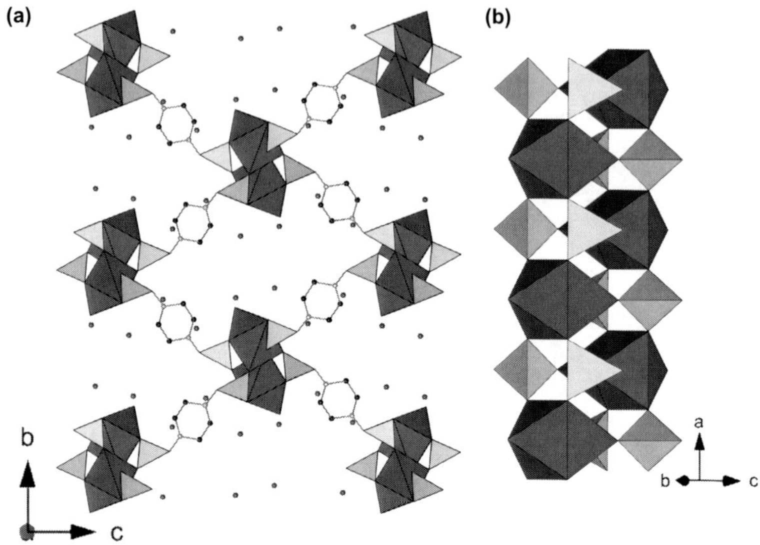

Figure 10.14 Channel (a) and chain (b) views of Ln($\mathbf{L}'$H)$\cdot x$H$_2$O (Ln = La–Nd; $\mathbf{L}' = \mathbf{L}'$ or $\mathbf{L}''$), **16**. The inorganic chains, with isolated LnO_7 polyhedra, are also observed in structures **11** and **12**. (LnO_7 polyhedra shown in dark grey, representation otherwise as Figure 10.2.)

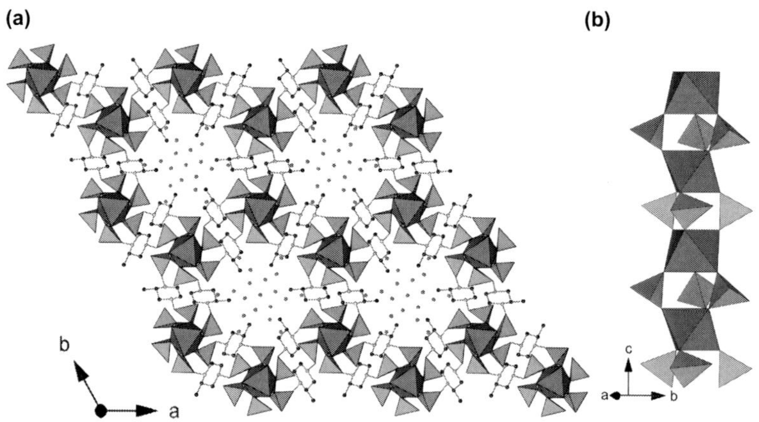

Figure 10.15 Channel (a) and chain (b) views of STA-13, M$_2$($\mathbf{L}'$H$_2$)$_3\cdot x$H$_2$O (M = Gd–Yb, Y, Sc), **17**, showing the favoured position of the 2-methyl group on the piperazine ring. Note that the methyl groups project into the pores and that the inorganic chains are similar to those observed in structures **7** and **13**. (Representation as Figure 10.7.)

and pre-organizing both the packing and the resulting H-bonding that direct the crystallization. As a footnote, it was found that using single enantiomers of the 2-methyl-derived ligand in place of the racemic compound, a form of structure **7** was prepared. It appears while a single enantiomer can be accommodated within **7**, for steric reasons both enantiomers cannot fit, and thus when using the racemic $H_4\mathbf{L}'$, rather than forming a conglomerate of crystals the true racemate STA-13 results.

10.8 Summary

Remarkable success has been achieved in the synthesis of crystalline open and microporous framework metal phosphonates using *N*,*N*′-piperazinebis(methylenephosphonic acid) ($H_4\mathbf{L}$) and related phosphonic acids with one or two methyl groups on the piperazinyl ring. Crystallization with these ligands tends to give framework bisphosphonates rather than layered metal phosphonates, with their structures based on a motif of inorganic metal phosphate chains running in one direction and cross-linked by the piperazine moieties of the ligands. The bisphosphonate MOFs of this family that have already been discovered show a great variety of different structure-related properties including permanent porosity (STA-13, 3 Å pore; MIL-91, 4 Å; STA-12, 9 Å); large breathing effects upon reversible adsorption and desorption of water (structures **7** and **8**); coordinatively unsaturated metal sites upon loss of coordinated water (STA-12); and cation exchange capacity (structure **8**).

The structural chemistry within this family is very varied, giving different structure types with different metal cations and even with the same cation in some cases, with individual structures also able to accommodate many different cations. The charge, size and electronegativity of cations all play an important role in phase selection and the synthesis pH can also be a controlling factor, particularly in determining the protonation state of the ligand. This structural diversity is an expression of the versatility of the ligand combined with the different possible types of metal phosphonate chains. By comparison, the terephthalic acid ligand that is such an integral part of many important carboxylate MOFs is quite rigid, with the many different structures arising solely from variation in its connectivity with metal oxide polyhedra, clusters and chains.

Primarily, ligand $H_4\mathbf{L}$ is distinguished by its wide range of coordinative behaviour (Table 10.2). The phosphonate group can bind strongly to metal cations via one, two or three phosphonate O atoms and may also bind via the N atom of the piperazinyl ring, if this is deprotonated. If the N atom is protonated, an H-bonding arrangement involving unbound P=O groups and protonated piperazinyl N atoms may additionally form. This opens up an important bonding possibility for phosphonate groups that are bidentate to metal cations, one that is not available to many other bisphosphonate ligands which do not possess auxiliary groups capable of H-bonding, for example $O_3PCH_2CH_2PO_3$, $O_3PCH_2C_6H_4CH_2PO_3$, etc. which have previously been used in the preparation of layered or framework phosphonates. This coordination mode, which increases the geometric bonding possibilities accessible to frameworks, is observed in MIL-91 and STA-13 (Figure 10.16).

Table 10.2

	Structure	Metal Charge	Metal Coord. & Connectivity		Structure Motif	Ligand Coordination					
						*No.**	*Symm.**	PO_3	*Ring N*	*N-H H-Bonding*	$-CH_2PO_3$ *Config.*
1	$VO(H_2O)\mathbf{L}H_2$	4+	MO_6	Isolated octahedra	Layered		Y	2		P=O	eq.
2	$M(H_2\mathbf{L})\cdot xH_2O$ (Mn,Co,Fe,Zn)	2+	MO_4	Isolated tetrahedra	Channel		Y	2		P=O	eq.
3	$Zn_2\mathbf{L}$	2+	MO_3N	Isolated tetrahedra	Layered		Y	3	1		eq.
4	$Cu_2(H_2O)_2\mathbf{L}$	2+	MO_4N	Isolated square pyramidal	Layered		Y	3	1		eq.
5	$M_2(H_2O)_2\mathbf{L}\cdot xH_2O$, (Mn,Fe, Co,Ni) *STA-12*	2+	MO_5N	Octahedra in edge-sharing chain	Channel		Y	2	1		ax.
6	$M(OH)(\mathbf{L}H_2)\cdot xH_2O$, (Al,V, Fe,In); $TiO(\mathbf{L}H_2).4.5H_2O$ - *MIL-91*	3+ 4+	MO_6	Octahedra in corner-sharing chain	Channel		Y	2		P=O	eq.
7	$M_2(\mathbf{L}H_2)_3\cdot xH_2O$, (Y, Gd)	3+	MO_6	Isolated octahedra	Channel	1	N	2/2		P=O/ H_2O	ax./eq. eq.
						2	Y	2		H_2O	
8	$MLn\mathbf{L}\cdot xH_2O$ (M=Na,K,Cs; Ln = La,Ce, Nd)	3+	MO_8	8-fold in edge-sharing chain	Channel		Y	3		H_2O	eq.
9	$La_2\mathbf{L}(\mathbf{L}H_2)\cdot 4.5H_2O$	3+	MO_7N	8-fold in edge-sharing chain	Channel	1	Y	2	1[#]	P=O	eq.
						2	Y	3			eq.
10	$Nd_2(\mathbf{L}H_2)_3\cdot 9H_2O$	3+	MO_8	8-fold in isolated clusters	Channel	1	Y	2		P=O	eq.
						2	Y	2		H_2O	eq.
						3	Y	2		P–O	eq.

11	$La_2(\mathbf{LH}_2)_2(\mathbf{LH}_4)Cl_2$	3+	MO_7	7-fold in isolated clusters	Channel	1	N	3		P=O/	eq./eq.
						2	Y	1		Cl	ax.
										P–O	
12	$La_2(\mathbf{LH}_2)_2(\mathbf{LH}_3)Cl$	3+	MO_7	7-fold in isolated clusters	Channel	1	Y	3		P=O	eq.
						2	N	3/3		Cl/	eq. ax./
						3	N	1/1		Cl/	eq.
										P-O	
13	$Zr_2\mathbf{L}_3H_4 \cdot 9H_2O$	4+	MO_6	Isolated octahedra	Channel		Y	2		H_2O	eq.
14	$ZrF_2(\mathbf{LH}_2)$	4+	MO_4F_2	Isolated octahedra	Layered		Y	2		P=O	eq.
15	$Co_4(\mathbf{L}')_{1.5}(OAc)1.5(OH,H_2O)_3 \cdot (NH_4)_{0.5},(H_2O)_{5.5}$	2+	$M_4O_{15}N_3$	Octahedra in isolated tetramer clusters	Channel		Y	3	1		eq.
16	$Ln(H\mathbf{L}') \cdot 4H_2O$, (Ln=La-Nd)	3+	MO_7	Isolated 7-fold	Channel		Y	3		H_2O	eq.
17	$M_2(H_2\mathbf{L}')_3 \cdot xH_2O$, (Sc, Y, Gd, Ho, Yb) *STA-13*	3+	MO_6	Isolated octahedra	Channel		Y	2		P=O	eq.

*'No.' is reference number of coordinatively, rather than crystallographically, distinct linker; Symm.' indicates whether linker possesses an inversion centre
#Long M-N bond: 2.87Å

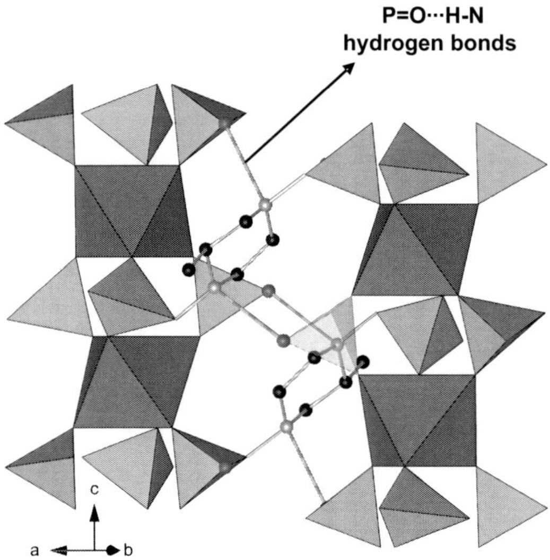

Figure 10.16 The H bonding scheme in STA-13 (**17**), showing hydrogen bonds (indicated) between terminal phosphonate P=O oxygen atoms and (protonated) piperazinyl N atoms. (Representation as Figure 10.7.)

The ability of the ligand to adopt different configurations also permits the crystallization of frameworks with different connectivities. Most importantly in this regard, methylenephosphonate groups can adopt equatorial or axial positions relative to the piperazinyl ring. This in turn makes it possible for the ligand to realize more of its coordinative potential. For example, whereas in the crystalline form of the acid and in structures **1–4**, **6**, **8–10** and **13–17** the methylenephosphonate groups are equatorial, their ability to adopt an axial configuration enables the remarkable STA-12 structure to form (Figure 10.17). The ligand itself is also flexible: rotation can occur around both the P–CH_2 bonds and also the H_2C–N bonds, allowing the phosphonate groups more freedom to coordinate to framework cations. The latter torsion is also found to allow the crystallization of structurally distorted variants of structure **8** to accommodate the steric demands of the 2-methyl and 2,5-dimethyl derivatives of the piperazine ligand. Furthermore, the different coordinative modes of the ligand are able to combine with a range of different types of inorganic chain, as shown in the structural diagrams throughout this chapter. For smaller cations these include chains of isolated tetrahedra and octahedra; chains of tetramers formed of edge-sharing octahedra; and edge-sharing spirals of octahedra. By contrast, for larger cations (such as the lighter lanthanides), isolated and edge-sharing chain motifs are again observed, as are chains of dimers, but cations are in all cases either 7- or 8-fold coordinated. In this way considerable structural variety arises, salient features of which are summarized in Table 10.2.

The flexibility of the ligand also manifests itself during structural transitions of this family of solids that occur as a consequence of reversible dehydration.

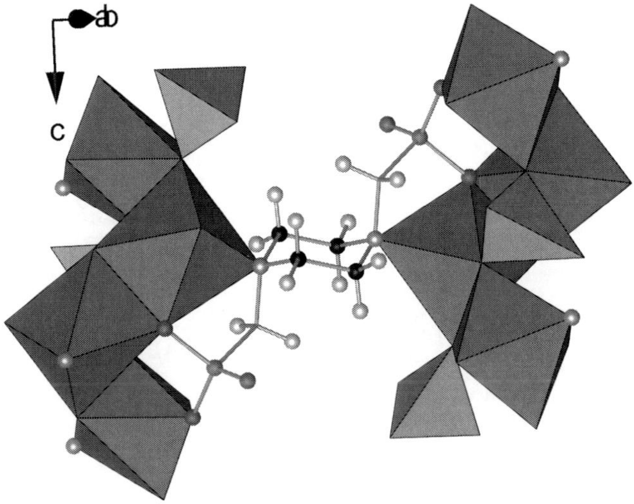

Figure 10.17 Portion of the STA-12 structure, showing the binding of the ligand (ball and stick representation) with the helical chains of MO_5N octahedra. (Representation as Figure 10.3.)

In STA-12, for example, loss of water coordinated to the metal cations leaves them with 5-fold MO_4N coordination, and hence also a Lewis acidic free metal site. Rather than remaining square pyramidal (as observed in the related CPO-27 structure, in which the dihydroxyterephthalate ligand is rigid[34]) the phosphonate ligand's flexibility allows relaxation of the coordination geometry closer to trigonal bipyramidal. This in turn reduces the strength of bonding of the Lewis acidic metal to basic molecules such as CO, as energy is expended in the required structural rearrangement upon chemisorption. Finally, in an extreme case, dehydration of structure **8** results in change of configuration of the piperazinyl ring from chair to boat as the channels close to enable better coordination of the metal cations.

10.9 Outlook

It is likely that the multi-faceted structural behaviour of metal phosphonate MOFs based on *N*,*N'*-piperazinebis(methylenephosphonate) and related ligands will profitably be extended in several directions in future. It should be possible, by ligand design and control of synthetic conditions, to prepare materials with larger pores, novel kinds of Lewis acidic sites and chiral internal surfaces. Towards the lattermost target, for example, the chiral *R*-2-methylpiperazine derivative of H_4**L** has already been prepared and included within STA-12(Ni), and future studies will determine whether it has been incorporated with its chirality retained.

Large pore MOFs have previously been prepared through the use of *isoreticular synthesis*, as exemplified by the IRMOF series of materials,[18,44] where

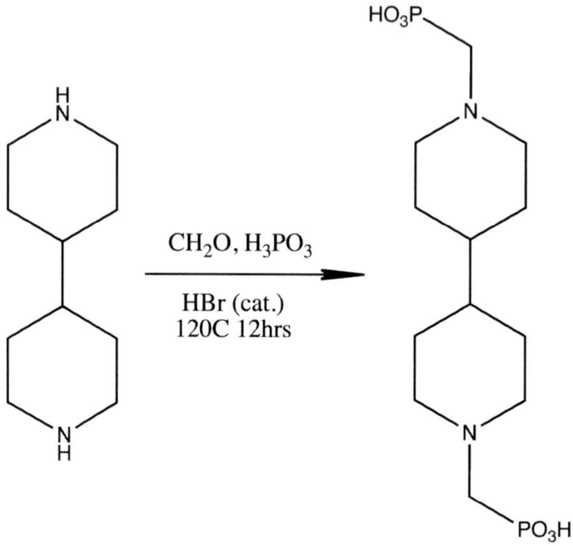

Scheme 10.3

ligand design is applied in the preparation of organic linkers of increasing length. Most recently this approach has been applied to the piperazine-derived bisphosphonates. By selection of a suitable substrate with a rigid core and terminal N-containing heterocycles, 4,4′-bipiperidine, and through the use of the same modified Mannich reaction used to prepare H_4**L**, it was possible to prepare *N*,*N*′-4,4′-bipiperidinebis(methylenephosphonic acid) (H_4**LL**) (Scheme 10.3). Reaction of this ligand with cobalt(II) acetate under similar conditions to those used in the synthesis of STA-12 (**5**) yielded a new material, referred to as STA-16 (**18**), with a network topology identical to that of STA-12, but with pores of approximately twice the free diameter (STA-12: 0.9 nm; STA-16: 1.8 nm; Figure 10.18a).[45] The structure of **18** has been confirmed by synchrotron powder X-ray diffraction studies and demonstrates permanent porosity to N_2 at 77 K. Non-local DFT calculations on the N_2 adsorption data indicate a pore size distribution with a single principal pore size of 1.85 nm, in agreement with crystallographic results (Figure 10.18b).[45] The topology of this material is particularly advantageous as, unlike many of the more open structured metal carboxylates, the relatively dense walls prevent interpenetration of the structure, ensuring consistent permanent porosity. This material is thought to represent the first example of isoreticular synthesis of a permanently porous phosphonate MOF and the synthesis and use of other ligands in the same strategy is likely to yield structures with still greater porosity.

Synthesis of the cobalt-tetramer-containing structure **15** points to two other possible synthetic directions: the use of mixed ligand systems and the synthesis of phosphonate MOFs based on clusters rather than chains. In fact these two strategies may operate best in tandem, as suggested by Tollis *et al.*[46] Such

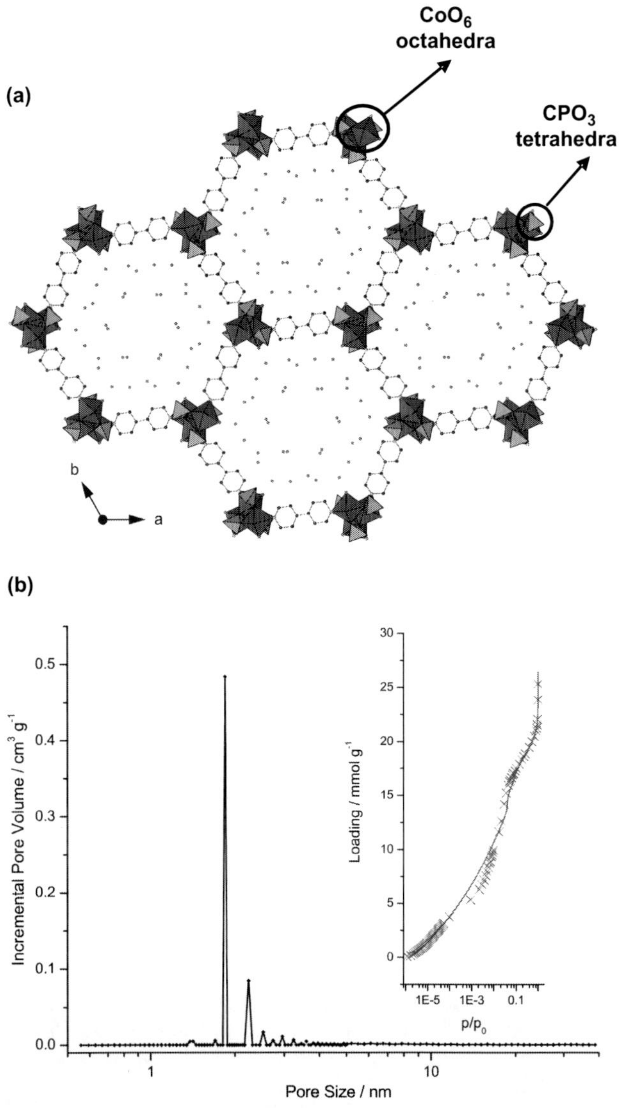

Figure 10.18 (a) Structure of STA-16(Co), $Co_2(H_2O)_2$**LL**$\cdot 11H_2O$, **18**, showing the 18 Å channels, viewed along the channel axis (CoO_6 octahedra and CPO3 tetrahedra are indicated, representation otherwise as Figure 10.7); (b) pore size distribution (PSD) derived from NLDFT calculation on N_2 adsorption isotherm at 77 K (assuming cylindrical pores with an oxide surface) with single maxima at 18.5 Å. Inset: log plot of adsorption isotherm (experimental – line) with isotherm calculated from the derived PSD (crosses). (Reproduced with permission from ref. 45.)

cluster-based solids with piperazine derived ligands may display 3D porosity, as does the MOF reported by Taylor *et al.*, based on Cu trimers and an adamantane tetraphosphonate-based ligand, for example.[47] Expansion of this and other families of phosphonate MOFs will lead to potential applications over the whole range currently being investigated for the better known carboxylate MOFs, and in some of these fields, metal phosphonates are likely to show greater stability and applicability.

References

1. K. Maeda, J. Akimoto, Y. Kiyozumi and F. Mizukami, *J. Chem. Soc., Chem. Commun.*, 1995, 1033.
2. K. Maeda, J. Akimoto, Y. Kiyozumi and F. Mizukami, *Angew. Chem., Int. Ed. Engl.*, 1995, **34**, 1199.
3. G. Alberti, M. Casciola, U. Costantino and R. Vivani, *Adv. Mater.*, 1996, **8**, 291.
4. A. Clearfield and Z. Wang, *J. Chem. Soc., Dalton Trans.*, 2002, 2937.
5. V. J. Carter, J. P. Kujanpää, F. G. Riddell, P. A. Wright, J. F. C. Turner, C. R. A. Catlow and K. S. Knight, *Chem. Phys. Lett.*, 1999, **313**, 505.
6. K. Maeda, Y. Kiyozumi and F. Mizukami, *J. Phys. Chem. B*, 1997, **101**, 4402.
7. J. Gonzalez, R. N. Devi, P. A. Wright, D. P. Tunstall and P. A. Cox, *J. Phys. Chem. B*, 2005, **109**, 21700.
8. C. Herdes, A. Valente, Z. Lin, J. Rocha, J. A. P. Coutinho, F. Medina and L. F. Vega, *Langmuir*, 2007, **23**, 7299.
9. S. R. Miller, E. Lear, J. Gonzalez, A. M. Z. Slawin, P. A. Wright, N. Guillou and G. Férey, *Dalton Trans.*, 2005, 3319.
10. G. Férey, *Chem. Soc. Rev.*, 2008, **37**, 191.
11. H. Li, M. Eddaoudi, M. O'Keeffe and O. M. Yaghi, *Nature*, 1999, **402**, 276.
12. S. S. Chui, S. M. Lo, J. P. H. Charmant, A. G. Orpen and I. D. Williams, *Science*, 1999, **283**, 1148.
13. K. Barthelet, J. Marrot, D. Riou and G. Férey, *Angew. Chem., Int. Ed. Engl.*, 2002, **41**, 281.
14. C. Serre, F. Millange, C. Thouvenot, M. Noguès, G. Marsolier, D. Louër and G. Férey, *J. Am. Chem. Soc.*, 2002, **124**, 13519.
15. S. Surble, C. Serre, C. Mellot-Draznieks, F. Millange and G. Férey, *Chem. Commun.*, 2006, 284.
16. G. Férey, C. Mellot-Draznieks, C. Serre, F. Millange, J. Dutour, S. Surblé and I. Margiolaki, *Science*, 2005, **309**, 2040.
17. S. Kitagawa, R. Kitaura and S. Noro, *Angew. Chem., Int. Ed. Engl.*, 2004, **43**, 2334.
18. O. M. Yaghi, M. O'Keeffe, N. W. Ockwig, H. K. Chae, M. Eddaoudi and J. Kim, *Nature*, 2003, **423**, 705.
19. J. R. Long and O. M. Yaghi (Eds.), *Chem. Soc. Rev.*, 2009, **38**, 1201.
20. K. Maeda, *Micropor. Mesopor. Mater.*, 2004, **73**, 47.

21. G. K. H. Shimizu, R. Vaidhyanathan and J. M. Taylor, *Chem. Soc. Rev.*, 2009, **38**, 1430.
22. J. A. Groves, S. R. Miller, S. J. Warrender, C. Mellot-Draznieks, P. Lightfoot and P. A. Wright, *Chem. Commun.*, 2006, 3305.
23. K. Moedritzer and R. R. Irani, *J. Org. Chem.*, 1966, **31**, 1603.
24. J. A. Groves, P. A. Wright and P. Lightfoot, *Inorg. Chem.*, 2005, **44**, 1736.
25. N. Choi, I. Khan, R. W. Matthews, M. McPartlin and B. P. Murphy, *Polyhedron*, 1994, **13**, 847.
26. V. P. Markhaeva, N. S. Erdyneev, G. P. Barchan and A. D. Garnovskii, *Izv. Sev.-Kavk. Nauchn. Tsentra Vyssh. Shk., Estestv. Nauki*, 1988, 106.
27. J. P. Mowat, J. A. Groves, M. T. Wharmby, S. R. Miller, Y. Li, P. Lightfoot and P. A. Wright, *J. Solid State Chem.*, 2009, **182**, 2769.
28. V. Soghomonian, R. Diaz, R. C. Haushalter, C. J. O'Connor and J. Zubieta, *Inorg. Chem.*, 1995, **34**, 4460.
29. R. LaDuca, D. Rose, J. R. D. DeBord, R. C. Haushalter, C. J. O'Connor and J. Zubieta, *J. Solid State Chem.*, 1996, **123**, 408.
30. J. A. Groves, P. A. Wright and P. Lightfoot, *Dalton Trans.*, 2005, 2007.
31. Y. Wang, S. Bao, W. Xu, J. Chen, S. Gao and L. Zheng, *J. Solid State Chem.*, 2004, **177**, 1297.
32. S. R. Miller, G. M. Pearce, P. A. Wright, F. Bonino, S. Chavan, S. Bordiga, I. Margiolaki, N. Guillou, G. Férey, S. Bourrelly and P. L. Llewellyn, *J. Am. Chem. Soc.*, 2008, **130**, 15967.
33. G. M. Pearce, *PhD Thesis*, University of St Andrews, 2009.
34. P. D. C. Dietzel, V. Besikiotis and R. Blom, *J. Mater. Chem.*, 2009, **19**, 7362.
35. R. D. Shannon, *Acta Crystallogr. A*, 1976, **32**, 751.
36. C. Serre, J. A. Groves, P. Lightfoot, A. M. Z. Slawin, P. A. Wright, N. Stock, T. Bein, M. Haouas, F. Taulelle and G. Férey, *Chem. Mater.*, 2006, **18**, 1451.
37. M. T. Wharmby, *unpublished results*, 2009.
38. J. A. Groves, N. F. Stephens, P. A. Wright and P. Lightfoot, *Solid State Sci.*, 2006, **8**, 397.
39. M. T. Wharmby, S. R. Miller, J. A. Groves, I. Margiolaki, S. E. Ashbrook and P. A. Wright, *Dalton Trans.*, 2010, **39**, 6389.
40. F. Hawthorne, *Acta Crystallogr. B*, 1994, **50**, 481.
41. M. Plabst, L. B. McCusker and T. Bein, *J. Am. Chem. Soc.*, 2009, **131**, 18112.
42. M. Taddei, F. Costantino and R. Vivani, *Inorg. Chem.*, 2010, **49**, 9664.
43. N. Guillou, Q. Gao, P. M. Forster, J. Chang, M. Noguès, S. Park, G. Férey and A. K. Cheetham, *Angew. Chem., Int. Ed. Engl.*, 2001, **40**, 2831.
44. M. Eddaoudi, J. Kim, N. Rosi, D. Vodak, J. Wachter, M. O'Keeffe and O. M. Yaghi, *Science*, 2002, **295**, 469.
45. M. T. Wharmby, J. P. S. Mowat, S. P. Thompson and P. A. Wright, *J. Am. Chem. Soc.*, 2011, **133**, 1266.
46. E. I. Tolis, M. Helliwell, S. Langley, J. Raftery and R. E. P. Winpenny, *Angew. Chem., Int. Ed. Engl.*, 2003, **42**, 3804.
47. J. Taylor, A. Mahmoudkhani and G. Shimizu, *Angew. Chem., Int. Ed. Engl.*, 2007, **46**, 795.

CHAPTER 11

Secondary Building Units and Framework Structures in Aluminium and Zinc Phosphates: The Connection Between

R. MURUGAVEL[a] AND C. N. R. RAO[b]

[a] Department of Chemistry, Indian Institute of Technology Bombay, Mumbai 400076, India; [b] Chemistry and Physics of Materials Unit, Jawaharlal Nehru Centre for Advanced Scientific Research, Jakkur, Bangalore-560064, India

11.1 Introduction

The chemistry of metal phosphates has remained an active area of research for a long time due to the relevance of these compounds in many biological systems. Phosphates are available in the human body as hydroxyapatite in bones, teeth and more importantly in nucleotides.[1] Many naturally occurring minerals also feature phosphates and this fact is of great importance to inorganic chemists as much effort is being directed towards the production of designer phosphate frameworks.[2] The term 'phosphate' refers to oxyanions of pentavalent phosphorus, PO_4^{3-}. Further, due to the occurrence of many phosphate-containing minerals in the environment and their utility in a wide rang of applications, it has remained a subject of much interest to inorganic chemists. Although phosphate-containing minerals are common in nature, they do not meet the complete need of the modern chemical industry in terms of their availability and properties such as larger pore size, high thermal stability and size-shape selectivity. The discovery of open framework aluminophosphates

Metal Phosphonate Chemistry: From Synthesis to Applications
Edited by Abraham Clearfield and Konstantinos Demadis

Published by the Royal Society of Chemistry, www.rsc.org

(ALPOs) by Flanigen and co-workers led to an outburst of activities towards the development of metal phosphates frameworks with tuneable porosity.[3] Using preformed molecular precursors, such extended solids can be built more rationally and with better control over the porosity. The chemistry of both molecular phosphates and extended phosphate frameworks has been recently reviewed.[2]

11.2 Phosphonates versus Phosphates

Phosphonic acids are essentially similar in their constitution to the phosphate monoesters, other than the presence of extra oxygen between phosphorus and the alkyl or aryl group in the latter class of compounds (Scheme 11.1). While the majority of the chapters in this book have dealt with metal phosphonates and their applications, the subject matter treated in this chapter is slightly different. The central phosphorus atom in the compounds described herein contains four oxygen atoms around them rather than three as in the case of metal phosphonates. The reaction chemistry of phosphonic acids with divalent or tetravalent metal ions is very well understood in terms of the layered solids isolated more often than any other structural type during the initial years of phosphonate chemistry,[4] although other structural types are becoming more prevalent in recent times.[5] Similarly, the reactions of phosphoric acid monoesters also lead to the isolation of a variety of structural types, including the double-4-ring secondary building units (SBUs) of zeolites. This reactivity difference between phosphonic acids[6] and phosphate monoester[7] towards zinc dication in the absence of any other co-ligands is schematically shown in Scheme 11.2. Similar building blocks can also be obtained directly starting from phosphoric acid under mild conditions.[2] More importantly, the high solubility of these SBUs in a variety of solvents render them as useful starting materials for the preparation of a variety of porous structures.

Thus, while monophosphonic acids normally yield layered solids that are amenable for intercalation type reactions, it is possible to isolate smaller soluble building blocks starting from phosphate esters and phosphoric acids and then effect transformations through a variety of reactions to build porous structures which may find applications similar to those of zeolites and MOFs.

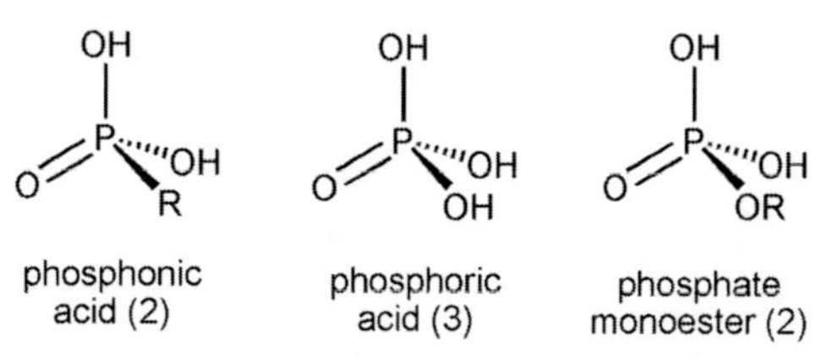

Scheme 11.1 Phosphorus-based acids (number of acidic protons).

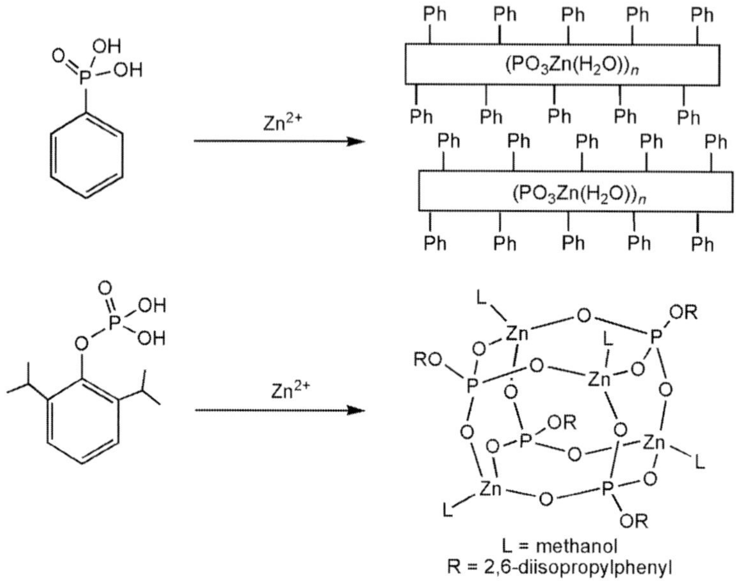

Scheme 11.2 Difference in reactivity between phosphonic and phosphoric acid towards a divalent metal ion.

11.3 Phosphate Monoesters as Building Blocks

The last two decades have seen a major impetus in research directed towards building main group and transition metal phosphate cage complexes which can be regarded as a soluble model of zeolite SBUs.[2,5] Phosphoric acid normally tends to form extended polymeric complexes and open framework metal phosphates. However, the formation of extended structures can be prevented if esters of phosphoric acid are used. Contributions from Tilley and co-workers[8] and our team[9] established that diesters of phosphoric acid, $(RO)_2P(O)(OH)$, form either mononuclear or dinuclear metal phosphates that are very useful precursors for fine particle metal phosphates by simple thermolysis.[2a] Phosphate monoesters, on the other hand, due to the presence of two acidic protons and one phosphoryl oxygen, tend to embrace more metal ions around them and form larger aggregates and form clusters that represent the SBUs of zeolites incorporating metal ions such as Ca,[10a] Sr,[10a] Ba,[10a] Ti,[10b] Mn,[10c] Fe,[10d] Cu,[10e] Zn,[7] Al,[10f,g] and Sn.[10h,i] Many of these complexes display unprecedented and interesting structural geometry. The chemistry of zinc[7] and aluminium phosphates[10f,g] is treated in detail here as representative examples for this chemistry.

11.3.1 Hierarchical Zinc Phosphates through Hydrogen Bonding

11.3.1.1 Zero-dimensional Cages

As noted earlier, in contrast to the well-known reaction of phosphonic acids, $RP(O)(OH)_2$, with divalent transition metal ions yielding layered metal phosphonates $[RPO_3M(H_2O)]_n$, the 2,6-di*iso*propyl-phenyl ester of phosphoric acid reacts with zinc acetate in methanol under ambient conditions to afford tetrameric zinc phosphate $[(ArO)PO_3Zn(MeOH)]_4$ (**1**).

Compound **1** serves as an ideal starting material for the synthesis of a large number of D4R zinc phosphate cubanes with a variety of ligands on zinc. Due to the well-known ability of zinc to form strong complexes with N-donor ligands, reactions of **1** with a large number of pyridine derivatives have been investigated. For example, the coordinated methanol in **1** can be readily exchanged by stronger Lewis basic ligands at room temperature. This strategy opens up a new avenue for building up D4R cubane-based supramolecular assemblies through strong inter-cubane hydrogen bonding interactions. Seventeen pyridinic ligands have been used to synthesize as many D4R cubanes $[(ArO)PO_3Zn(L)]_4$ (**2**–**18**) from **1** (Scheme 11.3) The ligands have been chosen in such a way that majority of them contain an additional functional group that could be used for non-covalent synthesis of extended structures. When the

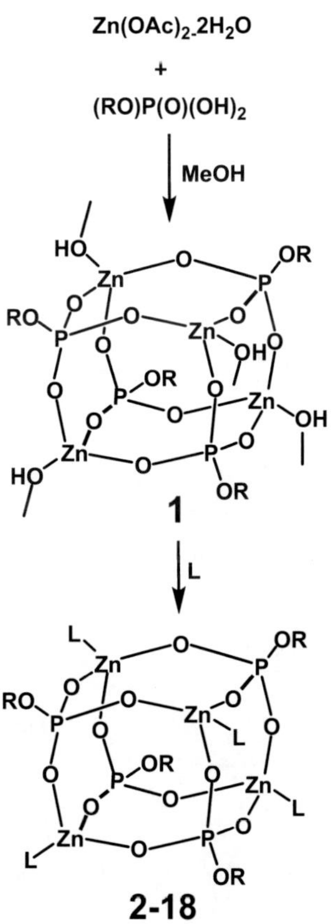

L =		
	2,4,6-trimethylpyridine	**2**
	pyridine	**3**
	2,6-dimethylpyridine	**4**
	2-aminopyridine	**5**
	2,6-diaminopyridine	**6**
	2-hydroxypyridine	**7**
	2-methylimidazole	**8**
	pyridine-2-methanol	**9**
	pyridine-3-methanol	**10**
	pyridine-4-methanol	**11**
	pyridine-2-ethanol	**12**
	3,5-dimethylpyrazole-2-ethanol	**13**
	N,N'-dimethylaminopyridine	**14**
	3-aminopyridine	**15**
	4-aminopyridine	**16**
	3-hydroxypyridine	**17**
	imidazole	**18**

Scheme 11.3 D4R cubanes derived from 2,6-diisopropylphenylphosphate.

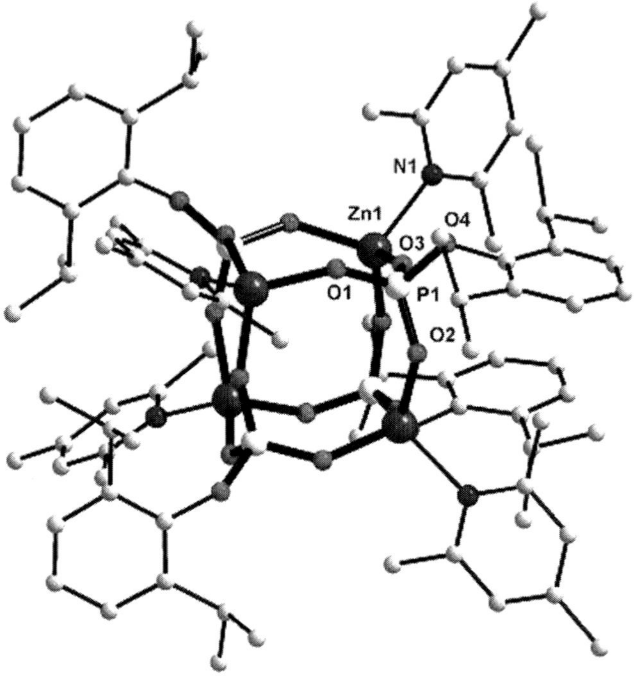

Figure 11.1 Molecular structure of $[(ArO)PO_3Zn(collidine)]_4$ (**2**).

ligand does not contain any other hydrogen bonding donor acceptor sites, zero-dimensional D4R cubanes has been obtained.

Addition of collidine to a solution of **1** leads to the isolation of $[(ArO)PO_3Zn(collidine)]_4$ (**2**).[7a] The presence of a single resonance in the ^{31}P NMR spectrum of **2** at $\delta = 4.7$ ppm indicates the presence of only one type of P atom in the molecule. A single-crystal X-ray diffraction study of **2** reveals the tetrameric nature of the zinc phosphate with a cubane-like structure (Figure 11.1). Compound **2** is the first structurally characterized metal phosphate with a core resembling D4R SBU of zeolites with tetrahedral metal and phosphorus centres.

11.3.1.2 One-dimensional Chains

The use of 2-aminopyridine, in place of collidine, leads to interesting results, however. The reaction as described above proceeds smoothly to yield $[(ArO)PO_3Zn(2apy)]_4$ (**5**), which once again is a D4R cubane (as shown in Scheme 11.4). The presence of NH_2 groups in ortho positions on the pyridine rings in **5** is the source of both intra- and intermolecular N–H···O hydrogen bonds. While each of the four NH_2 groups in **5** forms an intramolecular N–H···O interaction with one of the framework oxygen atoms, two of the NH_2 groups are also involved intermolecular N–H···O interactions with adjacent cubes, thus

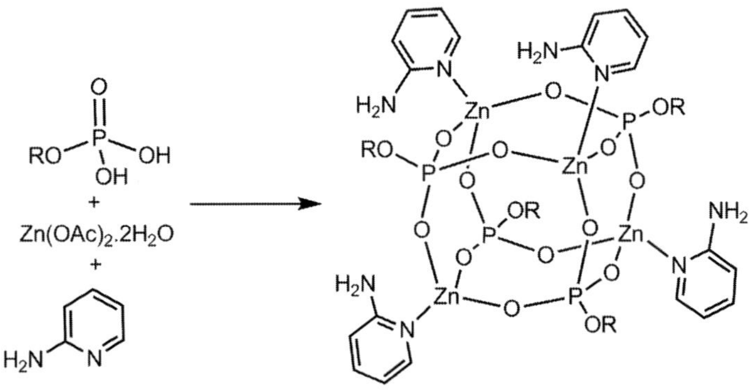

Scheme 11.4 Synthesis of $[(ArO)PO_3Zn(2apy)]_4$ (**5**).

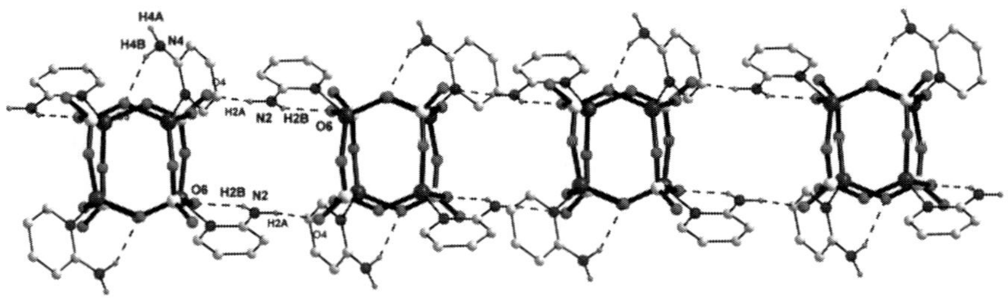

Figure 11.2 One-dimensional assembly of zinc phosphate cubanes in **5**.

resulting in a 1D assembly (Figure 11.2). The intercube distance of 6.812 Å is largely dictated by the strong intermolecular hydrogen bonding. Similarly, the use of pyridine, lutidine, and 2,6-diaminopyridine also results in the formation of linear or zigzag 1D assembly of D4R cubanes through strong intermolecular C–H···O or N–H···O interactions.[7a]

11.3.1.3 Two-dimensional Sheets

The construction of 2D assemblies of zinc phosphates has been achieved first by employing 2-hydroxy pyridine (2hpy) ligand in its reaction with **1** to yield $[(ArO)PO_3Zn(2hpy)]_4$ (**7**). The 2hpy ligand in **7** coordinates to the zinc ion through the phenoxide oxygen rather than the pyridinic nitrogen (Scheme 11.5). The phenolic –OH has undergone an H-shift during coordination, thus producing a zwitter-ionic pyridinium phenoxide ligand.

The structure of **7** represents a different type of supramolecular aggregation. Since the zinc ion is bound through the phenoxide rather than the pyridinic nitrogen, the –NH group is farther away from the framework oxygen centres and hence prefers intermolecular over intramolecular NH···O hydrogen bonds.

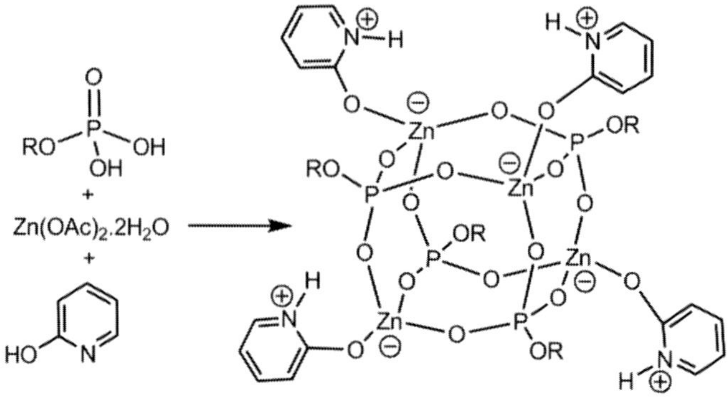

Scheme 11.5 Synthesis of [(ArO)PO_3Zn(2hpy)]$_4$ (**7**).

Thus, three out of the four –NH groups form intermolecular hydrogen bonds with the oxygen of the aryloxides of three different adjacent cubanes. The fourth –NH group is content with an intramolecular NH···O interaction with a framework oxygen. This results in the propagation of hydrogen bonds between the cubes only from three terminals which are approximately displaced from each other by 120°. Thus, every D4R cube has three neighbours, which results in the eventual formation of an extended 2D sheet of D4R cubes with well-defined hexagonal channels (Figure 11.3). Additional CH···O hydrogen bonds between pyridyl CH and framework oxygen centres stabilize the hexagonal arrangement, which results in a huge macrocycle in the centre of each hexagon. Similar construction of 2D assembly **8** has also been achieved by employing 2-methyl imidazole as the exo-cubane ligand.

11.3.1.4 Three-dimensional Frameworks

The isolation of 2D grid structures **7** and **8** led to two different important conclusions to increase the dimensionality of the self-assembly: (a) it is important to increase the number of intermolecular hydrogen bonds over the number of intramolecular hydrogen bonds and (b) it is imperative to engage the acidic protons of all the four ancillary ligands on four different pyridinic centres in intermolecular hydrogen bonding interactions. One of the best ways to achieve this result is not only to move the H-donor site from ortho- to either meta- or para- positions but also place a spacer of appropriate length in between the aryl carbon and the –NH or –OH group.

Complexes [(ArO)PO_3Zn(pyridine-3-methanol)]$_4$ (**10**) and [(ArO)PO_3Zn (pyridine-4-methanol)]$_4$ (**11**), due to the relative position of the –CH_2OH groups on the pridyl ring, exhibit for the first time the desired 3D self-assembly construction of zinc phosphates (Figure 11.4; top). The presence of –CH_2OH on the meta and para positions of the pyridinic ligand systems helps the complete elimination of intra-cubane H-bonding.

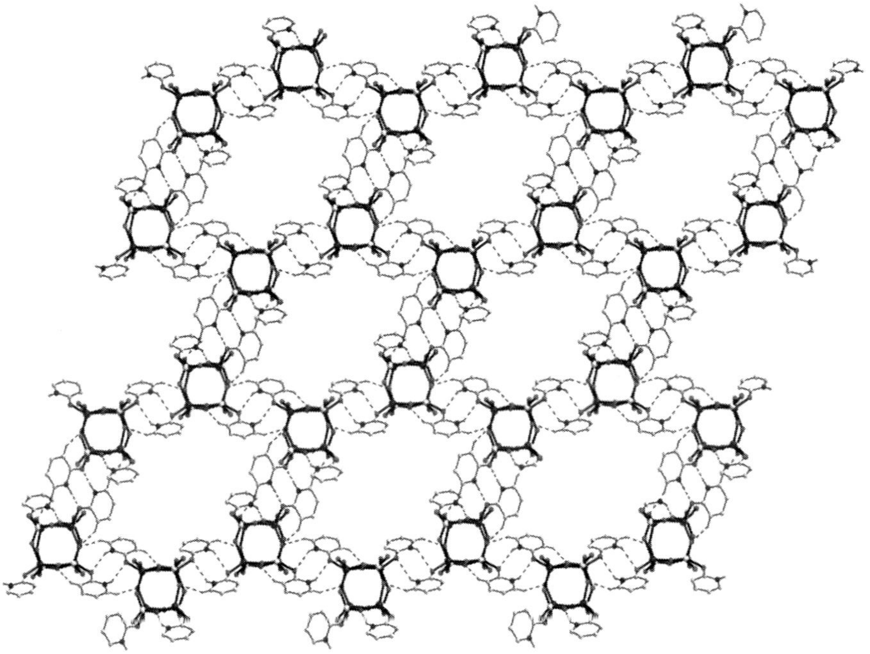

Figure 11.3 Two-dimensional assembly of zinc phosphate cubanes in **7**.

Since the use of pyridine-4-methanol was very useful in isolating the first examples of non-covalently linked porous zinc phosphate super structure, it was believed that the use of an even longer functional side chain on the pyridinic ligand would result in more interesting structures. Hence the reaction of **1** was carried out with readily available 3,5-dimethylpyrazole-2-ethanol and the cubane $[(ArO)PO_3Zn$ (3,5-dimethylpyrazole-2-ethanol)$]_4$ (**13**) was isolated. The molecular structure of **13** displays a unique 3D framework structure generated through the intermolecular interactions involving O–H···O hydrogen bonds, as shown in Figure 11.4 (bottom).

Strategies for arranging zinc phosphate SBUs into one-, two-, or three-supramolecular structures through simple non-covalent hydrogen-bonding interactions have thus been developed. These studies suggest that it is not only important to use an H-donor site on the pyridinic nitrogen, but also to place it on an appropriate position on the aryl ring (with a spacer). The structural determination of a variety of compounds further suggests that in order to build super structures with larger channels, it is important to reduce or eliminate the intramolecular hydrogen bonds. Further, the presence of functional groups on the surface of the clusters opens up new possibilities for further reactions with other metal alkyls (e.g. AlR_3) in order to build dendritic heterometallic super structures.

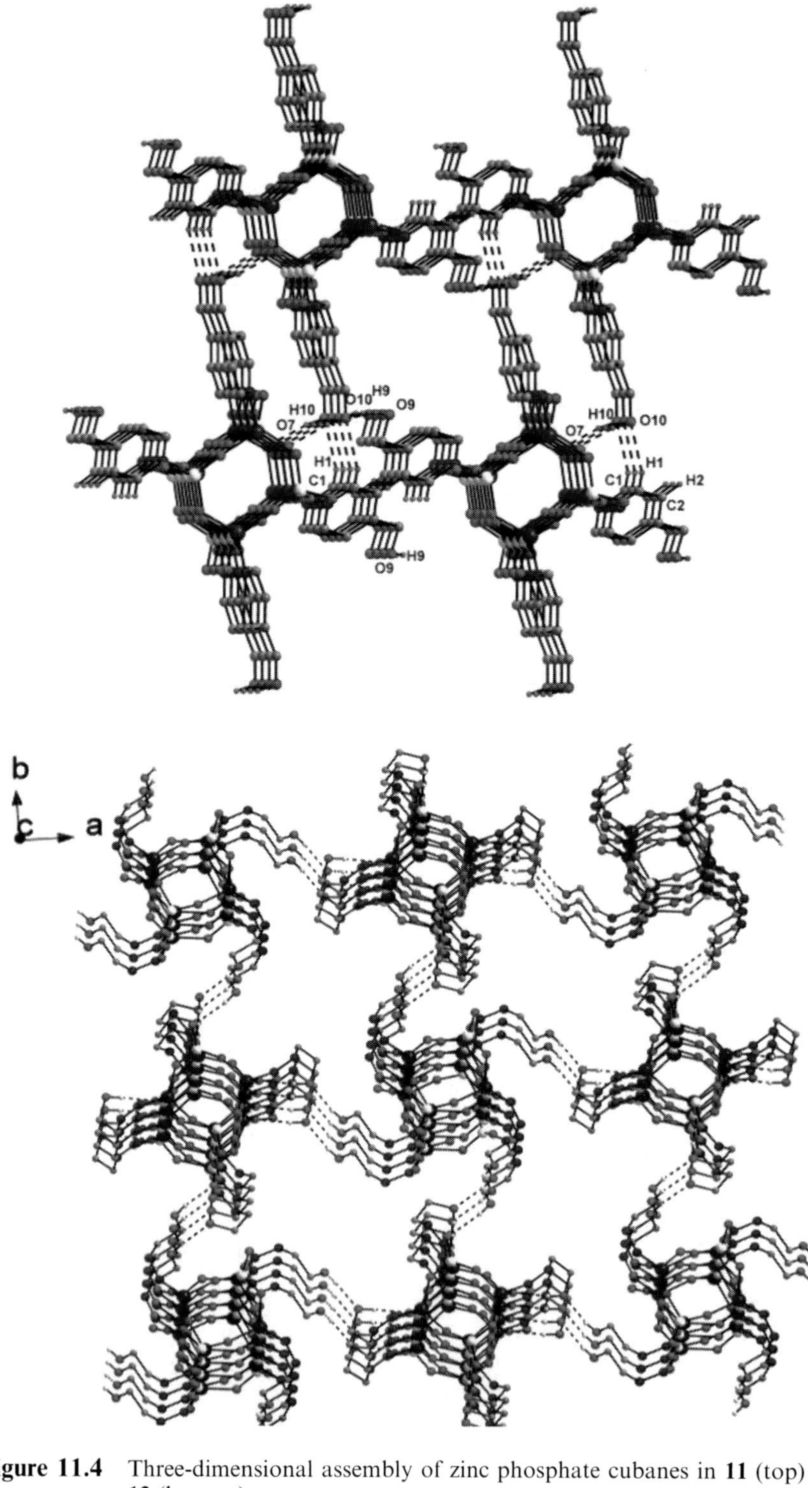

Figure 11.4 Three-dimensional assembly of zinc phosphate cubanes in **11** (top) and **13** (bottom).

11.3.2 Hierarchical Zinc Phosphates through Coordination

Although it would seem that the preparation and characterization of coordinatively linked D4R cubanes in the form of the 3D super structures is a fairly straightforward exercise, as shown in Scheme 11.6 (by the use of ditopic spacer), the isolation of fully characterizable products proved very difficult in practice. For example, addition of two equivalents of rigid ditopic spacer ligand 4,4′-bipyridine resulted in the formation of copious amounts of precipitate that is insoluble in almost all common organic solvents and water. Similarly, efforts to hierarchically build super structures using coordination chemistry, as described in Section 11.3.1, also failed since it was almost impossible to replace the coordinated methanol molecules from **1** in a stepwise fashion. This could be reasoned to the rather poor Lewis basic character of methanol compared to 4,4′-bipyridine.

In order to overcome this high reactivity of **1** in the presence of pyridines, compound **1** was transformed to relatively less reactive $[(ArO)PO_3Zn(DMSO)]_4$ (**19**) by the addition of excess DMSO to a solution of **1** in methanol.[11a] Large single crystals of **19** obtained from this reaction not only showed better solubility in many organic solvents, but also were found to be more amenable for stepwise replacement of coordinated DMSO by ditopic 4,4′-bipyridine.

For instance, the reaction of **19** with 2 equivalents of 4,4′-bipyridine is expected to yield once again an insoluble 3D framework. However, its reaction with 1.5 equivalents of 4,4′-bipyridine should produce a 2D sheet-like structure where each of the D4R cubanes still retains one of the coordinated DMSO molecules. The diffusion of 1.5 equivalents of 4,4′-bipyridine into a solution of **19** indeed produces $[((ArO)PO_3)_4Zn_4(4,4'\text{-bpy})_{1.5}(DMSO)]$ (**20**) in good yields (Scheme 11.7). The three free ends of the 4,4′-bpy ligands shown in Scheme 11.7 coordinate to three more D4R neighbours, leading to formation of a 2D sheet-like structure shown in Figure 11.5.

Although the synthesis of a true 3D coordinatively linked phosphate was initially difficult, as noted above, its eventual synthesis was finally achieved by

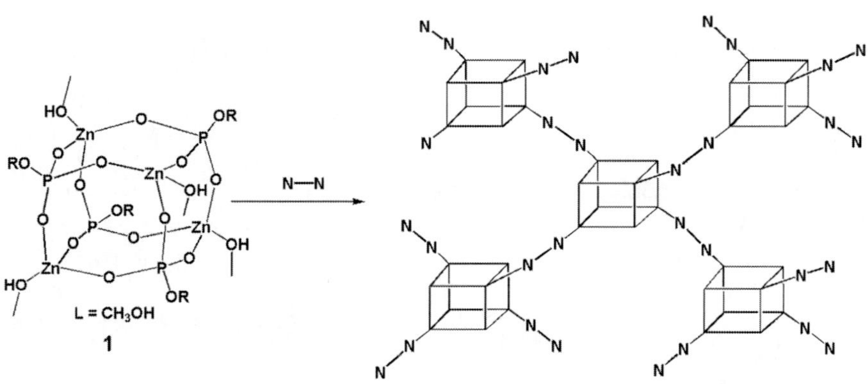

Scheme 11.6 Strategy for D4R coordination 3D framework structures.[6c]

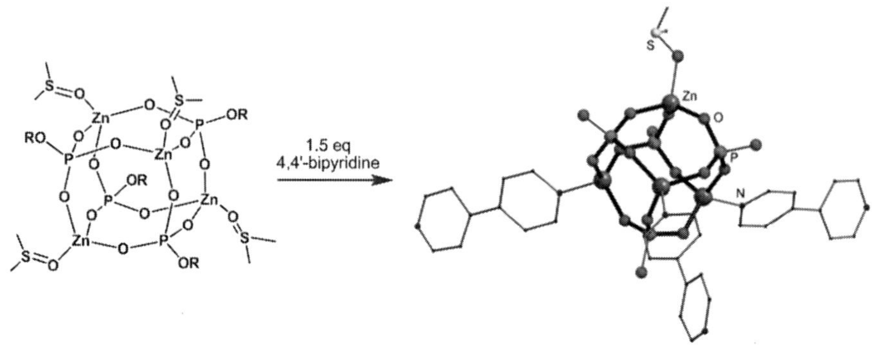

Scheme 11.7 Synthesis of $[((ArO)PO_3)_4Zn_4(4,4'\text{-bpy})_{1.5}(DMSO)]$ (**20**).[11b]

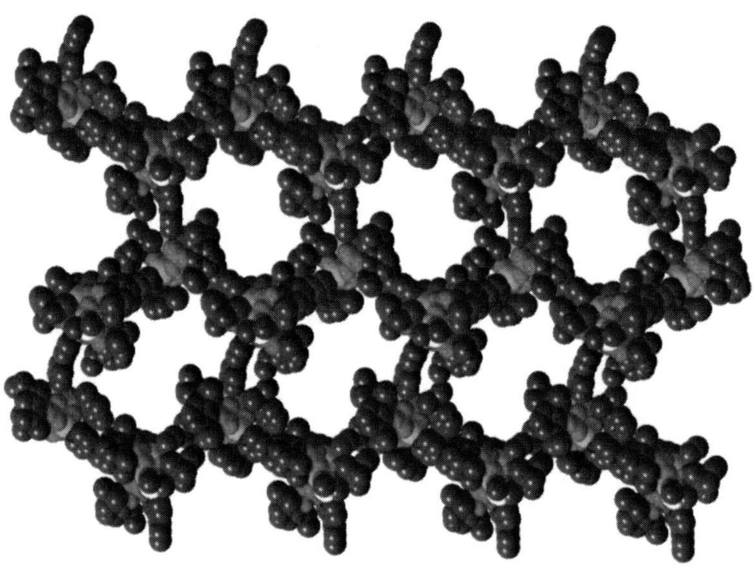

Figure 11.5 Coordinatively linked 2D assembly of zinc phosphate **20**.

doing reactions at infinite dilution and very slow diffusion rates, by simulating hydrothermal conditions at room temperature and normal pressure. One of the two structural isomers of the first 3D coordination framework based of D4R cubanes, $[(ArO)PO_3Zn(4,4'\text{-bpy})]_4$ (**21**). In the structure of **21**, each of the D4R cubanes are surrounded by four other D4R cubanes through four 4,4′-bpy ligands. The inter-cubane distances in the framework results in the formation of a $(D4R)_3$ units which then hexamerize to result in the final micro-mesoporous structure shown in Figure 11.6.

The successful synthesis of compounds such as **20** and **21** clearly demonstrates that even high molecular weight SBUs can be transformed into 3D

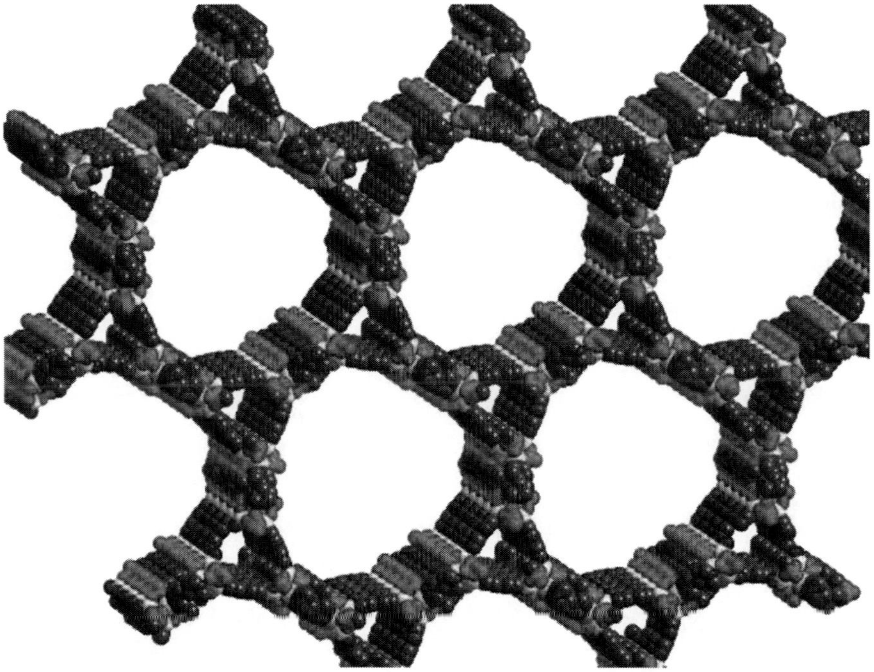

Figure 11.6 Covalent coordinatively linked 3D zinc phosphate **21**.

framework structures either by synthesis design (changing the precursor to **19** in case of the synthesis of **20**) or by very careful manipulation of the reaction conditions (as in the case of **21** and its isomers).[11b]

11.4 Molecular Zinc Phosphates from H_3PO_4 and Their Conversion to Higher Dimensional Structures

While all the above molecular phosphates were prepared starting from a phosphate monoester, there are a few examples of molecular phosphates synthesized from phosphoric acid in aqueous medium. Although under hydrothermal conditions the reactions of phosphoric acid with metal ions generally result in extended open framework structures, it is possible to isolate molecular zero-dimensional metal phosphates.[12,13] A zero-dimensional tin phosphate comprising a four-membered ring was isolated a few years ago.[12] Similar molecular zinc phosphates were isolated by carrying out the reaction between zinc oxide, phosphoric acid, and an amine at 80 °C.[13a] Another four-membered ring zinc phosphate was obtained by the reaction between a zinc salt, phosphoric acid, and the amine at room temperature over an extended period of time.[13b] This four-membered ring phosphate could also be obtained by the reaction of the amine phosphate with the zinc salt. What is interesting about these four-membered ring metal phosphates is their close resemblance to the S4R secondary building unit.

It has been shown recently that the four-membered ring zinc phosphate $[C_6N_2H_{18}][Zn(HPO_4)(H_2PO_4)_2]$ (**22**) transforms to 1D, 2D, and 3D structures under relatively mild conditions (Figure 11.7).[14,15] An especially noteworthy recent observation is the transformation of a four-membered ring zinc phosphate to a linear chain structure at room temperature on reaction with

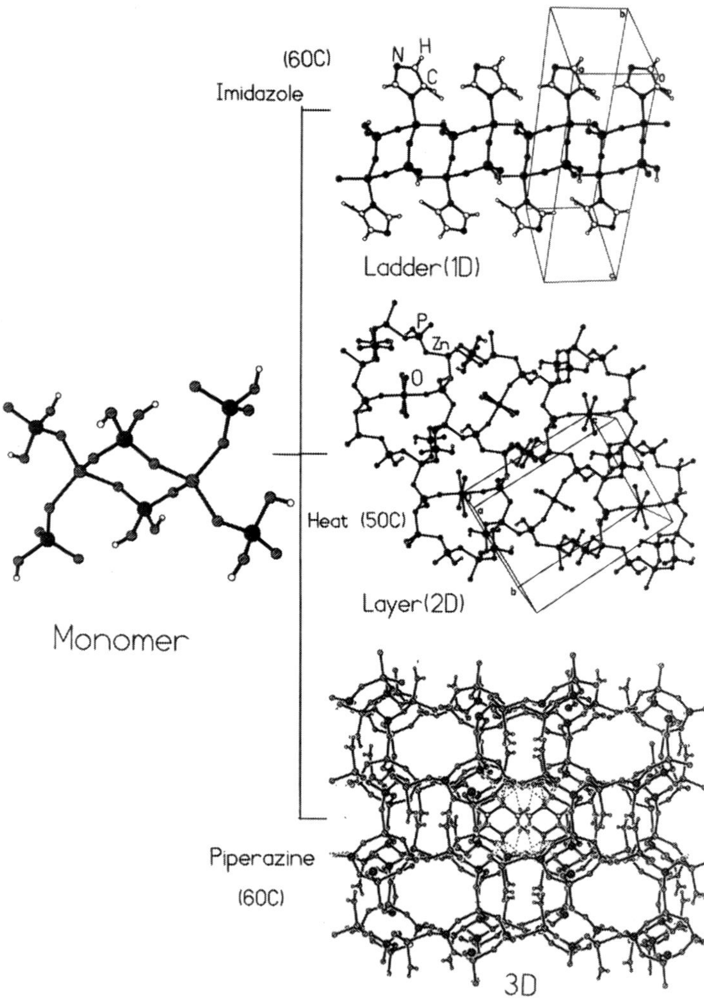

Figure 11.7 Transformation of a zero-dimensional zinc phosphate to 1D (ladder), 2D (layer), and 3D structures. Reproduced from ref. 2b with permission from American Chemical Society.

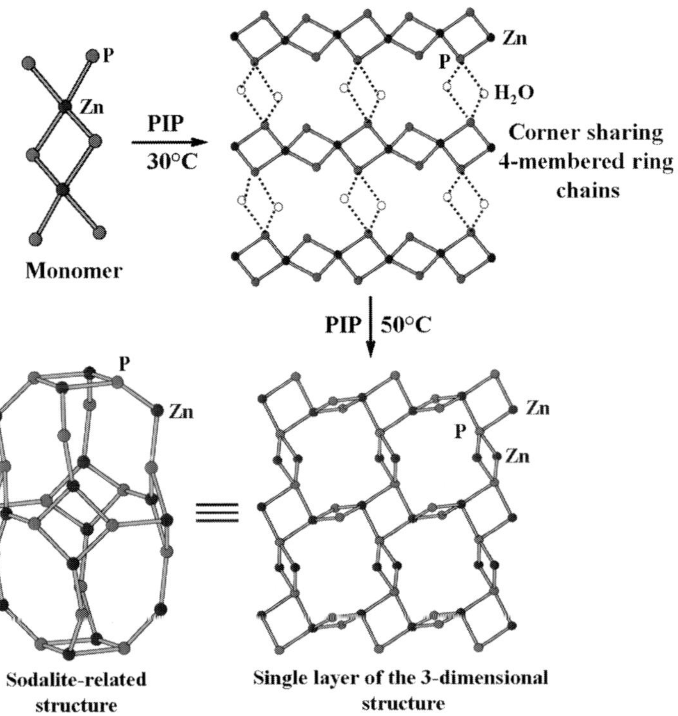

Figure 11.8 Transformation of a S4R SBU to a linear chain, and the transformation of the later to a 3D sodalite-related structure. Reproduced from ref. 2b with permission from American Chemical Society.

piperazine. The linear chain then transforms to a sodalite-related structure with a $4^6 8^8$ cage at 50 °C (Figure 11.8).[16] A transformation of metal squarates to sodalite structures has been observed recently,[17] showing thereby that the four-membered ring motif can generally act as a SBU.

11.5 Other Transformations within Framework Zinc Phosphates

Interestingly, other than **22**, 1D phosphates with linear chains or ladder structures can also be transformed to 2D and 3D structures, while the layered 2D structures transform to 3D structures.[18,19] These studies show that the formation of complex 3D structures may involve a progressive building up process, as indeed verified by synchrotron studies.[20] Acid degradation of 3D zinc phosphates with channel structures have been found to yield low-dimensional structures, showing thereby the reversibility of the 1D ↔ 2D ↔ 3D transformation process.[21] Some of the details of these transformations are described below.

1D metal phosphates can possess either linear chain or ladder structures. The linear chain zinc phosphates do not have pendant phosphate groups, and hence were found to be quite unreactive, unlike the 1D ladder structures with pendant phosphate groups.[14,18] Hence it is easier to carry out dimensionality transformations from the ladder structures under different conditions. On heating the ladder structure, $[C_6N_4H_{22}]_{0.5}[Zn(HPO_4)_2]$ (**23**), in water (**23**:H_2O = 1:100) at 150 °C for 100 h, the 3D structure $[C_6N_4H_{22}]_{0.5}[Zn_3(PO_4)_2(HPO_4)]$ (**24**) was obtained. The structure of **24** is built up of ZnO_4 and PO_4 tetrahedra sharing vertexes, forming a 3D structure (Figure 11.9), the connectivity between the ZnO_4 and PO_4 units giving rise to 16-membered 1D channels along the *bc* plane. However, by heating **23** in the presence of a base such as piperazine (PIP) or triethylenetetramine (TETA) at 150 °C in water (**23**:amine:H_2O = 1:5:100) for relatively short periods (24 h), another 3D structure, $[C_6N_4H_{22}]_{0.5}[Zn_2(PO_4)_2]$ (**25**) (Figure 11.9) was obtained. The structure of **25** is also built up of ZnO_4 and PO_4 tetrahedra, but the connectivity between these units creates eight-membered channels along all the crystallographic directions.

The same reaction carried out at a slightly higher temperature (165 °C) gave a new 3D structure, $[C_2N_2H_{10}]$ $[Zn_2(PO_4)_2]$ (**26**) (Figure 11.9). The connectivity between ZnO_4 and PO_4 tetrahedra in **26** results in four-membered rings, which are connected to each other via oxygens. The connectivity between the four-membered rings is such that they form an edge-shared four-ring ladder. The ladders are then connected together within and out of the plane, forming eight-membered channels along the *a* axis. Along the *c* axis, the connectivity between the tetrahedra gives rise to another eight-membered channel. It is noteworthy that all the three 3D structures, **24**, **25**, and **26**, obtained by the transformation of **23** contain ladder-like features of the starting material. By heating the ladder structure **23** in the presence of PIP at 165 °C, one can also obtain a 2D layer compound, $[C_4N_2H_{10}][Zn_2(PO_4)_2]$ (**27**).

An example for the transformation of 2D layered zinc phosphates to 3D structures is the conversion of the layer structure $[C_6N_4H_{22}]_{0.5}[Zn_2(HPO_4)_3]$ (**28**)[21b] (Figure 11.10) in water at 150 °C (**28**:H_2O = 1:200) to give the 3D structure **24** with 16-membered channels. It must be recalled that we could obtain this 3D structure from the ladder structure, **23**, as well. Similarly heating the tubular layer phosphate obtained with triethylenetetramine (TETA), $[C_6N_4H_{22}]_{0.5}[Zn_3(PO_4)_2(HPO_4)]$ (**29**)[21b] (Figure 11.10), at 150 °C in water (**29**:H_2O = 1:100), produced the 3D structure, **25**.

The chemistry described above in Sections 11.4 and 11.5 presents reasons to believe that the zero-dimensional monomers, comprising four-membered rings, are the most basic structural units of the open-framework phosphates and that after an optimal precursor state, such as the ladder structure, is formed, further building may occur spontaneously. Evidence for the occurrence of self-assembly in the formation of complex structures is provided by the presence of the structural features of the 1D starting material in the final products. These observations constitute the beginning of our understanding of the building-up principle of such complex structures.

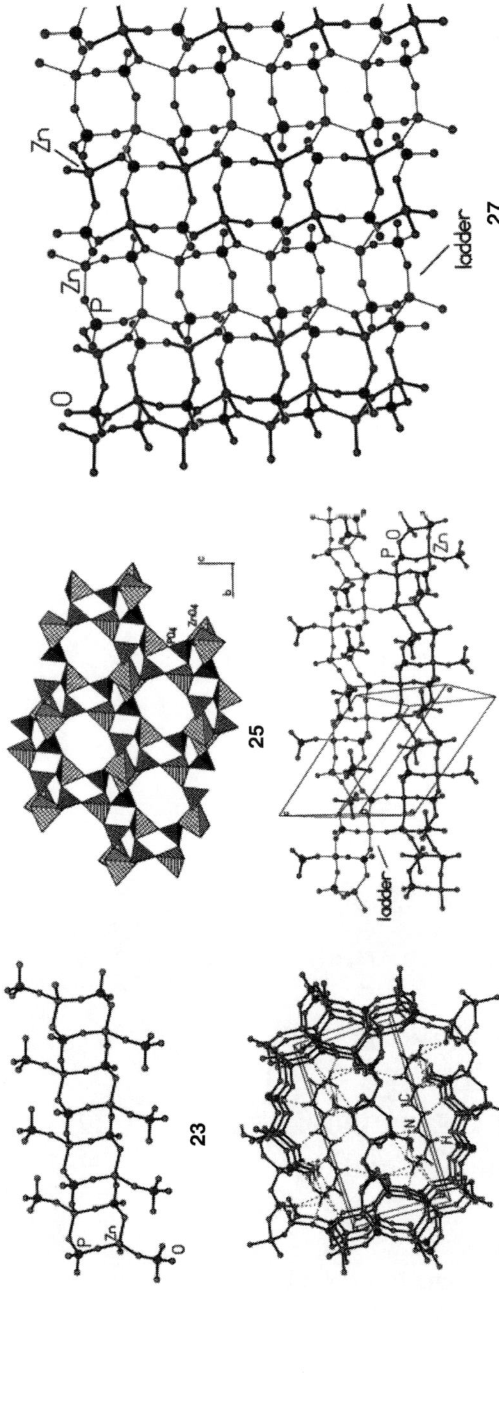

Figure 11.9 Structures of **23**–**27**. Compound **23** is a 1D layer phosphate, which either heated alone or in the presence of added amines transforms to 3D framework phosphates **24**–**26** and 2D layered phosphate **27**. Details in the text and also in ref. 14.

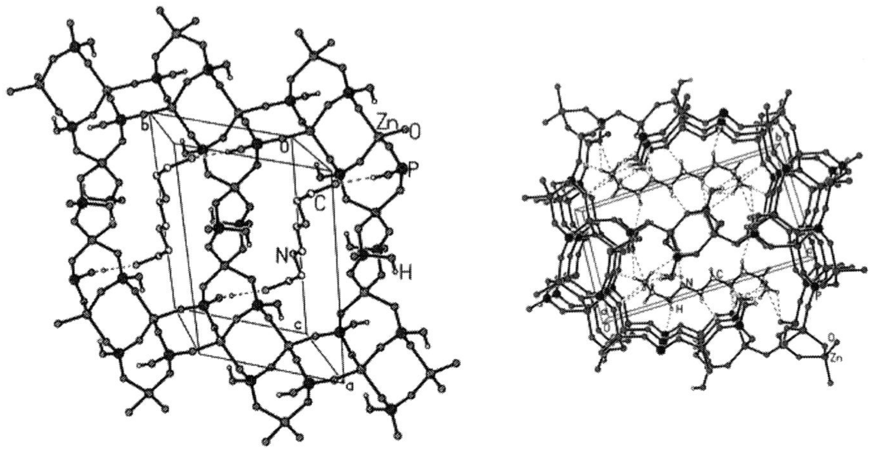

Figure 11.10 Structures of **28** and **29**. Compounds **28** and **29** transform to **24** and **25**, respectively.

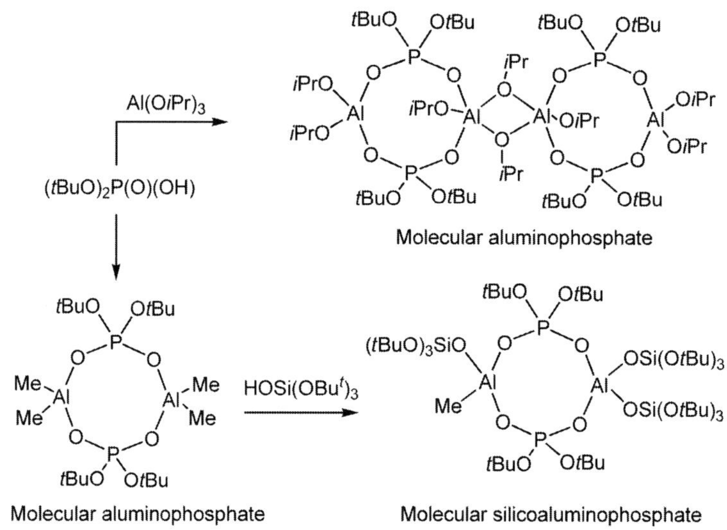

Scheme 11.8 Synthesis of molecular aluminophoshate precursors.

11.6 Framework Aluminophosphates from a Molecular Precursor

In the case of aluminium phosphates, Ozin *et al.*[22] have pointed out the possibility of the transformation of a chain structure to other chains or to layered and 3D open framework structures through a hydrolysis–condensation self-assembly pathway. We have recently carried out a few interesting molecule–material transformations in aluminophosphates. Thus, the centrosymmetric, tetrameric molecular aluminophosphate (described by Tilley *et al.*[23] for the preparation of fine particle AlPOs and SAPOs; see Scheme 11.8) transforms to a layered structure[24] of composition $[NH_3(CH_2)_2NH_3][Al_2(OH)_2(PO_4)_2(H_2O)]2H_2O$ on reaction with ethylenediamine in aqueous medium at 150 °C, as shown in Figure 11.11. Interestingly, this structure can also be obtained as the transformation product of a 1D ladder compound of composition $[NH_3(CH_2)_2NH_3][AlP_2O_8H]$.[25] This layered structure transforms to another layered structure by loss of water molecules, followed by

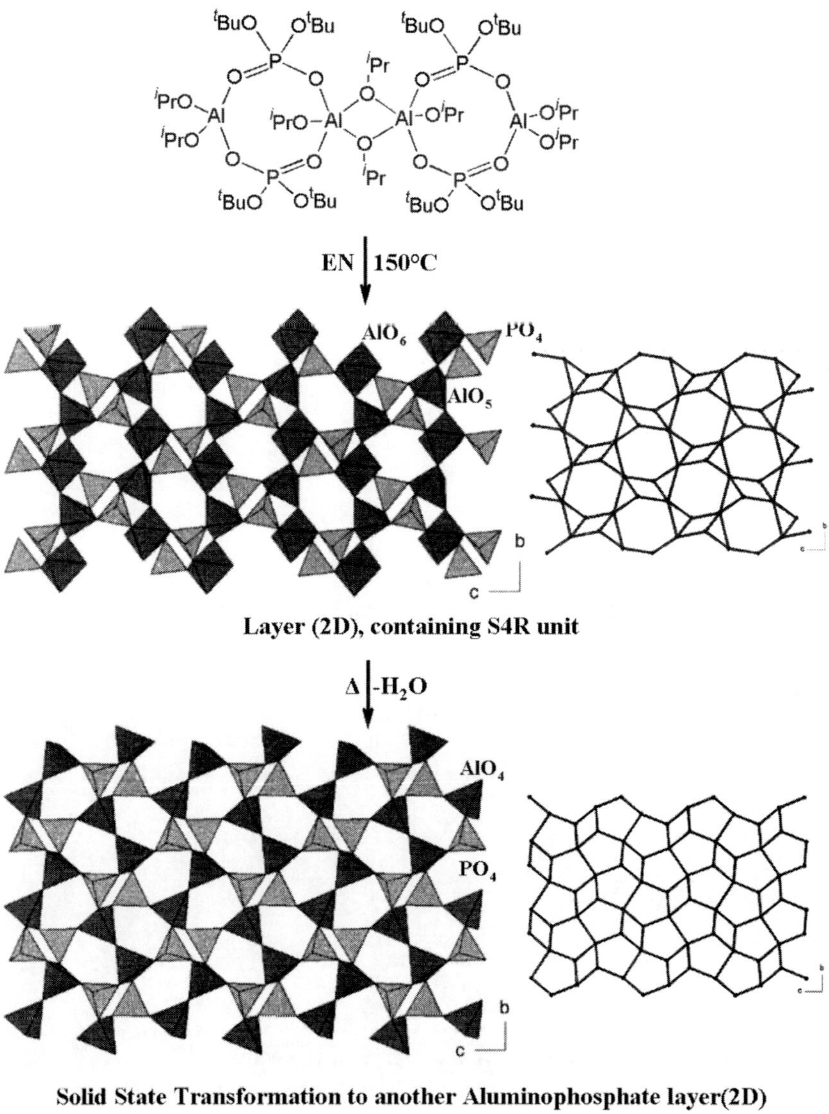

Figure 11.11 Transformations of a S4R aluminophosphate.

reorganization of the aluminophosphate layer (Figure 11.11). The molecular aluminium phosphate compound gives rise to a 3D structure with channels on reacting with piperazine in aqueous medium at 150 °C. These studies reveal that other types of aluminophosphates could be obtained by exploring reactions of this molecular aluminium phosphate precursor and similar compounds in various other media.

11.7 Conclusions

The discussion in the previous sections clearly testifies to the importance of the transformations of molecular compounds to materials. Studies of such transformations are only making a beginning, and there appears to be a great future for such investigations. While the isolation of organic soluble molecular zinc phosphates with a D4R core is an important milestone in this direction, the utility of this precursor under ambient conditions (room temperature and normal pressure) to build hierarchical structures either by using hydrogen bonding or by coordination interactions showcases the potential of the building-block approach to extended and porous solids. Similarly, the recent synthesis of a sodalite-related structure from a molecular zinc phosphate is noteworthy, besides the transformation of the centrosymmetric, tetrameric, molecular aluminophosphate to open framework structures. Finally, it would be rewarding to pursue investigations on the transformation of various molecular precursors including those involving aluminophosphate and silicate motifs.

References

1. H. Goldwhite, *Introduction to Phosphorus Chemistry*, Cambridge University Press, Cambridge, UK, 1981.
2. (a) R. Murugavel, A. Choudhury, R. Pothiraja, M. G. Walawalkar and C. N. R. Rao, *Chem. Rev.*, 2008, **108**, 3549; (b) R. Murugavel, M. G. Walawalkar, M. Dan, H. W. Roesky and C. N. R. Rao, *Acc. Chem. Res.*, 2004, **37**, 763.
3. S. T. Wilson, B. M. Lok, C. A. Messina, T. R. Cannan and E. M. Flanigen, *J. Am. Chem. Soc.*, 1982, **104**, 1146.
4. A. Clearfield, *Prog. Inorg. Chem.*, 1998, **47**, 371.
5. M. G. Walawalkar, H. W. Roesky and R. Murugavel, *Acc. Chem. Res.*, 1999, **32**, 117.
6. K. J. Martin, P. J. Squattrito and A. Clearfield, *Inorg. Chim. Acta*, 1989, **155**, 7.
7. (a) R. Murugavel, S. Kuppuswamy, R. Boomishankar and A. Steiner, Angew. Chem., *Int. Ed. Engl.*, 2006, **45**, 5536; (b) R. Murugavel, S. Kuppuswamy, N. Gogoi, R. Boomishankar and A. Steiner, *Chem. Eur. J.*, 2010, **16**, 994; (c) R. Murugavel, S. Kuppuswamy, N. Gogoi, R. Boomishankar and A. Steiner, *Inorg. Chem.*, 2010, **49**, 2153.
8. C. G. Lugmair and T. D. Tilley, *Monatsh. Chem.*, 2006, **137**, 557, and references cited therein.

9. R. Pothiraja, M. Sathiyendiran, R. J. Butcher and R. Murugavel, *Inorg. Chem.*, 2005, **44**, 6314.
10. (a) R. Murugavel and S. Kuppuswamy, *Inorg. Chem.*, 2008, **47**, 7686; (b) R. Murugavel, S. Kuppuswamy and S. Randoll, *Inorg. Chem.*, 2008, **47**, 6028; (c) R. Murugavel, S. Kuppuswamy, N. Gogoi, A. Steiner, R. Boomishankar and K. G. Suresh, *Chem. Asian J.*, 2009, **4**, 143; (d) R. Murugavel, N. Gogoi, K. G. Suresh, S. Layek and H. C. Verma, *Chem. Asian. J.*, 2009, **4**, 923; (e) R. Murugavel, S. Kuppuswamy, A. N. Maity and M. P. Singh, *Inorg. Chem.*, 2009, **48**, 183; (f) R. Murugavel and S. Kuppuswamy, *Angew. Chem., Int. Ed. Engl.*, 2006, **45**, 7022; (g) R. Murugavel and S. Kuppuswamy, *Chem. Eur. J.*, 2008, **14**, 3869; (h) R. Murugavel, S. Shanmugan and S. Kuppuswamy, *Eur. J. Inorg. Chem.*, 2008, 1508; (i) R. Murugavel and S. Shanmugan, *Organometallics*, 2008, **27**, 2784.
11. (a) S. Kuppuswamy, Ph.D. thesis, IIT Bombay 2008; (b) R. Murugavel, N. Gogoi, A. C. Kalita, S. Kuppuswamy, unpublished results.
12. S. Ayyappan, A. K. Cheethan, S. Natarajan and C. N. R. Rao, *J. Solid State Chem.*, 1998, **139**, 207.
13. (a) W. T. A. Harrison and L. Hannooman, *J. Solid State Chem.*, 1997, **131**, 363; (b) S. Neeraj, S. Natarajan and C. N. R. Rao, *J. Solid State Chem.*, 2000, **150**, 417.
14. C. N. R. Rao, S. Natarajan, A. Choudhury, S. Neeraj and A. A. Ayi, *Acc. Chem. Res.*, 2001, **34**, 80.
15. A. A. Ayi, A. Choudhury, S. Natarajan, S. Neeraj and C. N. R. Rao, *J. Mater. Chem.*, 2001, **11**, 1181.
16. M. Dan, D. Udayakumar and C. N. R. Rao, *Chem. Commun.*, 2003, 2212.
17. S. Neeraj, M. L. Noy, C. N. R. Rao and A. K. Cheetham, *Solid State Sci.*, 2002, **4**, 1231.
18. (a) A. Choudhury, S. Neeraj, S. Natarajan and C. N. R. Rao, *J. Mater. Chem.*, 2001, **11**, 1537; (b) A. Choudhury, S. Neeraj, S. Natarajan and C. N. R. Rao, *J. Mater. Chem.*, 2002, **12**, 1044.
19. R. I. Walton, F. Millange, A. Le Bail, T. Loiseau, C. Serre, D. O'Hare and G. Férey, *Chem. Commun.*, 2000, 203.
20. (a) R. I. Walton, A. J. Norquist, S. Neeraj, S. Natarajan, C. N. R. Rao and D. O'Hare, *Chem. Commun.*, 2001, 1990; (b) R. J. Francis, S. O'Brein, A. M. Fogg, P. S. Halasyamani, D. O'Hare, T. Loiseau and G. Ferey, *J. Am. Chem. Soc.*, 1999, **121**, 1002.
21. (a) A. Choudhury and C. N. R. Rao, *Chem. Commun.*, 2003, 366; (b) A. Choudhury, S. Natarajan and C. N. R. Rao, *Inorg. Chem.*, 2000, **39**, 4295.
22. S. Oliver, A. Kuperman and G. A. Ozin, *Angew. Chem., Int. Ed. Engl.*, 1998, **37**, 46.
23. C. G. Lugmair, T. D. Tilley and A. L. Rheingold, *Chem. Mater.*, 1999, **11**, 1615.
24. K. O. Kongshaug, H. Fjellvag and K. P. Lillerud, *J. Mater. Chem.*, 1999, **9**, 1591.
25. I. D. Williams, J. Yu, Q. Gao, J. Chen and R. Xu, *Chem. Commun.*, 1997, 1273.

CHAPTER 12

Structural Studies of Paramagnetic Molecular Phosphonates

KANDASAMY GOPAL, SHOAIB ALI AND RICHARD E. P. WINPENNY

School of Chemistry, The University of Manchester, Oxford Road, Manchester M13 9PL, UK

12.1 Introduction

The report that a dodecametallic manganese cage shows slow relaxation of magnetization[1] led to a huge increase in the study of polymetallic complexes containing paramagnetic metal ions; the $\{Mn_{12}\}$ cage in question is the first and the best studied 'single molecule magnet' (SMM). Since the first reports of SMMs in the early 1990s, many further SMMs have been discovered.[2] The key parameters for an SMM is a high ground state spin, S, and that that spin should have easy-axis type anisotropy, parameterized as a large negative zero-field splitting D. As attempts were made to design better SMMs, it became apparent that control of both spin and anisotropy were vital – very high spin compounds with no spin anisotropy are not SMMs.[3]

Phosphonates have been studied for many years as ligands that readily produce 2D structures (see early chapters in this book), and we reasoned that they could be ideal for making anisotropic metal cage complexes if we could prevent the formation of extended lattices. Several possible strategies for making molecular paramagnetic phosphonates suggested themselves. Here we discuss results from two of these paths. We restrict ourselves largely to structural descriptions, only highlighting magnetic properties where they are worth further investigation.

Metal Phosphonate Chemistry: From Synthesis to Applications
Edited by Abraham Clearfield and Konstantinos Demadis

Published by the Royal Society of Chemistry, www.rsc.org

12.1.1 Nomenclature

Phosphonic acids of general formula R-PO_3H_2 have been used in synthetic inorganic chemistry for many years; the coordination networks made using these ligands are reviewed in earlier chapters of this book. The R- group is an organic component and can be anything from as simple as methyl (Me-PO_3H_2) to bulky 2,4,6-tri*iso*propylphenylphosphonic acid, (2,4,6-*i*-Pr_3-C_6H_2-PO_3H_2) or triphenylmethylphosphonic acid (Ph_3C-PO_3H_2, tritylphosphonic acid). The phosphonic acid molecule can be doubly deprotonated to form the di-anion, R-PO_3^{2-}, which now has three identical oxygen donor atoms to bind to many metal centres. In principle a phosphonate ligand can bind up to nine metal ions through the nine lone pairs of the three oxygen donor atoms.

In order to describe the coordination mode of such high denticity ligands, we use Harris notation.[4] Harris notation describes the binding mode as [X.Y_1Y_2...Y_n] where X is the overall number of metal bound by the whole ligand, and each value of Y refers to the number of metal atoms attached to the different donor atoms. Therefore, it will be three values of Y for each R-PO_3^{2-} unit. The ordering of Y is listed by the Cahn–Ingold–Prelog priority rules; some of the common binding modes are given in Figure 12.1.

12.2 Strategies for Making Molecular Phosphonates

When compared to layered metal phosphonates, molecular phosphonates were few in number until recently. The ability of phosphonates to form coordination polymers, which are (in the main part) insoluble, is also the reason fewer polymetallic molecular species existed. To overcome this difficulty, five general methods are widely used for the synthesis of molecular metal phosphonates.

12.2.1 Solvothermal/Hydrothermal Method

Zubieta and co-workers have shown that under solvothermal/hydrothermal conditions, the insoluble polymers of metal phosphonates were going into the

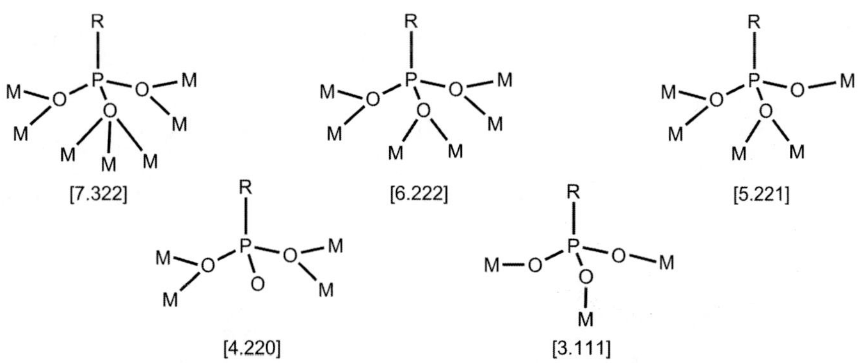

Figure 12.1 Examples of some common binding modes for the phosphonate ligands with the Harris notation for each mode below.

solution that gives the molecular cages of phosphonates. By this strategy they have reported many numbers of vanadium phosphonate cages with different nuclearities. This work has been reviewed[5] in 1995 and 2003 and we do not review this work here.

12.2.2 Alkyl Elimination Method

Roesky and co-workers, and others, have demonstrated the use of metal alkyls in constructing molecular phosphonates, largely of the main group metals and metalloids.[6] This strategy is very successful in isolating a variety of metal phosphonates containing boron, aluminium, gallium, indium, bismuth, titanium, zinc and cadmium. This work has been reviewed in 1999,[6] and covered in the previous chapter of this book, and as the compounds are all diamagnetic we do not discuss the work further here.

12.2.3 Bulky Phosphonic Acid Method

The use of bulky organo groups on phosphonic acids have been successfully employed to prepare a small number of molecular metal phosphonates. With $Ph_3C\text{-}PO_3H_2$, a series of tetranuclear transition metal phosphonates ($V^{III}{}_4$, $Mn^{II}{}_2Mn^{III}{}_2$, $Co^{II}{}_4$ and $Cu^{II}{}_4$)[7] was made. The cobalt complex $[Co(O_3PCPh_3)(py)]_4$ **1** is particularly interesting as it was the first example of a cage which resembled the D4R secondary building unit of zeolites (Figure 12.2); since that report other cubanes have been reported with cobalt(II) and group 10 metals using *t*-butylphosphonate,[8] and with 2,4,6-*i*-$Pr_3\text{-}C_6H_2\text{-}PO_3H_2$, a $\{Cu^{II}{}_4\}$ phosphonate has been reported.[9] It appears that the bulky units on the phosphonates restrict the nuclearity of the cages by preventing oligomerization and this approach seems to favour tetrametallic cages, at least for d-block metals.

12.2.4 Ancillary Ligand or Co-ligand Method

Chandrasekhar and co-workers introduced the idea of using an ancillary ligand or co-ligand for the construction of molecular metal phosphonates.[10] The co-ligands restrict oligomerization by occupying some of the metal coordination sites and hence enhance the solubility of the resultant metal phosphonate cages. There are a number of co-ligands used in this regard, such as pyrazoles, pyridine, 2,2′-bipyridine, 1,10-phenanthroline, 6-chloro-2-pyridinol and different carboxylates.

12.2.5 Cluster Expansion Method

An alternative approach involves the reaction of preformed polymetallic cages with phosphonates. Initially this work used oxo-centred trimetallic carboxylates[11,12] such as mixed valent $Mn^{III}{}_2Mn^{II}$ or homovalent $Fe^{III}{}_3$ complexes of

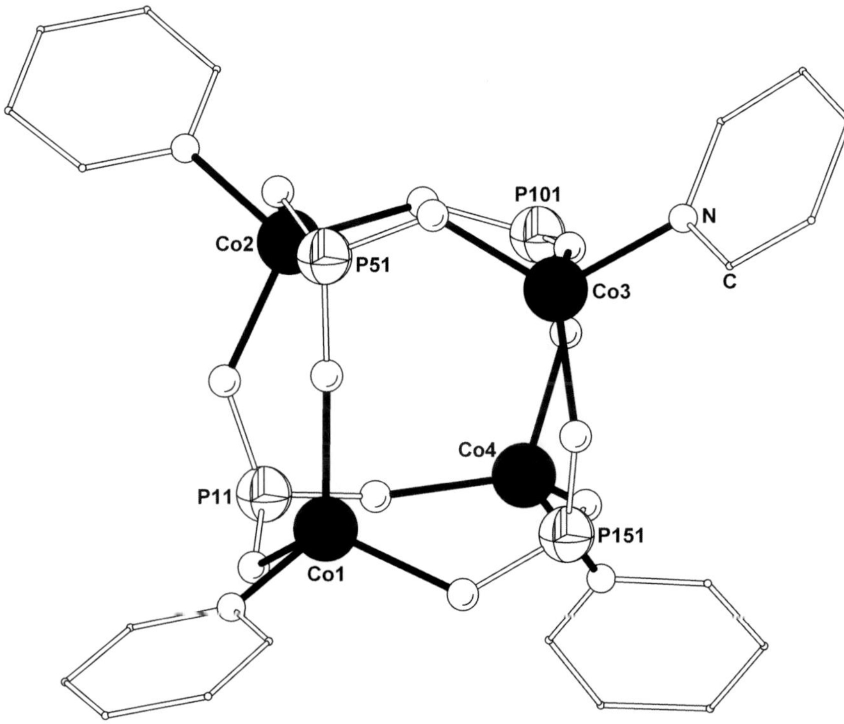

Figure 12.2 The structure of compound **1** in the crystal. Lightly shaded circles are oxygen atoms, other atoms are as labels. The organic side-chains of phosphonates and carboxylates are excluded from all figures for clarity.

general formula $[M_3O(O_2C\text{-}R)_6(OH_2)_3]X$, where R = -H, -Me, -Ph, *t*-Bu; M = Fe and X = NO_3^-, Cl^-; M = Mn and X = NO_3^-, Cl^-, ClO_4^- or nothing). The idea was that the phosphonate might displace the bridging carboxylate, and that the additional oxygen atom of the phosphonate ligand could then be used to bridge to a further metal centre leading to cluster expansion. The results often show that the triangular metal building block is retained, at least in part, in the final product. The difference between this approach and the co-ligand approach is simply that here the co-ligand is already present in the cage complex prior to reaction with phosphonates. In more recent work with V^{III} complexes,[13] isolation of the trimetallic vanadium carboxylates proved difficult, and we found that the best route was to add the carboxylates and phosphonates sequentially rather than isolating the intermediate compound.

12.3 Survey of Paramagnetic Molecular Paramagnetic Phosphonates

The first complex reported,[10] where a co-ligand was used to make large molecular phosphonates phosphonic acids, was the Cu_{12} phosphonate cage,

$[Cu^{II}_{12}(\mu_4\text{-}Cl)_4(\mu_3\text{-}Cl)_2(O_3P\text{-}t\text{-}Bu)_6(HO_3P\text{-}t\text{-}Bu)_2(DMPz)_4(DMPz\text{-}H)_6]$ **2** reported in 2000 by Chandrasekhar and Kingsley (Figure 12.3). It was synthesized by reaction of anhydrous $CuCl_2$, *t*-Bu-PO_3H_2 and 3,5-dimethylpyrazole (DMPz-H, as extra ligand called co-ligand) in the presence of Et_3N as base. The Chandrasekhar group has expanded this copper chemistry dramatically, making $\{Cu^{II}_2\}$, $\{Cu^{II}_3\}$, $\{Cu^{II}_4\}$, $\{Cu^{II}_6\}$ and $\{Cu^{II}_{16}\}$ cages.[9,14] The simplicity of the approach, and the potential to add any co-ligand, has led to a great deal of related work.

12.3.1 Cobalt Phosphonate Cages

We extended the co-ligand method, using 6-chloro-2-hydroxypyridine (chp-H) as co-ligand with phosphonates in appropriate mole ratios to prepare cobalt(II) cages; the unusual magnetic behaviour of 1D-, 2D- and 3D-cobalt phosphonates is discussed in Chapter 8 by Prof Zheng, and we hoped for similarly unusual magnetic behaviour. The first reaction we reported[11] gave Co_{13} cages $[Co^{II}_{13}(\mu_3\text{-}OH)_3(O_3P\text{-}Ph)_2(chp)_{19}(H_2O)_2(EtOAc)_2]$ **3** and $[Co^{II}_{13}(\mu_3\text{-}OH)_3(O_3P\text{-}Ph)_2(chp)_{19}(H_2O)_2(chp\text{-}H)_2]$ **4** with nearly identical core structures, with the only difference that EtOAc in **3** is replaced by chp-H in **4**; this difference is purely due to the solvent used for recrystallization. The structures are highly

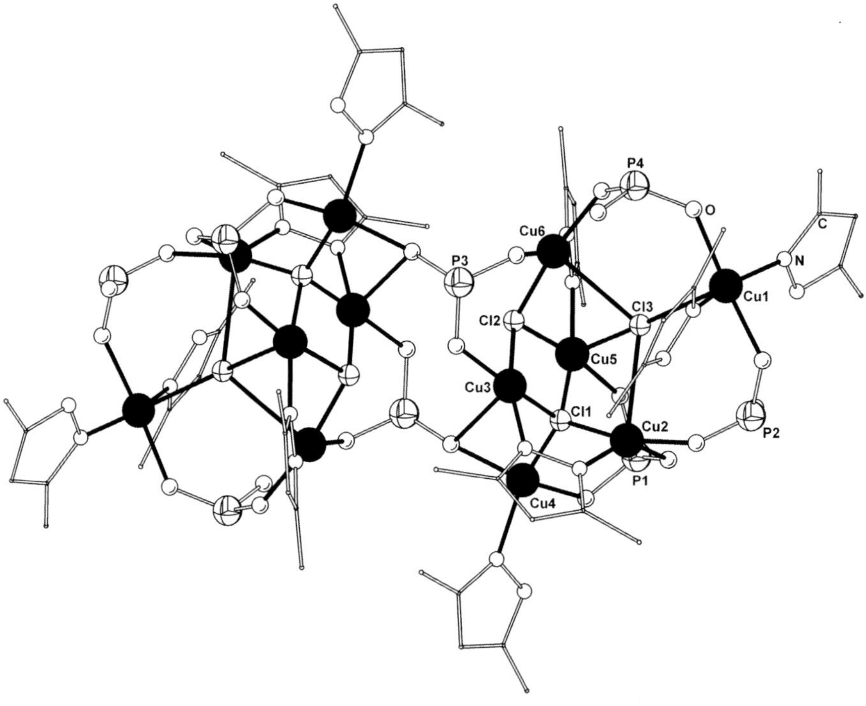

Figure 12.3 The structure of **2** in the crystal.

irregular and the phosphonates have bonding modes of [5.221] and [4.211] (Figure 12.4). The cage also contains three μ_3-OH ligands and all the cobalts are six coordinate. The magnetic susceptibility studies for compound **3** show a decline in χT value as the temperature is lowered, probably due to the usual crystal field effects of Co^{II} but anti-ferromagnetic exchange may also be present. This work showed features we have found for the molecular cobalt(II) phosphonates – extremely irregular and unpredictable structures, and an equally unpredictable dependence on reaction conditions, such as the solvent or counter-ions present.

If NaOMe is used to deprotonate the ligands prior to cobalt coordination, a hexanuclear cobalt cluster, $[Co^{II}_6Na_8(O_3P\text{-}Ph)_4(chp)_{12}(MeCN)_4]$ **5** results,[15] in which alkali metal ions are also included in the cage. Compound **5** contains four P atoms at alternate corners of a cube, with Na atoms at the other corners (Figure 12.5). The six Co^{II} ions are at each face of the central cube forming an octahedron of cobalt centres. The rest of the four Na atoms lie above four of

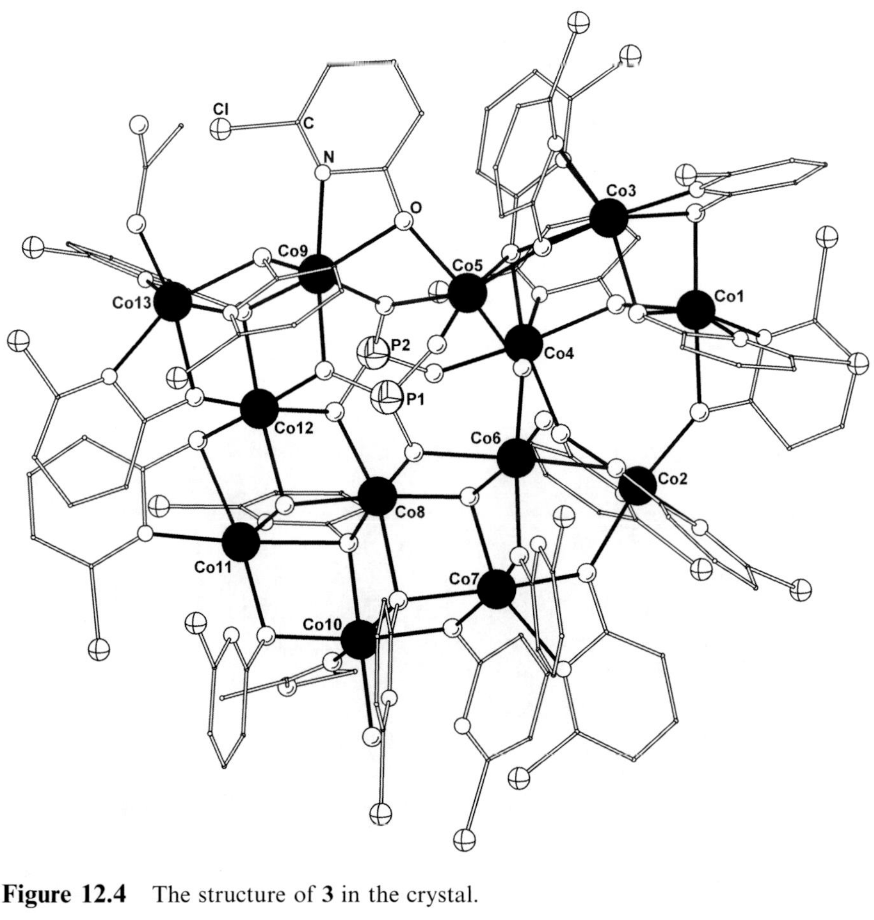

Figure 12.4 The structure of **3** in the crystal.

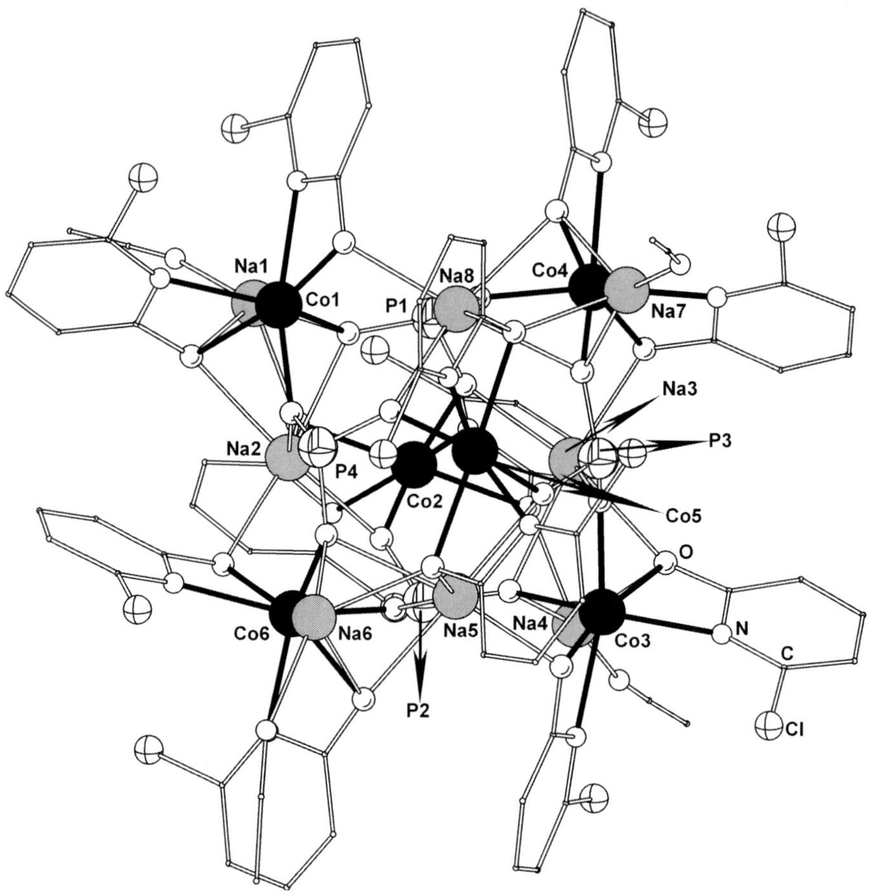

Figure 12.5 The structure of **5** in the crystal.

the edges of the cube. The binding mode of all the phosphonates in the complex is [7.322], which is unusual.

Similar reactions,[15] but using different combinations of cobalt salts and phosphonic acids, leads to the isolation of three further sodium-containing cobalt phosphonate cages, $[Co^{II}_{14}Na_4(O_3P\text{-}Ph)_6(chp)_{20}(chp\text{-}H)_2(OH_2)_8]$ **6**, $[Co^{II}_{13}Na_6(OH)_2(O_3P\text{-}CH_2Ph)_5(chp)_{20}(OH_2)(MeCN)(Me\text{-}OH)]$ **7** and $[Co^{II}_{15}Na(OH)_3(O_3P\text{-}Et)_3(chp)_{22}(chp\text{-}H)_2(OH_2)_3(Me\text{-}CN)_3]$ **8**. However, in the *t*-Bu-PO_3H_2 reaction we have isolated the Na free cage $[Co_{12}(F)_3(OH)_4(O_3P\text{-}t\text{-}Bu)(chp)_{15}]$ **9** in which the F^- ions came from decomposition of BF_4^- anion. These structures are all irregular; in **6** the cobalt ions form a distorted ring, with the sodium centres within the ring (Figure 12.6). Complex **9** can be deconstructed as a fragment of a cuboctahedron, but the simplest conclusion is that the structures are difficult to describe.

When we changed from alkali metal-based bases to Et_3N, eleven new cobalt cages were isolated by changing alkyl substituents on phosphonates and using

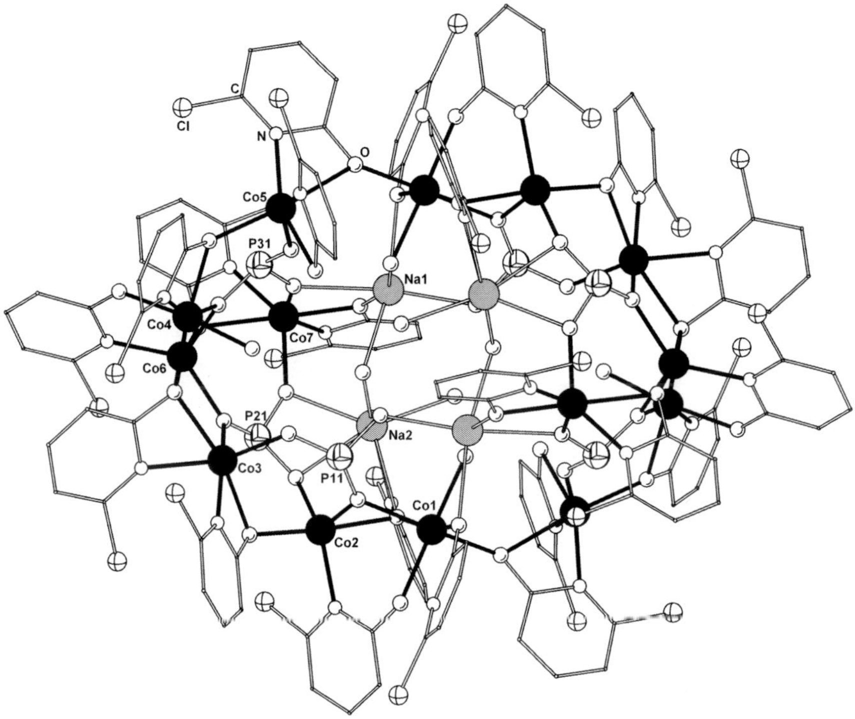

Figure 12.6 The structure of **6** in the crystal.

different cobalt salts.[16] The most common structure contains a tetradecanuclear cage of general formula $[Co^{II}_{14}(OH)_2X_2(O_3P\text{-}R)_2(chp)_{20}(OH_2)_2]$ where X = OH (**10** and **11**) or F (**12**–**14**), and R = -CH$_2$Ph, **10**; -Et, **11**;-Me, **12**; -*n*-octyl, **13**; -H **14** (Figure 12.7). These cages all have a stepped-layered structure, with two $\{Co_7\}$ fragments lying in a plane, but with a step between these two planes.

Other cages can be made by subtle variations in the reaction,[16,17] including: $[Co^{II}_{12}(OH)_2(HBO_3)(O_3P\text{-}CH_2Nap)(chp)_{18}(OH_2)]$ **15**, $[Et_3NH][Co^{II}_8(O_3P\text{-}Ph)_2(chp)_{10}(chp\text{-}H)_2(NO_3)_3]$ **16**, $[Co^{II}_{18}Co^{III}_2(O)_2(OH)_4(O_3P\text{-}Ph)_4(chp)_{26}(OH_2)_6]$ **17**, $[Co^{II}_6(O_3P\text{-}t\text{-}Bu)_2(chp)_8(chp\text{-}H)_2(OH_2)_2]$ **18**, $[Co^{II}_{10}(OH)_2(O_3P\text{-}CH_2CH_2\text{-}PO_3)(chp)_{14}]$ **19** and $[Co_{16}(OH)_6(O_3P\text{-}C_6H_9)_2(chp)_{22}(OH_2)_4]$ **20**. Compound **17** again contains planes of cobalt centres as in the tetradecanuclear cages, but the other cages have very different structures. Compound **15** is very irregular, containing distorted heterocubanes as fragments of the structure. The crystal structure of **16** shows that two P and four Co^{II} atoms lie on the vertices of a central trigonal prism. The rest of four Co^{II} atoms lie above the triangular faces of the prism. The bonding mode of phosphonates is [3.111] (Figure 12.8).

Compound **18** contains a planar array of six Co^{II} centres and consists of two $\{Co_3\}$ sections,[16] each bound to a 3.111 bridging phosphonate, with these two sections held together by bridging chp ligands The decanuclear cage **19** was

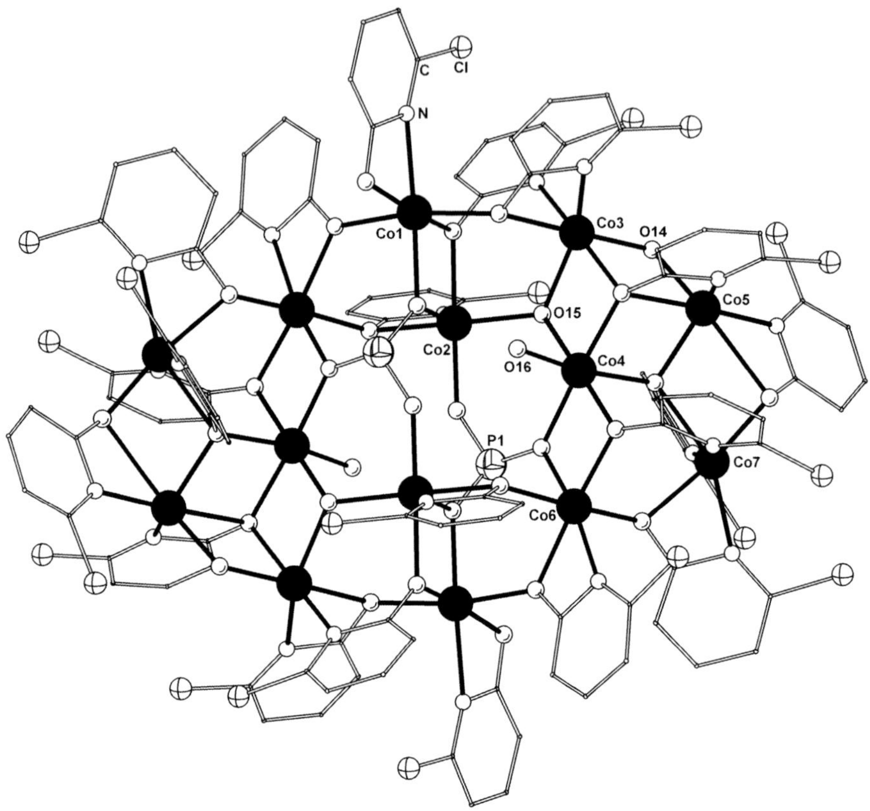

Figure 12.7 The structure of **10** in the crystal.

prepared by using ethylene diphosphonic acid, $H_2O_3P\text{-}CH_2CH_2\text{-}PO_3H_2$ in the place of mono-phosphonic acids. The structure shows that ten Co^{II} and two P lie on the vertices of an octahedron that is capped on four faces and two edges, giving an elongated hexagonal bipyramid. Both phosphonate end groups of the ethylene diphosphonate show the [5.221] binding mode which gives the overall binding mode of [8.221221], i.e. eight metal centres are bound to the diphosphonate with each oxygen bound to either two or one metal. The complex **20** has a cyclic structure,[17] with two 4.211 cyclohexylphosphonates holding together two $\{Co^{II}_8\}$ fragments.

If we consider the various cobalt–phosphonate cages we find that the structures are diverse and difficult to rationalize.[11,15–17] Compounds **3**, **4**, **10**–**14** and **17** contain planar arrays that can be related to cobalt hydroxide. In the other compounds it appears that the coordinative flexibility of cobalt(II) and the ligands, both phosphonates and chp, show multiple coordination modes in these complexes, and lead to little control in the types of structure observed.

Magnetic studies of these compounds have been performed.[15–17] In general there is a fall in the product $\chi_m T$ with temperature, and in the majority of cases

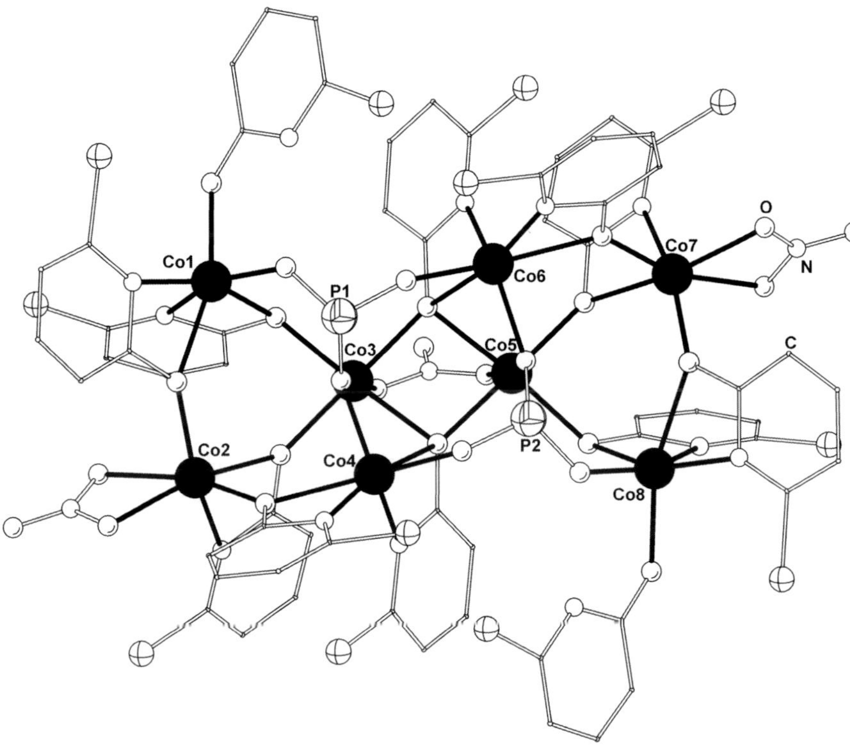

Figure 12.8 The structure of **16** in the crystal.

it appears that the ground state is diamagnetic. In some cases – compounds **7**, **8**, **9**, **15**, **16** and **18** – a low temperature maximum is found in the plot of $\chi_m T$ against *T*; this maximum may indicate a ground state with a significant spin; however, in no case was saturation in magnetization seen up to 6 T external fields and assigning a spin ground state is impossible.

A.C. susceptibility measurements were performed on the compounds that show the low temperature maximum. For **8**, **15** and **18** no peak was seen in the out-of-phase susceptibility (χ''), while for **9** there was a rise in χ'' at low temperature, but no maximum. Compounds **7** and **16** show frequency dependent peaks in χ'', which is often thought to be the signature for SMMs. However, analysis of the data in both cases give energy barriers that are unreasonably high for SMMs, and pre-exponential factors that are absurdly small. While there is clearly unusual behaviour, it appears only a fraction of the sample is undergoing slow relaxation in any of these complexes.[15,16]

12.3.2 Iron Phosphonate Cages

Oxo-centred tri-iron(III) carboxylates, $[Fe^{III}{}_3(\mu_3\text{-}O)(O_2C\text{-}R)_6(OH_2)_3]X$, are well known and seemed ideal candidates for ligand displacement reactions with phosphonates. Variation of -R, X and the reaction conditions, especially solvent, allows a variety of polynuclear cages to be made. In the majority of cases, the oxo-centred triangle building blocks are retained as part of the final structure.

A hexa-iron cluster $[Fe^{III}_6(\mu_3\text{-}O)(\mu\text{-}OH)_3(O_3P\text{-}Ph)_4(O_2C\text{-}Me)_3(py)_9](NO_3)_2$ **21** was obtained by the reaction of $Ph\text{-}PO_3H_2$ with $[Fe^{III}_3(\mu_3\text{-}O)(O_2C\text{-}Me)_6(OH_2)_3](NO_3)$ in pyridine (py).[12] The structure contains two iron triangles linked through the phosphonates (Figure 12.9). One triangle is clearly related to the precursor containing a μ_3-oxide and three carboxylates bridging the three Fe^{III} centres; however, three of the carboxylates in the original structure have been replaced by phosphonates, resulting in the formation of new cages of higher nuclearity. Each phosphonate shows the binding mode of [3.111].

Following the similar synthetic procedure, we isolated six other iron phosphonate cages by varying the precursor as well as the reaction condition.[12] These cages are $[Fe^{III}_4(\mu_3\text{-}O)(Cl)(O_3P\text{-}Ph)_3(O_2C\text{-}Ph)_3(py)_5]$ **22**, $[Fe^{III}_7(\mu_3\text{-}O)_2(O_3P\text{-}Ph)_4(O_2C\text{-}Me)_9(py)_6]$ **23**, $[Fe^{III}_7(\mu_3\text{-}O)_2(O_3P\text{-}Ph)_4(O_2C\text{-}Ph)_9(py)_6]$ **24**, $[Fe^{III}_7(\mu_3\text{-}O)_2(O_3P\text{-}Ph)_3(O_2C\text{-}Ph)_8(NO_3)(py)_6]$ **25**, $[Fe^{III}_{14}(\mu_3\text{-}O)_4(O_2)_2(O_3P\text{-}Ph)_8(O_2C\text{-}t\text{-}Bu)_{12}(OH_2)_{12}](NO_3)_2$ **26** and $[Fe^{III}_6(\mu_3\text{-}O)_2(O_2)(O_3P\text{-}Ph)_2(O_2C\text{-}t\text{-}$

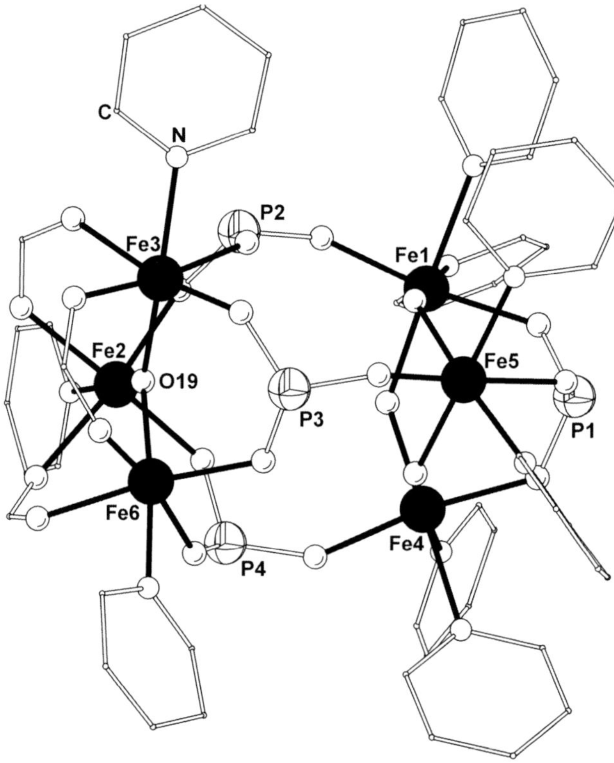

Figure 12.9 The structure of **21** in the crystal.

$Bu)_8(OH_2)_2]$ **27**. In all the structures, the triangular $\{Fe^{III}_3(\mu_3\text{-}O)\}$ core of the starting material is retained and some of the carboxylates displaced by phosphonates. The structure of **22** can be described as derivatives of $\{Fe^{III}_3(\mu_3\text{-}O)\}$-carboxylate triangle from which three carboxylate groups have been displaced by three phosphonates, of which the third oxygen of each phosphonate unit binds to a single Fe^{III} atom (Figure 12.10a). The cages **23**–**25** have identical metal cores, built from two $\{Fe^{III}_3(\mu_3\text{-}O)\}$ triangles in which two carboxylates on one edge are replaced by phosphonates, which both bridged to a single Fe^{III} atom (Figure 12.10b). The phosphonates in the structures of **22**–**25** show the binding mode of [3.111].

Though we did not use any peroxide in the reaction, structure **26** contains two bridging peroxides (Figure 12.11).[12] The structure contains two distorted trigonal prisms, made up of linked Fe_3 triangles, in which two phosphonates bind to two rectangular faces of the prism with the [4.211] mode. The final rectangular face of each prism is occupied by a 4.22-bridging peroxide. The two trigonal prisms are linked through two $\{Fe^{III}(OH_2)_4\}$ units *via* four [3.111] phosphonates. Deliberate addition of H_2O_2 yields cage **27**,[12] which contains the isolated trigonal prism from **26**.

In an another reaction the precursor $[Fe^{III}_3(\mu_3\text{-}O)(O_2C\text{-}t\text{-}Bu)_6(H_2O)_3](t\text{-}Bu\text{-}CO_2).2t\text{-}Bu\text{-}CO_2H$ was reacted with $Ph\text{-}PO_3H_2$ giving the nonanuclear iron cage, $[Fe^{III}_9(\mu_3\text{-}O)_4(O_3P\text{-}Ph)_3(O_2C\text{-}t\text{-}Bu)_{13}]$ **28**.[18] The structure of the cage contains a vertex-sharing triangle of oxo-centred $\{Fe^{III}_3(\mu_3\text{-}O)\}$ triangles that are connected to the final three Fe^{III} centres through three phosphonates with a binding mode of [5.221] (Figure 12.12). The Fe and P sites lie on the vertices of an $\{Fe_9P_3\}$ icosahedron. Each P centre has five Fe^{III} nearest neighbours, and each Fe^{III} site has five nearest neighbours (either four Fe and one P, or three Fe and two P). The Fe_9 core therefore describes a tridiminished icosahedron, which has non-crystallographic three-fold symmetry. Magnetic susceptibility studies of **28**, combined with EPR data, suggest a strong antiferromagnetic interaction between the metal centres with a degenerate pair of $S = 1/2$ states as the lowest in energy.

Recently, we have reported[19] further new multinuclear iron phosphonate cages $[Fe^{III}_7(\mu_3\text{-}O)_2(O_3P\text{-}t\text{-}Bu)_4(O_2C\text{-}t\text{-}Bu)_8(py)_8](NO_3)$ **29**, $[Fe^{III}_7(\mu_3\text{-}O)_2(O_3P\text{-}Ph)_4(O_2C\text{-}Me)_8(py)_8]$ **30**, $[Fe_9(\mu_3\text{-}O)_2(\mu_2\text{-}OH)(O_3P\text{-}Ph)_6(O_2C\text{-}t\text{-}Bu)_{10}(OH_2)_5(MeCN)]$ **31**, $[Fe^{III}_6(\mu_3\text{-}O)_2(O_2)(O_3P\text{-}C_{10}H_{17})_2(O_2C\text{-}t\text{-}Bu)_8(OH_2)_2]$ **32** and $Fe^{III}_6(\mu_3\text{-}O)_2(O_2)(O_3P\text{-}t\text{-}Bu)_2(O_2C\text{-}t\text{-}Bu)_8(py)_2]$ **33** synthesized from the reactions of oxo-centred iron-carboxylate triangles and phosphonic acid in different conditions. The core of the cage **29** is very similar to that of **23**–**25**, but the core of the heptametallic cage **30** is different than that of **23**–**25**. It contains two $\{Fe^{III}_3(\mu_3\text{-}O)\}$ triangles that are centrally bridged by an Fe^{III} atom *via* a phosphonate in one side and three phosphonates in the other side (Figure 12.13). In many ways the structure resembles the tetrametallic cage **22** linked to the precursor triangle. All the phosphonates show the binding mode of [3.111].

The structure of **31** contains two $\{Fe^{III}_3(\mu_3\text{-}O)\}$ triangles linked by a single Fe^{III} atom and a $Fe^{III}\text{-}(OH)\text{-}Fe^{III}$ *via* three phosphonate units in each side (Figure 12.14). In each triangle, the water molecules present in two of the Fe^{III}

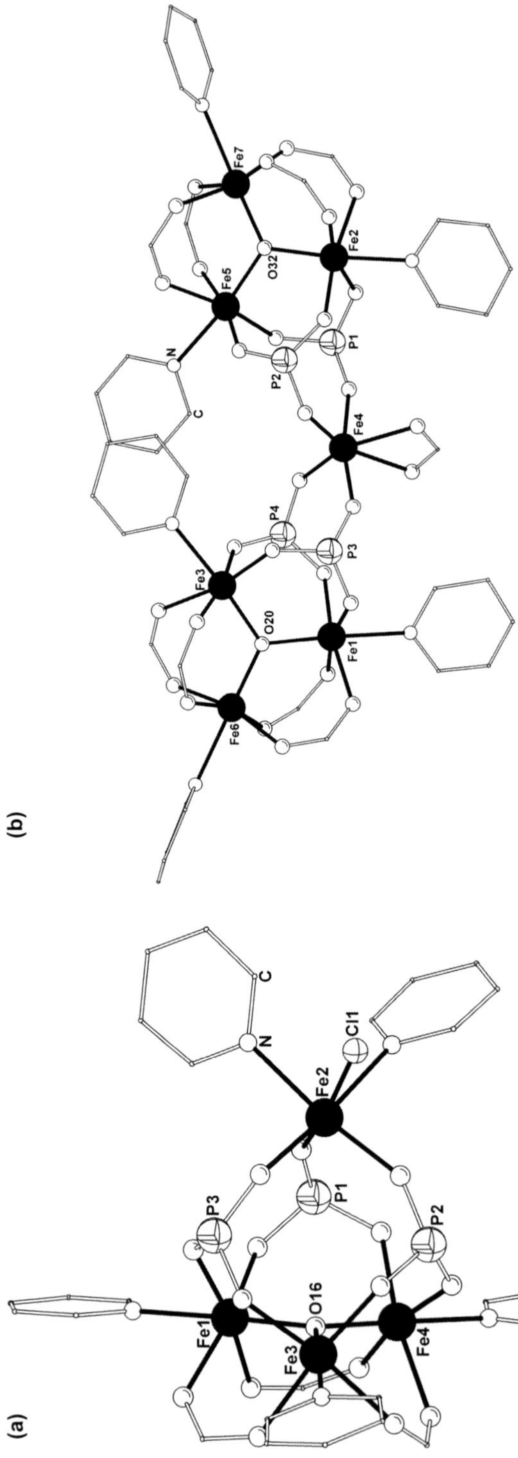

Figure 12.10 The structures of (a) compound **22** and (b) compound **23**.

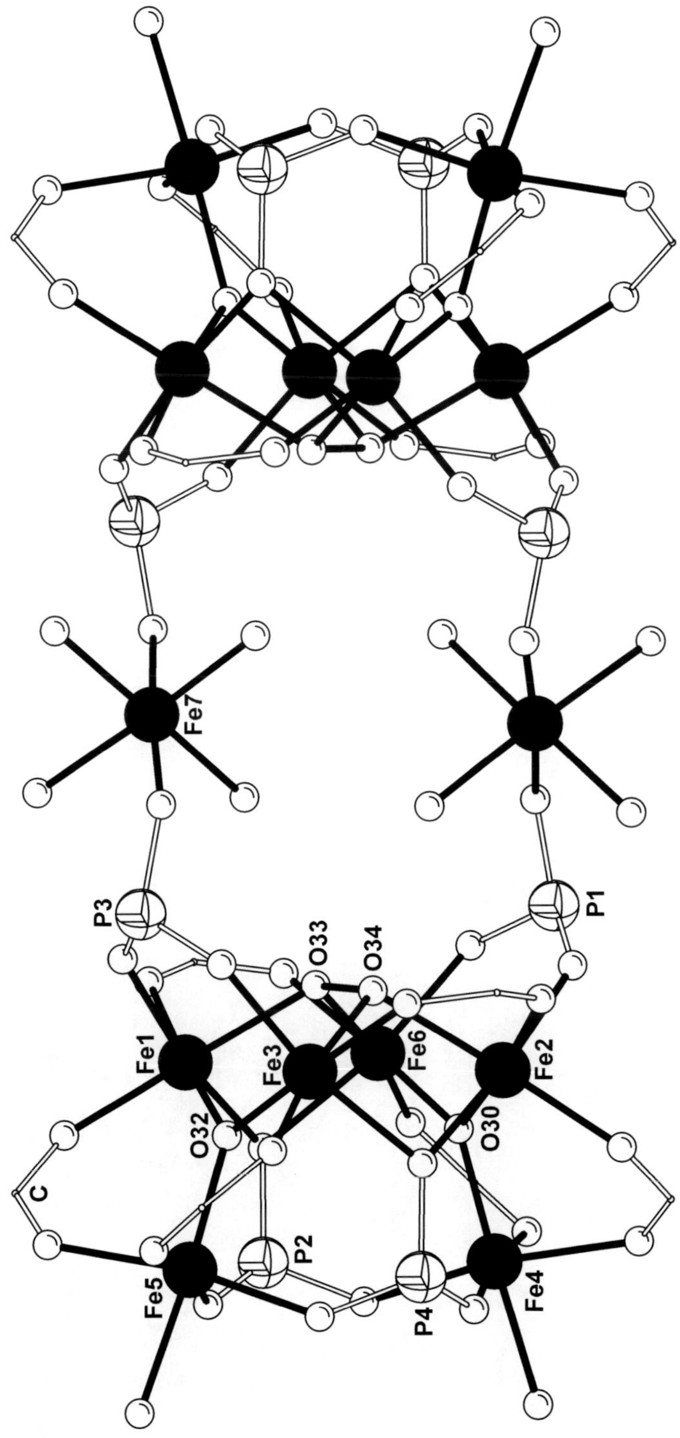

Figure 12.11 The structure of **26** in the crystal.

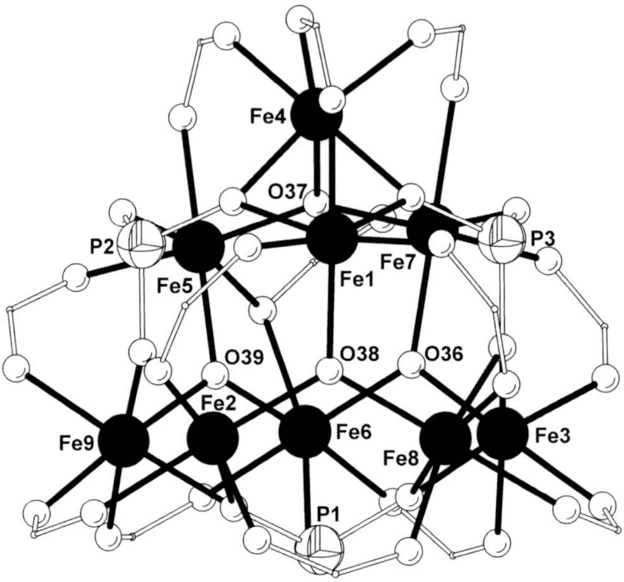

Figure 12.12 The structure of **28** in the crystal.

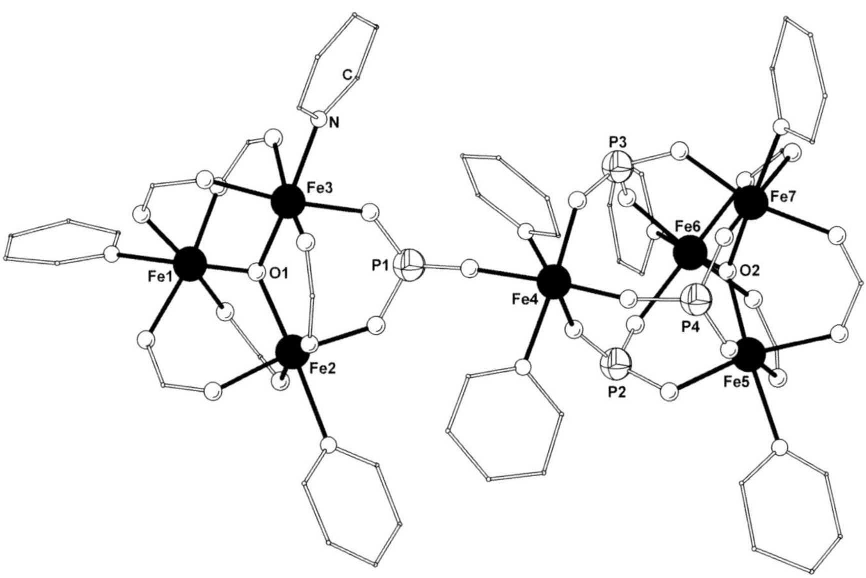

Figure 12.13 The structure of **30** in the crystal.

atoms of the precursor are retained, but the third one has been replaced by a carboxylate group, which is further involved in bridging to the Fe^{III}-(OH)-Fe^{III} subunit. Two of six phosphonates show the binding mode of [3.111] and the

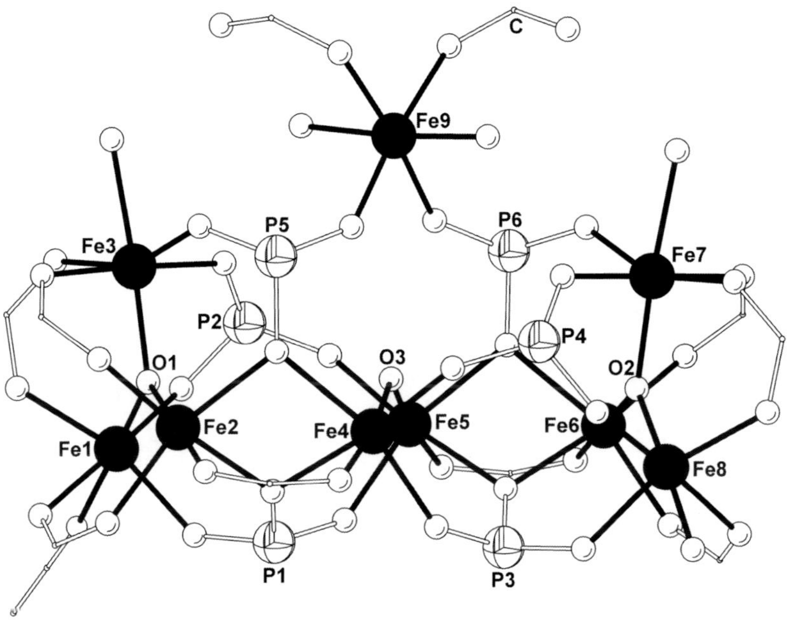

Figure 12.14 The structure of **31** in the crystal.

remaining four show [4.211] binding modes. The core of the cages **32** and **33** are similar to that of **27** with different peripheral substitutions.

Presence of bases in the reaction of oxo-centred iron-carboxylate triangles with phosphonic acids influences the final cage formation.[20] Thus, when Et_3N was the base, cages $[Fe^{III}_8(\mu_3\text{-}O)_2(\mu_3\text{-}OH)(\mu_2\text{-}OH)_2(O_3P\text{-}CH_2Ph)_3(O_2C\text{-}t\text{-}Bu)_{11}(py)_3]$ **34** and $[Fe^{III}_{10}(\mu_3\text{-}O)_2(\mu_3\text{-}OH)_4(\mu_2\text{-}OH)_4(O_3P\text{-}CH_2Ph)_4(O_2C\text{-}t\text{-}Bu)_{10}(pip)_2]$ **35** (pip = piperidine) were synthesized by reacting benzylphosphonic acid ($PhCH_2\text{-}PO_3H_2$) and $[Fe^{III}_3(\mu_3\text{-}O)(O_2C\text{-}t\text{-}Bu)_6(OH_2)_3]^+Cl^-$ or $[Fe^{III}_2Fe^{II}(\mu_3\text{-}O)(O_2C\text{-}t\text{-}Bu)_6(pip)_3]$. The cage $[Fe^{III}_6Li_5(\mu_3\text{-}O)_2(O_3P\text{-}t\text{-}Bu)_6(O_2C\text{-}t\text{-}Bu)_8(py)_4(MeOH)_2]$ **36** was isolated in the presence of LiOMe as base in the reaction of $[Fe^{III}_3(\mu_3\text{-}O)(O_2C\text{-}t\text{-}Bu)_6(OH_2)_3]^+Cl^-$ and $t\text{-}Bu\text{-}PO_3H_2$. The structure of **34** contains a triangular $[Fe^{III}_3(\mu_3\text{-}O)(O_2C\text{-}t\text{-}Bu)_5]^{2+}$ unit that is linked to a 'butterfly' of $[Fe^{III}_4(\mu_3\text{-}O)(\mu_3\text{-}OH)(O_2C\text{-}t\text{-}Bu)_4]^{5+}$ unit *via* a [3.111] phosphonate and hydroxo bridges (Figure 12.15). The butterfly unit is linked to a separate Fe^{III} centre (Fe8) by two phosphonates (with binding mode of [4.211] and [3.111]), one carboxylate and one hydroxo unit giving a $[Fe^{III}_8(\mu_3\text{-}O)(\mu_3\text{-}OH)(\mu_2\text{-}OH)_2(O_2C\text{-}t\text{-}Bu)_{10}]^{7+}$ core. There is a resemblance to **30** with an additional Fe^{III} site.

The complex **35** contains a $[Fe^{III}_{10}(\mu_3\text{-}O)_2(\mu_3\text{-}OH)_4(\mu_2\text{-}OH)_4]$ core that can be described as containing four edge-sharing $\{Fe^{III}_3(\mu_3\text{-}OH)\}$ triangles giving two $\{Fe^{III}_4(\mu_3\text{-}OH)_2\}$ butterflies, with the wing-tip Fe sites of these butterflies also involved in two $\{Fe^{III}_3(\mu_3\text{-}O)\}$ triangles. Peripheral ligation to the core is provided by ten $Me_3CCO_2^-$ groups in the familiar [2.11] bridging modes and four

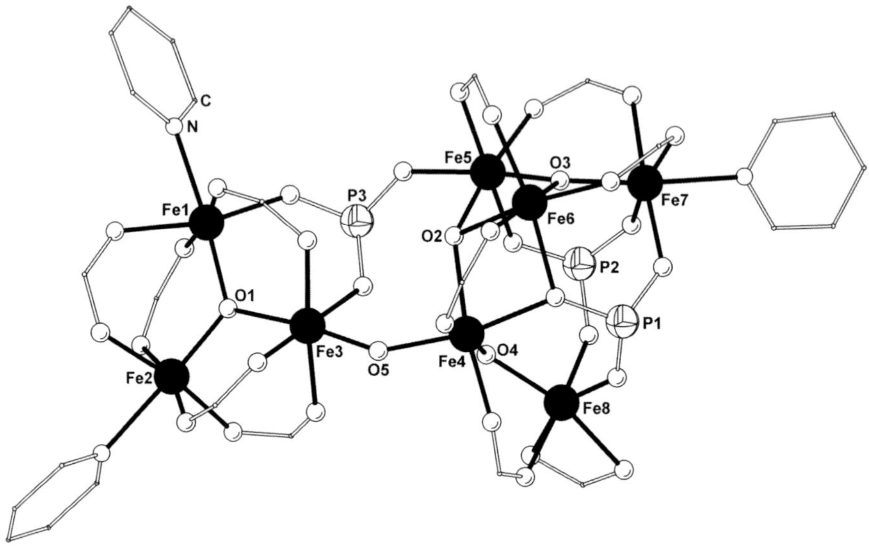

Figure 12.15 The structure of **34** in the crystal.

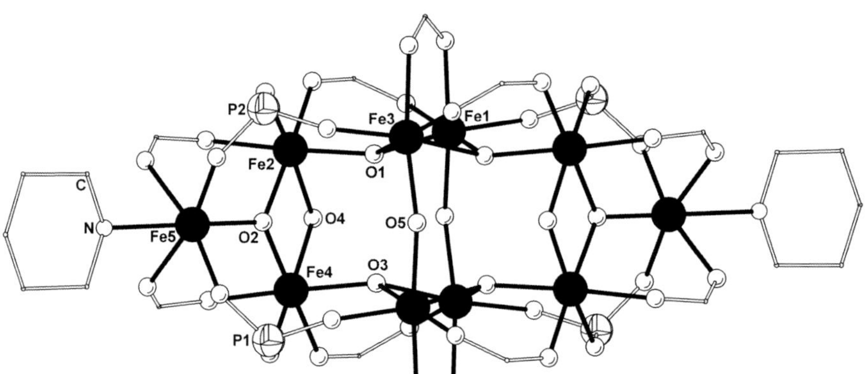

Figure 12.16 The structure of **35** in the crystal.

$PhCH_2PO_3^{2-}$ groups which show the 3.111 bridging mode (Figure 12.16). The structure of **36** consists of two iron triangles of the precursor bridged by phosphonates and Li^{I} atoms in the centre (Figure 12.17). Four of the six phosphonates have the [4.211] binding mode and the other two show the [5.221] binding mode.

Clearfield and co-workers used $\{Fe^{III}{}_3(\mu_3\text{-}O)\}$-carboxylate triangles and camphylphosphonic acid ($C_{10}H_{17}PO_3H_2$) to synthesize three new Fe^{III} phosphonate cages.[21] The cages reported are $[Fe^{III}{}_9(\mu_3\text{-}O)_4(O_3P\text{-}C_{10}H_{17})_3(O_2C\text{-}t\text{-}Bu)_{13}]$ **37**, $[Fe^{III}{}_9(\mu_3\text{-}O)_2(\mu\text{-}OH)(O_3P\text{-}C_{10}H_{17})_6(O_2C\text{-}Ph)_{10}(OH_2)_6]$.7MeCN **38**

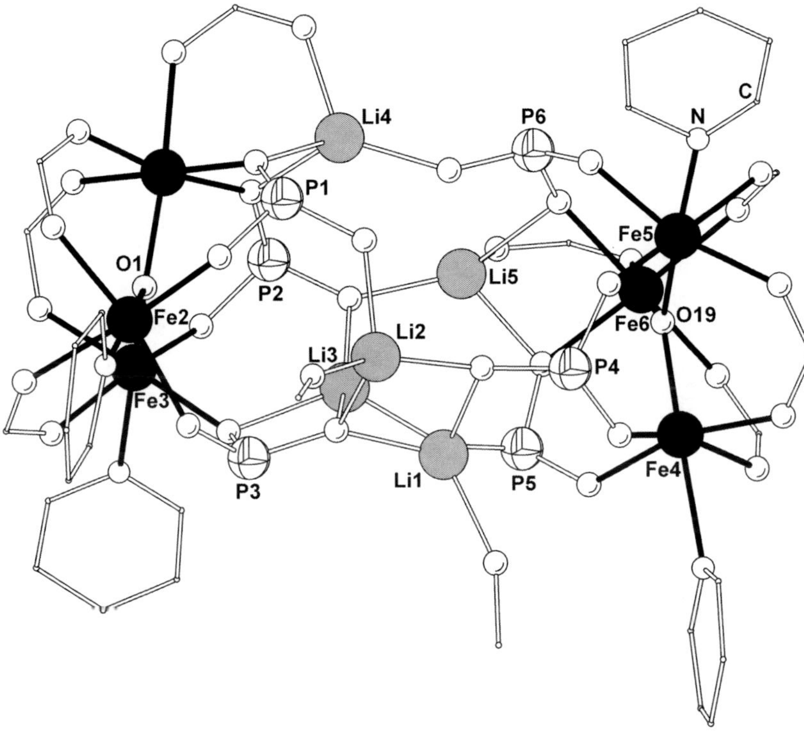

Figure 12.17 The structure of **36** in the crystal.

and $[Fe^{III}_6(\mu_3\text{-}O)_2(O_2)(O_3P\text{-}C_{10}H_{17})_2(O_2C\text{-}t\text{-}Bu)_8(OH_2)_2]$.4MeCN **39**. The presence of Et_3N in the above reaction gives the dodecanuclear cage, $[Fe^{III}_{12}(\mu_3\text{-}O)_4(\mu\text{-}O)_4(HO_3P\text{-}C_{10}H_{17})_6(O_2C\text{-}Ph)_{14}]$ **40**.[21] The core of the cages **37**, **38** and **39** are very similar to those of **28**, **31** and **27**, respectively, with different peripheral substitutions. The central Fe^{III}_{12} core in the structure of **40** can be described as two Fe^{III}_6 sub-units are linked *via* four oxo (μ_3-O) and two [4.211] phosphonates (Figure 12.18). Each Fe^{III}_6 sub-unit is butterfly shaped, made by two $\{Fe^{III}_3(\mu_3\text{-}O)\}$ triangles, which are connected by another two phosphonates with binding mode of [4.211].

The cages **37** and **38** can also be synthesized in a one-pot reaction with $FeCl_3$, $C_{10}H_{17}PO_3H_2$ and t-Bu-CO_2H/C_6H_5-CO_2H in the presence of excess Et_3N.[19] Similar syntheses with pyridine instead of Et_3N result in two new tetra-iron phosphonate cages: $[Fe^{III}_4(\mu_3\text{-}O)(O_3P\text{-}C_{10}H_{17})_3(O_2C\text{-}t\text{-}Bu)_4(py)_4]$.3MeCN **41** and $[Fe^{III}_4(\mu_3\text{-}O)(O_3P\text{-}C_{10}H_{17})_3(O_2C\text{-}Ph)_4(py)_4]$.3py.2MeCN **42**.[19] The core of the compounds **41** and **42** are similar to that of **22** with minor changes at the fourth Fe^{III} centre. Another dodecanuclear phosphonate cage, $[Fe^{III}_{12}(\mu_3\text{-}O)_4(\mu\text{-}O)_4(HO_3P\text{-}C_6H_4\text{-}4\text{-}t\text{-}Bu)_6(O_2C\text{-}CHPh_2)_{14}]$ **43** has been made by solvothermal methods using $FeCl_3$ as a precursor, adding diphenylacetic acid during the reaction as a co-ligand with the 4-t-butylphenylphosphonate.[22] The core of the cage **43** is similar to that of **40**.

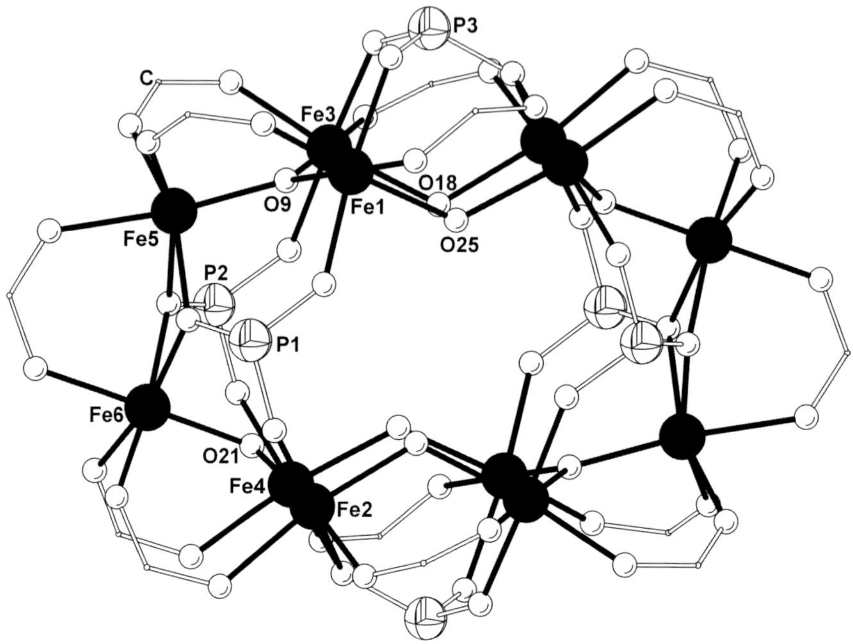

Figure 12.18 The structure of **40** in the crystal.

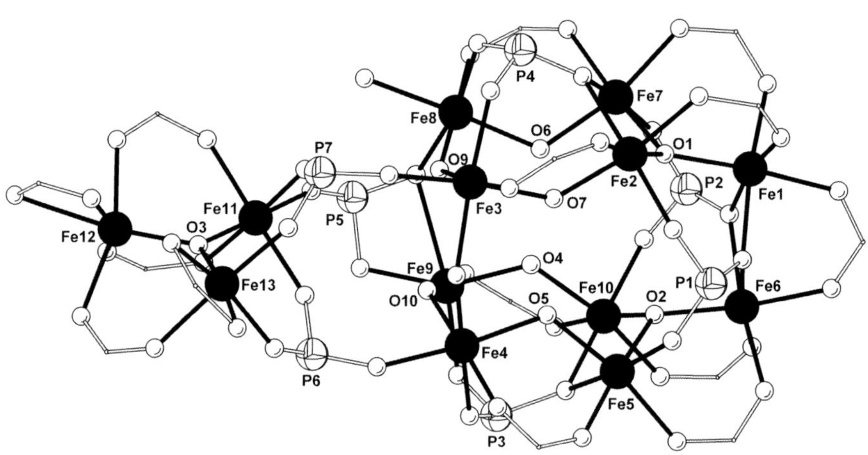

Figure 12.19 The structure of **44** in the crystal.

A one-pot reaction involving $FeCl_3$, *t*-Bu-PO_3H_2 and *t*-Bu-CO_2H in the presence of Et_3N leads to the formation of a new cage $[Et_3NH]_2[Fe^{III}{}_{13}(\mu_3\text{-}O)_3(\mu_2\text{-}OH)_7(O_3P\text{-}t\text{-}Bu)_7(O_2C\text{-}t\text{-}Bu)_{14}(OH_2)]$ **44**. The structure of **44** contains a $\{Fe^{III}{}_3(\mu_3\text{-}O)\}$ triangular unit linked to a $\{Fe^{III}{}_{10}(\mu_3\text{-}O)_3(\mu_2\text{-}OH)_7\}$ unit by a [4.211] phosphonate and two [3.111] phosphonates (Figure 12.19). The

$\{Fe^{III}_{10}(\mu_3\text{-}O)_3(\mu_2\text{-}OH)_7\}$ unit consisting of two $\{Fe^{III}_3(\mu_3\text{-}O)\}$ triangles are connected by two [4.211] phosphonates that are bridged to a central $\{Fe^{III}_4(\mu_2\text{-}OH)_3\}$ ring *via* four μ_2-hydroxides and another two [4.211] phosphonate groups. There is a clear relationship between the structures of **44** and **40** which can be seen from Figs. 12.18 and 12.19; the ten Fe^{III} centres on the right of both figures lie in similar positions in the structures.

Yao *et al.* reported[23,24] three new Fe^{III} phosphonate cages, $[Fe^{III}_4(\mu_3\text{-}O)(Cl)(O_3P\text{-}C_6H_9)_3(O_2C\text{-}Me)_3(py)_5]$ **45**, $[Fe^{III}_7(\mu_3\text{-}O)_2(O_3P\text{-}C_6H_9)_4(O_2C\text{-}Ph)_9(py)_6]$ **46** and $[Fe^{III}_9(\mu\text{-}O)_2(\mu\text{-}OH)_7(O_3P\text{-}C_6H_9)_8(py)_{12}].6H_2O$ **47**, synthesized using cyclohexylphosphonic acid ($C_6H_9\text{-}PO_3H_2$). The core of the cages **45** and **46**[23] are very similar to that of **22** and **24** respectively, with different peripheral substitutions. The structure of **47**[24] contains a ring of $\{Fe^{III}_9(\mu\text{-}O)_2(\mu\text{-}OH)_7\}$ capped by six phosphonates in the sides and two phosphonates above and below the ring, each with binding mode of [3.111] (Figure 12.20). Compound **47** is the first homometallic compound that could be described as a nine-metal ring, albeit the structure is a very distorted ring. Magnetic studies suggest an $S = 1/2$ ground state, as would be expected for a frustrated spin system of odd-electron centres.[24]

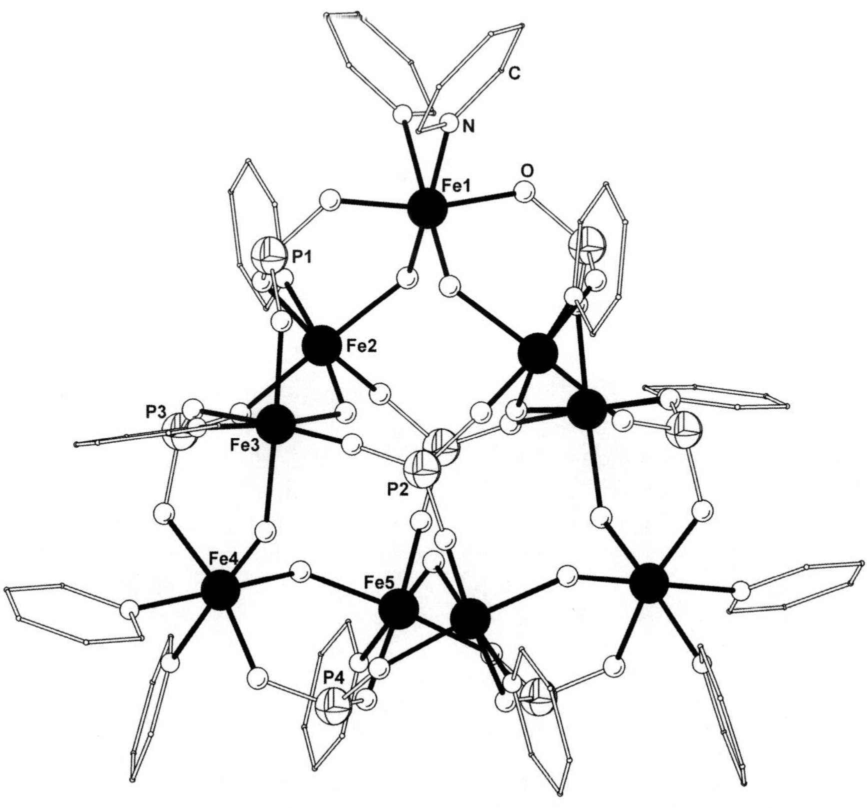

Figure 12.20 The structure of **47** in the crystal.

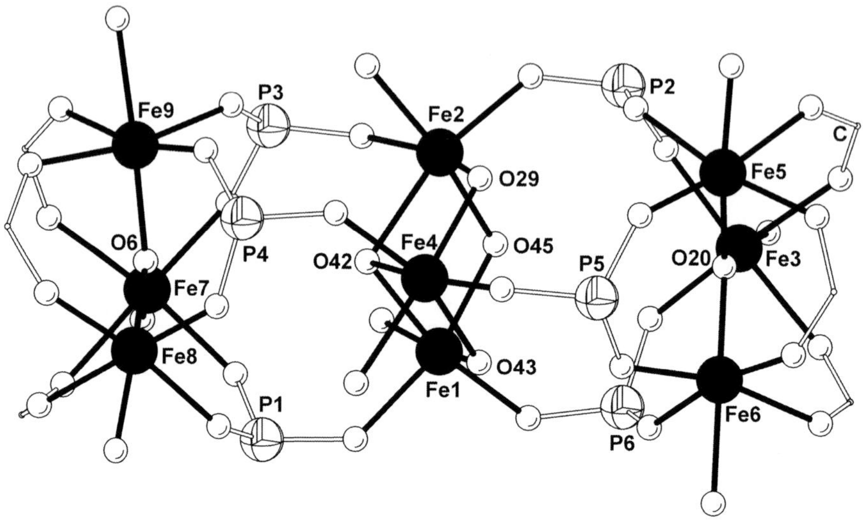

Figure 12.21 The structure of **49** in the crystal.

The reaction of $\{Fe^{III}_3(\mu_3\text{-}O)\}$-carboxylate triangles with phosphonic acids under solvothermal condition also shows potential.[22] By this method two new nonanuclear iron cages were synthesized, $[Fe_9(\mu_3\text{-}O)_4(O_3P\text{-}C_6H_3\text{-}3,5\text{-}Me_2)_3(O_2C\text{-}t\text{-}Bu)_{13}]$ **48** and $[Fe_9(O)_3(OH)_3(O_3P\text{-}CHPh_2)_6(O_2C\text{-}t\text{-}Bu)_6(OH_2)_9]$ **49**. The core of the cage **48** is similar to that of **28**, while cage **49** consists of three $\{Fe^{III}_3(\mu_3\text{-}O)\}$ triangles, stacked on top of one another, each of the two outer triangles is linked to the central triangle by three phosphonates (Figure 12.21).

Recently, Murugavel and co-workers[25] synthesized two new Fe^{III} phosphonate cages in the reaction of $[Fe^{III}_3(\mu_3\text{-}O)(O_2C\text{-}Ph)_6(OH_2)_3]^+(Cl^-)$ or $FeCl_3$ and *t*-butylphosphonic acid ($t\text{-}Bu\text{-}PO_3H_2$) in pyridine. The cages are $[Fe^{III}_4(\mu_3\text{-}O)(Cl)(O_3P\text{-}t\text{-}Bu)_3(O_2C\text{-}Ph)_3(py)_3]\cdot 3.5py$ **50** and $[Fe^{III}_{10}(\mu\text{-}OH)_8(HPO_4)(O_3P\text{-}t\text{-}Bu)_8(HO_3P\text{-}t\text{-}Bu)_4(py)_8]\cdot 4py\cdot 5H_2O$ **51**. The core of the cage **50** is very similar to that of **22** with different peripheral substitutions. The structure of the cage **51** consists of two phosphonate-bridged $\{Fe^{III}_5(\mu\text{-}OH)_4\}$ half-rings that are centrally bridged by a phosphate and four [3.111] dianionic phosphonate groups (Figure 12.22). In addition, each $\{Fe^{III}_5(\mu\text{-}OH)_4\}$ half-ring has two dianionic and two mono-anionic bridging phosphonates with binding modes of [3.111] and [2.110] respectively. The presence of phosphate in the middle is due to decomposition of $t\text{-}Bu\text{-}PO_3H_2$ which is quite unusual.

12.3.3 Manganese Phosphonate Cages

Initially, we reported[11] the synthesis of a hexanuclear Mn^{III} phosphonate cage, $[Mn^{III}_6(\mu_3\text{-}O)_2(O_3P\text{-}Ph)_2(O_2PH\text{-}Ph)_2(O_2C\text{-}Ph)_8(py)_2]$ **52** by reacting the mixed-valence manganese-carboxylate triangle $[Mn^{III}_2Mn^{II}(\mu_3\text{-}O)$

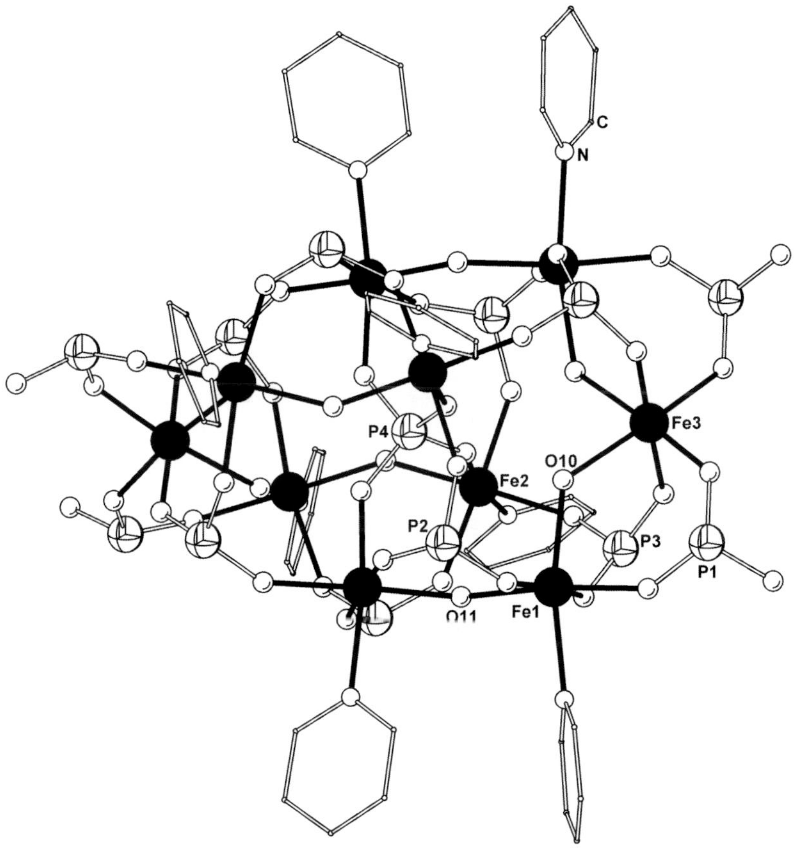

Figure 12.22 The structure of **51** in the crystal.

$(O_2C\text{-}Ph)_6(py)_2(OH_2)]$ and phenylphosphinic acid ($Ph\text{-}PHO_2H$), which is similar to Fe^{III}-phosphonate synthesis discussed previously. The structure of the cage **52** shows that the product retains the precursor of oxo-centered triangles of manganese centres $\{Mn_3(\mu_3\text{-}O)\}$ and are linked by two phosphonate and two phosphinate ligands (Figure 12.23). The bridging phosphonates show the binding mode of [4.211] whereas the phosphinates show the [2.11] binding mode. The same product **52** was obtained when using a homo-valent Mn-triangle $[Mn^{III}{}_3(\mu_3\text{-}O)(O_2C\text{-}Ph)_6(py)_3](ClO_4)$. Recently, we have synthesized[26] an all phosphonate version of cage **52** by using $Ph\text{-}PO_3H_2$ in the place of $Ph\text{-}PHO_2H$ in the above reaction. The formula of this cage is $[Mn^{III}{}_6(\mu_3\text{-}O)_2(O_3P\text{-}Ph)_2(HO_3P\text{-}Ph)_2(O_2C\text{-}Ph)_8(py)_2]$ **53**.

Four further Mn-phosphonate cages were made[27] by reacting $[Mn^{III}{}_2Mn^{II}(\mu_3\text{-}O)(O_2C\text{-}t\text{-}Bu)_6(py)_3]$ and $Ph\text{-}PO_3H_2$ or $Ph\text{-}PO_3Na_2$ in the presence of different bases. The cages are $[Mn_7(\mu_3\text{-}O)_3(O_3P\text{-}Ph)_3(O_2C\text{-}t\text{-}Bu)_8(py)_3]$ **54**, $[Mn_9Na(\mu_3\text{-}O)_4(\mu_4\text{-}O)_2(O_3P\text{-}Ph)_2(O_2C\text{-}t\text{-}Bu)_{12}(OH_2)_2(OH_2)_{0.67}(py)_{0.33}]$ **55**, $[Mn_{13}O_8(O_3P\text{-}Ph)_4(O_2C\text{-}t\text{-}Bu)_{10}(OMe)_8]$ **56** and $[Mn_{18}O_8(O_3P\text{-}Ph)_{14}$

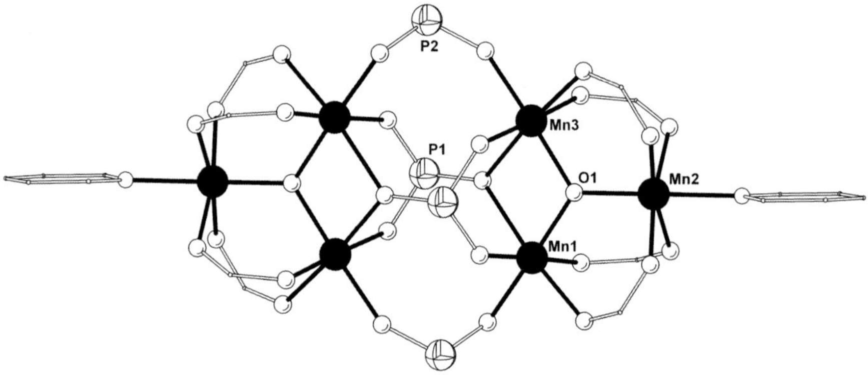

Figure 12.23 The structure of **52** in the crystal.

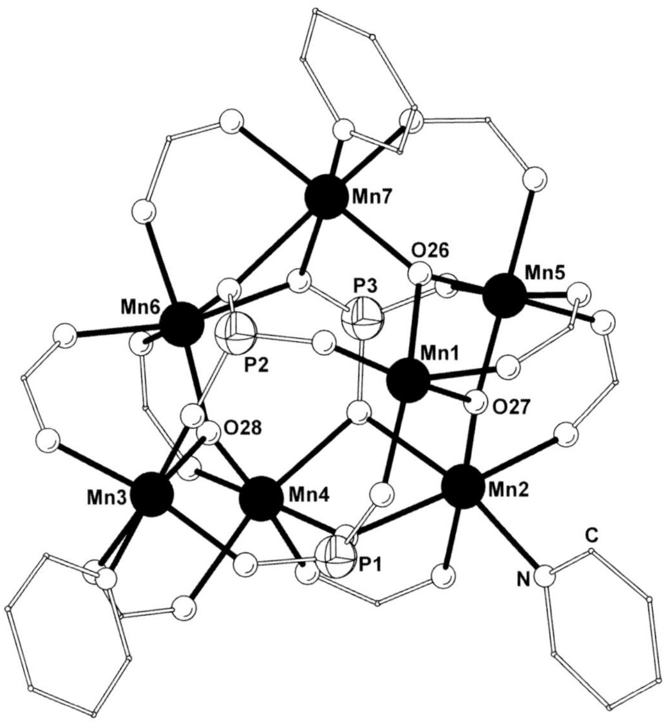

Figure 12.24 The structure of **54** in the crystal.

$(O_2C\text{-}t\text{-}Bu)_{12}(H_2O)_2(py)_6]$ **57**. The structure of **54** contains a $\{Mn_4(\mu_3\text{-}O)_2\}$ butterfly connected to a $\{Mn_3(\mu_3\text{-}O)\}$ triangle similar to the precursor through phosphonates and carboxylates (Figure 12.24). The BVS analysis of the structure reveals that it contains three different oxidation states of manganese ions and gives the complete formula of $[Mn^{II}{}_2Mn^{III}{}_4Mn^{IV}(\mu_3\text{-}O)_3$

$(O_3P\text{-}Ph)_3(O_2C\text{-}t\text{-}Bu)_8(Py)_3]$. One of the phosphonates show the binding mode of [5.221] and the remaining two show the [4.211] mode.

The structure of the cage **55** can be described as fragments of an icosahedron (Figure 12.25). Eight of the Mn atoms (except Mn2), two P atoms and the Na atom lie at the vertices of a distorted icosahedron that is missing one vertex which would complete the pentagonal ring of Mn3, Mn5, Mn6 and Mn7. While Mn2 may appear to be in this ring, it does not lie on the surface of an icosahedron but lies near the centre of the polyhedron. The BVS analysis of the structure reveals that all the manganese ions are in +3 oxidation state (all Mn^{III}). Both the phosphonates show the binding mode of [5.221].

The structure of the cage **56** contains two trigonal prisms of manganese ions and one Mn atom (Mn3) between the two prisms (Figure 12.26). Each trigonal prism contains two $\{Mn_3(\mu_3\text{-}O)\}$-triangular units bridged by two [4.211] phosphonates which occupy two square faces of the prism. The BVS analysis of the structure reveals that all the manganese ions are in +3 oxidation state (all Mn^{III}).

The structure of the octadecanuclear compound **57** shows that the cage retains most of $\{Mn_3(\mu_3\text{-}O)\}$ triangles from the precursor (Figure 12.27). At the centre of the cluster there are two $\{Mn_6(\mu_3\text{-}O)_3\}$ fragments that are linked centrally by two [5.221] and two [3.111] phosphonates. Each $\{Mn_6(\mu_3\text{-}O)_3\}$ fragment consists of three vertex-sharing triangles bridged by two [3.111] phosphonates. In addition, there are two $\{Mn_3(\mu_3\text{-}O)\}$ triangular units present

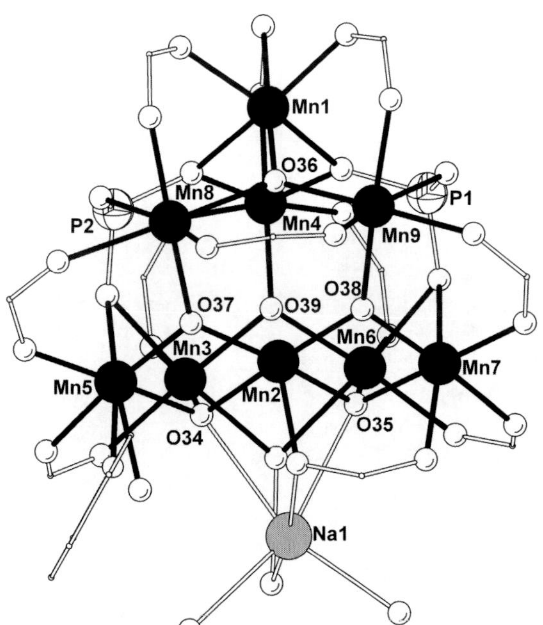

Figure 12.25 The structure of **55** in the crystal.

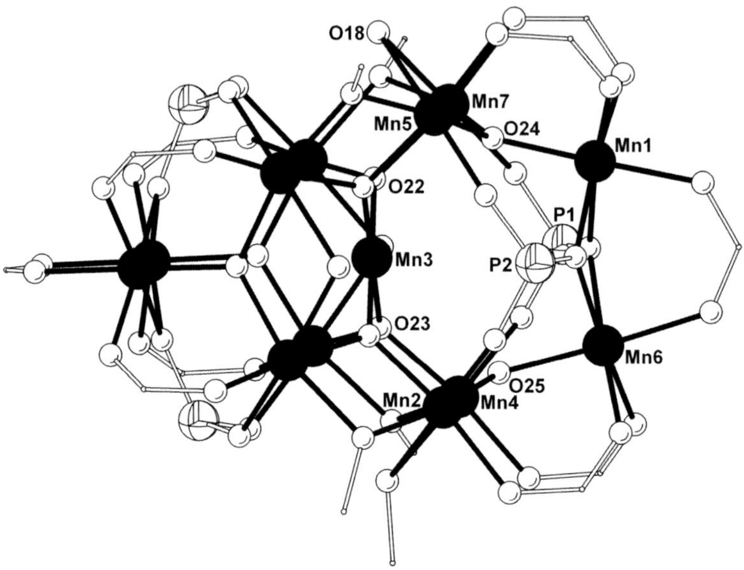

Figure 12.26 The structure of **56** in the crystal.

in each end of the periphery and are connected to the central core by three different phosphonates with binding mode of [5.221], [4.211] and [3.111]. The cage is mixed-valent that contains sixteen Mn^{III} ions and two Mn^{IV} ions assigned by BVS analysis of the structure.

Three structurally related mixed-valent Mn_{20} phosphonate cages have been reported;[28] $[Et_3NH]_2[Mn^{III}{}_{18}Mn^{II}{}_2(\mu_4\text{-}O)_8(\mu_3\text{-}O)_4(\mu_3\text{-}OH)_2(O_3PCH_2Ph)_{12}(O_2C\text{-}t\text{-}Bu)_{10}(Py)_2]$ **58**, $K_4[Mn^{III}{}_{16}Mn^{II}{}_4(\mu_4\text{-}O)_4(\mu_3\text{-}O)_6(O_3PCH_2Ph)_{14}(O_2CPh)_{12}(HO_2C\text{-}Ph)_{0.5}(MeCN)_2]$ **59** and $Na_6[Mn^{III}{}_{14}Mn^{II}{}_6(\mu_4\text{-}O)_4(\mu_3\text{-}O)_4(OH)_4(O_3P\text{-}CH_2Ph)_{14}(O_2CPh)_{12}(HO_2CPh)_2(OH_2)_4(MeCN)_4]$ **60** are formed from reacting mixed-valent or homo-valent Mn-triangles with benzylphosphonic acid, varying the reaction conditions. The structure of the cage **58** shows that many μ_3-oxo-centered triangles from the precursor are retained in the product, and form a layered structure held together by phosphonates and carboxylates (Figure 12.28). Out of twelve phosphonates, two of them show the [5.221] binding mode, six show the binding mode of [4.211] and the remaining four show the [3.111] binding mode. The core of the other two compounds **59** and **60** are similar to that of cage **58**. In all the cages the oxidation state of the manganese sites were assigned by BVS analysis. The magnetic susceptibility studies for compound **58** show that room temperature value of χT, in a χT versus T plot, is nearly 60 cm^3 K mol^{-1} and increases as the temperature is lowered, reaching a maximum value of 210 cm^3 K mol^{-1} and then a sharp decrease is observed at 10 K. This behaviour could be due to zero field splitting or intermolecular interactions and suggests a high spin ground state of $S = 19 \pm 1$. The detailed investigation shows that the molecule undergoes slow relaxation of magnetization and is a 'single molecule magnet'.

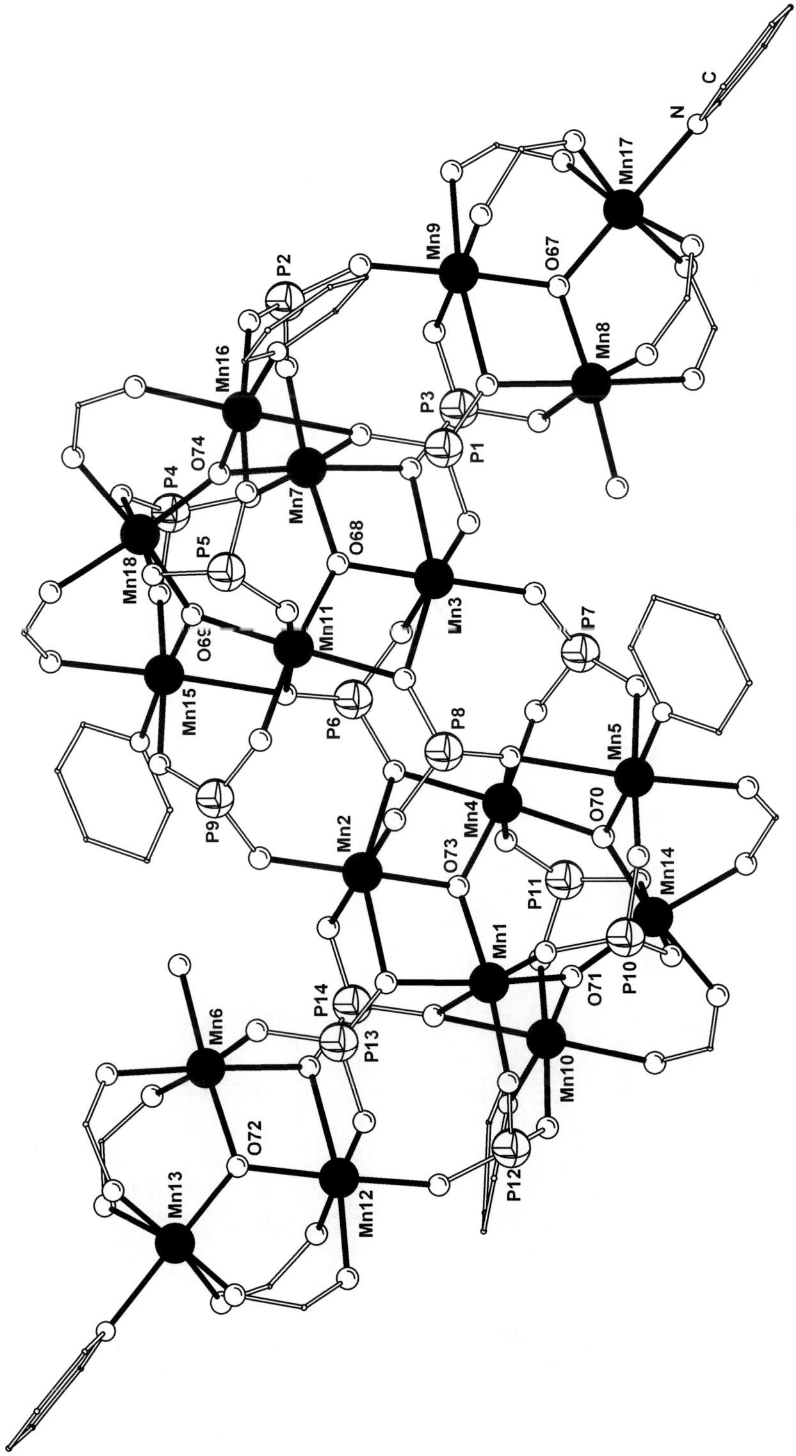

Figure 12.27 The structure of **57** in the crystal.

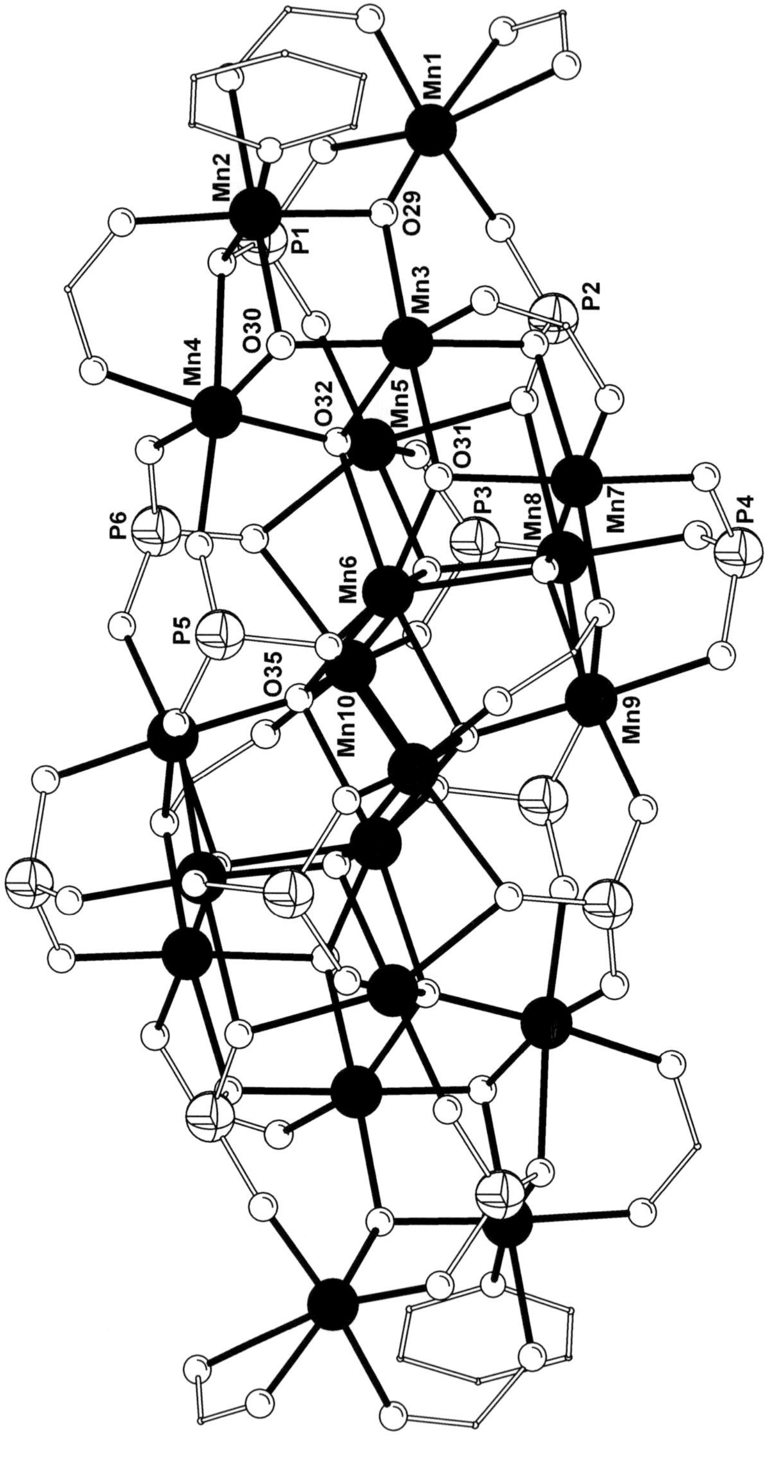

Figure 12.28 The structure of **58** in the crystal.

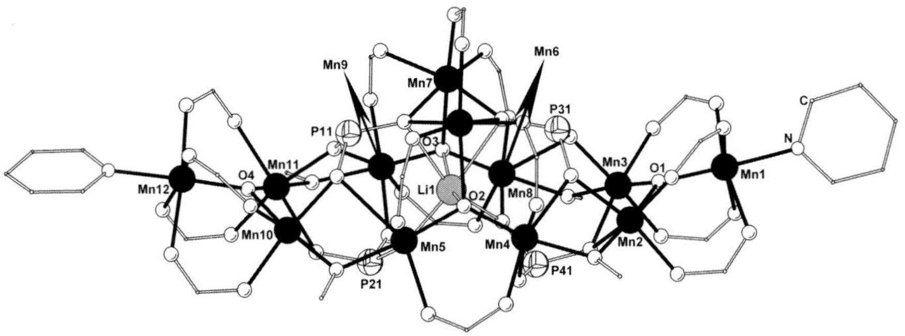

Figure 12.29 The structure of **61** in the crystal.

When LiOMe is used as base in a similar reaction, a dodecanuclear cage results, $[Mn^{III}_{10}Mn^{II}_2Li(\mu_3\text{-}O)_4(O_3P\text{-}CH_2Ph)_4(O_2C\text{-}Ph)_{15}(OMe)_4(py)_2]$ **61**.[26] The structure of this cage consists of a central trigonal prism of $\{LiMn_6(\mu_3\text{-}O)_2\}$ unit and two peripheral $\{Mn_3(\mu_3\text{-}O)\}$ triangles present in two sides of the prism (Figure 12.29). The trigonal prism is made up of two $\{Mn_3(\mu_3\text{-}O)\}$ triangles and these are bridged by a lithium atom in the centre *via* μ_3-oxygen. Two of the square faces of the prism are each bridged to one edge of a peripheral $\{Mn_3(\mu_3\text{-}O)\}$ triangle by two phosphonates with a binding mode of [5.221].

These results also show that by making minor changes in reaction conditions, such as changing solvent or using different bases, results in the formation of metal cages of different nuclearities. In all the above products the same Mn carboxylate triangle was used.

Li-Min Zheng and co-workers reported[29,30] four manganese phosphonate cages, $[Mn^{II}_8Mn^{III}_2(\mu_4\text{-}O)_2(O_3P\text{-}C_{10}H_7)_4(O_2C\text{-}Ph)_{10}(py)_4(OH_2)_2]$ **62**, $[Mn^{II}_4Mn^{III}_9(\mu_3\text{-}O)_4(\mu_3\text{-}OH)_2(O_3P\text{-}C_{10}H_7)_{10}(O_2C\text{-}Ph)_5(py)_8(OH_2)_6]$ **63**, $[Mn^{II}Mn^{III}_{12}(\mu_4\text{-}O)_6(\mu\text{-}OH)_6(O_3P\text{-}C_6H_{11})_{10}(py)_6]$ **64** and $[Mn^{II}_4Mn^{III}_{18}(\mu_3\text{-}O)_{12}(O_3P\text{-}C_6H_{11})_8(O_2C\text{-}Me)_{22}(OH_2)_6(py)_2]$ **65** synthesized from different reaction conditions. The structure of **62**[29] consists of a central $\{Mn^{III}_2Mn^{II}_4(\mu_4\text{-}O)_2\}$ bitetrahedral core connected to two Mn^{II} ions in each side by two phosphonates with a binding mode of [5.221] (Figure 12.30). The structure of **63** consists of a cyclic core with four distinct metal fragments.[29] Two $\{Mn_4(\mu_3\text{-}O)_2\}$ butterflies are linked by a $\{Mn_3(\mu_3\text{-}OH)_2\}$ fragment *via* μ_3-hydroxides and six phosphonates (Figure 12.31). Of six phosphonates, four of them show the [3.111] binding mode and the remaining two show the binding mode of [4.211]. At the other side of the cage, the $\{Mn_4(\mu_3\text{-}O)_2\}$ fragments are bridged by a $\{Mn_2(O_2C\text{-}Ph)\}$ unit. The magnetic studies show that **63** is an SMM, with a comparatively low *S* ground state.

The structure of **64**[30] consists of a close packed metallic core in which the manganese ions are bridged by six μ_4-oxo and six μ-hydroxo bridges (Figure 12.32). The central Mn^{II} ion (Mn7) is surrounded by six edge-sharing partial cubane units that are bridged by ten phosphonates. Of the ten phosphonates, four of them show the [3.111] binding mode and the remaining six show the binding mode of [4.211]. Similar to **63**, cage **65** consists of a cyclic core with four

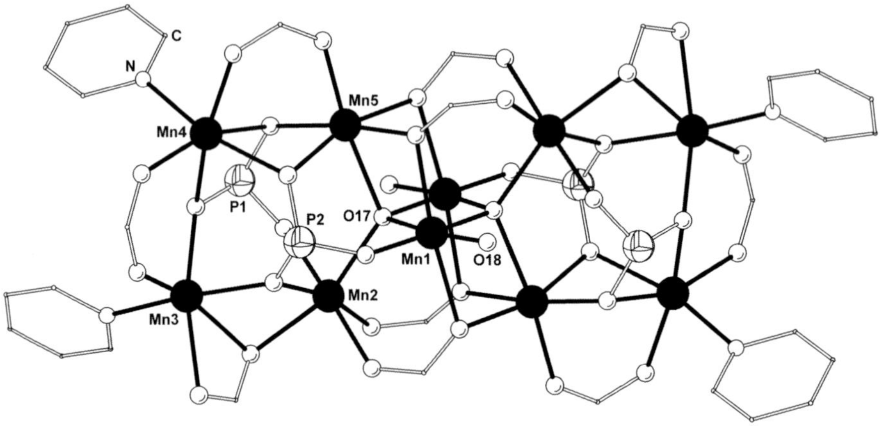

Figure 12.30 The structure of **62** in the crystal.

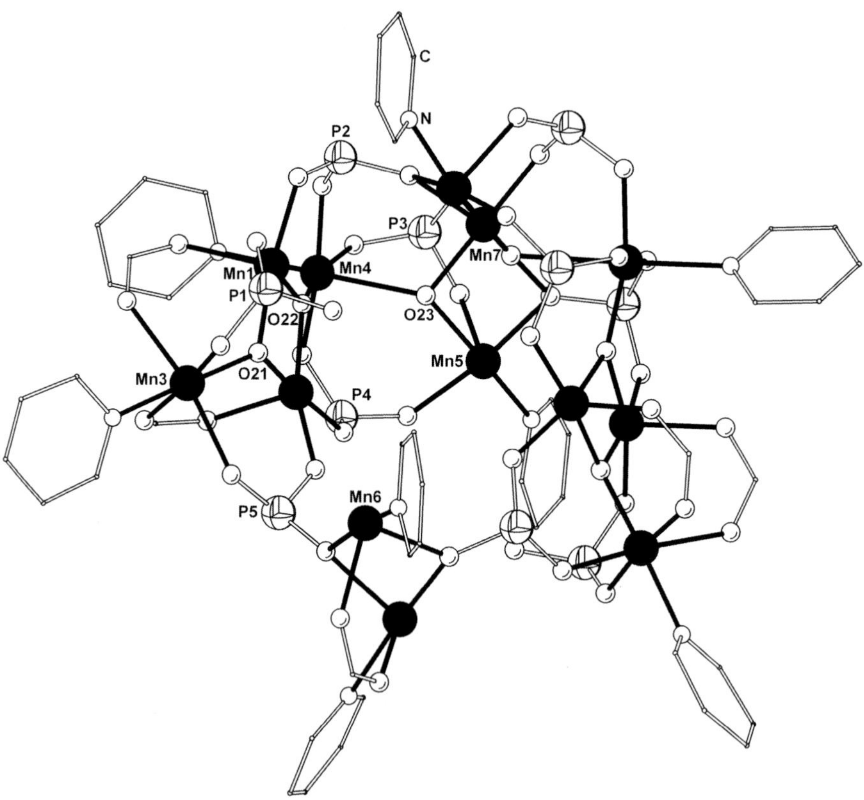

Figure 12.31 The structure of **63** in the crystal.

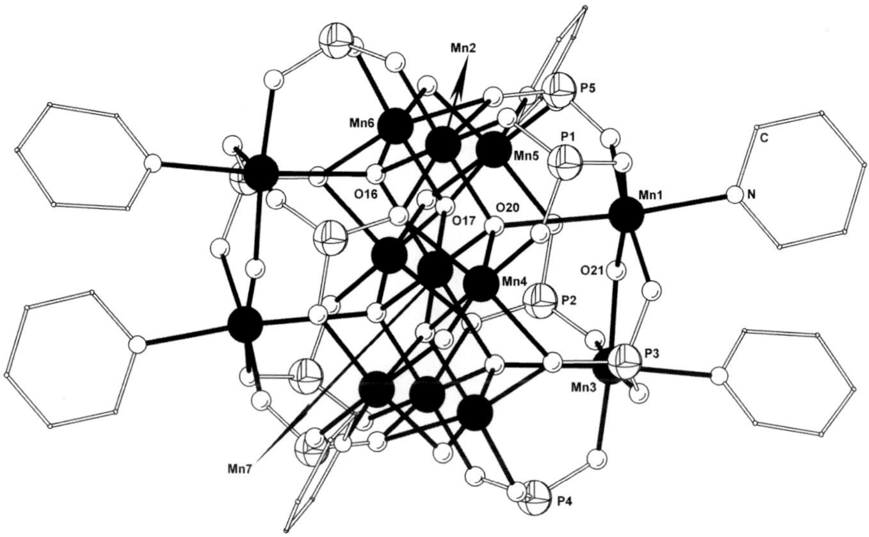

Figure 12.32 The structure of **64** in the crystal.

distinct metal fragments.[30] Two sets of $\{Mn_9(\mu_3\text{-}O)_6\}$ butterflies are bridged by two sets of $\{Mn_2(O_2C\text{-}Ph)\}$ fragments in both the sides *via* two carboxylates and two [4.211] phosphonates (Figure 12.33). The $\{Mn_9(\mu_3\text{-}O)_6\}$ butterfly consists of two $\{Mn_3(\mu_3\text{-}O)\}$ triangles that are centrally bridged by $\{Mn_3(\mu_3\text{-}O)_4\}$ units *via* its μ_3-oxygen. Each butterfly also contains two [5.221] phosphonate bridges.

Konar and Clearfield have reported[31] two high-nuclearity manganese phosphonate cages, $[Mn^{III}{}_{10}Mn^{II}{}_9(\mu_4\text{-}O)_6(\mu_3\text{-}O)_2(O_3P\text{-}C_6H_3\text{-}3,5\text{-}Me_2)_6(O_2C\text{-}t\text{-}Bu)_{10}(OMe)_{16}]$ **66** and $[Mn^{III}{}_{10}Mn^{II}{}_6(\mu_4\text{-}O)_6(\mu_3\text{-}O)_2(O_3P\text{-}CHPh_2)_8(O_2C\text{-}t\text{-}Bu)_6(OMe)_4(HOMe)_2(OH_2)(py)_2]$ **67** by reacting mixed-valent compound $[Mn^{III}{}_2Mn^{II}(\mu_3\text{-}O)(O_2C\text{-}t\text{-}Bu)_6(py)_3]$ and 3,5-dimethylphenylphosphonic acid ($3,5\text{-}Me_2\text{-}C_6H_3\text{-}PO_3H_2$) or diphenylmethylphosphonic acid ($Ph_2CH\text{-}PO_3H_2$). The structure of the cage **66** contains at its centre a $\{Mn^{III}{}_6Mn^{II}(\mu_4\text{-}O)_6\}$ hexagon with six Mn^{III} ions at the corners bonded to a Mn^{II} at the centre through six μ_4-oxygens (Figure 12.34). This central core is bonded to two pairs of $\{Mn_3(\mu_3\text{-}O)\}$ triangles above and below by two sets of three phosphonate units with the binding mode of [6.222].

The structure of **67** has an irregular metal core that can be described as a combination of two subunits, an isovalent $\{Mn_3(\mu_3\text{-}O)\}$ triangle that is bonded to a larger $\{Mn_{13}(\mu_4\text{-}O)_6(\mu_3\text{-}O)\}$ unit through three phosphonates with three different binding modes of [6.222], [5.221] and [4.211] (Figure 12.35). The $\{Mn_{13}(\mu_4\text{-}O)_6(\mu_3\text{-}O)\}$ has a central Mn^{II} (Mn2 in Figure 12.35) bonded to five μ_4-oxides, which are bridged to a total of 10 Mn atoms. It also has four [3.111] and one [5.221] phosphonate bridges.

Two homovalent Mn^{III} cages were reported by Chen *et al.*,[32] using *t*-butylphosphonate. $[Mn_9O_6(O_3P\text{-}t\text{-}Bu)_2(O_2CMe)_{11}(HO_2CMe)(H_2O)]$ **68** contains two oxo-centred Mn_3 triangles linked to a linear Mn_3 unit through oxides and

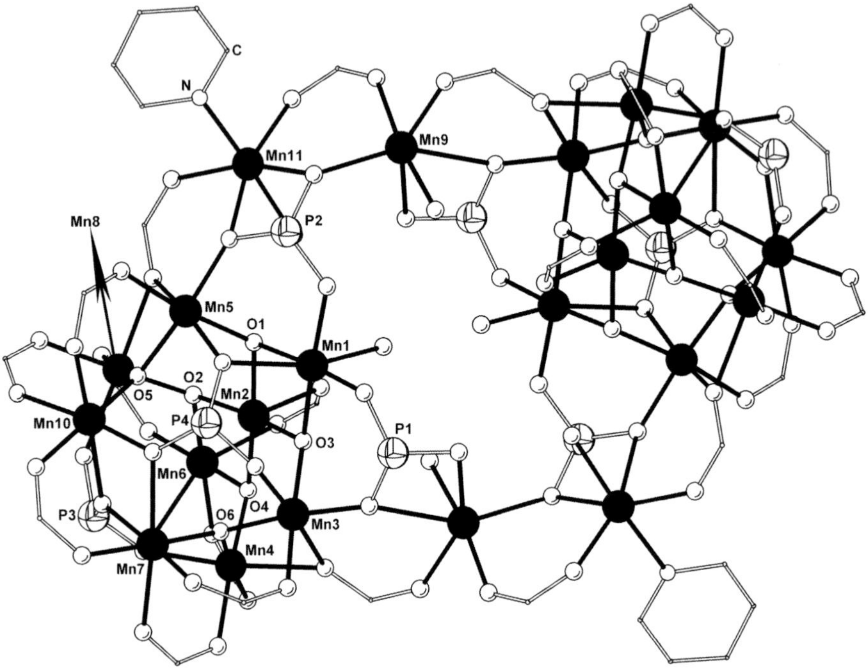

Figure 12.33 The structure of **65** in the crystal.

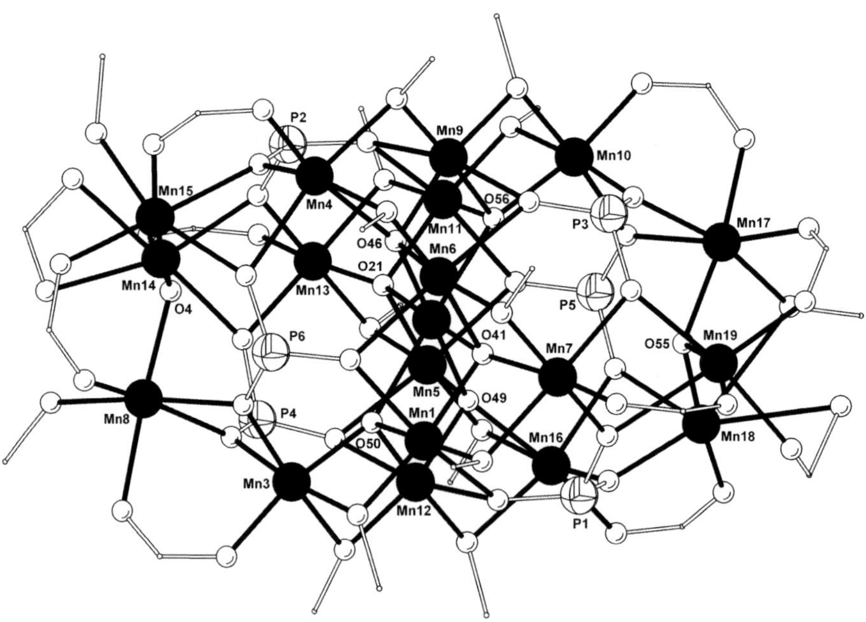

Figure 12.34 The structure of **66** in the crystal.

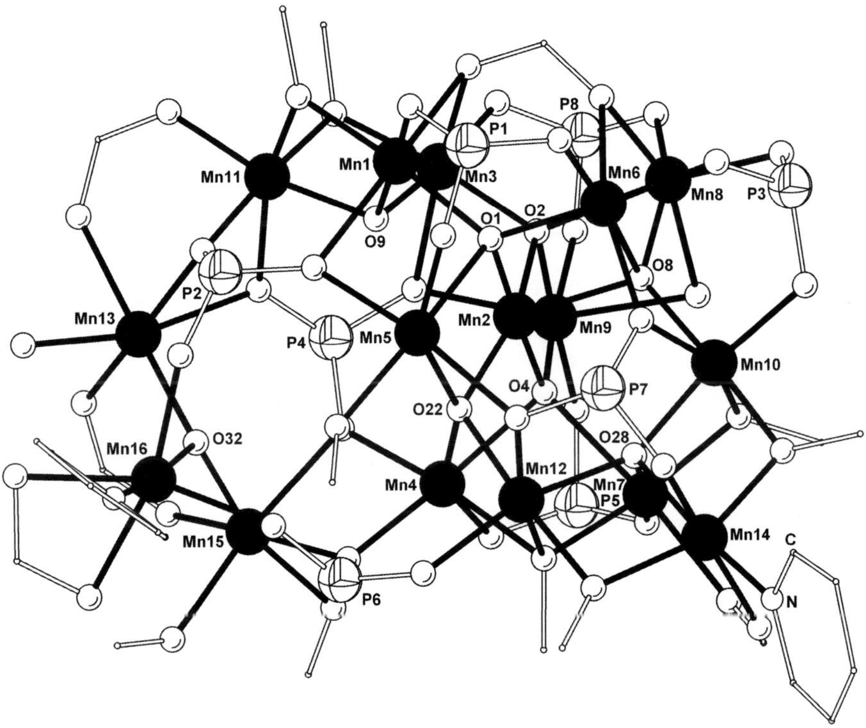

Figure 12.35 The structure of **67** in the crystal.

[5.221] bridging phosphonates. [N*n*-Bu_4][$Mn_{13}O_6(OH)_2(O_3P$-*t*-$Bu)_{10}(N_3)_6$ $(HO_2CMe)_2(H_2O)_2$] **69** has a centred hexagon disc of seven Mn ions at the middle of the structure, with a dimer and a separate isolated Mn^{III} above the disc, and the same fragments below the disc.

Zheng and co-workers have reported[33,34] five new manganese phosphonate cages prepared by the co-ligand method. The cages isolated are [$Mn^{III}_4(\mu_3$-$O)_2(O_3P$-$C_6H_{11})_2(O_2C$-$Me)_4(phen)_2$] **70**, [$Mn^{III}_4(\mu_3$-$O)_2(O_3P$-$C_6H_{11})_2(O_2C$-$Ph)_4(bpy)_2$] **71**, [$Mn^{III}_{12}(\mu_4$-$O)_3(\mu_3$-$O)_5(O_3$-$PC_6H_9)_7(O_2C$-$Me)_6(bpy)_3$] **72**, [$Mn^{III}_{12}(\mu_4$-$O)_3(\mu_3$-$O)_5(O_3$-$PC_6H_9)_7(O_2C$-$Ph)_6(bpy)_3$] **73** and [$Mn^{III}_{12}(\mu_4$-$O)_3$ $(\mu_3$-$O)_5(O_3$-$PC_6H_9)_7(O_2C$-$Ph)_6(phen)_3$] **74** made from the conventional reactions of *n*-Bu_4NMnO_4, manganese(II) acetate or benzoate and different phosphonic acids in the presence of 1,10-phenanthroline (phen) or 2,2′-bipyridine (bpy) as a co-ligand. The structure[32] of the cage **70** contains a {$Mn^{III}_4(\mu_3$-$O)_2$} butterfly core in which the Mn^{III} centres are bridged by [3.111] phosphonates above and below (Figure 12.36). Each phen ligand is coordinated to a single Mn^{III} atom at both ends. The structure of the cage **71** is similar to that of **70** with bpy coordination. Further examples of this tetranuclear core have also been reported, varying the phosphonate and the carboxylate:[35,36] [Mn^{III}_4 $(\mu_3$-$O)_2(O_3P$-$Me)_2(O_2C$-$Me)_4(bpy)_2$].4$MeCO_2H$ **75**,[35] [$Mn^{III}_4(\mu_3$-$O)_2(O_3P$-$Et)_2$ $(O_2C$-$Me)_4(bpy)_2$].7H_2O **76**,[35] [$Mn^{III}_4(\mu_3$-$O)_2(O_3P$-*n*-$Bu)_2(O_2C$-$Me)_4(bpy)_2$].

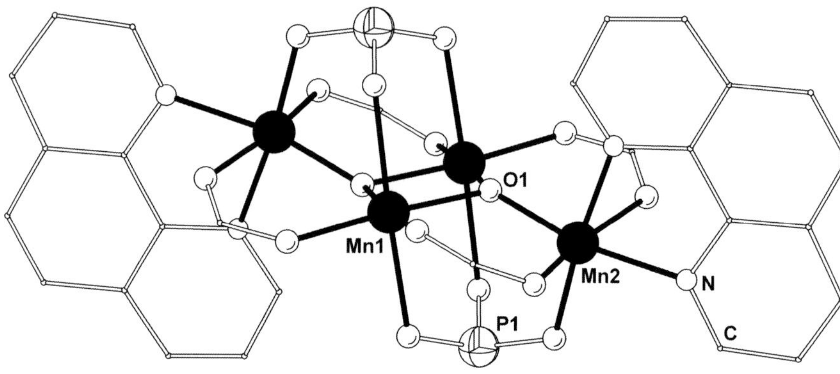

Figure 12.36 The structure of **70** in the crystal.

$4H_2O.(CO_2H)_2$ **77**,[35] $[Mn^{III}_4(\mu_3\text{-}O)_2(O_3P\text{-}Me)_2(O_2C\text{-}Ph)_4(phen)_2].4Ph\text{-}CO_2H.2MeCN$ **78**,[35] $[Mn^{III}_4(\mu_3\text{-}O)_2(O_3P\text{-}t\text{-}Bu)_2(O_2C\text{-}Me)_4(bpy)_2]$ **79**[36] and $[Mn^{III}_4(\mu_3\text{-}O)_2(O_3P\text{-}t\text{-}Bu)_2(O_2C\text{-}Ph)_4(bpy)_2]$ **80**.[36]

The structure of compound **72** contains a $\{Mn^{III}_{12}(\mu_4\text{-}O)_3(\mu_3\text{-}O)_5(\mu\text{-}O_3P)_3\}$ core, which can be described as three distinct parallel layers (Figure 12.37).[34] The bottom layer consists of a $\{Mn^{III}_3(\mu_3\text{-}O)\}$ triangle containing a [3.111] phosphonate unit, which is connected to the middle layer that contains six Mn^{III} ions through three μ_4-oxygens and three phosphonates with the binding mode of [3.111]. The middle layer consists of a $\{Mn^{III}_6(\mu_4\text{-}O)_3(\mu_3\text{-}O)_4\}$ fragment, which can be described as three edge-sharing partial-cubane units. The upper layer consists of three $\{Mn^{III}(bpy)\}$ fragments and is connected to the middle layer *via* three μ_3-oxygens and three phosphonates with the binding mode of [4.211]. The structure of the other dodecanuclear cages **73** and **74** are very similar to that of **72**.

Chen and co-workers have reported[37] two high nuclearity manganese phosphonate cages: $[Mn^{III}_6Mn^{II}_9(\mu_4\text{-}O)_6(O_3P\text{-}Me)_2(O_2C\text{-}Me)_{18}(OH_2)_{12}]^{2+}[Me\text{-}PO_3H^-]_2\cdot 12H_2O$ **81** and $[Mn^{III}_{20}Na_4(\mu_3\text{-}O)_{12}(O_3P\text{-}t\text{-}Bu)_{12}(O_2C\text{-}Me)_{16}(OH_2)_{12}]\cdot 5H_2O$ **82**. The structure of the cage **81** contains three vertex-sharing $\{Mn^{III}_2Mn^{II}_4(\mu_4\text{-}O)_2\}$ bitetrahedral units in a cyclic arrangement and are centrally bridged by two phosphonates of [6.222] binding mode (Figure 12.38).

The core of complex **82** is composed of two $\{Mn^{III}_4(\mu_3\text{-}O)_2\}$ butterflies and four oxo-centred Mn_3 triangles in a cyclic arrangement (Figure 12.39). The butterflies are on two sides of the cage, linked to a triangle at each end. The triangles are arranged in stacked pairs, forming distorted trigonal prisms similar to those found in iron(III) cages **26** and **27**. Four sodium ions are found within the cavity of the cage. The phosphonates within the structure show the [4.211] mode.

Two pentanuclear cages have also been reported[36] from similar reactions involving carboxylates and phen. $[Mn^{III}_5(\mu_3\text{-}O)_3(O_3P\text{-}t\text{-}Bu)_2(O_2C\text{-}Me)_5(phen)_2(OH_2)]$ **83** and $[Mn^{III}_5(\mu_3\text{-}O)_3(O_3P\text{-}t\text{-}Bu)_2(O_2C\text{-}Ph)_5(phen)_2]$ **84** have

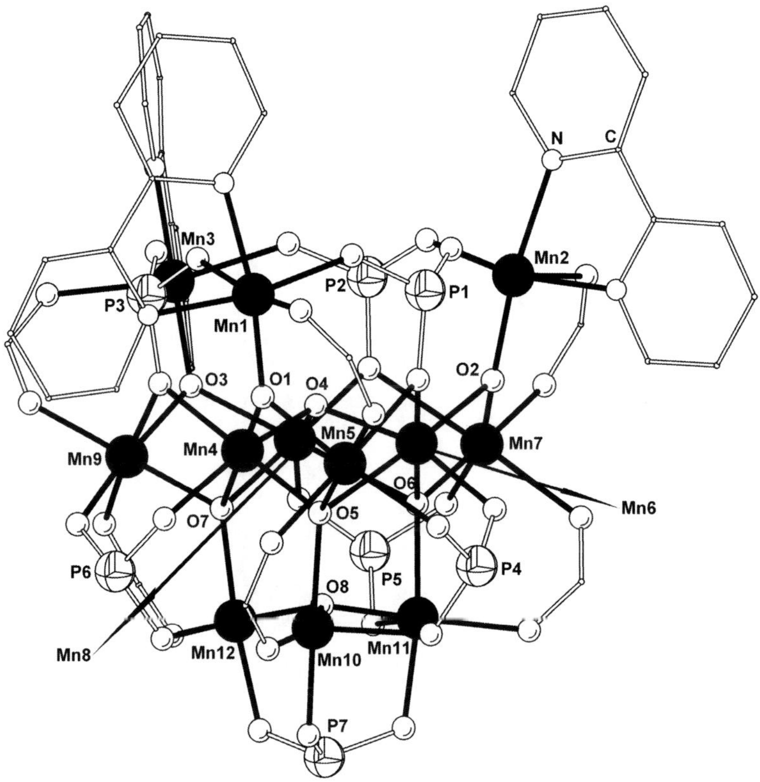

Figure 12.37 The structure of **72** in the crystal.

basket-like cores in which the five Mn^{III} ions are bridged by three μ_3-oxygens and two phosphonates with the binding mode of [4.211] and [3.111] (Figure 12.40).

Recently, we have isolated[38] a decanuclear cage $[Mn^{III}_4Mn^{II}_6K_3(O_3P\text{-}Ph)_2(O_2C\text{-}Ph)_{10}(tea)_2(tea\text{-}H)_4](PhCO_2) \cdot 6MeCN$ **85** from the conventional or solvothermal reaction of $KMnO_4$, $Ph\text{-}PO_3H_2$ and $Ph\text{-}CO_2H$ in the presence of triethanolamine (tea-H_3) as co-ligand. The structure of the cage **85** has a planar cyclic core comprising six Mn^{II} and four Mn^{III} ions (Figure 12.41). Three potassium ions present in the middle of the structure are capped by different oxygen and nitrogen donors. The two phosphonates show the binding mode of [5.221].

12.3.4 Vanadium(III) Phosphonate Cages

As we mentioned at the beginning of the chapter, beautiful high-valent vanadium(IV/V) phosphonate cages were reported by Zubieta and co-workers several years ago.[5] More recently, low-valent vanadium(III) and mixed-valence vanadium(III/IV) phosphonate cages have been reported.[13] These include six new cages, $[V^{III}_4(\mu_4\text{-}O)(\mu\text{-}OH)_2(O_3P\text{-}CPh_3)_4(py)_4]$ **86**, $[V^{III}_5(\mu_3\text{-}OH)(O_3P\text{-}t\text{-}$

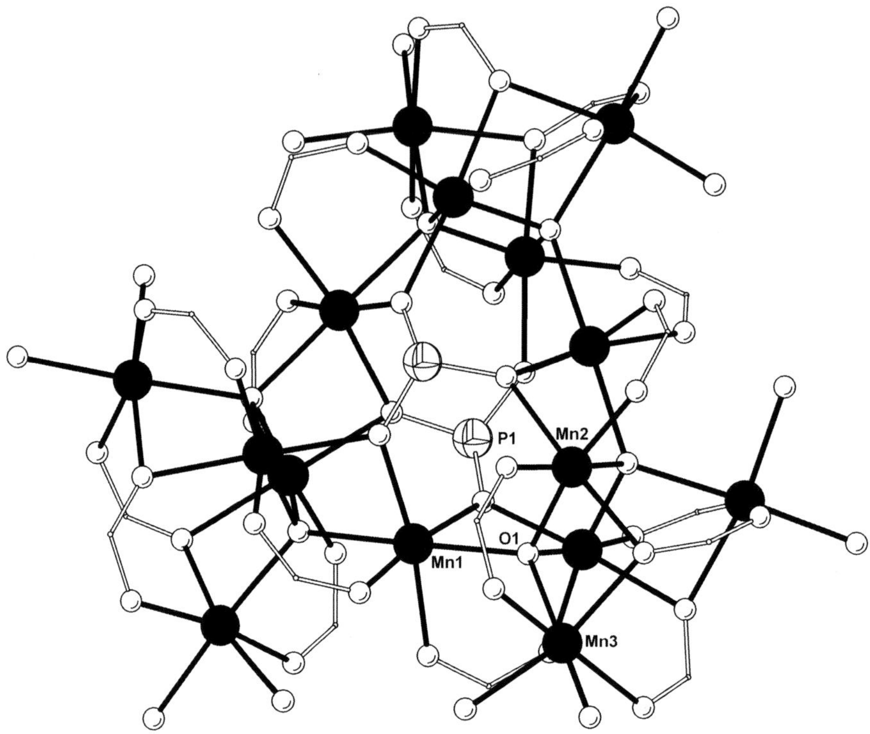

Figure 12.38 The structure of **81** in the crystal.

Bu)$_6$Cl$_2$(py)$_6$] **87**, [$V^{III}{}_6(\mu_3$-O)$_2$(O$_3$P-*t*-Bu)$_2$(HO$_3$P-*t*-Bu)$_2$(O$_2$C-*t*-Bu)$_8$(HO$_2$C-*t*-Bu)$_2$] **88**, [$V^{III}{}_8(\mu_4$-O)$_2$(μ-OH)$_2$(O$_3$P-CH$_2$Np)$_8$(EtOH)$_8$](Cl)$_2$ **89**, [$V^{III}{}_9(\mu_3$-O)$_4$(O$_3$P-*t*-Bu)$_3$(O$_2$C-*t*-Bu)$_{13}$] **90** and [$V^{III}{}_{12}$(V^{IV}O)(μ_3-OH)$_4$(μ_2-OH)$_8$(μ_2-OEt)$_4$(O$_3$P-*t*-Bu)$_8$(O$_2$C-Ph)$_4$(EtOH)$_4$](Cl)$_2$ (**109**) from the reactions of solid VCl_3 or [VCl_3(THF)$_3$] and phosphonic acids under ambient pressure or solvothermal conditions.[13]

The structure of the tetranuclear cage **86** is based on a μ_4-oxide-centred tetrahedron of V^{III} ions (Figure 12.42). It shows a D_{2d} distortion from a perfect tetrahedron, with opposite edges of the tetrahedron having a μ-hydroxide group inserted into the V---V edge. Each face of the tetrahedron is capped by a bridging phosphonate with binding mode of [3.111]. The structure of the pentanuclear cage **87** has the V^{III} sites on the vertices of a trigonal bipyramid, and the [3.111] bound phosphonates occupy the six triangular faces of this polyhedron (Figure 12.43). The structure of the hexanuclear cage **88** is made of two {$V^{III}{}_3(\mu_3$-O)} triangles that are centrally connected by two di-anionic [4.211] phosphonates and two mono-anionic [2.110] phosphonates. Overall, the core of the structure resembles the hexanuclear manganese phosphonate cage **53** (see Figure 12.23).

The structure of complex **89** contains a central {$V^{III}{}_6(\mu_3$-O)$_2$} bitetrahedral core connected on each side to an isolated V^{III} ion by an μ-hydroxide and two

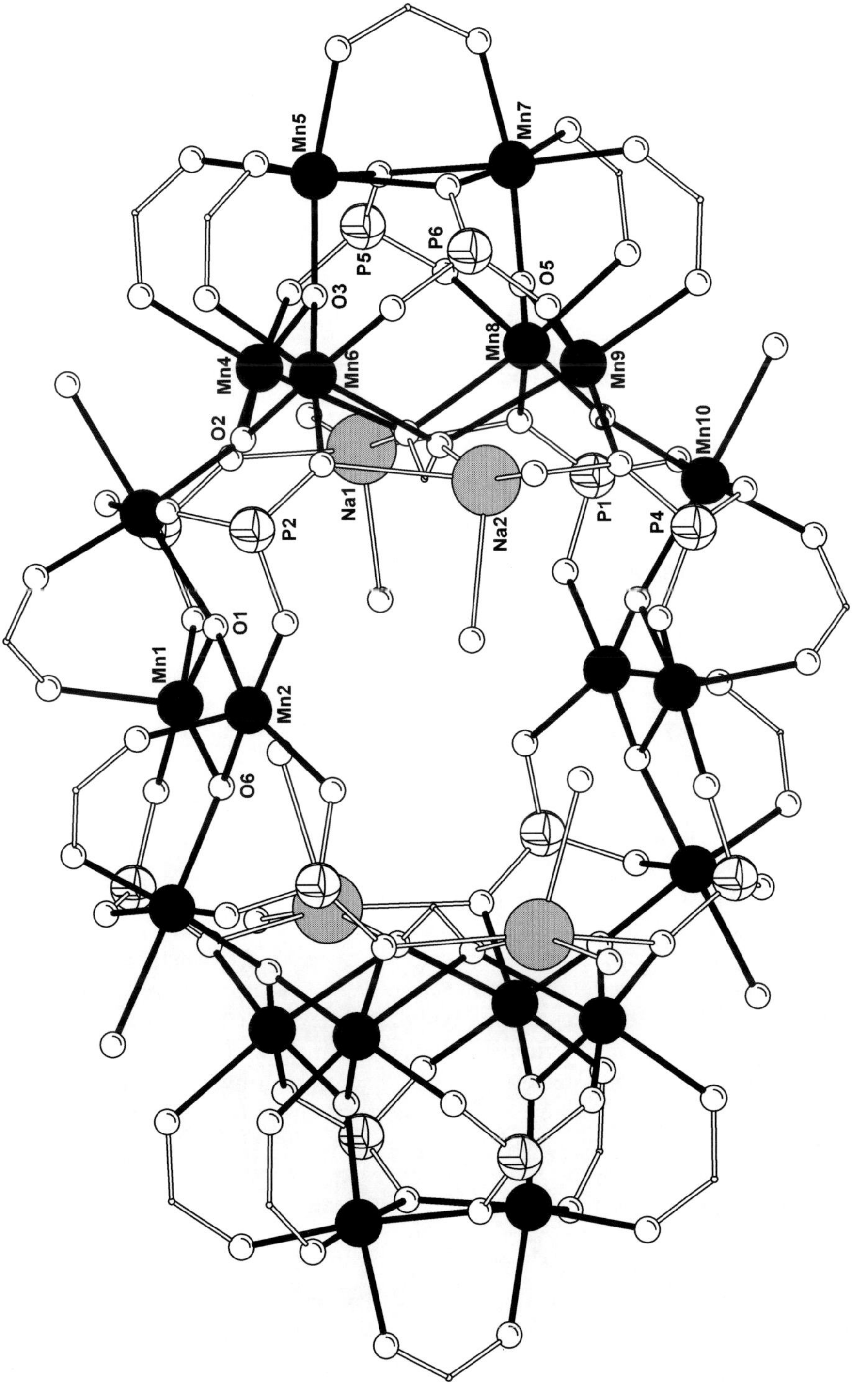

Figure 12.39 The structure of **82** in the crystal.

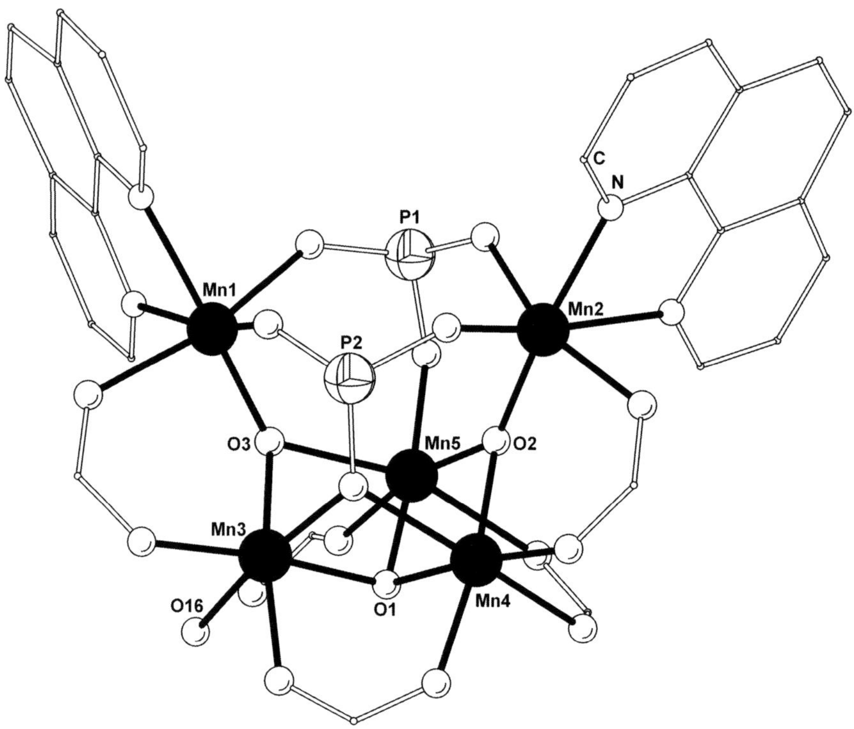

Figure 12.40 The structure of **83** in the crystal.

phosphonates adopt a [4.211] binding mode (Figure 12.44). The central $\{V^{III}{}_6(\mu_3\text{-}O)_2\}$ bitetrahedral core itself contains four phosphonate bridges with the binding mode of [3.111].

The structure of the cage **90** is analogous to the Fe-phosphonate compounds **28**, **37** and **48** (see Figure 12.12). The core of **90** is a tridiminished icosahedron of V^{III} sites with three phosphonates adopting a [5.221] binding mode. It contains of two fragments, a oxo-centred $\{V^{III}{}_3(\mu_3\text{-}O)\}$ triangle linked *via* phosphonates to a $\{V^{III}{}_6(\mu_3\text{-}O)_3\}$ triangle of oxo-centred triangles.

The structure of **91** is composed of four hydroxo-centred $\{V^{III}{}_3(\mu_3\text{-}OH)\}$ triangles and each one is connected to two others by an μ-hydroxide and [4.211] phosphonate bridges (Figure 12.45). Also, each hydroxo-centred $\{V^{III}{}_3(\mu_3\text{-}OH)\}$ triangle contain one phosphonate cap with a [3.111] binding mode. The resultant $\{V^{III}{}_{12}(\mu_3\text{-}OH)_4(\mu\text{-}OH)_4\}$ core is in a cyclic arrangement with a vanadyl ion, $\{V^{IV}O\}$ bound to the centre of the cage, linked *via* four other μ-hydroxide units.

Use of pre-formed polymetallic starting materials such as $[V_2OCl_4(THF)_6]$ or the oxo-centred triangle, $[V^{III}{}_3(\mu_3\text{-}O)(O_2C\text{-}t\text{-}Bu)_6(Py)_3][ClO_4]$, leads to two further new cages including: $[V^{III}{}_6(\mu_3\text{-}O)_2(\mu\text{-}Cl)_4(O_3P\text{-}t\text{-}Bu)_4(Cl)_2(py)_8]$ **92** and $[V^{III}{}_8(\mu_3\text{-}O)_2(O_3P\text{-}CH_2Ph)_6(HO_3P\text{-}CH_2Ph)_8(py)_8]$ **93**.[39] The structure of the cage **92** is composed of two oxo-centred $\{V^{III}{}_3(\mu_3\text{-}O)\}$ triangles connected by

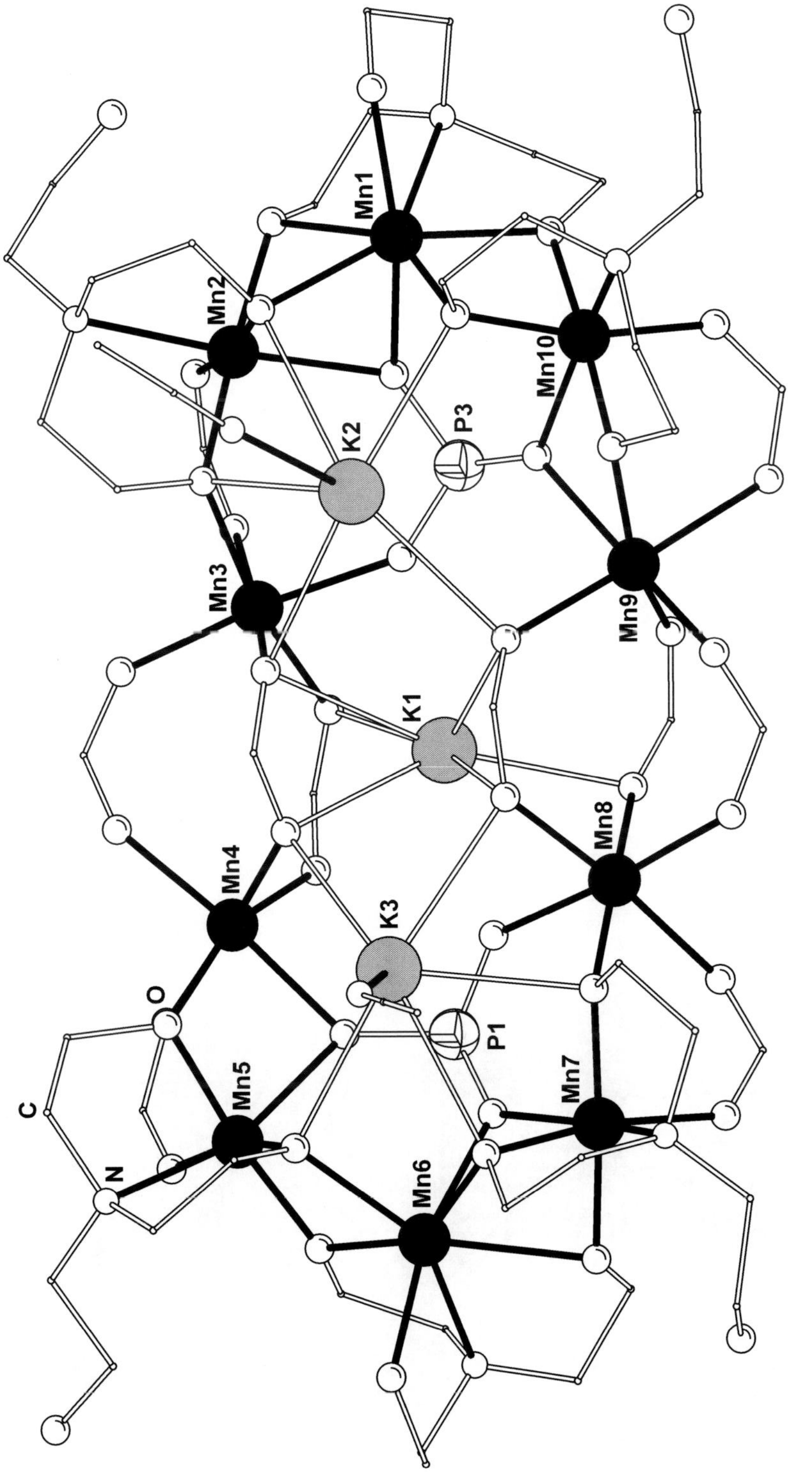

Figure 12.41 The structure of **85** in the crystal.

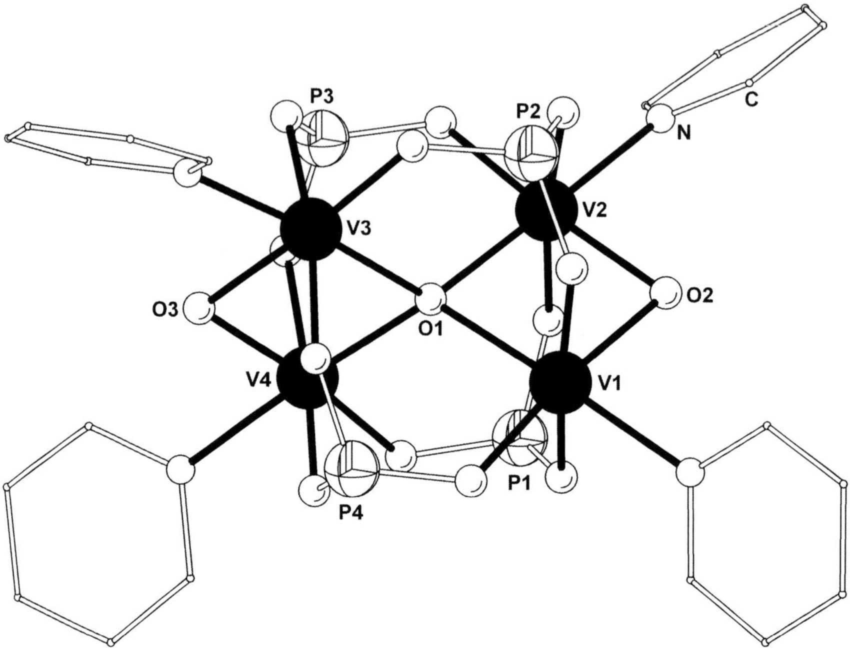

Figure 12.42 The structure of **86** in the crystal.

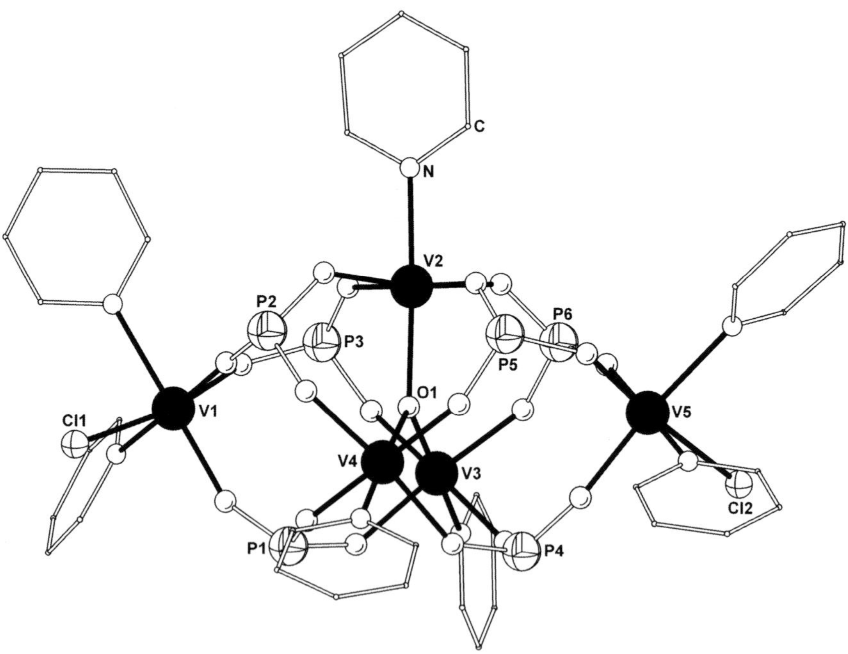

Figure 12.43 The structure of **87** in the crystal.

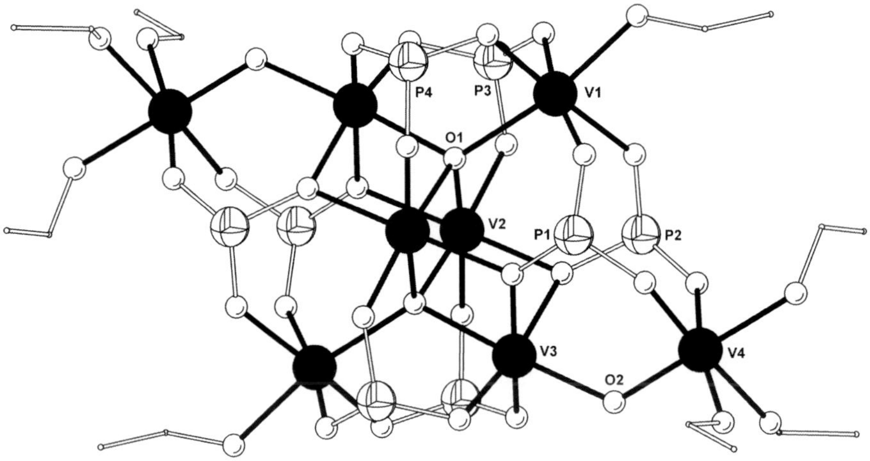

Figure 12.44 The structure of **89** in the crystal.

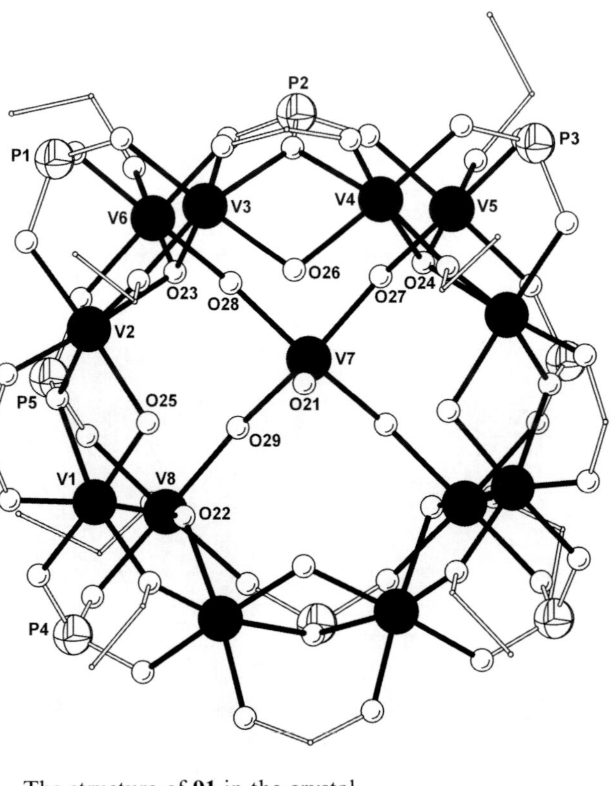

Figure 12.45 The structure of **91** in the crystal.

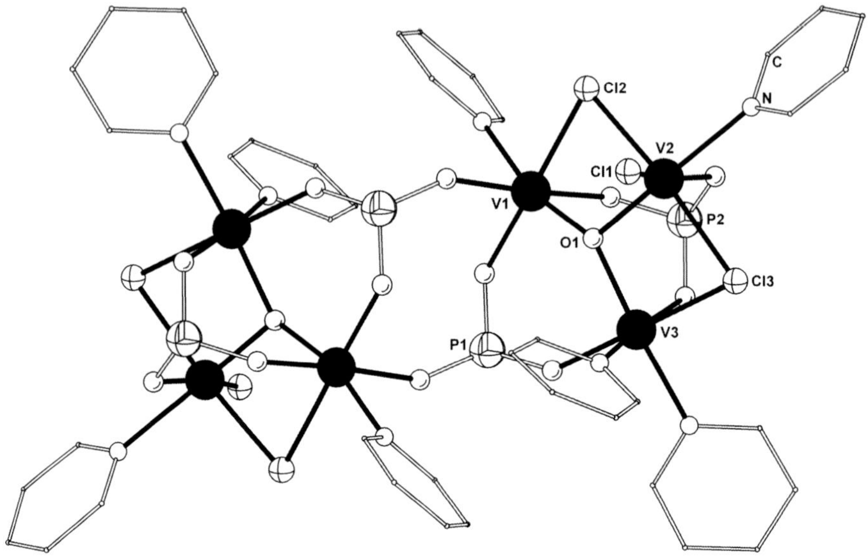

Figure 12.46 The structure of **92** in the crystal.

two [4.211] phosphonate bridges (Figure 12.46). Each triangle is capped by a [3.111] phosphonate and two of the three edges are bridged by μ-chlorides.

The structure of the cage **93** is composed of two tetranuclear $\{V^{III}_4(\mu_3\text{-}O)\}$ fragments that are linked by two mono-anionic phosphonates with the binding mode of [2.110] (Figure 12.47). Each tetranuclear $\{V^{III}_4(\mu_3\text{-}O)\}$ fragment is made of an oxo-centred $\{V^{III}_3(\mu_3\text{-}O)\}$ triangle which is connected to the fourth V^{III} site through three [3.111] phosphonate bridges. In addition, each triangle is capped by three mono-anionic phosphonates with the binding mode of [2.110] which is very unusual for oxo-centred triangles of metal ions. Unlike reactions of oxo-centred iron and manganese triangles, no carboxylate is retained in the final product. This suggests that V^{III} may be more susceptible to substitution.

Konar and Clearfield have reported[40] chloride encapsulation in two dodecanuclear cages, $[Et_3NH^+]_2[(V^VO)_8(V^{IV}O)_4(\mu\text{-}O)_8(O_3P\text{-}CHPh_2)_8(OH_2)_{12}][Cl^-]_2 \cdot 7H_2O \cdot 4MeCN$ **94** and $[Py\text{-}H^+]_5[(V^VO)_{12}(\mu\text{-}O)_{12}(O_3P\text{-}CHPh_2)_6(OH_2)_6][Cl^-]_5 \cdot 6Py \cdot 10H_2O \cdot 3MeCN$ **95**. The structure of the cage **94** reveals that the structure consists of a pair of Et_3NH^+ cations and a neutral mixed-valent vanadium (V^V and V^{IV}) cluster that encapsulates two chloride anions (Figure 12.48). It can be described as consisting of two tetranuclear cationic $\{V^V_4O_8\}^{4+}$ caps, bridged symmetrically to four independent $\{V^{IV}O(OH_2)_3\}^{2+}$ units through two [3.111] phosphonates. The compound could also be described as a combination of two positively charged $\{V^V_4O_8\}^{4+}$ caps bridged by four hypothetical negatively charged complex anions, $\{VO(OH_2)_3(O_3P\text{-}CHPh_2)_2\}^{2-}$. The two encapsulated chloride anions are located inside the cluster, very near to the positively charged $\{V^V_4O_8\}^{4+}$ caps.

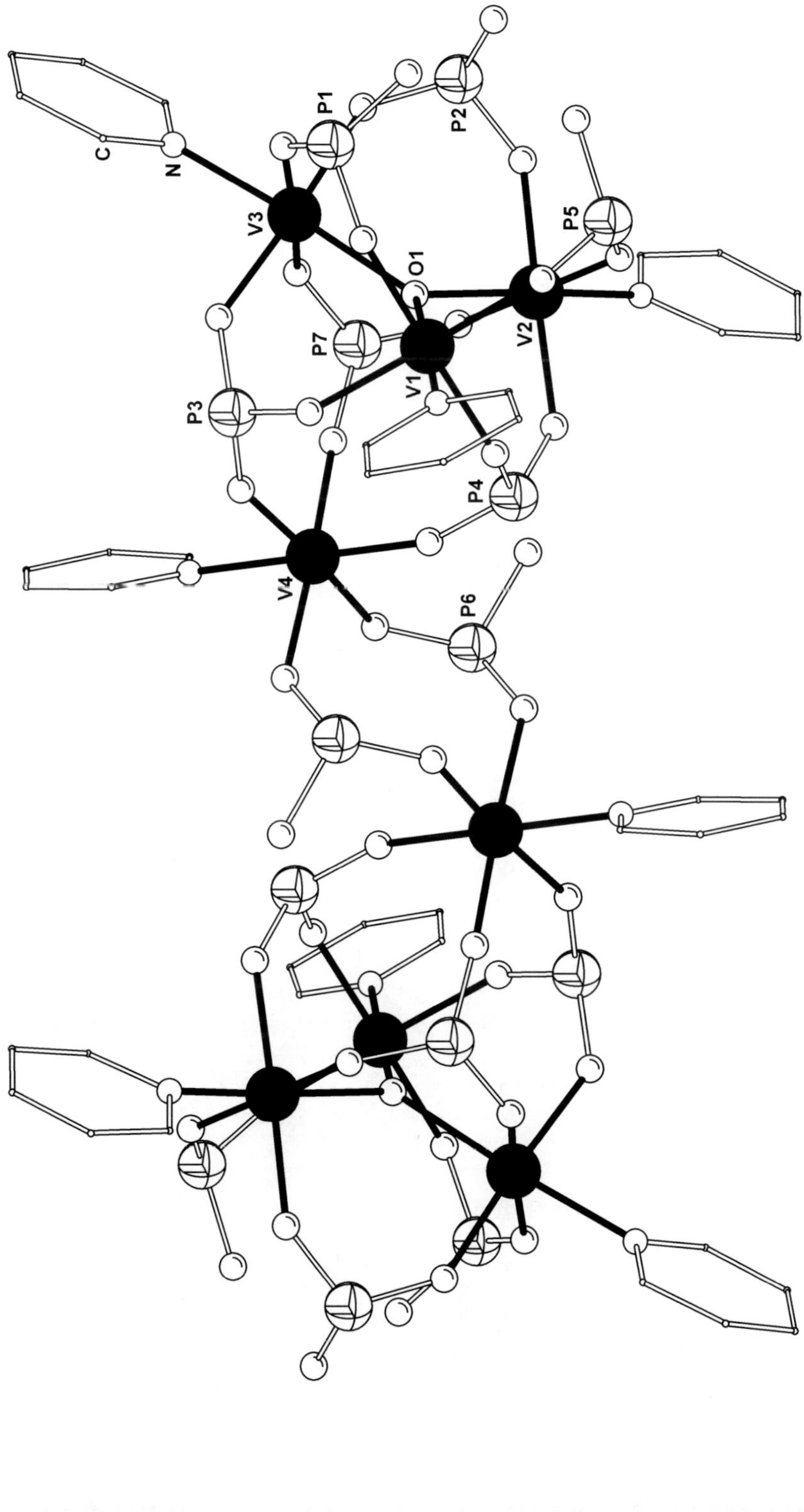

Figure 12.47 The structure of **93** in the crystal.

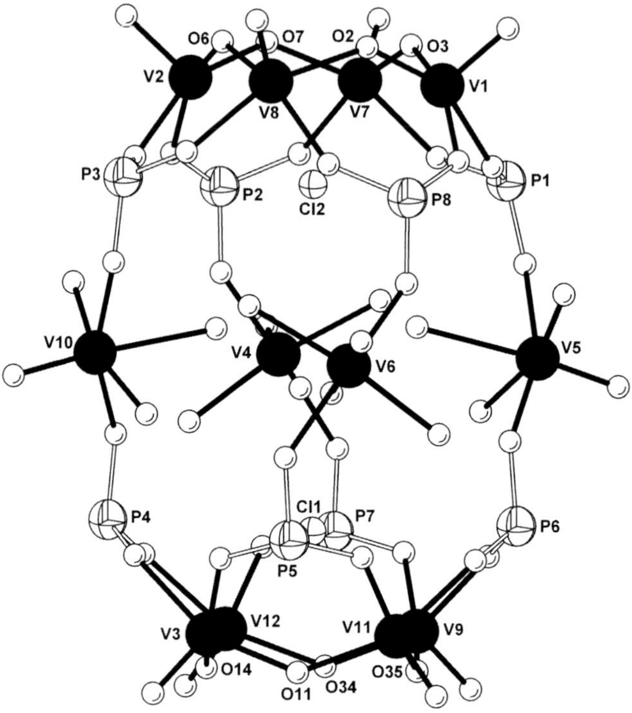

Figure 12.48 The structure of **94** in the crystal.

The structure of the cage **95** is composed of two open bowl-shaped neutral $\{(V^VO)_6(\mu\text{-}O)_6(O_3P\text{-}CHPh_2)_3(OH_2)_3\}$ cages that are bridged by three chloride anions *via* strong O–H...Cl hydrogen bonds (Figure 12.49). In addition, each bowl encapsulates one chloride anion in its centre. Each bowl cage consists of six $\{V^VO\}$ units arranged in a trigonal prism with six of the edges containing μ-oxide bridges. Three of the faces of the prism are capped by three phosphonates with the binding mode of [3.111]. Each of the three V^V sites at the edge are bonded with a water molecule that pointed toward the open face of another cage. There is a mirror plane between the two open bowls, three chloride anions are positioned on that plane and each of the chloride anions is strongly hydrogen-bonded to stick the two bowls face-to-face.

12.3.5 Nickel(II) Phosphonate Cages

Only one paper on nickel phosphonate cages has thus far appeared.[41] Three multinuclear nickel phosphonate cages were reported, made from reaction of a dinickel pivalate complex, $[Ni^{II}_2(\mu\text{-}OH_2)(O_2C\text{-}t\text{-}Bu)_4(HO_2C\text{-}t\text{-}Bu)_4]$, with either phenyl- or *t*-butyl-phosphonate. The cages $[Ni^{II}_8(\mu_3\text{-}OH)_4(\mu\text{-}OH_2)_2(O_3P\text{-}CH_2Ph)_2(O_2C\text{-}t\text{-}Bu)_8(HO_2C\text{-}t\text{-}Bu)_6]$ **96**, $[Ni^{II}_9(\mu_3\text{-}OH)_3(\mu\text{-}OH_2)(O_3P\text{-}CH_2Ph)$

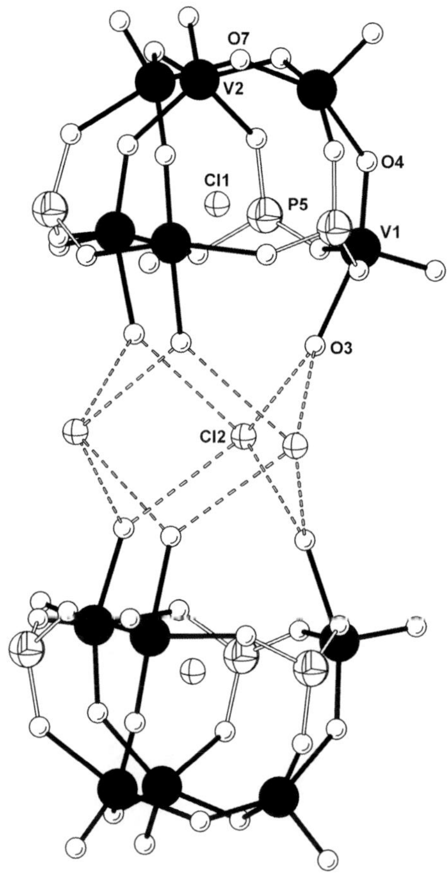

Figure 12.49 The structure of **95** in the crystal. The O-H···Cl hydrogen bondings are shown in dotted lines.

$(HO_3P\text{-}CH_2Ph)(O_2C\text{-}t\text{-}Bu)_8(chp)_4(HO_2C\text{-}t\text{-}Bu)_2(OH_2)]$ **97** and $[Ni^{II}_{12}(\mu_3\text{-}OH)_4(O_3P\text{-}Ph)_4(O_2C\text{-}t\text{-}Bu)_{12}(L)_6]$ **98** (where L = disordered mixture of HO_2C-Me, MeCN, H_2O and or HO_2C-*t*-Bu) were isolated. The structure of the octa-nuclear cage **96** consists of two puckered $\{Ni^{II}_4(\mu_3\text{-}OH)_2\}$ butterfly cores that are connected by two μ-water and two [5.221] phosphonates (Figure 12.50). The structure of the cage **97** contains eight Ni^{II} ions arranged in a similar, but not identical core to that of **96**. An additional Ni^{II} ion is found in the cage which leads to one of the phosphonates binding in a [6.222] mode and the other in a [5.221] mode (Figure 12.51).

The core of complex **98** is made of four $\{Ni^{II}_3(\mu_3\text{-}OH)\}$ triangles that are connected to each other by four phosphonates with a binding mode of [6.222], to give a spherical cage (Figure 12.52). The core can be thought of as an ε-Keggin ion with the central tetrahedral atom absent.

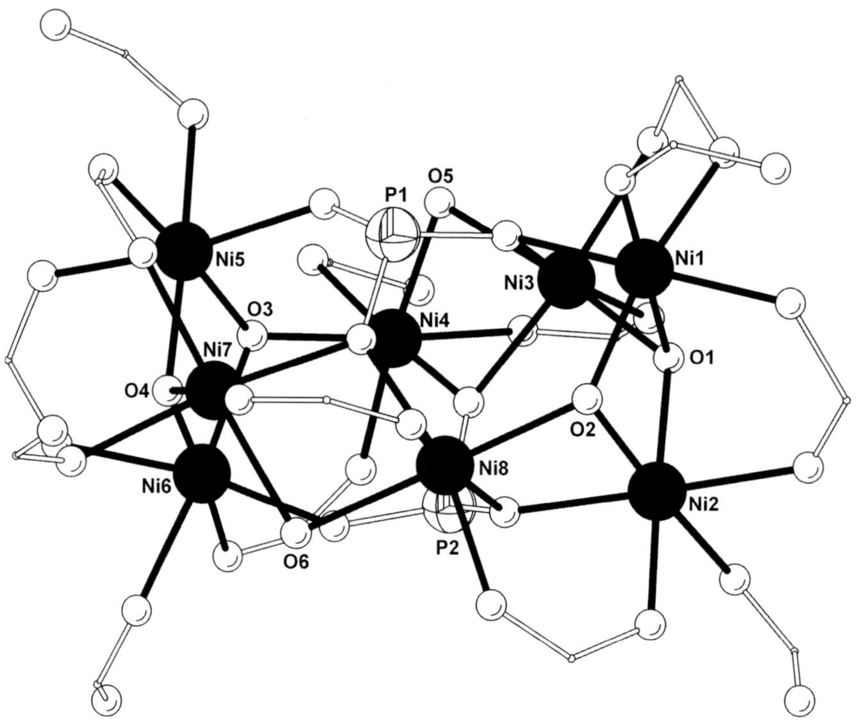

Figure 12.50 The structure of **96** in the crystal.

12.3.6 Heterometallic Phosphonate Cages

It is also possible to isolate heterometallic phosphonate cages, including the combination of 3d–4f and 3d–3d metal ions. We have reported[42] three new cages based on Cu^{II}–Ln^{III}. The formula of these cages are $[Cu^{II}_{24}Dy^{III}_{8}(\mu_3\text{-}OH)_{24}(\mu\text{-}OH)_{18}(\mu_9\text{-}NO_3)(O_3P\text{-}CPh_3)_6(HO_3P\text{-}CPh_3)_6(O_2C\text{-}Me)_{12}(HO_2C\text{-}Me)_6(OH_2)_6]$ **99**, $(Me_4N)_2K_2[Cu^{II}_{24}Gd^{III}_{8}(\mu_3\text{-}OH)_{24}(\mu\text{-}OH)_{18}(\mu_9\text{-}NO_3)(O_3P\text{-}CPh_3)_6(HO_3P\text{-}CPh_3)_6(O_2C\text{-}Me)_{12}(HO_2C\text{-}Me)_{12}]$ **100**, $[Me_4N]_3[Cu^{II}_{16}Y^{III}_{12}(\mu_3\text{-}OH)_{22}(\mu\text{-}OH)_2(\mu_7\text{-}NO_3)_2(O_3P\text{-}CPh_3)_{12}(O_2C\text{-}Me)_{22}(OH_2)_6]$ **101**. The structure of **99** consists of eight Dy^{III} centres at the vertices of a cube with twelve Cu^{II} centres inside and twelve Cu^{II} centres outside the cube (Figure 12.53). The twelve inner Cu^{II} ions are arranged in the form of a cuboctahedron which are connected by one NO_3^- and twenty-four OH^- bridges. Each Dy^{III} caps a triangular face of the Cu^{II}_{12} cuboctahedron; hence forming a $Dy^{III}Cu^{II}_3$ tetrahedron with each Cu^{II} atom being part of two such tetrahedra. Further, each square face of the Dy^{III} cube is capped externally by a dimer of $\{Cu^{II}_2(\mu\text{-}OH)\}$ unit; hence twelve Cu^{II} (six dimers) surround the Dy^{III}_8 cube. Of the twelve phosphonates present, six are dianionic with the binding mode of [3.111], while the other six being monoanionic with a [2.110] binding mode. The structures of **99** and **100** are very similar in the arrangement of the metal atoms with slight

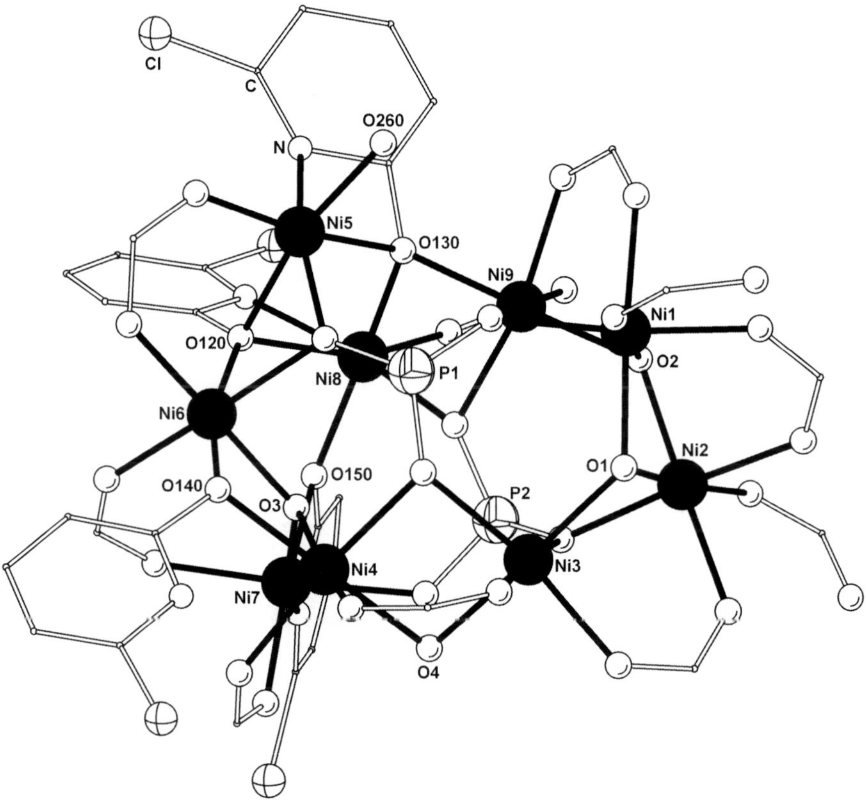

Figure 12.51 The structure of **97** in the crystal.

difference in the binding modes of ligands (carboxylates and phosphonates). A.c. susceptibility and micro-SQUID measurements show that **99** is an SMM with a very small energy barrier ($U_{\rm eff} \approx 4.6$ K) and $\tau_0 = 2.1 \times 10^{-8}$ s. Compound **100** is not an SMM, but as the magnetization saturates rapidly to a value of 68 Bohr magnetons, it may be useful in low temperature magnetic cooling.[42]

The structure of complex **101**[42] consists of two sets of $[\mathrm{Cu^{II}}_8\mathrm{Y^{III}}_6(\mu_3\text{-OH})_{11}(\mu\text{-OH})(\mu_7\text{-NO}_3)(\mathrm{O_3P\text{-}CPh_3})_6(\mathrm{O_2C\text{-}Me})_{10}(\mathrm{OH_2})_3]$ clusters that are centrally bridged by two acetate units (Figure 12.54). Each set of cluster is a very irregular arrangement and contains eight $\mathrm{Cu^{II}}$ and six $\mathrm{Y^{III}}$ atoms connected together by $\mathrm{OH^-}$ and $\mathrm{NO_3}^-$ bridges. It contains six dianionic phosphonates, of which four exhibit the [3.111] binding mode while the two other exhibit the binding mode of [4.211].

Two heterometallic $\mathrm{Cr^{III}}$–$\mathrm{Co^{II}}$ cages have also been reported:[43] $[\mathrm{Co^{II}}_4\mathrm{Cr^{III}}_4(\mu_3\text{-OH})_4(\mu\text{-OEt})_4(\mathrm{O_3P\text{-}}t\text{-Bu})_2(\mathrm{O_2C\text{-}}t\text{-Bu})_6(\mathrm{OEt})_2(\mathrm{HO_2C\text{-}}t\text{-Bu})_4]$ **102** and $[\mathrm{Co^{II}}_8\mathrm{Cr^{III}}_4(\mu_3\text{-OMe})_2(\mu\text{-OMe})_{10}(\mathrm{O_3P\text{-}}t\text{-Bu})_4(\mathrm{O_2C\text{-}}t\text{-Bu})_8(\mathrm{HOMe})_6]$ **103**. The structure of the cage **102** can be described as two $\{\mathrm{Co^{II}}_2\mathrm{Cr^{III}}_2(\mu\text{-OH})_2\}$ butterfly

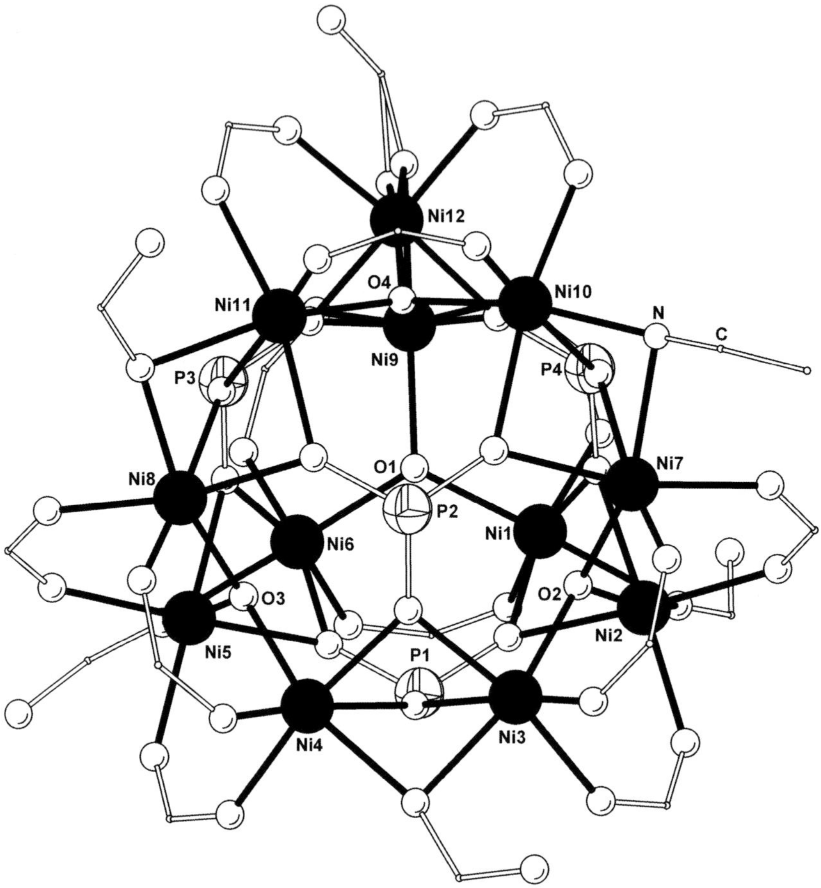

Figure 12.52 The structure of **98** in the crystal.

cores are bridged together by two μ-ethoxide and two phosphonates that adopt an [5.221] binding mode (Figure 12.55a). The structure of the cage **103** features two edge-sharing $\{Co^{II}{}_4Cr^{III}{}_2\}$ hexagons with the outside edge containing a $[Co^{II}Cr^{III}{}_2(\mu_3\text{-}OMe)(\mu\text{-}OMe)_3]$ cube fragment with a vertex missing (Figure 12.55b). There are two phosphonate ligands, one located in the centre of each hexagon with the binding mode of [6.222]. Two unusual [3.220] phosphonates are disordered with two [3.22] pivalate ligands, these chelate and bridge the Co^{II} sites of the shared edge of the two edge-sharing $\{Co^{II}{}_4Cr^{III}{}_2\}$ hexagons. Magnetic studies in both cases suggest low-spin ground states.

Wang *et al.* have reported[44] two Ce–Mn cage complexes using phosphonates; The smaller complex has the formula $[Ce^{IV}{}_2Mn^{IV}{}_6O_6(OH)_5(O_3P\text{-}t\text{-}Bu)_6(O_2C\text{-}t\text{-}Bu)_3]$ **104**, has three-fold symmetry and contains three equivalent distorted $\{Ce_2Mn_2O_2(OH)_2\}$ cubanes sharing the Ce edge (Figure 12.56). The phosphonate groups adopt the [3.111] bridging mode. The larger cage,

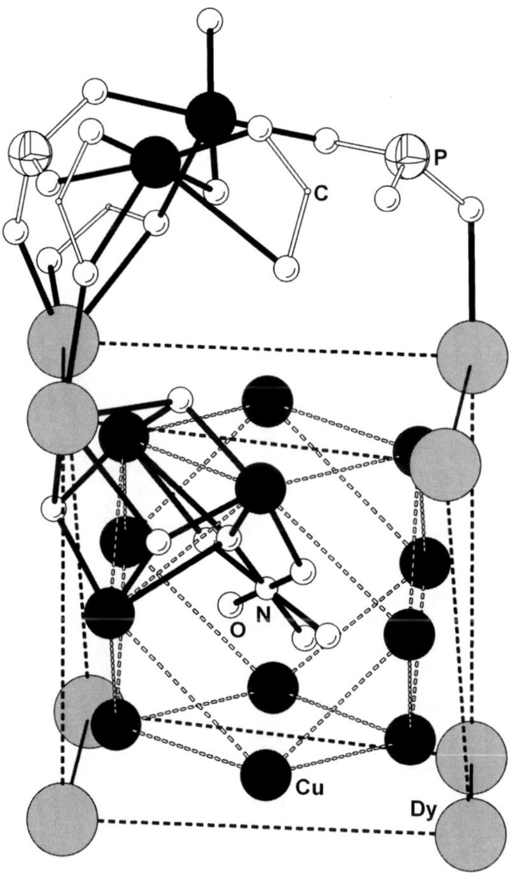

Figure 12.53 A view of the $Dy^{III}{}_8$ cube encapsulating a $Cu^{II}{}_{12}$ cuboctahedron and one of the six outer $\{Cu^{II}{}_2(\mu\text{-}OH)\}$ dimers present outside the six face of the cube in **99**.

$[Ce_{22}Mn_{12}O_{34}(O_3PMe)_{12}(O_2CMe)_{33}(OMe)_6(NO_3)(H_2O)_{12}]_n$ **105** is a 1D chain, containing $\{Ce_{11}Mn_6\}$ sub-units bridged by nitrates and carboxylates. The $\{Ce_{11}Mn_6\}$ sub-units have non-crystallographic three-fold symmetry, containing three $\{Ce^{IV}{}_2Mn_2O_4\}$ cubanes arranged around a $\{Ce^{IV}{}_3Ce^{III}{}_2\}$ trigonal bipyramid, with the unit held together by six [4.221] methylphosphonates (Figure 12.57).

Three Co–Gd complexes have also been reported.[45] $[Co^{II}{}_8Gd^{III}{}_8(\mu_3\text{-}OH)_4(NO_3)_4(O_2C\textit{t}\text{-}Bu)_{16}(O_3P\textit{t}\text{-}Bu)_8]$ **106**, $[Co^{II}{}_8Gd^{III}{}_4(O_3P\textit{t}\text{-}Bu)_6(O_2C\textit{t}\text{-}Bu)_{16}]$ **107** and $[Co^{II}{}_4Gd^{III}{}_6(O_2C\textit{t}\text{-}Bu)_{14}(O_3PCH_2Ph)_6(MeCN)_2]$ **108** have structures that can be related to molecular grids. In **106** the centre of the molecule features a [2×2] $\{Co_4\}$ square grid in which six-coordinate cobalt(II) ions are bridged by four [2.210] nitrates. The phosphonate ligands show a [4.122] coordination mode that bridge between the inner $\{Co_4\}$ square and an outer $\{Gd_8Co_4\}$ frame

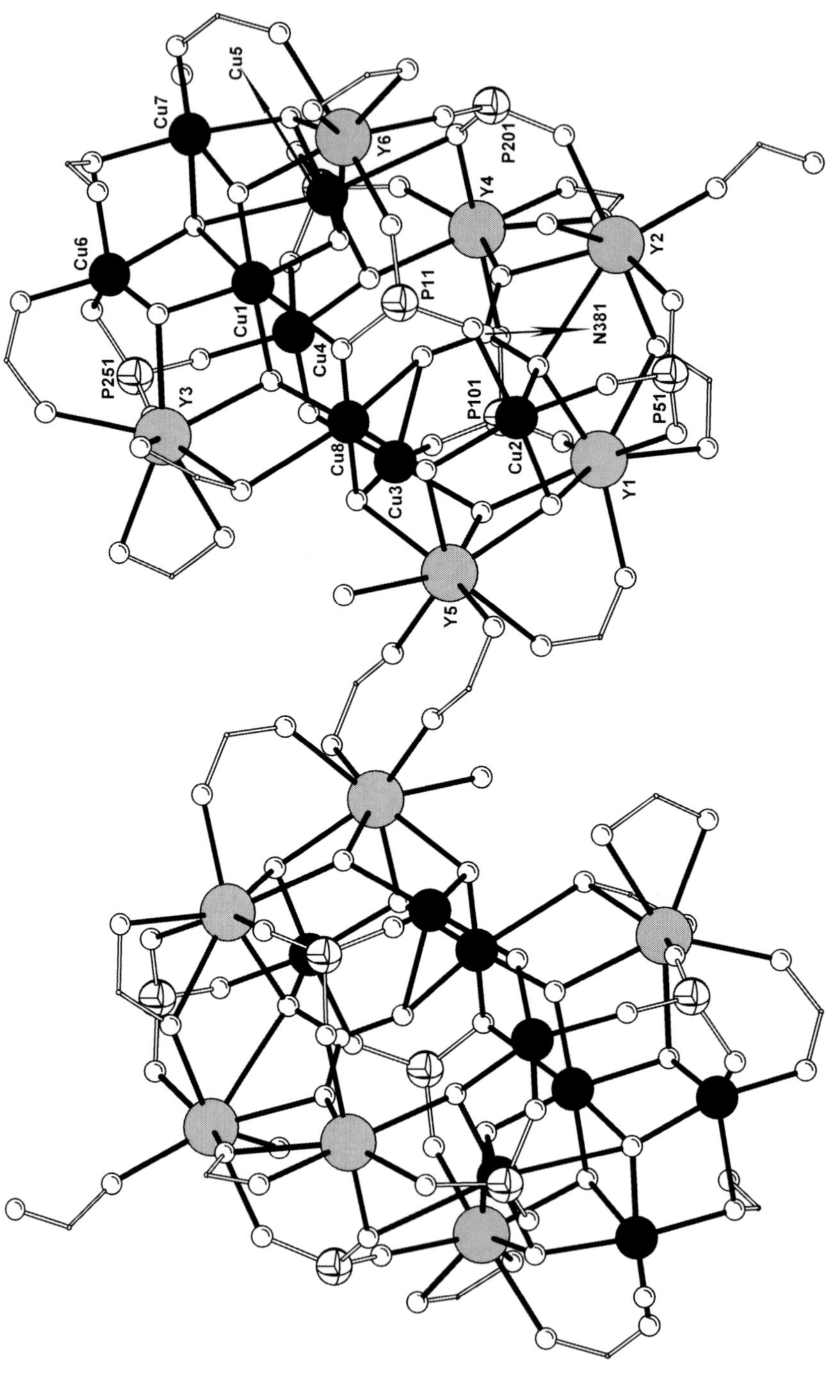

Figure 12.54 The structure of **101** in the crystal.

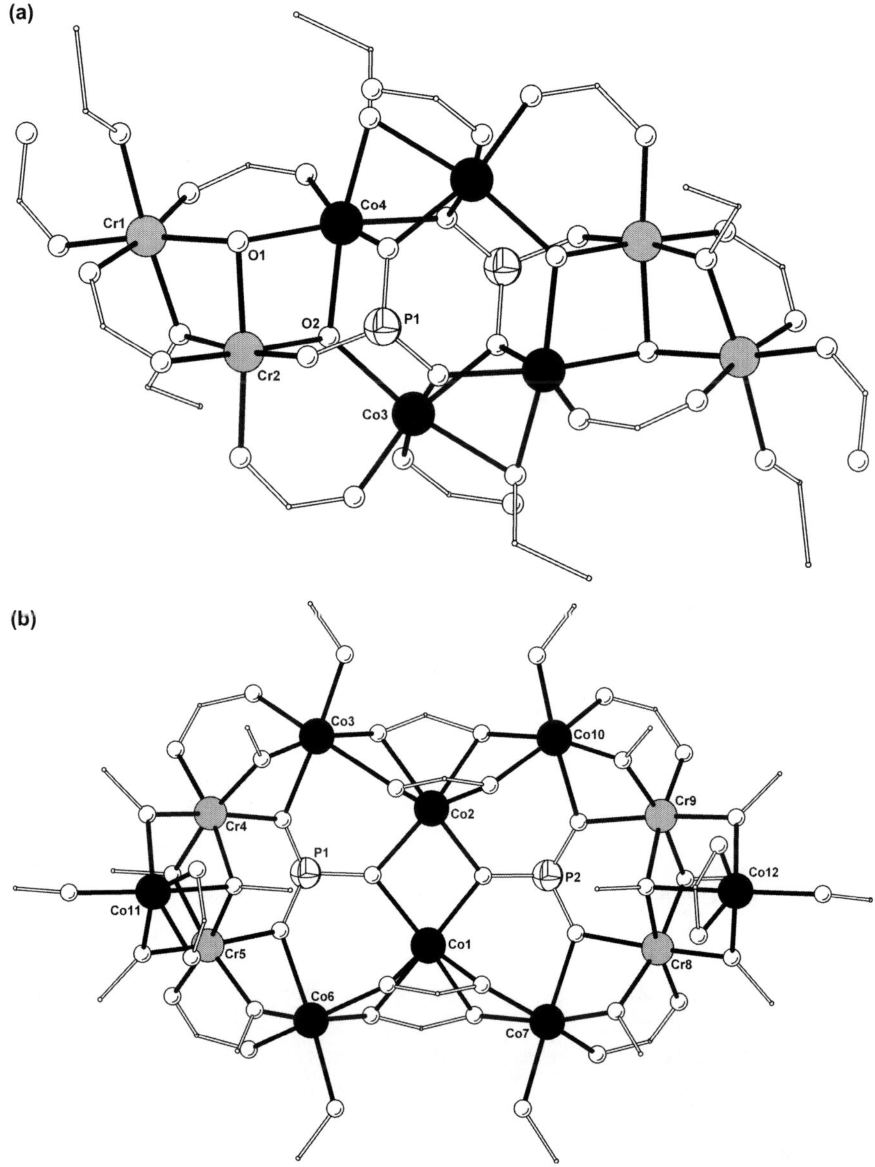

Figure 12.55 The structures of (a) **102** and (b) **103**.

(Figure 12.58). Compared with a regular [4×4] grid the central $\{Co_4\}$ square is rotated through 45° with respect to the outer $\{Gd_8Co_4\}$ frame. Compound **107** features a [4×3] grid structure. The outer rows of four metals are symmetry equivalent, and each contains two five-coordinate terminal Co^{II} ions which are triply bridged to the two central Gd^{III} ions by one [2.21] carboxylate, one [2.11]

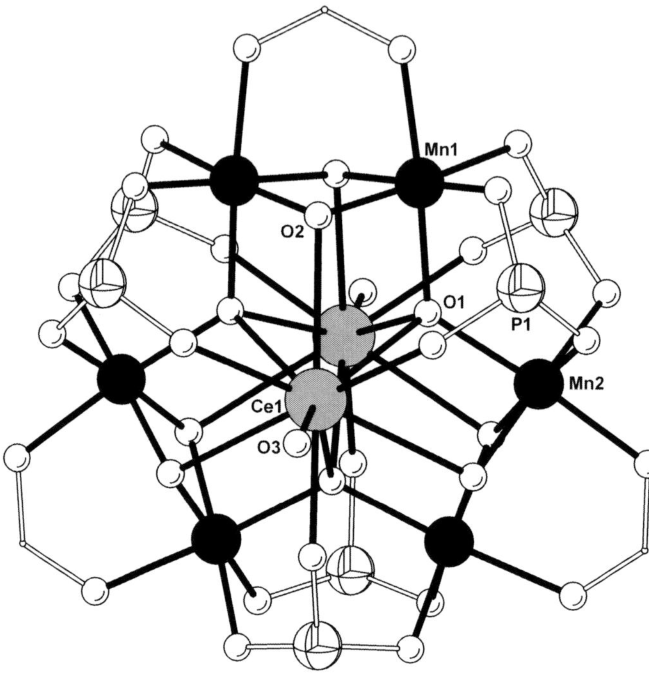

Figure 12.56 The structure of **104** in the crystal.

carboxylate and one oxygen from a [4.221] phosphonate. The central row of the grid contains four four-coordinate Co^{II} ions and they are bridged by 4.221 and 4.222 phosphonates.

Compound **108** has a pair of octahedral Co^{II} ions in the centre, which are surrounded by six Gd^{III} ions and two tetrahedral Co^{II} ions. If the central Co^{II} dimer were regarded as a 'node', this molecule would be a [3×3] grid. The central Co^{II} dimer is part of a Co_2O_2 ring, with the oxygen atoms coming from two [5.222] phosphonates. Each Gd···Gd edge is triply bridged by an oxygen from a [4.221] phosphonate and two [2.11] pivalates. The corner tetrahedral Co^{II} ions are doubly bridged to the adjacent Gd ions by one [2.11] pivalate and one O-atom from a [4.221] phosphonate. These three compounds show very large magneto-caloric effects.

12.4 Conclusions

The growth in the number of paramagnetic molecular phosphonates has been spectacular. The two routes discussed here – the co-ligand approach pioneered by Chandresekhar and the cluster expansion route we introduced – have both led to some remarkably beautiful new cage complexes. However, it is clear we are only scratching the surface, e.g. the cluster expansion route has largely concentrated

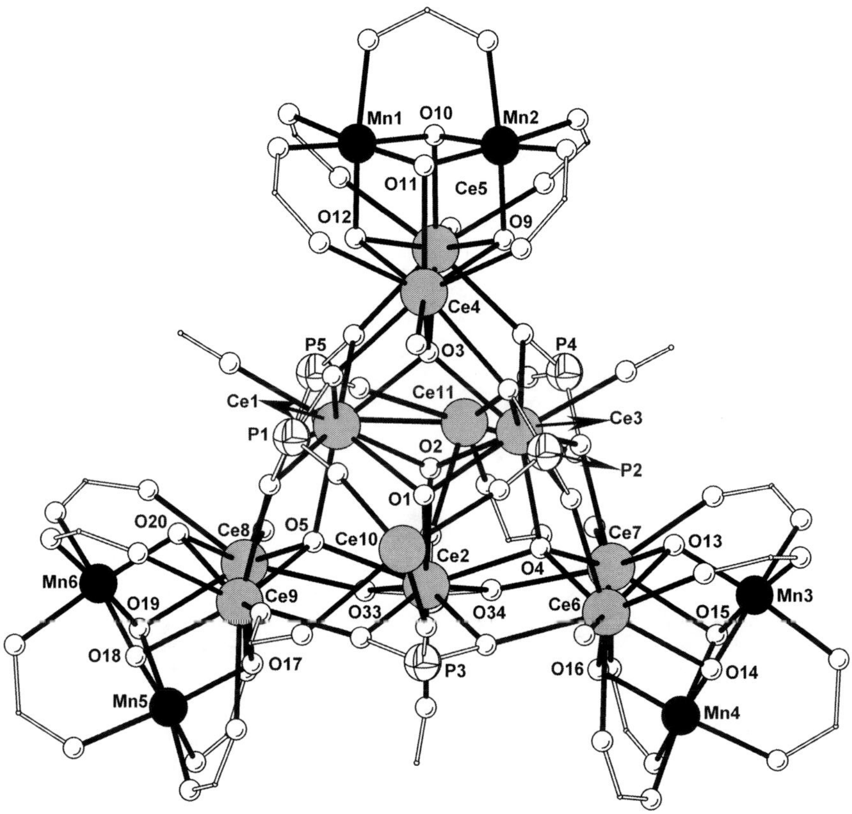

Figure 12.57 The structure of **105** in the crystal.

on using oxo-centred metal carboxylate triangles in preference to the many other carboxylate cages in the chemical literature. Similarly, the reports of heterometallic cage complexes with phosphonates are all from 2010 or later, and it seems clear from these early papers that very large cages could easily be made from 3d–4f mixtures by varying the phosphonates and reaction conditions.

The structures are extremely diverse, and trends are far from obvious. It is noticeable that there are parallels between the structures of Fe^{III} cages and V^{III} cages, but fewer parallels between these cages and those of manganese. Possibly this is because the manganese cages are frequently mixed-valent. The cobalt(II) cages are complex and frequently indescribable.

Magnetic studies on the existing compounds show that the paramagnetic molecular phosphonates will be as interesting as the equivalent carboxylate complexes. Several of the manganese phosphonates are SMMs, and the heterometallic Co–Gd cages look promising for magnetic cooling. The complexity of the cobalt phosphonate magnetism is as challenging as the magnetism of many other cobalt cage complexes.

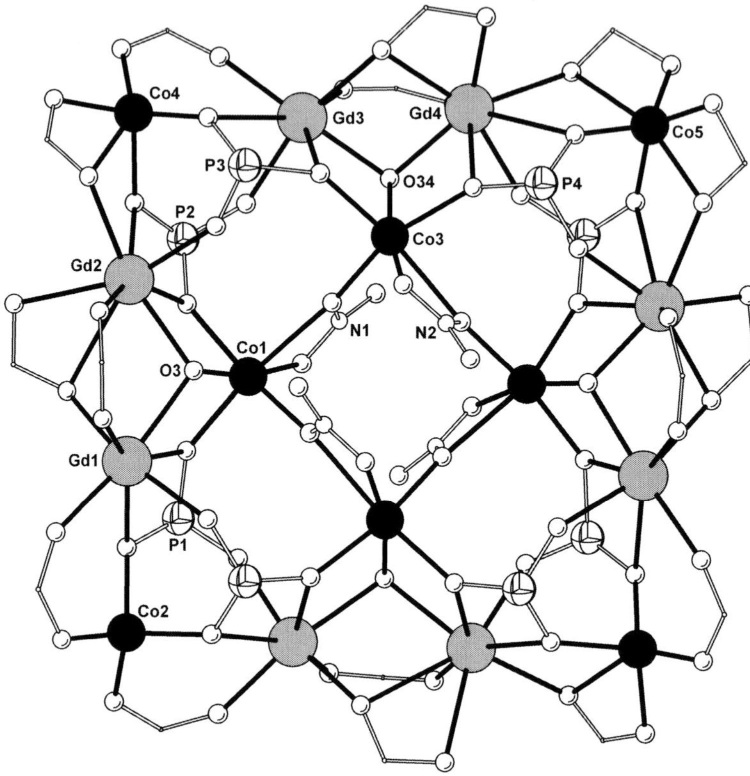

Figure 12.58 The structure of **106** in the crystal.

Acknowledgements

The synthetic work performed in Manchester that is described in this review was performed by Euan Brechin, Vangelis Tolis, Stuart Langley, Maheswaran Shanmugam, Baskar Viswanathan, Barbara Breeze, Marc Kloth, Carolina Sanudo, Sumit Khanra and Yan-Zhen Zheng; some of this work was jointly supervised by David Collison and Eric McInnes. The structures were collected, solved and refined by Chris Muryn, Robin Pritchard and Simon J. Teat (Advanced Light Source, Berkeley). We also collaborated with Sanjit Konar and Abraham Clearfield on some of the studies of iron(III) phosphonates, when we found that both groups were making equivalent structures but with different peripheries. Magnetic measurements on our compounds were done in Manchester by Floriana Tuna, and in collaboration with Roberta Sessoli (Florence), Talal Mallah (Orsay) and Wolfgang Wernsdorfer (Grenoble). G. K. has been funded by a Marie Curie Incoming International Fellowship, S. A. by the Higher Education Commission of

Pakistan and R. E. P. W. is supported by a Royal Society Wolfson Merit Award. Our work has more generally been supported by the EPSRC(UK) and the E.C. through networks including MAGMANet, MolNanoMag, QueMolNa and MolSPINQIP.

References

1. (a) A. Caneschi, D. Gatteschi, R. Sessoli, A. L. Barra, L. C. Brunel and M. Guillot, *J. Am. Chem. Soc.*, 1991, **113**, 5873; (b) R. Sessoli, H.-L. Tsai, A. R. Schake, S. Wang, J. B. Vincent, K. Folting, D. Gatteschi, G. Christou and D. N. Hendrickson, *J. Am. Chem. Soc.*, 1993, **115**, 1804; (c) R. Sessoli, D. Gatteschi, A. Caneschi and M. A. Novak, *Nature*, 1993, **365**, 141.
2. G. Aromí and E. K. Brechin, *Structure Bonding*, 2006, **122**, 1.
3. For example: (a) R. Shaw, R. H. Laye, L. F. Jones, D. M. Low, C. Talbot-Eeckelaers, Q. Wei, C. J. Milios, S. J. Teat, M. Helliwell, J. Raftery, M. Evangelisti, M. Affronte, D. Collison, E. K. Brechin and E. J. L. McInnes, *Inorg. Chem.*, 2007, **46**, 4968; (b) A. M. Ako, I. J. Hewitt, V. Mereacre, R. Clérac, W. Wernsdorfer, C. E. Anson and A. K. Powell, *Angew. Chem., Int. Ed. Engl.*, 2006, **45**, 4926.
4. R. A. Coxall, S. G. Harris, D. K. Henderson, S. Parsons, P. A. Tasker and R. E. P. Winpenny, *Dalton Trans.*, 2000, 2349.
5. (a) M. I. Khan and J. Zubieta, *Prog. Inorg. Chem.*, 1995, **43**, 1; (b) R. C. Finn, J. Zubieta and R. C. Haushalter, , 2003, **51**, 421.
6. M. G. Walawalkar, H. W. Roesky and R. Murugavel, *Acc. Chem. Res.*, 1999, **32**, 117.
7. V. Baskar, M. Shanmugam, E. C. Sanudo, M. Shanmugam, D. Collison, E. J. L. McInnes, Q. Wei and R. E. P. Winpenny, *Chem. Commun.*, 2007, 37.
8. R. Murugavel and S. Shanmugan, *Dalton Trans.*, 2008, 5358.
9. V. Chandrasekhar, L. Nagarajan, R. Clérac, S. Ghosh and S. Verma, *Inorg. Chem.*, 2008, **47**, 1067.
10. V. Chandrasekhar and S. Kingsley, *Angew. Chem., Int. Ed. Engl.*, 2000, **39**, 2320.
11. E. K. Brechin, R. A. Coxall, A. Parkin, S. Parsons, P. A. Tasker and R. E. P. Winpenny, *Angew. Chem., Int. Ed. Engl.*, 2001, **40**, 2700.
12. E. I. Tolis, M. Helliwell, S. Langley, J. Raftery and R. E. P. Winpenny, *Angew. Chem., Int. Ed. Engl.*, 2003, **42**, 3804.
13. S. Khanra, M. Kloth, H. Mansaray, C. A. Muryn, F. Tuna, E. C. Sañudo, M. Helliwell, E. J. L. McInnes, Richard and E. P. Winpenny, *Angew. Chem., Int. Ed. Engl.*, 2007, **46**, 5568.
14. (a) V. Chandrasekhar, P. Sasikumar and R. Boomishankar, *Dalton Trans.*, 2008, 5189; (b) V. Chandrasekhar and L. Nagarajan, *Dalton Trans.*, 2009, 6712; (c) V. Chandrasekhar, T. Senapati and E. C. Sanudo, *Inorg. Chem.*, 2008, **47**, 9553; (d) V. Chandrasekhar, T. Senapati, E. C. Sanudo and R. Clérac, *Inorg. Chem.*, 2009, **48**, 6192; (e) V. Chandrasekhar, R. Azhakar,

T. Senapati, P. Thilagar, S. Ghosh, S. Verma, R. Boomishankar, A. Steiner and P. Kögerler, *Dalton Trans.*, 2008, 1150.
15. S. Langley, M. Helliwell, R. Sessoli, S. J. Teat and R. E. P. Winpenny, *Dalton Trans.*, 2009, 3102.
16. S. Langley, M. Helliwell, R. Sessoli, S. J. Teat and R. E. P. Winpenny, *Inorg. Chem.*, 2008, **47**, 497.
17. Y.-S. Ma, Y. Song, X.-Y. Tang and R.-X. Yuan, *Dalton. Trans.*, 2010, **39**, 6262.
18. E. I. Tolis, L. P. Engelhardt, P. V. Mason, G. Rajaraman, K. Kindo, M. Luban, A. Matsuo, H. Nojiri, J. Raftery, C. Schröder, G. A. Timco, F. Tuna, W. Wernsdorfer and R. E. P. Winpenny, *Chem. Eur. J.*, 2006, **12**, 8961.
19. S. Khanra, S. Konar, A. Clearfield, M. Helliwell, E. J. L. McInnes, E. Tolis, F. Tuna and R. E. P. Winpenny, *Inorg. Chem.*, 2009, **48**, 5338.
20. S. Khanra, M. Helliwell, F. Tuna, E. J. L. McInnes and R. E. P. Winpenny, *Dalton Trans.*, 2009, 6166.
21. S. Konar, N. Bhuvanesh and A. Clearfield, *J. Am. Chem. Soc.*, 2006, **128**, 9604.
22. S. Konar and A. Clearfield, *Inorg. Chem.*, 2008, **47**, 5573.
23. H.-C. Yao, Y.-Z. Li, L.-M. Zheng and X.-Q. Xin, *Inorg. Chim. Acta*, 2005, **358**, 2523.
24. H.-C. Yao, J.-J. Wang, Y.-S. Ma, O. Waldmann, W.-X. Du, Y. Song, Y.-Z. Li, L.-M. Zheng, S. Decurtins and X.-Q. Xin, *Chem. Commun.*, 2006, 1745.
25. R. Murugavel, N. Gogoi and R. Clérac, *Inorg. Chem.*, 2009, **48**, 646.
26. M. Shanmugam, M. Shanmugam, G. Chastanet, R. Sessoli, T. Mallah, W. Wernsdorfer and R. E. P. Winpenny, *J. Mater. Chem.*, 2006, **16**, 2576.
27. M. Shanmugam, G. Chastanet, T. Mallah, R. Sessoli, S. J. Teat, G. A. Timco and R. E. P. Winpenny, *Chem. Eur. J.*, 2006, **12**, 8777.
28. S. Maheswaran, G. Chastanet, S. J. Teat, T. Mallah, R. Sessoli, W. Wernsdorfer and R. E. P. Winpenny, *Angew. Chem., Int. Ed. Engl.*, 2005, **44**, 5044.
29. Y.-S. Ma, Y. Song, Y.-Z. Li and L.-M. Zheng, *Inorg. Chem.*, 2007, **46**, 5459.
30. J.-T. Li, Y.-S. Ma, S.-G. Li, D.-K. Cao, Y.-Z. Li, Y. Song and L.-M. Zheng, *Dalton Trans.*, 2009, 5029.
31. S. Konar and A. Clearfield, *Inorg. Chem.*, 2008, **47**, 3489.
32. M. Wang, C. Ma, H. Wen and C. Chen, *Dalton Trans.*, 2008, 4612.
33. Y.-S. Ma, H.-C. Yao, W.-J. Hua, S.-H. Li, Y.-Z. Li and L.-M. Zheng, *Inorg. Chim. Acta*, 2007, **360**, 1645.
34. H.-C. Yao, Y.-Z. Li, Y. Song, Y.-S. Ma, L.-M. Zheng and X.-Q. Xin, *Inorg. Chem.*, 2006, **45**, 59.
35. M. Wang, C. Ma, D. Yuan, M. Hu, C. Chen and Q. Liu, *New J. Chem.*, 2007, **31**, 2103.
36. M. Wang, C.-B. Ma, D.-Q. Yuan, H.-S. Wang, C.-N. Chen and Q.-T. Liu, *Inorg. Chem.*, 2008, **47**, 5580.

37. M. Wang, C. Ma, H. Wen and C. Chen, *Dalton Trans.*, 2009, 994.
38. C. Dendrinou-Samara, C. A. Muryn, F. Tuna and R. E. P. Winpenny, *Eur. J. Inorg. Chem.*, 2010, 3097.
39. S. Khanra, L. Batchlelor, M. Helliwell, F. Tuna, E. J. L. McInnes and R. E. P. Winpenny, *J. Mol. Struct.*, 2008, **890**, 157.
40. S. Konar and A. Clearfield, *Inorg. Chem.*, 2008, **47**, 3492.
41. B. A. Breeze, M. Shanmugam, F. Tuna and R. E. P. Winpenny, *Chem. Commun.*, 2007, 5185.
42. V. Baskar, K. Gopal, M. Helliwell, F. Tuna, W. Wernsdorfer and R. E. P. Winpenny, *Dalton. Trans.*, 2010, **39**, 4747.
43. Y.-Z. Zheng, B. A. Breeze, G. A. Timco, F. Tuna and R. E. P. Winpenny, *Dalton Trans.*, 2010, **39**, 6175.
44. M. Wang, D.-Q. Yuan, C.-B. Ma, M.-J. Yuan, M.-Q. Hu, N. Li, H. Chen, C.-N. Chen and Q.-T. Liu, *Dalton. Trans.*, 2010, **39**, 7276.
45. Y.-Z. Zheng, M. Evangelisti, F. Tuna and R. E. P. Winpenny, *Chem. Sci.*, 2011, **2**, 99.

CHAPTER 13

Application of Metal Phosphonates to Biotechnologies

BRUNO BUJOLI, PASCAL JANVIER AND MARC PETIT

University of Nantes, UMR CNRS 6230 CEISAM, UFR Sciences et Techniques, 2 Rue de la Houssinière, BP 92208, 44322 Nantes Cedex 03, France

13.1 Introduction

Among organic-inorganic hybrid materials, the development of metal organic frameworks (MOFs) is a fast expanding field, which mainly includes carboxylate-based[1–4] and phosphonate-based architectures. For the latter category, also termed metal phosphonates, it has been demonstrated since the late 1970s that phosphonic acids, RPO_3H_2 (R is an organic component), have the ability to coordinate a wide range of cations, leading to metal organic frameworks. The cohesion of these networks is ensured by the formation of PO–metal bonds which can be highly covalent, according to the nature of the metal.[5]

At the beginning, the main efforts were focused on structural studies, investigating the dimensionality (0D to 3D) and type of networks accessible as a function of the size of the phosphonic acid, the nature, number and location of functional groups possibly present on the organic backbone, the nature of the metal, and the conditions used for the synthesis (i.e. solvent, temperature, pressure, pH, additives, stoichiometry, . . .).

During the last two decades, this unique property has quite naturally led to the use of phosphonic acids as a route to functional materials by surface modification of metals and metal oxides, as well as any type of solid having surface exposed metal centers. This chapter will highlight how metal phosphonate chemistry can be used for the design of novel materials in the area of biotechnologies.

Metal Phosphonate Chemistry: From Synthesis to Applications
Edited by Abraham Clearfield and Konstantinos Demadis

Published by the Royal Society of Chemistry, www.rsc.org

13.2 Surface Modification of Implants

13.2.1 Derivatization with Bioactive Molecules to Favor Interaction of the Implant with Cells

Titanium and its alloys (Ti-6Al-4V) are of widespread use as orthopedic implants, since they are biocompatible and mechanically resistant. However, a stable implant to bone tissue binding is highly desirable to ensure the long-term stability of the implanted biomaterial. In this context, numerous strategies have been developed to improve adhesion of bone to the metal.[6] For example, di-, tri-, and tetra-phosphonic acids were chemically bound to titanium discs and the bioactivity of the resulting surfaces was investigated by probing their ability to promote calcium phosphate growth upon incubation in a physiological solution. Compared to the bare titanium surface, the phosphonate-coated analogs were found to induce higher calcium phosphate deposition, which however remained relatively low.[7] Hence, the current research is mainly focused on the modification of the native oxide surface of the implant with bioactive molecules. For this purpose, silanization has long been the "gold standard" method, although competition between surface attachment and silane cross-linking often takes place, and stability problems due to hydrolysis under physiological conditions can occur.

Accordingly, the use of functional phosphonic acids was recently reported for the derivatization of this type of implants, and proved to be superior to the siloxane approach, in particular in terms of chemical robustness.[8] For instance, 11-hydroxyundecyl phosphonic acid and 12-carboxydodecyl phosphonic acid were reported to spontaneously assemble on the surface,[9] and subsequent activation allowed to bind cell-adhesive peptides (such as the sequence arginine–glycine–aspartate [RGD]) to make the surface bioactive and favor cell proliferation[10,11] (Figure 13.1). In order to increase the specificity of the cells attachment, the immobilization of bone morphogenic proteins (i.e. BMP-2) according to a similar strategy has also been considered to enhance local bone growth.[12]

On the other hand, the direct binding of peptides functionalized with terminal polyphosphonic acid[13,14] anchors was also reported. The resulting advantages are: (i) a simpler and direct procedure which makes it of more practical use than the previous multistep process; (ii) an ability of polyphosphonic acids, in particular gem-bisphosphonate groups,[15] to form monolayers on titanium with higher

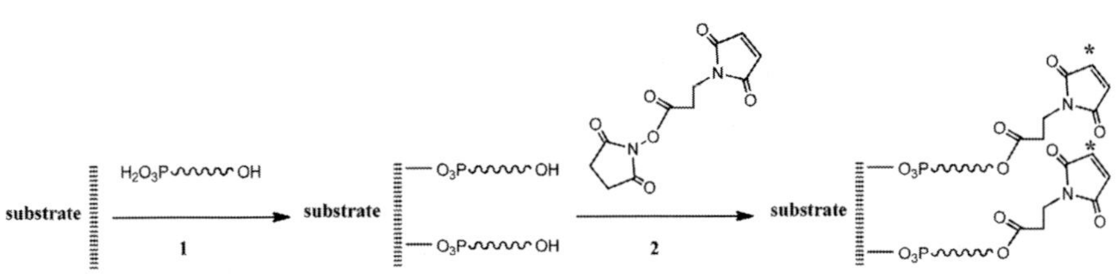

Figure 13.1 Activation of titanium substrates for the covalent binding of peptides: step (1) grafting of a functional phosphonate primer; step (2) activation of the terminal functional groups. * denote the resulting activated sites for the attachment of peptides, for example via a cysteine residue.

affinity and chemical stability when compared to monophosphonic acid binding heads. In parallel, the higher mean molecular area of the bisphosphonate headgroup results in a decrease of the packing of the lateral chain of the monolayer, when compared to the case of monophosphonates.[16]

Finally, alternative routes using α,ω-diphosphonic acid-based self-assembled monolayers (SAM) are currently under development, in which the PO_3 head groups are bound to the titanium, while the PO_3 tail groups exposed at the SAM surface can serve (i) to adsorb positively charged proteins (i.e. collagen I) via electrostatic interactions[17] and (ii) to complex zirconium(IV) alkoxides which can be further derivatized to bind RGD peptides through (maleimido)alkyl carboxylate intermediates.[18]

13.2.2 Derivatization with Terminal Functions Capable of Inhibiting Bacterial Adhesion

In the biomaterials area, many studies are devoted to the surface modification of implants for the prevention of nosocomial infections. In contrast with the previous paragraph, various coatings capable of inhibiting bacterial adhesion and biofilm formation have been developed according to two main strategies:

- passive surfaces for which the hydrophobic/hydrophilic balance is tuned to inhibit bacterial adhesion, with for example quite good results obtained using hydrophilic terminal chains such as polyethylene glycol[19]
- active surfaces which promote degradation of bacteria due to their sterile character (i.e. electric dipoles such as quaternary ammonium salts)[20] or their ability to release antibacterial agents (antibiotics, silver ions, nitrogen monoxide).[21]

In the case of metal or metal alloy implants, the use of phosphonate anchors for binding self-assembled monolayers on these surfaces has thus been investigated. For instance, phosphonate monolayers functionalized with solver thiolate species were bound onto titanium and stainless steel substrates (Figure 13.2a), and were shown to strongly decrease bacterial adhesion and biofilm formation, compared to the non-coated metal surface.[22,23] According to a similar strategy, SAM based on gem-bisphosphonic acids bearing terminal quaternary ammonium groups (Figure 13.2b) were found to exhibit a

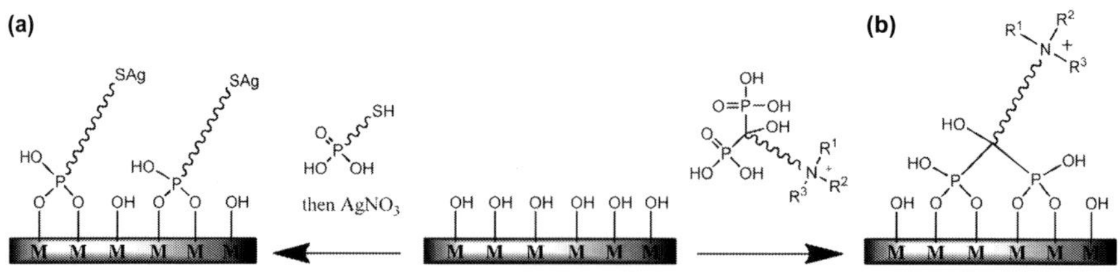

Figure 13.2 Some examples of biofouling coatings on metal oxide substrates, using functional phosphonic acids.

biofouling behavior[24] and the higher affinity of the bisphosphonate group for oxidized metal surface as compared to monophosphonate allowed the formation of monolayers under much milder conditions.

13.2.3 Derivatization with Drugs for the Subsequent Use of the Implants as Local Drug Delivery Systems

In the two previous sections, various coatings grafted via phosphonate and bisphosphonate anchors were described. Given that some gem-bisphosphonates (BPs) are widely used in medicine for the treatment of various osseous pathologies, many attempts to graft these drugs onto implants were reported in the literature, in order to design local drug delivery systems. Indeed, systemic administration of bisphosphonates (oral or intravenous) is commonly prescribed to prevent fractures of the hip, wrist, or vertebral bodies, related to the bone degradation resulting from osteoporosis. Bisphosphonates (Figure 13.3) are synthetic carbon-substituted analogs of pyrophosphates, which have a high affinity for the mineralized bone matrix and contribute to restore the bone remodeling balance and reinforce the bone micro-architecture.

The affinity of BPs for bone is enhanced when R_1 is a hydroxyl group, and the activity of the drug strongly depends on the R_2 side chain,[25–28] the more potent molecules being aminobisphosphonates.

Recent studies have demonstrated the advantages of a local bisphosphonate delivery as a promising improvement to prevent the side effects of systemic treatments (*per os* or intravenously) and to promote local bone reinforcement on weakened implanted sites. One of the main strategies consists in using a calcium phosphate-coated metal implant as a local delivery vehicle.

In recent years, hydroxyapatite (HAP; $Ca_{10}(PO_4)_6(OH)_2$) and tricalcium phosphate (TCP; $Ca_3(PO_4)_2$) coatings obtained by plasma spraying or electrochemical deposition have been used to improve the fixation of metallic implants to bone. Attempts to decrease the peri-implant osteolysis via the local elution of bisphosphonates were investigated by placing hip titanium implants coated with HAP in a bisphosphonate solution (Zoledronate). Hydroxyapatite thus acts as a carrier for the drug and experiments on osteoporotic rats have

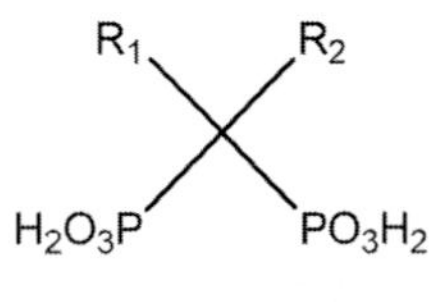

$R_1 = R_2 = Cl$, clodronate
$R_1 = OH$, $R_2 = (CH_2)_2NH_2$, pamidronate
$R_1 = OH$, $R_2 = (CH_2)_3NH_2$, alendronate
$R_1 = OH$, $R_2 = (CH_2)_2N(CH_3)$(pentyl), ibandronate
$R_1 = OH$, $R_2 = CH_2$-imidazole, zoledronate
$R_1 = OH$, $R_2 = CH_2$-3-pyridine, risedronate

Figure 13.3 Structure of selected examples of gem-bisphosphonate anti-osteoporotic drugs.

shown an increase of the mechanical fixation, with a dose/response effect.[29,30] A similar approach was reported with tantalum implants coated with HAP.[31] Finally, thin films of bisphosphonate (Alendronate)/HAP composites have been prepared by matrix-assisted pulsed laser evaporation (MAPLE) on titanium substrates, and resulted in an enhanced bioactivity due to the local availability of alendronate, capable of promoting bone formation and preventing bone resorption.[32]

Not surprisingly, many studies have been focused on the use of pure HAP as bisphosphonate drug carrier, given that HAP-based bioceramics are widely used as bone-graft substitutes in various orthopedic applications, since they are biocompatible, osteoconductive, resorbable, and thus appropriate for the filling of osseous defects. The local elution of the BP from the bisphosphonate grafted-HAP was found to promote substantial bone augmentation around the implantation site, thus giving evidence that the BP/calcium phosphate combination does not impair the healing process by inhibiting bone remodeling.[33–38]

Unlike the case of HAP, less interest has been devoted to the potential combination of BPs with other calcium phosphates (CaPs) of biological interest, such as BCP (biphasic calcium phosphate – a mixture between HAP and β-TCP), pure β-TCP or calcium-deficient apatite (CDA). Two types of reaction can result from a reaction of a bisphosphonate aqueous solution with CaPs: (i) precipitation of crystalline bisphosphonate complex onto the surface of BCP and β-TCP resulting from a partial dissolution of the CaP[39,40] or (ii) reversible chemical binding of the bisphosphonate to the CaP through a chemical exchange between phosphate groups from the CaP (CDA) with phosphonate groups from the BP (Figure 13.4).[41,42]

The latter mechanism was also found to take place in human bone on the basis of NMR spectroscopy investigations.[25,26] Then, injectable formulations made of a suspension of Zoledronate-loaded CDA granules in a cellulose-based hydrogel have been injected in the femur of osteoporotic rats and ewes. These combined systems were found to promote new bone formation and reinforce the existing trabecular micro-architecture in osteoporotic sites.[43]

This approach has also been extended to the case of apatitic calcium phosphate cements (CPCs) to address indications where mechanical properties are

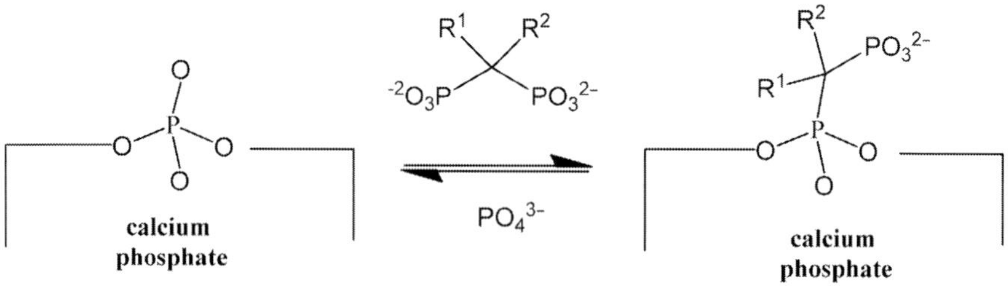

Figure 13.4 Binding mechanism of gem-bisphosphonates on calcium deficient apatites. (Reproduced with permission from reference 40. Copyright ACS publications.)

required, in particular for biomaterials for load-bearing applications. Due to the cement-setting retardant properties of phosphonates (i.e. BPs are efficient calcium scavengers), the incorporation of BPs in CPCs is challenging, and as a result only very low doses of BP (~0.004–0.01 wt%, with respect to the solid phase) could be incorporated to minimize inhibition of the setting.[44] Higher doses, however, could be incorporated either by addition of a gelatin additive (BP in the liquid phase)[45] or by addition of BP-loaded CDA (BP grafted onto the CDA),[46] the cement paste having the advantage to be injectable in the latter case.

In addition, it is interesting to note that the strong affinity of the gem-bisphosphonate motif for calcium phosphate discussed in this section also resulted in attempts to use it as a targeting function for calcified tissues, in order to design bone-seeking medicinal agents for imaging and therapy.[47–49] For this purpose, the drug of interest is functionalized with a gem-bisphosphonate moiety.

13.3 Functional Metal Phosphonate Surfaces

13.3.1 Efficient and Oriented Immobilization of Biological Probes for the Design of Microarrays

Biological microarrays have emerged as a convenient and powerful tool for highly parallel, high throughput experimentation in molecular biological research.[50–56] These biosensors typically consist of biological probes (i.e. ss- or ds-DNA, proteins, antibodies, ...) that are bound to a surface and available for subsequent complexation by targets, leading to a signal that is detected by an imaging technology, most often fluorescence. Usual applications include, for example, gene mapping, DNA sequencing, study of protein function and expression, and diagnosis of genetic diseases.

Glass substrates are preferred for arrays, since they are cheap and their flat and non-porous character allows using small volumes of solution during incubation with the targets. A significant challenge is attachment of the probes on the chip surface. Immobilization by non-specific adsorption is often associated with problems such as a high background signal and significant loss of the protein probes during stringent washes. Furthermore, random orientations of the immobilized probes affect their activities, which decrease the sensitivity of detection. Therefore, the current trend is to form a surface-bound monolayer of functional groups that are available to react only with a specific group on the probe terminus, resulting in covalent linkage. Organosilane-coating protocols are commonly used to fix the active groups to the glass surface, with now well-established technologies for the preparation of oligonucleotide microarrays (Figure 13.5), most often involving attachment of DNAs modified with primary amine tethers.[57,58]

Recently, a fundamentally different route for covalently attaching probes to surfaces for array applications was proposed, based on the derivatization of the

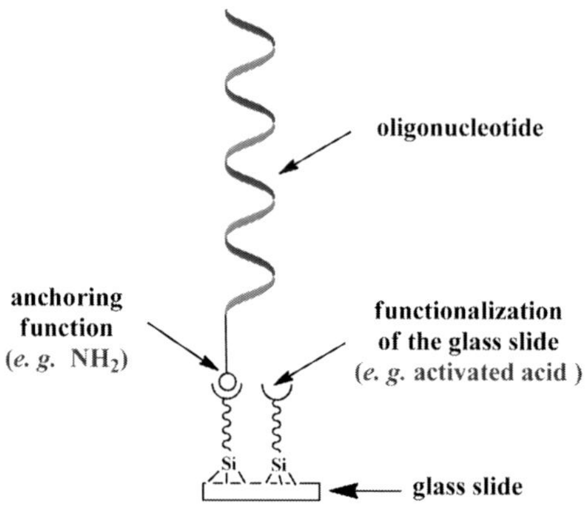

Figure 13.5 Scheme illustrating covalent immobilization of oligonucleotide probes onto glass substrates using organosilane chemistry.

glass with a zirconated phosphonate monolayer to generate a reactive surface. Probe attachment proceeds then via a highly specific iono-covalent linkage between a terminal phosphate group on the probes and the inorganic ions on the glass surface (Figure 13.6).

An advantage over other methods currently in use is that phosphate is a naturally occurring function that can be introduced with enzymatic routes, offering the possibility of avoiding chemical modifications which can result in denaturation of the probes, especially in the case of fragile species, such as proteins and antibodies. Different routes have been reported to prepare zirconium phosphonate coatings on various substrates, such as gold, glass, or metal oxides, for a wide range of applications.[59–69] Concerning the field of biological microarrays, an early study by Mallouk and Bard[70,71] has reported the use of an aluminum phosphonate thin film deposited on a gold electrode for the electrogenerated chemiluminescence detection of ss- and ds-DNAs immobilization. Adsorption of the probes was assumed to occur via the interaction of the surface exposed Al^{III} binding sites with the phosphodiester groups of the DNA backbone. In this context, analogous films with Zr^{IV} and La^{III} centers were found to be less efficient for binding DNA.

A significant breakthrough was achieved by Nonglaton *et al.*[72] who proposed the introduction of a free phosphate group on the 5′-end of oligonucleotides which should dominate over the backbone phosphate diester groups for binding to a metal phosphonate surface. For this purpose, smooth and uniform films generated on hydrophobic glass slides using Langmuir–Blodgett methods were investigated.[73–78] The reactive surface is made of an octadecylphosphonic acid (ODPA) Langmuir monolayer, which upon exposure to a solution of Zr^{4+} ions that strongly cross-link the PO_3H_2 acid group directed away from the support, provides a well-defined interface of zirconium phosphonate sites. With this system, the specific binding of the probe terminus to the surface was evidenced from competition studies between probes of the same sequence, one of

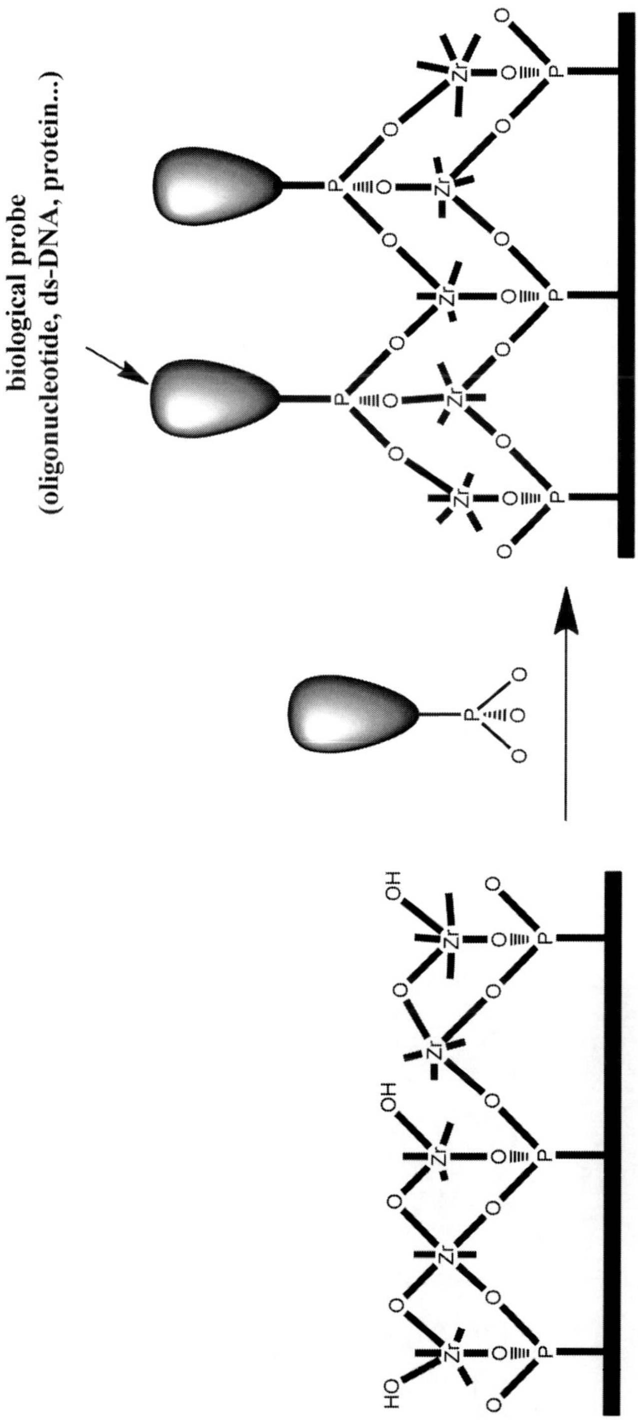

Figure 13.6 Adsorption onto a zirconated phosphonic acid surface of phosphate-containing biological probes.

which containing a 5′-phosphate group, showing that unmodified oligonucleotides binds very poorly on this surface. Accordingly, this methodology was optimized and successfully used to form efficient oligonucleotide arrays (Figure 13.7a).[72,79,80]

More recently, Monot *et al.* have shown that the same approach can be extended to phosphate-terminated ds-DNA, and probes immobilized in this way can be used to detect protein targets. The efficacy of the protein capture was found to be influenced by several features, in particular the presence of a polyguanine segment between the protein interaction domain and terminal phosphate groups,[81] as well as the number and location of the phosphate linker in the duplex probes (Figure 13.7b). Using a ds-DNA/protein interaction model, the sensitivity and specificity of the arrays were observed to be the best when two phosphate groups are introduced on the two ends of one of the two strands of the ds-DNA probes.[82]

Finally, Cinier *et al.* investigated the potential of the zirconium phosphonate surface for the design of protein microarrays usable in proteomics. It was found that unmodified as well as oligo-histidine-tagged proteins are poorly adsorbed on this surface. Therefore, bifunctional adaptors containing a multivalent phosphonic acid anchor at one extremity and one (I) or two (II) nitrilotriacetate (NTA) groups at the other were synthesized (Figure 13.8).

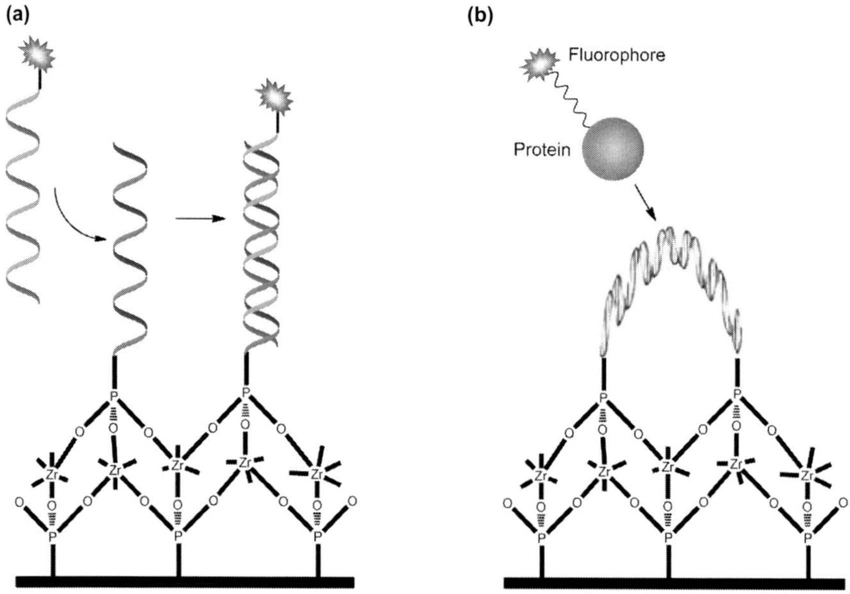

Figure 13.7 Phosphate-terminated oligonucleotide (a) or ds-DNA (b) probes attach specifically via coordinate covalent bonding to a zirconated monolayer of octadecylphosphonic acid (ODPA) deposited onto a hydrophobic slide. The resulting arrays can be used for the analysis of (a) DNA–DNA or (b) protein–(ds-DNA) interactions.

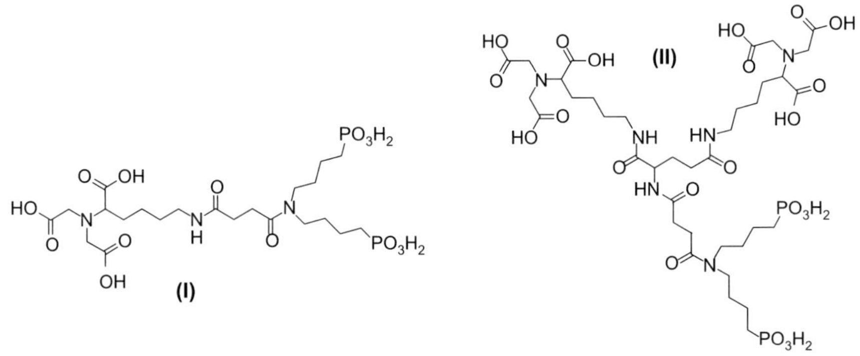

Figure 13.8 Synthesis of mono-NTA and bis-NTA bifunctional adaptors for the design of zirconium phosphonate-based protein microarrays.

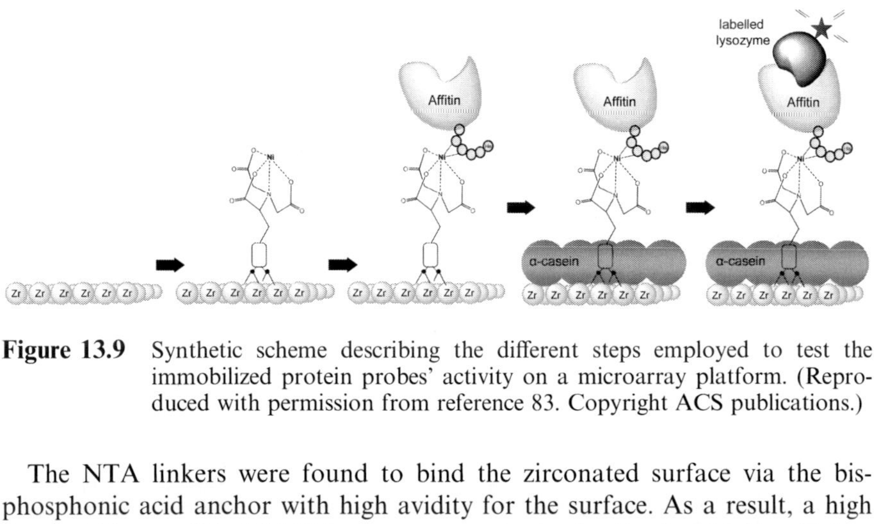

Figure 13.9 Synthetic scheme describing the different steps employed to test the immobilized protein probes' activity on a microarray platform. (Reproduced with permission from reference 83. Copyright ACS publications.)

The NTA linkers were found to bind the zirconated surface via the bisphosphonic acid anchor with high avidity for the surface. As a result, a high density of accessible Ni-NTA groups was obtained, allowing highly efficient protein binding (affitin capture agents) and control of their orientation on the surface (Figure 13.9).

Proteins immobilized in this way retain their ability to capture protein target with high sensitivity, and the modified zirconium phosphonate surfaces were shown to compete favorably with commercially available substrates designed for protein microarrays.[83]

13.3.2 Purification of Complex Biological Media upon Selective Interaction with Specific Target Molecules

Mass spectrometry has been established as the method of choice to characterize phosphorylation of proteins by kinases and phosphatases, which is one of the major regulatory mechanisms of important cellular functions. However, due to the low abundance of phosphorylated proteins, isolation and enrichment steps are often required prior to analysis. Among the various techniques developed for phosphopeptide enrichment in biological samples, IMAC (immobilized metal ion affinity chromatography) is commonly employed, consisting in the specific binding of the phosphate residues of phosphopeptide to metal complexes decorating the stationary phase.[84] In this context, Fe^{III} complexed to iminodiacetate (IDA) or nitrilotriacetate (NTA) ligands are the most frequently used (Figure 13.10).

Using a concept similar to that described in the previous section, the design of zirconium phosphonate(ZrP)-modified substrates for phosphopeptide capture has recently emerged, taking advantage of the strong interaction between Zr^{4+} ions present on the surface and the peptide phosphate groups (Figure 13.11).

The resulting ZrP-coated adsorbents were found to be more specific than the conventional Fe^{3+}-IMAC materials for binding phosphopeptides selectively in complex biological mixtures. In this context, the ZrP coating of various organic and inorganic substrates was reported, including polymer beads,[85] porous silicon wafers,[86] magnetic nanoparticles,[87] and polymer-based capillary column stationary phases.[88] In most cases, substrates were first functionalized with terminal primary amine groups, which allowed subsequent ZrP derivatization of the surface by reaction with $POCl_3$, and then incubation of the substrate in a $ZrOCl_2$ solution (Figure 13.12).

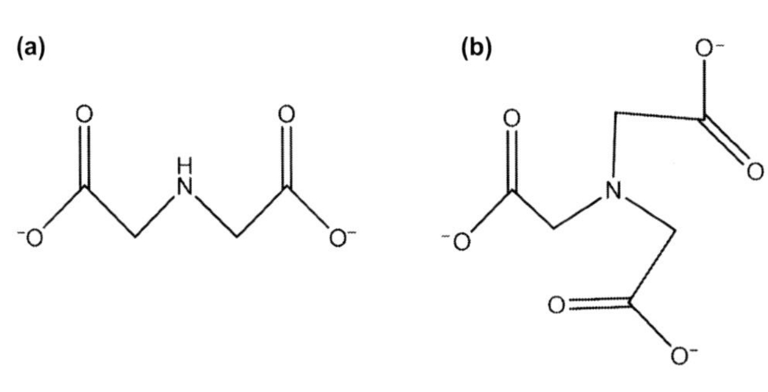

Figure 13.10 Structure of usual metal-binding polycarboxylate employed in IMAC: (a) IDA, (b) NTA.

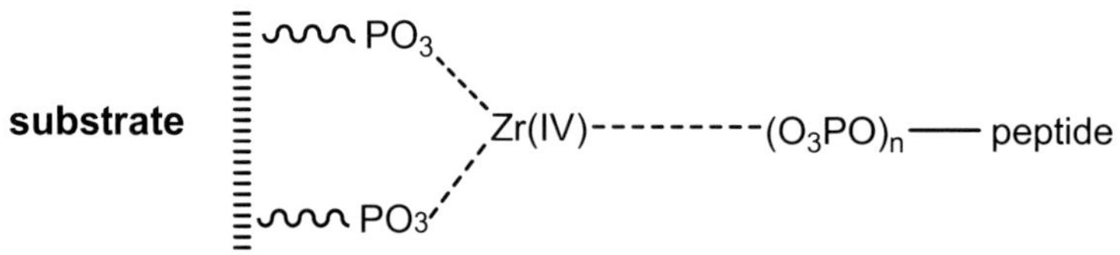

Figure 13.11 Schematic interaction of phosphopeptides with ZrP-coated materials.

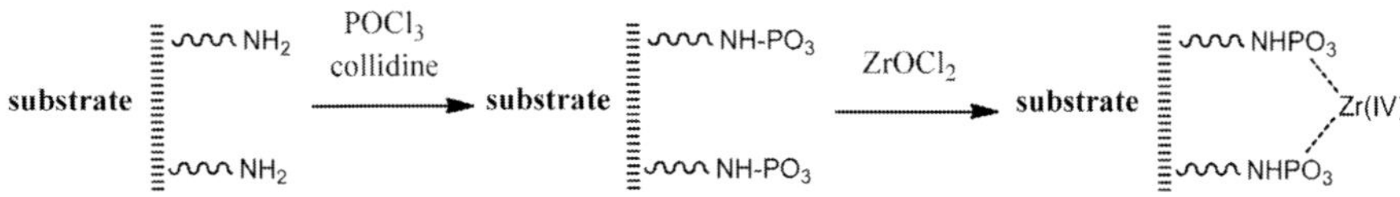

Figure 13.12 Scheme for the preparation of Zr-P coatings on aminated substrates.

In parallel, ion-exchange chromatography is a popular method for protein separations. In this context, zirconia has received considerable interest as stationary phase due to its high stability towards a wide range of mobile phases as well as high temperatures and pressures, being thus most often superior to silica and polymer phases. However, modification of zirconia's surface is necessary to tune its hard Lewis acidity, due to the presence of coordinatively unsaturated zirconium species, which makes it not appropriate for the chromatography of proteins. Accordingly, various modifier species were investigated, including EDPA, a phosphonate analog of EDTA (ethylene diamine-*N*,*N*,*N*′,*N*′-tetra (methylene phosphonic) acid, $[(H_2O_3PCH_2)_2NCH_2CH_2N(CH_2PO_3H_2)_2]$). Given the well-documented coordinating ability of phosphonate groups towards zirconium ions,[5] the zirconia surface modification is likely to occur via the formation of covalent zirconium–O–P bonds, thereby attenuating the overall acidity of the inorganic support. This results in a biocompatible stationary phase useful for small and large-scale purification of biomolecules,[89–92] including monoclonal antibodies, some commercially available versions of this system being now available at Zirchrom®. In addition, once EDTPA has been immobilized, alkylation at the nitrogen centers of the phosphonate ligand result in modified separation selectivities.

In a different area, solid supported bilayer lipid membranes are of particular interest since they can be used as models to investigate many features of cell membranes, including their interaction with proteins. In this context, among different other routes, phosphonate chemistry has also been used to prepare stable lipid bilayers, partially anchored to a solid support, taking advantage of strong binding of the phosphate and phosphonate groups present in the bilayer to surface metal ions, such as zirconium (zirconium phosphonate monolayer transferred on glass)[93] and aluminum (aluminum oxide sputtered on gold).[94] This results in an inner layer which shows low fluidity in contrast to the outer layer, and it was shown that such supported lipid bilayers can be considered as viable membrane mimics.

13.4 Surface Modification of Nanoparticles and Their Bioconjugation

The exceptional binding properties of phosphonic acids to non-siliceous oxide surfaces has naturally led to the capping of nanoparticles (NPs) using phosphonate-terminated molecules, to stabilize them or make them functional.[95–99]

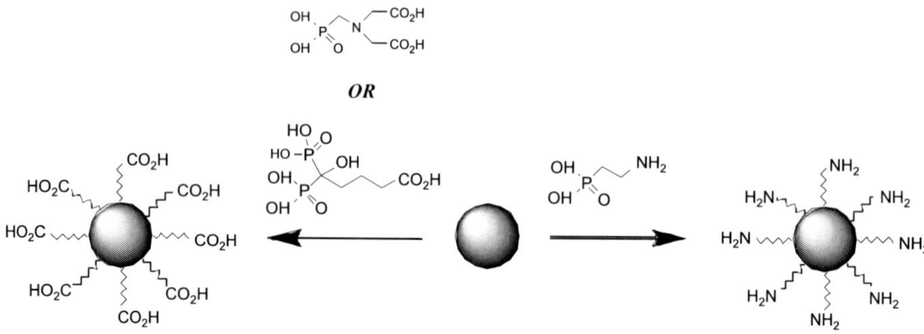

Figure 13.13 Schematic strategy for the design of potential nanomagnetic platforms towards diagnosis and therapy. Bioconjugation can be performed using pendant primary amine groups (CO_2H platform) or carboxylic acid, aldehyde, and isocyanate groups (NH_2 platform) present on the biomolecule [drug, protein, peptide...].

For example, the biofunctionalization of NPs has received increasing interest, in particular those based on superparamagnetic iron oxide, which play an important role as magnetic resonance imaging (MRI) contrast agents. For this purpose, the NPs have to be coated, most often using a thick layer of polymer (commonly dextran or polyethylene glycol), to prevent their aggregation, and inhibit *in vivo* adsorption of proteins and subsequent activation of the opsonization process. As an alternative, other routes are currently being investigated, including the covalent grafting of biologically active species via phosphonic or bisphosphonic acid anchors. Thus, ammonium-terminated bisphosphonates were found to bind strongly to iron oxide NPs, leading to stable species at neutral pH.[100] More recently, biological functional ligands were bound to NPs by bioconjugation, after prior derivatization of the surface with terminal functional groups, such as primary amine (from 2-aminoethyl phosphonic acid[101]) or carboxylic acid groups (from *N*-phosphonomethyl iminodiacetic acid (PMIDA)[102,103] or 5-hydroxy-5,5′-bis(phosphono)pentanoic acid)[104] (Figure 13.13).

For example, the resulting ultrasmall superparamagnetic iron oxide nanocore could thus be coupled with folic acid, to selectively target cancer cells overexpressing the folate receptor and detect them by magnetic resonance imaging.[102]

Interestingly, in some cases the phosphonate coating could be performed in a single step concomitantly to the generation of the NP, via the alkali-mediated co-precipitation of Fe^{2+} and Fe^{3+} in the presence of the selected phosphonic acid (i.e. PMIDA).[103]

13.5 Conclusions

Contrary to the case of polymer coatings or layer by layer deposition of ionic species, grafting appropriate functional molecules onto inorganic substrates can provide a better control of the arrangement of the organic component at the surface. This feature is of particular importance when designing functional surfaces in the biomaterials area. Grafting of organic molecules onto inorganic surfaces in a controlled and stable manner requires the presence of functional groups which vary according to the chemical nature of the substrate to be modified, the most popular ones for many years now being for example [thiol/gold] and [SiH_3, $Si(OR)_3$ or $SiCl_3$/silica or glass].

In this context, if we exclude silicon-based materials (e.g. glass or silica), the phosphonic acid functionality is far superior to alkoxysilane groups for this purpose since the PO–Si bond can be cleaved easily by protic solvents.[105]

Additional advantages of phosphonic acids include: (i) a high compatibility with other organic functional groups; (ii) high robustness and stability; (iii) surface modification achievable using both ester [$RP(O)(OR)_2$] and acid [$RP(O)(OH)_2$] forms thus often making them highly soluble in many solvents including water. For this reason, the use of metal phosphonates in the field of biotechnologies is now rapidly expanding.

Acknowledgements

For part of the work described in this chapter, financial support has come from the Region Pays de la Loire (Bioregos I program), ANR (RNTS 2005 program – grant 5A0659), the Graftys company, and the Direction Générale de l'Armement (DGA – grants n° 04.04.025 and n° 2008.34.0010).

References

1. A. K. Cheetham, C. N. R. Rao and R. K. Feller, *Chem. Commun.*, 2006, 4780.
2. A. U. Czaja, N. Trukhan and U. Muller, *Chem. Soc. Rev.*, 2009, **38**, 1284.
3. M. Eddaoudi, D. B. Moler, H. L. Li, B. L. Chen, T. M. Reineke, M. O'Keeffe and O. M. Yaghi, *Acc. Chem. Res.*, 2001, **34**, 319.
4. D. J. Tranchemontagne, J. L. Mendoza-Cortes, M. O'Keeffe and O. M. Yaghi, *Chem. Soc. Rev.*, 2009, **38**, 1257.
5. A. Clearfield, *Progr. Inorg. Chem.*, 1998, **47**, 371.
6. R. Beutner, J. Michael, B. Schwenzer and D. Scharnweber, *J. R. Soc. Interface*, 2010, **7**, S93.
7. C. Viornery, Y. Chevolot, D. Leonard, B. O. Aronsson, P. Pechy, H. J. Mathieu, P. Descouts and M. Gratzel, *Langmuir*, 2002, **18**, 2582.
8. B. M. Silverman, K. A. Wieghaus and J. Schwartz, *Langmuir*, 2005, **21**, 225.
9. S. Tosatti, R. Michel, M. Textor and N. D. Spencer, *Langmuir*, 2002, **18**, 3537.
10. E. S. Gawalt, M. J. Avaltroni, M. P. Danahy, B. M. Silverman, E. L. Hanson, K. S. Midwood, J. E. Schwarzbauer and J. Schwartz, *Langmuir*, 2003, **19**, 200.

11. S. Pallu, C. Bourget, R. Bareille, C. Labrugere, M. Dard, A. Sewing, A. Jonczyk, M. Vernizeau, M. C. Durrieu and J. Amedee-Vilamitjana, *Biomaterials*, 2005, **26**, 6932.
12. N. Adden, L. J. Gamble, D. G. Castner, A. Hoffmann, G. Gross and H. Menzel, *Langmuir*, 2006, **22**, 8197.
13. J. Auernheimer and H. Kessler, *Bioorg. Med. Chem. Lett.*, 2006, **16**, 271.
14. J. Auernheimer, D. Zukowski, C. Dahmen, M. Kantlehner, A. Enderle, S. L. Goodman and H. Kessler, *ChemBioChem*, 2005, **6**, 2034.
15. J. Beuvelot, D. Portet, G. Lecollinet, M. F. Moreau, M. F. Basle, D. Chappard and H. Libouban, *J. Biomed. Mater. Res. B Appl. Biomater.*, 2009, **90B**, 873.
16. G. Lecollinet, N. Delorme, M. Edely, A. Gibaud, J. F. Bardeau, F. Hindre, F. Boury and D. Portet, *Langmuir*, 2009, **25**, 7828.
17. Y. X. Ni, B. Feng, J. X. Wang, X. Lu, S. X. Qu and J. Weng, *J. Mater. Sci.*, 2009, **44**, 4031.
18. M. P. Danahy, M. J. Avaltroni, K. S. Midwood, J. E. Schwarzbauer and J. Schwartz, *Langmuir*, 2004, **20**, 5333.
19. X. P. Qian, S. J. Metallo, I. S. Choi, H. K. Wu, M. N. Liang and G. M. Whitesides, *Anal. Chem.*, 2002, **74**, 1805.
20. B. Gottenbos, H. C. van der Mei, F. Klatter, P. Nieuwenhuis and H. J. Busscher, *Biomaterials*, 2002, **23**, 1417.
21. E. M. Hetrick and M. H. Schoenfisch, *Chem. Soc. Rev.*, 2006, **35**, 780.
22. J. Amalric, P. H. Mutin, G. Guerrero, A. Ponche, A. Sotto and J. P. Lavigne, *J. Mater. Chem.*, 2009, **19**, 141.
23. P. H. Mutin, G. Guerrero and J. Amalric, 2007, Patent WO2007080291.
24. B. Denizot, F. Hindre and D. Portet, 2006, Patent WO2006/053910.
25. S. Mukherjee, C. Huang, F. Guerra, K. Wang and E. Oldfield, *J. Am. Chem. Soc.*, 2009, **131**, 8374.
26. S. Mukherjee, Y. C. Song and E. Oldfield, *J. Am. Chem. Soc.*, 2008, **130**, 1264.
27. G. H. Nancollas, R. Tang, R. J. Phipps, Z. Henneman, S. Gulde, W. Wu, A. Mangood, R. G. G. Russell and F. H. Ebetino, *Bone*, 2006, **38**, 617.
28. J. Robinson, I. Cukrowski and H. M. Marques, *J. Mol. Struct.*, 2006, **825**, 134.
29. B. Peter, O. Gauthier, S. Laib, B. Bujoli, J. Guicheux, P. Janvier, G. H. van Lenthe, R. Muller, P. Y. Zambelli, J. M. Bouler and D. P. Pioletti, *J. Biomed. Mater. Res. A*, 2006, **76A**, 133.
30. B. Peter, D. P. Pioletti, S. Laib, B. Bujoli, P. Pilet, P. Janvier, J. Guicheux, P. Y. Zambelli, J. M. Bouler and O. Gauthier, *Bone*, 2005, **36**, 52.
31. J. D. Bobyn, K. McKenzie, D. Karabasz, J. J. Krygier and M. Tanzer, *J. Bone Joint Surg.*, 2009, **91**, 23.
32. A. Bigi, E. Boanini, C. Capuccini, M. Fini, I. N. Mihailescu, C. Ristoscu, F. Sima and P. Torricelli, *Biomaterials*, 2009, **30**, 6168.
33. E. Boanini, M. Gazzano, K. Rubini and A. Bigi, *Adv. Mater.*, 2007, **19**, 2499.

34. E. Boanini, P. Torricelli, M. Gazzano, R. Giardino and A. Bigi, *Biomaterials*, 2008, **29**, 790.
35. H. Denissen, E. van Beek, C. Lowik, S. Papapoulos and A. Vandenhooff, *Bone Miner.*, 1994, **25**, 123.
36. H. Denissen, E. van Beek, R. Martinetti, C. Klein, E. vanderZee and A. Ravaglioli, *J. Periodont. Res.*, 1997, **32**, 40.
37. Y. Gao, S. J. Zou, X. G. Liu, C. Y. Bao and J. Hu, *Biomaterials*, 2009, **30**, 1790.
38. M. Yoshinari, Y. Oda, T. Inoue, K. Matsuzaka and M. Shimono, *Biomaterials*, 2002, **23**, 2879.
39. S. Josse, C. Faucheux, A. Soueidan, G. Grimandi, D. Massiot, B. Alonso, P. Janvier, S. Laib, P. Pilet, O. Gauthier, G. Daculsi, J. Guicheux, B. Bujoli and J.-M. Bouler, *Biomaterials*, 2005, **26**, 2073.
40. H. Roussiere, F. Fayon, B. Alonso, T. Rouillon, V. Schnitzler, E. Verron, J. Guicheux, M. Petit, D. Massiot, P. Janvier, J. M. Bouler and B. Bujoli, *Chem. Mater.*, 2008, **20**, 182.
41. S. Josse, C. Faucheux, A. Soueidan, G. Grimandi, D. Massiot, B. Alonso, P. Janvier, S. Laib, O. Gauthier, G. Daculsi, J. Guicheux, B. Bujoli and J. M. Bouler, *Adv. Mater.*, 2004, **16**, 1423.
42. H. Roussiere, G. Montavon, S. Laïb, P. Janvier, B. Alonso, F. Fayon, M. Petit, D. Massiot, J. M. Bouler and B. Bujoli, *J. Mater. Chem.*, 2005, **15**, 3869.
43. E. Verron, O. Gauthier, P. Janvier, P. Pilet, J. Lesoeur, B. Bujoli, J. Guicheux and J.-M. Bouler, *Biomaterials*, 2010, **31**, 7776.
44. S. Panzavolta, P. Torricelli, B. Bracci, M. Fini and A. Bigi, *J. Inorg. Biochem.*, 2009, **103**, 101.
45. S. Panzavolta, P. Torricelli, B. Bracci, M. Fini and A. Bigi, *J. Inorg. Biochem.*, 2010, **104**, 1099.
46. V. Schnitzler, F. Fayon, C. Despas, I. Khairoun, C. Mellier, T. Rouillon, D. Massiot, A. Walcarius, P. Janvier, O. Gauthier, G. Montavon, J.-M. Bouler and B. Bujoli, *Acta Biomater.*, 2011, **7**, 759.
47. C. Rill, Z. I. Kolar, G. Kickelbick, H. T. Wolterbeek and J. A. Peters, *Langmuir*, 2009, **25**, 2294.
48. T. Vitha, V. Kubicek, P. Hermann, L. V. Elst, R. N. Muller, Z. I. Kolar, H. T. Wolterbeek, W. A. P. Breeman, I. Lukes and J. A. Peters, *J. Med. Chem.*, 2008, **51**, 677.
49. S. F. Zhang, G. Gangal and H. Uludag, *Chem. Soc. Rev.*, 2007, **36**, 507.
50. J. L. DeRisi, V. R. Iyer and P. O. Brown, *Science*, 1997, **278**, 680.
51. V. Knezevic, C. Leethanakul, V. E. Bichsel, J. M. Worth, V. V. Prabhu, J. S. Gutkind, L. A. Liotta, P. J. Munson, E. F. Petricoin and D. B. Krizman, *Proteomics*, 2001, **1**, 1271.
52. B. Li, L. X. Jiang, Q. F. Song, J. X. Yang, Z. L. Chen, Z. B. Guo, D. S. Zhou, Z. M. Du, Y. J. Song, J. Wang, H. X. Wang, S. Y. Yu, J. Wang and R. F. Yang, *Infect. Immun.*, 2005, **73**, 3734.
53. J. C. Miller, H. P. Zhou, J. Kwekel, R. Cavallo, J. Burke, E. B. Butler, B. S. Teh and B. B. Haab, *Proteomics*, 2003, **3**, 56.

54. M. Schena, D. Shalon, R. W. Davis and P. O. Brown, *Science*, 1995, **270**, 467.
55. A. Sreekumar, M. K. Nyati, S. Varambally, T. R. Barrette, D. Ghosh, T. S. Lawrence and A. M. Chinnaiyan, *Cancer Res.*, 2001, **61**, 7585.
56. S. Steller, P. Angenendt, D. J. Cahill, S. Heuberger, H. Lehrach and J. Kreutzberger, *Proteomics*, 2005, **5**, 2048.
57. M. Beier and J. D. Hoheisel, *Nucleic Acids Res.*, 1999, **27**, 1970.
58. L. M. Demers, S. J. Park, T. A. Taton, Z. Li and C. A. Mirkin, *Angew. Chem., Int. Ed. Engl.*, 2001, **40**, 3071.
59. H. Byrd, E. P. Suponeva, A. B. Bocarsly and M. E. Thompson, *Nature*, 1996, **380**, 610.
60. J. C. Horne and G. J. Blanchard, *J. Am. Chem. Soc.*, 1996, **118**, 12788.
61. J. C. Horne, Y. Huang, G. Y. Liu and G. J. Blanchard, *J. Am. Chem. Soc.*, 1999, **121**, 4419.
62. H. E. Katz and M. L. Schilling, *Chem. Mater.*, 1993, **5**, 1162.
63. H. E. Katz, M. L. Schilling, C. E. D. Chidsey, T. M. Putvinski and R. S. Hutton, *Chem. Mater.*, 1991, **3**, 699.
64. H. Lee, L. J. Kepley, H. G. Hong, S. Akhter and T. E. Mallouk, *J. Phys. Chem.*, 1988, **92**, 2597.
65. H. Lee, L. J. Kepley, H. G. Hong and T. E. Mallouk, *J. Am. Chem. Soc.*, 1988, **110**, 618.
66. G. A. Neff, M. R. Helfrich, M. C. Clifton and C. J. Page, *Chem. Mater.*, 2000, **12**, 2363.
67. T. M. Putvinski, M. L. Schilling, H. E. Katz, C. E. D. Chidsey, A. M. Mujsce and A. B. Emerson, *Langmuir*, 1990, **6**, 1567.
68. J. L. Snover, H. Byrd, E. P. Suponeva, E. Vicenzi and M. E. Thompson, *Chem. Mater.*, 1996, **8**, 1490.
69. K. J. Hoekstra and T. Bein, *Chem. Mater.*, 1996, **8**, 1865.
70. X. H. Xu and A. J. Bard, *J. Am. Chem. Soc.*, 1995, **117**, 2627.
71. X. H. Xu, H. C. Yang, T. E. Mallouk and A. J. Bard, *J. Am. Chem. Soc.*, 1994, **116**, 8386.
72. G. Nonglaton, I. O. Benitez, I. Guisle, M. Pipelier, J. Leger, D. Dubreuil, C. Tellier, D. R. Talham and B. Bujoli, *J. Am. Chem. Soc.*, 2004, **126**, 1497.
73. I. O. Benitez, B. Bujoli, L. J. Camus, C. M. Lee, F. Odobel and D. R. Talham, *J. Am. Chem. Soc.*, 2002, **124**, 4363.
74. H. Byrd, J. K. Pike and D. R. Talham, *Chem. Mater.*, 1993, **5**, 709.
75. H. Byrd, J. K. Pike and D. R. Talham, *J. Am. Chem. Soc.*, 1994, **116**, 7903.
76. C. N. Nixon, K. Le Claire, F. Odobel, B. Bujoli and D. R. Talham, *Chem. Mater.*, 1999, **11**, 965.
77. M. A. Petruska, G. E. Fanucci and D. R. Talham, *Chem. Mater.*, 1998, **10**, 177.
78. A. P. Wu and D. R. Talham, *Langmuir*, 2000, **16**, 7449.
79. S. M. Lane, J. Monot, M. Petit, B. Bujoli and D. R. Talham, *Colloids Surf. B Biointerfaces*, 2007, **58**, 34.
80. B. Bujoli, S. M. Lane, G. Nonglaton, M. Pipelier, J. Leger, D. R. Talham and C. Tellier, *Chem. Eur. J.*, 2005, **11**, 1981.

81. S. M. Lane, J. Monot, M. Petit, C. Tellier, B. Bujoli and D. R. Talham, *Langmuir*, 2008, **24**, 7394.
82. J. Monot, M. Petit, S. M. Lane, I. Guisle, J. Leger, C. Tellier, D. R. Talham and B. Bujoli, *J. Am. Chem. Soc.*, 2008, **130**, 6243.
83. M. Cinier, M. Petit, M. N. Williams, R. M. Fabre, F. Pecorari, D. R. Talham, B. Bujoli and C. Tellier, *Bioconjug. Chem.*, 2009, **20**, 2270.
84. J. D. Dunn, G. E. Reid and M. L. Bruening, *Mass Spectrom. Rev.*, 2010, **29**, 29.
85. S. Feng, M. L. Ye, H. J. Zhou, X. G. Jiang, X. N. Jiang, H. F. Zou and B. L. Gong, *Mol. Cell. Proteomics*, 2007, **6**, 1656.
86. H. J. Zhou, S. Y. Xu, M. L. Ye, S. Feng, C. Pan, X. G. Jiang, X. Li, G. H. Han, Y. Fu and H. Zou, *J. Proteomics Res.*, 2006, **5**, 2431.
87. J. Y. Wei, Y. J. Zhang, J. L. Wang, F. Tan, J. F. Liu, Y. Cai and X. H. Qian, *Rapid Commun. Mass Spectrom.*, 2008, **22**, 1069.
88. J. Dong, H. Zhou, R. Wu, M. Ye and H. Zou, *J. Separ. Sci.*, 2007, **30**, 2917.
89. A. M. Clausen and P. W. Carr, *Anal. Chem.*, 1998, **70**, 378.
90. A. M. Clausen, A. Subramanian and P. W. Carr, *J. Chromatogr. A*, 1999, **831**, 63.
91. A. Subramanian, P. W. Carr and C. V. McNeff, *J. Chromatogr. A*, 2000, **890**, 15.
92. A. Subramanian and S. Sarkar, *J. Chromatogr. A*, 2002, **944**, 179.
93. R. M. Fabre and D. R. Talham, *Langmuir*, 2009, **25**, 12644.
94. R. F. Roskamp, I. K. Vockenroth, N. Eisenmenger, J. Braunagel and I. Koper, *ChemPhysChem*, 2008, **9**, 1920.
95. G. Chouhan, D. S. Wang and H. Alper, *Chem. Commun.*, 2007, 4809.
96. A. G. Hu, G. T. Yee and W. B. Lin, *J. Am. Chem. Soc.*, 2005, **127**, 12486.
97. C. A. Traina and J. Schwartz, *Langmuir*, 2007, **23**, 9158.
98. M. A. White, J. A. Johnson, J. T. Koberstein and N. J. Turro, *J. Am. Chem. Soc.*, 2006, **128**, 11356.
99. C. Yee, G. Kataby, A. Ulman, T. Prozorov, H. White, A. King, M. Rafailovich, J. Sokolov and A. Gedanken, *Langmuir*, 1999, **15**, 7111.
100. D. Portet, B. Denizot, E. Rump, J. J. Lejeune and P. Jallet, *J. Colloid Interface Sci.*, 2001, **238**, 37.
101. M. Das, D. Mishra, T. K. Maiti, A. Basak and P. Pramanik, *Nanotechnology*, 2008, **19**, 415101.
102. M. Das, P. Dhak, S. Gupta, D. Mishra, T. K. Maiti, A. Basak and P. Pramanik, *Nanotechnology*, 2010, **21**, 125103.
103. M. Das, D. Mishra, P. Dhak, S. Gupta, T. K. Maiti, A. Basak and P. Pramanik, *Small*, 2009, **5**, 2883.
104. Y. Lalatonne, C. Paris, J. M. Serfaty, P. Weinmann, M. Lecouvey and L. Motte, *Chem. Commun.*, 2008, 2553.
105. C. E. McKenna, M. T. Higa, N. H. Cheung and M.-C. McKenna, *Tetrahedron Lett.*, 1977, **18**, 155.

CHAPTER 14

Structural Diversity in Metal Phosphonate Frameworks: Impact on Applications

KONSTANTINOS D. DEMADIS AND NIKOLETA STAVGIANOUDAKI

Crystal Engineering, Growth & Design Laboratory, Department of Chemistry, University of Crete, Heraklion, Crete, GR-71003, Greece

14.1 Introduction

The emerging fields of supramolecular chemistry, crystal engineering, and materials chemistry have made long strides forward with impressive growth during the last decades. The chemistry of phosphonate ligands has undoubtedly played an important role in widening these areas of research.[1–42] Phosphonate ligands have attracted considerable attention in the context of fundamental research, but they have also been extensively used in several other technologically/industrially significant areas, such as water treatment,[43–46] oilfield drilling,[47–54] minerals processing,[55,56] corrosion control,[57–62] metal complexation and sequestration,[63,64] dental materials,[65–68] enzyme inhibition,[69–71] bone targeting,[72–75] cancer treatment,[76–78] etc. There exists a plethora of metal phosphonate materials whose crystal structures exhibit attractive features that depend on several variables, such as nature of M^{n+} (metal oxidation state, ionic radius, and coordination number in particular), number of phosphonate groups on the ligand backbone, presence of other functional moieties (carboxylate, sulfonate, amine, hydroxyl), and, naturally, on process variables (reactant ratio and concentration, temperature, pressure, etc.). The synthesis of metal phosphonates is usually carried out in aqueous solutions (or in mixtures of water and a polar organic solvent such as alcohols, DMF, acetone), so it is

Metal Phosphonate Chemistry: From Synthesis to Applications
Edited by Abraham Clearfield and Konstantinos Demadis

Published by the Royal Society of Chemistry, www.rsc.org

not surprising that water is commonly found in their lattice, participating in extensive hydrogen bonding, which is predominant in these architectures resulting in 1D, 2D, and 3D supramolecular networks.[79–81] Lastly, it is worth noting that the vast majority of metal phosphonates are coordination polymers and clusters (to a lesser extent), although there are reports of metal phosphonate molecular complexes.

The structural literature of metal phosphonate materials is fairly large. However, in this chapter we will present a selective overview, focusing on some interesting structural features of metal phosphonate materials that contain metal ions in a variety of oxidation states combined with a diverse selection of phosphonate ligands. We will also explore possible relationships between structural features and properties, with an eye towards applications. The structures of the phosphonate ligands (and their "mixed" analogs) mentioned in this chapter are shown in Figure 14.1.

14.2 The P–O Bond in "Free" Phosphonic Acids and "Metal-coordinated" Phosphonates

The phosphonate group is a strongly anionic moiety.[82] Thus, it tends to form strong bonds with metal ions M^{n+} ($n = 1$–4). The phosphonic acid moiety, $R\text{-}PO_3H_2$ and its two deprotonated forms, $R\text{-}PO_3H^-$ and $R\text{-}PO_3^{2-}$ exhibit an interesting, pH-dependent behavior (Figure 14.2). Usually the first proton is very acidic, while the second is a couple orders of magnitude less acidic.[83–89]

There is a plethora of published crystal structures of "free" (not metal-coordinated) phosphonic acids and their deprotonated forms. It is beyond the scope of this chapter to compile an exhaustive analysis of all the structural features. Only some relevant comments will be made here, that are of significance to the discussion on metal phosphonates.

Polyanchuk *et al.*[90] have determined the structure of ethylenediamine-tetrakis(methylphosphonic acid) dihydrate, EDTMP (an EDTA analog) from single crystal data (see Figure 14.3). It is worthwhile to have a closer look at the geometrical features of the phosphonate groups.

Due to the high basicity of the N atoms, an internal protonation has occurred. Two protons originating from two diametrically opposite phosphonic acid groups have protonated the N atoms. Hence, the EDTMP molecule is best described as a zwitter-ion with two phosphonic acid groups and two monoprotonated phosphonate groups. The P–O bond distances in the two phosphonic acid groups (P(2) will be used as the example) are 1.529(3) Å, 1.547(3) Å, and 1.476(3) Å, i.e. two "long" and one "short" bond lengths. It is obvious that the "short" bond, 1.476(3) Å, belongs to the P=O phosphoryl group, and the two "long" bonds to the two P–O(H) groups. Similarly, the P–O bond distances in the two monoprotonated phosphonate groups (P(1) will be used as the example) are 1.557(3) Å, 1.486(3) Å, and 1.507(3) Å, i.e. one "long", one "short", and one "intermediate" bond lengths. Again, the "short" bond, 1.486(3) Å, belongs to the P=O phosphoryl group, and the "long" bond,

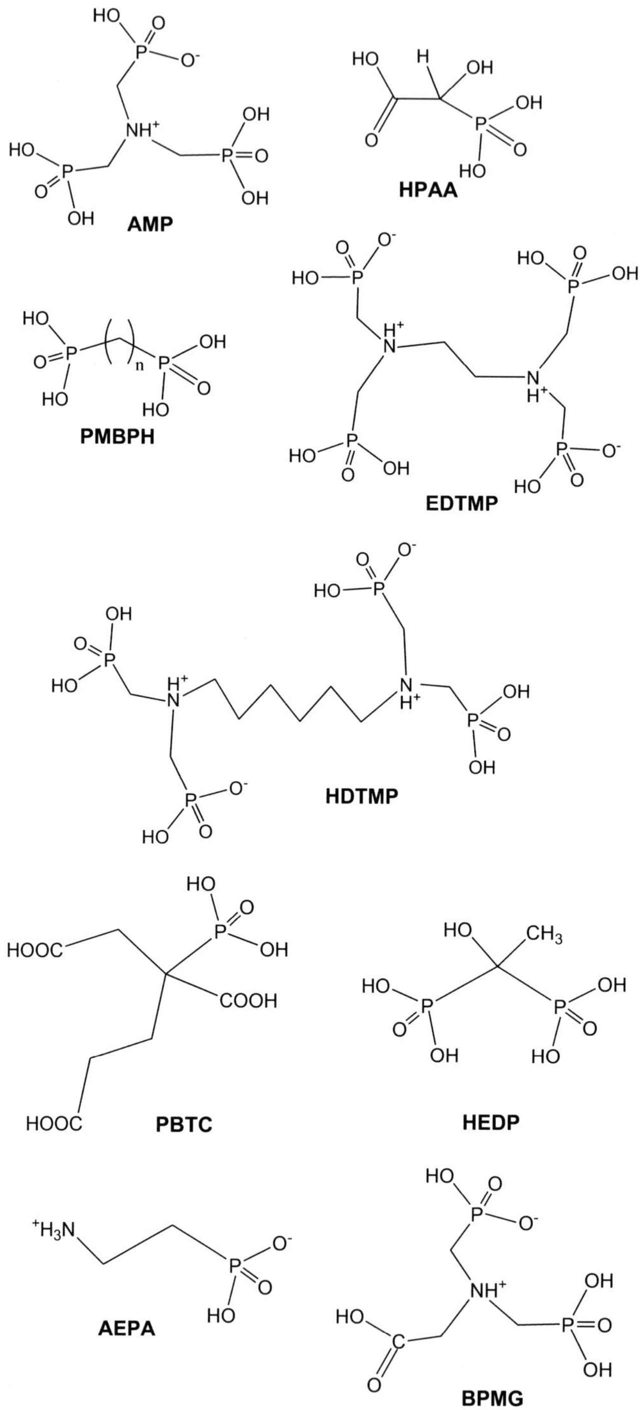

Figure 14.1 Schematic structures of several mono- and poly-phosphonic acids.

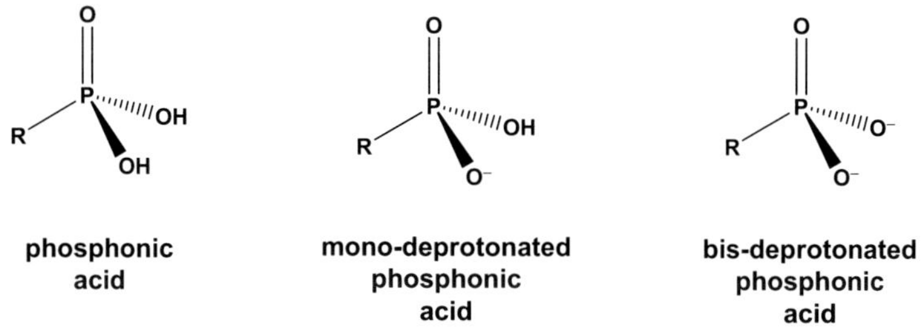

Figure 14.2 The three forms of the phosphonic acid group.

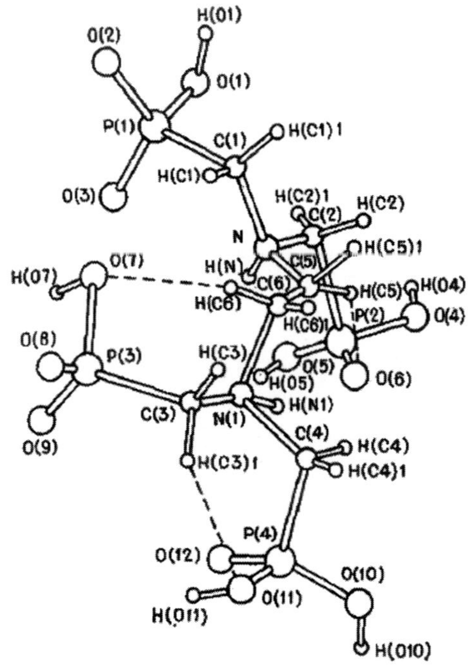

Figure 14.3 The crystal structure of EDTMP dihydrate. (Reproduced with permission from ref. 90.)

1.557(3) Å, to the P–O(H) group. It is therefore reasonable to assume that the "intermediate" bond length be assigned to the monoprotonated –P–O$^-$ moiety. The same observations can be noted for the three phosphonic acid groups in the structures of benzene-1,3,5-triphosphonic acid (Figure 14.4),[91] 1,3,5,7-tetrakis(4-phosphonophenyl)adamantane (Figure 14.5),[92] and 1,2,4-phosphobutane tricarboxylic acid monohydrate (Figure 14.6).[93]

The same observations can be noted for the monoprotonated phosphonate groups in the ionic tetraphosphonate structures of (en)(HDTMP), (NH_4)(EDTMP), $Zn(H_2O)_6$(TDTMP) (en = ethylenediamine, see Figure 14.1 for phosphonate names and structures).[94]

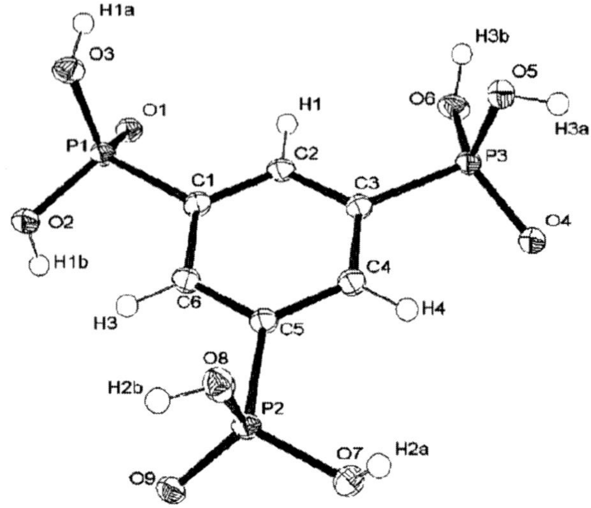

Figure 14.4 The molecular structure of benzene-1,3,5-triphosphonic acid. (Reproduced with permission from ref. 91.)

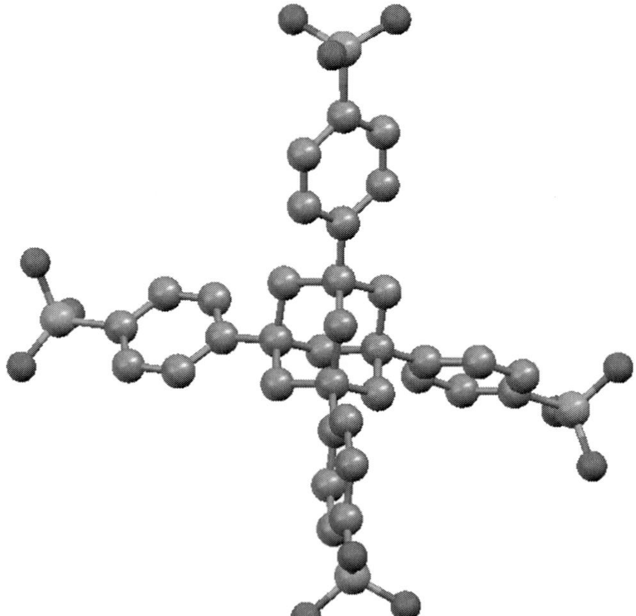

Figure 14.5 The molecular structure of 1,3,5,7-tetrakis(4-phosphonophenyl)adamantane. (Reproduced with permission from ref. 92.)

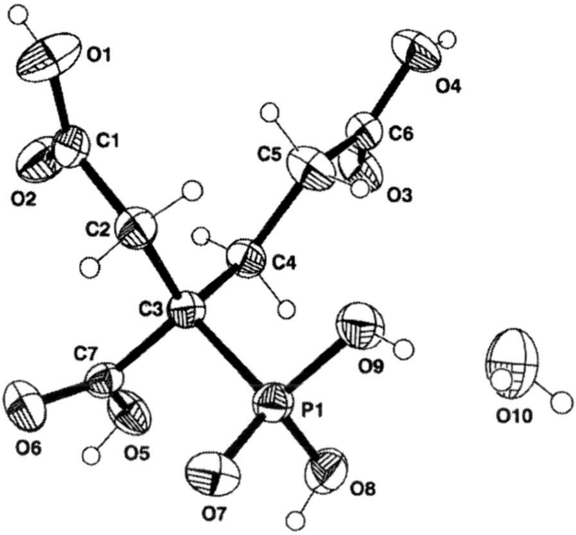

Figure 14.6 The molecular structure of 1,2,4-tricarboxylic acid monohydrate. (Reproduced with permission from ref. 93.)

Upon coordination of the phosphonate oxygens with a metal ion there are a number of structural changes related to the P–O bond lengths. We will demonstrate this by using a number of examples, starting with 1D $[Cu(HO_3PCH(OH)CO_2)(H_2O)_2]\cdot H_2O$, Figure 14.7.[95] This material possesses a monodeprotonated phosphonic acid group that is coordinated to a metal ion in a monodentate fashion. The three P–O bond lengths are P(1)–O(4) 1.4931(18) Å, P(1)–O(6) 1.5212(18) Å, P(1)–O(5) 1.5664(19) Å. It is obvious that the 1.5664(19) Å bond distance belongs to the P–O(H) acid group. The shortest distance, 1.4931(18) Å, is assigned to the P=O groups, whereas the remaining "intermediate" distance, 1.5212(18) Å, is the P–O(Cu) moiety. This situation resembles that of a monoprotonated, but not metal-coordinated $-PO_3H^-$ moiety, discussed above.

The AMP ligand (AMP = amino-tris(methylenephosphonate) in the layered 2D material $[Ca(AMP)(H_2O)]\cdot 3.5H_2O$ coordinates to Ca^{2+} in a monodentate fashion, but in one occasion the phosphonate group is terminal and in the second it is bridging (Figure 14.8).[43]

Focusing on the Ca–O–P–O–Ca moiety, we note the following bond lengths: P(2)–O(5) 1.4942(15) Å, P(2)–O(4), 1.5102(14) Å, P(2)–O(6) 1.5684(15) Å. Again, the longest P–O bond length, 1.5684(15) Å, belongs to the uncoordinated and protonated P–O(H) moiety. However, the other two P–O bond lengths tend to merge. This occurs because the anionic charge is now more effectively delocalized over the Ca–O–P–O–Ca moiety and it becomes almost impossible to distinguish between the P=O and the $P-O^-$ groups.

In the structure of Sr(AMP) (which is anhydrous) one of the AMP arms coordinates to two Sr^{2+} centers in a chelating-bridging fashion. The

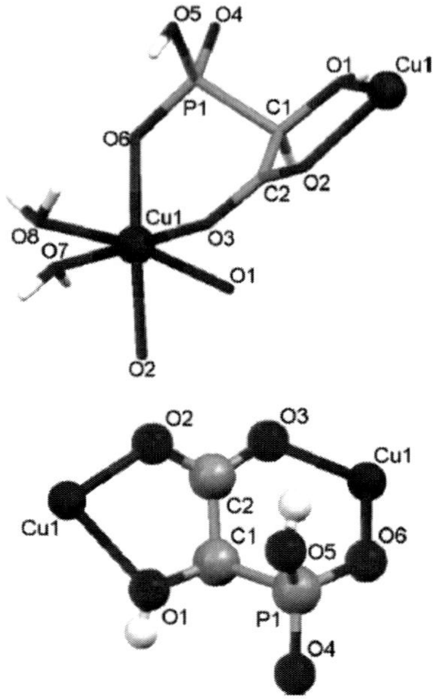

Figure 14.7 Asymmetric unit of $[Cu(HO_3PCH(OH)CO_2)(H_2O)_2]\cdot H_2O$ (upper) and the ligand chelating/bridging coordination mode (lower). (Reproduced with permission from ref. 95.)

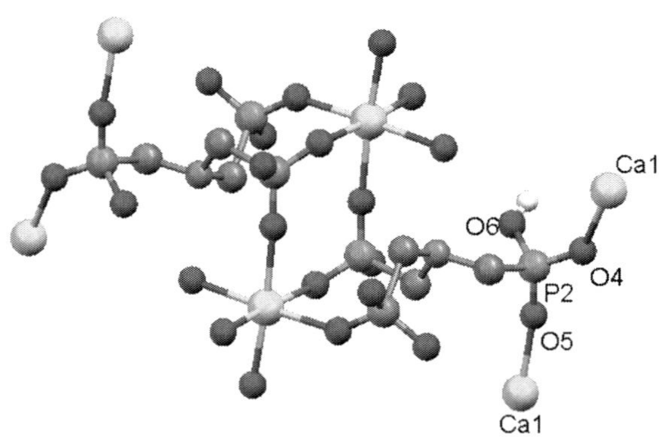

Figure 14.8 Structure of $[Ca(AMP)(H_2O)]\cdot 3.5H_2O$.

P–O(H) group is not coordinated (Figure 14.9). Similar bond lengths are observed.[96]

In the structure of Ba-AMP the P–O(H) group is now coordinated to a Ba center, creating a 4-membered chelate ring with the P–O group (Figure 14.10).[96]

There are a few examples of metal phosphonate structures in which the mono-deprotonated phosphonic acid group coordinates to three different metal ions.[97] For instance, in the structure of Ba-CEPA (CEPA = carboxyethylphosphonic acid) a structural motif is observed that is shown in Figure 14.11.

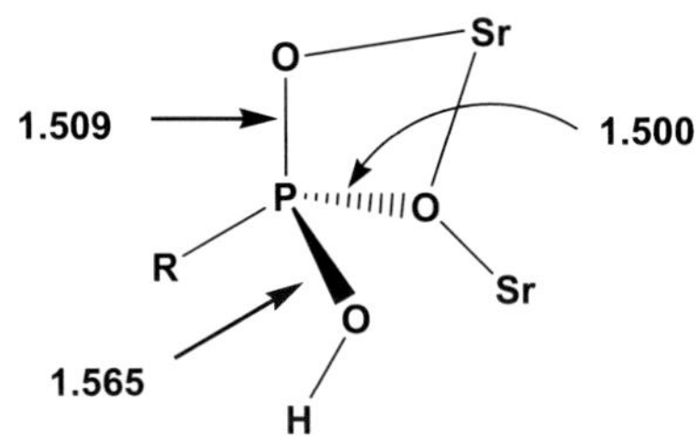

Figure 14.9 Portion of the structure of Sr-AMP. Distances are in Å.

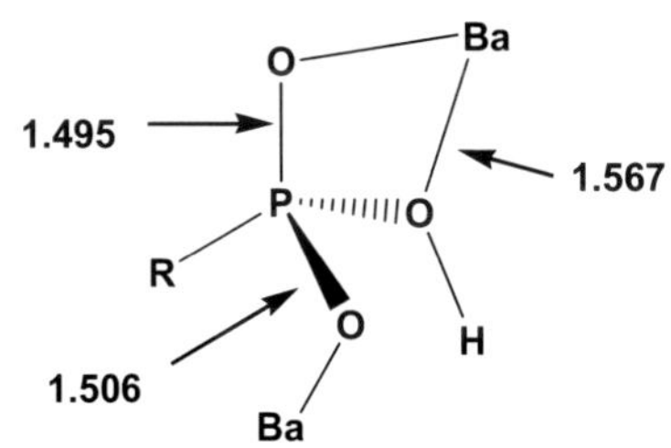

Figure 14.10 Portion of the structure of Ba-AMP. Distances are in Å.

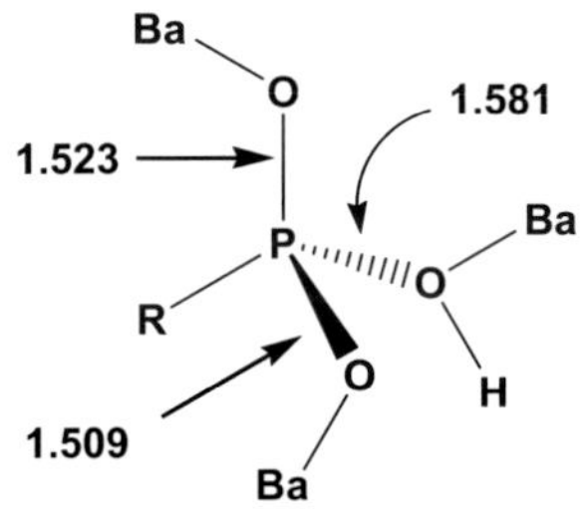

Figure 14.11 Portion of the structure of Ba-CEPA. Distances are in Å.

A slight elongation of all P–O bonds is observed. Still, the P–O(H) bond (although the O is now coordinated to a Ba) is the longest one, 1.581 Å. A similar situation is observed in the structure of Ba-HPAA (HPAA = hydroxyphosphonoacetic acid).[98,99]

Let us now take a look at the structural changes on the P–O bond lengths when the phosphonate group is doubly deprotonated. In the structures of $[M(phen)_3]\cdot C_6H_5PO_3\cdot 11H_2O$ (M = Co, Ni, Cu, phen = 1,10-phenanthroline; Figure 14.12), the phenylphosphonate is found in its doubly deprotonated state, and it is not coordinated.[100]

Thus, this is a good example to assess the changes induced in the P–O bond lengths upon complete deprotonation. Thus, the P–O bond lengths are 1.544 Å, 1.497 Å, and 1.505 Å. One of the three P–O bonds (1.544 Å) is still longer than

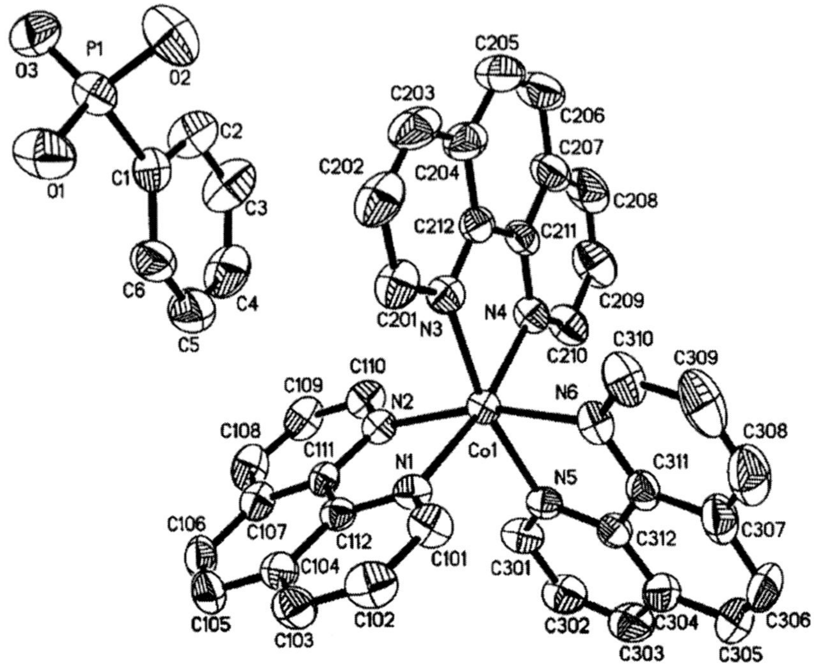

Figure 14.12 The structure of $[M(phen)_3] \cdot C_6H_5PO_3 \cdot 11H_2O$. (Reproduced with permission from ref. 100.)

the other two, whose bond lengths tend to merge. It should be noted that the $PhPO_3^{2-}$ dianion participates in an extensive network of hydrogen bonds with the lattice water molecules.

Interaction of strontium acetate with dippH$_2$ (dippH$_2$ = 2,6-diisopropylphenyl phosphate) in methanol at room temperature leads to the isolation of the ionic phosphate $[\{Sr_2(\mu\text{-}H_2O)_4(H_2O)_{10}\}\{dipp\}_2] \cdot 4CH_3OH$.[101] Interestingly, the P–O bond lengths in the uncoordinated dipp dianion are nearly identical, P1–O1 1.509(7) Å, P1–O2 1.503(6) Å, P1–O3 1.494(6) Å (Figure 14.13).

Upon metal coordination of the doubly deprotonated phosphonic acid group, elongation of the P–O bond lengths is observed. For example, in the structure of $Ba_2[O_3P(CH_2)_3PO_3] \cdot 3H_2O$ the P–O bond distances are: P(1)–O(3) 1.532(11) Å, P(1)–O(2) 1.536(10) Å, P(1)–O(1) 1.540(10) Å (Figure 14.14).[102]

All phosphonate groups are also deprotonated in the barium phosphonate $[Ba_3(O_3PCH_2NH_2CH_2PO_3)_2(H_2O)_4] \cdot 3H_2O$.[103] The P–O bond lengths range from 1.507(3) Å to 1.530(3) Å (Figure 14.15).

14.3 The Metal–Oxygen Bond in Metal Coordinated Phosphonates

In this section we will attempt to see whether there are systematic correlations between the oxidation state of the metal ion and the M–O bond distance.

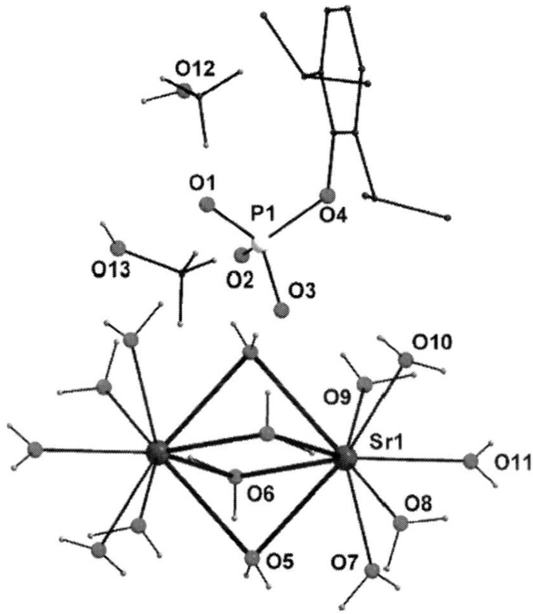

Figure 14.13 The structure of $[\{Sr_2(\mu\text{-}H_2O)_4(H_2O)_{10}\}\{dipp\}_2] \cdot 4CH_3OH$. (Reproduced with permission from ref. 101.)

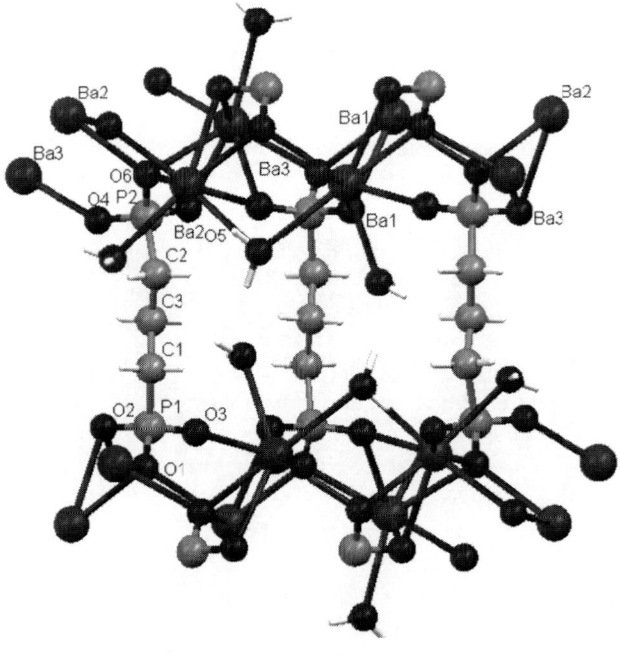

Figure 14.14 The structure of $Ba_2[O_3P(CH_2)_3PO_3] \cdot 3H_2O$. The diphosphonate ligand acts as a pillar connecting the Ba/O inorganic layer. Structure drawn from the cif file taken from ref. 102.

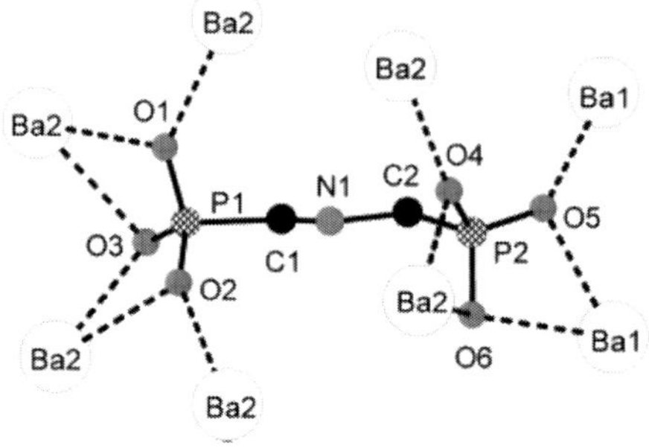

Figure 14.15 The partial structure of $[Ba_3(O_3PCH_2NH_2CH_2PO_3)_2(H_2O)_4] \cdot 3H_2O$, showing the coordination mode of the fully deprotonated phosphonate groups. (Reproduced with permission from ref. 103.)

Correlations are meaningful for metal environments with the same coordination numbers, as a word of caution. The range of Na–O bond distances in the compound Na-N-(phosphonomethyl)glycine is 2.401(2) Å to 2.361(2) Å.[104] Similarly, the Na–O bond distances in the material $Na_2(AMP) \cdot 1.5H_2O$ range from 2.336(5) Å to 2.489(5) Å.[105] The Li–O bond distances in the compound *catena*-poly[lithium-μ_3-ethylenediphosphonato] are much shorter, 1.962(4) Å to 1.905(3) Å. This is an indication that the ionic radius plays a significant role, as will be discussed later in this section. It is noteworthy that the P=O bond coordinates to the Li ion, as reported by the authors.[106] Incorporation of a larger monovalent cation in the material $[Ag_4(O_3PCH_2CH_2PO_3)]$ creates a wider range of Ag–O bond distances, 2.203(12) Å to 2.530(10) Å.[107] A similar observation can be made for the material Ag-N-(phosphonomethyl)glycine, in which the Ag–O distances are 2.21 Å to 2.53 Å.[104]

We will now discuss in a systematic way in the next paragraphs the M–O bond distances, where M = alkaline earth metal cation. It is useful to take a closer look at the M–O bond distances, where M = transition metal cation. A good starting point is the copper phosphonate materials, such as the 1D compound $[Cu(R,S\text{-}HO_3PCH(OH)CO_2)(H_2O)_2] \cdot H_2O$. The Cu–O distance is 1.9687(18) Å. The phosphonate group is monodentate in this compound.[95] In a similar material, $Cu_{1.5}(O_3PCH_2CO_2) \cdot H_2O$ (Figure 14.16), the phosphonate group is pentadentate and the Cu–O distances range from 1.909(3) Å to 1.972(3) Å.[95] It is interesting to note that one of the phosphonate oxygens participates in a Cu–μ_2-O–Cu bridge; however, the Cu–O distances seem to be unaffected. Lastly, there is a "long" Cu–O distance (2.284(3) Å), but this dramatic elongation is due to the Jahn–Teller effect (common in a d^9 O_h coordination environment).

Similar Cu–O bond distances are noted for the material $Cu(HO_3PCH_2CO_2) \cdot H_2O$.[95] 1D material $[Cu(R,S\text{-}HO_3PCH(OH)CO_2)(H_2O)_2] \cdot H_2O$ upon complete dehydration transforms to anhydrous 3D material $Cu(R,S\text{-}HO_3PCH(OH)CO_2)$, in which the Cu–O bond distances range from 1.994(4) Å to

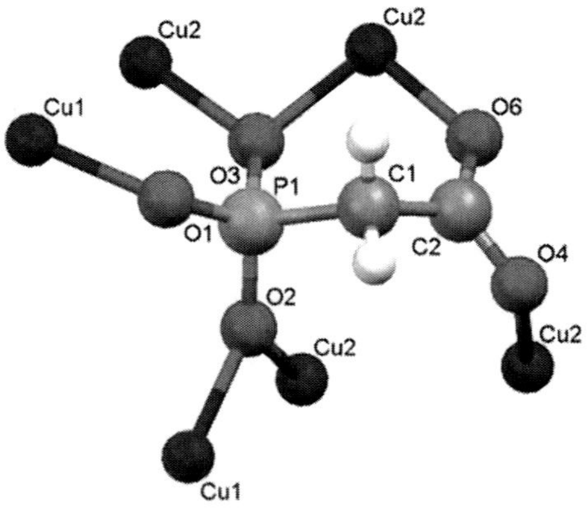

Figure 14.16 Portion of the structure of $Cu_{1.5}(O_3PCH_2CO_2)\cdot H_2O$ that shows the coordination mode of phosphonoacetate. (Reproduced with permission from ref. 95.)

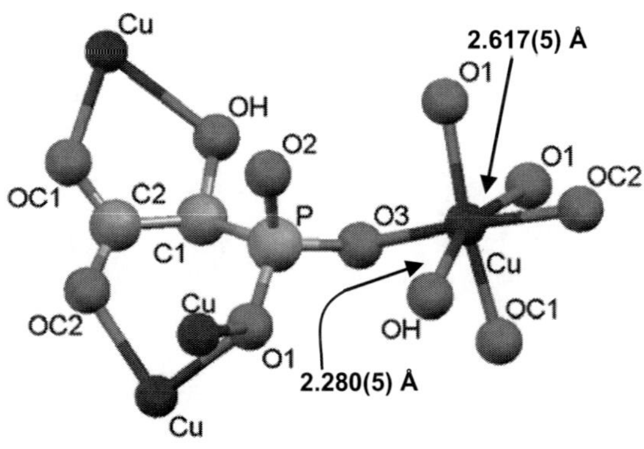

Figure 14.17 Portion of the structure of $[Cu(R,S\text{-}HO_3PCH(OH)CO_2)(H_2O)_2]\cdot H_2O$ that shows the coordination mode of the hydroxyphosphonoacetate. (Reproduced with permission from ref. 108.)

1.988(4) Å (Figure 14.17). As before, there is a long Cu–O distance, 2.617(5) Å, due to the Jahn–Teller effect.[108]

1D phases with compositions $[M\{HO_3PCH(OH)CO_2\}(H_2O)_2]\cdot 2H_2O$ (M = Mg, Co, and Zn) have been isolated.[109] These materials are isostructural. The Zn material and its various hydrated forms display a range of Zn–O distances from 2.042 Å to 2.064 Å. In the molecular Co^{II} phosphonate $[Co(H_2O)_6][Co(H_2O)_4(HDTMP)_2]\cdot 12H_2O$ the Co–O distance is 2.106(2) Å.[110] The Co–O bond lengths in the 3D material poly[cesium(I) [μ_4-ethylenediphosphonato)cobalt(II)]], $\{Cs[Co(C_2H_5O_6P_2)]\}_n$, fall in the range from 1.934(3) Å to 1.965(2) Å. The Co is found in a tetrahedral environment.[111] These Co–O bond lengths are comparable to those found in other cobalt(II) phosphonates containing Co^{II} in a tetrahedral environment.[23,112–117]

In 1D linear $[Fe(H_2PMIDA)(H_2O)_2]$ (PMIDA = phosphonomethyl-imino-diacetic acid) there are two Fe–O bond distances, 2.051(2) Å and 2.182(2) Å.[118] This paper furnishes useful crystallographic information on Fe–O and Fe–N bond distances, based on statistical analyses of a CSD search. In the 3D material $(NH_3C_6H_4NH_3)Fe_2(HEDP)\cdot 2H_2O$, HEDP = $C(CH_3)(OH)(PO_3)_2{}^{2-}$, the Fe–O bond distances are 2.050(2) Å to 2.150(2) Å. Here, there seems to be a wide range of bond lengths.[119]

The 3D material $Mn[(HO_3PCH_2)_2N(H)(CH_2)_4(H)N(CH_2PO_3H)_2]$ is built from alternating corner-linked $[MnO_6]$ and $[PO_3C]$ polyhedra forming a 2D net of eight rings. These layers are connected to a pillared structure by the diaminobutane groups. Magnetic susceptibility data confirms the presence of Mn^{2+} ions.[120] The Mn–O bond distances are in the range 2.138(2)–2.219(2) Å.

In the 2D zigzag material Zn-AMP the Zn–O distances are 2.049 Å and 2.122 Å.[121] In a different, 3D Zn-AMP material, $Zn_2[HO_3PCH_2NH(CH_2PO_3)_2]$, containing tetrahedral Zn centers, the Zn–O bond distance is 1.928(7) Å to 1.964(8) Å.[122] In Zn-HDTMP the Zn–O distances are in the range 2.022(3)–2.067(3) Å.[123] In the 1D Zn-HPAA material the Zn–O distance is 2.042(2) Å.[109]

In the 1D material Ni-HPAA the Ni–O distance is 2.0322(18) Å.[124]

In the chromium(II) methyl phosphonate dihydrate, $[Cr(CH_3PO_3)(H_2O)]\cdot H_2O$, the Cr–O bond distances are 2.031 Å, and there is a "longer" one corresponding to a μ_2-O phosphonate, of 2.084 Å.[125] Similar distances are reported for the organometallic Cr^{II} phosphonate $[\eta^6\text{-}1,2\text{-}C_6H_4(OCH_3)(P(O)(OCH_2CH_3)_2)]Cr(CO)_3$.[126]

In Al-AMP phosphonate material, $Al[(HO_3PCH_2)_3N]\cdot H_2O$, the Al–O bond distances were reported to be in the range 1.841(14) Å to 2.011(14) Å.[127] In the framework of Al^{III}-*N,N′*-piperazinebis(methylenephosphonate) the Al–O bond distances are 1.98(2) Å and 2.05(2) Å.[35] The Al–O bond distances in the layered phosphonate $Al(OH)(O_3PCH_3)\cdot H_2O$ are in a "tighter" range 1.870(1) Å to 1.881(1) Å.[128] Similarly, the Al–O distances in the material $(C_5H_5NH)\{AlF[O_3PCH_2PO_2(OH)]\}\cdot 0.61H_2O$ fall into the range 1.876(3) Å to 1.898(2) Å.[129] For Fe^{III} phosphonates, the Fe–O distances are a bit longer than the corresponding Al–O ones. For example, in the iron diphosphonate $Fe^{III}(HO_3PCH_2CH_2PO_3)(H_2O)$ the Fe–O distances range from 1.980(3) Å to 1.989(3) Å.[19] Similar Fe–O distances are observed in $[Fe_4O(t\text{-}BuPO_3)_3(O_2CPh)_3(py)_3Cl]\cdot 3.5py$ (py = pyridine), 1.913(9) Å to 1.964(5) Å.[130]

In the framework of Ti^{IV} *N,N′*-piperazinebis(methylenephosphonate) (MIL-91(Ti)) the Ti–O bond distances are 2.16(2) Å, 2.01(2) Å.[35] In the "mixed" Sn^{IV} oxalate/methylphosphonate material $Sn_2(O_3PCH_3)(C_2O_4)$ a Sn–O bond distance of 2.092 Å has been noted.[131] Contrast this to a Sn^{II} compound, tin(II)-phenylbis(phosphonate), $Sn_2(O_3PC_6H_4PO_3)$, where the Sn–O distances are only a bit longer, 2.095(4) Å to 2.145(2) Å.[132] Another Sn^{II} polymeric compound, poly[tin(II)-μ-phenylphosphonato], $[Sn(C_6H_5O_3P)]_n$, shows a Sn–O distance range of 2.097(3) Å to 2.133(3) Å.[133] Similarly, the Sn–O distances in the Sn^{II} compound $Sn_2[O_3PCH_2C_6H_4CH_2PO_3]$, are 2.11(1) Å to 2.19(1) Å.[134]

The reader is referred to Chapter 9, which describes in detail structural aspects of Zr-phosphonate chemistry. Only a few representative examples will be mentioned here. In the linear chain zirconium organophosphonate $(NH_4)Zr[F_2][H_3\{O_3PCH_2NH(CH_2CO_2)_2\}_2]\cdot 3H_2O\cdot NH_4Cl$, the Zr–O bonds are found in the range 2.058(2) Å to 2.092(2) Å.[135]

Interestingly, the Zr–O distances in the compound $Zr_2(O_3PCH_2CH_2$-bipyridinium-CH_2-$CH_2PO_3)$-F_6-$2H_2O$ are slightly longer, 2.130(7) Å to 2.157(6) Å.[136] In a similar compound, $Zr_2(O_3PCH_2CH_2$-viologen-$CH_2CH_2PO_3)$ $F_6\cdot 2H_2O$, the Zr–O distances are from 2.03(1) Å to 2.09(1) Å.[137] Similar Zr–O distances are noted for the zirconium diphosphonate fluoride, $ZrHF(O_3PCH_2)_2NHC_3H_6CO_2)$ that contains the 4-[bis(phosphonomethyl) amino]butanoic acid ligand, 2.01(1) Å to 2.16(1) Å.[138] Additionally, the Zr–O distances in $ZrF(O_3PCH_2)_2NHCH_2C_6H_5$ fall in the range 2.051(7) Å to 2.112(6) Å.[139] The layered zirconium chloromethylphosphonate, $Zr(O_3PCH_2Cl)_2$, displays Zr–O distances from 1.991(8) Å to 2.085(8) Å.[140] Recently, two interesting Zr^{IV} tetraphosphonate compounds were reported, $Zr(HPO_3CH_2)_2N$-C_4H_8-$N(CH_2PO_3H)_2\cdot 4H_2O$ and $Zr(O_3PCH_2)_2N$-C_6H_{10}-$N(CH_2PO_3)_2Na_2H_2\cdot 5H_2O$, in which the Zr–O distances fall in the ranges 2.093–2.119 Å (for the former) and 2.014–2.083 Å (for the latter).[41] Similar Zr–O distances were reported for a family of layered diphosphonate compounds, $ZrF(O_3PCH_2)_2NHC_nH_{2n+1}$ ($n = 1$–10).[42]

The previous examples point to an interesting observation, that there are no systematic trends between the metal ion oxidation state and the M–O bond distance. Therefore, it will be informative to seek similar trends between the metal ion ionic radius and the M–O bond distance.

Our discussion on the effect of the metal ionic radius on the M–O bond distance will focus on divalent metal ions, for the convenience of comparisons and also due to the plethora of divalent metal phosphonate structures. A good starting example is the series of M-AMP materials (M = Mg, Ca, Sr, Ba). Apparently, there is a steady increase of the M–O bond distance as the metal ion increases in size.[43,96] This is clearly shown in Figure 14.18, where M–O bond distances (shortest and longest) are plotted versus the ionic radii.

An identical trend is observed for the series M-HPAA materials (M = Mg, Ca, Sr, Ba), as shown in Figure 14.19.[98,99,109]

Bond distances versus ionic radii of the first row divalent transition metal ions are plotted in Figure 14.20. It is obvious that there is no discernible trend.

In contrast, there is a clear trend in the M–O bond distances, when M is a monovalent alkali metal ion (Figure 14.21). M–O bond distances were taken from references 106 (Li), 104 (Na), 109 (K), 141(Rb), and 97 (Cs).

14.4 Metal–Phosphonate versus Metal–Carboxylate Bonding

Among the battery of anionic ligands available for the construction of inorganic–organic hybrids, polycarboxylates are predominant.[142–150]

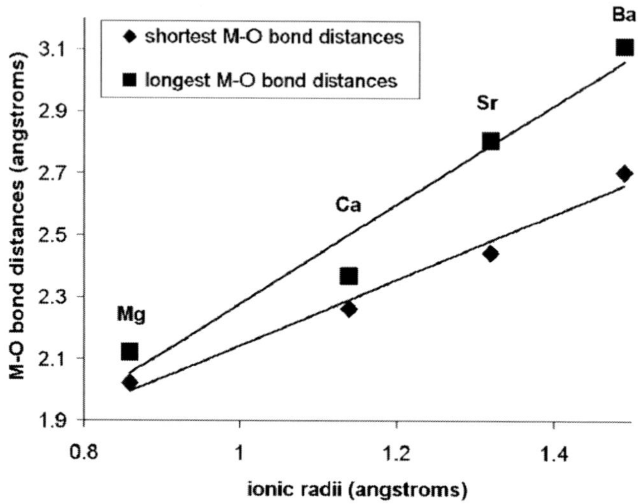

Figure 14.18 Correlation of M–O bond distances (M = Mg, Ca, Sr, and Ba) with the corresponding ionic radii in a series of metal-AMP hybrids.

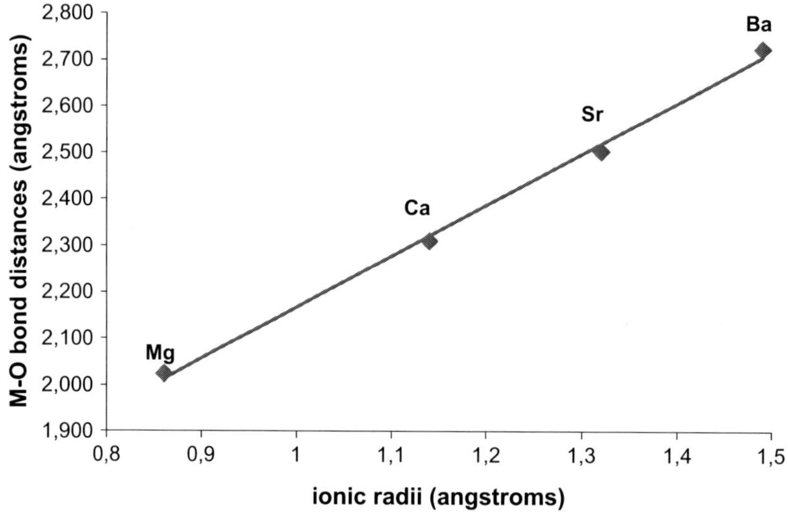

Figure 14.19 Correlation of M–O bond distances (M = Mg, Ca, Sr, and Ba) with the corresponding ionic radii in a series of metal-HPAA hybrids.

Polyphosphonates have also attracted significant interest.[3,6,40,64] because they exhibit a number of similarities, but also differences, to the carboxylates.

(a) Phosphonate building blocks possess three O atoms linked to the phosphorus atom in the coordinating moiety, compared to two O atoms in the

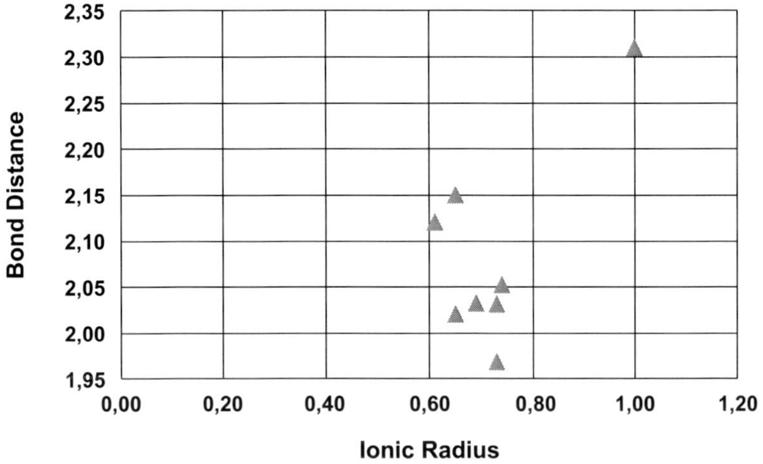

Figure 14.20 Correlation of M–O bond distances (M = first row divalent transition metal ions) with the corresponding ionic radii in a various metal(II)-phosphonate hybrids.

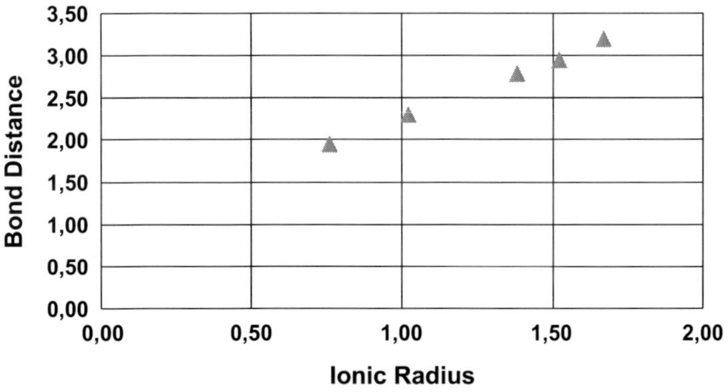

Figure 14.21 Correlation of M–O bond distances (M = alkali metal ions) with the corresponding ionic radii in a various metal(I)-phosphonate hybrids.

case of carboxylates. This increases the possibilities for access to novel structures.

(b) The phosphonic acid moiety can be doubly deprotonated in two well-defined successive steps, depending on solution pH.[82] Carboxylic acid ligands can only be deprotonated once (Figure 14.22). Again, this allows access to a variety of potential novel phosphonate-containing structures, by simply varying the pH.

(c) The phosphonate group can be (potentially) doubly esterified, in contrast to the carboxylate group that can only be monoesterified.[151–154] Introduction of at least one phosphonate ester in the building block is expected

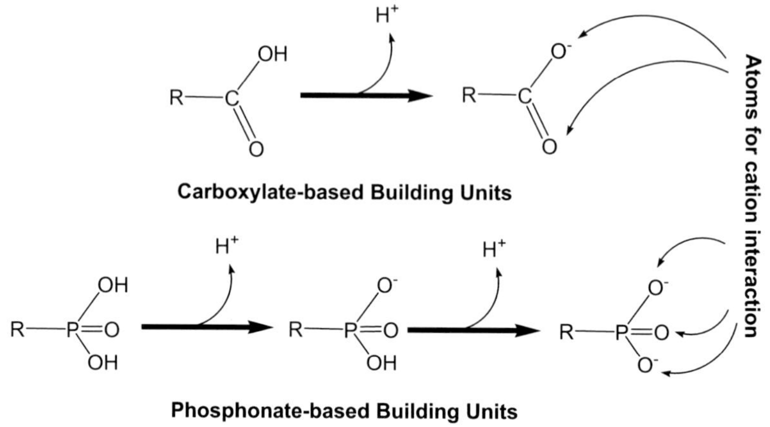

Figure 14.22 Structural and functional differences between carboxylic and phosphonic acids.

to enhance solubility (in the case of very insoluble materials), or by virtue of its hydrolysis,[95] to yield structural diversity in the end material.

(d) Synthesis of metal phosphonate materials can be carried out via a number of different routes that do not necessarily give products with the same structure. There is hence a greater potential of structural diversity in the products derived. Several of these methods lend themselves to a combinatorial approach allowing high-throughput screening of candidate materials to be achieved.[28] In this context, a recent review was published on "non-carboxylate" MOFs.[155]

14.5 Synthetic Considerations

14.5.1 pH

Recently, an important review was published on the effect of pH on the assembly of metal–organic architectures.[156] It is clear that pH plays an important role in determining the final structures of the inorganic–organic hybrids. Significant progress has been made in synthetic endeavors; however, it appears that prediction of any specific effects caused by pH changes remains elusive. Such effects are unknown *a priori*, and there is often more than one effect that comes into play, resulting in complex and diverse structures of such materials. Furthermore, the assembly of the solid state structures of such architectures is governed by a number of intermolecular forces in addition to the pH influence. Frequently, the balance between said forces is so subtle that a convincing explanation of some structures observed is argumentative. An example experienced in our laboratories is the formation of two Sr-HPAA materials, namely $Sr[(HPAA)(H_2O)_3]\cdot H_2O$ and $Sr(HPAA)(H_2O)_2$.[98,99] These two materials are synthesized under identical conditions, except that the former forms at pH 2.0, whereas the latter forms at pH 2.7. Further progress and a deeper understanding of the diverse effects of pH on synthesis depend on further systematic studies. For instance, isolation of possible intermediates should provide useful insight. Computational studies should also shed some light on possible pH effects. Combined, the results of experimental and computational studies will allow further and more thoughtful design in the context of crystal engineering with the goal of gaining access to structurally and functionally more sophisticated and useful materials.[157–160]

14.5.2 Temperature

It has been noted that by changing the reaction conditions, it is sometimes possible to synthesize a number of different materials from the same combination of metal salt and phosphonate anion. Examples include the families of aluminum phenylphosphonates and benzylphosphonates reported by Cabeza[161] and Zakowsky,[26] respectively, and the copper methylphosphonates reported by Bujoli.[162] It is quite common that different products are formed from a particular set of reagents have the different compositions, and are not polymorphs of one another. More often the materials will contain different metal:phosphonate ratios, arising from charge variations on the phosphonate anion backbone, which, in turn, are dependent on reaction mixture pH. Other variations include the number of water molecules (either metal-coordinated or in the lattice) in the structure. This is often a result of differences in the coordination number, environment of the metal atoms, and hydrogen bonding schemes.

The hydrothermal synthetic procedure was carried out using different Cu:P ratios and at different temperatures.[95] Figure 14.23 shows which products are obtained from each of the Cu:P ratios used at the different reaction temperatures.

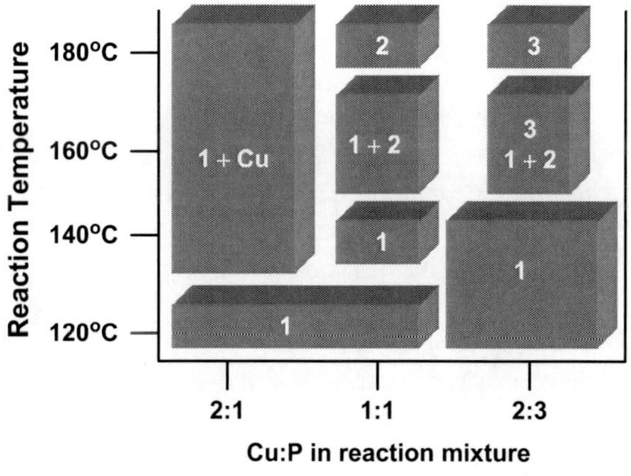

Figure 14.23 Products formed from reactions between copper acetate and diethylphosphonoacetic acid, under hydrothermal conditions at different temperatures and with varying reactant ratios. α-$Cu_{1.5}(O_3PCH_2CO_2)\cdot H_2O$ (**1**), β-$Cu_{1.5}(O_3PCH_2CO_2)\cdot H_2O$ (**2**), $Cu(HO_3PCH_2CO_2)\cdot H_2O$ (**3**). (Reproduced with permission from ref. 95.)

The initial experiment carried out in the investigation of this system was carried out using a 1:1 mole ratio of copper acetate to DPAA (or Cu:P ratio) at 160 °C. The products appeared to consist of three phases; large dark green block crystals, large bright blue plates, and small turquoise plates. Crystals were separated by hand and investigated by single crystal methods, which showed the bright blue and turquoise materials to be the same phase. Hence the products of this reaction are biphasic, with both phases having a composition of $Cu_{1.5}(O_3PCH_2CO_2) \cdot H_2O$, but with different structures (see below).

Treatment of the 1:1 Cu:P reaction mixture at 140 °C resulted in the formation of the blue material (hereafter referred to as the α-phase) as a single phase product. Another single crystal measurement was carried out on this material, confirming it to be identical to the blue phase obtained at 160 °C. The single phase nature of this material was confirmed by comparison of an observed powder X-ray diffraction pattern with one generated from the single crystal data. The same result was obtained for a synthesis carried out at 120 °C. Isolation of the material as a single phase allowed CHN analyses to be carried out. These results confirmed the elemental composition determined during the single crystal analysis.

Treatment of the same 1:1 reaction mixture at 180 °C results in the formation, as a single phase product, of the green phase obtained at 160 °C. Again this was confirmed by additional single crystal analysis and comparison with powder diffraction patterns. CHN analysis confirms the composition of the β-phase to be the same as that of the α-phase and corresponds to the stoichiometry obtained from the single crystal analysis.

Treatment of a reaction mixture with a Cu:P ratio of 2:3 (taking into account the stoichiometry of the previously obtained products) at 160 °C again gave rise to a mixture of products. In this case however, the products contained, $Cu_{1.5}(O_3PCH_2CO_2) \cdot H_2O$ (**1**), $Cu_{1.5}(O_3PCH_2CO_2) \cdot H_2O$ (**2**) and bright blue, block shaped crystals. Single crystal analysis of the new product (see below) showed the product to have a composition of $Cu(HO_3PCH_2CO_2) \cdot H_2O$ (**3**). When the reaction temperature was raised to 180 °C this material was obtained as a single phase, allowing a CHN analysis to be carried out, and again confirming the composition determined from the single crystal analysis. Using the same 2:3 Cu:P reactant ratio at a reaction temperature of 140 °C or below results in the formation of **1** only.

A further change in the relative amounts of the starting reagents such that the Cu:P ratio was 2:1 did not yield any new materials. At temperatures of 140 °C and above, the hydrothermal reaction gave rise to a mixture of copper metal and **1**. The quantity of copper metal obtained increased as the reaction temperature was raised.

14.6 Auxiliary Ligands

A secondary ligand is occasionally introduced into the synthesis mixture in the hope of gaining access to novel products. There are cases where this ligand (often an amine-based molecule) acts as a template for the formation of end products. There are cases reported where this ligand ends up in the final structure. For example, $(NH_4)Zn[O_3PCH(OH)CO_2]$ forms from a mixture of zinc acetate, NH_4Cl, and HPAA in water (final pH 4.12) at 190 °C.[163]

Often, the auxiliary ligands used are bidentate chelates, such as 2,2′-bpy or 1,10-phen, although there are reports of 4,4′-bpy utilization as a bridge. The reaction of copper acetate with phosphonoacetic acid in the presence of phen yields a binuclear complex in which phen acts as a chelating ligand, whereas the two Cu ions are bridged by two phosphonoacetate ligands.[164] Two pillared metal phosphonates, $[Mn_3(4,4'\text{-bpy})(HPAA)_2]\cdot(4,4'\text{-bipy})_{0.5}$ and $[Co_3(4,4'\text{-bpy})(H_2O)_2(HPAA)_2]\cdot(4,4'\text{-bpy})_{0.5}$ have been synthesized hydrothermally, and structurally characterized. The Mn^{II} and Co^{II} ions are found in octahedral coordination environment and bridged by the HPAA ligand through –OH, –COO and $–PO_3$ groups into a hybrid layer. The layers are further pillared by the coordinated 4,4′-bpy into a 3D neutral framework.[165] In the presence of ethylenediamine template agents, a Fe^{II} phosphonate, $[NH_3CH_2CH_2NH_3][Fe_2(HPAA)_2(H_2O)_2]\cdot 2H_2O$ has been synthesized under hydrothermal conditions and characterized by single-crystal X-ray diffraction.[166] The Fe^{II} cation is octahedrally coordinated by six oxygen atoms from the three ligands and one coordinated water molecule to form a 2D layered structure with a 1D channel system in the *a* axis direction. In this case ethylenediamine does not coordinate to the Fe^{II}. A new Co^{II} phosphonate, $[NH_3CH_2CH_2NH_3]_{0.5}[Co(HPAA)(H_2O)]\cdot H_2O$, has been hydrothermally synthesized and structurally characterized. It has a honeycomb-like layer structure which is composed of CoO_6 octahedra and μ_3-HPAA ligands with the protonated ethylenediamine and lattice water molecules stabilized between the layers. As before, the ethylenediamine does not coordinate to the Co^{II}.[167] In the presence of ethylenediamine, a layered antimony(III) phosphonate, $[NH_2CH_2CH_2NH_2][Sb_2(HPAA)_2]$ has been synthesized by hydrothermal reaction at 140 °C and characterized by single-crystal X-ray diffraction.[168] In this case, although the ethylenediamine is not protonated in the crystal lattice, it does not coordinate to the Sb^{III}. Two new metal phosphonatobenzenesulfonates using 1,10-phenanthroline as auxiliary ligand, namely, $[Cu(HL)(phen)]\cdot 0.5H_2O$ and $[Y(L)(phen)(H_2O)_2]\cdot 2H_2O$ ($H_3L = m\text{-}HO_3S\text{-}C_6H_4\text{-}PO_3H_2$, phen = 1,10-phenanthroline), have been synthesized by hydrothermal reactions.[169] Both compounds feature 1D chain-like structures. The 1D chain structure in the first features $Cu_2(PO_3)_2$ dimers fused by the bridging of pairs of bidentate phosphonate groups, which are further interconnected via pairs of monodentate sulfonate groups. In contrast, the 1D chain in the latter features an edge-shared four-membered ring which is composed of two Y^{III} ions and two tridentate phosphonate groups. The different 1D chain structures in them are mainly due to the various coordination modes of the 3-phosphonobenzesulfonic acid ligand. The reaction of copper perchlorate with monophosphonic acids of the type RPO_3H_2 (R = cyclopentyl, isopropyl, trichloromethyl) in the presence of chelating nitrogen ligands bpya (bpya = 2,2-bipyridylamine) or bpy afforded dinuclear copper phosphonates that feature the $[Cu_2(\mu_2\text{-}RPO_3)_2]$ core.[170] Similar reaction with copper salts and monophosphonates in the presence of phen and acetic acid yield higher nuclearity clusters, such as $[Cu_4(\mu\text{-}CH_3COO)_2(\mu_3\text{-}C_6H_{11}PO_3)_2(2,2'\text{-bpy})_4](CH_3COO)_2$.[171]

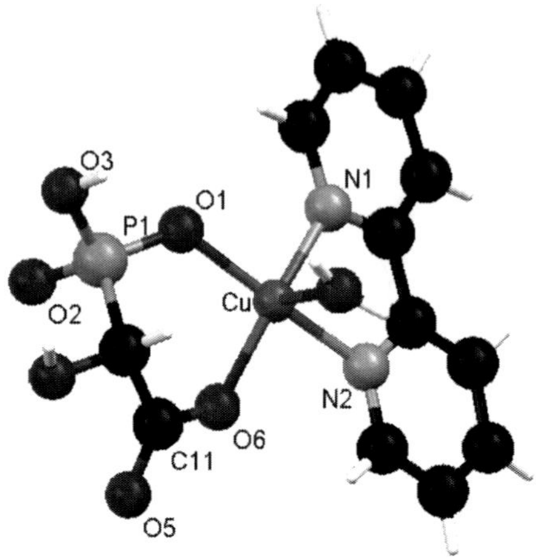

Figure 14.24 The molecular structure of the mononuclear complex, [Cu(*R*,*S*-HO_3P-CH(OH)CO_2)(bpy)(H_2O)] · H_2O. The lattice water has been omitted for clarity.

In an effort to diverge the synthesis end material from [Cu(*R*,*S*-HO_3P-CH(OH)CO_2)(H_2O)$_2$] · H_2O, we utilized bpy in the reaction mixture. The end product was a mononuclear complex, [Cu(*R*,*S*-HO_3PCH(OH)CO_2)(bpy)-(H_2O)] · H_2O (Figure 14.24).[172]

Chapter 5 in this book offers an in-depth discussion on metal phosphonate materials that contain an auxiliary ligand.

14.7 Applications

Organophosphonate anions (see Figure 14.1) function as linkers connecting inorganic centers to yield hybrid frameworks. Tailored design of the linker molecules is easy because phosphonic acids are prepared in relatively simple procedures, or occasionally are commercially available. Chemical and thermal stability of phosphorus–carbon bonds in phosphonates is considered to be high. Generally, metal phosphonate solids are prepared either at room temperature or under mild hydrothermal conditions. Such experimental conditions cause little or no decomposition of organic moieties in several cases. Therefore, phosphonates are suitable to design inorganic–organic nano-hybrid materials. Such merits of phosphonates have been recognized for layered materials since Alberti and co-workers reported the first metal phosphonate in 1978.[173] Structural similarities between metal organophosphonates and the corresponding phosphates were revealed through studies on layered phosphonates[174] and tailored design of the materials became subsequently possible. A detailed review on metal phosphonate materials by Clearfield is available,[5] along with reviews on layered metal phosphonates and their use.[175,176]

During the last decade, possibilities of incorporating organic molecules into ordered nanoporous inorganic solids were extensively explored. As surveyed in a review by Maeda,[177] organically modified mesoporous silicas were realized both by post-modification of the purely inorganic materials using organosilylation agents and by direct incorporation of organosiloxane during assembly of silicate frameworks. When these approaches are applied for microporous materials, however, the post-modification has difficulties in leaving accessible pores owing to their much smaller pore sizes. Direct crystallization needs careful selection of a suitable system. For example in the silicate system, partial substitution of the framework oxygen atoms in several known zeolite structures by methylene groups was attained recently.[178] Considerable parts of the Si–C bond of methylene-bridged organosilane used as the silicon source, however, were cleaved under strongly basic hydrothermal conditions. Thus, chemical and thermal stabilities of organophosphonic acids and moderately mild synthetic conditions of metal phosphonates tempted many researchers to explore open-framework solids. Structural and chemical properties of metal phosphonates are variable depending on many synthetic parameters such as phosphonate source, metal source, metal/P ratio, solvent, concentration, pH, and reaction temperature.

In the following sections we will discuss several properties and applications of metal phosphonates and we will make an attempt to correlate structure with function.

14.8 Corrosion Protection by Metal Phosphonate Thin Film Formation

Corrosion is a wide and highly interdisciplinary field of science. Corrosion inhibition by chemical means (i.e. with the use of corrosion inhibitors) is a significant part of corrosion science, whose scientific literature is vast. Inhibition approaches vary according to the peculiarities of the system, its metallurgy being one of the most significant. In this section we briefly review the state-of-the-art of the field of phosphonates as corrosion inhibitors, providing some characteristic examples of contributions originating from our laboratory.

Phosphonate-based corrosion inhibitors are effective in decreasing metallic corrosion in nearly neutral conditions by forming poorly soluble compounds with the metal ion existing in the solution. These metal phosphonate "complexes" precipitate onto the metallic surface to form a protective layer. Such inhibitors (often called interphase inhibitors) for cooling water treatment technology in the last decades comprise different types of phosphonic acids.[179] Widely used phosphonic acids are 1-hydroxyethane-1,1-diphosphonic acid (HEDP), amino-tris(methylenephosphonic acid) (AMP), hydroxyphosphonoacetic acid (HPAA), etc. Phosphonates are introduced into the system to be protected in the acid form or as alkali metal soluble salts, but readily form more stable complexes with other metal cations found in the process stream (most commonly Ca, Mg, Sr, or Ba), depending on the particular application. It should be emphasized that research in this area has been stimulated by the need to develop inhibitor formulations that are free from carcinogenic chromates,[180] nitrates, nitrites, inorganic phosphorus compounds, etc. Phosphonates when blended with certain metal cations and polymers reduce the optimal inhibitor concentration needed for inhibition due to synergistic effects.[181] Synergism is one of the important effects in the inhibition process and serves as the basis for the development of all modern corrosion inhibitor formulations.[182]

In spite of the significant body of literature, evidence about the molecular identity of the thin protective metal-phosphonate films lags behind. In this section, the corrosion inhibition performance of several metal-phosphonate materials is reported and a more systematic look at their corrosion inhibition mechanisms at the molecular level is presented. These exhibit dramatically different anticorrosion efficiencies, which are linked to their molecular structure. Herein we will only mention a few examples to illustrate the point. The reader is referred to a recent review published by our group for further details.[183] These metal phosphonates are the following coordination polymers: Zn-AMP, $\{Zn(AMP)\cdot 3H_2O\}_n$; Zn-HDTMP, $\{Zn(HDTMP)\cdot H_2O\}_n$; Ca-HPAA, $Ca_3(HPAA)_2(H_2O)_{14}$; M (Sr, Ba)-HPAA, $\{M(HPAA)(H_2O)_2\}_n$.

In the following sections the corrosion rate (CR) is calculated from the equation:

$$CR = \frac{534.57 \times (\text{mass loss})}{(\text{area})(\text{time})(\text{metal density})}$$

Units: CR in mm/year, mass loss in mg, area in cm^2, time in hours, metal density = 7.85 g cm^{-3} (for carbon steel, but this varies according to the particular metallurgy of the system).

14.8.1 Inhibitory Thin Films by the Material $\{Zn(AMP)\cdot 3H_2O_n\}$

Synergistic combinations of 1:1 molar ratio Zn^{2+} and AMP are reported to exhibit superior inhibition performance than either Zn^{2+} or AMP alone.[184] However, no mention is made regarding the identity of the inhibitor species involved in corrosion inhibition. Therefore, corrosion experiments were carried out in order to verify the literature results and prove that the protective material acting as a corrosion barrier is an organic–inorganic hybrid composed of Zn and AMP.[121] A synergistic combination of Zn^{2+} and AMP in a 1:1 ratio (under identical conditions used to prepare crystalline Zn-AMP) offers excellent corrosion protection for carbon steel. Although differentiation between the "control" and "Zn-AMP" protected specimens is evident within the first hours, the corrosion experiment is left to proceed over a 3-day period. Based on mass loss measurements the corrosion rate for the "control" sample is 2.5 mm/year, whereas for the Zn-AMP protected sample it was 0.9 mm/year, a 270% reduction in corrosion rate. The filming material is collected and subjected to FT-IR, XRF, and EDS studies. These show that the inhibiting film is a material containing Zn (from added Zn^{2+}) and P (from added AMP) in an approximately 1:3 ratio, as expected. Fe was also present apparently originating from

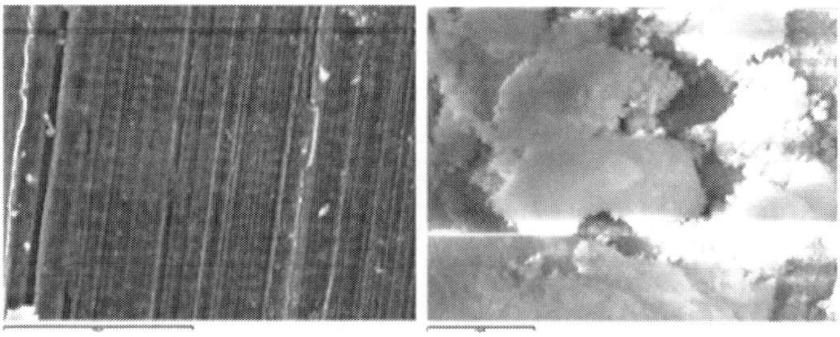

Figure 14.25 SEM images of a bare carbon steel surface (left, bar = 60 m) and a Zn-AMP protected steel surface (right, bar = 10 m). Deposition of an anticorrosive Zn-AMP material is obvious.

Figure 14.26 The anticorrosive effect of Zn-HDTMP films on carbon steel. The upper specimen is the "control", no inhibitor present. Corrosion inhibition in the lower specimen by a 1 mM Zn^{2+}/HDTMP synergistic combination is obvious.

the steel specimen. FT-IR showed multiple bands associated with the phosphonate groups that closely resemble those of an authentically prepared Zn-AMP material. For comparison, EDS and XRF spectra of a "protected" and an "unprotected" region show the presence of Zn and P in the former, but complete absence in the latter. A characteristic example of a Zn-AMP film is shown in Figure 14.25 and is compared to a "bare" iron metal surface.

14.8.2 Inhibitory Thin Films by the Material $\{Zn(HDTMP)\cdot H_2O_n\}$

A combination of Zn^{2+} and HDTMP in a 1:1 ratio (under identical conditions used to prepare crystalline Zn-HDTMP) offers excellent corrosion protection for carbon steel (Figure 14.26).[123]

Although differentiation between the "control" and "Zn-HDTMP" protected specimens is profound within the first hours, the corrosion experiment is left to proceed over a 3-day period. Based on mass loss measurements the corrosion rate for the "control" sample is 7.28 mm/year, whereas for thc Zn-HDTMP protected sample 2.11 mm/year, a ~170% reduction in corrosion rate. The filming material is collected and subjected to FT-IR, XRF, and EDS studies.

These show that the corrosion inhibiting film is a material containing Zn^{2+} (from externally added Zn^{2+}) and P (from added HDTMP) in an approximate 1:4 ratio. Fe was also present apparently originating from the carbon steel specimen. FT-IR of the filming material showed multiple bands associated with the phosphonate groups in the 950–1200 cm^{-1} region that closely resemble those of the authentically prepared Zn-HDTMP material (Figure 14.27). For comparison, EDS and XRF spectra of a "protected" and an "unprotected" region show presence of Zn and P in the former, but complete absence in the latter.

The Zn-HDTMP material was prepared from first principles and its crystal structure was determined.[123] It is shown in Figure 14.28.

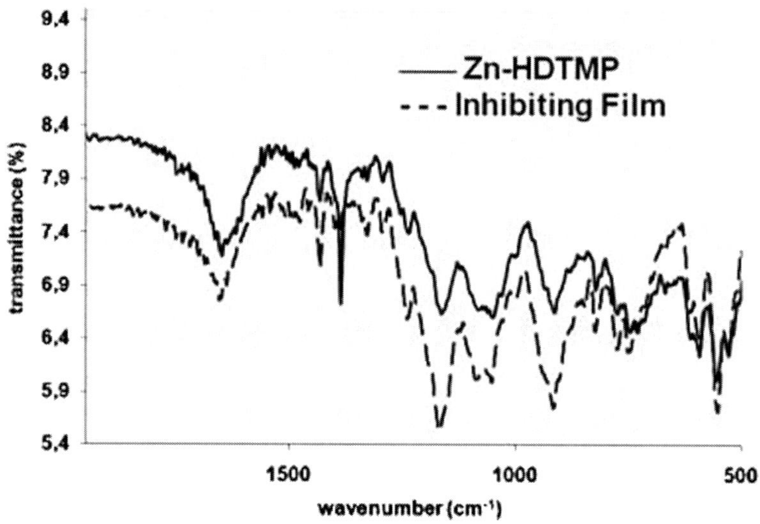

Figure 14.27 FT-IR spectra of "genuine" Zn-HDTMP and of the corrosion inhibiting film formed in situ from a 1:1 Zn^{2+}:HDTMP synergistic combination.

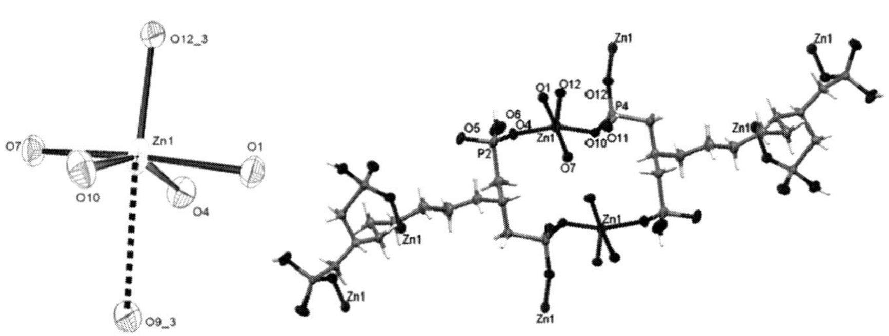

Figure 14.28 Structure of the coordination polymer Zn(HDTMP)(H_2O): the Zn^{2+} coordination environment (left) and portion of the structure showing the extensive 18-membered ring (right). (Reproduced with permission from ref. 123.)

14.8.3 Inhibitory Thin Films by the Molecular Trimer $Ca_3(HPAA)_2(H_2O)_{14}$

The effectiveness of corrosion protection by synergistic combinations of Ca^{2+} and HPAA in a 1:1 ratio is dramatically pH-dependent (Figure 14.29).[185]

At pH 2.0 mass loss from the steel specimens is profound, resulting in high corrosion rates (353×10^{-3} mm/year). However, specimen **2** (Figure 14.29) appears relatively clean from corrosion products, presumably because HPAA (either free or metal-bound) at the surface acts as a Fe-oxide dissolving agent. At pH 7.3 corrosion rates are appreciably suppressed (4×10^{-3} mm/year) in the presence of combinations of Ca^{2+} and HPAA (specimen **4**), reaching nearly quantitative inhibition.

The inhibiting film is fairly uniform and contains Ca^{2+} and P (from externally added Ca^{2+} and HPAA), in an approximate 3:2 molar ratio (by EDS), suggesting a ratio of three Ca^{2+} and two $HPAA^{3-}$ ligands. FT-IR of the filming material showed multiple bands (in the 950–1200 cm^{-1} region) associated with the phosphonate groups and two bands at 1590 and 1650 cm^{-1} (assigned to the $\nu_{C=O}$ stretches), that closely match those of $Ca_3(HPAA)_2(H_2O)_{14}$, prepared from first principles. Its molecular structure is shown in Figure 14.30. An XRD powder pattern of the Ca-HPAA anticorrosion film deposited on a carbon steel substrate is identical to that of a crystalline sample of $Ca_3(HPAA)_2(H_2O)_{14}$. This is unequivocal proof that the protective film on the steel surface is $Ca_3(HPAA)_2(H_2O)_{14}$.

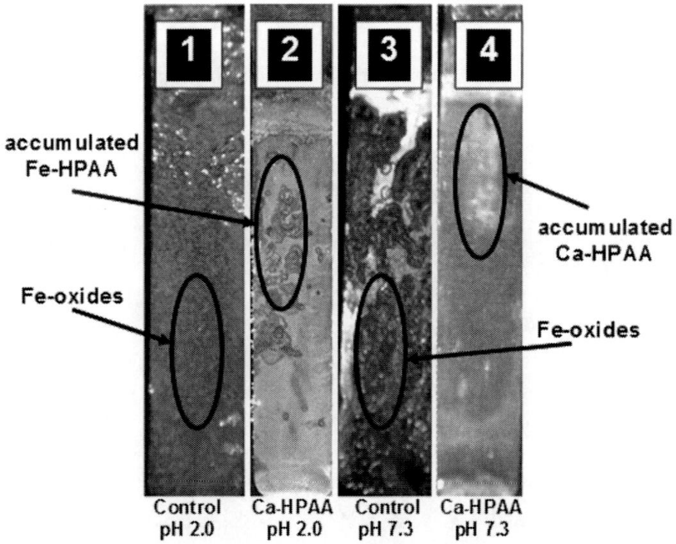

Figure 14.29 The anticorrosion protection of Ca-HPAA films on carbon steel. The effect of Ca-HPAA is especially demonstrated in specimen **4**. (Reproduced with permission from ref. 185.)

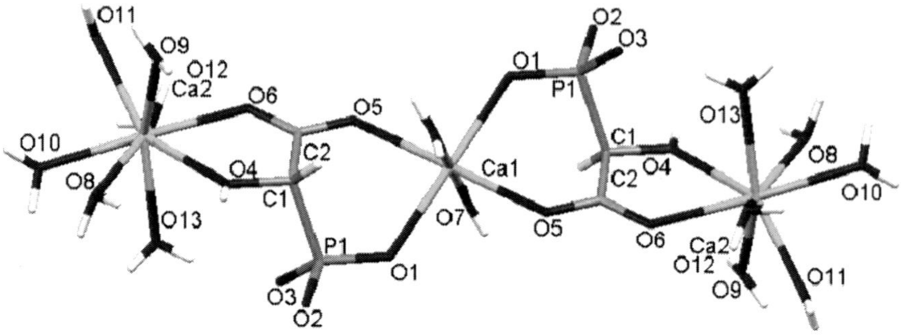

Figure 14.30 The molecular structure of the Ca-HPAA trimer. (Reproduced with permission from ref. 185.)

14.8.4 Inhibitory Thin Films by the Materials $\{M(HPAA)(H_2O)_2\}_n$ (M = Sr, Ba)

Initial experiments were focused on exposure of carbon steel specimens to synergistic combinations of M^{2+} (Sr or Ba) and HPAA in oxygenated aqueous solutions, in a 1:1 ratio (under identical conditions used to prepare crystalline M-HPAA, at pH 2.0). Although the visual effect was at first encouraging, quantification of the corrosion rates demonstrated that they were actually much higher than the "control".[98,99] Explanations for this lack of anti-corrosion performance at low pH regions could be that the HPAA added first reacts preferentially with the Fe-oxide layer (formed almost instantaneously upon exposure of the carbon steel surface to oxygenated water) before it interacts with soluble Sr^{2+} or Ba^{2+}. Another possibility is that Sr-HPAA or Ba-HPAA compounds that may form in solution never reach the steel surface because they undergo bulk precipitation. We have discounted this scenario, based on the following arguments. Indeed, we have observed white precipitates formed in the bulk in our corrosion experiments (pH 2.0) whose FT-IR, however, is distinctly different from those of authentically prepared Sr-HPAA or Ba-HPAA. These FT-IR spectra are the same as that of a Fe-HPAA material prepared at pH 2.0 using a Fe:HPAA ratio of 1:1, whose composition is consistent with the formula $Fe(HPAA)\cdot H_2O$. A reasonable assumption is that HPAA at the surface acts as an Fe oxide dissolving agent. We have observed a similar behavior in similar experiments with M^{2+} (M = Ca, Zn) and 2-phosphonobutane-1,2,4-tricarboxylic acid (PBTC).[186] Due to the ineffectiveness of the metal-HPAA materials to act as corrosion inhibitors at pH 2.0, no further experiments were pursued at that pH.

Hence, corrosion experiments were set up at higher pH (pH 7.3). In general, corrosion rates are lower as pH increases. This was confirmed in our "control" experiments (reduction of the corrosion rate by half). At higher pH (7.3) and in the presence of Sr^{2+} or Ba^{2+} and HPAA combinations, corrosion rates are dramatically suppressed and corrosion inhibition reaches almost 100%. Although differentiation between the "control" and "metal-HPAA"-protected specimens is profound within the first hours, the corrosion experiments were left to proceed over a 6-day period. Anti-corrosion inhibitory activity was based on mass loss measurements. To further characterize the protective film, the corrosion specimens and film material were subjected to SEM, FT-IR, XRF, and EDS studies (Figure 14.31).

SEM images reveal a fairly uniform inhibiting film. This coating was found (by EDS, Figure 14.31, right) to contain M^{2+} (Sr or Ba from externally added salts) and P (from added HPAA), in an approximate 1:1 molar ratio. Fe was also present, apparently originating from the carbon steel specimen.

Furthermore, a complementary study of the inhibiting film was pursued by FT-IR spectroscopy. Figure 14.32 shows comparative FT-IR spectra of the

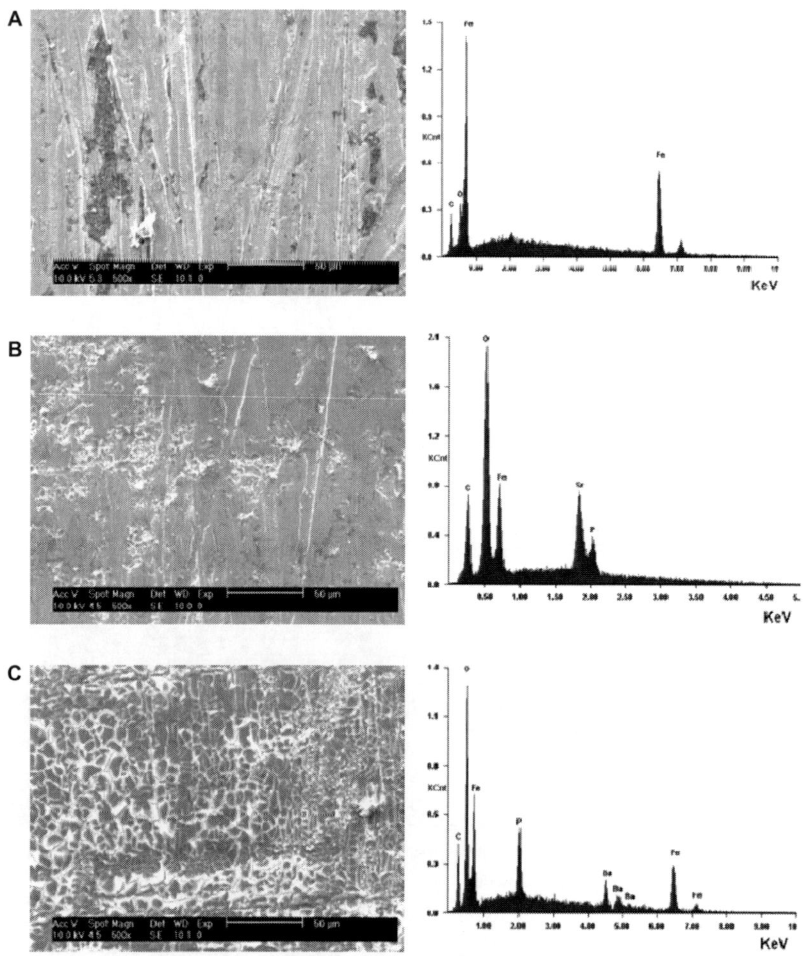

Figure 14.31 Morphology of corroded and metal-HPAA-protected steel surfaces by SEM: control (A), Sr-HPAA (B), and Ba-HPAA (C) at pH 7.3. Occasional film cracking is due to drying. Identification of film components (Fe, C, O, Sr, Ba, and P) was possible by EDS: control (A), Sr-HPAA (B), and Ba-HPAA (C). (Reproduced with permission from ref. 99.)

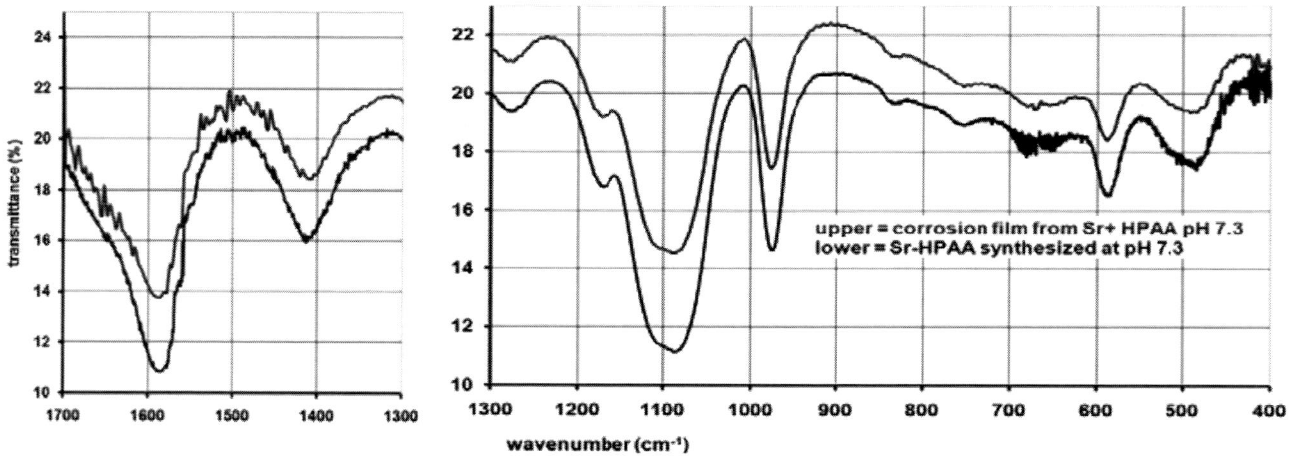

Figure 14.32 FT-IR of the anti-corrosion protective film formed by combination of Sr^{2+} and HPAA at pH 7.3 and, for comparison, of Sr-HPAA synthesized at pH 7.3. Left: ν(C=O) asymmetric and symmetric stretching vibrations. Right: vibrations associated with the phosphonate group. (Reproduced with permission from ref. 99.)

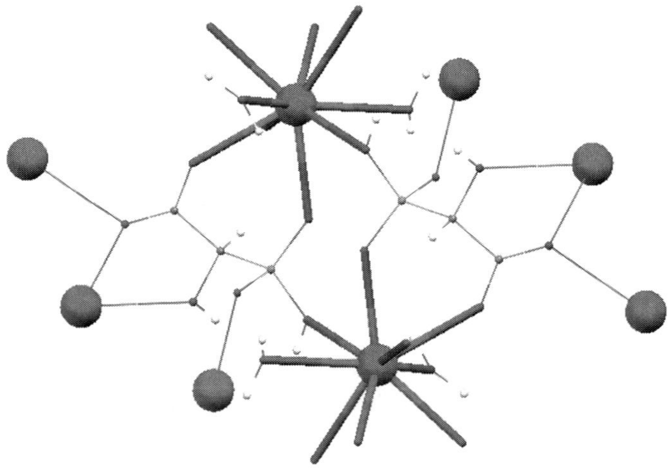

Figure 14.33 Coordination function of the HPAA ligand in $M[(HPAA)(H_2O)_3] \cdot H_2O$, with metal ions as exaggerated spheres.

filming material (from a corrosion experiment with Sr^{2+} and HPAA at pH 7.3) and a Sr-HPAA material that was synthesized at pH 7.3. A similar FT-IR spectrum was obtained for the Ba^{2+}/HPAA system. It is obvious that there is an excellent agreement between the two spectra.

The materials $\{M(HPAA)(H_2O)_2\}_n$ (M = Sr, Ba) are isostructural, and their structure is depicted in Figure 14.33.

The anti-corrosion coatings composed of Sr-HPAA or Ba-HPAA function as corrosion inhibitors by reducing the cathodic current. This results in lower corrosion rates. The films prevent oxygen diffusion towards the steel surface. This phenomenon is well-known for phosphonate additives.[187]

14.8.5 Inhibitory Thin Films by other Structurally Characterized Metal Phosphonate Materials

Similar studies such as that mentioned above have been carried out. These included the metal phosphonates $\{Ca(PBTC)(H_2O)_2 \cdot 2H_2O\}_n$,[186] $\{M(PMIDA)\}_n$ (M = Ca, Sr, Ba),[172] $\{M[(EDTMP)(H_2O)_2] \cdot H_2O\}_n$ (M = Ca, Sr),[188] and $\{Ba[(HDTMP)(H_2O)_6] \cdot 2H_2O\}_n$.[189] The reader is referred to the above original papers for further details.

14.8.6 A "Holistic" Look at Metal Phosphonate Materials as Corrosion Inhibitors

Metallic corrosion is a phenomenon affected by several factors. It is well established that pH is one of the major factors affecting corrosion rates.

The more acidic the fluid (water) in contact with the metal surface, the more aggressive the corrosion. Therefore, pH plays a profound role in corrosion inhibition as well. In Figure 14.34, the performance of several metal phosphonate corrosion inhibitors has been systematically gathered and plotted. There are a number of important observations to be discussed. At higher pH regions (~7) most metal phosphonate coatings perform well. At low pH regions (<3) corrosion inhibition is more challenging. This may have to do with incomplete formation of the protective coating. In addition, the high concentration of H^+ and the resulting high corrosion rates may be prevailing over the formation of the metal phosphonate protective coating.

The results with Ca-PBTC and Zn-PBTC require further discussion. Corrosion rates in the presence of inhibitor are *higher* than those for the control (no inhibitor). This, at a first glance, is contrary to results obtained with several other inhibitors. This may be explained by several arguments. First, the metal-phosphonate film may not be robust, but porous in its microscopic nature. This, as mentioned before, would lead to localized attack and metal pitting. Such phenomena have not been observed upon examination of the metal specimens after the corrosion experiments. Second, the metal phosphonate (Ca, or Zn-PBTC) is too soluble to deposit onto the metal surface, so it does not form a protective and anticorrosion thin film. This argument would be consistent with literature data on metal-PBTC complex formation constants (4.4 for Ca-PBTC and 8.3 for Zn-PBTC) that are considered to be very low.[64] The difference in complex formation constants between Ca- and Zn-PBTC would be consistent with the fact that Zn-PBTC is a more effective corrosion inhibitor than Ca-PBTC, as long as both inhibitors form films (albeit unstable) on the metal surface. If film formation does not take place, then corrosion rates in the presence of Ca-PBTC or Zn-PBTC would be the same as the control, which is not the case. Therefore, the results obtained with Ca-PBTC and Zn-PBTC, indicate that these materials are soluble and due to their acidic nature they actually act as metal dissolvers rather than corrosion inhibitors.

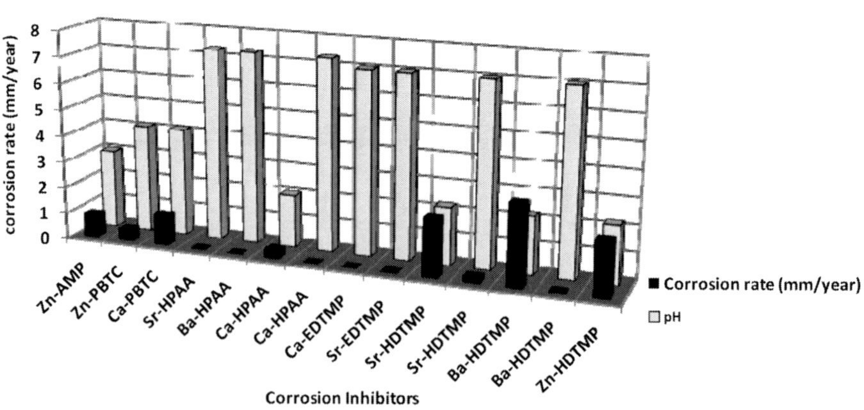

Figure 14.34 Corrosion rates of metal phosphonate-protected surfaces as a function of pH.

The Zn-HDTMP material by virtue of its long chain linker between the two amino-bis(methylenephosphonate) moieties might be thought of as a porous material. However, porosity measurements on this and the other phosphonates show absence of any porous structure. Therefore, differences in porosity cannot be invoked to explain the various anticorrosion properties of these metal-phosphonate materials.

Lastly, the ability of a metal-phosphonate corrosion inhibitor to adhere onto the metal surface plays a vital role in corrosion efficacy. Bulk precipitation of a metal-phosphonate complex will lead to loss of active inhibitor to precipitation, leading to insufficient levels for thin-film formation. Surface adherence of the inhibitor films is a property that cannot be precisely predicted. However, it is a necessary condition for acceptable inhibition. In addition, the metal-phosphonate protective layer has to be robust and uniform.

Corrosion is a vast scientific field; however, it presents several facets that touch upon economics and people.[190] For example, the overall demand for corrosion inhibitors after a rather constant rise of ~4.4%/year, reached $ 1.6 billion in 2006 in the USA. [191] The petroleum refining sector was expected to have a ~$ 400 million share in this. Corrosion is an economical burden for several industry sectors. Research on the subject has been active for several decades.[192,193] The solution to this complicated issue requires a multidisciplinary approach that unifies researchers from a diverse list of scientific and technological disciplines: chemistry, chemical engineering, electrochemistry, materials science, engineering, and many others.

In this section, our contribution to advancing solutions for corrosion issues relevant to industrial problems lies with the study of the corrosion event and its inhibition *at the molecular level*. In this context, we have shown that conveniently synthesized and structurally characterized organic–inorganic hybrid polymeric materials can act as protective corrosion inhibitors.

An ideal phosphonate corrosion inhibitor of the "complexing type" is required to possess the following features:

(a) it must be capable of generating metal-phosphonate thin films on the surface to be protected
(b) it should not form very soluble metal complexes, because these will not eventually "deposit" onto the metal surface, but will remain soluble in the bulk
(c) it should not form sparingly soluble metal complexes because these may never reach the metal surface to achieve inhibition, but may generate undesirable deposits in the bulk or on other critical system surfaces
(d) its metal complexes generated by controlled deposition on the metal surface must create dense thin films with robust structure. If the anticorrosion film is non-uniform or porous, then uneven oxygen permeation may create sites for localized attack, leading to pitting of the metal surface.

14.9 Gas Storage

Hybrid organic–inorganic microporous materials, particularly metal carboxylate frameworks, are of great importance in solid-state and materials chemistry. Many microporous, thermally stable to 700 K, 3D connected framework structures have been discovered that have pores up to ca. 20 Å in free diameter, and have remarkable adsorption properties, including very high surface areas, structural flexibility, and uncoordinated metal adsorption sites. MOF-5,[194] MIL-100,[195] and HKUST-1[196] are among the widely known carboxylate MOFs. Other families of porous MOFs including those based on amines, such as zeolitic imidazoles,[197] mixed amine-carboxylates,[198] and amino acids[199] have also attracted attention.

Metal phosphonates were among the first microporous organic–inorganic hybrid adsorbents to be investigated. They offer an alternative set of chemical and structural possibilities but none of them is being possessed of pores larger than 6 Å resulting to low uptake of adsorbates. In metal phosphonate chemistry there is a rich variety of phosphonate building units that can be used as reagents and form strong bonds with a range of metal cations when they bind through one, two, or three oxygen atoms per phosphonate groups. Due to the fact that the O_3P–C bond is stable to elevated temperatures, there is a great prospect for the preparation of thermally stable phosphonate frameworks that would be competitive with metal carboxylate MOFs in applications.

Efforts for the synthesis of structures with larger pores and adsorption capacities using *N*,*N*′-piperazinebis(methylenephosphonic acid) (H_4L) were successfully accomplished, resulting in the first large pore metal phosphonates which followed up the syntheses of divalent metal piperazinebis (methylenephosphonate)s of Co(II), Mn(II), Fe(II), and Ni(II) among which the nickel and cobalt bis(phosphonate)s showed remarkable pore volume and thermal stability.[200]

Additional reactions of Fe(II), Co(II), Ni(II) acetates with H_4L at pH values below 6.5, resulted in the formation of structure type $[Mn(H_2O)_2\ L]\cdot 5.4\ H_2O$ materials. Their framework is based on inorganic columns of helical chains of edge-sharing NiO_5N octahedra and comprises the first porous solids in which the ligand coordinates through both oxygen and nitrogen atoms. The arrangement of inorganic chains and organic linkers leaves a structure with a hexagonal array of channels with a free diameter taking into account van der Waals radii of hydrogen atoms, of ca. 10 Å. Decomposition of materials occurs above 400 °C leading to the conversion of unit cell from rhombohedral to primitive triclinic for cobalt and nickel versions. Using the 2-methylpiperazine version of the bis-phosphonic acid in the syntheses gives the same structure materials for nickel and cobalt versions. After dehydration of Ni_2L' the symmetry remains rhombohedral resulting to higher absorption of nitrogen in regard to Ni_2L.

Using cobalt acetate in reactions with H_4L' gives a novel structural material where the inorganic unit is composed of chains of tetramers of octahedral cobalt cations linked by three phosphonate units. Coordination of outer cobalt anions by three phosphonate oxygens, nitrogen of the piperazine ring, and two remaining oxygens from acetate groups and water molecules sets up inorganic columns which are linked to three other identical columns of piperazine units in chair conformation. Upon dehydration the structure remains crystalline, without change in symmetry and indicates channels with free dimension of around 7 Å, showing unremarkable nitrogen adsorption. Both phosphonate structures offer potential as adsorbents. Due to the fact that they can be prepared with a racemic mixture of *N*,*N*′-2-methylpiperazinebis(phosphonic acid)s there are great possibilities for chiral separations, while the use of the 2-methylpiperazine derivative of the acid is both preventing distortion of structures upon dehydration and increasing nitrogen absorption. Moreover, accessible metal sites show Lewis acidity resulting in the catalytic activity of materials as well as the nature of inorganic chains in these solids suggest their interesting magnetic properties.

Recently, porous divalent metal phosphonates have been prepared comprising Fe, Co, and Ni, materials labeled as STA-12, being the first fully crystalline phosphonate MOFs with pores approaching 1 nm and specific pore volume approaching values observed for large pore zeolites.[201] The nickel version of STA-12 is highly porous containing channels filled with physisorbed and chemisorbed water molecules coordinated to the nickel cations when fully hydrated. The framework of the as-prepared solid is built from helical chains of edge-sharing NiO_5N octahedra, linked into a honeycomb arrangement via the ligands. The ligand coordinates to the metal atoms of the inorganic chain via phosphonate oxygen atoms and the piperazine ring N atom, a structural feature which imparts thermal stability up to 650 K. The structure of Ni-STA-12 at different levels of hydration reveals changes in the local structure of the framework during reversible dehydration and rehydration. H_2 and CO are excellent probes of adsorption sites at low temperatures, while CO_2, CH_3OH, CD_3CN, and CH_4 are useful probes at room temperature in terms of the fully dehydrated sample. All measurements were carried out at 303 K in order to evaluate the adsorption properties of the material for potential storage and separation applications. These greenhouse gases have been chosen both with respect to their relevance in several applications and also to permit a comparison between interactions of the adsorbent with molecules possessing a significant quadrupole moment (CO_2) and without any permanent moment (CH_4). During reversible dehydration, significant structural rearrangement arises with a symmetry change from rhombohedral to triclinic resulting in a small reduction in pore size. Rehydration has been found to be rapid, whereas dehydration takes place in two main steps including the loss of physisorbed water after gentle heating to give a rhombohedral structure while chemisorbed water is lost completely above 373 K to give a triclinic unit cell. The fully dehydrated structure displays remarkable structural complexity including NiO_4N coordination polyhedra intermediate between tetragonal pyramids and trigonal bypyramids, linked via either edge- or corner-sharing. The structure has adapted to water loss by movement of the phosphonate tetrahedra and transformation of the nickel coordination to NiO_4N polyhedral, resulting in the formation of a number of adsorption sites that are different from those observed in other MOFs or zeolites and include coordinatively unsaturated nickel sites and accessible P=O groups on one in three phosphonates as well as sites on the organic linkers. Interactions of these sites with adsorbed probe molecules confirm the presence of Lewis acid sites of moderate strength. In the partially dehydrated form, free and hydrogen bonded hydroxyl groups are present instead. Both types of samples exhibit permanent porosity in the adsorption of gases but higher heats and capacity are observed in the fully dehydrated solid. Additional adsorption data of H_2 and CO at 77 K show strong adsorption of the former, and strong interactions for the latter suggesting that the structure can adjust to accommodate the CO at all nickel sites but in terms of H_2 only on the accessible nickel cation sites. The different types of nickel cations act as Lewis acid sites with different strengths of interaction, depending on their local geometry, and there is also evidence that that P=O groups interact weakly. This unusual array of adsorption sites is expected to show unusual adsorption selectivity for separation and purification whereas the structure can show characteristic catalytic activities.

A tetrahedral phosphonate ligand 1,3,5,7-tetrakis(4-phosphonatophenyl)adamantine (LH_8) was designed to prevent the formation of simple layers and direct the formation of an open framework upon metal complexation.[202] The tetrahedral ligand forms large diamondoid motifs rather than densely packed arrangement complexes with Ti^{4+} and V^{3+}, which show nitrogen sorption, as well as the crystal structure of diamondoid solid $[Cu_3(H_3L)(OH)(H_2O)_3]\cdot H_2O\cdot MeOH$ which displays permanent porosity as confirmed by CO_2 sorption analysis. All diamondoid nets have a topology which readily enables interpenetration to fill potential void space but despite this, an open channel structure is still formed. The clusters are formed by an "anti-SBU" approach and the geometry of L is structure determining as it directs the topology and geometry of the resultant metal cluster. Magnetic measurements on $[Cu_3(H_3L)(OH)(H_2O)_3]\cdot H_2O\cdot MeOH$ indicate that antiferromagnetic exchange interactions occur between the copper centers within the clusters.

Layered salts of zirconium phosphate (ZrP) yields pillared scaffolds where the surface phosphate groups in γ-ZrP layers can be replaced by phosphonates and hypophosphite which leads to drastic changes in the porosity. The incorporation of rigid 4,4′-terpenyldiphosphonate into γ-ZrP followed by further exchange of residual surface phosphate groups with hypophosphite gives a much more porous, non-polar solid.[203] The exchange reaction with 4,4′-terpenyldiphosphonate leads to material **I** where some of the diphosphonates do not form pillars but are bonded only by one end to the organic layers. Treatment of this material with hypophosphorous acid gives a product with similar hydrogen and carbon content (material **II**) which does not involve loss of terpenyldiphosphonate but shows reduction in crystallinity and interlayer distance. Hypophosphite is not incorporated into the matrix contrary to phosphate and phosphonate groups. The reaction is not topotactic, as occurs in the case of pillared γ-ZrP with polyethylenoxadiphosphonates, but in turn directs the formation of a mixed phase leading to the production of a much more porous solid which takes up hydrogen in amounts of 74 cm^3 g^{-1} at 650 Torr.

Pillared clays are a new class of microporous solids prepared from the intercalation of polyoxocations in the interlayer region of layered aluminosilicates, possessing microporosity. As this fact is not restricted to aluminosilicates, many other porous pillared materials can be designed, showing initial interest in structures of different layered metal phosphonates, arsenates, and phosphates, derived from various combinations of tetrahedral and octahedral building blocks. Up to this point three main layered structures are known: α-layered structure, γ-layered structure, and layered compounds of general formula $ZrPO_4LL'$.[204] All layered structures can be pillared both with organic and inorganic pillars to obtain microporous and mesoporous compounds. Moreover, some engineered zirconium diphosphonate compounds in which special pillars are employed result in materials with particular characteristics such as topochemical reactivity, electron transfer, and nonlinear optical properties whereas photochromism, via the photoreduction of pillars containing viologen is promising in solar energy storage application.

As it concerns α-layered structure, the zirconium bis(monohydrogenphosphate) monohydrated (hereafter α-ZrP) is the most representative member. In this compound the metal shares six oxygens with six different O_3P–OH groups where a monolayer of metal atoms, bridged by monohydrogenphosphate groups lying alternately above and below it, share three of their oxygens with the tetravalent metal. The forces holding the layers together are of van der Waals type and the intercalated water molecules form hydrogen bonding with the P–OH groups belonging to only one side of the layers. In the γ-layered structure the zirconium atoms lie in two different planes and are bonded by tetrahedral PO_4 and H_2PO_4 groups. The PO_4 group shares all four oxygens with zirconium atoms while the H_2PO_4 shares two oxygens with two different zirconium atoms. Adjacent layers are linked by hydrogen bonds involving the water molecules and the –$P(OH)_2$ groups. The compounds of this class can be formulated as $ZrPO_4LL'$ in which L is a neutral ligand and L′ is an anionic monovalent ligand (see Chapter 1).

Although α- and γ-layers have the same chemical composition, their structural characteristics are quite different, such as distances between active sites, thickness and flexibility of the layers, and interlayer hydrogen bonds, affecting the mechanism of intercalation, the exfoliation, and the distribution of the intercalated species in the interlayer region which impacts the porosity in the interlayer of layered materials. The driving force for the intercalation of pillars, or their precursors, is their interaction with some active sites present on the surface of the lamellae which have polar character. The reciprocal value which corresponds to the surface density of the active sites on one side of a layer is usually called *free area*. As it concerns α-ZrP, the active sites are the acid –OH groups covalently bonded to phosphorus whereas in the case of γ-ZrP there are two acid –OH groups bonded to the same phosphorus but only one is active for intercalation of large pillars. The effect of distance between active sites is established by the interaction of a divalent pillar with two active sites belonging to adjacent layers inducting a lateral distance which corresponds to the distance between the two active sites. If the latter distance is too short, the pillars tend to be overcrowded in the interlayer region and no microporosity can be obtained unless special precautions are taken. Moreover, the tendency towards formation of solid solutions in the interlayer region is higher in solids with rigid layers. In the case of pillaring with polyoxocations, the more rigid the layers are the greater the possibility of the formation of stable solid solutions in which the polyoxocations are laterally spaced by small intercalated species that can then be removed to obtain or to increase the microporosity in the interlayer region. In some layer compounds interlayer hydrogen bonds are present, a fact that has to be taken into consideration when pillaring with large molecular species must be obtained from colloidal dispersion of the exfoliated layered compound. As it concerns layered metal phosphates, γ-ZrP exhibits the greater thickness and the greater layer rigidity; as a result it exhibits the greater tendency to the formation of solid solutions.

If in the synthesis of α-ZrP, phosphoric acid is replaced by either phenylphosphonic or diphosphonic acid, then the zirconium phosphonate or diphosphonate which results respectively has the same inorganic network as α-ZrP. A meaningful difference, however, is that in the case of phenylphosphonic acid the monovalent organic radicals pendant in the interlayer region whereas in diphosphonates, inorganic layers are covalently joined by divalent ones. A significant finding for these materials is that the surface area of those that possess a low degree of crystallinity is largely due to intercrystal mesoporosity, a fact related to the high flexibility of the α-layers. The problem is mitigated by the use of pillars with cross sections larger than 0.24 nm^2. One of these pillars, 3,3′,5,5′-tetramethylbiphenyldiphosphonic acid, was prepared exhibiting a high phosphate percentage, a good degree of crystallinity, and interlayer microporosity. Additionally, large excess of phosphorous acid in the synthesis of zirconium phosphate-diphosphonates induces the formation of a high percentage of mesopores with high specific surface area and possibility of modification of the diameter of the mesopores by changing the ratio and concentrations of the reagents. Organic pillared compounds of γ-ZrP, unlike those of α-ZrP, that can be prepared only by direct synthesis, have been obtained only by topotactic replacement of $O_2P(OH)_2^-$ lying in the external sides of the γ-lamellae with $O_2(OH)P\text{-}R\text{-}P(OH)O_2^-$ group. Reaction takes place with both monophosphonic and diphosphonic acids and is homologous to the replacement of monohydrogen phosphate with phosphonate groups in α-ZrP. However, the topotactic replacement of the monohydrogen phosphate groups with phosphonate groups in α-ZrP can take place only on the surface of the microcrystals, while in γ-ZrP the replacement occurs in the interlayer region. A variety of organic derivatives has been obtained using aliphatic and aromatic phosphinic and phosphonic acids, X-ray diffraction patterns of which prove that the γ-layer framework remains unaltered as well as that the interlayer groups influence the interlayer distance. Accommodation of the organic moieties in the interlayer region by means of simple structural models derived on the basis of the structure of the original γ-ZrP indicate that *a* and *b* dimensions are sufficiently large to accommodate one linear alkyl chain per each zirconium and as a result full conversion can be expected for n-alkylphosphonic acid. Though, a different situation is found for phenylphosphonic acid as the rigidity of the aromatic ring causes a steric hindrance with the adjacent –OH groups. In the case of phenylphosphonic group, the steric hindrance due to hydrogen atoms is markedly smaller and 100% of replacement is observed whereas partial replacement is perceived when using dimethylphosphonic acid. All these derivatives can find practical application as catalysts or solid electrolytes and especially the compound containing benzenesulfonic acid groups that exhibit a very high protonic conduction.

When diphosphonic acid with alkyl chains is used for topotactic reaction with exfoliated γ-ZrP, compounds of composition $ZrPO_4(H_2PO_4)_{1-x}(HO_3P$-$(CH_2)_m$-$PO_3H)_{x/2} \cdot nH_2O$ result, in which the alkyl chains act as pillars. The pillaring does not occur by non-rigid R organic radicals and is of significant interest because of the variation of the conformation of the pillars. Pillared compounds with $m = 4$, 6, 10, 12 are revealed that the interlayer distance of hydrate samples depend linearly on the m value. For anhydrous samples, the linear dependence of the interlayer distance from m is found at very high pillaring percentages while is almost independent for low pillaring percentages. The effect of alteration of elongation and shortening of the interlayer distance is called *accordion effect* and is reversible.[205] The compounds at low pillaring percentages act as layer solids with chained adjacent layers. If the lateral distance is large enough to permit collapse of the chains the interlayer distance depends on the species intercalated. However, the maximum interlayer distance of chained layer solids is fixed by the length of the alkyl chain joining the layers. As a result these pillared solids behave as intermediates between layer solids and pillared compounds with rigid pillars. Due to the fact that solid solutions are formed at low pillaring percentages and very long alkyl chains can be used as non-rigid pillars, interlayer spaces as large as 2 × 2 × 3 units are possible. Such interlayer spaces can be pertained as molecular vessels in which selective reactions can be carried out.

The incorporation of rigid 4,4′-biphenyldiphosphonic (BPDP) and phosphorous acids (in different ratios) into aluminum salts results in amorphous microporous materials designated as UAM-150, UAM-151, and UAM-152.[206] H_2 adsorption data reveal its augmentation as the amount of phosphorous acid increases whereas the H_2 intake displayed by material UAM-152 is close to the highest reported for organic–inorganic materials at 77 K and atmospheric pressure. Moreover, the last two materials show lower C % contents and inferior TGA losses at high temperature, which strongly support that phosphorous acid incorporates into the matrix. Further measurements indicate adsorption in wide micropores and on external surface and prove that the incorporation of H_3PO_4 to the crystals led to a substantial increase in porosity. Pore size distribution analysis for material UAM-150 shows mainly micropores and ultra micropores and only the latter for material UAM-152, a fact that intimates strong interactions between H_2 and the material due to increased van der Waals contact areas associated with very small pore sizes. Finally the fact that UAM-152 adsorbed 1.64% H_2 shows a relatively limited surface area although most of the molecular systems adsorbing this amount of H_2 or even lower have areas even well above this value. The authors stated that a large specific surface area is not a prerequisite in design of porous materials for efficient hydrogen storage.

14.10 Ion Absorption

Bein *et al.* have recently published a significant paper on the exceptional ion-exchange selectivity of a flexible open-framework sodium-lanthanum(III) tetrakis-phosphonate, $NaLa[(HO_3P)_2CH\text{-}C_6H_4\text{-}CH(PO_3H)_2]\cdot 4H_2O$.[207] The exchange of the Na^+ ions in this material with alkaline earth, alkaline, and selected transition metal ions was studied. In aqueous media, the Na^+ ions hosted in the channels of the structure can be exchanged with other monovalent ions with ionic radii ranging from 0.76 Å (Li^+) to 1.52 Å (Rb^+), while divalent ions in the same size range are rejected.

Two open framework and microporous materials, $[Pb_7(HEDTP)_2(H_2O)]\cdot 7H_2O$ and $[Zn_2(H_4EDTP)]\cdot 2H_2O$ [$H_8EDTP = N,N,N',N'$-ethylenediamine-tetrakis(methylenephosphonic acid)], have been synthesized and structurally characterized.[208] Both materials adsorbed >96% of Fe^{3+} in aqueous solutions. They were also found to adsorb Fe^{3+} ions selectively over other metal ions, such as Ca^{2+}, Cr^{2+}, Mn^{2+}, Cu^{2+}, Zn^{2+}, Cd^{2+}.

Another (non-phosphonate) coordination polymer material, $[Ba_2(L)(H_2O)_5]Cl$, where L is the trianion of 1,3,5-tris(sulfomethyl)benzene, was found to exchange the chloride ions of the structure for fluoride ions in a facile manner (80% yield).[209] This exchange was quantified by elemental analyses, gravimetric determination, and ^{19}F NMR spectroscopy. Confirmation of retention of structure is provided by standardized powder X-ray diffraction experiments. This last point is notable as the F^- analog of the structure is not attainable by a direct synthesis.

Several new tin(IV) nitrilotris(methylene)triphosphonates of different composition have been synthesized by a gel method in granular form (as spherical beads).[210] The ion exchange behavior of tin(IV) nitrilotris(methylene)triphosphonates towards alkali, alkaline earth, and some transition metal ions in individual and complex solutions were studied. It was found that these exchangers contain highly acidic adsorption sites, which enables them to operate efficiently even at pH close to 0. High affinity of the tin(IV) nitrilotris(methylene)triphosphonates for lead, cadmium, and copper in acid solutions makes them promising for treatment of some technological solutions and waste.

A novel hybrid ion-exchange material, zirconium-hydroxy ethylidene diphosphonate [ZrHEDP], has been synthesized by the sol-gel technique.[211] No crystal structure is available for this material. ZrHEDP was investigated for its possible use as an ion exchanger. The Na^+ ion-exchange capacity of the material was determined, and kinetics and thermodynamics of exchange for the metal ions Cu^{II}, Ni^{II}, Zn^{II}, and Mn^{II} towards ZrHEDP were studied. The sorption of the metal ions followed the order $Zn^{II} > Cu^{II} > Mn^{II} \approx Ni^{II}$.

Mesoporous surface-phosphonated titania, hierarchically meso-/macroporous titania phosphonate, and titanium phosphonate hybrid adsorbents have been prepared by a simple template-free nanoparticle-assembly procedure with the use of different kinds of organophosphonic acids as the coupling molecules.[212,213] The phosphonates used were 1-hydroxy ethylidene-1,1-diphosphonic acid (HEDP), ethylenediamine-tetra(methylene phosphonic acid) (EDTMP), and diethylenetriamine-penta(methylene phosphonic acid) (DTPMP). With the increase of the molar ratio of the original added phosphonic acid and tetrabutyl titanate, the structural phase transformed from crystalline phosphonated titania to semicrystalline titania-phosphonated hybrid to amorphous titanium phosphonate, and simultaneously the nanoarchitecture also changed from disordered mesoporous to hierarchically meso-/macroporous structure. The selective complexation affinity with different metal ions on different hybrid adsorbents depends on the nature and positions of organic functional groups in the synthesized samples. Mesoporous organophosphonated titanias exhibited unique selective complexation affinity for Cd^{2+} with the sequence of $Pb^{2+} < Cu^{2+} < Cd^{2+}$, and titania phosphonate showed a selective complexation affinity for Cu^{2+} with the sequence of $Cd^{2+} < Pb^{2+} < Cu^{2+}$, while hierarchical meso-/macroporous titanium phosphonates with intra-framework ethylenediamine groups exhibited a large capacity for all the three kinds of ion adsorption with a preference sequence of $Cd^{2+} < Cu^{2+} < Pb^{2+}$ and effective regeneration ability. Also, a β-cyclodextrin-added derivative of macroporous titanium organophosphonate was prepared.[214] The obtained macroporous titanium phosphonate materials exhibited selective absorption of Cu^{2+} ions compared to Cd^{2+} and Pb^{2+}.

Phosphonic acids have also been immobilized (grafted) onto polymeric matrices. Although these materials are not "technically" metal phosphonates, a brief reference to some of these may be useful to the reader. These polymeric resins have been used successfully for the removal of metal ions from aqueous solutions. Popa *et al.* reported the preparation and characterization of some chelating resins, phosphonate grafted on polystyrene-divinylbenzene supports.[215] The resins were prepared by an Arbuzov-type reaction between chloromethyl polystyrene-divinylbenzene copolymers and triethylphosphite, yielding the phosphonate ester copolymer (resin A). This can be hydrolyzed by HCl to yield the phosphonate/phosphonic acid copolymer (resin B). The total sorption capacity of the phosphonate ester-functionalized resin (A) and phosphonate/ phosphonic acid-functionalized resin (B) for divalent metal ions such as Ca^{2+}, Cu^{2+}, and Ni^{2+} was studied in aqueous solutions. Resin A retains ~3.25 mg Ca^{2+}/g copolymer , 2.75 mg Cu^{2+}/g copolymer, but retains no Ni^{2+} at $pH = 1$. On the other hand, resin B retains 8.46 mg Ca^{2+}/g copolymer , 7.17 mg Cu^{2+}/g copolymer, and no Ni^{2+} at $pH = 1$. Efficient Ni^{2+} retention was observed at $pH = 7$ only for the phosphonate/phosphonic acid-functionalized resin (B) at the level of 19 mg Ni^{2+}/g polymer B. Polymer A was incapable of retaining Ni^{2+} at $pH = 7$.

Separation of hafnium and zirconium was achieved by applying resins with –PO_3H_2 groups and defined structure.[216] Diphonix® resin is a commercially available chelating cation exchanger containing diphosphonic, sulfonic, and carboxylic acid groups bonded to the polymer matrix. The best medium for separation of hafnium and zirconium was 0.5 M sulfuric acid. Decrease in temperature from 22 °C to 5 °C lowered the degree of metal separation, while lower flow rates through the column increased separation ability. Under the best examined conditions, a 10-fold decrease of initial content of hafnium in zirconium could be achieved, with a 45 % efficiency of zirconium recovery. These results prove that Diphonix® resin is very promising for the separation of zirconium from hafnium.

14.11 Intercalation

The earliest reported data on the intercalation chemistry described the insertion of sulfuric acid into lamellar graphite in 1841. The field of intercalation chemistry expanded dramatically over the last four decades, as more and more new inorganic layered materials were discovered as potential "hosts" to "guest" molecules. As with several other scientific fields, potential applications were the drivers of intercalation development: adsorption, ion exchange, catalysis, proton and ion conductivity, etc. Host materials have to possess high thermal stability, resistance to chemical oxidation, and selectivity to ions and molecules. Most importantly, they must be able to expand their interlamellar space when guest molecules are to be inserted.[217,218]

Layered metal phosphonates have been extensively used for intercalation applications (for reviews, see references 219 and 220). Most lamellar metal mono-phosphonate compounds have the general formula $M(O_3PR)_x \cdot nH_2O$, for various metals (M), where R is an aliphatic or aromatic group. Molecules such as *n*-alkylmonoamines,[221–224] *n*-alkyldiamines,[225] aromatic amines (pyridine and the likes),[226] dendritic polyamines,[227] etc.

Ruiz and Airoldi have reported on the interaction of *n*-alkylmonoamines, $R\text{-}(CH_2)_n\text{-}NH_2$ ($n = 0$–6) with zirconium phenylphosphonate solid, $Zr(O_3PC_6H_5)_2$.[228] It may be expected that in neat amine (very high pH), a replacement of the phosphonate by OH^- likely takes place. Thus, a subsequent metal–N interaction may be possible leading to the amine taken up by the solid. In case of the longer chain length of 9.79 Å for pentylamine, when bonded to opposing zirconium acid centers inside of the same lamella, an expansion could be expected, as observed for the subsequent *n*-alkylmonoamines.

Lazarin and Airoldi studied the intercalation of pyridine (py) and a-, b-, and c-picolines (a-, b-, and c-pic) into the crystalline structure of hydrated barium phenylphosphonate, $Ba[(HO)O_2PC_6H_5]_2 \cdot 2H_2O$,[226] which results in the formation of an intercalated material of the type:

$$Ba[(HO)O_2PC_6H_5]_2 \cdot 2H_2O \cdot \lambda\chi AA \cdot (2-\lambda\chi)H_2O(s) + \lambda\chi H_2O$$

The isotherms of intercalation present an increase in number of moles of amines intercalated with time or with the concentration of the supernatant. In all cases, intercalation causes a total saturation of the inorganic layered structure showing a decrease in the sequence py > c-pic > b-pic > a-pic.

Loss of the coordinated water in zinc phenylphosphonate monohydrate results in the dehydrated, also layered, zinc phenylphosphonate, $Zn(O_3PC_6H_5)$. When $Zn(O_3PC_6H_5) \cdot H_2O$ is in contact with liquid amine, the water molecule is replaced by amine producing the thermally stable $Zn(O_3PC_6H_5) \cdot (RNH_2)$ were the amine occupies the same coordination site as the water molecule in the monohydrate.[229] The rate of intercalation of the amine into the dehydrated zinc phenylphosphonate depends on the size of the amine – the longer the alkyl chain, the slower the reaction. The interlayer spacing of zinc phenylphosphonate is 14.339(5) Å but layer expansion is observed for pentyl and longer chain amines and specifically an increase of about 4.2 Å occurred upon intercalation of heptylamine. Although the basic structures are similar, the interlayer distances do not uniformly increase as the size of the amine alkyl chain increases. This may be explained by the fact that the Zn coordination environment changes from six-coordinated to four-coordinated (tetrahedral), so "small" amines inserted do not protrude beyond the phenyl group within the interlayer space.

Dehydration of the copper phosphonates $Cu(O_3PC_6H_5) \cdot H_2O$ and $Cu(O_3PCH_3) \cdot H_2O$ leads to the formation of materials with evident amine adsorption where copper is in an unusual five-coordinate environment, occupying a water molecule bonded in the equatorial position of a distorted square pyramidal arrangement.[175] High dehydration temperatures are expected for these compounds, which are lower than the Co and Zn derivatives, a fact indicative of their relative instability. Anhydrous copper phosphonates are less crystalline than their monohydrates due to the increase in the interlayer spacing upon dehydration which may be caused by a reversible structural rearrangement due to the change in the coordination environment around Cu after dehydration. Like Co and Zn methylphosphonates, anhydrous copper methylphosphonate is quite reactive with primary amines forming amine intercalates through both solid–liquid and solid–gas reactions.[228] Using solid–liquid method, the reaction begins instantaneously and reaches completion within a few hours followed by a phase change while drying in air. As the sample dries, the crystalline intercalated phase or phases converts to a final stable crystalline layered phase. The time required to complete the phase transformation during drying is proportional to the size of the primary amines.

Lima and Airoldi have studied the intercalation of crystalline calcium phenylphosphonate[230] and calcium methylphosphonate[231] with *n*-alkylmonoamines. The chemical formula of the anhydrous calcium phosphonate[232] was previously proposed as $Ca(O_3PC_6H_5)_2$. Their investigations focused on the hydrated compound, $Ca(HO_3PC_6H_5)_2 \cdot 2H_2O$, in which two water molecules are coordinated to the inorganic backbone. When basic polar molecules are intercalated, these water molecules are gradually replaced.[232] During the intercalation process the guest molecules are oriented inside the free lamellar cavity, causing an increase in the interlamellar distance to accommodate them inside the interlayer space.[233] Thus, intercalation changes both guest and host properties. Calorimetric data, based on calorimetric titration[234] and the breaking thin-glass ampoules,[235] on intercalation between hydrated $Ca(HO_3PC_6H_5)_2 \cdot 2H_2O$ and anhydrous $Ca(O_3PC_6H_5)_2$ crystalline calcium phenylphosphonates and *n*-alkylmonoamines, $CH_3(CH_2)_nNH_2$ ($n = 0$–4, 7), establish correlations between the physical properties of the host inorganic matrix and the guest molecules. The variation in enthalpy increases as the aliphatic amine chain length increases, as expressed by negative G values. For the guest molecule *n*-butylamine intercalated in the matrix in aqueous solution, the enthalpy values show an increase from the hydrated to anhydrous hosts, clearly a favorable condition of intercalation. This fact suggested that immersion of the anhydrous compound in water does not replace the solvent molecules in the lattice, as indicated by distinguishable enthalpy values. The enthalpy values obtained from calorimetric titration are linearly correlated with the number of carbon atoms on the aliphatic *n*-alkylmonoamines intercalated in water into $Ca(HO_3PC_6H_5)_2 \cdot 2H_2O$.

Inorganic titanium phenylphosphonate and titanium phenylarsonate, $M(O_3XC_6H_5)_2$, ($M = Ti$, $X = P$, As) matrices can also be used in the synthesis of such compounds. Intercalation of ethyl and propylamines, $H_3C(CH_2)_nNH_2$ ($n = 1$, 2) in ethanol or acenonitrile solvents causes a small change in lamellar space in both inorganic matrices.[236] Thus, an evaluation of the capacity of intercalation of these two inorganic supports can be made. The most important effects (largest absolute magnitudes) are the type of inorganic material, the solvent, and the material mass.

Poojary and Clearfield have studied the intercalation of propyl-, butyl-, and pentylamine intercalates into both the hydrated and anhydrous zinc phenylphosphonates and were able to solve the intercalated structures from powder data.[237] In all three intercalates the zinc atom exhibits a slightly distorted tetrahedral geometry. The coordination sites are occupied by three oxygen atoms from three different phosphonate groups and a nitrogen atom of the amine. Among the six bond angles around the metal atom, four have normal tetrahedral values while one is larger and the other is smaller possibly due to the spacial requirements for the positioning of the amine group. It may be noted that the distortion in the bond angles is less for the butylamine compound as compared to the other two intercalates. The phosphonate tetrahedron and the phenyl group show very regular bond parameters in all three compounds. The bond parameters involving the carbon atoms of the amine are also normal, although the values have larger errors which increase with the number of carbon atoms in the chain.

The structure of $Cd(O_3PCH_3) \cdot H_2O$ is composed by neutral slabs where Cd atoms are octahedrally coordinated by five oxygen atoms of phosphonate groups and a water molecule.[238] This structure is similar to the one of $Cd(O_3PC_6H_5) \cdot H_2O$ as it concerns coordination environment around the Cd atoms and the plane arrangement of phosphonate groups and water molecule. A significant difference is that the CdO_6 octahedron in $Cd(O_3PC_6H_5) \cdot H_2O$

possesses a mirror symmetry whereas the CdO_6 octahedron in $Cd(O_3PCH_3)\cdot H_2O$ does not. Moreover, the stacking of the layers is different. Cadmium phenylphosphonate is isomorphous with manganese and zinc phenylphosphonates as they show the same unusual disorder of the phenyl rings whose positions are disordered among orientations. As a result, they reveal similarities in their dehydration and intercalation chemistry. It is interesting to note that $Cd(O_3PCH_3)\cdot H_2O$ crystallizes in the space group $Pmn2_1$ as divalent organophosphonates with the structural formula $M^{II}(O_3PR)\cdot H_2O$, while $Cd(O_3PCH_3)\cdot H_2O$ has no counterpart among the inorganic phosphates. Dehydration of the latter is likely to proceed topotactically and the compounds are formed with vapor and non-aqueous solution of amines including long-chain alkyl amines. Products can be formulated as $Cd(O_3PCH_3)(RNH_3)$ and their stability depends on the volatility of the amine involved. The intercalation compounds of lower molecular weight amines are not stable at room temperature while the compounds of higher amines are. Cadmium methylphosphonate intercalates amines with no branching at the α-carbon as the four methyl groups surrounding the site of the water molecule restrict the accessibility of the site sterically. Cadmium phenylphosphonate also reacts with amines due to structural changes associated with the reaction process which may proceed through a dissolution recrystallization mechanism. Finally, it is interesting that weakly basic alkanethiols intercalate into the dehydrated cadmium phenylphosphonates with layer expansion.

Besides amines, other molecules can also be intercalated into the metal phosphonate structure. For example, Mao *et al.* have reported the intercalation of 1,3,5-benzenetricarboxylate (H_3BTC) into the structure of lead bisphosphonate, where the phosphonate is $C_6H_5CH_2N(CH_2PO_3H_2)_2$ (H_4L).[239] Hydrothermal reaction of H_4L with lead(II) carbonate in the presence of butanedicarboxylic acid results in the materials $Pb_3L(H_2L)\cdot 1.5H_2O$ with different layered structures. When H_3BTC is used instead of butanedicarboxylic acid, a novel lead(II) biphosphonate, with the molecular formula $Pb_3(HL)_2\cdot 2(H_3BTC)\cdot 2H_2O$, is isolated with a porous 3D network.

Seven new ethylenediphosphonates with the formula $[A\{HO_3P(CH_2)_2PO_3H_2\}]$ (A = Li, Na, K, Rb, Cs, Tl, and NH_4) have been synthesized and characterized by single-crystal X-ray diffraction, infrared spectroscopy, and thermal studies.[240] They have four types of 3D structures, with variations in the coordination of the A^+ ions. All of them are solid Brönsted acids that undergo room temperature, acid–base intercalation reactions with ammonia.

The 2D layered material $Ca(HDTMP)\cdot 2H_2O$ exhibits interesting NH_3 absorption properties.[241] The basic building block for this phosphonate consists of an eight-membered ring formed by Ca^{2+} chelation by two phosphonate oxygens (from two different phosphonate groups) and the organic ligand. The latter is located *within* the layers, which are held together by H-bonding. Thermal treatment and subsequent exposure to NH_3 and/or H_2O vapors led to four new compounds showing high storage capacity of guest species between the layers, up to 10 H_2O/NH_3 molecules, and a maximum unit cell volume increase of 55% (Figure 14.35).

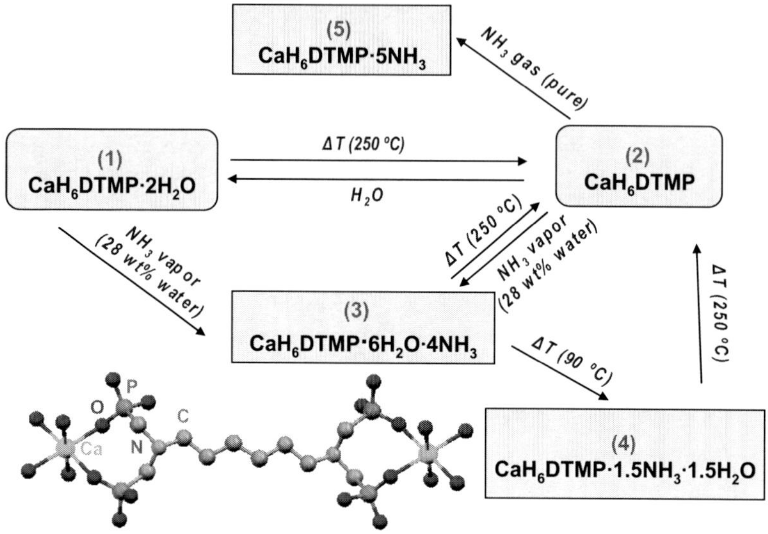

Figure 14.35 Interconversions between the reported calcium tetraphosphonate hybrids due to incorporation/removal of host molecules. The basic structural unit for this family is also shown. (Reproduced with permission from ref. 240.)

The structural analysis revealed that uptake/removal of guest species (H_2O and NH_3) induces significant changes in the framework, not only by changing the interlayer distances, but also through important conformational changes of the organic ligand (Figure 14.36). Anisotropic breathing motion could be quantified by the changes of unit cell dimensions and ligand arrangements in four crystalline derivatives. Complete characterization revealed the existence of interconversion reactions between the different phases upon gas uptake and release.

Due to the intense interest lately in ammonia storage, we will add a word on its storage in covalent organic frameworks (COFs). These are porous crystalline materials composed of light elements linked by strong covalent bonds. A number of these materials contain a high density of Lewis acid boron sites that can strongly interact with Lewis basic guests, which makes them ideal for the storage of corrosive chemicals such as ammonia. We found that a member of the covalent organic framework family, COF-10, shows the highest uptake capacity (15 mol kg^{-1}, 298 K, 1 bar) of any porous material, including microporous 13X zeolite (9 mol kg^{-1}), Amberlyst 15 (11 mol kg^{-1}) and mesoporous silica, MCM-41 (7.9 mol kg^{-1}).[242] Notably, ammonia can be removed from the pores of COF-10 by heating samples at 200 °C under vacuum. In addition, repeated adsorption of ammonia into COF-10 causes a shift in the interlayer packing, which reduces its apparent surface area to nitrogen. However, owing to the strong Lewis acid–base interactions, the total uptake capacity of ammonia and the structural integrity of the COF are maintained after several cycles of adsorption/desorption.

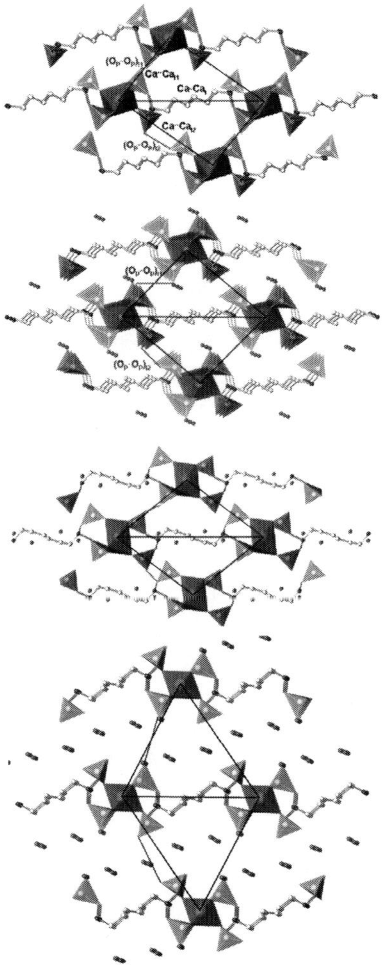

Figure 14.36 Structures of the hybrids showing the breathing with the guest species uptake. From top to bottom: (2), (1), (4), and (3). For the numbering scheme, see Figure 14.35. (Reproduced with permission from ref. 240.)

References

1. A. Clearfield, *Chem. Mater.*, 1998, **10**, 2801.
2. K. Maeda, *Microporous Mesoporous Mater.*, 2004, **73**, 47.
3. K. D. Demadis, in *Solid State Chemistry Research Trends*, ed. R. W. Buckley, Nova Science Publishers, New York, 2007, p. 109.
4. C. V. K. Sharma and A. Clearfield, *J. Am. Chem. Soc.*, 2000, **122**, 4394.
5. A. Clearfield, *Prog. Inorg. Chem.*, 1998, **47**, 371.
6. A. Vioux, L. Le Bideau, P. Hubert Mutin and D. Leclercq, *Top. Curr. Chem.*, 2004, **232**, 145.
7. A. H. Mahmoudkhani and V. Langer, *Phosphorus Sulfur Silicon*, 2002, **177**, 2941.

8. A. H. Mahmoudkhani and V. Langer, *J. Mol. Struct.*, 2002, **609**, 97.
9. A. N. Lazar, A. Navaza and A. W. Coleman, *Chem. Commun.*, 2004, 1052.
10. A. H. Mahmoudkhani and V. Langer, *Cryst. Growth Des.*, 2002, **2**, 21.
11. Z.-Y. Du, A. V. Prosvirin and J.-G. Mao, *Inorg. Chem.*, 2007, **46**, 9884.
12. Z.-Y. Du, H.-B. Xu and J.-G. Mao, *Inorg. Chem.*, 2006, **45**, 6424.
13. B.-P. Yang and J.-G. Mao, *Inorg. Chem.*, 2005, **44**, 566.
14. C. Lei, J.-G. Mao, Y.-Q. Sun, H.-Y. Zeng and A. Clearfield, *Inorg. Chem.*, 2003, **42**, 6157.
15. A. K. Cheetham, G. Ferey and T. Loiseau, *Angew. Chem. Int. Ed. Engl.*, 1999, **38**, 3268.
16. P. M. Forster and A. K. Cheetham, *Top. Catal.*, 2003, **24**, 79.
17. C. A. Merrill and A. K. Cheetham, *Inorg. Chem.*, 2007, **46**, 278.
18. S. H. Jhung, J. W. Yoon, J.-S. Hwang, A. K. Cheetham and J.-S. Chang, *Chem. Mater.*, 2005, **17**, 4455.
19. C. A. Merrill and A. K. Cheetham, *Inorg. Chem.*, 2005, **44**, 5273.
20. M. M. Gomez-Alcantara, A. Cabeza, M. Martinez-Lara, M. A. G. Aranda, R. Suau, N. Bhuvanesh and A. Clearfield, *Inorg. Chem.*, 2004, **43**, 5283.
21. A. Cabeza, O. Y. Xiang, C. V. K. Sharma, M. A. G. Aranda, S. Bruque and A. Clearfield, *Inorg. Chem.*, 2002, **41**, 2325.
22. A. Cabeza, M. A. G. Aranda and S. Bruque, *J. Mater. Chem.*, 1999, **9**, 571.
23. A. Turner, P.-A. Jaffres, E. J. MacLean, D. Villemin, V. McKee and G. B. Hix, *J. Chem. Soc. Dalton Trans.*, 2003, 1314.
24. G. B. Hix, D. S. Wragg, P. A. Wright and R. E. Morris, *J. Chem. Soc. Dalton Trans.*, 1998, 3359.
25. A. Turner, B. M. Kariuki, M. Tremayne and E. J. MacLean, *J. Mater. Chem.*, 2002, **12**, 3220.
26. N. Zakowsky, G. B. Hix and R. E. Morris, *J. Mater. Chem.*, 2000, **10**, 2375.
27. N. Stock and T. Bein, *Angew. Chem. Int. Ed. Engl.*, 2004, **43**, 749.
28. S. Bauer and N. Stock, *Angew. Chem. Int. Ed. Engl.*, 2007, **46**, 6857.
29. P. M. Forster, N. Stock and A. K. Cheetham, *Angew. Chem. Int. Ed. Engl.*, 2005, **44**, 7608.
30. S. Bauer, T. Bein and N. Stock, *J. Solid State Chem.*, 2006, **179**, 145.
31. B. A. Breeze, M. Shanmugam, F. Tuna and R. E. P. Winpenny, *Chem. Commun.*, 2007, 5185.
32. S. Langley, M. Helliwell, J. Raftery, E. I. Tolis and R. E. P. Winpenny, *Chem. Commun.*, 2004, 142.
33. V. Baskar, M. Shanmugam, E. C. Sañudo, M. Shanmugam, D. Collison, E. J. L. McInnes, Q. Wei and R. E. P. Winpenny, *Chem. Commun.* 2007, 37.
34. A. Harrison, D. K. Henderson, P. A. Lovatt, A. Parkin, P. A. Tasker and R. E. P. Winpenny, *Dalton Trans.* 2003, 4271.
35. C. Serre, J. A. Groves, P. Lightfoot, A. M. Z. Slawin, P. A. Wright, N. Stock, T. Bein, M. Haouas, F. Taulelle and G. Ferey, *Chem. Mater.*, 2006, **18**, 1451.

36. G. Ferey, C. Mellot-Draznieks, C. Serre and F. Millange, *Acc. Chem. Res.*, 2005, **38**, 217.
37. C. Serre, C. Lorentz, F. Taulelle and G. Ferey, *Chem. Mater.*, 2003, **15**, 2328.
38. K. Barthelet, M. Nogues, D. Riou and G. Ferey, *Chem. Mater.*, 2002, **14**, 4910.
39. G. Alberti, M. Casciola, U. Costantino and R. Vivani, *Adv. Mater.*, 1996, **8**, 291.
40. R. Vivani, G. Alberti, F. Costantino and M. Nocchetti, *Microporous Mesoporous Mater.*, 2008, **107**, 58.
41. R. Vivani, F. Costantino, U. Costantino and M. Nocchetti, *Inorg. Chem.*, 2006, **45**, 2388.
42. U. Costantino, M. Nocchetti and R. Vivani, *J. Am. Chem. Soc.*, 2002, **124**, 8428.
43. K. D. Demadis and S. D. Katarachia, *Phosphorus Sulfur Silicon*, 2004, **179**, 627.
44. K. D. Demadis and P. Lykoudis, *Bioinorg. Chem. Appl.*, 2005, **3**, 135.
45. K. D. Demadis, *Phosphorus Sulfur Silicon*, 2006, **181**, 167.
46. K. D. Demadis and E. Mavredaki, *Env. Chem. Lett.*, 2005, **3**, 127.
47. S. J. Dyer, C. E. Anderson and G. M. Graham, *J. Pet. Sci. Eng.*, 2004, **43**, 259.
48. J. E. Oddo and M. B. Tomson, *Appl. Geochem.*, 1990, **5**, 527.
49. J. J. Xiao, A. T. Kan and M. B. Tomson, *Langmuir*, 2001, **17**, 4668.
50. S. J. Friedfeld, S. He and M. B. Tomson, *Langmuir*, 1998, **14**, 3698.
51. V. Tantayakom, H. S. Fogler, P. Charoensirithavorn and S. Chavadej, *Cryst. Growth Des.*, 2005, **5**, 329.
52. F. H. Browning and H. S. Fogler, *AIChE J.*, 1996, **42**, 2883.
53. R. Pairat, C. Sumeath, F. H. Browning and H. S. Fogler, *Langmuir*, 1997, **13**, 1791.
54. V. Tantayakom, H. S. Fogler, F. F. de Moraes, M. Bualuang, S. Chavadej and P. Malakul, *Langmuir*, 2004, **20**, 2220.
55. A.-L. Penard, F. Rossignol, H. S. Nagaraja, C. Pagnoux and T. Chartier, *Eur. J. Ceram. Soc.*, 2005, **25**, 1109.
56. M. J. Pearse, *Min. Eng.*, 2005, **18**, 139.
57. I. Sekine, T. Shimode and M. Yuasa, *Ind. Eng. Chem. Res.*, 1992, **31**, 434.
58. B. Mosayebi, M. Kazemeini and A. Badakhshan, *Br. Corr. J.*, 2002, **37**, 217.
59. Yu I. Kouznetsov, *Prot. Met.*, 2001, **37**, 434.
60. Yu V. Balaban-Irmenin, A. M. Rubashov and N. G. Fokina, *Prot. Met.*, 2006, **42**, 133.
61. J. L. Fang, Y. Li, X. R. Ye, Z. W. Wang and Q. Liu, *Corrosion*, 1993, **49**, 266.
62. A. Paszternák, S. Stichleutner, I. Felhősi, Z. Keresztes, F. Nagy, E. Kuzmann, A. Vértes, Z. Homonnay, G. Pető and E. Kálmán, *Electrochim. Acta*, 2007, **53**, 337.
63. *Biogeochemistry of Chelating Agents*, ed. B. Nowack, J. M. Van Briessen, J. M. ACS Symposium Series, ACS, Washington DC, 2003, Vol. 910.

64. T. P. Knepper, *Trends Anal. Chem.*, 2003, **22**, 708.
65. K. Miyazaki, T. Horibe, J. M. Antonucci, S. Takagi and L. C. Chow, *Dent. Mater.*, 1993, **9**, 46.
66. M. Atai, M. Nekoomanesh, S. A. Hashemi and S. Amani, *Dent. Mater.*, 2004, **20**, 663.
67. J. W. Nicholson and G. Singh, *Biomaterials*, 1996, **17**, 2023.
68. H. Tschernitschek, L. Borchers and W. Geurtsen, *J. Prosth. Dent.*, 2006, **96**, 12.
69. F. Cheng and E. Oldfield, *J. Med. Chem.*, 2004, **47**, 5149.
70. C. Temperini, A. Innocenti, A. Guerri, A. Scozzafava, S. Rusconi and C. T. Supuran, *Bioorg. Med. Chem. Lett.*, 2007, **17**, 2210.
71. E. Davini, C. Di Leo, F. Norelli and P. Zappelli, *J. Biotechnol.*, 1993, **28**, 321.
72. M. Bottrill, L. Kwok and N. J. Long, *Chem. Soc. Rev.*, 2006, **35**, 557.
73. I. G. Finlay, M. D. Mason and M. Shelley, *Lancet Oncol.*, 2005, **6**, 392.
74. V. Kubicek, J. Rudovsky, J. Kotek, P. Hermann, L. Vander Elst, R. N. Muller, Z. I. Kolar, H. T. Wolterbeek, J. A. Peters and I. Lukeš, *J. Am. Chem. Soc.*, 2005, **127**, 16477.
75. H. Kung, R. Ackerhalt and M. Blau, *J. Nucl. Med.*, 1978, **19**, 1027.
76. S. S. Padalecki and T. A. Guise, *Breast Cancer Res.*, 2001, **4**, 35.
77. V. Stresing, F. Daubiné, I. Benzaid, H. Mönkkönen and P. Clézardin, *Cancer Lett.*, 2007, **257**, 16.
78. R. Layman, K. Olson and C. Van Poznak, *Hematol. Oncol. Clinics North Am.*, 2007, **21**, 341.
79. A. Clearfield, *Curr. Opin. Solid State Mater. Sci.*, 2002, **6**, 495.
80. A. Clearfield, *Curr. Opin. Solid State Mater. Sci.*, 1996, **1**, 268.
81. E. Barouda, K. D. Demadis, S. Freeman, F. Jones and M. I. Ogden, *Cryst. Growth Des.*, 2007, **7**, 321.
82. K. Popov, H. Rönkkömäki and L. H. J. Lajunen, *Pure Appl. Chem.*, 2001, **73**, 1641.
83. M. Dyba, M. Jezowska-Bojczuk, E. Kiss, T. Kiss, H. Kozłowski, Y. Leroux and D. El Manouni, *J. Chem. Soc. Dalton Trans.*, 1996, 1119.
84. E. Gumienna-Kontecka, J. Jezierska, M. Lecouvey, Y. Leroux and H. Kozłowski, *J. Inorg. Biochem.*, 2002, **89**, 13.
85. E. Gumienna-Kontecka, R. Silvagni, R. Lipinski, M. Lecouvey, F. C. Marincola, G. Crisponi, V. M. Nurchi, Y. Leroux and H. Kozłowski, *Inorg. Chim. Acta*, 2002, **339**, 111.
86. E. Matczak-Jon, B. Kurzak, A. Kamecka and P. Kafarski, *Polyhedron*, 2002, **21**, 321.
87. B. Boduszek, M. Dyba, M. Jezowska-Bojczuk, T. Kiss and H. Kozłowski, *J. Chem. Soc. Dalton Trans.*, 1997, 973.
88. B. Kurzak, A. Kamecka, K. Kurzak, J. Jezierska and P. Kafarski, *Polyhedron*, 1998, **17**, 4403.
89. E. Matczak-Jon, B. Kurzak, A. Kamecka, W. Sawka-Dobrowolska and P. Kafarski, *J. Chem. Soc. Dalton Trans.*, 1999, 3627.

90. G. V. Polyanchuk, L. M. Shkol'nikova, M. V. Rudomino, N. M. Dyatlova and S. S. Makarevich, *J. Struct. Chem.*, 1985, **26**, 586.
91. G. B. Hix, V. Caignaert, J.-M. Rueff, L. Le Pluart, J. E. Warren and P.-A. Jaffrès, *Cryst. Growth Des.*, 2007, **7**, 208.
92. K. M. E. Jones, A. H. Mahmoudkhani, B. D. Chandler and G. K. H. Shimizu, *CrystEngComm*, 2006, **8**, 303.
93. K. D. Demadis, R. G. Raptis and P. Baran, *Bioinorg. Chem. Appl.*, 2005, **3**, 119.
94. K. D. Demadis, E. Barouda, H. Zhao and R. G. Raptis, *Polyhedron*, 2009, **28**, 3361.
95. S. Lodhia, A. Turner, M. Papadaki, K. D. Demadis and G. B. Hix, *Cryst. Growth Des.*, 2009, **9**, 1811.
96. K. D. Demadis, S. D. Katarachia, H. Zhao, R. G. Raptis and P. Baran, *Cryst. Growth Des.*, 2006, **6**, 836.
97. K. D. Demadis, A. Panera, unpublished results.
98. K. D. Demadis, M. Papadaki, R. G. Raptis and H. Zhao, *J. Solid State Chem.*, 2008, **181**, 679.
99. K. D. Demadis, M. Papadaki, R. G. Raptis and H. Zhao, *Chem. Mater.*, 2008, **20**, 4835.
100. J. Yang, J.-F. Ma, G.-L. Zheng, L. Li, F.-F. Li, Y.-M. Zhang and J.-F. Liu, *J. Solid State Chem.*, 2003, **174**, 116.
101. R. Murugavel, S. Kuppuswamy and S. Randoll, *Inorg. Chem.*, 2008, **47**, 6028.
102. M. Tuikka, M. Haukka and M. Ahlgren, *Solid State Sci.*, 2007, **9**, 535.
103. S. Bauer, H. Muller, T. Bein and N. Stock, *Inorg. Chem.*, 2005, **44**, 9464.
104. D. S. Sagatys, C. Dahlgren, G. Smith, R. C. Bott and J. M. White, *J. Chem. Soc. Dalton Trans.*, 2000, 3404.
105. H. S. Martínez-Tapia, A. Cabeza, S. Bruque, P. Pertierra, S. García-Granda and M. A. G. Aranda, *J. Solid State Chem.*, 2000, **151**, 122.
106. C.-Y. Cheng and K.-J. Lin, *Acta Crystallogr.*, 2006, **C62**, m363.
107. R. Fu, S. Xia, S. Xiang, S. Hu and X. Wu, *J. Solid State Chem.*, 2004, **177**, 4626.
108. K. D. Demadis, M. Papadaki, M. A. G. Aranda, A. Cabeza, P. Olivera-Pastor and Y. Sanakis, *Cryst. Growth Des.*, 2010, **10**, 357.
109. R. M. P. Colodrero, P. Olivera-Pastor, A. Cabeza, M. Papadaki, K. D. Demadis and M. A. G. Aranda, *Inorg. Chem.*, 2010, **49**, 761.
110. G.-L. Zheng, J.-F. Ma and J. Yang, *J. Chem. Res.* 2004, 387.
111. S.-J. Fu, C.-Y. Cheng and K.-J. Lin, *Acta Crystallogr.*, 2007, **C63**, m368.
112. W. R. Gemmill, M. D. Smith and B. A. Reisner, *J. Solid State Chem.*, 2005, **178**, 2658.
113. O. H. S. Gustschke, D. J. Price, A. K. Powell and P. T. Wood, *Angew. Chem. Int. Ed. Engl.*, 1999, **38**, 1088.
114. A. Distler, D. L. Lohse and S. C. Sevov, *J. Chem. Soc. Dalton Trans.*, 1999, 1805.
115. R. LaDuca, D. Rose, J. R. D. DeBord, R. C. Haushalter, C. J. O'Connor and J. Zubieta, *J. Solid State Chem.*, 1996, **123**, 408.

116. P. Rabu, P. Janvier and B. Bujoli, *J. Mater. Chem.*, 1999, **9**, 1323.
117. D. L. Lohse and S. C. Sevov, *Angew. Chem. Int. Ed. Engl.*, 1997, **36**, 1619.
118. F.-N. Shi, F. A. Almeida Paz, P. Girginova, J. Rocha, V. S. Amaral, J. Klinowski and T. Trindade, *J. Mol. Struct.*, 2006, **789**, 200.
119. P. Yin, S. Gao, Z.-M. Wang, C.-H. Yan, L.-M. Zheng and X-Q. Xin, *Inorg. Chem.*, 2005, **44**, 2761.
120. N. Stock, M. Rauscher and T. Bein, *J. Solid State Chem.*, 2004, **177**, 642.
121. K. D. Demadis, S. D. Katarachia and M. Koutmos, *Inorg. Chem. Commun.*, 2005, **8**, 254.
122. J.-G. Mao, Z. Wang and A. Clearfield, *New J. Chem.*, 2002, **26**, 1010.
123. K. D. Demadis, C. Mantzaridis, R. G. Raptis and G. Mezei, *Inorg. Chem.*, 2005, **44**, 4469.
124. J. Li, D.-P. Dong, C.-Y. Huang, Z.-G. Sun and Y.-Y. Zhu, *Acta Crystallogr.*, 2007, **E63**, m2348.
125. E. M. Bauer, C. Bellitto, P. Imperatori, G. Righini, M. Colapietro, G. Portalone and C. J. Gomez-Garcia, *Inorg. Chem.*, 2010, **49**, 7472.
126. L. R. Olano, H. T. Tran, J. T. Mague and D. A. Knight, *J. Chem. Crystallogr.*, 2003, **33**, 497.
127. A. Cabeza, S. Bruque, A. Guagliardi and M. A. G. Aranda, *J. Solid State Chem.*, 2001, **160**, 278.
128. K. Maeda, Y. Hashiguchi, Y. Kiyozumi and F. Mizukami, *Bull. Chem. Soc. Jpn.*, 1997, **70**, 345.
129. H. G. Harvey, M. P. Attfield, *Solid State Sci.*, 2006, **8**, 404.
130. R. Murugavel, N. Gogoi and R. Clerac, *Inorg. Chem.*, 2009, **48**, 646.
131. B. Adair, S. Natarajan and A. K. Cheetham, *J. Mater. Chem.*, 1998, **8**, 1477.
132. A. Subbiah, N. Bhuvanesh and A. Clearfield, *J. Solid State Chem.*, 2005, **178**, 1321.
133. D. E. Lansky, P. Y. Zavalij and S. R. J. Oliver, *Acta Crystallogr.*, 2001, **C57**, 1051.
134. N. Stock, N. Guillou, T. Bein and G. Férey, *Solid State Sci.*, 2003, **5**, 629.
135. B. Zhang, D. M. Poojary and A. Clearfield, *Inorg. Chem.*, 1998, **37**, 249.
136. D. M. Poojary and A. Clearfield, *J. Organomet. Chem.*, 1996, **512**, 237.
137. D. M. Poojary, L. A. Vermeulen, E. Vicenzi, A. Clearfield and M. E. Thompson, *Chem. Mater.*, 1994, **6**, 1845.
138. R. Vivani, U. Costantino and M. Nocchetti, *J. Mater. Chem.*, 2002, **12**, 3254.
139. R. Vivani, F. Costantino, M. Nocchetti and G. D. Gatta, *J. Solid State Chem.*, 2004, **177**, 4013.
140. D. M. Poojary, C. Bhardwaj and A. Clearfield, *J. Mater. Chem.*, 1995, **5**, 171.
141. P. P. Charpin, M. Lance, M. Nierlich, D. Vigner, M.-R. Lee, J.-P. Silvestre and N. Q. Dao, *Acta Crystallogr.*, 1988, **C44**, 990.
142. A. M. Beaty, B. A. Helfrich, G. A. Hogan and B. A. Reed, *Cryst. Growth Des.*, 2006, **6**, 122.

143. K. D. Demadis, Z. Anagnostou and H. Zhao, *ACS-Appl. Mater. Interf.*, 2009, **1**, 35.
144. S. M. Hawxwell, H. Adams and L. Brammer, *Acta Crystallogr.*, 2006, **B62**, 808.
145. K. M. Fromm, *Coord. Chem. Rev.*, 2008, **252**, 856.
146. R. Robson, *Dalton Trans.*, 2008, 5113.
147. A. U. Czaja, N. Trukhan and U. Muller, *Chem. Soc. Rev.*, 2009, **38**, 1284.
148. M. Kurmoo, *Chem. Soc. Rev.*, 2009, **38**, 1353.
149. G. Ferey, *Dalton Trans.*, 2009, 4400.
150. N. L. Rosi, M. Eddaoudi, J. Kim, M. O'Keeffe and O. M. Yaghi, *CrystEngComm*, 2002, **4**, 401.
151. K. D. Demadis, J. D. Sallis, R. G. Raptis and P. Baran, *J. Am. Chem. Soc.*, 2001, **123**, 10129.
152. J. Jokiniemi, S. Peräniemi, J. J. Vepsäläinen and M. Ahlgreìn, *CrystEngComm*, 2008, **10**, 1011.
153. P. A. Turhanen, K. D. Demadis, S. Peräniemi and J. J. Vepsäläinen, *J. Org. Chem.*, 2007, **72**, 1468.
154. (a) H. Kumpulainen, R. Saari, M. Lehtonen, J. Rautio, T. Järvinen and J. Vepsäläinen, *Tetrahedron Lett.*, 2006, **47**, 2003; (b) R. Murugavel, A. Choudhury, M. G. Walawalkar, R. Pothiraja and C. N. R. Rao, *Chem. Rev.*, 2008, **108**, 3549.
155. S. Natarajan and P. Mahata, *Curr. Opin. Solid State Mater. Sci.*, 2009, **13**, 46.
156. L.-S. Long, *CrystEngComm*, 2010, **12**, 1354.
157. R. G. Lin, L. S. Long, R. B. Huang and L. S. Zheng, *Inorg. Chem. Commun.*, 2007, **10**, 1257.
158. S. J. Dalgarno, J. L. Atwood and C. L. Raston, *Cryst. Growth Des.*, 2006, **6**, 174.
159. C. Liu, F. Luo, W. P. Liao, D. Q. Li, X. F. Wang and R. Dronskowski, *Cryst. Growth Des.*, 2007, **7**, 2282.
160. S. J. Dalgarno, M. J. Hardie and C. L. Raston, *Cryst. Growth Des.*, 2004, **4**, 227.
161. A. Cabeza, M. A. G. Aranda, S. Bruque, D. M. Poojary, A. Clearfield and J. Sanz, *Inorg. Chem.*, 1998, **37**, 4168.
162. J. Le Bideau, C. Payen, P. Palvadeau and B. Bujoli, *Inorg. Chem.*, 1994, **33**, 4885.
163. R. Fu, H. Zhang, L. Wang, S. Hu, Y. Li, X. Huang and X. Wu, *Eur. J. Inorg. Chem. 2005*, 3211.
164. Q.-H. Zhao, L. Dua and R.-B. Fang, *Acta Crystallogr.*, 2006, **E62**, m219.
165. R. Fu, S. Hu and X. Wu, *Dalton Trans.*, 2009, 9843.
166. Z.-G. Sun, D.-P. Dong, J. Li, L.-Y. Cui, Y.-Y. Zhu, J. Zhang, Y. Zhao, W.-S. You and Z.-M. Zhu, *J. Coord. Chem.*, 2007, **60**, 2541.
167. Y.-Y. Zhang, M.-H. Zeng, Y. Qi, S.-Y. Sang and Z.-M. Liu, *Inorg. Chem. Commun.*, 2007, **10**, 33.
168. J. Li, L. Meng, Z.-G. Sun, L.-Y. Cui, J. Zhang, Y.-Y. Zhu, D.-P. Dong, H. Chen, W.-S. You and Z.-M. Zhu, *Inorg. Chem. Commun.*, 2007, **10**, 535.

169. Z.-Y. Du, J.-J. Huang, Y.-R. Xie and H.-R. Wen, *J. Mol. Struct.*, 2009, **919**, 112.
170. V. Chandrasekhar, T. Senapati and R. Clérac, *Eur. J. Inorg. Chem. 2009*, 1640.
171. V. Chandrasekhar, R. Azhakar, T. Senapati, P. Thilagar, S. Ghosh, S. Verma, R. Boomishankar, A. Steiner and P. Kogerler, *Dalton Trans.*, 2008, 1150.
172. K. D. Demadis and D. Varouchas, *unpublished results.*
173. G. Alberti, U. Constantino, S. Allulli and N. Tomassini, *J. Inorg. Nucl. Chem.*, 1978, **40**, 1113.
174. G. Cao, H.-G. Hong and T. E. Mallouk, *Acc. Chem. Res.*, 1992, **25**, 422.
175. Y. P. Zhang and A. Clearfield, *Inorg. Chem.*, 1992, **31**, 2821.
176. M. E. Thompson, *Chem. Mater.*, 1994, **6**, 1168.
177. K. Maeda and F. Mizukami, *Catal. Surv. Jpn.*, 1999, **3**, 119.
178. K. Yamamoto, Y. Sakata, Y. Nohara, Y. Takahashi and T. Tatsumi, *Science*, 2003, **300**, 470.
179. G. Gunasekaran, R. Natarajan, V. S. Muralidharan, N. Palaniswamy and B. V. Appa Rao, *Anti-Corr. Meth. Mater.*, 1997, **44**(4), 248.
180. S. M. Cohen, *Corrosion*, 1995, **51**, 71.
181. S. Rajendran, B. V. Appa Rao and N. Palaniswamy, *Anti-Corr. Meth. Mater.*, 1999, **46**(1), 23.
182. E. Kalman, I. Lukovits and G. Palinkas, *ACH-Models in Chemistry*, 1995, **132**, 527.
183. M. Papadaki and K. D. Demadis, *Commun. Inorg. Chem.*, 2009, **30**, 89.
184. B. P. Boffardi, in *Reviews on Corrosion Inhibitor Science and Technology*, ed. A. Raman, P. Labine, NACE International, Houston, TX, 1993, p. 6.
185. K. D. Demadis, M. Papadaki and I. Cisarova, *ACS-Appl. Mater. Interf.*, 2010, **2**, 1814.
186. K. D. Demadis, P. Lykoudis, R. G. Raptis and G. Mezei, *Cryst. Growth Des.*, 2006, **6**, 1064.
187. M. Benabdellah, A. Dafali, B. Hammouti, A. Aouniti, M. Rhomari, A. Raada, O. Sehaji and J. J. Robin, *Chem. Eng. Commun.*, 2007, **194**, 1328.
188. K. D. Demadis, E. Barouda, N. Stavgianoudaki and H. Zhao, *Cryst. Growth Des.*, 2009, **9**, 1250.
189. K. D. Demadis, E. Barouda, R. G. Raptis and H. Zhao, *Inorg. Chem.*, 2009, **48**, 819.
190. J. Menke, *NACE Conference 1999*, Corrosion/99, Paper number 503.
191. Editorial, *Chem. Process.* 2002, December, 11.
192. K. D. Demadis, B. Yang, P. R. Young, D. L. Kouznetsov, D. G. Kelley, in *Advances in Crystal Growth Inhibition Technologies*, ed. Z. Amjad, Plenum Publishing Corporation, New York, 2000, Chapter 16, p. 215.
193. K. D. Demadis, in *Compact Heat Exchangers and Enhancement Technology for the Process Industries*, ed. R. K. Shah, Begell House Inc. New York, 2003, p. 483.
194. H. Li, M. Eddaoudi, M. O'Keeffe and O. M. Yaghi, *Nature*, 1999, **402**, 276.

195. G. Ferey, C. Mellot-Draznieks, C. Serre and F. Millange, *Acc. Chem. Res.*, 2005, **38**, 217.
196. S. S. Y. Chui, J. P. H. Charmant, A. G. Orpen and I. D. Williams, *Science*, 1999, **283**, 1148.
197. K. S. Park, Z. Ni, A. P. Côté, J. Y. Choi, R. D. Huang, F. J. Uribe-Romo, H. K. Chae, M. O'Keeffe and O. M. Yaghi, *Proc. Natl. Acad. Sci. USA*, 2006, **103**, 10186.
198. R. Matsuda, R. Kituara, S. Kitagawa, Y. Kubota, R. V. Belosludov, T. C. Kobayashi, H. Sakamoto, T. Chiba, M. Takata, Y. Kawazoe and Y. Mita, *Nature*, 2005, **436**, 238.
199. (a) R. Vaidhyanathan, D. Bradshaw, J. N. Rebilly, J. P. Barrio, J. A. Gould, N. G. Berry and M. J. Rosseinsky, *Angew. Chem. Int. Ed. Engl.*, 2006, **45**, 6495; (b) I. Imaz, M. Rubio-Martinez, J. An, I. Sole-Font, N. L. Rosi and D. Maspoch, *Chem. Commun.*, 2011, **47**, 7287.
200. J. A. Groves, S. R. Miller, S. J. Warrender, C. Mellot-Draznieks, P. Lightfoot and P. A. Wright, *Chem. Commun.*, 2006, 3305.
201. S. R. Miller, G. M. Pearce, P. A. Wright, F. Bonino, S. Chavan, S. Bordiga, I. Margiolaki, N. Guillou, G. Ferey, S. Bourrelly and P. L. Liewellyn, *J. Am. Chem. Soc.*, 2008, **130**, 15967.
202. J. M. Tylor, A. H. Mahmoudkhani and G. K. H. Shimizu, *Angew. Chem. Int. Ed. Engl.*, 2007, **46**, 795.
203. E. Brunet, H. M. H. Alhendawi, C. Cerro, M. J. de la Mata, O. Juans and J. C. Rodriguez-Ubis, *Angew. Chem. Int. Ed. Engl.*, 2006, **45**, 6918.
204. G. Alberti, R. Vivani, F. Marmottini and P. Zappelli, *J. Porous Mater.*, 1998, **5**, 205.
205. A. Clearfield and Z. K. Wang, *J. Chem. Soc. Dalton Trans.*, 2002, 2937.
206. E. Brunet, C. Cerro, O. Juanes, J. C. Rodriguez-Ubis and A. Clearfield, *J. Mater. Sci.*, 2008, **43**, 1155.
207. M. Plabst, L. B. McCusker and T. Bein, *J. Am. Chem. Soc.*, 2009, **131**, 18112.
208. J. Wu, H. Hou, H. Han and Y. Fan, *Inorg. Chem.*, 2007, **46**, 7960.
209. S. A. Dalrymple and G. K. H. Shimizu, *Chem. Eur. J.*, 2002, **8**, 3010.
210. A. I. Bortun, L. N. Bortun, A. Clearfield, S. A. Khainakov and J. R. García, *Solv. Extr. Ion Exch.*, 1998, **16**, 651.
211. A. Jayswal and U. Chudasama, *Turk. J. Chem.*, 2008, **32**, 63.
212. T.-Y. Ma, X.-J. Zhang and Z.-Y. Yuan, *J. Mater. Sci.*, 2009, **44**, 6775.
213. X.-J. Zhang, T.-Y. Ma and Z.-Y. Yuan, *J. Mater. Chem.*, 2008, **18**, 2003.
214. T.-Y. Ma, X.-J. Zhang, G.-S. Shao, J.-L. Cao and Z.-Y. Yuan, *J. Phys. Chem. C*, 2008, **112**, 3090.
215. A. Popa, C.-M. Davidescu, N. Petru, I. Gheorghe, A. Katsaros and K. D. Demadis, *Ind. Eng. Chem. Res.*, 2008, **47**, 2010.
216. M. Smolik, A. Jakóbik-Kolon and M. Porański, *Hydrometallurgy*, 2009, **95**, 350.
217. K. Peeters, P. Grobet and E. F. Vansant, *J. Mater. Chem.*, 1996, **6**, 239.
218. R. Hoppe, G. Alberti, U. Contantino, C. Dionigi, G. S. Ekloff and R. Vivani, *Langmuir*, 1997, **13**, 7252.

219. S. L. Suib, *Chem. Rev.*, 1993, **93**, 803.
220. G. Cao, H.-G. Hong and T. Mallouk, *Acc. Chem. Res.*, 1992, **25**, 420.
221. A. M. Lazarin and C. Airoldi, *J. Inclusion Phen. Macrocyclic Chem.*, 2005, **51**, 33.
222. C. B. A. Lima and C. Airoldi, *Solid State Sci.*, 2002, **4**, 1321.
223. B. Zhang, D. M. Poojary, A. Clearfield and G. Peng, *Chem. Mater.*, 1996, **8**, 1333.
224. P. Gendraud, M. E. de Roy and J. P. Besse, *Inorg. Chem.*, 1996, **35**, 6108.
225. T. Kijima, S. Watanabe and M. Machida, *Inorg. Chem.*, 1994, **33**, 2586.
226. A. M. Lazarin and C. Airoldi, *J. Chem. Thermodyn.*, 2005, **37**, 243.
227. T. Kijima, K. Ohe, F. Sasaki, M. Yada and M. Machida, *Bull. Chem. Soc. Jpn.*, 1998, **71**, 141.
228. V. S. O. Ruiz and C. Airoldi, *Thermochim. Acta*, 2004, **420**, 73.
229. Y. Zhang, K. J. Scott and A. Clearfield, *J. Mater. Chem.*, 1995, **5**, 315.
230. C. B. A. Lima and C. Airoldi, *Thermochim. Acta*, 2003, **400**, 51.
231. C. B. A. Lima and C. Airoldi, *Int. J. Inorg. Mater.*, 2001, **3**, 907.
232. G. B. Hix and K. D. M. Harris, *J. Mater. Chem.*, 1998, **8**, 579.
233. F. Fredoueil, D. Massiot, P. Janvier, F. Gingl, M. B. Doeuff, M. Evain, A. Clearfield and B. Bujoli, *Inorg. Chem.*, 1999, **38**, 1831.
234. C. Airoldi and L. N. H. Arakaki, *J. Colloid Interf. Sci.*, 2002, **249**, 1.
235. R. F. de Farias and C. Airoldi, *Thermochim. Acta*, 2001, **378**, 113.
236. V. S. O. Ruiz, S. L. P. Dias, Y. Gushikem, R. E. Bruns and C. Airoldi, *J. Solid State Chem.*, 2004, **177**, 675.
237. D. M. Poojary and A. Clearfield, *J. Am. Chem. Soc.*, 1995, **117**, 11278.
238. G. Cao, V. M. Lynch and L. N. Yacullot, *Chem. Mater.*, 1993, **5**, 1000.
239. Z.-M. Sun, J.-G. Mao, B.-P. Yang and S.-M. Ying, *Solid State Sci.*, 2004, **6**, 295.
240. K. P. Rao and K. Vidyasagar, *Eur. J. Inorg. Chem.*, 2006, 813.
241. R. M. P. Colodrero, A. Cabeza, P. Olivera-Pastor, A. Infantes-Molina, E. Barouda, K. D. Demadis and M. A. G. Aranda, *Chem. Eur. J.*, 2009, **15**, 6612.
242. C. J. Doonan, D. J. Tranchemontagne, T. G. Glover, J. R. Hunt and O. M. Yaghi, *Nature Chem.*, 2010, **2**, 235.

CHAPTER 15

Metal Organophosphonate Proton Conductors

GEORGE K. H. SHIMIZU, JARED M. TAYLOR AND KARL W. DAWSON

Department of Chemistry, University of Calgary, Calgary, Alberta T2N 1N4, Canada

15.1 Introduction

Ion conductors are ubiquitous in modern and next-generation energy materials. Proton exchange membrane fuel cells require, as a pre-eminent challenge, more effective and efficient electrolyte membranes if hydrogen-powered vehicles are to become an affordable reality.[1] Better proton ionomers are also necessary for direct alcohol fuel cells.[2] With many alternative forms of energy generation being developed that do not provide on-demand supply (e.g. solar, wind), better energy storage materials are also urgently needed and, in this role, better Li ion battery materials are at the forefront.[3] For hybrid electric vehicles, global use of this green technology (depending on the source of the electricity) may exhaust lithium supplies for the batteries and so other ionomers (e.g. Na) would be needed. The scope of this chapter is potentially very broad and many possible topics for inclusion have been reviewed elsewhere.[4–9] As such, rather than focus on themes which have been well discussed in other works, this chapter will deal specifically with the opportunities and challenges of employing metal phosphonate structures as proton-conducting materials.

Within the themes mentioned above, the vast majority of research involving metal phosphonates as ionomers has focussed on proton conduction.

Metal Phosphonate Chemistry: From Synthesis to Applications
Edited by Abraham Clearfield and Konstantinos Demadis

Published by the Royal Society of Chemistry, www.rsc.org

The main, and most topical, application for proton ionomers is as low and intermediate temperature fuel cell membranes and so a brief overview of these challenges will be presented. This will be followed by a presentation of the *metal* phosphonate materials that have been examined as proton conductors. For the purposes of discussion, the materials will be categorized into: (1) proton conducting phosphonates where the metal phosphonate backbone acts a scaffolding for a secondary functionality (primarily hybrid inorganic–organic layered systems with pendant acidic groups); (2) metal phosphonate materials where the phosphonate is a part of the proton conduction pathway; (3) polymer composite materials of proton conducting metal phosphonates. The final sections will provide a comment on future prospects.

Regardless of the ion being transported there are a number of common desirable qualities for a conductor. Ion conductivity is proportional to the charge of the ion, the mobility of the ion and the number of charges.[10] Given a specific ion of interest, the charge of the ion is not a variable so performance can be improved by either increasing the number of charges and their mobility or both. With respect to ion mobility, a discussion of the possible mechanisms is merited. Broadly, ion mobility can be categorized as occurring either by diffusion of the ion or by transfer mediated by a carrier. Diffusion of ions in solids is most efficiently realized when there are ion-binding pockets of similar energy separated by a small activation barrier. For proton conduction, direct diffusion involves also transporting a solvation/hydration sphere and is typically a higher energy process than a carrier-mediated transfer, particularly when the carriers are water molecules. The H_2O-mediated proton transfer process involves water becoming hydronium, then transferring one of the other protons to the next water molecule in the carrier chain. Individual carriers must then reorient to accept additional protons. This mechanism, known as a Grotthuss transfer,[11] typically has activation energies of 0.1–0.4 eV. This hopping mechanism can occur along surfaces (for example, between sulfonic acid/sulfonate groups) or with amphiprotic carriers other than water. For example, the imidazole/imidazolium couple can also exhibit Grotthuss-like transfer[12] but obviously with a higher rotational energy barrier than water or through any combination of available proton donor/acceptor sites provided a reasonable activation barrier is present.

With a myriad of possible building blocks and manners of linking them into potentially conducting matrices, insight into the design of better ionomers at a molecular level is invaluable. This would logically stem from molecular insights into existing conductors through both experiment and computational modelling.[13] In this regard, crystalline materials offer the ability for structural determination by X-ray diffraction methods and direct visualization of potential ion transfer pathways. At the same time, materials with a more fluid and dynamic structure would intuitively seem better suited to enabling transport phenomena in the solid state. Thus, a dichotomy in the generation of ion conductors is that some of the most desirable features in a material (crystallinity and dynamics) are seemingly in conflict. This point can be further extended to the concept of preorganization of an ion transfer pathway, where a perfectly arranged network of acceptors and donors is clearly favored, versus the adaptability of a material where *facile* motion of components would augment transport. The ability to design binding sites of similar energy, continuously linked in multiple dimensions via low energy pathways, is fundamental to the generation of better ion conductors.

15.2 Proton Conducting Metal Phosphonates

The most attractive target for a new proton conductor would be to address the challenges of improving the efficiency of proton exchange membrane fuel cells (PEMFCs). As such, as a prelude to discussion on proton conducting phosphonates, a brief introduction on challenges for PEMFCs will be presented. One of the major obstacles to widespread commercialization of hydrogen fuel cell powered vehicles is the cost efficiency and one of the major hurdles to overcome to address the efficiency is to improve the durability and performance of the proton conducting electrolyte membrane. Core properties of a proton electrolyte would be conductivity in or above the range of 0.1 S cm^{-1}, electronically insulating, and a structure that would minimize fuel crossover between the half-cells. Requirements for fuel cell systems for automotive applications[14] have driven the need for a high-performance membrane that operates under conditions of low relative humidity and temperatures exceeding the boiling point of water. The benefits of higher temperature include better kinetics of the electrode reactions, diminished CO poisoning of Pt catalysts (where the CO would be present in hydrogen from its reforming preparation), easier management of water as only steam would be present above 100 °C, and possibly better compatibility with the hydrogen storage material. These conditions put a severe dependence on hydration of the fuel cell membrane electrode assembly (MEA), limiting conductivity in conventional membrane materials. Membranes must also have high mechanical strength. In macromolecular membranes, some swelling or contraction with variation in hydration is typical but this can often cause the contact with the electrode to be compromised. The durability of the material is obviously critical as well and many existing materials rely upon a perfluoro polymer backbone to provide chemical robustness. Performance requirements for the membrane would span the full operating range of temperature from start-up (sub-zero temperatures) to full power (up to 120–130 °C) and over lifetimes of 5000 h for automotive applications. The "conventional" polymers that have been studied for PEMFCs are Nafion-like perfluorosulfonic acid (PFSA) polymers which, as mentioned, possess the fluoro backbone for mechanical strength and pendant sulfonic acids groups which segregate the material into hydrophobic and hydrophilic domains. More recent work has concerned alternative polymers such as acid-doped polybenzimidazole,[15] a system capable of proton conduction above 150 °C. Other recent developments have concerned radiation-induced graft polymerization on fluorinated aliphatic polymers,[16] and more easily prepared sulfonated polymers, in particular polysulfones,[17] polyetherketones,[18] and polyimides.[19]

Another general consideration for proton conducting materials is the extent to which the solid is ordered. While order is not a requirement of the membrane (indeed many may argue that order would more than likely translate to a static structure and hence disfavor ion conduction), order of a solid would permit X-ray diffraction studies and provide much greater insight to proton conduction pathways. This in turn would enable establishment of structure/activity relationships and ultimately better design of materials. Design can be achieved both through chemical intuition and by using the structural models as firm handholds to calibrate or validate computational methods.[20] Regarding order and dynamics, these two points would not intuitively be highly compatible; that said, they are also not mutually exclusive in network solids.[21–24] Efficient ion conduction is enabled by domains of similar affinity for given ions separated by a small activation barrier; this point is not at all in conflict with crystallinity. Given that, for protons, the most efficient transfer involves the Grotthuss hopping mechanism, a *translationally* static structure optimally arranged would be a highly desirable conducting material provided it allowed for sufficient *rotational* movement of components in place, again a feature not necessarily in conflict with long range order. An excellent illustration of this would be to compare models of conduction for macromolecular proton conductors with the enhanced precision of models for ion conduction in ordered oxides.[25] While this may appear as a comparison between two disparate species, the domain of metal phosphonates transcends this gap, albeit with more similarities to the solid acids. In $CsHSO_4$, CsH_2PO_4 and related salts, the oxoanion can exhibit dynamics in the solid state at elevated temperature.[26] This rotational motion augments proton mobility and is obviously enabled by the spherical shape of the anion and the coordinative flexibility of the Cs ion. The crystallinity of salts such as $CsHSO_4$ has led to much more precise models for proton conduction than exist for macromolecular ion conductors.

This leads to a discussion of the advantages of metal phosphonates as potential proton conducting materials. Metal phosphonates offer a number of intriguing possibilities that are discussed below. Any single material could not incorporate all the features but the intent is to outline prospects and opportunities. The first point is that the organophosphonate group can coordinate to metal ions as a hydrogen phosphonate.[27] This is a key feature as network structures can be prepared while still offering acidic sites for proton exchange. A metal organophosphonate could show similar properties to $CsHSO_4$ with respect to ion dynamics, enhancing conduction, but to a diminished degree owing to the fact that rotation of the oxo anion component in the phosphonate would have fewer rotational degrees of freedom. The second point is that the organophosphonate component of a metal phosphonate can be readily varied with respect to its size, shape and denticity, offering structural variation well beyond hydrogen sulfate or phosphate salts.[28,29] Beyond size and shape, given some degree of structural pliancy is potentially beneficial to ion conduction; the flexibility of the linker can also be varied.[30] The third is that metal phosphonates are largely crystalline materials and so X-ray diffraction can provide meaningful insights into structure and hence activity. Activation energies can be rationalized based on the distribution of H-bond donors and acceptors, diffusion pathways for transport, dimensionality of proton transfer pathways, *etc*. A fourth point is that the metal phosphonate structure can be made porous.[31–36] The potential benefit of this is that the structure could be loaded with guest molecules that could act as proton carriers. Water is an obvious candidate but, given the interest in higher temperature proton conductors, loading pores with amphiprotic heterocycles such as imidazole would be an attractive prospect.[37,38] This is somewhat of a double edged sword in that porosity of an electrolyte membrane would facilitate fuel crossover. It has been shown though that in metal organic framework materials, heterocycle-loaded micropores can be made gas tight.[37] A fifth point is that some metal organophosphonate backbones are chemically highly robust (e.g. $Zr(RPO_3)_2$) and can be modified via sulfonation reactions (see below) to yield proton conductors where the metal phosphonate serves as a skeleton upon which to anchor the active proton hopping sites. Again, the regular backbone serves to position proton hopping sites at defined intervals in space. This final option actually encompasses the large majority of research done on proton conducting metal phosphonates.

The remainder of this chapter will be broken into three sections concerning proton conduction in metal phosphonates: (i) proton conductors where the metal phosphonate serves as a scaffolding for actual proton conducting sites; (ii) proton conductors where the phosphonate group itself is an integral component of the proton transfer pathway; (iii) polymer composites of metal phosphonate proton conductors.

15.3 Metal Phosphonates as Scaffolds for Proton Conducting Sites

By far the largest body of work on proton conducting metal phosphonates has concerned robust metal phosphonate architectures which either contain, or are modified post-synthetically to incorporate, proton conducting sites. A survey of the published works follows. To provide a historical perspective of the development of the field, we have chosen to present the work in this section largely chronologically.

Zr *m*-sulfophenylphosphonate (m-SPP) has been widely incorporated into proton conducting materials. In this chapter, unless stated otherwise, the use of sulfophenyl phosphonate or $O_3PC_6H_4SO_3H$ can be assumed to refer to the sulfonic acid group being in the 3-position, meta to the phosphonate. In 1992, Alberti *et al.*[39] prepared a mixed zirconium sulfophenylphosphonate and hydroxymethyl phosphonate (HMP) material for proton conductivity. $Zr(O_3P\text{-}C_6H_4\text{-}SO_3H)_{0.73}(O_3P\text{-}CH_2\text{-}OH)_{1.27}\cdot nH_2O$ was prepared by mixing an aqueous solution of the phosphonic acids with $ZrOCl_2$, heating at 80 °C for 4 days then direct precipitation in the presence of HF. The sample was characterized by

TGA, ^{31}P NMR (after dissolving in HF), and PXRD. Ligand ratios were determined by NMR and the sample was observed to decompose past 200 °C. The material was a layered solid with individual layers resembling the alpha zirconium phosphate phase by assumption that a pure Zr phosphonate cannot form the gamma phase structure. Based on the d-spacing in the first peak in the PXRD, the structure was regarded as one with sulfonate groups opposite hydroxyl groups in adjacent layer (this is for dehydrated material). The solid absorbed significant quantities of water with varying relative humidity, with $n=2.4$ at 22% RH, to $n=7.5$ at 90% RH. The PXRD showed d-spacing increases with increasing water as expected. Impedance analysis was done on pressed pellets with Pt black electrodes with the pellets equilibrated for 3 days at set humidity. All measurements were taken by decreasing temperature down to –20 °C from room temperature to avoid dehydration. For the mixed ligand material, conductivity was measured as 1.6×10^{-2} S cm^{-1} at 90% RH, and room temperature. This is an excellent value for proton conduction albeit at low temperature and high humidity. This conductivity value drops to 10^{-5} S cm^{-1} at 20% RH. Arrhenius behavior was observed with a low activation energy (21 kJ mol^{-1}) at 90% RH and 62 kJ mol^{-1} at 20% RH. For comparison, pure $Zr(O_3P\text{-}CH_2\text{-}OH)_2$ was measured, and the conductivity was a very low 10^{-8} S cm^{-1}. EMF measurements on the material, using a half blocking electrode configuration, were performed and confirmed that the observed charge mobility was purely protonic in nature.

A follow-up study from Alberti *et al.*[40] concerned two other mixed phosphonate zirconium systems, $Zr(m\text{-}O_3P\text{-}C_6H_4\text{-}SO_3H)_x(O_3PR)_y \cdot nH_2O$, where $R=CH_2OH$ (in a different ratio) or C_2H_5. Based on the PXRD patterns, both solids were again assumed to be the alpha Zr phosphate structure and also assumed to be interdigitated. From the ratio of phosphorus peaks in the ^{31}P-NMR, the ratios of the two phosphonate groups present, x and y, were found to be $x=0.85$ and $y=1.15$ for $R=C_2H_5$ and $x=0.97$ and $y=1.03$ for $R=CH_2OH$. Water content for each material was examined by TGA, and interlayer distance measured through PXRD. Proton conductivity was measured by AC impedance analysis in the presence and absence of water in the temperature range 100–200 °C (Figure 15.1). For the $R=C_2H_5$ material, anhydrous conductivity was measured between 100 °C and 180 °C and Arrhenius behavior was observed. Conductivity was measured as 1.2×10^{-5} S cm^{-1} at 180 °C with an activation energy of 62 kJ mol^{-1}. When the ethylphosphonate derivative was held above 200 °C, conductivity was slowly lost, due to decomposition of the organic portion (likely desulfonation). AC/DC measurements were performed to determine if there was any electronic conductivity and showed this contribution was negligible. The conductivity of pure $Zr(O_3PCH_2OH)_2$ was measured to be 10^{-9} S cm^{-1} at 140 °C, while the conductivitiy of $Zr(O_3PCH_2CH_3)_2$ was so low it could not be measured, affirming that the observed conductivity was integrally connected to the SO_3H group. The stability of the $R=CH_2OH$ derivative was increased by the presence of water in the atmosphere at high temperature. The decomposition pathway was thought to be through reaction between the SO_3 and OH groups forming water

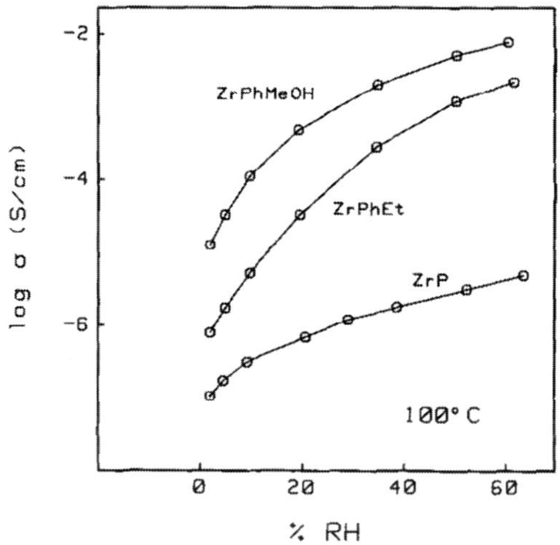

Figure 15.1 Log of conductivity as a function of humidity for ZrPhMeOH, ZrPhEt and alpha-Zr phosphate (ZrP). (Reprinted from G. Alberti *et al.*, Protonic conductivity of layered zirconium phosphonates containing -SO_3H groups. II. Ac conductivity of zirconium alkyl-sulphophenyl phosphates in the range 100–200 °C, in the presence or absence of water vapour, *Solid State Ionics*, 1992, **58**, 339, with permission from Elsevier.)

and S–O–C bridges so it would be suppressed by higher hydration. Conductivity of $R = CH_2OH$ can be restored by treatment under 15% RH at 140 °C for 9 days. Anhydrous conductivity of the $R = CH_2OH$ form was not actually reported, but an Arrhenius plot was given which showed slightly lower conductivity than the $R = C_2H_5$ analogue. In hydrated conditions, the conductivity of the $R = CH_2OH$ was about an order of magnitude higher than the $R = C_2H_5$ which is quite intuitive given the polar, proton accepting hydroxyl groups. Water content of both phases was roughly the same at any given level of relative humidity and temperature.

During roughly this same time period, other potential proton transferring groups, beyond sulfonic acids, were being appended to the Zr phosphonate framework. $Zr(O_3P\text{-}(CH_2)_n\text{-}COOH)_2$ ($n = 1$–5) (ZrPCn) were prepared and studied for proton conductivity also by Alberti *et al.*[41] $ZrOCl_2$ was employed as the zirconium source with the same reaction conditions as for the mixed phosphonate above. Composition was determined by elemental analysis and TGA. PXRD confirmed the expected layered motif, that of the alpha $Zr(PO_4)$ structure, and the layer spacing. The material, given the rather polar interlayer region, was isolated as an anhydrous solid, with no weight loss on TGA until decomposition at 300 °C. Depending on the length of the carbon chain in the carboxylic acid (*n* even or odd), a variation in hydrogen bonding between the layers causes the spacing to vary (Figure 15.2). Also a reversible phase transition is noted in the ZrPC2 and C4, attributed to a breaking hydrogen bond. Conductivity was measured to be low in general, but higher for odd *n* values.

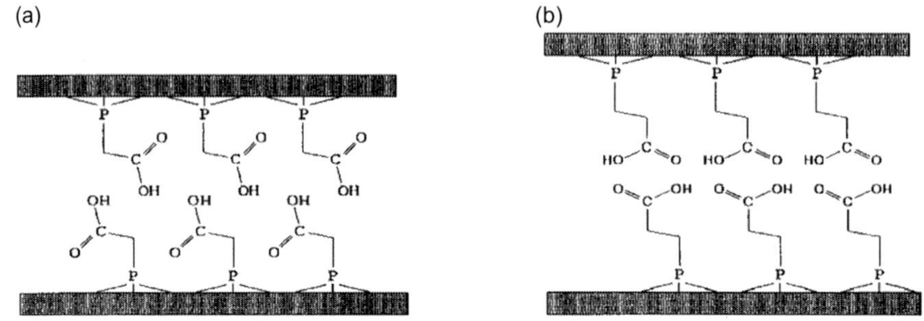

Figure 15.2 Postulated arrangements of the carboxyalkyl chains in the interlayer region of $Zr(O_3P\text{-}(CH_2)_n\text{-}COOH)_2$ ($n = 1$, 2) showing the even/odd dependence. (Reprinted from G. Alberti *et al.*, Proton conductivity of zirconium carboxy n-alkyl phosphonates with an α-layered structure, *Solid State Ionics*, 1991, **46**, 61, with permission from Elsevier.)

The specific values ranged from $\sim 10^{-10}$ S cm^{-1} for even values of n and 10^{-9} to 10^{-7} S cm^{-1} for odd values of n. The activation energy values that were measured ranged from moderate to high being 77 kJ mol^{-1} for $n = 1$, 104 kJ mol^{-1} for $n = 2$ and 132 kJ mol^{-1} for $n = 3$.

Amine groups were also appended to the Zr phosphonate framework. In 1995, Casciola *et al.*[42] reported layered Zr alkylphosphonates with amine groups in the omega terminus of the alkyl chain. These layered solids were examined with intercalated n-alkyl carboxylic acids for high temperature proton conduction. The layered Zr phosphonate with the pendant amine was prepared as for the other systems already discussed. The amine was formed as a hydrogen chloride salt, $Zr(O_3P(CH_2)_2NH_3Cl)_2$ (ZrAPHCl), then deprotonated with 0.1 M NaOH solution to form $Zr(O_3P(CH_2)_2NH_2)_2$, ZrAP. Intercalation was by mixing solid ZrAP with an aqueous solution of the carboxylic acid. Interlayer distance measured by PXRD, and again assumed to be the layered alpha Zr phosphate phase. The unit cell was extracted from the powder pattern and the density was predicted based on a predicted structure. The composition and stability of the network were assessed with TGA, with stability only up to 200 °C for the ZrAP, and up to 400 °C for hydrochloride salt, ZrAPHCl. Intercalation was done in 0.1 M solutions of the carboxylic acids with the family of acids from formic to heptanoic acid studied. The formulae were found to be $Zr(O_3P(CH_2)_2NH_2)_2 \cdot xRCOOH \cdot nH_2O$, where x ranged from 1.6 to 1.8. As expected, intercalation increased interlayer distance and it did so linearly with chain length. ZrAPHCl showed no conductivity, but ZrAP showed 2×10^{-6} S cm^{-1} at 200 °C in a pure form, with an activation energy of 20 kcal mol^{-1}. It was postulated that the proton conductivity in ZrAP may have been due to residual HCl, allowing jumps from protonated to non-protonated amino groups whereas the hydrochloride salt would have been saturated. A reasonable possibility in this work was that the conductivity could have been due to Cl^- ion conduction, but no measurements were performed to (in)validate this scenario. Impedance analyses were not performed on any of the carboxylic acid intercalated species.

In 1996, Stein *et al.*[43] reported the proton conductivity of Zr/Ti sulfophenylphosphonates and, in a very different approach, Zr/Ti sulfophenylphosphonates interstratified with Zr/Th pyridyl phosphonates (Figure 15.3). The sulfophenylphosphonate components of these materials were first formed from the phenyl phosphonic acid, and then post synthetically sulfonated using fuming sulfuric acid. Ti salts were more resistant to sulfonation, so reaction conditions were slightly harsher. Ultimately, the formulae for the sulfonated complexes were, for Zr: $Zr(O_3PC_6H_4SO_3H)_2 \cdot 3.6H_2O$; for Ti: $Ti(O_3POH)_{0.25}(O_3PC_6H_5)_{0.12}(O_3PC_6H_4SO_3H)_{1.63} \cdot 3.64H_2O$; these were confirmed by TGA and elemental analyses and corroborated evidence from PXRD, ^{31}P-MAS-NMR and IR spectra. Proton conductivity of the pure Zr sulfophenylphosphonate ranged from 10^{-6} to 10^{-2} S cm^{-1} at relative humidities of 20–85% at 5 °C. Notably, the activation energy was 0.28 eV (27.0 kJ mol^{-1}) which is in the range for typical Grotthuss transfer of protons. The Ti complexes gave conductivities of from 10^{-6} to 10^{-1} S cm^{-1} over the same range.

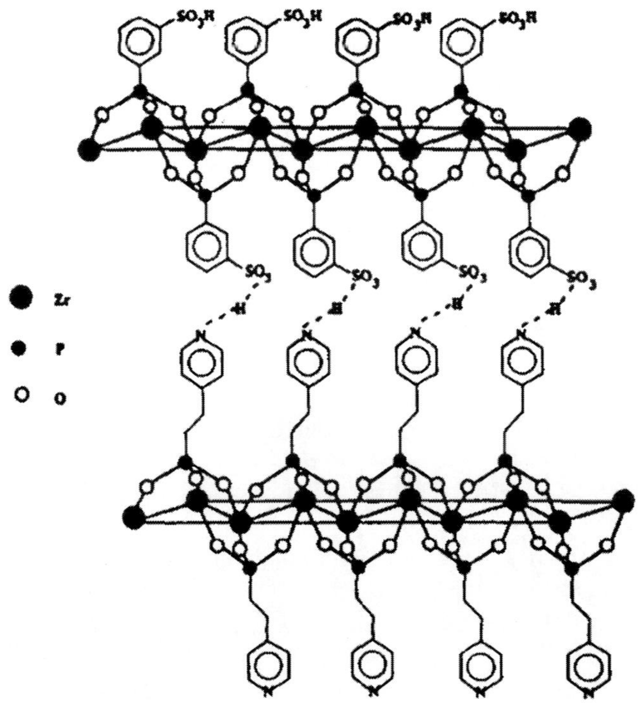

Figure 15.3 Schematic representation of the inter-stratified pyridine-sulfophenyl phosphonate materials formed from exfoliated layers of pure compounds. (Reprinted from E.W. Stein *et al.*, Conductivity of group IV metal sulfophosphonates and a new class of interstratified metal amine-sulfophosphonates, *Solid State Ionics*, 1996, **83**, 113, with permission from Elsevier.)

It was also reported that Zr phosphonates could be prepared mixed with phosphate by increasing the SO_3 percentage in the fuming sulfuric acid during sulfonation. The conductivity of materials with the same metal, but with increasing phosphate content, was found to be lower. This finding may not necessarily be a general phenomenon as there would be an ideal placement of proton donors and acceptors. An interesting aspect of this study was the mixing of different lamellae that were each able to exfoliate (separation of layers in some solvent to release individual sheets). The sulfonated materials were observed to spontaneously exfoliate in water. Coupling this with the observation that metal pyridyl phosphonates also exfoliated when protonated offered the potential to make metal phosphonates with layers of differing composition and chemistry, in this case, proton accepting properties. Mixing dispersions of the sulfonated phosphonates with the pyridylphosphonates caused spontaneous precipitation. The mixed phase composition was confirmed by ^{31}P-MAS-NMR and elemental analyses indicated a ratio of S/N of 1.02 (*i.e.* close to the ideal alternating arrangement as shown in Figure 15.3). The conductivity of the mixed phases was lower than the pure sulfophenyl derivates across the full range of analysis giving values from 10^{-6} to 10^{-1} S cm^{-1}. This result is not surprising in that the interstratified complexes would have stronger acids interspersed with more basic sites. The net result would be that the protons would be more localized on the N sites which would lead to lower proton conductivity. Nonetheless, the illustration that a mixed species could be prepared was very significant. A mixed phase zirconium phosphonate polyimine, $(Zr(O_3PCH_2NHCH_2CH_2N_2)_2$, was also prepared then interstratified with Zr sulfophenylphosphonate. Variable temperature conductivity was performed (Figure 15.4) at 30% RH. This study showed that the activation energy increased to 0.41 eV in the interstratified phase, compared to pure Zr sulfophenylphosphonate (0.28 eV). Again, the addition of more basic sites was likely localizing the protons.

In 1996, Alberti *et al.*[44] reported a study on a series of sulfonated Zr layered networks. The actual organic pendant was varied in this study, as shown in Figure 15.5. A methylene spacer was added between the P atom and the arene, which would diminish the acidity of the sulfonic acid, while fluorine substituents were added to the aryl ring to simultaneously increase the acidity of the SO_3 group.

These phosphonates were inserted into gamma Zr phosphate through a topotactic reaction, forming $Zr(PO_4)(F\cdot H_2O)_{0.2}(O_2P(OH)_2)_{0.8-x}(O_2P(OH)R)_x\cdot nH_2O$ from $Zr(PO_4)(F\cdot H_2O)_{0.2}(O_2P(OH)_2)_{0.8}\cdot 2H_2O$. The fluoride is present from the synthesis of the zirconium phosphate as ammonium fluoride is needed in high concentration. All compounds were characterized by TGA, PXRD, ^{31}P NMR (run in 3 M HF), ion chromatography for Cl^- content, and gas sorption. Crystallinity of the materials was observed to decrease after the topotactic reaction, indicating exchange rather than recrystallization. As well as the first peak in the PXRD occurred at lower 2-theta values consistent with the larger pillar. Interlayer distance depended on water content as expected. Figure 15.6 shows the structural model for the mixed phosphate/phosphonate species.

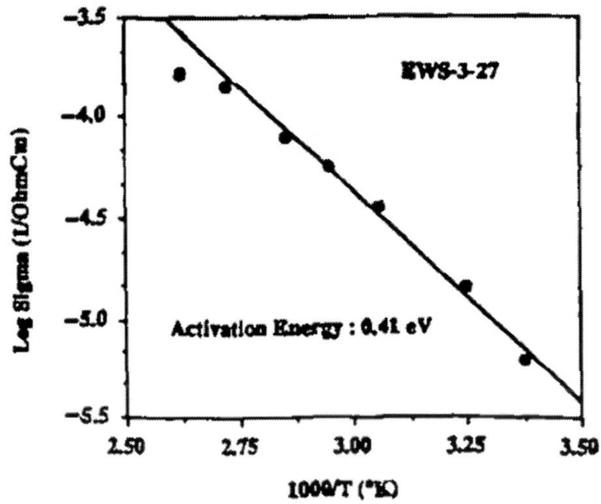

Figure 15.4 Plot of log of conductivity versus $1000/T$ for the inter-stratified complex with $Zr(O_3PCH_2NHCH_2CH_2N_2)_2$. (Reprinted from E.W. Stein *et al.*, Conductivity of group IV metal sulfophosphonates and a new class of interstratified metal amine-sulfophosphonates, *Solid State Ionics*, 1996, **83**, 113, with permission from Elsevier.)

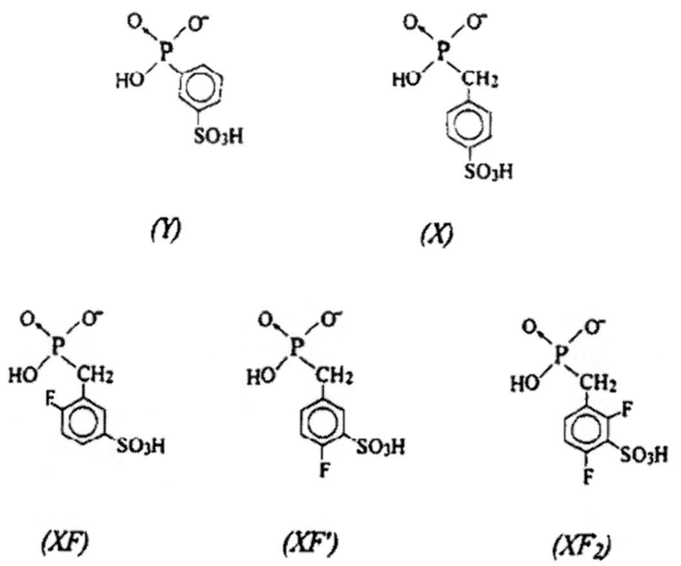

Figure 15.5 The different anions and their abbreviation in brackets for a study varying sulfonic acid acidity in a family of gamma-Zr sulfoarylphosphonate/phosphates. (Reprinted from G. Alberti *et al.*, Protonic conductivity of layered zirconium phosphonates containing $-SO_3H$ groups. III. Preparation and characterization of γ-zirconium sulfoaryl phosphonates, *Solid State Ionics*, 1996, **84**, 97, with permission from Elsevier.)

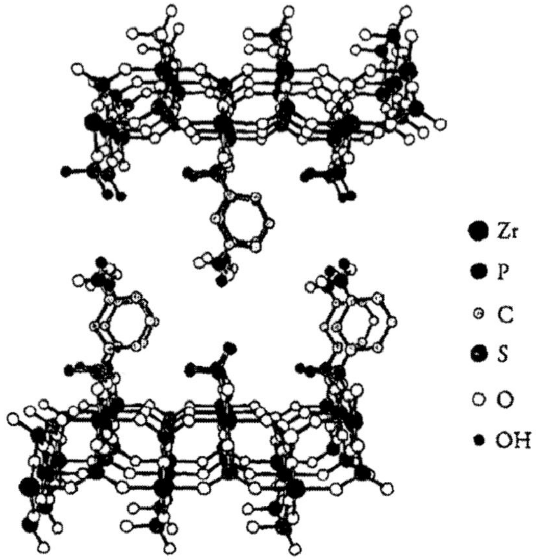

Figure 15.6 Model for the gamma-Zr m-sulfophenylphosphonate/phosphate complex. (Reprinted from G. Alberti *et al.*, Protonic conductivity of layered zirconium phosphonates containing $-SO_3H$ groups. III. Preparation and characterization of γ-zirconium sulfoaryl phosphonates, *Solid State Ionics*, 1996, **84**, 97, with permission from Elsevier.)

Conductivity was highly dependent on humidity (Figure 15.7). It was found that the unaltered meta-sulfophenyl phosphonic acid was still the best conductor, as the phosphonate acted as a strong electron withdrawing group, more so than even the difluoroarylsulfonate as estimated by Hammett indicators. High conductivity, confirmed as arising solely from protons, of 0.01 S cm^{-1} was measured at 90% RH for meta-sulfophenyl phosphonic acid at 20 °C, with a very low activation energy of 4–9 kJ mol^{-1}. Conductivity is still lower than the alpha Zr phosphonate phase by 1.5–2 times. The authors also used the measured surface area (2 m^2 g^{-1}) to determine that surface conductivity was roughly 3.5×10^{-3} S cm^{-1}, confirming that bulk transport was more significant than surface transport in these systems.

Jaimez, Hix and Slade reported[45] in 1997, the synthesis, characterization, porosity and H^+ conductivity of Ti^{4+} salt of *N*,*N*-(diphosphonomethyl)glycine. The phosphonate was prepared by reaction between glycine, phosphorus trichloride and paraformaldehyde in water. The Ti salts were prepared by reaction of the phosphonic acid with $TiCl_3$ in water, where the Ti^{3+} was slowly air oxidized to Ti^{4+}; this was done to obtain a measure of kinetic control and obtain a more crystalline sample. The network was studied by TGA/DTA, EA, IR, ^1H and ^{31}P solution NMR, ^{31}P solid state NMR (SS NMR), PXRD, N_2 adsorption and impedance analysis. PXRD showed that the material was amorphous but, based on the other data, the authors suggested that the compound contains one phosphonate that is binding to the Ti, and another "free"

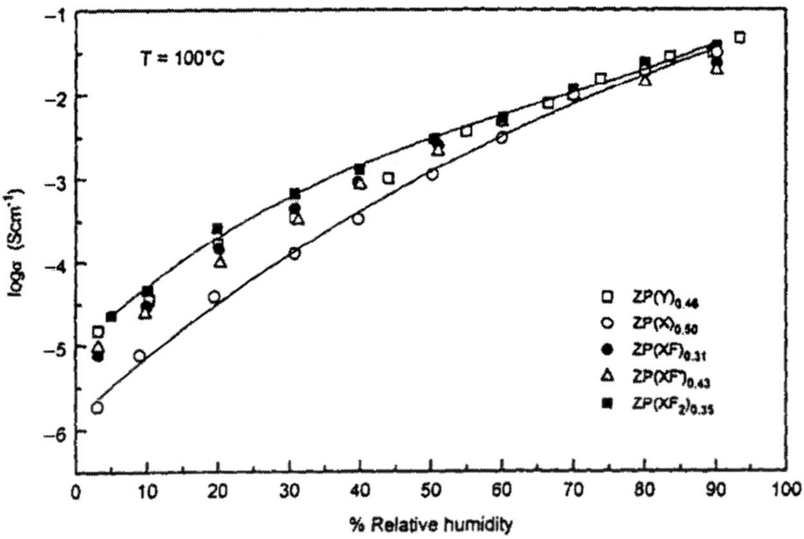

Figure 15.7 Log of conductivity as a function of relative humidity for the gamma-Zr phosphonate/phosphates noted in Figure 5. (Reprinted from G. Alberti *et al.*, Protonic conductivity of layered zirconium phosphonates containing $-SO_3H$ groups. III. Preparation and characterization of γ-zirconium sulfoaryl phosphonates, *Solid State Ionics*, 1996, **84**, 97, with permission from Elsevier.)

phosphonate, with some of the "free" phosphonate bridging Ti centers (ratio of 1.4:1 for Ti coordinated to free phosphonate) by SS NMR. The formula determined by TGA/DTA was $Ti[HO_2CCH_2N(CH_2PO_3)_2]_{0.28}[HO_2CCH_2N(CH_2PO_3H_2)CH_2PO_3]_{1.43} \cdot 0.6H_2O$ for a sample dried at 50 °C, which agreed with NMR results. The surface area was determined to be 119 $m^2\ g^{-1}$ by N_2 sorption indicating a significant degree of porosity to the structure. DC conductivity indicated purely ionic conduction and AC impedance indicated that the conductivity was highly humidity dependent. Arrhenius plots were taken for conductivity at room temperature and at zero (10^{-7} S cm^{-1}) and 100% relative humidity (3.7×10^{-4} S cm^{-1}). No measurable conductivity was observed over 130 °C.

Jaimez, Hix and Slade[46] also reported a titanium(IV) phosphonomethyliminodiacetic acid and its proton conductivity. The phosphonate group employed in this study was $H_2O_3PCH_2N(CH_2CO_2H)_2$ (PMIDA) and the Ti complex was stated as being layered and similar, though not identical, to gamma zirconium phosphate. The PXRD analysis showed an amorphous compound but ^{31}P SS NMR, IR, TGA, elemental analysis and N_2 sorption were employed to assign a formula of $[Ti_2(PO_4)\{O_3PCH_2N(CH_2CO_2H)_2\}\{O_3PCH_2N(CH_2CO_2H)(CH_2CO_2)\}(H_2O)] \cdot nH_2O$, thus a mixed carboxylic acid/carboxylate. The BET surface area of the solid was 28 $m^2\ g^{-1}$, which would indicate largely a densely packed material. Proton conductivity in this system was measured at different temperatures varying relative humidity from

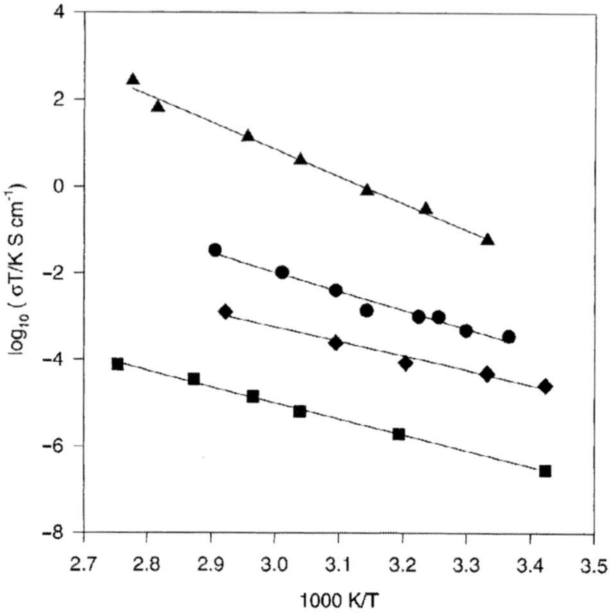

Figure 15.8 Arrhenius plot for conductivity in Ti(PMIDA) at, top to bottom, 100, 60, 40, 0% RH. (Reprinted from E. Jaimez *et al.*, A phosphate–phosphonate of titanium(IV) prepared from phosphonomethyliminodiacetic acid: characterisation, n-alkylamine intercalation and proton conductivity, *Solid State Ionics*, 1997, **97**, 195, with permission from Elsevier.)

0, 40, 60 to 100% (Figure 15.8). The highest conductivity measured was 8×10^{-1} S cm^{-1} at 100% RH and 360 K.

In an application of isomorphous replacement to fine tune properties of a solid material, Viviani and co-workers reported,[47] in 1998, a study of a series of barium salts (and doped derivatives) with 3,5-disulfophenyl phosphonic acid (H_4Y). The authors prepared four different salts, $Ba_2Y \cdot H_2O$, $Ba_{1.8}H_{0.4}Y \cdot 1.2H_2O$, $BaKHY \cdot 0.5H_2O$ and BaH_2Y by mixing the H_4Y, $Ba(OH)_2$ and KOH in water, then evaporating to dryness. Formulae were determined from initial reagent ratios and agreed with ultimate analyses based upon TGA, density, ^{1}H and ^{31}P NMR used for characterization. PXRD of Ba_2Y, $Ba_{1.8}H_{0.4}Y$ and BaKHY gave similar unit cells after indexing, showing isomorphous replacement as a means of tuning the acidity. Conductivity measurements at 100 °C and 70% RH were performed, with Ba_2Y showing less than 10^{-8} S cm^{-1} under these conditions. Impressively, partial replacement with H^+ and K^+ made the conductivity value jump to 10^{-2} S cm^{-1}. The order of increasing conductivity in the series was $Ba_2Y < Ba_{1.8}H_{0.4}Y < BaKHY < BaH_2Y$ with the best result being that BaH_2Y showed 0.12 S cm^{-1} at 80% RH and 100 °C. At 20 °C, BaH_2Y showed a jump from 2×10^{-4} S cm^{-1} to 3×10^{-3} S cm^{-1} from 53–75% RH, and increasing water content from 2 to 3 molecules per formula unit. $H_2(p_1,RH_1)$, Pt|$Ba_{(4-k-h)/2}K_kH_hY$|a-$Zr(HPO_4)_2 \cdot H_2O$|Pt, $H_2(p_2,RH_2)$ was used as an electrochemical cell in EMF measurements to test for the nature of the proton conduction and it was found that BaKHY was a mixed K^+/H^+ conductor. IR spectroscopy was used to determine if the H^+ was on the phosphonate, or occurring as H_3O^+, and in BaKHY it appeared to be on the phosphonate. However, in BaH_2Y, it may have been as hydronium, but the analysis was inconclusive due to overlap with other IR regions.

In 2001, Alberti *et al.* reported[48] a Ti mixed phosphate/sulfophenylphosphonate of formula, $Ti(HPO_4)_1(O_3PC_6H_4SO_3H)_{0.85}(OH)_{0.3} \cdot nH_2O$, (TiP(SPP)). This material showed very high proton conductivity data but was not nearly as robust as some other layered M^{IV} phosphonates as TGA showed the material was only stable to 200 °C. ^{31}P SS NMR gave two resonances indicating only one environment for PO_4H and one for the phosphonate. There was also a weak resonance at 0 ppm, indicating presence of some free H_3PO_4, which grew with increasing % RH, indicating some phosphate hydrolysis. However, at 90% RH, H_3PO_4 made up only 0.2% of the total phosphorus. Conductivity versus number of water per sulfonate is shown in Figure 15.9. Between 5–90% RH, conductivity rose by three orders of magnitude to ~0.1 S cm^{-1} with the number of water molecules per sulfonic acid increasing from 2 to 11.

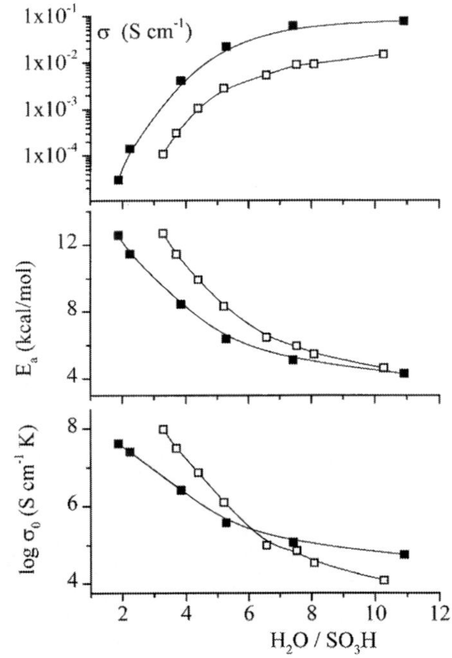

Figure 15.9 Conductivity, activation energy and log of conductivity at 20 °C as a function of hydration number in TiP(SPP), filled squares, and $Zr(O_3PC_6H_4SO_3H)_{0.73}(O_3PCH_2OH)_{1.27} \cdot nH_2O$, empty squares. (Reprinted from G. Alberti *et al.*, Preparation, characterization and proton conductivity of titanium phosphate sulfophenylphosphonate, *Solid State Ionics*, 2001, **145**, 249, with permission from Elsevier.)

Conductivity values higher than 0.01 S cm^{-1} were obtained for % RH greater than 50% (> 6 water per sulfonic acid). Conductivity is an order of magnitude higher than the corresponding zirconium phosphonate, and the authors suggested that the presence of H_3PO_4 might have increased the carrier concentration and hence the conductivity in this material compared to the Zr phase. An outstanding value for conductivity of 0.25 S cm^{-1} was obtained at 100 °C and 90% RH. This manuscript also dealt with hybrid membranes with Nafion and this will be discussed in the composite section of this chapter.

In 2003, Jang *et al.* reported[49] a cerium(IV) sulfophenylphosphonate where the layer structure was post-synthetically sulfonated using 30% fuming H_2SO_4. C/H/N/S elemental analysis values agreed with the formulation $Ce(O_3PC_6H_4SO_3H)_2$. PXRD showed a layered structure that was fairly crystalline, and was isomorphous with the Zr analogue. Water content, as synthesized, was measured at 3.86 mol equivalents and this varied with relative humidity. TGA/DTA showed that this solid was not stable above 200 °C. AC impedance analysis showed very promising results for conductivity. Proton conduction was observed to be humidity dependent and increased by 1 order of magnitude between 40 to 100% RH. At 100 % RH and 100 °C, conductivity values in excess of 10^{-2} S cm^{-1} were obtained. Figure 15.10 shows the plots of conductivity versus relative humidity at different temperatures. The results even at 40% RH at elevated temperatures are very good. Conductivity decreases past 175 °C probably because of decomposition of the sulfonate group but this is above the target operating range for PEMFCs. The activation energy was measured for the temperature range of 100–175 °C, with a relative water vapor pressure of 1, and E_a was 0.21 eV, consistent with a Grotthuss mechanism.

In 2010, Zima *et al.* reported[50] Zr sulfophenylphosphonates where the sulfonation was in the 4-position, para to the phosphonate group. Synthesis of the ligand was slightly more complicated and is shown in Figure 15.11. EDX analysis on the Zr complex gave a Zr:P ratio of 1:2, consistent with a formula of

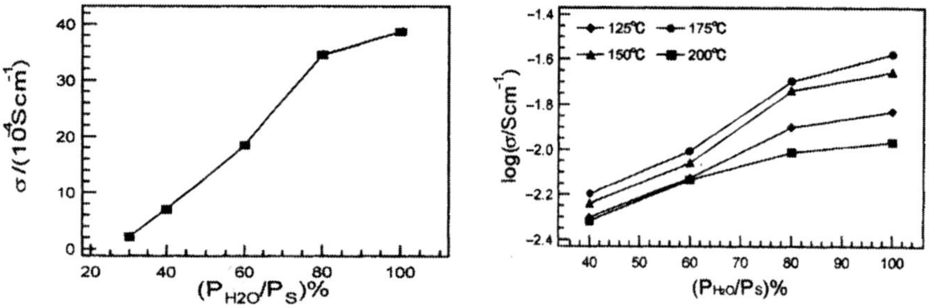

Figure 15.10 Left: plot of conductivity versus relative humidity for $Ce(O_3PC_6H_4SO_3H)_2 \cdot 3.86H_2O$ at 25 °C. Right: plot of log of conductivity versus relative humidity for $Ce(O_3PC_6H_4SO_3H)_2 \cdot 3.86H_2O$ from 125 to 200 °C. (Reprinted from M. Y. Jang *et al.*, Electrochemistry, 2003, **71**, 691, with permission from The Electrochemical Society of Japan.)

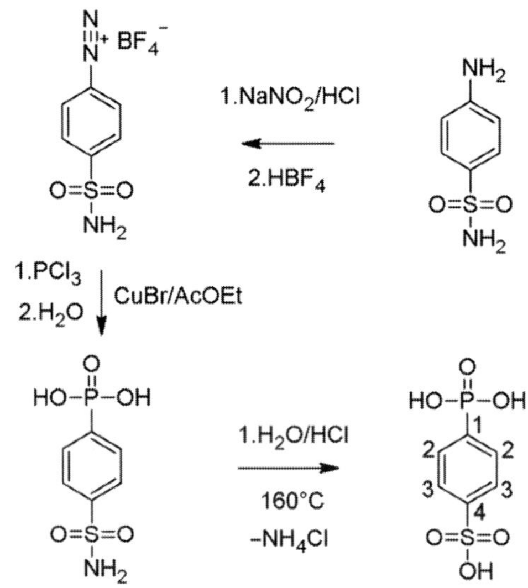

Figure 15.11 Synthesis of 4-sulfophenylphosphonic acid. (Reprinted from, V. Zima *et al.*, Synthesis and characterization of new zirconium 4-sulfophenylphosphonates, *Solid State Ionics*, 2010, **181**, 705, with permission from Elsevier.)

$Zr(HO_3SC_6H_4PO_3)_2 \cdot 2H_2O$, called ZrL_2, where the hydration number was determined by TGA. A mixed phosphate/phosphonate phase of formula $Zr(HPO_4)_{0.7}(HO_3SC_6H_4PO_3)_{1.3} \cdot 2H_2O$ (ZrPL) was also prepared in this paper by adding phosphoric acid to the reaction mixture. Water content in both solids varied with relative humidity. PXRD indicated crystalline materials with similar patterns, although the pattern of ZrPL had more disorder (slightly broadened peaks and a noisier baseline). ^{1}H-NMR experiments were performed to determine the proton dynamics in the system. Acidic protons in ZrL_2 are shown to be highly dynamic, with fast chemical exchange and are all in mutual contact. These experiments also show that the number of labile protons is increased in ZrPL as compared to ZrL_2. Conductivity was measured as a function of temperature at 75% RH (Figure 15.12) and as a function of % RH at 100 °C (Figure 15.13).

In ZrPL, conductivity goes from 0.028 to 0.063 S cm^{-1} from 50 to 90% RH, which matches almost exactly with the 3-sulfophenylphosphonate, $(Zr(HPO_4)_{0.7}(m\text{-}HO_3SC_6H_4PO_3)_{1.3})$, of similar composition. In comparison, the ZrL_2 was less conductive and more sensitive to changes in % RH. The authors show that the ZrPL phase retains water better than the ZrL_2 phase and this was ascribed as the reason conductivity was higher. The activation energy in ZrPL was calculated to be 14.4 kJ mol^{-1} and the activation energy of ZrL_2 was not calculated because it was found that the water content at constant humidity would vary as a function of temperature in this material, making the activation energy calculated invalid.

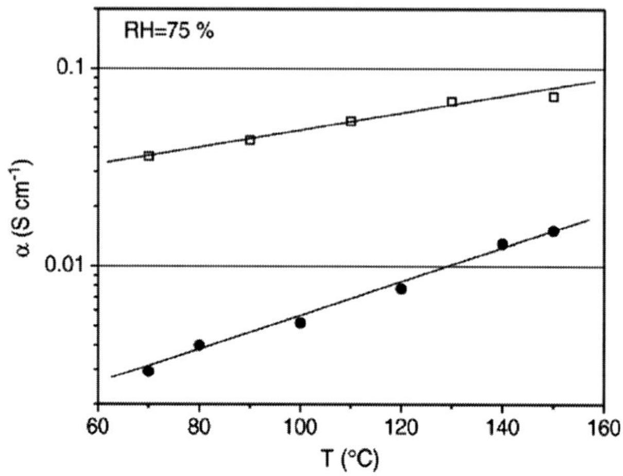

Figure 15.12 Conductivity as a function of temperature at 75% RH for Zr phosphate 4-sulfophenylphosphonate (empty squares) and Zr 4-sulfophenylphosphonate (filled squares). (Reprinted from V. Zima *et al.*, Synthesis and characterization of new zirconium 4-sulfophenylphosphonates, *Solid State Ionics*, 2010, **181**, 705, with permission from Elsevier.)

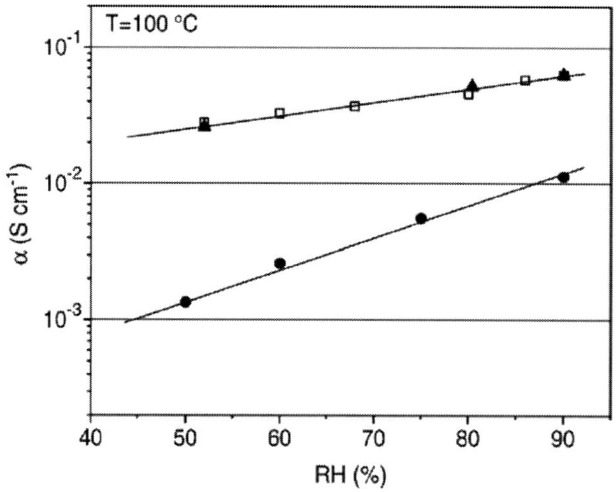

Figure 15.13 Conductivity at 100 °C as a function of relative humidity for Zr phosphate 4-sulfophenylphosphonate (squares), Zr 4-sulfophenylphosphonate (circles), and Zr phosphate 3-sulfophenylphosphonate (triangles). (Reprinted from V. Zima *et al.*, Synthesis and characterization of new zirconium 4-sulfophenylphosphonates, *Solid State Ionics*, 2010, **181**, 705, with permission from Elsevier.)

15.4 Metal Phosphonates with Phosphonate Groups as Proton Conducting Units

In listing the potential advantages of metal phosphonate groups for proton conducting materials, one significant factor was that the functional group could coordinate as a hydrogen phosphonate and offer an acidic proton. This was recognized over two decades ago by Clearfield *et al.*,[51] who reported a linear chain uranyl phenylphosphonate, $[UO_2(HO_3PC_6H_5)_2(H_2O)]_2 \cdot 8H_2O$, that demonstrated proton conductivity. The material, Figure 15.14, had a 1D chain structure with four unique phenyl phosphonates and two unique UO_2 groups, each with a single coordinated water. Two phosphonate groups bridged each uranyl unit through U–O–P–O–U linkages, with the uncoordinated P–O

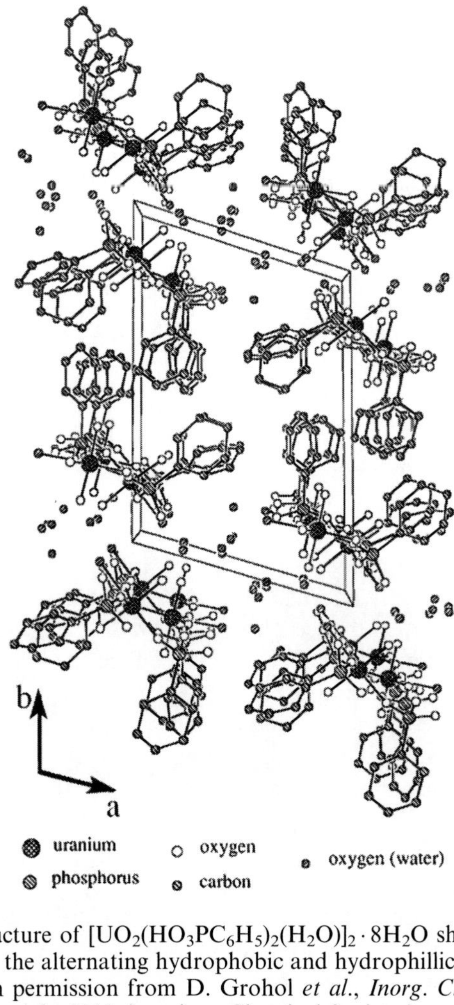

Figure 15.14 Structure of $[UO_2(HO_3PC_6H_5)_2(H_2O)]_2 \cdot 8H_2O$ showing the 1D chains and the alternating hydrophobic and hydrophillic domains. (Reprinted with permission from D. Grohol *et al.*, *Inorg. Chem.*, 1996, **35**, 5264. Copyright 2011 American Chemical Society.)

containing the acidic proton. Phenyl groups all protruded from the same side of the chain, pointing towards the phenyl groups from an adjacent chain, forming a pseudo-layered structure with alternating hydrophobic and hydrophilic regions. Water molecules filled the space between the hydrophilic portions of the chains. Uncoordinated water molecules were easily lost upon exposing the crystals to low humidity. The material was quite stable, with uncoordinated water lost by 100 °C, the two coordinated waters at 260–335 °C and decomposition by 450–600 °C. ^{31}P-MAS-NMR corroborated the crystal structure. Interestingly, dehydration of the sample was only reversible in crystals, a ground up powder could not be reversibly dehydrated, as indicated by PXRD. Impedance measurements were also done from 20 to 80% RH at room temperature and showed conductivity values that ranged from 7×10^{-7} to 3×10^{-3} S cm^{-1} with increasing humidity. The activation energy was determined to be 0.36 eV. The authors believed the Grotthuss mechanism for proton conduction was in operation at high humidity but a shift to a combined hopping/diffusion mechanism could occur upon dehydration. In keeping with this postulate, the uranyl phosphonate showed strong dependence of conductivity to relative humidity. While coordination of a hydrogenphosphonate augments acidity, it does compromise structural robustness. This point that can be addressed through design (3D versus 2D or 1D architectures) and so it is far from a catastrophic shortcoming.

A more recent work by Goward *et al.*[52] in 2008 examined the protonic mobility in rubidium methane hydrogenphosphonate using solid state NMR. The rationale for using this material was to have the methyl group disrupt the hydrogen bonding in the system (in comparison to RbH_2PO_4) and possibly lower the superprotonic phase transition temperature relative to the hydrogen phosphate. $Rb(MePO_3H)\cdot 2H_2O$ was synthesized and the structure was determined by single crystal X-ray diffraction at 100 K (Figure 15.15). DSC on the material indicated two endothermic phase transitions at 327 K and 340 K with no corresponding mass loss. ^{1}H-SS-NMR and ^{87}Rb-SS-NMR was performed to determine proton and structural dynamics within the material. NMR indicated that the phase transition around 320 K was acid protons moving to water protons. In the dehydrated sample, NMR indicates that the protons are static, and no phase transition is observed. No AC impedance analyses were performed on this sample but the results on the high temperature phase indicate promise for proton conduction. The proton transfer would likely still have challenges much above the 330 K temperature studied as crystalline water was integral to the pathway. Such studies though provide meaningful insights to proton transfer mechanisms and design of better conductors at all temperatures.

In this similar vein, Taylor *et al.*[53] reported, in 2010, Zn_3(1,3,5-benzenetriphosphonate) $(H_2O)_2\cdot 2H_2O$, a layered network where the interlayer region was lined with phosphonate oxygen atoms and Zn-ligated water molecules (Figure 15.16). In this solid, the trigonal shape of the organic core prevented the formation of a hybrid inorganic–organic motif. Although the three phosphonate groups are fully deprotonated, they can serve to act as H-bond acceptors

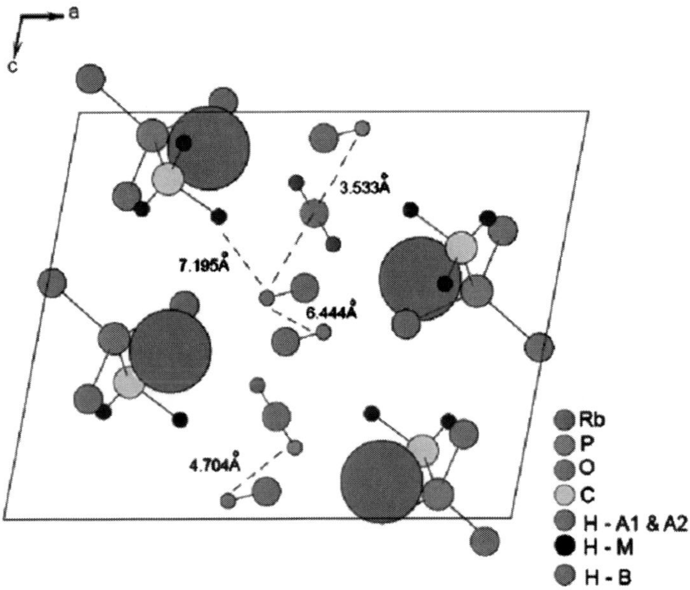

Figure 15.15 Crystal structure of $Rb(MeHPO_3)$ showing a potential proton conduction pathway through crystalline water molecules. A1 and A2 represent protons of different acidities by ^1H SSNMR. (Reprinted with permission from M. Vijawakumar *et al.*, *J. Phys. Chem. C*, 2008, **122**, 5221. Copyright 2011 American Chemical Society.)

in the interlayer. Free water molecules were located crystallographically (with good positional if not rotational order) and they were observed to form a continuous H-bond pathway along with the coordinated water molecules. This solid was stable as a tetrahydrate below 30 °C, confirmed by a room temperature X-ray structure, but would dehydrate to trihydrate and monohydrate phases with heating. To assess proton conductivity and develop an Arrhenius relationship, impedance measurements were carried out in the range of 5–25 °C. While absolute conductivity values were modest (5×10^{-5} S cm^{-1} at 25 °C and 98% RH), the activation barrier was determined to be 0.17 eV, a very low value indicating a facile Grotthuss mechanism. ^2H-SS-NMR studies were also carried out which showed that all deuterons (in the D_2O analogue) were fully exchanging even at –20 °C.

With the ever growing interest in new proton conducting materials, the prospects for new hydrogen phosphonate based metal complexes appears very promising. While these solids would always be less robust than their fully deprotonated analogues, the fact that layered motifs are less likely to be default structures, offers options for design of 3D materials. Moreover, as shown by the final example above, even relatively simple modifications to the ligand backbone can provide macroscopic changes to structure and topology. These are powerful options, coupled with the crystallinity of meta phosphonates, for the design of better proton conductors. To ultimately address the true materials

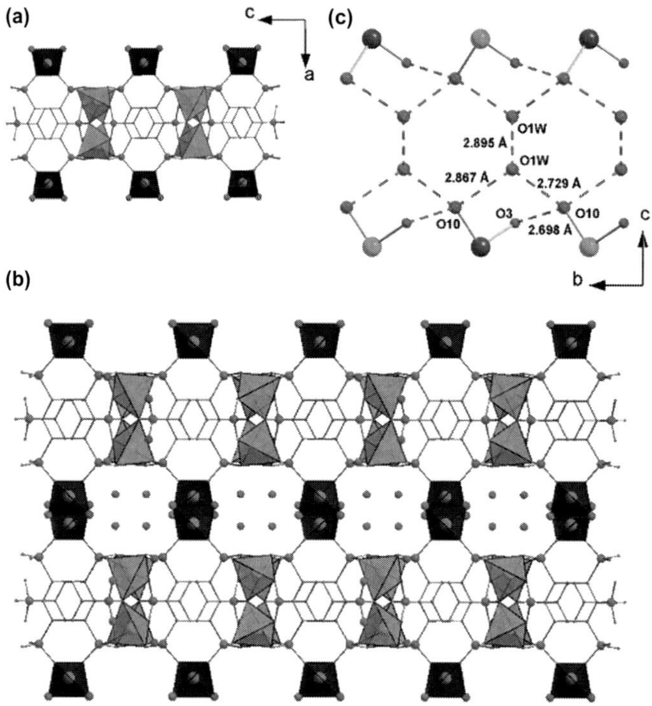

Figure 15.16 Structure of Zn_3(1,3,5-benzenetriphosphonate)$(H_2O)_2 \cdot 2H_2O$, showing: (a) single layer where light tetrahedra are fully phosphonate-ligated Zn centers and dark tetrahedra represent aquo-ligated Zn centers; (b) layers stacked to show the channels and included water molecules; (c) a cross-section of the channel with H-bonds between channel and coordinated water molecules. (Reprinted with permission from J. M. Taylor *et al.*, *J. Am. Chem. Soc.*, 2010, **132**, 14055. Copyright 2011 American Chemical Society.)

challenges of fuel cell operation, most pure metal phosphonates would likely need to be incorporated into some polymer matrix to enable good conduction across all temperature ranges, easier processing into a thin membrane, enhance durability and moderate electrode contacts. The next section of this chapter deals with the research of metal phosphonate polymer composites.

15.5 Polymer Composites of Proton Conducting Metal Phosphonates

Although metal phosphonates show promise for applications in proton conduction due to their conductivity at temperatures above 100 °C at lower levels of humidification, their powdered morphology makes it difficult to form membrane electrode assemblies (MEAs) and fuel cell stacks for practical applications. On the other end of the scale, polymer membranes, such as

Nafion®, show good material properties and proton conductivity below 100 °C and at high levels of humidification, but dehydration of the membranes, coupled with fuel crossover at elevated temperatures, limits their potential. Based on these observations, several research groups have investigated forming composite membranes that incorporate the desirable mechanical properties of polymer membranes with the conductive properties of metal phosphonates. The polymer matrices chosen can either be proton conducting themselves, such as Nafion® and sulfonated poly(ether ketone) (s-PEK), or they can be non-conducting matrices, such as polyvinylidenedifluoride (PVDF), which simply act as a binder agent for the metal phosphonates. Choosing an inherently conducting polymer has the advantage of giving a larger total conductive area, with the polymer conducting at lower temperatures, while the phosphonate additive can conduct at higher temperatures. The polymer, unfortunately, may not be compatible with the metal phosphonate, or might not be stable to the higher desired operating temperatures, which may make the non-conducting polymers a more suitable choice. The majority of the work in phosphonate composite materials has investigated composites based on zirconium sulfophenylphosphonate as they are very robust materials with high levels of proton conductivity at elevated temperatures; however, there are two other metal phosphonate composite systems which have been investigated and will be discussed first: zirconium tricarboxybutylphosphonate (ZrPBCT) and titanium phosphate sulfophenylphosphonate [TiP(SPP)].

ZrPBCT composites with polybenzimidazole (PBI) have been studied by Yamazaki *et al.*[49,54] When unmodified, polybenzimidazole[15] is a non-conducting polymer with good thermal stability and mechanical properties, but it can become an ionic conductor by phosphorylation or sulfonation of the polymer in concentrated acid. PBI composites containing various amounts of ZrPBCT (30, 50 and 70 weight %) were made by dispersing ZrPBCT particles in a solution of the polymer, then casting the mixture and allowing the solvent to evaporate under elevated temperatures. The characterization of the resulting membranes demonstrates the challenges of characterizing these composite materials. Powder X-ray diffraction is an obvious method to demonstrate that the crystalline metal phosphonate is present and unchanged in the composite membrane; however, zirconium phosphonates are notorious for being amorphous in character due to their poor solubility and rapid precipitation during formation. Coupled with the fact that polymers tend to show large amorphous humps at low angles (10–30° in 2 theta), where the peaks from metal phosphonates tend to be most intense, it can become difficult to identify the presence of the metal phosphonate in a composite, especially if the sample is thin or the concentration of the metal phosphonate in the polymer is low. In the case of the ZrPBCT composites, the PXRD data is fairly inconclusive due to the largely amorphous character of the materials obtained. Similar issues are also observed in the infrared spectroscopy patterns, where the absorption peaks that were assigned to the PO_3 vibrations are apparently absent in the spectrum obtained from the composite materials. Conductivity measurements at 25 °C show that the conductivity increases as the concentration of ZrPBCT increases in the samples, but still remains about an order of magnitude lower than in pure ZrPBCT. Variable humidity measurements show this conductivity is humidity dependent, and variable temperature measurements at 100% RH show that the conductivity increases with temperature up to about 90 °C, at which point it remains approximately constant up to 200 °C, at $\sim 5 \times 10^{-3}$ S cm^{-1} for all membranes. Despite the fact that the 70 wt% loaded sample showed the best conductivity, it was also very brittle and not mechanically stable, thus the 50 wt% sample was used for post-synthetic phosphorylation or sulfonation of the membranes in order to try to improve the conductivity of the samples.[54] The phosphorylated and sulfonated samples showed the same general trends, with a slight increase in conductivity to 8.13×10^{-3} S cm^{-1} at 200 °C and 100% RH for the sulfonated sample (c.f. 3.82×10^{-3} S cm^{-1} non-sulfonated).

A proton conducting composite of titanium phosphate sulfophenylphosphonate [TiP(SPP)] with Nafion® was published in 2001 by Alberti *et al.*[48] TiP(SPP), $Ti(O_3POH)(O_3PC_6H_4SO_3H)_{0.85}(OH)_{0.3}$ (5–30 wt%) was loaded into Nafion® 1100 by casting a suspension of the two and evaporating the solvents at elevated temperatures. Membranes with greater than 20 wt% TiP(SPP) were found to become brittle and began to lose mechanical stability. This poor mechanical stability was reflected in the conductivity data, which showed a rapid decrease in conductivity for the highly doped membranes. One likely reason for this would be the poor interfacial contact between the TiP(SPP) and Nafion®, resulting in a large grain boundary effect. At 100 °C and 80% RH, the composite membranes showed a comparable, although slightly lower, level of conductivity than the unmodified Nafion®. Examination of the humidity dependence at 100 °C with the 20 wt% loaded composite showed that the conductivity was slightly worse below 80% RH than Nafion®, but was improved slightly above 80% RH. While the composite membrane was stable up to 100% RH, giving a maximum conductivity of 0.15 S cm^{-1}, this improvement in conductivity relative to the unmodified Nafion® was attributed to loss of interfacial contact between the unmodified Nafion® and the electrodes due to large amounts of membrane swelling, while the doped membrane was less susceptible to swelling.

Apart from the previous two examples, the remaining examples of composite membranes all involve the use of zirconium sulfophenylphosphonate (ZrSPP) or zirconium phosphate sulfophenylphosphonate [ZrP(SPP)], again where the SO_3 group is in the 3-position. In 2004, ZrSPP composites were reported by Kim *et al.*[55] where crystalline ZrSPP was used as a 6.5 and 12.5 wt% additive in Nafion®. The conductivity measurements were made at an unspecified relative humidity from room temperature to between 100 and 160 °C. Conductivity values were very similar below 90 °C, with conductivity values in the range of 5×10^{-2} and 1×10^{-1} S cm^{-1}. At 90 °C, the unmodified Nafion® membrane began to lose conductivity, while the 6.5 wt% and 12.5 wt% composites did not begin to lose conductivity until 120 and 140 °C, respectively. The results do show an improved conductivity range over the unmodified Nafion® which is an uncommon, and interesting, result, but the lack of any humidification data makes it difficult to compare to other results. There is also no data provided on the long-term stability of the membrane, or on the mechanical stability (brittleness, thermal/humidity induced expansion and contraction), which makes its practical and industrial relevance difficult to determine. Nonetheless, the expanded conductivity range is an important finding.

ZrP(SPP) composite membranes have been examined in several different papers by Alberti and Casciola, starting in 2000.[56] Amorphous ZrP(SPP) was made with two different ratios of P to SPP, 1:1 and 0.5:1.5, and was loaded into sulfonated poly(ether ether ketone) (s-PEEK) at 40 wt%. Conductivity measurements at 100% RH at 100 and 150 °C give values around 0.03 S cm^{-1}, which is marginally higher than the unmodified membrane, although lower than the ZrP(SPP) alone (0.07–0.1 S cm^{-1}). A similar trend was observed when variable humidity conductivity measurements were made at 100 °C, with the composites being slightly better than s-PEEK, but almost an order of magnitude worse than ZrP(SPP) alone. This paper exemplifies one of the difficulties in making composite metal phosphonate proton conductors, in that, despite very good conductivity being observed in the unadulterated metal phosphonate, the composite must, by design, not be 100% the phosphonate additive, and thus, a lower level of conductivity is expected. In this case, the 40% composite shows approximately only 40% of the conductivity of the parent material. Despite this, in comparison to the pure s-PEEK membrane, the composites showed comparable or higher conductivities under all humidity and temperatures measured.

In order to try to improve the conductivity of the composites, gels of $ZrP_{0.7}(SPP)_{1.3}$ were made with DMF in order to try to reduce settling of the ZrP(SPP) during drying of the membranes.[57] The hypothesis was that this should make the membranes more homogenous by improving the dispersion of the ZrP(SPP) within the polymer, which should in turn improve the conductivity by creating highly conducting pathways throughout the entire membrane. Composites (10 and 20 wt%) were made with s-PEK and their conductivity was measured from 80 to 150 °C at 90% relative humidity and compared with the individual components. Compared to s-PEK, which conducted just below 1×10^{-3} S cm^{-1}, the 10 and 20 wt% composites showed, respectively, 5- and 10-fold increases in conductivity below 110 °C. Past 110 °C, the conductivity began to decrease, which was attributed to a loss of mechanical integrity of the membrane–electrode assembly due to expansion of the various components. Measurements made up to 150 °C at 70% RH showed temperature independent conductivity for the 20 wt% composite, although the conductivity was lower ($0.9–1.1 \times 10^{-3}$ S cm^{-1}). Calculations based on the composition and conductivities of the individual components, in comparison to the results obtained, indicated that such an increase in conductivity would require about 40% of the metal phosphonate in the composite to form complete conduction pathways through the membrane. In order to prove that continuous conduction pathways could be formed with only a 20 wt% loading of ZrP(SPP), a 20 wt% composite was made with PVDF, a non-conducting polymer, and the conductivity was found to be $\sim 10^{-3}$ S cm^{-1}. Based on their calculations, they concluded that only 5% of the filler ($\sim$1 wt%) formed complete conduction pathways, assuming that only non-tortuous pathways are produced. This represented a minimum value, since the pathways would be tortuous to some extent. In comparison to the case with s-PEK, it appeared that there was some cooperative effect between the polymer and the metal phosphonate which led to the apparent higher amount of pathways forming (40% *vs.* 5%). These results also conclusively showed that composites with non-conducting polymers can be conductive to an appreciable degree.

Due to the fact that these non-conducting polymers are less hygroscopic, and thus less prone to expansion in high humidity environment, Alberti and Casciola continued their investigation into PVDF, making composites with $ZrP_1(SPP)_1$.[58] They were also interested in any differences between 20 wt% composites formed with either crystalline or amorphous ZrP(SPP). The crystalline ZrP(SPP), being a layered compound with high aspect ratios, did not form homogeneous membranes as observed from TEM and SEM analysis (Figures 15.17 and 15.18).

The amorphous ZrP(SPP) created a much more homogeneous membrane with a good dispersion of the metal phosphonate throughout the polymer. Based on these observations, they determined that the amorphous ZrP(SPP) composites would give better results, and thus the crystalline samples were not investigated further. After deciding to use the amorphous ZrP(SPP), membranes with varying amounts of ZrP(SPP) (5–25 wt%) were produced. To determine whether the initial assumption, that the hydrophobic PVDF would reduce membrane swelling, was correct, water uptake experiments were performed on circular membranes in order to determine the amount of swelling that would occur. The change in diameter for these circular membranes, at the equivalent of 90% relative humidity, was found to be less than 2.5%, which corroborates their hypothesis that the PVDF would prevent swelling. Conductivity measurements between 80 and 100 °C and at 90% RH showed that

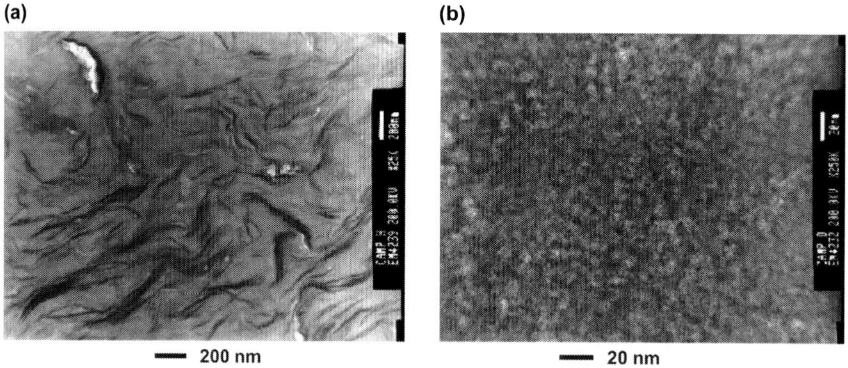

Figure 15.17 TEM images of membranes made of PVDF and (a) 20% α-layered Zr(SPP) and (b) 20% amorphous Zr(SPP). (Reprinted from M. Casciola *et al.*, Nanocomposite membranes made of zirconium phosphate sulfophenylenphosphonate dispersed in polyvinylidene fluoride: Preparation and proton conductivity, *Solid State Ionics*, 2005, **176**, 2985, with permission from Elsevier.)

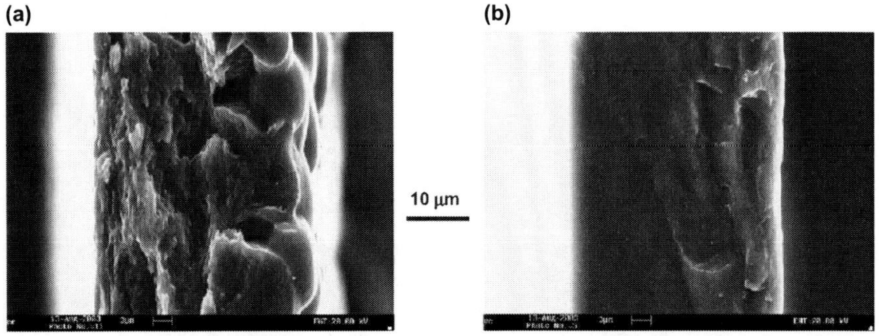

Figure 15.18 SEM images of membranes made of PVDF and (a) 20% α-layered Zr(SPP) and (b) 20% amorphous Zr(SPP). (Reprinted from M. Casciola *et al.*, Nanocomposite membranes made of zirconium phosphate sulfophenylenphosphonate dispersed in polyvinylidene fluoride: Preparation and proton conductivity, *Solid State Ionics*, 2005, **176**, 2985, with permission from Elsevier.)

the conductivity increased with filler loading: from $\sim 5 \times 10^{-5}$ S cm^{-1} at 5 wt% up to $\sim 1 \times 10^{-3}$ S cm^{-1} at 25 wt%; however, the increase was not linear and the conductivity showed a sigmoidal dependence on loading, which the authors attributed to a lack of mechanical stability of the membrane at higher loadings. They also showed that the conductivity was humidity dependant, but that the relative change in conductivity for all the samples was the same irrespective of the loading in the polymer, in comparison to their conductivity at 90% RH.

In order to solve the problem of mechanical stability, in 2005, Alberti *et al.* developed a "precursor solution" process to deposit higher loadings of the ZrP(SPP) in both PVDF and s-PEK, while still maintaining mechanical integrity.[59] A precursor solution of zirconyl propionate (2×10^{-3} mol) in DMF (or NMP), was slowly mixed at room temperature with a solution of phosphoric acid (2×10^{-3} mol) and sulfophenylphosphonic acid (2×10^{-3} mol) in DMF (or NMP). These precursor solutions were stable at room temperature, without any signs of precipitate formation after 2 days. This precursor solution was then dissolved in 2-propanol, mixed with solutions of Nafion® or s-PEK, cast, and dried at elevated temperatures to yield membranes with ZrP(SPP) concentrations up to 40 wt%. TEM (see Figure 15.19) and SEM analysis showed very homogeneous membranes were obtained, that were both transparent and flexible. Conductivity measurements at 100 °C with varying relative humidity (Figure 15.20) shows increased conductivity with increased loading, with a 33 wt% ZrP(SPP) in s-PEK showing the highest conductivity at $\sim 9 \times 10^{-3}$ S cm^{-1} at 90% relative humidity.

In the most recent paper, in 2009, from Casciola *et al.*,[60] Nafion® composite membranes with several loadings (20, 27, 35 and 40 wt%) of zirconium phosphate (ZrP) were prepared via the precursor method, and were then post-synthetically modified using 4-sulfophenylphosphonic acid to give ZrP(SPP)/Nafion® composite membranes. ^{31}P-MAS-NMR measurements indicate that

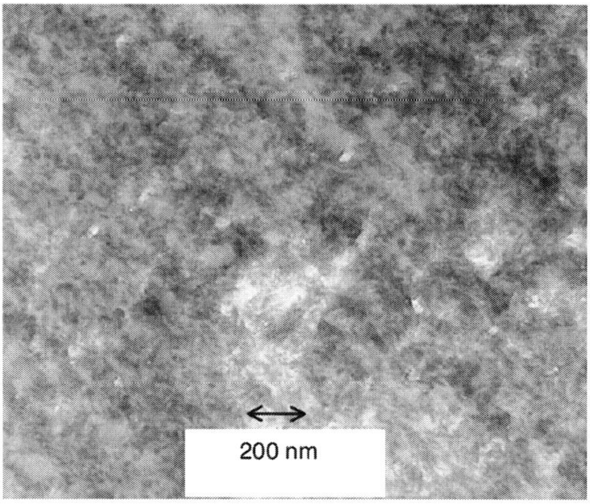

Figure 15.19 TEM image of a PVDF membrane with 40% ZrP(SPP). (Reprinted from G. Alberti *et al.*, New preparation methods for composite membranes for medium temperature fuel cells based on precursor solutions of insoluble inorganic compounds, *Fuel Cells*, 2005, **5**, 366–374. Copyright Wiley-VCH Verlag GmbH & Co. KGaA. Reproduced with permission.)

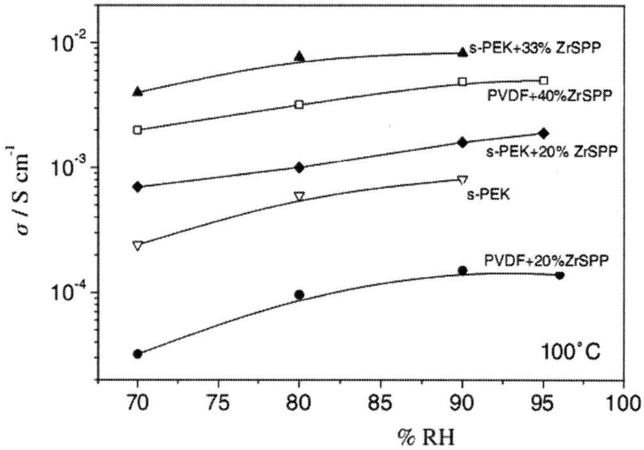

Figure 15.20 Conductivity at 100 °C as a function of relative humidity. Here ZrSPP = $Zr(O_3POH)(O_3PC_6H_4SO_3H)$. (Reprinted from G. Alberti *et al.*, New preparation methods for composite membranes for medium temperature fuel cells based on precursor solutions of insoluble inorganic compounds, *Fuel Cells*, 2005, **5**, 366–374. Copyright Wiley-VCH Verlag GmbH & Co. KGaA. Reproduced with permission.)

the SPP groups are coordinating to two (2.9 ppm) and three (–5.4 ppm) zirconium atoms, from which they conclude that the ZrP filler has been successfully modified to form $ZrP_x(SPP)_y$. ICP spectrometry showed that $x+y$ varied between 1.92 and 1.95, which is consistent with the expected value of 2, and indicates once again that successful exchange of P for SPP has occurred. Integration of the NMR peak areas showed that the amount of SPP varied from $y=0.32$ in the 20 wt% loaded sample down to $y=0.17$ in the 40 wt% loaded sample. The decrease in SPP exchange at higher loadings was attributed to the particles becoming larger at higher loadings, and thus having less total surface area available for exchange and slower migration of the SPP to the center of the particles. Conductivity, as a function of relative humidity at 100 °C, shows that as the filler loading increases, the conductivity of the composite membranes decrease; however, the lowest loadings give slightly better conductivity than unmodified Nafion®. Conductivity as a function of temperature was also examined (at 90% RH), and the composite membranes began to lose conductivity at ~140 °C, while Nafion® lost conductivity starting at 130 °C. This 10 °C difference was attributed to the increased mechanical stability of the composite membranes at the higher temperatures.

15.6 Conclusions and Future Prospects

This chapter has presented a survey of the past and the prospects for metal phosphonate proton conductors. The need for better proton conducting materials is pressing and metal phosphonates have inherent qualities that make them intriguing candidates. Metal phosphonates can form highly robust frameworks that can be modified with acidic sites. They often offer crystalline architectures which can facilitate deeper understanding of proton conduction pathways and also serve as firm handholds to validate computational methods. Metal phosphonates can be incorporated into polymer composites where, examples have shown, synergistic benefits to a wider range of conduction and mechanical processability can be obtained. One area that seems to be underrepresented is the chemistry of hydrogenphosphonates as this group would be highly relevant to better proton conductors. Additionally, frameworks that are 3D versus layered would offer advantages for application. Three-dimensional structures would offer more routes for proton percolation (as well as more routes for fuel crossover), greater pore variation to allow for more possible carriers, and lower aspect ratios to favor incorporation into composites. While the focus has been on fuel cell application, any efficient ion conductor would have application as potentiometric and amperometric gas sensors, such as those observed with layered Zr phosphates.[61] There is ample opportunity in this field both with respect to applications and with respect to possible candidate metal phosphonates.

References

1. J. Zhang, Z. Xie, J. Zhang, Y. Tang, C. Song, T. Navessin, Z. Shi, D. Song, H. Wang, D. P. Wilkinson, Z. S. Liu and S. Holdcroft, *J. Power Sources*, 2006, **160**, 872.
2. K. D. Kreuer, *J. Membr. Sci.*, 2001, **185**, 29.
3. M. S. Whittingham, Y. N. Song, S. Lutta, P. Y. Zavalij and N. A. Chernova, *J. Mater. Chem.*, 2005, **15**, 3362.
4. A. Clearfield, *Dalton Trans.*, 2008, 6089.
5. G. Alberti and M. Casciola, *Annu. Rev. Mater. Sci.*, 2003, **33**, 129.
6. Q. F. Li, R. H. He, J. O. Jensen and N. J. Bjerrum, *Chem. Mater.*, 2003, **15**, 4896.
7. S. J. Hamrock and M. A. Yandrasits, *Polym. Rev.*, 2006, **46**, 219.
8. D. J. Jones and J. Rozière, *Adv. Polym. Sci.*, 2008, **215**, 219.
9. L. Carrette, K. A. Friedrich and U. Stimming, *ChemPhysChem*, 2000, **1**, 162.
10. J. T. S. Irvine, D. C. Sinclair and A. R. West, *Adv. Mater.*, 1990, **2**, 132.
11. P. Colomban and A. J. Novak, *Mol. Struct.*, 1988, **177**, 277.
12. K. D. Kreuer, *Chem. Mater.*, 1996, **8**, 610.
13. K. A. Mauritz and R. B. Moore, *Chem. Rev.*, 2004, **104**, 4535.
14. C. Wieser, *Fuel Cells*, 2004, **4**, 245.
15. J. S. Wainright, J. T. Wang, R. F. Savinell, M. Litt, H. Moaddel and C. Rogers, *Proc. Electrochem. Soc.*, 1994, **94**, 255.
16. M. Rikukawa and K. Sanui, *Prog. Polym. Sci.*, 2000, **25**, 1463.
17. R. Nolte, K. Ledjeff, M. Bauer and R. Mülhaupt, *J. Membr. Sci.*, 1993, **83**, 211.
18. B. Bauer, D. J. Jones, J. Rozière, L. Tchicaya, G. Alberti, M. Casciola, L. Massinelli, A. Peraio, S. Besse and E. Rammuni, *J. New Mater. Electrochem. Syst.*, 2000, **3**, 93.
19. S. Besse, P. Capron, O. Diat, G. Gebel, F. Jousse, D. Marsacq, M. Pinéri, C. Marestin and R. Mercier, *J. New Mater. Electrochem. Syst.*, 2002, **5**, 109.
20. K. D. Kreuer, S. J. Paddison, E. Spohr and M. Schuster, *Chem. Rev.*, 2004, **104**, 4637.
21. S. Horike, S. Shimomura and S. Kitagawa, *Nature Chem.*, 2009, **1**, 695.
22. M. P. Suh and Y. E. Cheon, *Aus. J. Chem.*, 2006, **59**, 605.
23. J. P. Zhang, Y. Y. Lin, W. X. Zhang and X. M. Chen, *J. Am. Chem. Soc.*, 2005, **127**, 14162.
24. K. Takaoka, M. Kawano, M. Tominaga and M. Fujita, *Angew. Chem., Int. Ed. Engl.*, 2005, **44**, 2151.
25. L. Malavasi, C. A. J. Fisher and M. S. Islam, *Chem. Soc. Rev.*, 2010, **39**, 4370.
26. S. M. Haile, C. R. I. Chisholm, K. Sasaki, D. A. Boysen and T. Uda, *Faraday Discuss.*, 2007, **134**, 17.
27. A. Clearfield, *Curr. Opin. Solid State Mater. Sci.*, 1996, **1**, 268.
28. G. K. H. Shimizu, R. Vaidhyanathan and J. M. Taylor, *Chem. Soc. Rev.*, 2009, **38**, 1430.

29. G. Alberti, M. Casciola, U. Costantino and R. Vivani, *Adv. Mater.*, 1996, **8**, 291.
30. G. K. H. Shimizu, *J. Solid State Chem.*, 2005, **178**, 2519.
31. K. Maeda, *Micropor. Mesopor. Mater.*, 2004, **73**, 47.
32. J. G. Mao, *Coord. Chem. Rev.*, 2007, **251**, 1493.
33. R. Murugavel, M. G. Walawalkar, M. Dan, H. W. Roesky and C. N. R. Rao, *Acc. Chem. Res.*, 2004, **37**, 763.
34. D. L. Lohse and S. C. Sevov, *Angew. Chem., Int. Ed. Engl.*, 1997, **36**, 1619.
35. K. Maeda, Y. Kiyozumi and F. Mizukami, *Angew. Chem., Int. Ed. Engl.*, 1994, **33**, 2335.
36. Q. Gao, N. Guillou, M. Nogues, A. K. Cheetham and G. Férey, *Chem. Mater.*, 1999, **11**, 2937.
37. J. A. Hurd, R. Vaidhyanathan, V. Thangadurai, C. I. Ratcliffe, I. M. Moudrakovski and G. K. H. Shimizu, *Nature Chem.*, 2009, **1**, 705.
38. S. Bureekaew, S. Horike, M. Higuchi, M. Mizuno, T. Kawamura, D. Tanaka, N. Yanai and S. Kitagawa, *Nature Mater.*, 2009, **8**, 831.
39. G. Alberti, M. Casciola, R. Palombari and A. Peraio, *Solid State Ionics*, 1992, **58**, 339.
40. G. Alberti, M. Casciola, U. Costantino, A. Peraio and E. Montoneri, *Solid State Ionics*, 1992, **50**, 315.
41. G. Alberti, U. Costantino, M. Casciola, R. Vivani and A. Peraio, *Solid State Ionics*, 1991, **46**, 61.
42. M. Casciola, U. Costantino, A. Peraio and T. Rega, *Solid State Ionics*, 1995, **77**, 229.
43. E. W. Stein, A. Clearfield and M. A. Subramanian, *Solid State Ionics*, 1996, **83**, 113.
44. G. Alberti, L. Boccaili, M. Casciola, L. Massinelli and E. Montoneri, *Solid State Ionics*, 1996, **84**, 97.
45. E. Jaimez, G. B. Hix and R. C. T. Slade, *J. Mater. Chem.*, 1997, **7**, 475.
46. E. Jaimez, G. B. Hix and R. C. T. Slade, *Solid State Ionics*, 1997, **97**, 195.
47. F. Adani, M. Casciola, D. J. Jones, L. Massinelli, E. Montoneri, L. Roziere and R. Vivani, *J. Mater. Chem.*, 1998, **8**, 961.
48. G. Alberti, U. Costantino, M. Casciola, S. Ferroni, L. Massinelli and P. Staiti, *Solid State Ionics*, 2001, **145**, 249.
49. M. Y. Jang, Y. S. Park and Y. Yamazaki, *Electrochemistry*, 2003, **8**, 691.
50. V. Zima, J. Svoboda, K. Melanova, L. Benes, M. Casciola, M. Sganappa, J. Brus and M. Trchova, *Solid State Ionics*, 2010, **181**, 705.
51. D. Grohol, M. A. Subramanian, D. M. Poojary and A. Clearfield, *Inorg. Chem.*, 1996, **35**, 5264.
52. M. Vijayakumar, J. W. Traer, J. F. Britten and G. R. Goward, *J. Phys. Chem. C*, 2008, **122**, 5221.
53. J. M. Taylor, R. K. Mah, I. L. Moudrakovski, C. I. Ratcliffe, R. Vaidhyanathan and G. K. H. Shimizu, *J. Am. Chem. Soc.*, 2010, **132**, 14055.
54. Y. Yamazaki, M. Y. Jang and T. Taniyama, *Sci. Tech. Adv. Mater.*, 2004, **5**, 455.

55. Y. Kim, M. Song, K. Kim, S. Park, S. Min and H. Rhee, *Electrochimica Acta*, 2004, **50**, 645.
56. B. Bonnet, D. Jones, J. Roziere, L. Tchicaya, G. Alberti, M. Casciola, L. Massinelli, B. Bauer, A. Peraio and E. Ramunni, *J. New Mat. Electrochem. Syst.*, 2000, **3**, 87.
57. G. Alberti, M. Casciola, E. D'Alessandro and M. Pica, *J. Mater. Chem.*, 2004, **14**, 1910.
58. M. Casciola, G. Alberti, A. Ciarletta, A. Cruccolini, P. Piaggio and M. Pica, *Solid State Ionics*, 2005, **176**, 2985.
59. G. Alberti, M. Casciola, M. Pica, T. Tarpanelli and M. Sganappa, *Fuel Cells*, 2005, **5**, 366.
60. M. Casciola, D. Capitani, A. Donnadio, V. Frittella, M. Pica and M. Sganappa, *Fuel Cells*, 2009, **9**, 381.
61. G. Alberti, F. Cherubini and R. Palombari, *Sensors and Actuators B*, 1995, **24–25**, 270.

CHAPTER 16

Luminescent Metal Phosphonate Materials

GARY B. HIX

School of Science and Technology, Nottingham Trent University, Nottingham NG11 8NS, UK

16.1 Introduction

Luminescence is the emission of light by a material in response to it having absorbed some energy. Luminescent processes and materials are finding increasing usage in everyday life, from the ubiquitous technologies underlying most television sets to anti-counterfeiting measures that have been introduced into the banknotes of many modern currencies. There are many common luminescent materials, but the number of distinct materials used in the commercial applications is still rather limited. Most are based on inorganic materials and they are often based on ceramic or refractory oxides. Organic–inorganic hybrid materials such as metal phosphonates provide the research community with copious new opportunities for developing new materials with properties that differ from those offered by materials currently in use commercially.

The type of energy used to excite the material, resulting in the luminescent response, gives rise to the name of the process. Hence in *photoluminescence*, a material is excited by the absorption of photons or light, in *electroluminescence* the excitation is electrical energy, and in *cathodoluminescence* a beam of electrons or cathode rays provides the energy. The word *luminescence* also covers a number of processes which result in the emission of light by a material, and the terms used to describe the process are often used interchangeably leading to

Metal Phosphonate Chemistry: From Synthesis to Applications
Edited by Abraham Clearfield and Konstantinos Demadis

Published by the Royal Society of Chemistry, www.rsc.org

some confusion. *Fluorescence* is a process which is usually characterized by a short lifetime of the excited state ($\sim 10^{-7}$ s or less), whilst in *phosphorescence* the decay of the excited state is longer. The most apparent consequence of the different processes is that a fluorescent emission will cease almost immediately upon removal of the stimulating energy source and a phosphorescent material will exhibit an *afterglow*. Perhaps the best known application of such materials is in CRT television/monitor screens. The phosphors in these devices are usually based around Y_2O_2S doped with either Eu or Tb, and ZnS doped with AgCl, which are excited by a cathode ray, giving rise to red, green and blue emissions respectively.

Electroluminescence usually requires the application of large applied voltages, under which conditions metal phosphonates are rarely stable. Similarly, the bombardment of electrons associated with cathodoluminescence will also result in decomposition of materials such as metal phosphonates, either by direct abrasion of the material or by thermal decomposition due to intense localized heating. The photoluminescence process is somewhat less demanding in terms of the stability of the material being studied, but even in this case, the application of laser light sources can cause significant charring of the sample if the beam is not sufficiently attenuated. Due to these constraints, the discussion of luminescent processes and materials in this chapter will be restricted to photoluminescent phenomena.

Photoluminescent materials generally take the form of a host lattice that has been doped with small quantities of *activator* ions. Occasionally, materials are co-doped with *sensitiser* ions, and the resulting materials are known as *phosphors*. Various types of ions can act as activator or sensitiser, e.g. Eu^{3+}, Tb^{3+}, Zn^{2+}, Mn^{2+}.

It is almost always the case that the light emitted is of lower energy (or longer wavelength) than that of the excitation source. In many studies reported in the literature it can be seen that the excitation source is in the UV region, with the emission being in the visible or IR regions. The increase in wavelength is sometimes referred to a *Stokes shift*, synonymous with the term used in the study of Raman spectroscopy. The UV light is absorbed by either the activator ions or the sensitiser ions in the material. In the latter case, energy is transferred from the sensitiser ions to the activator ions. In either case the result is that the activator ions are raised to a higher energy electronic state. The excited state then decays to a lower energy configuration releasing the excess energy as light. This process is shown schematically in Figure 16.1.[1] During the decay process, however, there are a number of possible pathways, some of which involve non-radiative intersystem crossing (ISC), and this is one of the causes of the observed Stokes shift.

As will become clear throughout this chapter, luminescent processes are not solely the domain of the lanthanide elements. Similar properties can be observed for materials based on transition metals or indeed main group metals. Furthermore, luminescent properties, and fluorescence in particular, can be observed for many organic compounds. Indeed the study of many lanthanide compounds is carried out in the form of complexes formed between the metal

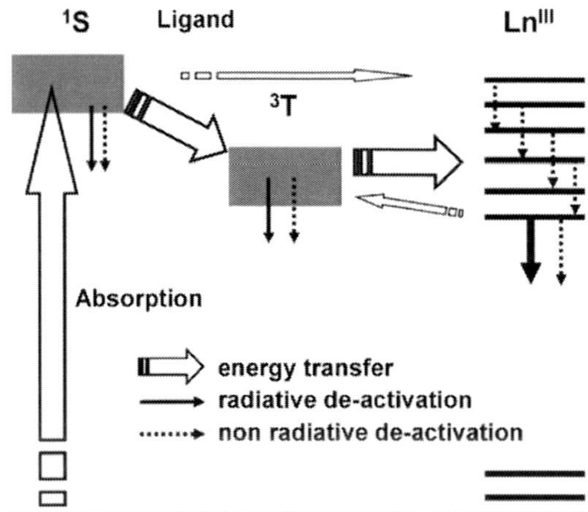

Figure 16.1 Jablonski diagram showing energy transitions in ligands and metal ions upon excitation (from *Chem. Soc. Rev.*, 2005, **34**, 1048).[1]

ions and photoactive organic ligands. The ability to combine active organic functional groups and metals in the solid state makes metal phosphonates ideal candidates when attempting to design new materials with specific properties.

16.1.1 Fluorescence of Ln^{3+} (Ln = Lanthanide) Ions

Many compounds of the Ln^{3+} ions fluoresce when exposed to UV radiation. The majority of electronic transitions are rearrangements of the f electron configuration, known as *f–f transitions*. These transitions are governed by selection rules in the same way that *d–d electronic transitions* are in transition metal compounds. In these latter compounds the selection rules are relaxed by *vibronic coupling* which relaxes the symmetry constraints imposed by the selection rules. However, the electrons in the 4f orbitals are essentially shielded and play little or no part in bonding to atoms around the Ln^{3+} ions. Thus the effects of vibronic coupling are much weaker, and hence the intensities of the associated electronic transitions are much weaker.

The fact that the 4f electrons do not participate in bonding also leads to an important point regarding the energies of the different f electron configurations. Crystal field effects ($\sim$100 cm^{-1} compared to $\sim$2000 cm^{-1} for the transition elements) are of much less significance compared to spin-orbit coupling effects. This means that a particular transition will always have the same energy (within a few nm) regardless of the nature of the ligand around the Ln^{3+} ion. The intensity of the absorption, however, will depend strongly upon the symmetry of the coordination environment of the Ln^{3+} ion.

Broader, more intense bands can be observed for the lanthanide ions, but these arise from f–d transitions, e.g. $[Xe]4f^n \rightarrow [Xe]4f^{n-1}5d^1$. The increase in intensity arises since the transition becomes allowed for all selection rules in the same manner as the LMCT effects that give rise to the intense colours of permanganate and dichromate solutions. Alongside the increase in intensity comes variability in the energy associated with the transition since the 5d electrons interact with the ligands (i.e. they are not shielded) and hence are far more prone to the effects of the ligand field. Ce^{3+} compounds can exhibit $[Xe]4f^1 \rightarrow [Xe]4f^0 5d^1$transitions from 49 737 cm^{-1} to 17 650 cm^{-1} depending on the ligand.

The luminescent properties of many lanthanide materials are enhanced by the so-called *antenna effect*. This is observed when the ligand connected to the Ln^{3+} ion contains an unsaturated organic group. This ligand may be part of a phosphonate anion or be present as a second ligand. Such ligands act as light gathering groups, harvesting UV light which is incident on the material, resulting in excitation of the electronic state of the organic group, e.g. a $\pi \rightarrow \pi^*$ transition. The excited state can then return to the ground state via a number of different routes as show in the Jablonski diagram in Figure 16.1. The excited singlet state can drop back to the ground state giving rise to a ligand-based fluorescence, or it can undergo an internal conversion (IC) to an excited triplet state.

16.2 Lanthanide Phosphonates

A large number of lanthanide phosphonates have been reported in the literature over the last two decades, a great many of which are described in the review by Mao along with their luminescent properties.[2] Since that review many more examples have been prepared and characterized. In terms of luminescence properties, it is clear that Eu and Tb are the most widely studied of the lanthanide ions. This is almost certainly due to the fact that these ions demonstrate strong luminescence in the visible part of the spectrum, and hence their activity can be established visually, simply by illuminating the samples with a low power laboratory UV lamp.

Mao's review[1] reveals that a large number of phosphonate anions have been combined with lanthanide ions to form solid materials, and since that review was published many more materials have been reported in the literature. The organic parts of the anions are wide-ranging and include simple and complex, aliphatic and aromatic moieties. The structures of materials described vary from layered, through pillared-layered, to 3D networks. In general, the coordination numbers of the lanthanide ions observed in phosphonate materials are the same as those observed in other lanthanide compounds; higher coordination numbers (CN = 8, 9) being observed for the bigger, early lanthanides and lower ones (CN = 6) for the smaller ions of the later lanthanides. It was noted that in the simple aryl and alkyl mono- and diphosphonic acid derivatives of the ions La–Tb reported, the lanthanide ion assumes an 8-fold coordination environment.[3–8] Despite the apparent predictable behaviour of the lanthanide ions, there exists the possibility of more than one local symmetry being associated with a coordination number, which can have an effect upon the observed luminescence.

16.2.1 Lanthanide Phosphonates Exhibiting Visible Photoluminescent Emission

Europium chemistry is dominated by the Eu^{3+} ionic state with an f^6 electronic configuration. Excitation by illumination with UV light results in population of the $^5D_0$ state, and luminescence occurs when this drops down to the ^7F_J (J = 0–4) states. These transitions take place in the range 580–700 nm and give rise to the characteristic red emission in the visible range which is associated with Eu.

Comparing the photoluminescence of $EuH(O_3PCH_2PO_3)$ ($\lambda_{ex} = 457.9$ nm)[9] with that of $EuH(O_3PC_6H_5)_2$ ($\lambda_{ex} = 378$ nm),[10] (Table 16.1) the positions of the emission bands are in good agreement. Furthermore Rosa noted that the emission characteristics of $EuH(O_3PCH_3)_2$ are also remarkably similar to that of the phenylphosphonate.[10] In all cases it can be noted that the $^5D_0 \rightarrow {}^7F_3$ electric dipole transition is not observed as it is forbidden by the selection rules.

In the layered material, $\{Eu_2(HO_3PCH_2NH_2CH_2CO_2)_2(H_2O)_8(ClO_4)_4\}$, which is based on *N*-phosphonomethylglycine, the Eu^{3+} ions are found in three crystallographically distinct environments.[11,12] The PL spectra support the crystallographic results indicating the presence of two unique Eu environments, since there are two peaks observed for the $^5D_0 \rightarrow {}^7F_0$ transition (578 and 579 nm). One is eight-coordinate and the other nine-coordinate. For CN = 8 there are three possible types of coordination polyhedra; dodecahedral (D_{2h}), square antiprism (D_{4h}) and bicapped trigonal prism (C_{2v}). The nine-coordinate centre can have either capped square antiprism (C_{4v}) or tricapped trigonal prismatic (D_{3h}) geometry. The relative intensities of the $^5D_0 \rightarrow {}^7F_0$ peaks and the number of Stark components leads to the conclusion that the Eu atoms have local symmetries of C_{2v} (CN = 8) and C_{4v} (CN = 9).

Table 16.1 Observed $^5D_0 \rightarrow {}^7F_J$ emission bands in the PL spectra of $EuH(O_3PCH_2PO_3)$ ($\lambda_{ex} = 457.9$ nm)[9], $EuH(O_3PC_6H_5)$[10,11], $Eu_2(HO_3PCH_2NH_2CH_2CO_2)_2(H_2O)_8(ClO_4)_4$[12], $[Eu(C_2O_4)\{MeNH(CH_2CO_2)(CH_2PO_3H)\}]\cdot 0.5H_2O$,[26] $[Eu((S)\text{-}HO_3PCH_2\text{-}NHC_4H_7\text{-}CO_2)_3]\cdot 2H_2O$[13] ($\lambda_{ex} = 248$ nm), and $Eu(O_3PC_6H_4CO_2)$[15] ($\lambda_{ex} = 266$ nm).

J	*EuH (O₃PCH₂-PO₃)*	*EuH(O₃-PC₆H₅)₂*	*Eu₂(HO₃-PCH₂NH₂-CH₂CO₂)₂-(H₂O)₈-(ClO₄)₄*	*[Eu(C₂O₄)-{MeNH(CH₂-CO₂)(CH₂-PO₃H)}]·0.5H₂O*	*[Eu((S)-HO₃PCH₂-NHC₄H₇-CO₂)₃]·2H₂O*	*Eu(O₃PC₆-H₄CO₂)*
0	580	580	578, 579	–	588	580
1	590–595	590	594	592	596	590
2	610–622	614–615	615	616	611	615
3	–	–	–	654*	652	650
4	700	700	695	699	700	700

*Peak observed in the data presented in the paper,[26] but not noted by the authors.

Excitation of the Eu^{III} compound $[Eu((S)\text{-}HO_3PCH_2\text{-}NHC_4H_7\text{-}CO_2)_3]\cdot 2H_2O$[13] at 248 nm, results in characteristic luminescent bands comprising the familiar red emission (Table 16.1, Figure 16.2). Of these transitions, the strongest band is that of $^5D_0 \rightarrow {}^7F_2$ transition, an electric dipole transition which is sensitive to chemical bonds in the vicinity of Eu^{III} ion, and the one mainly responsible for the bright red luminescence. This observation may also be made for all of those materials mentioned previously, regardless of phosphonate anion or coordination environment of the Eu ion. It is known that the intensity of the $^5D_0 \rightarrow {}^7F_2$ transition increases as the site symmetry of Eu^{III} decreases, whereas the $^5D_0 \rightarrow {}^7F_0$ transition is strictly forbidden in a symmetric environment. For this material, the second most intense band is observed for the $^5D_0 \rightarrow {}^7F_0$ transition at 588 nm. The prominence and singularity of this peak demonstrates the noncentrosymmetry of Eu^{III} and the presence of only one type of Eu^{III} site in the unit cell. The $^5D_0 \rightarrow {}^7F_1$ transition for Eu^{3+} is a magnetic dipole transition; its intensity varies with the crystal field strength ion. The intensity ratio of the $^5D_0 \rightarrow {}^7F_2$ transition to the $^5D_0 \rightarrow {}^7F_1$ can thus be used as a measure of the coordination state and the site symmetry of the rare earth ion.[14] For $[Eu((S)\text{-}HO_3PCH_2\text{-}NHC_4H_7\text{-}CO_2)_3]\cdot 2H_2O$, the intensity ratio $I(^5D_0 \rightarrow {}^7F_2)/I(^5D_0 \rightarrow {}^7F_1)$ was found to be around 4.6, a clear indicator that Eu^{III} ions are not located on an inversion centre.[13]

In all of the materials described above the intensity maximum is due to the $^5D_0 \rightarrow {}^7F_2$ transition seen around 610 nm, which results in the red emission characteristic of many Eu compounds. It would seem then that all Eu phosphonates have the same emission characteristics independent of the nature of the phosphonate anion, and the coordination environment of the Eu ion.

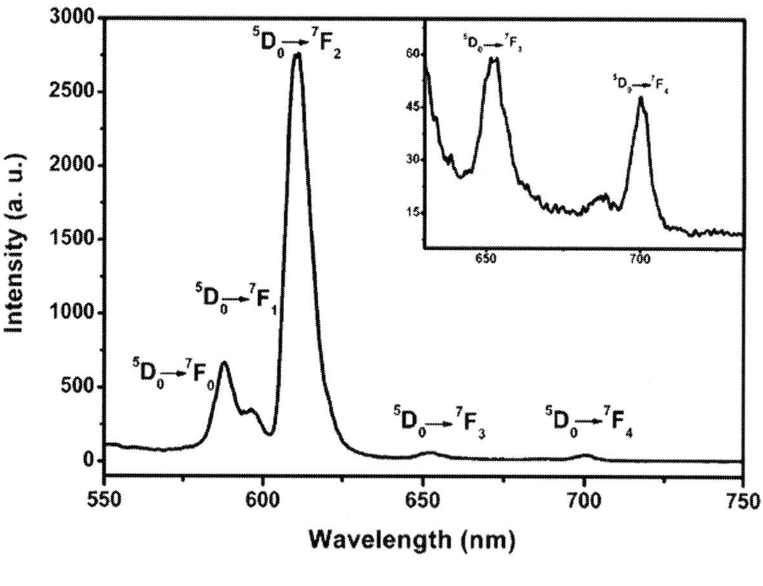

Figure 16.2 Emission spectrum of $[Eu((S)\text{-}HO_3PCH_2\text{-}NHC_4H_7\text{-}CO_2)_3]\cdot 2H_2O$. (Copied with permission from *Inorg. Chem.*, 2006, **45**, 4431. Copyright 2006 American Chemical Society.)

As such, it would be expected that luminescent emission from Eu phosphonates will always be red. This is not however, the case.

In the layered phosphonate, $Eu(O_3PC_6H_4CO_2)$, the local symmetry of the Eu is such that the ratio $I(^5D_0 \rightarrow {}^7F_2)/I(^5D_0 \rightarrow {}^7F_1) < 1$; that is to say, the $^5D_0 \rightarrow {}^7F_1$ transition is more intense (Figure 16.3). This arises due to the fact that the Eu is located on a mirror plane in the structure. The net result is that the observed emission is now at a lower wavelength, and is more orange in colour than red.[15]

Whilst it would therefore seem possible to change the luminescence emission colour, what must be remembered is that the positions of the various lines in the emission spectrum are essentially unchanged. All that may be possible is to vary the relative contributions of the lines to the emitted light by developing materials with multiple Eu environments and controlling occupancy of those sites. Hence variations from orange through to red are possible.

$[Ln(H_2cmp)(H_2O)]$ [where cmp = *N*-(carboxymethyl)iminodi(methylphosphonic acid) and $Ln^{3+} = Y^{3+}$, La^{3+}, Pr^{3+}, Nd^{3+}, Sm^{3+}, Eu^{3+}, Gd^{3+}, Tb^{3+}, Dy^{3+}, Ho^{3+} and Er^{3+}] contains a single Ln^{3+} centre in a highly distorted dodecahedral coordination environment with one water molecule in the coordination sphere.[16] Since the chemistry of the lanthanides is often very similar across the whole series it is often simple to 'dope' structures with different metals. Studies of the mixed lanthanide $[(Gd_{0.95}Eu_{0.05})(H_2cmp)(H_2O)]$ material show that Gd^{3+}-to-Eu^{3+} energy transfer occurs within the layers upon excitation at 272 nm. The Gd^{3+} ion has an f^7 configuration, and as such the f–f

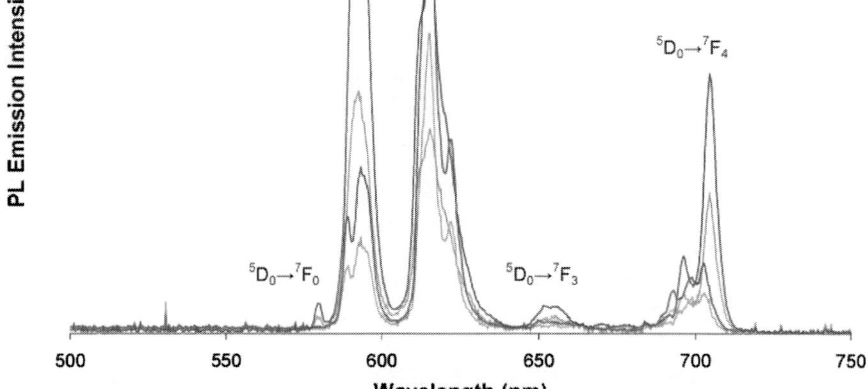

Figure 16.3 RT PL emission spectra for $Eu(p\text{-}O_3PC_6H_4CO_2)$ and $EuH(O_3PC_6H_5)_2$ ($\lambda_{ex} = 366$ nm) (from *Dalton Trans.*, 2009, 10614).[15]

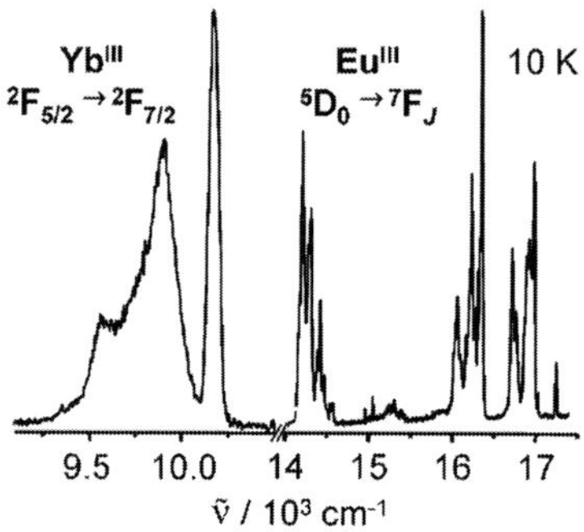

Figure 16.4 Luminescence spectra at 10 K [α_{ex}] 28 180 (left region) and 21 468 cm^{-1} (right region) of the polycrystalline $Na_6Eu_{9-x}Yb_x(L)_{16}$ (H_2L = 5-methyl-2,2-bipyridyl-6-phosphonic acid) cluster showing emission of both Yb^{III} and Eu^{III}. (Copied with permission from *Inorg. Chem.*, 2006, **45**, 3158. Copyright 2006 American Chemical Society.)

transitions, $^6P_{5/2,7/2} \rightarrow {}^8S_{7/2}$ involve a significant configurational change and are observed in the UV region. The emitted light from this transition excites the transitions of the Eu^{3+} ion, and the intensities of the luminescence in this material is greater than that observed for the isostructural Eu-only derivative. This effect is not restricted materials containing UV emitting ions. It has been shown that the green emission from Tb can enhance and complement the red emission from Eu in some lanthanide carboxylate MOFs.[17]

In a similar manner Bünzli showed that the cluster phosphonate $[Na_6Eu_9(L)_{16}](OH) \cdot 51H_2O$ (H_2L = 5-methyl-2,2-bipyridyl-6-phosphonic acid)[18] can be isostructurally synthesised to contain Yb, $[Na_6Eu_{9-x}Yb_x(L)_{16}](OH) \cdot 51H_2O$. The Yb^{3+} ion has an f^{13} configuration and emits in the IR region. The doped material containing both Eu and Yb thus exhibits emission in both the visible (red) and IR regions (Figure 16.4).

Terbium is also widely used in the synthesis of luminescent materials, likely due to the visible emission in the green part of the spectrum. The Tb^{3+} ion has an electronic configuration of f^8, and with regards to f–f transitions the term symbols are similar to those observed for Eu. Excitation populates the $^5D_4$ excited state and the luminescent transitions to ^7F_J (J = 6–3) fall in the region of 490–622 nm.

$Tb(HPMIDA)(H_2O)_2 \cdot H_2O$ [PMIDA = *N*-(phosphonomethyl)iminodiacetic acid) is one of seven isostructural lanthanide phosphonates reported by Mao *et al.*[19] The lanthanide ion is eight coordinate and exhibits four very strong characteristic emission bands in the visible region under excitation at 380 nm. These emission lines are seen at 489.5 nm ($^5D_4 \rightarrow {}^7F_6$), 546.5 nm ($^5D_4 \rightarrow {}^7F_5$), 582.0 nm ($^5D_4 \rightarrow {}^7F_4$) and 621.5 nm ($^5D_4 \rightarrow {}^7F_3$) (Figure 16.5). It is interesting to note in this material that the phosphonate anion does not contain an antenna species, *i.e.* it does not undergo any π–π* transitions, and as such the Tb^{3+} ion is excited directly. Furthermore, comparison of simple phenyl- and methyl-phosphonates of Tb show that there is no appreciable difference in the intensity

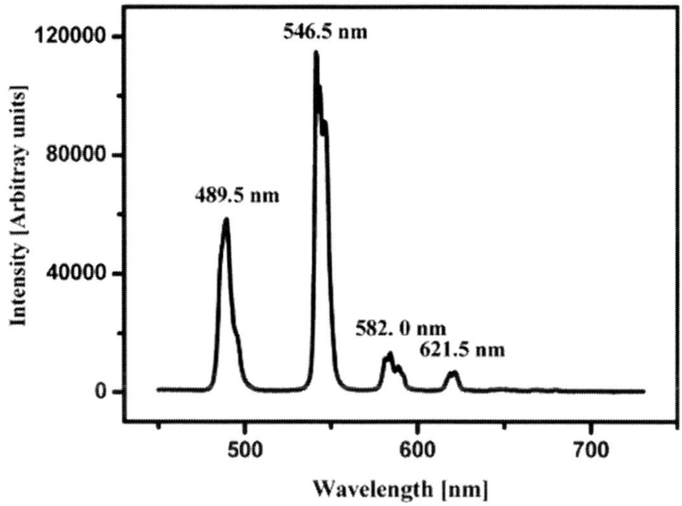

Figure 16.5 RT PL emission spectrum for $Tb(HPMIDA)(H_2O)_2 \cdot H_2O$ ($\lambda_{ex} = 380$ nm). (Adapted with permission from *Cryst. Growth Des.*, 2006, **6**, 2322. Copyright 2006 American Chemical Society.)

of the emitted light when an antenna moiety is present in the phosphonate anion. Similarly, illumination of Tb phenylphosphonate with UV light which is too low in energy to excite a transition in the organic moiety has little effect on the brightness of the emission. This implies that the observed emission arises almost totally from f–f transitions in the lanthanide ion.[20]

The positions of the transitions observed in the emission spectra of Tb phosphonates are, like those of Eu, not significantly altered by changing the nature of the phosphonate anion, varying by only a few nm. But, unlike the Eu-containing materials, there is little reported variation in the relative intensity of the peaks in the spectra, and it is the most intense peak centred around 546 nm (Figure 16.5) that yields the characteristic green colour. For example the PL spectrum of $[Tb(H_2L)_3] \cdot 2H_2O$ (where $H_3L = (S)\text{-}HO_3PCH_2\text{-}NHC_4H_7\text{-}CO_2H$) is shown in Figure 16.6.[13] In this material the Tb ion resides in a distorted octahedral environment, and whilst the spectrum shows some evidence of crystal field splitting, the relative intensities of the peaks are much the same as those observed for Tb in an eight-coordinate environment (Figure 16.6). This rather global assertion does however require some qualification, as at the present time there are few reports in the literature which describe Tb phosphonates with unusual coordination environments, and fewer still where both structure and photoluminescent behaviour have both been described.

As mentioned previously, Eu and Tb-containing materials make up the largest proportion of the photoluminescent materials reported. This stems from the fact that theirs are some of the brightest emissions of the lanthanides as well as being in the visible part of the spectrum and partly due to the fact that they are relatively inexpensive compared to other lanthanide salts.

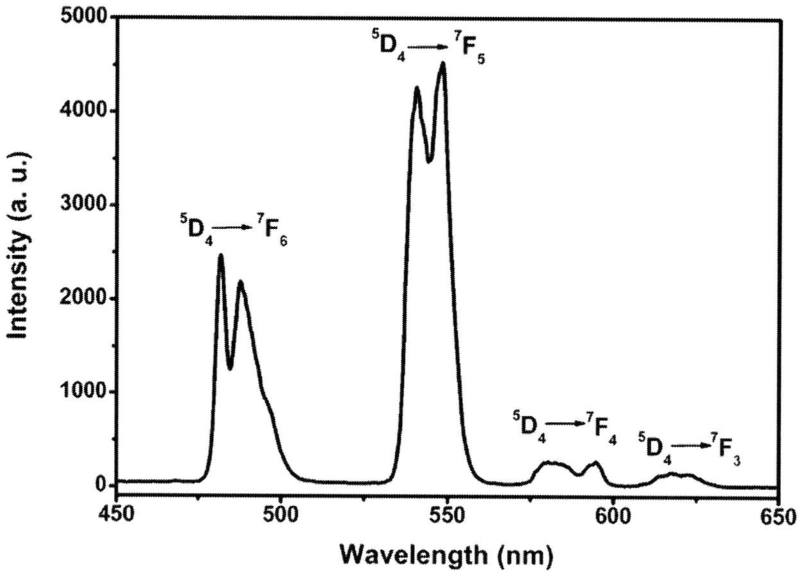

Figure 16.6 The emission spectrum of $[Tb(H_2L)_3] \cdot 2H_2O$ (where $H_3L = (S)$-HO_3PCH_2-NHC_4H_7-CO_2H) corresponding to the $^5D_4 \rightarrow {}^7F_J$ ($J = 6, 5, 4, 3$) transitions in solid state at room temperature, $\lambda_{exc} = 249$ nm. (Copied with permission from *Inorg. Chem.*, 2006, **45**, 4431. Copyright 2006 American Chemical Society.)

Of the remaining lanthanides, few exhibit emission in the visible region. Those that do, show only weak responses that are often easily missed by the naked eye. Dy compounds luminesce giving a visible, but very weak, yellow emission. Excitation of $[Dy_4(C_2O_4)_5(2\text{-pmpH})_2(H_2O)_7] \cdot 5H_2O$ [where 2-$pmpH_2$ = 2-pyridylmethylphosphonic acid] at 285 nm leads to the yellow luminescent response typical of Dy^{3+}.[21] The two bands at 480 and 587 nm are attributed to transitions of $^4D_{9/2} \rightarrow {}^6H_{15/2}$ and $^4D_{9/2} \rightarrow {}^6H_{13/2}$, respectively (Figure 16.7). The structure of this material contains four crystallographically distinct ions each of which is eight coordinate with differences in local symmetry. This accounts for the fact that the decay was best fit by a biexponential process with lifetimes of $t_1 = 0.92$ ms (84%) and $t_2 = 5.41$ ms (16%).

$[Dy_2(HL)(H_2L)(L)(H_2O)_2] \cdot 4H_2O$ (where $H_3L = 4$-HOOC-C_6H_4-CH_2 $NHCH_2PO_3H_2$) also displays yellow luminescence in the solid state when it is excited at 348 nm. The two emission bands in are seen at 478 nm and 576 nm for the same two transitions as the previous example, and the relative intensities of the peaks in the spectrum are very similar. It contains a single crystallographically unique Dy ion in an 8-fold coordination environment, which is best viewed as a distorted bicapped trigonal prism. Once again this demonstrates the lack of variation in response with respect to the presence of different phosphonate anions and coordination environment that was observed for Tb-based materials, although the caveat regarding the lack of combined structural and luminescent data must be noted.[22]

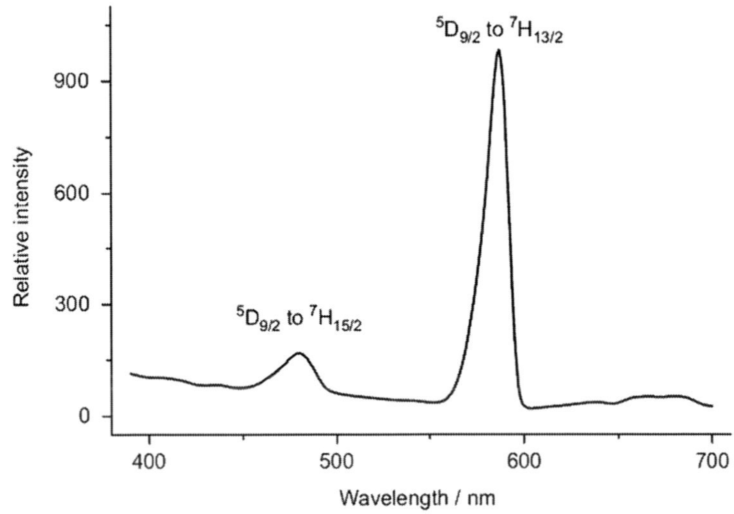

Figure 16.7 The solid state emission spectrum of $[Dy_4(C_2O_4)_5(2\text{-pmpH})_2(H_2O)_7]\cdot 5H_2O$ [where 2-pmpH$_2$ = 2-pyridylmethylphosphonic acid], $\lambda_{ex} = 285$ nm. (Reprinted from *Journal of Solid State Chemistry*, **183**, p.1159 T. H. Yang, D. K. Cao, Y. Z. Li and L. M. Zheng, Copyright 2010, with permission from Elsevier.)

Samarium is a little studied member of the lanthanide family when it comes to phosphonate derivatives. There are few reports of structural studies and fewer concerning the photophysical properties of materials. This is probably due to the fact that the emission spectra of Sm-containing materials are very weak, as noted by Sun *et al.*[23] $[Sm_3(HL)(C_2O_4)_4(H_2O)_4]\cdot 2H_2O$ [where $H_2L = O(CH_2CH_2)_2NCH_2PO_3H_2$)] is one of a seven lanthanide phosphonates containing *N*-morpholinomethylphonate anions. Excitation at a wavelength of 404 nm gives rise to three characteristic bands in the orange-red part of the visible spectrum, which are attributed to $^4G_{5/2} \rightarrow {}^6H_J$ ($J = 5/2, 7/2, 9/2$), $^4G_{5/2} \rightarrow {}^6H_{5/2}$ (563 nm), $^4G_{5/2} \rightarrow {}^6H_{7/2}$ (597 nm) and $^4G_{5/2} \rightarrow {}^6H_{9/2}$ (643 nm) transitions.

Praesodymium is little studied, but also luminesces in the visible region giving rise to an orange-red emission. Figure 16.8 shows the RT emission spectrum of $Na[Pr_3(H_2O)_4(C_2O_4)_4(CH_3PO_3)].2H_2O$[24] and unlike the emission spectra of other lanthanides, it contains a series of emission bands that cover the entire green to red wavelengths. The peaks observed at 480–540 nm and those at 550–570 nm can be assigned to the $^3P_0 \rightarrow {}^3H_4$ and $^3P_0 \rightarrow {}^3H_5$ transitions, respectively. Those peaks between 580 and 650 nm are a mixture of the $^1D_2 \rightarrow {}^3H_4$ and $^3P_0 \rightarrow {}^3H_6$ transitions, whereas those between 670 and 718 nm corresponded to the mixed emissions of $^1D_2 \rightarrow {}^3H_5$ and $^3P_0 \rightarrow {}^3F_2$. The observed transitions fall in to one of three types: $^3P_0 \rightarrow {}^3H_J$ (triplet–triplet), $^3P_0 \rightarrow {}^3F_J$ (triplet–triplet) and $^1D_2 \rightarrow {}^3H_J$ (singlet–triplet). Excitation at 473 nm raises Pr^{III} from the ground $^3H_4$ state to the excited $^3P_0$ state which can then relax to the $^1D_2$ state. Emission from either of these states can be observed. Attempts to

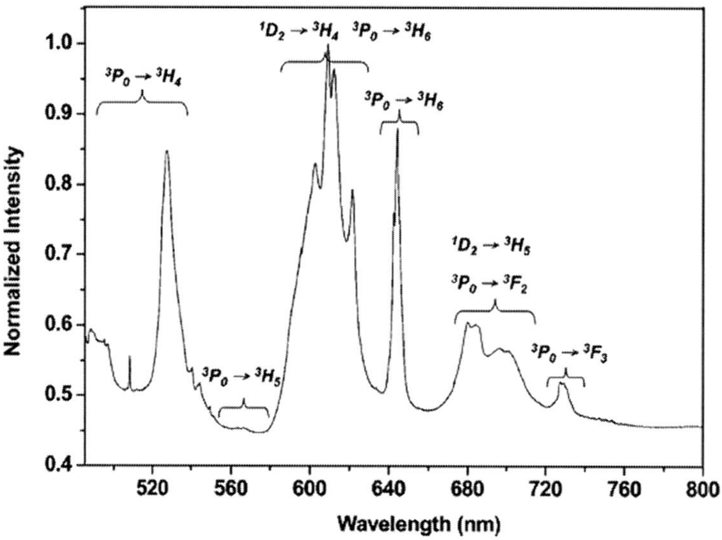

Figure 16.8 Photoluminescence spectrum of $Na[Pr_3(H_2O)_4(C_2O_4)_4(CH_3PO_3)] \cdot 2H_2O$ excited at 473 nm. (Copied with permission from *Chem. Mater.*, 2007, **19**, 3232. Copyright 2007 American Chemical Society.)

distinguish between the states, and hence assign the peaks in this complex spectrum, on the basis of excited state lifetimes proved unsuccessful.

16.2.2 Lanthanide Phosphonates Exhibiting Non-visible Photoluminescent Emission

The number of lanthanides exhibiting emission in the visible region is relatively small. Many of the others also undergo electronic transitions which give rise to photoluminescent emission phenomena. However, most occur in the IR region of the electromagnetic spectrum which gives rise numerous problems, not least of which is observing and recording the emission spectrum.

Under the excitation at 310 nm, $Nd(HL_1)(H_2L_2)_{0.5}(H_4L_2)_{0.5}(H_2O)] \cdot (H_2O)_{1.5}$ [where H_3L_1 = 5-sulfosalicylic acid and H_4L_2 = *N*,*N*′-piperazine(bis-methylenephosphonic acid)][25] displays the characteristic emission bands for neodymium(III) ion: a middle emission band at 892 nm ($^4F_{3/2} \rightarrow {}^4I_{9/2}$), a strong emission band at 1059 nm ($^4F_{3/2} \rightarrow {}^4I_{11/2}$) and a weak band at 1349 nm ($^4F_{3/2} \rightarrow {}^4I_{13/2}$) in the near-infrared (NIR) region (Figure 16.9). They are in agreement with the previously reported emission spectra of Nd^{III} compounds.[26–28]

As seen for other lanthanide ions, the presence of an aryl antenna group is not necessary for excitation to occur. The simple compound $Na[Nd_3(H_2O)_4(C_2O_4)_4(CH_3PO_3)].2H_2O$[24] containing only methylphosphonate, is excited at 514.5 nm and yields an emission spectrum almost identical to that shown in Figure 16.9.

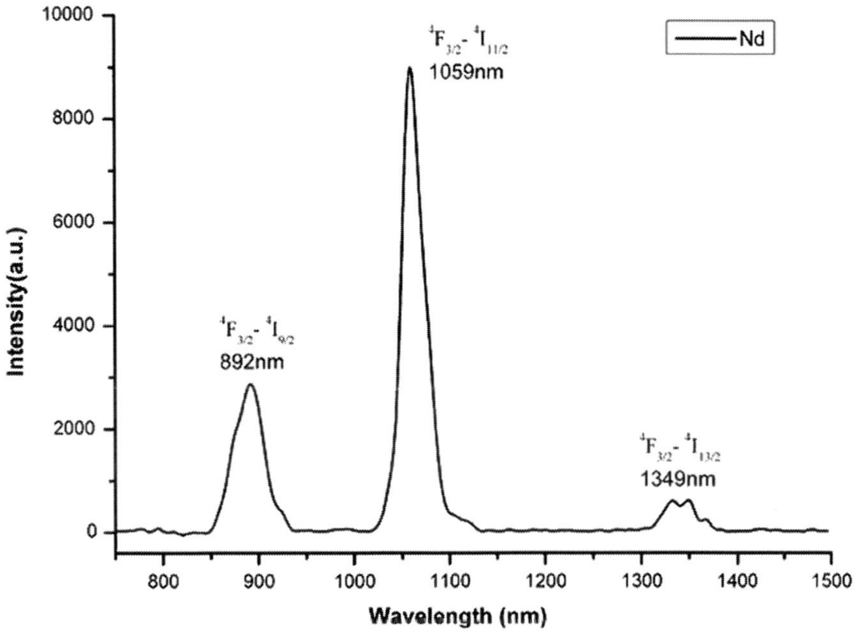

Figure 16.9 Solid State emission spectrum of $Nd(HL_1)(H_2L_2)_{0.5}(H_4L_2)_{0.5}(H_2O)] \cdot (H_2O)_{1.5}$ [where H_3L_1 = 5-sulfosaclicylic acid and H_4L_2 = *N,N'*-piperazine(bis-methylenephosphonic acid)]. $\lambda_{ex} = 310$ nm. (Reprinted from *Journal of Solid State Chemistry*, **182**, p. 2290, X. Li, Z, Xie, J. Lin and R. Cao, Copyright 2009, with permission from Elsevier.)

$Er(H_3L^2)(H_4L^2)$ [where H_5L^2 = 4-HOOC–C_6H_4–$CH_2N(CH_2PO_3H_2)_2$] and $Er(HL^3)(H_2L^3)(H_2O)$ [where $H_2L^3 = H_2O_3PCH_2N(CHO)(CH_2$–$C_6H_4$–$COOH)]$[29] are two related Er phosphonates which exhibit quite different photoluminescent behaviour. The former, anhydrous, material exhibits three strong peaks in the IR region (Figure 16.10a) at 1477, 1541 and 1561 nm under 520 nm excitation. These are attributed to the $^4I_{13/2} \rightarrow {}^4I_{15/2}$ transition for the Er^{3+} ion; the splitting of the transition into several sub-bands is due to the low symmetry of the coordination geometry around the Er^{3+} ion (C_1), an effect that has been seen in other Er phosphonates.[26] The latter, hydrated, compound shows no emission in the NIR region at all. This lack of NIR emission is almost certainly due to the presence of water in the material. The excitation spectrum ($\lambda_{em} = 330$ nm, Figure 16.10b) shows two strong absorbances at 324 and 408 nm. Subsequent excitation of the material at 408 nm shows two strong, broad emissions at 460 and 571 nm. The breadth and intensity of these bands leads to the conclusion that these are ligand centred bands, associated with π–π* transitions within the organic part of the anion.

Emission in the IR part of the spectrum is often difficult to observe, not only due to the fact that it cannot be detected by the naked eye. It is very often the

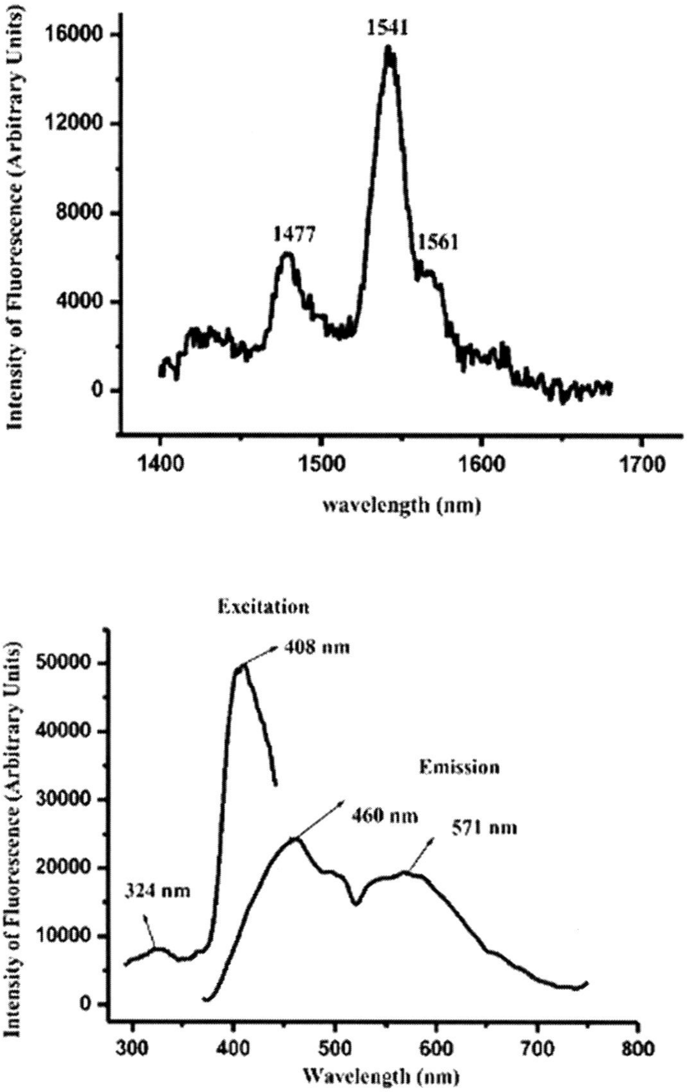

Figure 16.10 Solid state photoluminescent spectra of (a) $Er(H_3L^2)(H_4L^2)$ [where $H_5L^2 = 4 - HOOC–C_6H_4–CH_2N(CH_2PO_3H_2)_2$] ($\lambda_{ex} = 520$ nm) and (b) $Er(HL^3)(H_2L^3)(H_2O)$ [where $H_2L^3 = H_2O_3PCH_2N(CHO)(CH_2–C_6H_4–COOH)$] ($\lambda_{em} = 330$ nm, $\lambda_{ex} = 408$ nm). (S. F. Tang, J. L. Song and J. G. Mao, *European Journal of Inorganic Chemistry*, 2006, 2011. Copyright Wiley-VCH Verlag GmbH & Co. KGaA. Reproduced with permission.)

case that emissions cannot be detected in materials from which they might be expected. The root of this problem lies in the fact that the vast majority of phosphonate materials contain coordinated water molecules. Samarium-, erbium- and dysprosium-derived luminescence is very sensitive towards water, compared to europium or terbium luminescence. The explanation lies in the value of the energy gap between the excited luminescent states and the highest J levels of the ground states of these lanthanides. This energy gap is around 7500 cm^{-1} for Sm^{III} and 7800 cm^{-1} for Dy^{III}, both of which are significantly smaller than those of Eu^{III} and Tb^{III} which display values around 12 300 cm^{-1} and 14 800 cm^{-1}, respectively.[30] Since OH oscillators act independently ($\nu_{O\text{–}Hstr.symm} = 3657\ cm^{-1}$), it can be seen that, in the case of samarium and dysprosium, fewer water molecules are needed to bridge the gap than in the case of europium and terbium (*ca.* 2 versus 3–4). The higher water sensitivity explains why lifetimes of Dy^{III} or Sm^{III} luminescence are generally much shorter than those observed with Eu^{III} or Tb^{III}, that is, tens of microseconds versus hundreds of microseconds to milliseconds, and in cases where materials have high levels of hydration, the emissions are not observed at all. It has been reported, however, that the weak emissions exhibited by ions such as Sm^{III} and Dy^{III} can be enhanced by replacing the water molecules with D_2O, which of course has markedly different vibrational energy levels.[31]

16.2.3 Applications of Lanthanide Phosphonates

The emissive properties of lanthanides are well known, and there are already many applications where these properties are exploited. Such applications include phosphors for imaging and display purposes, phosphors for security printing (e.g. banknotes) and as luminescent markers in bioassays. Many of these applications are based on simple oxide or oxide-related materials. Lanthanide phosphonates offer some advantages over traditional oxides which might expand the range of applications in which they might be employed.

The ready manner in which substitution of anions can be carried out can lead to materials with multiple responses, although as already noted, there can be technological issues with different ions requiring different excitation wavelengths. The latter problem is not really of such huge significance given the relative ease with which one can purchase LEDs which can cover any region of the electromagnetic spectrum from UV to IR.

The use of luminescent materials in security is well known. We have all seen the Eu and Tb emissions when modern banknotes are illuminated with a UV lamp. This is an example of an overt response, which is one that is readily detectable by the naked eye and permits rapid verification of the validity of a document. These are, however, relatively simple to forge for someone with a rudimentary chemical knowledge, so the security industries are interested in covert responses, *i.e.* ones which cannot be discerned by the naked eye and require machine readers. In particular the IR emission properties are of interest. Many of these emissions can be excited by wavelengths which are also in the IR region, e.g. those of Yb^{3+}. This leads to the ability to include covert information in printed documents where such materials can be included in printing media. Measurement of the lifetimes of excited states gives another level of complexity which can be exploited in security applications, leading to an almost forensic ability to specifically identify the source of a particular material.

The presence of the organic functional group lends additional possibilities for applications of lanthanide phosphonates. The range of functional groups available is extraordinary, and one can almost select a functional group to give rise to materials which have particular properties. Many lanthanide phosphonates have open structures and contain solvent molecules. Upon removal of these molecules luminescent properties can be radically altered; for example the IR emission of many lanthanide phosphonates is impossible where water (or other solvent) molecules are present, or the relative intensity of emission peaks or excited state lifetimes can be altered as the coordination environment of the metal changes. Such observations lead us to the possibility of sensing applications.

The luminescent properties of the lanthanide phosphonates is relatively little studied in comparison to structure and applications such as catalysis. Whilst the energy levels in the metal ions are intransigent, and hence the emission wavelengths are do not change overmuch, there are other variations, such as lifetimes and the production of mixed ion materials, which can be exploited in finding new applications.

16.2.4 Lanthanide Phosphonates: Conclusions

Observing the apparent ease with which lanthanides ions can be substituted into various frameworks and the range of emission wavelengths available from blue to IR, it might seem as though it would be a straightforward matter to create a phosphor that will emit with any chosen colour simply by selecting an appropriate combination of two or more metal ions. There are a number of factors that need to be considered. Firstly, it is not always simple to isostructurally substitute different ions as there are significant differences in the sizes of the lanthanide ions at either end of the f block. Substitution works well where the ions are near neighbours in the lanthanide series or where substitution is at low levels, but is by no means guaranteed. Significant differences in size can lead to different structures and in some cases changes in the local coordination environment.

It should also be noted, for example, that in $[Na_6Eu_{9-x}Yb_x(L)_{16}](OH)\cdot 51H_2O$ ($H_2L = $ 5-methyl-2,2-bipyridyl-6-phosphonic acid)[17] the two different ions are excited by different wavelengths; $\lambda_{ex} = 355$ nm for Yb and $\lambda_{ex} = 466$ nm for Eu. This means that if the aim is to synthesize a material with a combination of emissions which appeared a certain colour to the eye, it would be necessary to consider that multiple excitation sources might be required; something which may or may not be easy to achieve.

16.3 Transition Metal (TM) Phosphonates

Luminescent properties of complexes of transition metals have long been studied, with those containing ruthenium being particularly well known. Somewhat ironically, ruthenium is one of the few cation forming elements of the periodic table which shows little tendency to form metal phosphonate materials, although metal phosphonates containing organic groups can be used to coordinate Ru species.[32] Photoluminescence in TM compounds is usually associated with ions with a d^5 or d^{10} electronic configuration. Unlike the lanthanide ions, the formation of the excited state involves exciting and electron between orbitals with different *angular momentum quantum numbers*, e.g. for Ag^+ the ground state is $4d^{10}$ and the excited state is $4d^9 5p^1$.

One significant difference in the photoluminescent properties of the transition and lanthanide metals is the line widths of the observed spectra. The energy levels of the d valence electrons in the transition metals vary far more in energy than those of the f electrons as they are more closely involved in bonding and hence much more susceptible to ligand field effects and vibronic coupling which will lower the local symmetry of the metal ion. In this instance the selection rules suggest that such transitions are more allowed than those for f–f transitions, and hence the intensity of observed luminescence tends to be greater.

The same underlying principles apply, however, and the presence of antenna groups in the phosphonate anion can lead to energy transfer between the organic and inorganic parts of these materials giving rise to luminescent behaviour.

16.3.1 Zinc Phosphonates

One of the metals that possesses all the properties required to demonstrate luminescent properties is zinc. It is also one of the most prolifically used cations in metal phosphonate synthesis. The sheer number of zinc-containing phosphonates precludes an exhaustive review being carried out in this chapter, so only a few examples are included which it is hoped will convey the breadth of luminescent behaviour that can be observed for these materials.

$[(NH_4)_2Zn_3(O_3P(CH_2)_3PO_3)_2]$ is a 3D framework material in which the Zn atoms are tetrahedrally coordinated (Figure 16.11).[33] Tetrahedral coordination is the most common observed for Zn ions. The tetrahedra are linked by propylene disphosphonate groups to form a porous material with the NH_4^+ ions located in an 8MR channel.

When excited at 270 nm at room temperature, $[(NH_4)_2Zn_3(O_3P(CH_2)_3PO_3)_2]$ emits a luminescence in the ultraviolet region with a narrow band profile (Figure 16.12, top). The peak band appears at 319 nm with a shoulder at 310 nm. The full width at half maximum (fwhm) is only about 35 nm. The lifetime of the excited state ($\lambda_{ex} = 270$ nm, monitored at 320 nm) is 5.3(2) ns, which is very short in comparison with those of the lanthanide materials discussed previously.

Extending the length of the alkyl chain by one $-CH_2-$ unit gives rise to the related butylenediphosphonate material, $[NH_4Zn(O_3P(CH_2)_4PO_3H)]$,[33] which is again based on ZnO_4 tetrahedra, though this time the structure is more like that of a pillared layered material than 3D framework (Figure 16.13).

In contrast to the previous material, when excited at 280 nm at room temperature, $[NH_4Zn(O_3P(CH_2)_4PO_3H)]$ gives off a blue-green luminescence with a

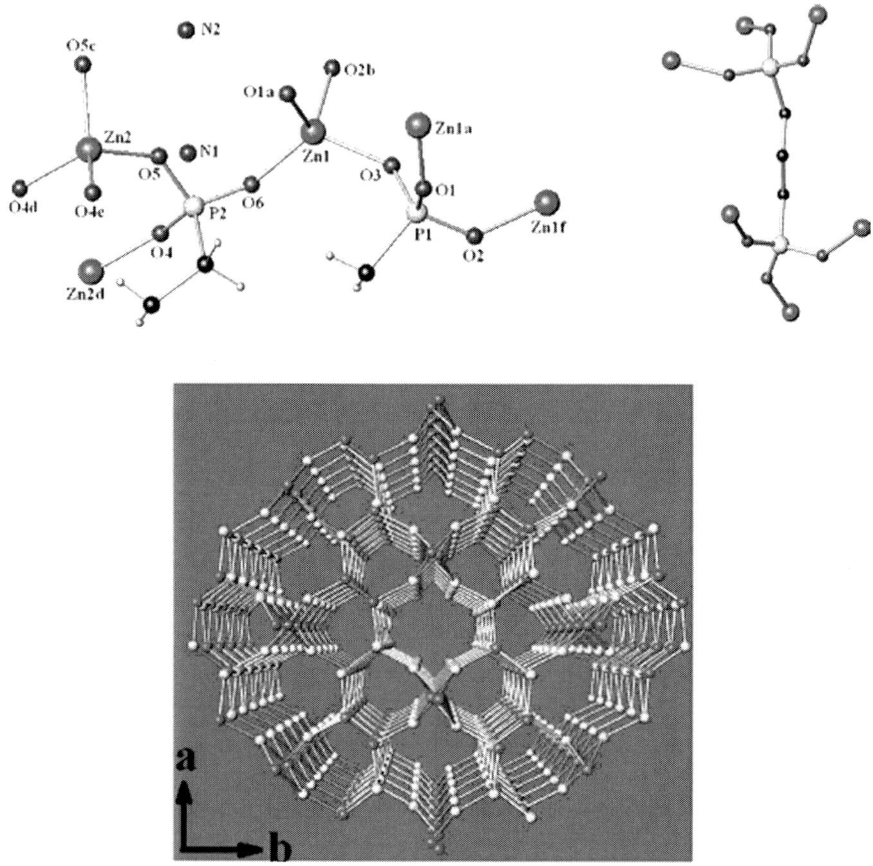

Figure 16.11 The structure and asymmetric unit of $[(NH_4)_2Zn_3(O_3P(CH_2)_3PO_3)_2]$. (Copied with permission from *Cryst. Growth Des.*, 2007, **7**, 1134. Copyright 2007 American Chemical Society.)

broad band profile (Figure 16.12, bottom). The peak band appears at 434 nm, which is blue-green in colour. The fwhm is about 215 nm, almost 6 times larger than the material containing the shorter alkyl chain, whilst the lifetime (λ_{ex} = 397 nm, monitored at 466 nm) is 55(3) ns, is around 10 times longer.[33] The authors offer no explanation for the observed luminescence properties. Given the short lifetimes of the excited state and, since the organic part of the phosphonate anion is saturated and there is no possibility of π –π* transitions, it is unlikely that ligand centred (LC) fluorescence is being observed. In both cases excitation will result in $3d^{10} \rightarrow 3d^9 4p^1$ transition. In the material with the shorter chain the transition back to the ground state occurs with less energy loss than the longer, possibly due to more vibrational excitation occurring in the latter. The main point to be taken is that UV emission is possible from Zn-containing materials, a phenomenon that is observed for ZnO. It is notable that

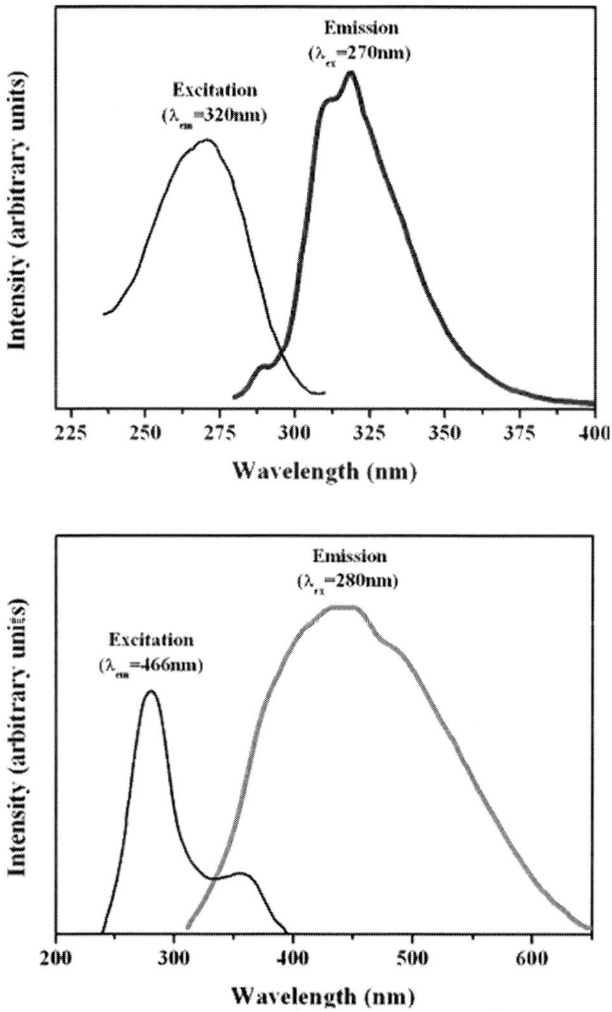

Figure 16.12 Top: solid state photoluminescence ($\lambda_{ex} = 270$ nm) and excitation ($\lambda_{ex} =$ 320 nm) spectra of $[NH_4Zn(O_3P(CH_2)_3PO_3H)]$. Bottom: solid state photoluminescence ($\lambda_{ex} = 280$ nm) and excitation ($\lambda_{ex} = 466$ nm) spectra of $[NH_4Zn(O_3P(CH_2)_4PO_3H)]$. (Copied with permission from *Cryst. Growth Des.*, 2007, **7**, 1134. Copyright 2007 American Chemical Society.)

in these examples there is no π system present in the phosphonate anion, hence the excitation process is one whereby the metal ion is excited directly.

$[(C_5H_{14}N_2)_{0.5}Zn_5(HO_3PCH(OH)CO_2)(O_3PCH(OH)CO_2)_3 \cdot 2H_2O] \cdot 2H_2O$ and $[(C_4H_{12}N_2)Zn_2(O_3PCH(OH)CO_2)_2]$[34] similarly do not contain π electron systems in either the phosphonate anion or the charge balancing cations. In the first of these Zn is present in octahedral (ZnO_6) and square pyramidal (ZnO_5) coordination environments, while in the latter coordination is purely tetrahedral (ZnO_4). The luminescent behaviour of the hydrated material $[(C_5H_{14}N_2)_{0.5}$

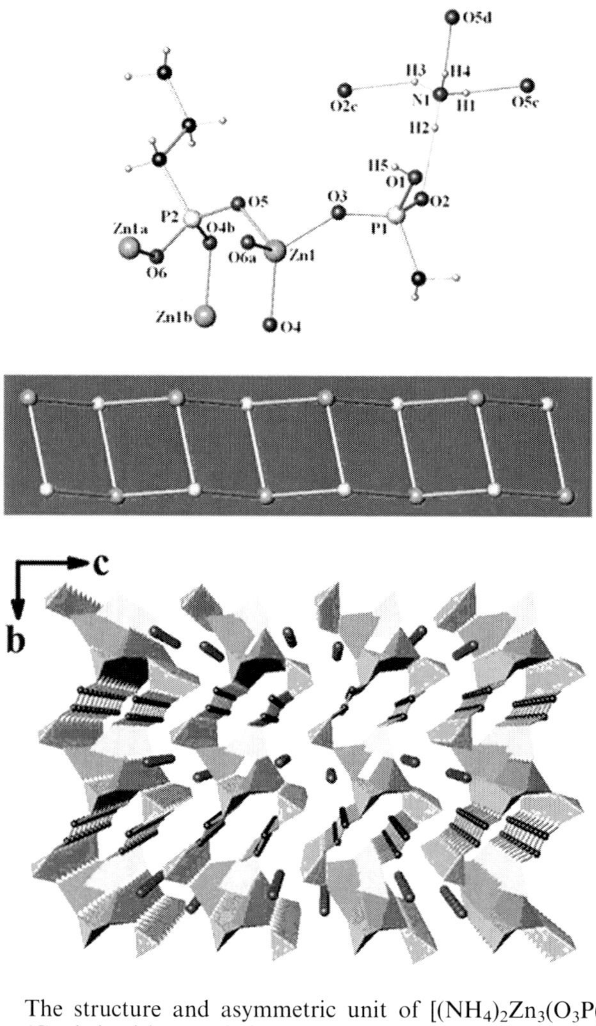

Figure 16.13 The structure and asymmetric unit of $[(NH_4)_2Zn_3(O_3P(CH_2)_3PO_3)_2]$. (Copied with permission from *Cryst. Growth Des.*, 2007, **7**, 1134. Copyright 2007 American Chemical Society.)

$Zn_5(HO_3PCH(OH)CO_2)(O_3PCH(OH)CO_2)_3 \cdot 2H_2O] \cdot 2H_2O$ is rather interesting. The as-prepared sample shows a weak purple luminescence around 425 nm, which the authors describe as being ligand centred. Heat treatment up to 250 °C removes water from the material, retaining the structure, and results in the luminescent response shifting to higher wavelengths. Treatment at 200 °C changes the emission spectrum to give a yellow band centred at 442 nm. The lifetime of ($\lambda_{ex,em} = 365$, 445 nm) is 2.1(1) ns, matching that of the parent material. A bright white emission can be seen with the naked eye when the solid, heated to 250 °C, is excited at 365 nm. This is ascribed to the fact that the 2-methylpiperazine and 2-hydroxyl(phosphono)acetate can give off purple-blue and yellow emissions respectively. The ratio of their intensities were alternated by shifting the wavelength of excited light, resulting in a form of reversible tuning of the luminescent colours. By increasing or decreasing the excitation wavelength, emission can be changed from green-yellow into purple-blue, bright white and blue-green. Increasing heat-treatment temperature to 300 °C results in the decomposition of the methylpiperazine cation and the luminescence becomes orange-red in colour. All the above results indicate that luminescent colours of this solid can be irreversibly transformed by simple heat-treatment.

The second material, $[(C_4H_{12}N_2)Zn_2(O_3PCH(OH)CO_2)_2]$, luminesces with a green-yellow emission, and due to the lack of water does not show the same variance in its emission characteristics.[34] The same phosphonic acid was used in the synthesis of four isostructural materials of composition $NaM(O_3P\text{-}CH(OH)CO_2)$ [M = Mn, Fe, Co and Zn).[35] The Zn derivative (along with the Mn and Fe derivatives) exhibit a bright blue-green emission which is ascribed to a ligand centred n–π* transition.

The presence of Zn in a phosphonate, however, does not guarantee a luminescent response. Zheng *et al.*[36] synthesized a series of zinc phosphonates containing ancillary ligating species: $[Zn_2(4\text{-cppH})_2(L^1)]$, $[Zn_3(4\text{-cpp})_2(L^2)_2]\cdot 2.5H_2O$, $[Zn_3(4\text{-cpp})_2(L^3)]\cdot H_2O$ where L^1 = 1,4-bis(imidazol-1-ylmethyl)benzene, L^2 = 1,3-bis(imidazol-1-ylmethyl)-benzene, L^3 = 1,2-bis(imidazol-1-ylmethyl)benzene and 4-$cppH_3$ = 4-carboxyphenylphosphonic acid. Of these three materials, the only one to display luminescence is the last, $[Zn_3(4\text{-}O_3PC_6H_4CO_2)_2$(1,2-bis(imidazol-1-ylmethyl)benzene)][36] with a violet emission at 433 nm (λ_{ex} = 335 nm). The emission in this material was ascribed to the 1,2-bis(imidazol-1-ylmethyl)benzene after the luminescent properties of the ancillary ligands showed this was the only one of the three which was active.

The related zinc phosphonate, $Zn_3(m\text{-}O_3PC_6H_4CO_2)_2$, however, shows a strong yellow luminescence at 578 nm on excitation at 265 nm (Figure 16.14).[37] Illumination of *m*-phosphonobenzoic acid alone shows a blue green response, and since this is not seen in the spectrum of the zinc salt, it is thought that the response comes solely from the Zn ions.

Rather interestingly, the Ag^+ derivative of the same phosphonic acid yields a material which exhibits a weak green luminescence centred at 502 nm, under the same excitation conditions (Figure 16.14).

16.3.2 Other Transition Metal Phosphonates

On the whole, the luminescent properties of transition metal phosphonates is little studied. Few examples of studies exist outside those containing Zn, where the luminescent response is a ligand-centred response. Such responses are usually blue or green in colour, being slightly shifted from the excitation wavelength.

More interesting observations can be made regarding Mn-containing phosphonates. It is known that Zn^{2+} ions can be excited in the phosphonates such as phenyl- or benzylphosphonates[37] and in these cases the luminescent response is notably brighter than that of ZnO. Mn^{II} derivatives of phenyl- and

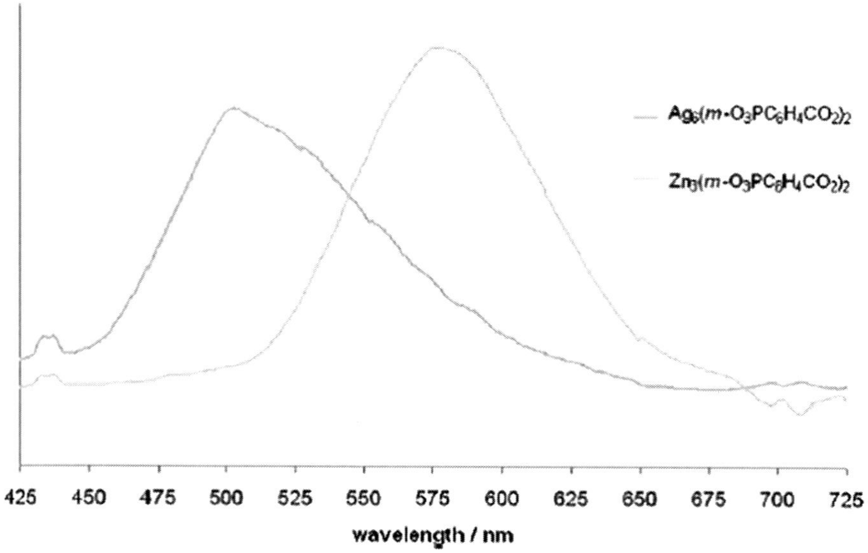

Figure 16.14 Photoluminescence spectra for zinc- and silver *m*-phosphonobenzoate ($\lambda_{ex} = 265$ nm) (from *Dalton Trans.*, 2010, **39**, 6024).[37]

benzylphosphonic acid can also be synthesized which are isostructural with the Zn-based materials. Materials containing Mn^{2+} have been shown to give a red luminescent response.[38] Again the versatility of metal phosphonate synthesis allows substitution of Mn into Zn frameworks. In a series of phenylphosphonates, it has been shown how the luminescent response of mixed metal phosphonates can be tuned by changing the relative content of the metals.[37] However, there is a large disparity in the intensity of the responses from Zn and Mn, which means that the emission from Zn is completed swamped by that of Mn if the Mn content is much above a fraction of a percent (Figure 16.15); in part it is thought the blue green emission from the Zn ions at 370 nm is exciting the Mn ions.

Substitution of Zn is not restricted to Mn. In the phenylphosphonate Zn can also be replaced by a small quantity (< 8%) of silver without disrupting the structure. The inclusion of silver in Zn phenylphosphonate results in a reduction of the blue component of the Zn emission spectrum around 370 nm which means that the response is dominated by the emission at 500 nm, giving a greener colour.

16.3.3 Transition Metal Phosphonates: Conclusions

Like the lanthanide phosphonates in the previous section, there is a temptation to believe that a transition metal phosphonate can be tailor-made to have a particular set of desired luminescent properties. Whilst is true that the relative ease of synthesis of these materials means that different combinations of metals

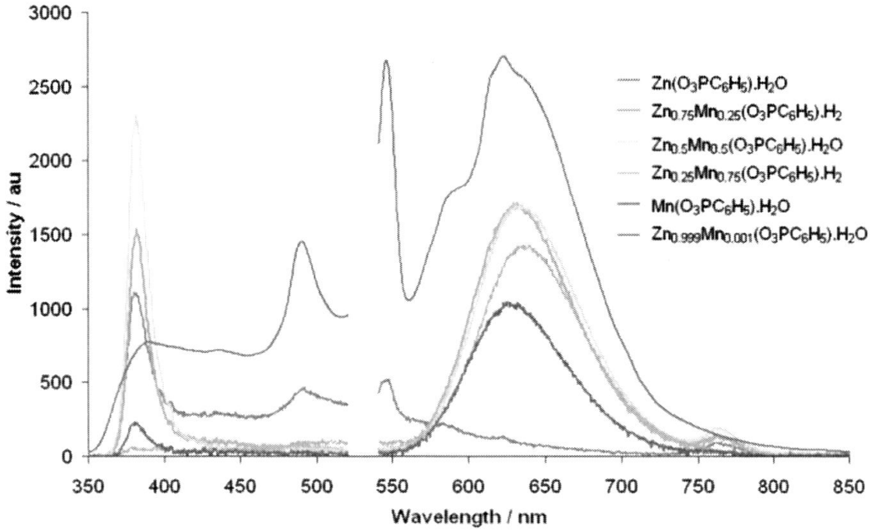

Figure 16.15 Emission spectra and images for a series of $Zn_{1-x}Mn_x$ phenylphosphonates (from *Dalton Trans.*, 2010, **39**, 6024).[37]

within the same structure is often simple to achieve, it is not always the case. Even then, where isostructural substitution is a simple matter the result may not be quite what is anticipated, due to the relative intensities of the different emissions, or the effects of one emission exciting a transition in the other ion(s) present.

It remains the case that luminescence is little studied in transition metal phosphonates when compared to lanthanide phosphonates, but there are several groups interested in this work. The difficulties not withstanding, this area of study is one with a great deal of interesting applications in the same areas as lanthanide-containing materials. One of the benefits of the transition metals is that their salts are relatively inexpensive and hence phosphors can be synthesized on the bulk scale. On the downside, the broad lines in the emission spectra and the relative ease with which the luminescent response undergoes significant change upon dehydration means that they are unlikely to find many applications in security applications.

16.4 Studying the Luminescent Properties of Metal Phosphonates

The visible, or more often, non-visible result of excitation of luminescent materials is the emission of light. Studying luminescent properties of materials is not simply a case of applying whatever excitation source is at hand and recording the emitted spectrum. As will be seen in the rest of this chapter, the study of lanthanide and other metal phosphonates uses a number of different excitation wavelengths, dependent on which metals and which phosphonates are contained in the material. Selection of the correct excitation wavelength can be crucial in order to observe an emission. The study of the photoluminescent properties of materials should also therefore include some form of absorption measurement. In the first and simplest instance this can take the form of diffuse reflectance UV–Vis spectroscopy, which will identify the wavelengths at which absorption of light occurs.

Once the absorption bands have been identified then these can be used to illuminate and excite the sample so that a photoluminescence spectrum can be recorded. Each absorption band identified in the diffuse reflectance measurement should be used to record the resulting emission spectra, as some transitions will be excited to a greater of lesser extent by different stimuli. This is often the most difficult part as it requires access to either a tunable laser source or a fluorimeter which is capable of studying solids, neither of which are commonplace pieces of equipment. For the sake of completeness, excitation spectra are often recorded. In these experiments the intensity of a particular emission peak is monitored as a function of the exiting wavelength. This allows clear identification of the most efficient excitation wavelength for particular emission band.

The study of IR emitting and absorbing materials brings further complication. Few standard absorption spectrometers and fluorimeters are capable of studying wavelengths in excess of 1000 nm, which causes obvious problems in studying the large number of materials whose most interesting properties exist beyond this wavelength. Those that can be used in these ranges often have much poorer resolution in the IR region ($\sim$5–10 nm) than in the UV–Vis part of the spectrum (often $<$1 nm). Where specialist equipment is available, it is usually much more expensive than standard spectrometers, which sadly leads many researchers to balance the relative gains against the cost in the study of these materials, and perhaps explains the relative dearth of examples in the literature.

Few chemistry laboratories, outside of specialist photochemical labs, are fully equipped to carry out the full suite of luminescent studies. It is often the case that the synthetic chemist must establish collaborations with other laboratories or make friends with colleagues in the physics department! Whichever is the case, it can be very rewarding to measure the properties of materials and conceive of new applications for them.

Acknowledgements

The author wishes to thank everyone who has contributed to metal phosphonate research at Nottingham Trent University and De Montfort University over the years. Special thanks go to Professor Paul-Alain Jaffrès of the Université de Brest, the numerous PhD and project students, and the many international and UK collaborators who have made all the work possible. Thanks are also due to EPSRC, The British Council, NTU and DMU for funding the work done within GBH's group.

References

1. J.-C. G. Bünzli and C. Piguet, *Chem. Soc. Rev.*, 2005, **34**, 1048.
2. J. G. Mao, *Coord. Chem. Rev.*, 2007, **251**, 1493.
3. G. Cao, V. M. Lynch, J. S. Swinnea and T. E. Mallouk, *Inorg. Chem.*, 1990, **29**, 2112.
4. R.-C. Wang, Y. Zhang, H. Hu, R. R. Frausto and A. Clearfield, *Chem. Mater.*, 1992, **4**, 864.
5. F. Serpaggi and G. Férey, *J. Mater. Chem.*, 1998, **8**, 2749.
6. E. Antic-Fidancev, F. Serpaggi and G. Férey, *J. Alloys Compd.*, 2001, **319**, 140.
7. F. Serpaggi, G. Férey and F. Antic-Fidancev, *J. Solid State Chem.*, 1990, **148**, 347.
8. O. R. Evans, H. L. Ngo and W. Lin, *J. Am. Chem. Soc.*, 2001, **123**, 10395.
9. F. Serpaggi, G. Férey and F. Antic-Fidancev, *J. Solid State Chem.*, 1999, **148**, 347.
10. I. L. V. Rosa, E. J. Nassar and O. A. Serra, *J. Alloys Compd.*, 1998, **275–277**, 315.
11. J. Legendziewicz, P. Gawryszewska, E. Gałdecka and Z. Gałdecki, *J. Alloys Compd.*, 1998, **275–277**, 356.
12. E. Gałdecka, Z. Gałdecki, P. Gawryszewska and J. Legendziewicz, *New J. Chem.*, 2000, **24**, 387.
13. Q. Yue, J. Yang, G.-H. Li, G.-D. Li and J.-S. Chen, *Inorg. Chem.*, 2006, **45**, 4431.
14. Q. H. Xu, L. S. Li, X. S. Liu and R. R. Xu, *Chem. Mater.*, 2002, **14**, 549.
15. J. M. Rueff, N. Barrier, S. Boudin, V. Dorcet, V. Caignert, P. Boullay, G. B. Hix and P. A. Jaffrès, *Dalton Trans.*, 2009, 10614.
16. L. Cunha-Silva, S. Lima, D. Ananias, P. Silva, L. Mafra, L. D. Carlos, M. Pillinger, A. A. Valente, F. A. Almeida Paz and J. Rocha, *J. Mater. Chem.*, 2009, **19**, 2618.
17. P. C. R. Soares-Santos, L. Cunha-Silva, F. A. Almeida Paz, R. A. Sa′ Ferreira, J. Rocha, T. Trindade, L. D. Carlos and H. I. S. Nogueira, *Cryst. Growth Des.*, 2008, **8**, 2505.
18. S. Comby, R. Scopelliti, D. Imbert, L. Charbonnière, R. Ziessel and J. C. G. Bünzli, *Inorg. Chem.*, 2006, **45**, 3158.
19. S. F. Tang, J. L. Song, X. L. Li and J. G. Mao, *Cryst. Growth Des.*, 2006, **6**, 2322.
20. G. B. Hix and L. Neeves, unpublished work.
21. T. H. Yang, D. K. Cao, Y. Z. Li and L. M. Zheng, *J. Solid State Chem.*, 2010, **183**, 1159.
22. T. H. Zhou, F. Y. Yi, P. X. Li and J. G. Mao, *Inorg. Chem.*, 2010, **49**, 905.
23. Y. Y. Zhu, Z. G. Sun, H. Chen, J. Zhang, Y. Zhao, N. Zhang, L. Liu, X. Lu, W. N. Wang, F. Tong and L. C. Zhang, *Cryst. Growth Des.*, 2009, **9**, 3228.
24. Y. L. Huang, M. Y. Huang, T. H. Chan, B. C. Chang and K. H. Lii, *Chem. Mater.*, 2007, **19**, 3232.

25. X. Li, Z. Xie, J. Lin and R. Cao, *J. Solid State Chem.*, 2009, **182**, 2290.
26. J. L. Song and J. G. Mao, *Chem. Eur. J.*, 2005, **11**, 1417.
27. Z. Y. Du, H. B. Xu and J. G. Mao, *Inorg. Chem.*, 2006, **45**, 9780.
28. Y. Yuan, T. Cardinaels, K. Lunstroot, K. V. Hecke, L. V. Meervelt, C. Görller-Walrand, K. Binnemans and P. Nockemann, *Inorg. Chem.*, 2007, **46**, 5302.
29. S. F. Tang, J. L. Song and J. G. Mao, *Eur. J. Inorg. Chem.*, 2006, 2011.
30. S. W. Magennis, S. Parsons and Z. Pikramenou, *Chem.-Eur. J.*, 2002, **8**, 5761.
31. R. Ferreira, P. Pires, B. de Castro, R. A. Sá Ferrerira, L. D. Carlos and U. Pischel, *New J. Chem.*, 2004, **28**, 1506.
32. I. Gillaizeau-Gauthier, F. Odobel, M. Alebbi, R. Argazzi, E. Costa, C. A. Bignozzi, P. Qu and G. J. Meyer, *Inorg. Chem.*, 2001, **40**, 6073.
33. R. Fu, S. Hu and X. Wu, *Cryst. Growth Des.*, 2007, **7**, 1134.
34. R. Fu, S. Hu and X. Wu, *Dalton Trans.*, 2009, 9440.
35. Z. Lai, R. Fu, S. Hu and X. Wu, *Eur. J. Inorg. Chem.*, 2007, 5439.
36. P. F. Wang, Y. Duan, D. K. Cao, Y. Z. Li and L. M. Zheng, *Dalton Trans.*, 2010, **39**, 4559.
37. R. Singleton, J. Bye, J. Dyson, G. Baker, R. M. Ranson and G. B. Hix, *Dalton Trans.*, 2010, **39**, 6024.
38. J. Escobal, J. L. Pizarro, J. L. Mesa, L. Lezama, R. Olazcuaga, M. I. Arriortua and T. Rojo, *Chem. Mater.*, 2000, **12**, 376.

CHAPTER 17

N-(Phosphonomethyl) iminodiacetic Acid in the Construction of Coordination Polymers

FILIPE A. ALMEIDA PAZ AND JOÃO ROCHA

University of Aveiro, Department of Chemistry, CICECO, Campus Universitário de Santiago, 3810-193 Aveiro, Portugal

17.1 Introduction

The design and construction of crystalline hybrid networks relies on the judicious choice of metal centers and organic ligands containing two or more coordination capable functional groups. Under certain reaction conditions, these inorganic and organic primary building blocks self-assemble into (1D, 2D, 3D) multi-dimensional crystalline architectures, usually known as coordination polymers.[1] In particular, the 3D architectures are commonly named metal-organic frameworks (MOFs). Because of the endless possible combinations of metal centers and ligands, there is no surprise in the exponential growth of papers published in the past decade in this field of research.[2]

Historically, research on coordination polymers was born among the crystal engineering field community that considered the hydrogen bond the main structure-directing force. Because the coordinative bond is considerably stronger than the typical hydrogen bonds, it ultimately emerged as a more convenient approach for designing robust and stable networks. Additionally, the use of

Metal Phosphonate Chemistry: From Synthesis to Applications
Edited by Abraham Clearfield and Konstantinos Demadis

Published by the Royal Society of Chemistry, www.rsc.org

transition metal centers also resulted in predictable coordination numbers and geometries for the network nodes. The combination of metallic centers and organic ligands, connected through strong coordinative bonds, allowed the isolation of novel extended compounds exhibiting interesting physicochemical properties, some typical of the organic component, others associated with the metal centers, such as photoluminescence[3] and magnetism.[4] This chapter is not focused on the general aspects of the synthesis and properties of coordination polymers. Several excellent reviews and book chapters addressing these matters are available in the literature and the reader is referred to them.[3–14]

It is common to find coordination polymers in which the organic bridges are solely, or partially, composed of rigid moieties like aromatic rings, such as 4,4′-bipyridine or terephthalic acid.[15] These rigid spacers provide a way of creating multi-dimensional scaffoldings while maintaining overall robustness. Flexible organic ligands are often avoided because they tend to promote the formation of both low-dimension complexes and dense structures. Flexible organic molecules also promote the formation of microcrystalline powders or poorly crystalline materials, which are not amenable to structural elucidation by X-ray diffraction techniques.

The most common multidentate, chelating and highly flexible organic ligand is, probably, ethylenediaminetetraacetic acid (EDTA, Scheme 17.1), a molecule containing six coordinating sites spanning over five highly flexible organic "arms". The result is that almost all metallic centers are easily sequestered by EDTA through the formation of several chelate rings. This binding capability makes this molecule very useful in laboratorial analyses (*e.g.* in the determination of metallic centers in natural waters),[16] as a food additive,[17] in medicine (*e.g.* as the chelating agent for lead and mercury poisoning),[18] and also in industry by removing undesirable metals from the reaction processes. Despite this, reports of crystal structures containing this molecule are not abundant. A search in the Cambridge Structural Database (CSD, Version 5.31 with four updates until August 2010)[19,20] reveals only 125 crystal structures out of which less than 15% are polymeric. Indeed, the ability of this molecule to sequester almost any type of metal center is also its weakness in the polymerization step, because it favors the formation of discrete complexes with no available coordinating sites to promote polymerization.

The strategy to design and construct 1D–3D coordination polymers with highly flexible ligands must consider the following: (i) the organic ligand should have a small number of highly flexible "arms" to reduce the sequestration degree although (ii) the molecule should keep a fair number of donor atoms to promote the formation of stable complexes; (iii) the terminal coordinating groups should be rich in donor atoms (such as carboxylates or phosphonates). We have, for example, employed with success molecules such as *N*-(phosphonomethyl)iminodiacetic acid or nitrilotris(methylenephosphonic acid) (Scheme 17.1, bottom) to form highly crystalline and photoluminescent layered networks having rare earth centres.[21–23]

In this chapter we shall focus on a slightly distinct flexible molecule, *N*-(phosphonomethyl)iminodiacetic acid (Scheme 17.1, top right), which has been

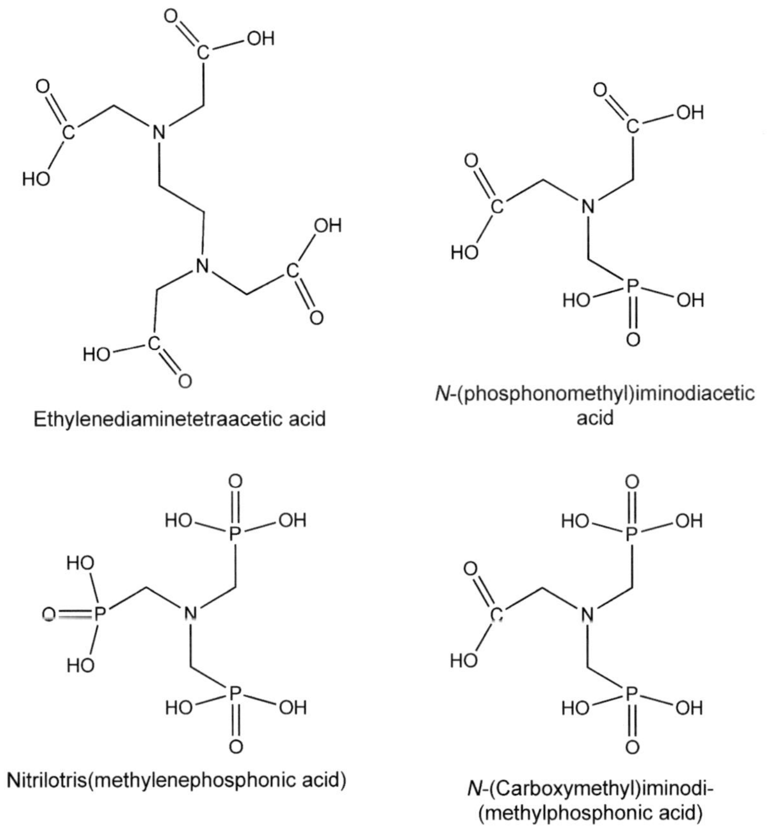

Scheme 17.1

employed in the construction of a number of multi-dimensional coordination polymers. This outstanding molecule may be envisaged as a fragment of EDTA in which the section starting at the second nitrogen atom was replaced by a phosphonic acid group. Here, we will show that this phosphonic acid group has important consequences on the overall connectivity of the ligand. In particular, this group effectively promotes the formation of physical bridges between metal centers, often leading to the self-assembly of dimeric units and to connections to a third metal, essential for the construction of MOFs.

17.2 *N*-(phosphonomethyl)iminodiacetic Acid

N-(phosphonomethyl)iminodiacetic acid is known in the literature and in industry as H_4pmida (Scheme 17.1, top right). Its worldwide importance resides in the fact that it is used in large scale as the main raw material to prepare glyphosate, a potent broad-spectrum herbicide. Glyphosate was developed by the Monsanto Company and appeared in the Roundup® herbicide in 1974.

Since the expiration of the original Monsanto patent, several other new attempts have appeared describing novel methods of controlling the formation of this widely employed herbicide. It is, thus, not surprising to still find much industrial research on this molecule, mainly focused on its physicochemical properties, such as solubility.[24,25]

The industrial production process of glyphosate is based on Mannich's condensation of iminodiacetic acid with paraformaldehyde and phosphorous acid, leading to an intermediate species that is finally converted into glyphosate by an oxidative decarboxylation reaction. This reaction usually occurs in the presence of a catalyst: under homogeneous conditions, mixtures of H_2O_2/H_2SO_4 or O_2/Co^{2+} are employed; in heterogeneous systems, activated carbons or Pt-C catalysts with oxygen (or air) are used instead as the oxidizing agent.

17.3 Literature Survey

17.3.1 Methodology and Notation

The main tool employed in the search for crystal structures containing residues of H_4pmida was the Cambridge Structural Database (Version 5.31 with four updates until August 2010).[19,20] The interface employed was ConQuest,[26] and all hits were restricted to crystal structures with the 3D coordinates properly determined and deposited. Variable bond types were allowed for the carboxylate and phosphonate groups. The same type of variable connection was enforced for the bonds between the oxygen atoms and the metal centers (defined as "any" within the periodic table). The search yielded a total of 52 crystal structures. The crystallographic details of the retrieved entries were inspected using the software package Mercury.[26,27] All representations given in this chapter have been created from the corresponding CSD entries (refcodes are indicated in each figure) using either Crystal Impact Diamond[28] or Chimera.[29]

Discrete complexes are depicted with a "0" in the top left corner (e.g., ^{0}[*composition*]) of the formula while polymeric structures indicate their dimensionality by a number (e.g., $_\infty{}^1$[*composition*] refers to a 1D coordination polymer).

17.3.2 Metallic Centers

Residues of H_4pmida (hereafter designated as $H_{4-x}pmida^{x-}$ to account for all possible species) can easily coordinate to a large number and type of metallic centers, spanning through all main blocks of the periodic table (see Scheme 17.2 for a graphical representation of all cations coordinated to these residues).

Alkali metals (s block) appear mainly as charge-balancing cations. The first reported examples in the literature emerge from Crans and co-workers and concern Na^+: $Na_8[V_2O_2(pmida)_2]_2 \cdot 16H_2O$ and $Na_4[V_2O_2(pmida)_2] \cdot 10H_2O$.[30] However, the most striking report is the series of structures reported by Wood and co-workers: $M_x[Fe(pmida)]_6 \cdot nH_2O$ where $M^+ = Na^+$ ($x = 11$), K^+

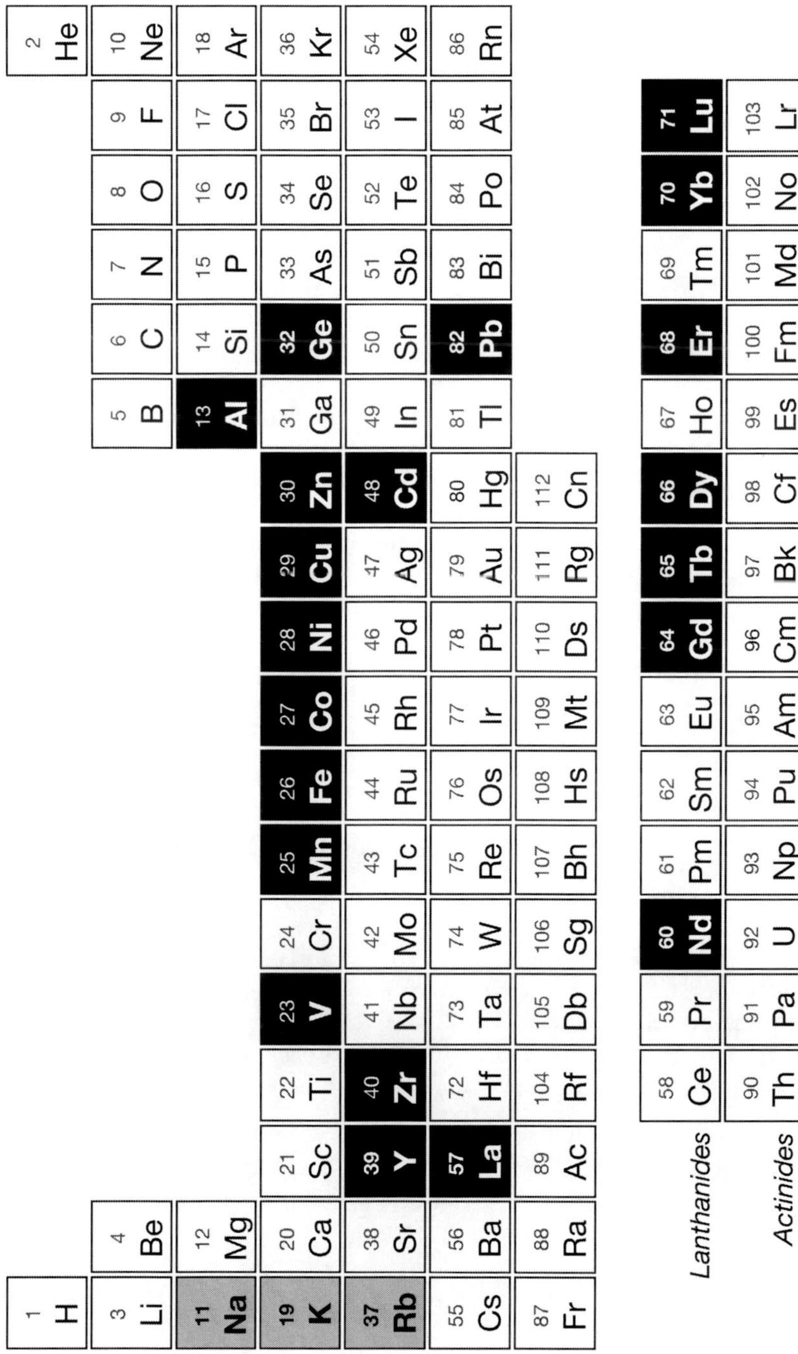
1 H
2 He
3 Li 4 Be 5 B 6 C 7 N 8 O 9 F 10 Ne
11 Na 12 Mg 13 Al 14 Si 15 P 16 S 17 Cl 18 Ar
19 K 20 Ca 21 Sc 22 Ti 23 V 24 Cr 25 Mn 26 Fe 27 Co 28 Ni 29 Cu 30 Zn 31 Ga 32 Ge 33 As 34 Se 35 Br 36 Kr
37 Rb 38 Sr 39 Y 40 Zr 41 Nb 42 Mo 43 Tc 44 Ru 45 Rh 46 Pd 47 Ag 48 Cd 49 In 50 Sn 51 Sb 52 Te 53 I 54 Xe
55 Cs 56 Ba 57 La 72 Hf 73 Ta 74 W 75 Re 76 Os 77 Ir 78 Pt 79 Au 80 Hg 81 Tl 82 Pb 83 Bi 84 Po 85 At 86 Rn
87 Fr 88 Ra 89 Ac 104 Rf 105 Db 106 Sg 107 Bh 108 Hs 109 Mt 110 Ds 111 Rg 112 Cn
Lanthanides
58 Ce 59 Pr 60 Nd 61 Pm 62 Sm 63 Eu 64 Gd 65 Tb 66 Dy 67 Ho 68 Er 69 Tm 70 Yb 71 Lu
Actinides
90 Th 91 Pa 92 U 93 Np 94 Pu 95 Am 96 Cm 97 Bk 98 Cf 99 Es 100 Fm 101 Md 102 No 103 Lr

Scheme 17.2

($x = 11$) and Rb^+ ($x = 10$).[31] Connectivity between $[Fe_2(pmida)_2]^{2-}$ units (with Fe^{3+}) leads to the formation of charged layers, as described in detail in Section 17.4.3. The alkali cations appear, thus, to balance the charge of the layers while they also establish bridges between adjacent nets. Another feature worth mentioning is that when the size of the charge-balancing cation increases, the overall crystal symmetry is also modified, with the Na^+ structure being the most symmetrical one (in space group *R*-1).

Elements of the p block have been rarely coordinated to $H_{4-x}pmida^{x-}$ residues, leading mainly to discrete binuclear complexes such as those with Al^{3+} ($[Al_2(pmida)_2(H_2O)_2]^{2-}$ and $[Al_2(pmida)_2(OH)_2]^{4-}$),[32] and Ge^{4+} ($[Ge_2(pmida)_2(OH)_2]^{2-}$).[33] Nevertheless, one of the most iconic structures with this organic ligand, and described in detail in Section 17.4.4, concerns the 3D porous material formulated as $[Pb_2(pmida)] \cdot 1.5H_2O$.[34]

The vast majority of structures are discrete complexes, or low-dimension polymers, with d-block cations, mainly those belonging to the first row. Remarkably, there is to date an extensive study of the structural consequences of the self-assembly of $H_{4-x}pmida^{x-}$ residues with V^{4+}, of which the use of the anionic $[V_2O_2(pmida)_2]^{4-}$ unit is a landmark.[35] As a consequence of the great focus from the research community on the first-row transition metals, some structures were also found to exhibit interesting magnetic properties such as the hexanuclear complex $[Co_6(pmida)_6]^{12-}$.[36] More recently some research groups have been studying the engineering of rare earth elements into multi-dimensional hybrid structures. In this context the reader is referred to the reports by Gu *et al.*[37] and Tang *et al.*[38] Remarkably, most likely due to the combined effect of the large coordination numbers of rare earth cations and the high connectivity of $H_{4-x}pmida^{x-}$ residues, all these structures are 3D frameworks.

17.3.3 Sequestering Modes of $H_{4-x}pmida^{x-}$ Residues

Residues of H_4pmida can bind to metal centers (Scheme 17.3) through several types of coordination modes. The existence of three oxygen-based functional groups, plus the central nitrogen atom, confers to $H_{4-x}pmida^{x-}$ a great ability to simultaneously coordinate to several cations. This creates physical links and, thus, expands the dimensionality of the structures. Scheme 17.3 summarizes the various types of sequestering coordination modes encountered for $H_{4-x}pmida^{x-}$ residues in most of the structures highlighted in the following subsections.

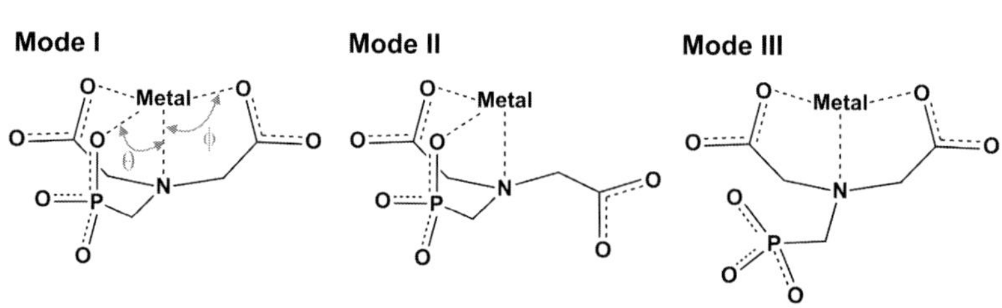

Scheme 17.3

The most common coordination fashion is best described as a sequestration of the cations by way of forming three five-membered chelate rings, in which the central nitrogen acts as the *pivot* donor atom (coordination mode I, Scheme 17.3). Despite being extremely stable, this sequestration still leaves, at least, two vacant coordinative sites in the sphere of the metallic centers. This permits the coordination of either auxiliary bridging ligands (*e.g.* 4,4′-bipyridine), or symmetry related carboxylate/phosphonate groups of neighboring $H_{4-x}pmida^{x-}$ residues in order to create extended structures. For this common coordination mode, it is possible to define two bite angles: θ for the N–metal–$O_{phosphonate}$ angles and φ for the corresponding N–metal–$O_{carboxylate}$ angle (shown in Scheme 17.3). For all structures with sequestrating mode I these angles were found in the $71.6 < \theta < 89.6°$ and $62.1 < \varphi < 86.4°$ ranges, respectively. It is possible to observe that for both θ and φ the lower values usually appear in structures with either small radii cations (*e.g.* Y^{3+}, Er^{3+}, Lu^{3+})[38] or in densely packed compounds such as $[Pb_2(pmida)] \cdot 1.5H_2O$ reported by Clearfield and co-workers.[34] The larger values of θ usually appear for structures in which the chelated carboxylate groups further establish *syn,anti*-bridges[39] with neighboring metallic centers. A striking example is the 86.4° bite angle for the bridge between Cu^{2+} and Nd^{3+} in the recently reported $[Cu_2Nd_2(pmida)_2(C_2O_4)(H_2O)_6] \cdot 3.5H_2O$ compound.[37] Conversely, the large values of θ usually appear in structures where the phosphonate group bridges several metal centers, *e.g.* in $[Cu_2Co_2(pmida)_2(H_2O)_6] \cdot 3H_2O$ this tetrahedral group establishes connections among four distinct metal centers, leading to a bite angle of *ca.* 89.2°.[37]

The release of one or two of the chelating pendant arms of $H_{4-x}pmida^{x-}$ leads to slightly distinct coordination fashions. Nevertheless, it is important to emphasize that these modes are significantly less common, with only a handful of structures exhibiting them. For example, one carboxylate-containing arm can appear free (mode II, Scheme 17.3b), but this was only observed for the 1D coordination polymer $_{\infty}{}^{1}[Fe(H_2pmida)(H_2O)_2]$.[40] Another possibility is the phosphonate moiety not being coordinated to the sequestrated metal (mode III, Scheme 17.3c), as observed in the dimeric anionic unit $[Al_2(pmida)_2(OH)_2]^{4-}$.[32]

17.3.4 Other Coordination Modes of $H_{4-x}pmida^{x-}$ Residues

The aforementioned *N*-(carboxymethyl)iminodi (methylphosphonic acid) or nitrilotris(methylenephosphonic acid) usually appear in the crystal structures with the central nitrogen atom protonated.[21–23] The occurrence of this structural feature for residues of $H_{4-x}pmida^{x-}$ would lead to coordination modes strikingly distinct because the sequestration could not occur. Indeed, three uncommon structures have been reported with, at least, one $H_{4-x}pmida^{x-}$ ligand having the central nitrogen atom protonated, which clearly leads to remarkably distinct molecular arrangements.

The most striking material is $[Mn(H_2O)(H_2pmida)]$,[41] where H_2pmida^{2-} does not encapsulate any cation and the chelating flexible arms are instead extended and, therefore, able to establish physical links between multiple neighboring

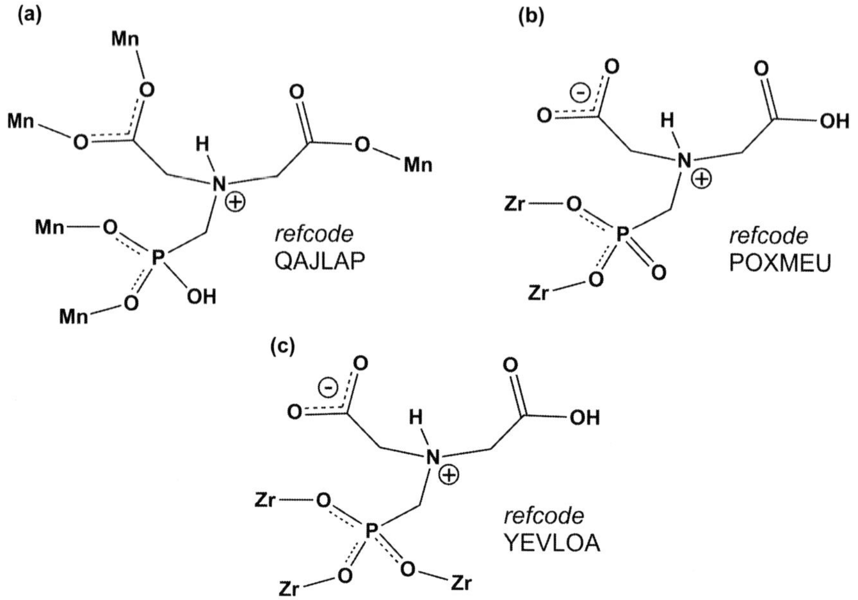

Scheme 17.4

Mn^{2+} centers (see Scheme 17.4a). This results in high connectivity leading to a 3D framework.

The ability of the phosphonate group to establish strong connections between metal centers is also clearly shown in the 1D and 2D (respectively) Zr^{4+} structures from the group of Clearfield:[42,43] $(NH_4)[ZrF_2\{H_3(Hpmida)_2\}]\cdot 3H_2O\cdot NH_4Cl$ (further details in Section 17.4.2) and $[Zr(PO_4)(H_2pmida)(Hpmida)(H_2O)_2]$ (further details in Section 17.4.3). In these structures only the phosphonate moiety of $H_{4-x}pmida^{x-}$ is bound to the metal centers, with the two pendant uncoordinated carboxylate groups being engaged in hydrogen bonding interactions (see Schemes 17.4b and 17.4c).

Inspection of the employed synthetic methods which led to the protonation of the central nitrogen atom is not very clear because of the small number of reported structures exhibiting this unusual feature. Nevertheless in the three preparations a strong base is always added at the beginning to deprotonate H_4pmida (either tetramethylammonium hydroxide or ammonium hydroxide). In the structures reported by Clearfield, a series of acids are also added afterwards (HCl, HF and phosphoric acid) and prior to the hydrothermal synthesis. Thus, it is feasible to assume that in the latter structures (POXMEU and YEVLOA) protonation most certainly arises from a strong acidic medium; on the other hand, in the structure of Fan *et al.* (QAJLAP)[41] protonation might be a strict requirement in order to balance the crystal charge. The use of a strong acidic medium in the structures of Clearfield is also well shown in the existence of a protonated carboxylic acid group.

17.4 Compounds with $H_{4-x}pmida^{x-}$

17.4.1 Discrete Complexes

The performed search for crystallographic details of compounds with $H_{4-x}pmida^{x-}$ residues showed that the vast majority of known structures are discrete complexes. This kind of predominance is very much similar to that found for ethylenediaminetetraacetic acid (see Section 17.1) and also for nitrilotriacetic acid (a search in the CSD for compounds with residues of this molecule resulted in a total 356 structures, of which only about one third are truly polymeric).

17.4.1.1 Monomers

In general, $H_{4-x}pmida^{x-}$ residues tend to aggregate the available metal centers in the reactive medium into high-nuclearity complexes rather than into simple, mononuclear-based compounds. Nevertheless, there are a handful of mononuclear complexes reported in the literature and fully characterized in the solid state (all with first row d-block elements). There are three common features among these compounds. First, all structures exhibit the typical coordination mode I for the sequestering $H_{4-x}pmida^{x-}$ residues, as this is the most effective way to chelate to cations. Second, in all structures the necessary charge-balancing protons are always, at least, located in the phosphonate group of $H_{4-x}pmida^{x-}$, which is a consequence of the stronger acidic nature of the two carboxylic groups. Third, all crystal structures are rich in hydrogen bonding interactions (*e.g.* of the O – H···O and N – H···O kinds), which were revealed to be the main supramolecular force directing crystal packing.

The most simple mononuclear complex was described by Ma *et al.*,[44] having Zn^{2+} coordinated to $Hpmida^{3-}$ and to two water molecules: $(NH_4)^0[Zn(Hpmida)(H_2O)_2]$ (Figure 17.1a). It is well known from coordination chemistry that these two water molecules may be displaced by, for example, the use of chelating agents. The other two known complexes, both with Cu^{2+}, can thus be obtained by using phenanthroline and diethylenetriamine in the

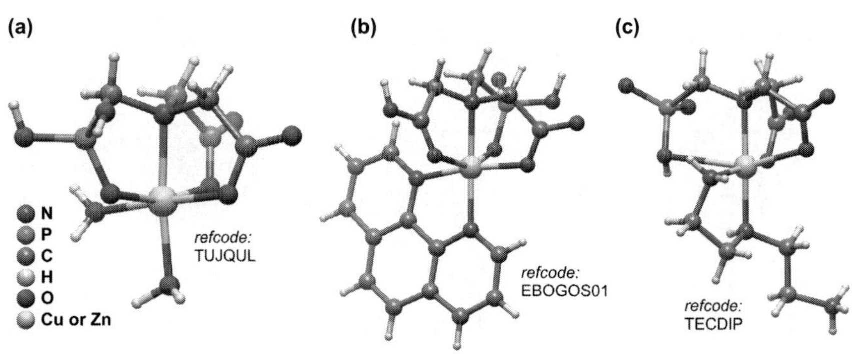

Figure 17.1 Ball-and-stick representations of the discrete (a) $^0[Zn(Hpmida)(H_2O)_2]$, (b) $^0[Cu(H_2pmida)(phen)]$, and (c) $^0[Cu(Hpmida)(Hdet)]$ complexes.

reactive medium. In one compound, H_2pmida^{2-} residues are coordinated to Cu^{2+} which is then chelated by a phenanthroline (phen) molecule: $^0[Cu(H_2pmida)(phen)] \cdot 3H_2O$ (Figure 17.1b). This compound was first described in 2004,[45] but its crystal structure was redetermined at low temperature more recently.[46] Indeed, the latter crystallographic study allowed the unequivocal identification of the protonated groups of H_2pmida^{2-} (the phosphonate and one carboxylate groups contain, each, one hydrogen rather than being located both in the phosphonate moiety, as proposed by Pei *et al.*). Consequently, a complete description of the supramolecular structure of this compound could be achieved. Complexes close pack, mediated by strong $O-H \cdots O$ hydrogen bonds, aggregating into supramolecular ladders which are strongly π–π stacked to each other *via* the *N*,*N*-chelated phenanthroline molecules. The second known discrete mononuclear complex[47] contains a flexible (i.e. non-aromatic) *N*,*N*-chelated moiety, diethylenetriamine (det): $^0[Cu(Hpmida)(Hdet)] \cdot 1.5H_2O$ (Figure 17.1c). In this structure it is possible to observe proton transfer from one carboxylic acid group to the more basic pendant amine moiety of the ligand det.

17.4.1.2 Dimers

This organic ligand, in its various deprotonated states, has a great tendency to induce the formation of discrete binuclear complexes with a range of metallic centers, mainly those belonging to the d block. Figures 17.2 and 17.3 depict the dimeric units usually encountered and their representation in space filling mode (i.e. by using van der Waals radii for all elements), with the objective to show their molecular shape and size.

In general terms, the formation of these dimeric species is driven by the high connectivity of the phosphonate moiety which, while chelating to one metal center *via* coordination mode I (Scheme 17.3a), also establishes a bridge with a neighboring metal by donating one oxygen atom to its coordination sphere (Figures 17.3 and 17.4). This connectivity can be envisaged as a head-to-head connection between similar monomers, imposing average intermetallic distances of about 5.23, 4.61, 4.86 and 4.99 Å for V $\cdots$ V,[30] Ge $\cdots$ Ge,[33] Al $\cdots$/ Al[32] and Fe $\cdots$ Fe,[31] respectively. The exception to this coordination fashion is the case of one Al^{3+} dimer, $[Al_2(pmida)_2(OH)_2]^{4-}$ (Figure 17.2d),[32] for which $H_{4-x}pmida^{x-}$ adopts the rarely encountered mode III (Scheme 17.3c). In this complex, direct intermetallic bridges are mainly formed by μ_2-hydroxyl groups, which imposes a close proximity between metal centers: Al $\cdots$ Al distance of *ca.* 2.89 Å.

An interesting crystallographic detail concerns the symmetry elements that drive the formation of these dimeric units. In general, dimers are formed by inversion symmetry, with the crystallographic operation being located at the center of gravity of the units, as well shown from a visual inspection of the V^{4+}, Al^{3+} and $Fe^{2+/3+}$ moieties depicted in Figure 17.2. Thus, the two coordination polyhedra of the metal centers are also inverted in relation to each other. The

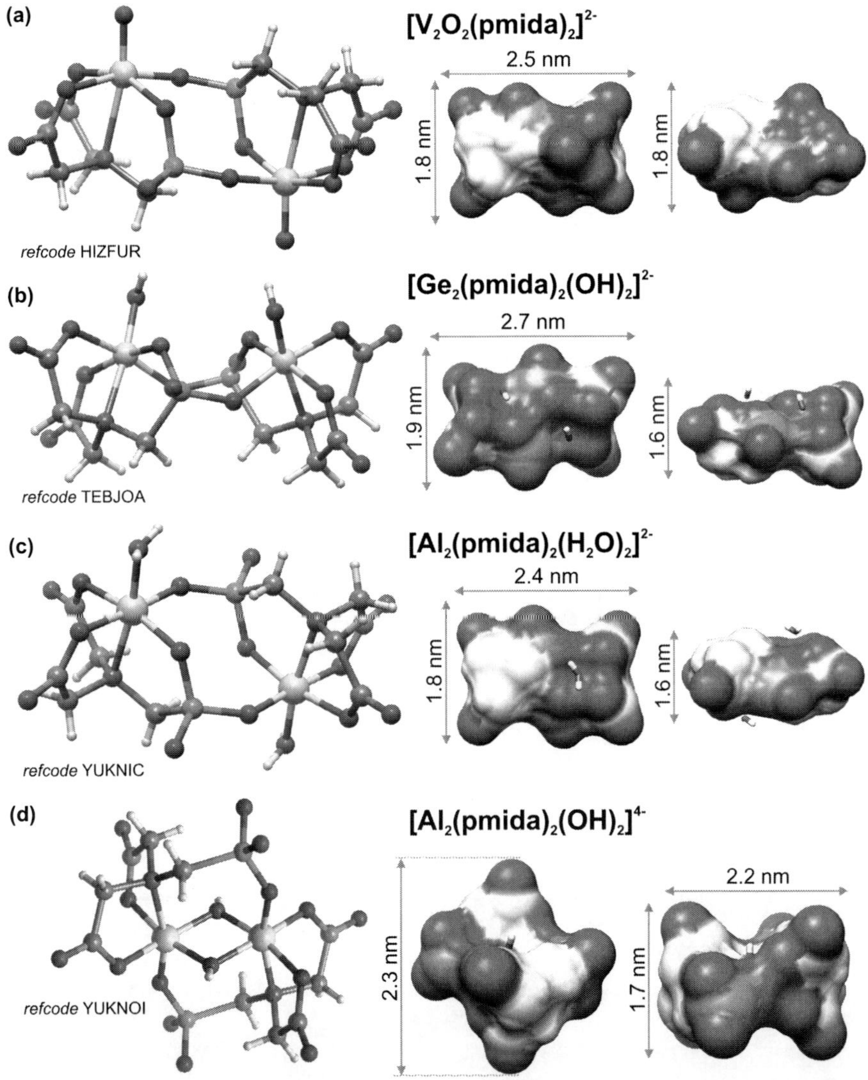

Figure 17.2 Symmetric dimeric units with $H_{4-x}pmida^{x-}$ residues for which the coordination spheres of the metallic centers (V^{4+}, Ge^{4+}, and Al^{3+}) are complete. The molecular surfaces represented on the left have been calculated by assuming a probe molecule with a radius of 1.4 Å (vertex density 10.0).

anionic $[Ge_2(pmida)_2(OH)_2]^{2-}$ unit deviates from this general crystallographic trend, being instead assembled by a two-fold rotation axis. Concomitantly, in this unit the two $pmida^{4-}$ ligands are located on the same side while, on the other, two hydroxyl groups coordinate to Ge^{4+} and point outwards of the dimer (Figure 17.2b).

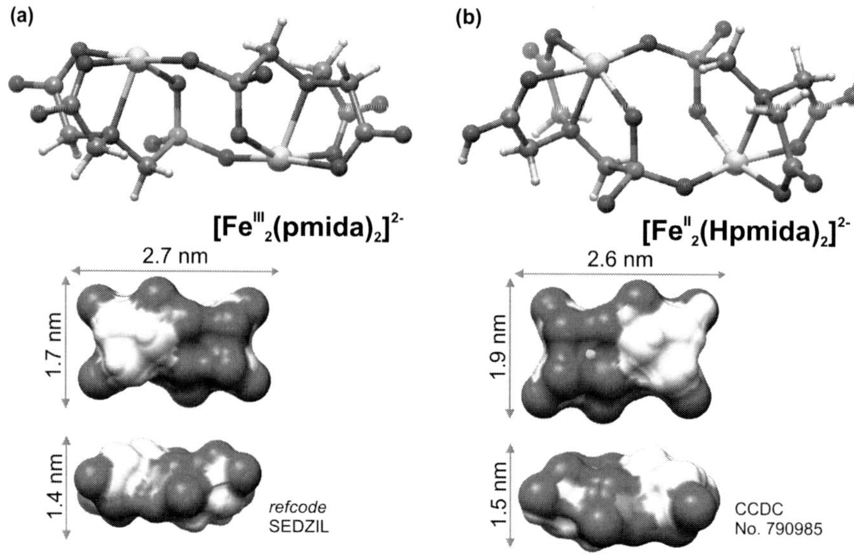

Figure 17.3 Centrosymmetric dimeric units with H_{4-x}pmida^{x-} residues connected to $Fe^{2+/3+}$ centers having incomplete coordination spheres for these metallic centers. The molecular surfaces represented below have been calculated by assuming a probe molecule with a radius of 1.4 Å (vertex density 10.0).

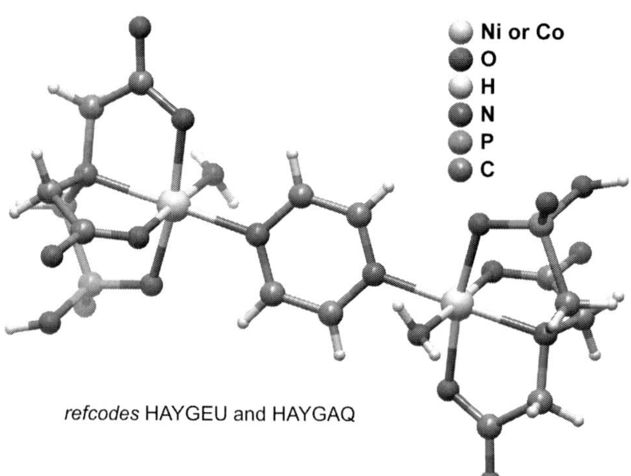

Figure 17.4 Ball-and-stick representation of the centrosymmetric anionic $[M_2(Hpmida)_2(pyr)(H_2O)_2]^{2-}$ dimeric unit described by Shi *et al.* (where $M^{2+} = Ni^{2+}$ or Co^{2+}; pyr stands for pyrazine).[53]

These dimers can be divided into two small distinct families concerning the features of the coordination sphere of the metallic centers which, as depicted from the empirical formulae, always contains one pmida^{4-} ligand: while for the structures with V^{4+}, Ge^{4+} and one with Al^{3+} the sixth coordination site is occupied by oxido, hydroxyl or water groups, for the $Fe^{2+/3+}$ moieties this site is completely available to further coordination. As shown in the space filling representations, even though for the three former units these terminal coordinating groups are located in the molecular peripheries, they cannot be easily displaced and replaced by other ligands. Because $[Al_2(pmida)_2(H_2O)_2]^{2-}$ has one water molecule occupying this terminal position, ligand replacement is, thus, more probable (water is a more labile ligand than a charged species). This arises as the main reason why Georgantas *et al.*[32] were able to isolate and report the two distinct units depicted in Figures 17.2c and 17.2d. The authors investigated some of the chemical conditions which tend to promote the formation of each species and found that the pH of the medium was of crucial importance: while at low pH values $[Al_2(pmida)_2(H_2O)_2]^{2-}$ exists predominantly in solution, when the pH is raised to 7–8 the phosphonate group leaves its role as the intermetallic bridge leading to the formation of $[Al_2(pmida)_2(OH)_2]^{4-}$ with the concomitant drastic reduction of the Al···Al distance by about 2 Å. The exact mechanism by which this transformation occurs remains unclear, but these studies helped to show that all dimeric units presented in this chapter, in particular that with the labile water molecules, might have, according to the conditions, some structural flexibility and dynamics in solution.

The incomplete coordination spheres of $Fe^{2+/3+}$ (Figure 17.3) make these units highly "reactive". In opposition to that observed for all other dimers depicted in Figure 17.2, these $Fe^{2+/3+}$ units could not be isolated individually, being always impregnated into polymeric matrices usually promoted by connections among themselves, as will be explained in more detail in subsequent sections (see, for example, those concerning the formation of 2D and 3D polymers). The anionic $[V_2O_2(pmida)_2]^{2-}$, $[Ge_2(pmida)_2(OH)_2]^{2-}$, $[Al_2(pmida)_2(H_2O)_2]^{2-}$, $[Al_2(pmida)_2(OH)_2]^{4-}$ units have, thus, the structural features which can render them as potential secondary buildings units (SBUs) to be employed in the construction of modular MOFs:[48,49] (i) they are chemically stable, both in solution and in the solid state (for the case of the Al^{3+} dimers the pH has to be controlled very closely); (ii) they can be isolated as individual units in the solid state; (iii) they all exhibit anchoring sites amenable for coordination to a multitude of metallic centers (in total, four carboxylate and two phosphonate moieties); (iv) they exhibit some structural rigidity. All this potential has already been put into practice from the reports of several research groups who have used, for example, the $[V_2O_2(pmida)_2]^{2-}$ unit to construct a handful of remarkable networks. The reader is referred to the detailed explanation of the concept in Section 17.5.2.

The structural versatility of $[V_2O_2(pmida)_2]^{2-}$ is further shown in a number of different crystallographic reports scattered in the literature in which this unit acts as a charge-balancing, bulky but structurally robust anion: in compounds $(4,4'\text{-bpyH})_2[M(4,4'\text{-bpy})(H_2O)_4][V_2O_2(pmida)_2]\cdot 2H_2O$ (where 4,4′-bpy stands for 4,4′-bipyridine and $M^{2+} = Mn^{2+}$ and Co^{2+}) reported by Shi *et al.*;[50] in $[Zn(H_2O)_6]_2[V_2O_2(pmida)_2]\cdot 2H_2O$ described by Gao *et al.*;[51] and the salt $(4,4'\text{-bpyH}_2)[V_2O_2(pmida)_2]\cdot 2H_2O$ by Fernandes *et al.*[52] Because of these combined features, the use of anionic dimeric units in the self-assembly of modular MOFs should not be discarded. In this context, it is important to stress that the molecular sizes and geometries of these units are convenient for their use as building bricks: on the one hand, all units have nanometric sizes, being *ca.* 2.2–2.7 nm long, *ca.* 1.7–1.9 nm wide and *ca.* 1.4–1.8 nm in height (Figures 17.2 and 17.3); on the other, the molecular surfaces are rich in oxygen donor atoms. A special note should be given, once again, to the $[Al_2(pmida)_2(OH)_2]^{4-}$ dimer which has a considerably distinct shape from the remaining moieties, with the distance between opposite carboxylate groups reaching *ca.* 2.3 nm (Figure 17.2d). The unit itself has a small cavity ending with the bridging hydroxyl groups that might act, in a supramolecular framework, as hydrogen bonding recognition sites.

There are also molecular dimers with $H_{4-x}pmida^{x-}$ residues in which the metallic bridge is not achieved by any functional group of this chelating ligand. This situation is rare and, to date, there is only one report by Shi *et al.* describing such a compound.[53] The centrosymmetric dimeric $[M_2(Hpmida)_2(pyr)(H_2O)_2]^{2-}$ unit (where $M^{2+} = Ni^{2+}$ or Co^{2+}; pyr stands for pyrazine) has the $Hpmida^{3-}$ ligand chelated to M^{2+} through the typical coordination mode I (see Section 17.3.3). As observed in other situations, the charge-balancing proton is located at the phosphonate group (see Figure 17.4). The remaining two coordinative positions are occupied by a water molecule and a bridging pyrazine ligand which effectively leads to the formation of a linear dimer, very much resembling a rotaxane. The intermetallic distances imposed by this bridge are *ca.* 6.93 Å and 6.83 Å for Co···Co and Ni···Ni, respectively.

17.4.1.3 Tetramers and Tetranuclear Complexes

While in the majority of the dimeric units described in the previous section the bridges *via* $H_{4-x}pmida^{x-}$ occur by way of the phosphonate moieties (head-to-head connections), in the rare examples of tetrameric units resembling "molecular squares" the connections are instead established through *syn,anti*-bridging carboxylate groups, in what can be seen a head-to-tail connection among adjacent monomers. One can find in the literature two descriptions of such type of compounds, both with the $H_{4-x}pmida^{x-}$ residues sequestrating the d-block metals *via* coordination mode I: $(H_3O)_8[Ni_4O_2(Hpmida)_4(H_2O)_2]$[54] and $(H_3O)_6[Co_4(H_2O)_4(Hpmida)_2(pmida)_2]\cdot 2H_2O$.[55] It is important to emphasize that in the report of the former complex the same research group also described the formation of a ${}_{\infty}^{1}[Ni(pmida)(H_2O)]^{2n-}$ anionic 1D polymer (see Section 17.4.2) which could be obtained by the occurrence of an *anti,anti*-bridge for the carboxylate group. These two compounds are based on rare centrosymmetric tetrameric units (Figures 17.5a and 17.5b, respectively), which share a number of similar geometrical features: (i) the *syn,anti*-bridges impose very similar intermetallic separations, found in the *ca.* 5.34–5.49 Å and 5.38–5.48 Å ranges for Ni···Ni and Co···Co, respectively; (ii) the $H_{4-x}pmida^{x-}$ residues are located in the periphery of the complexes; (iii) $H_{4-x}pmida^{x-}$ protonation only occurs in the phosphonate moieties; (iv) their molecular sizes (based on van der

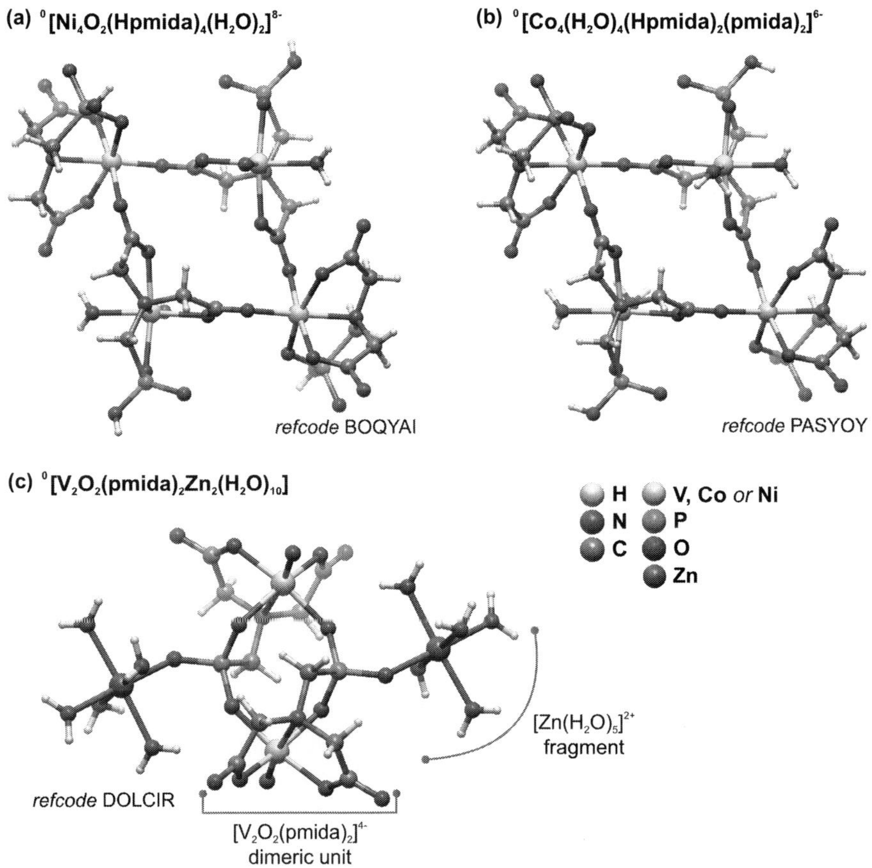

Figure 17.5 Centrosymmetric tetrameric units with $H_{4-x}pmida^{x-}$ residues.

Waals radii) are of about 3.6 × 1.8 nm (longest size and height); (v) half of the metallic centers have coordination spheres decorated by water molecules. Concerning this last structural feature, it is interesting to note that for the Ni^{2+} compound one water molecule was replaced by an oxido group which led, on the one hand, to the need of complete protonation of all $Hpmida^{3-}$ moieties and, on the other, to the protonation of the two uncoordinated water molecules (hence the distinct empirical formulae for the two structures). Noteworthy, the report of Fan and Song (Ni^{2+} "molecular square")[54] is of considerable better crystallographic quality, and the authors were able to discern the locations of all hydrogen atoms.

A completely distinct type of tetranuclear complex can also be obtained by using as a molecular canvas the anionic $[V_2O_2(pmida)_2]^{4-}$ unit described in the previous section. This unit has several possible sites for coordination and it is relatively easy to find in the literature a small family of structures in which coordination to a second metallic center occurs *via* the free oxygen atoms of the bridging phosphonate groups. Figure 17.5c depicts the centrosymmetric neutral $[V_2O_2(pmida)_2Zn_2(H_2O)_{10}]$ complex,[51,56] with similar crystal structures described in the literature for Mn^{2+} and Co^{2+}.[57,58] We note that this coordination does not always have to necessarily occur in the solid state. Proof of that concerns the structure reported by Gao *et al.* and mentioned in the previous section, $[Zn(H_2O)_6]_2[V_2O_2(pmida)_2]\cdot 2H_2O$,[51] in which the aforementioned V^{4+} anionic unit co-crystallizes with the hexaaquocomplex of zinc(II). These mixed-metal tetranuclear complexes are significantly more distorted (*e.g.* for the Zn^{2+}-based compound the Zn···V distances range from *ca.* 5.31 to 6.00 Å), but the two external Zn^{2+} centers are prone to ligand replacement in order to extend the dimensionality of the structures. This feature has important structural consequences and is further explored in the following sections by giving examples of compounds based on the $[V_2O_2(pmida)_2]^{4-}$ unit.

17.4.1.4 High Nuclearity Complexes

Considering the coordination modes described in Scheme 17.3 for $H_{4-x}pmida^{x-}$ residues, in particular mode I which is a "molecular trap" for metallic centers, it is somehow surprising to find in the literature reports of discrete complexes formed by a large number of metals. Because in these structures there are no direct connections between the metallic centers, as required by the definition of Cotton and Wilkinson[59] for a cluster of metals, these compounds can only be defined as high nuclearity complexes. In the genesis of these complexes one can usually find the phosphonate as a coordinating pivot group, either for establishing connections between metal centers or to aggregate a large number of cations in a small region of space.

Wood and co-workers reported in the late 1990s the hexameric anionic $^0[Co_6(pmida)_6]^{12-}$ molecular wheel (Figure 17.6a) exhibiting canted antiferromagnetism.[36] In this structure the Co^{2+} centers are coordinated to $pmida^{4-}$ *via* coordination fashion I but, remarkably, these cations only exhibit a five-coordination number (overall geometry resembling a distorted trigonal bipyramid, which is chemically feasible for Co^{2+} centers): the fifth coordination site is occupied by a phosphonate bridging oxygen atom from a neighboring monomer; the bridge imposes an intermetallic Co···Co separation of about 5.52 Å. This molecular wheel is assembled by head-to-tail connections between adjacent monomers (as observed for the formation of the "molecular squares" described in the previous section), but in this case they can occur either in clockwise or anti-clockwise fashions. The crystal structure is formed by a racemic mixture of both wheels (distributed in alternating layers).

The most remarkable high nuclearity complex based on $H_{4-x}pmida^{x-}$ residues is $^0[Co_{15}(pmida)_6(H_2O)_{24}]^{6+}$ (Figure 17.6b), discovered by Gu and Sevov in 2009.[37] This is a true nanocluster with average sizes of about 3.6 × 3.0 nm (diameter × height), and is composed of 15 Co^{2+} metallic centers which can be divided into three different layers (an "onion"-type cluster): the central Co^{2+} cation acts as the "aggregator", being coordinated to the six $pmida^{4-}$ anions

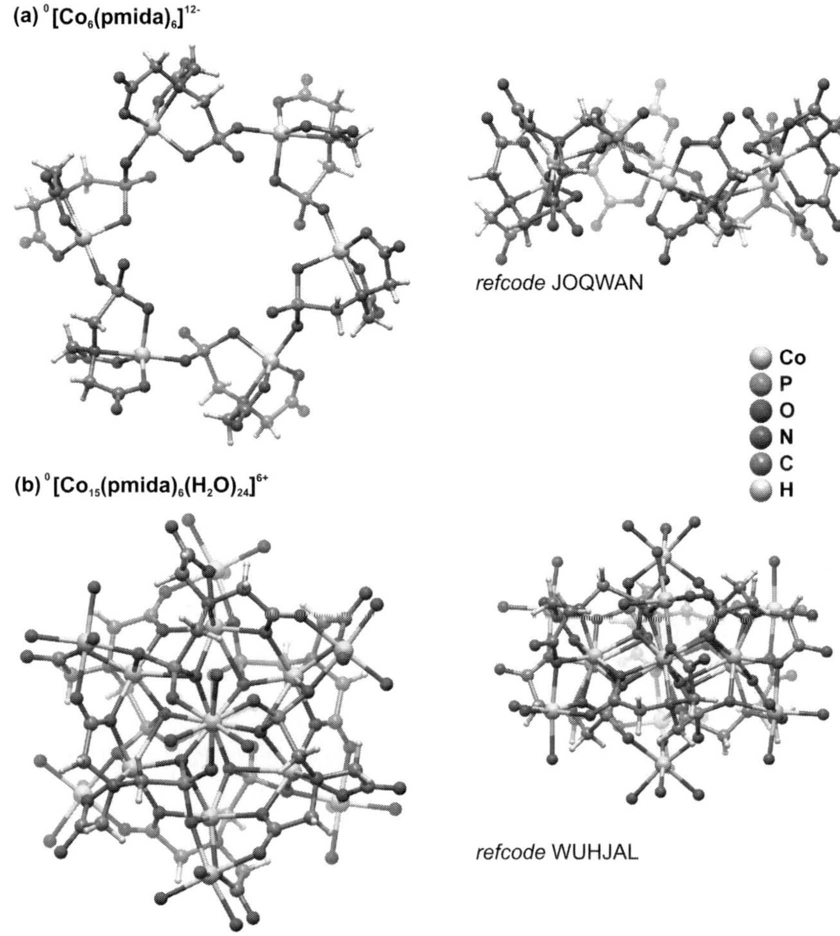

Figure 17.6 Ball-and-stick representation of high nuclearity complexes with $H_{4-x}pmida^{x-}$ residues.

which, in turn, encapsulate (*via* coordination mode I) six Co^{2+} cations composing the middle layer; the peripheral layer is formed by eight metallic centers, mainly coordinated to the carboxylate groups. One important feature regarding this cluster concerns the presence of 24 coordinated water molecules (three for each Co^{2+} of the peripheral layer), which can be displaced by bridging ligands to hierarchically form higher dimensional structures. Indeed, the same research group was able to design a 3D framework using this cluster as a building unit (see Section 17.5.1) while using the partially deprotonated *cis* enantiomer of 1,2,3,4,5,6-cyclohexanehexacarboxylic acid (H_6E) to directly coordinate to the outer layer Co^{2+} cations. This nanocluster was isolated in the solid state as a salt while co-crystallizing with the fully deprotonated *trans* enantiomer: $[Co_{15}(pmida)_6(H_2O)_{24}][\textit{trans}\text{-E}] \cdot xH_2O$.[37]

17.4.2 1D Polymers

Linear polymers can be deconstructed into individual monomers to illustrate how extended coordination-based structures can be self-assembled by way of specific functional groups. In this sense, 1D polymers are the best model for that purpose. One can find in the literature two interesting examples of 1D coordination polymers containing solely sequestrating $H_{4-x}pmida^{x-}$ residues, which helps to illustrate how the overall topology is influenced by the way individual monomers interconnect. In $_{\infty}{}^{1}[Fe(H_2pmida)(H_2O)_2]$ (Figure 17.7a)[40] H_2pmida^{2-} appears encapsulating Fe^{2+} through the rarely encountered coordination mode II (Scheme 17.3b), with the coordination sphere of this metal center further comprising a total of two water molecules. Connections between nearby units occur *via* the protonated phosphonate group, imposing Fe···Fe intermetallic distances of *ca.* 6.33 Å. This polymer is truly unique because it has a pendant arm with an uncoordinated carboxylic acid group, engaged in hydrogen bonding interactions. Three years later, besides the tetranuclear cluster with Ni^{2+} (details in Section 17.4.2.2), the research groups of Fan and Song also described a charged (anionic) 1D polymer formulated as $(NH_4)_2[Ni(pmida)(H_2O)]$ (Figure 17.7b).[54] The basic building chelating unit of $_{\infty}{}^{1}[Ni(pmida)(H_2O)]^{2n-}$ is in this case based on the commonly encountered coordination mode I. The polymeric nature comes from *anti,anti*-bridges *via* one carboxylate group. Due to the small size of this bridging-capable functional group, the shortest intermetallic distance decreases to *ca.* 5.78 Å, a value slightly shorter than that created by a typical phosphonate bridge. Moreover, contrasting with the previous polymer, $_{\infty}{}^{1}[Ni(pmida)(H_2O)]^{2n-}$ grows in a zigzag fashion (Figure 17.7b), most likely to minimize steric hindrance between nearby monomers.

The first discovered 1D coordination polymer with sequestrating $H_{4-x}pmida^{x-}$ residues precedes the two materials mentioned above, being published in 2004 by Song *et al.*, and containing 4,4′-bipyridine (4,4′-bpy) as an auxiliary bridging ligand: $[Co_2(pmida)_2(H_2O)_5(4,4'\text{-bpy})]\cdot 4H_2O$.[60] In this structure the phosphonate group promotes instead the formation of Co^{2+} dimers (separation of *ca.* 6.24 Å). The $_{\infty}{}^{1}[Co_2(pmida)(H_2O)_5(4,4'\text{-bpy})]$ polymer is, thus, formed by bridges between these dimeric units through 4,4′-bpy: this auxiliary molecule establishes a physical and directional link between the pmida-sequestered and non-sequestered Co^{2+} centers as depicted in Figure 17.7c, promoting the formation of a zigzag polymer. Because of the combined effect of this undulated feature of $_{\infty}{}^{1}[Co_2(pmida)(H_2O)_5(4,4'\text{-bpy})]$, and also the fact that it is neutral (in opposition to $_{\infty}{}^{1}[Ni(pmida)(H_2O)]^{2n-}$ which requires two ammonium cations per Ni^{2+} in their vicinity), the close packing is not very effective. Large channels running through the crystal, and parallel to the crystallographic *a* axis, are filled with water molecules, as represented in Figure 17.8.

In Section 17.3.4 it was briefly mentioned that $H_{4-x}pmida^{x-}$ residues can also adopt different coordination modes other than those involving sequestration of the metallic center. A striking example comes from Clearfield and co-workers

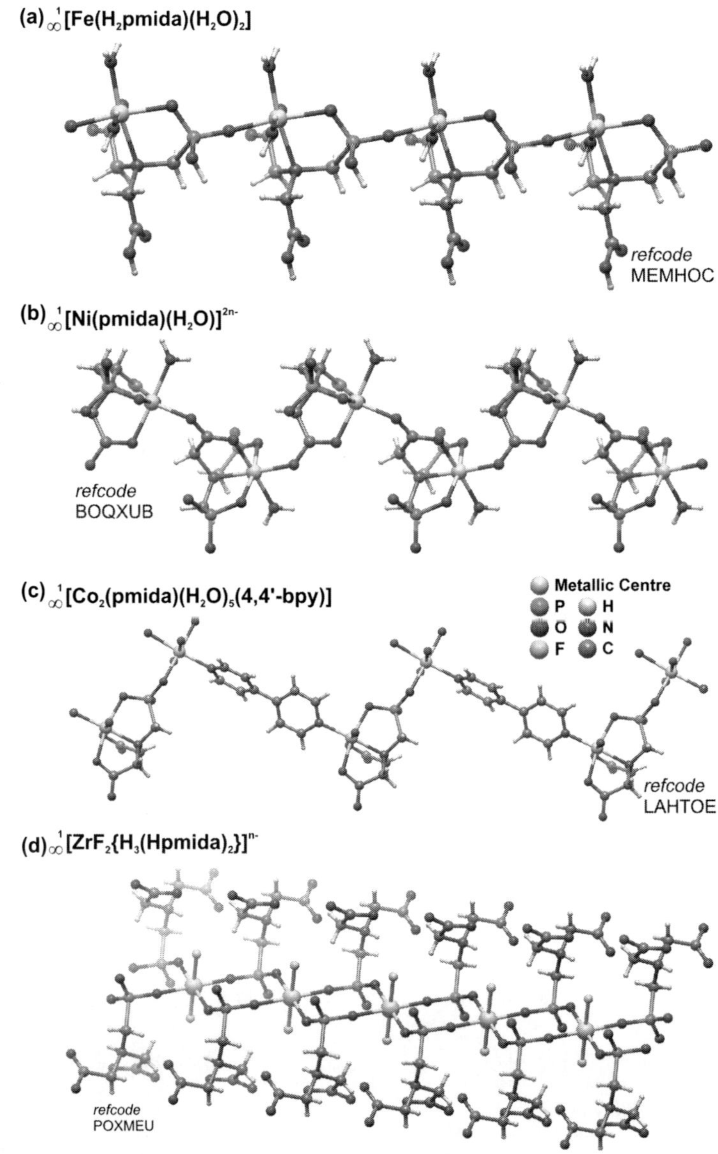

Figure 17.7 Ball-and-stick representation of 1D coordination polymers containing sequestrating and non-sequestrating $H_{4-x}pmida^{x-}$ residues.

who, in 1998, reported a rare linear coordination-based anionic chain in which a non-sequestrating $H_{4-x}pmida^{x-}$ residue is connected to Zr^{4+} metallic centers: $(NH_4)[ZrF_2\{H_3(Hpmida)_2\}]\cdot 3H_2O\cdot NH_4Cl$.[43] The organic ligand appears with the central nitrogen atom protonated, thus preventing chelation. The slightly distorted octahedral coordination sphere of Zr^{4+} is only composed of

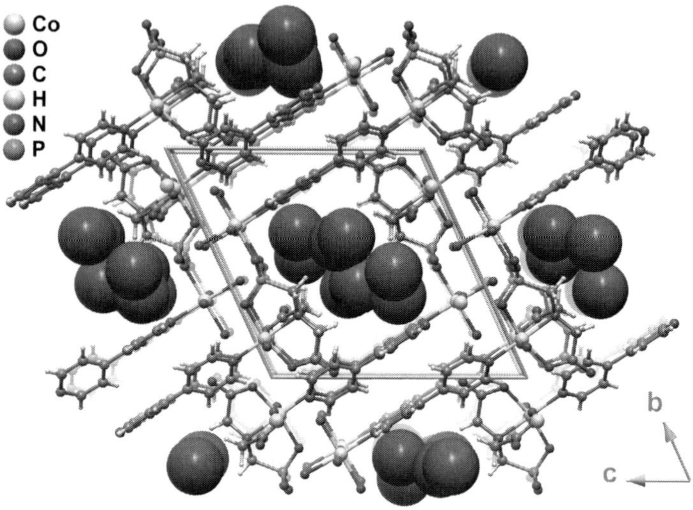

Figure 17.8 Close packing of $_\infty{}^1[Co_2(pmida)_2(H_2O)_5(4,4'\text{-bpy})]$ 1D polymers forming a porous crystalline structure having the channels housing confined water molecules (represented in space filling mode).

phosphonate groups and fluoride anions, with the pendant carboxylates of $_\infty{}^1[ZrF_2\{H_3(Hpmida)_2\}]^{n-}$ (Figure 17.7d) being engaged in hydrogen bonds, just like that observed for $_\infty{}^1[Fe(H_2pmida)(H_2O)_2]$. It was suggested in this report that the steric hindrance of the $H_{4-x}pmida^{x-}$ residue (due to uncoordinated, hydrogen-bonded carboxylate moieties) prevented the coalescence of the structure into layers: to promote the formation of the stable α-phase of zirconium phosphate the ligand needs to necessarily occupy an area smaller than about 24 $Å^2$; because H_2pmida^{2-} occupies a larger area, a 1D polymer is formed instead. It is worth noting that Clearfield and co-workers were able to further design functional layered materials by adding a controlled amount of hydrogenophosphate anions as auxiliary spacers (details in the next section).

17.4.3 2D Layers

Polymerization in two dimensions of space does not seem to be an easy experimental task using a simple combination of only one metal center and $H_{4-x}pmida^{x-}$ residues: the combined effect of the ability of these ligands to sequester metallic centers (which, as mentioned, removes available coordination sites), and the intensive research performed on d-block elements (with coordination numbers typically up to 6), seems to create some structural limitation in the self-assembly process. In this sense, only a handful of 2D materials are known, with all being depicted in Figures 17.9 and 17.10. Moreover, it is also possible to conclude that structural diversity is somehow limited. The exception is the $M_x[Fe(pmida)]_6 \cdot nH_2O$ family of plane nets [where $M^+ = Na^+$

($x = 11$), K^+ ($x = 11$) and Rb^+ ($x = 10$)] reported by Wood and co-workers.[31] In these materials the alkali metals only balance the crystal charge and do not participate directly in the self-assembly of the polymer. Compounds are based on the previously described centrosymmetric dimeric anionic $[Fe_2(pmida)_2]^{2-}$ units (mainly with Fe^{3+}), which interconnect directly *via* the vacant coordinative site of iron and the phosphonate group to form a maple leaf plane lattice: ${}_{\infty}^{2}[Fe(pmida)]_{6n}{}^{11n-}$ (Figure 17.10a). These inter-dimer phosphonate bridges impose Fe···Fe separations of *ca.* 5.73 Å. As described in Section 17.3.2, the Na^+-based structure of this family is the most symmetrical. The inclusion of

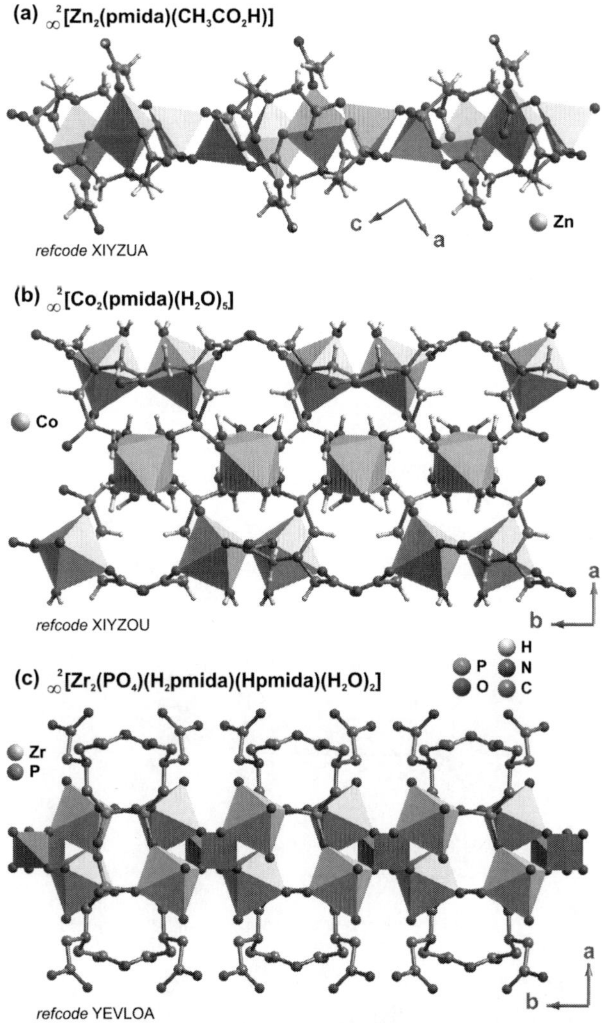

Figure 17.9 Mixed ball-and-stick and polyhedral side representations of 2D coordination polymers containing $H_{4-x}pmida^{x-}$ residues and only type of metallic center in their composition (see Scheme 17.2 for a graphical listing).

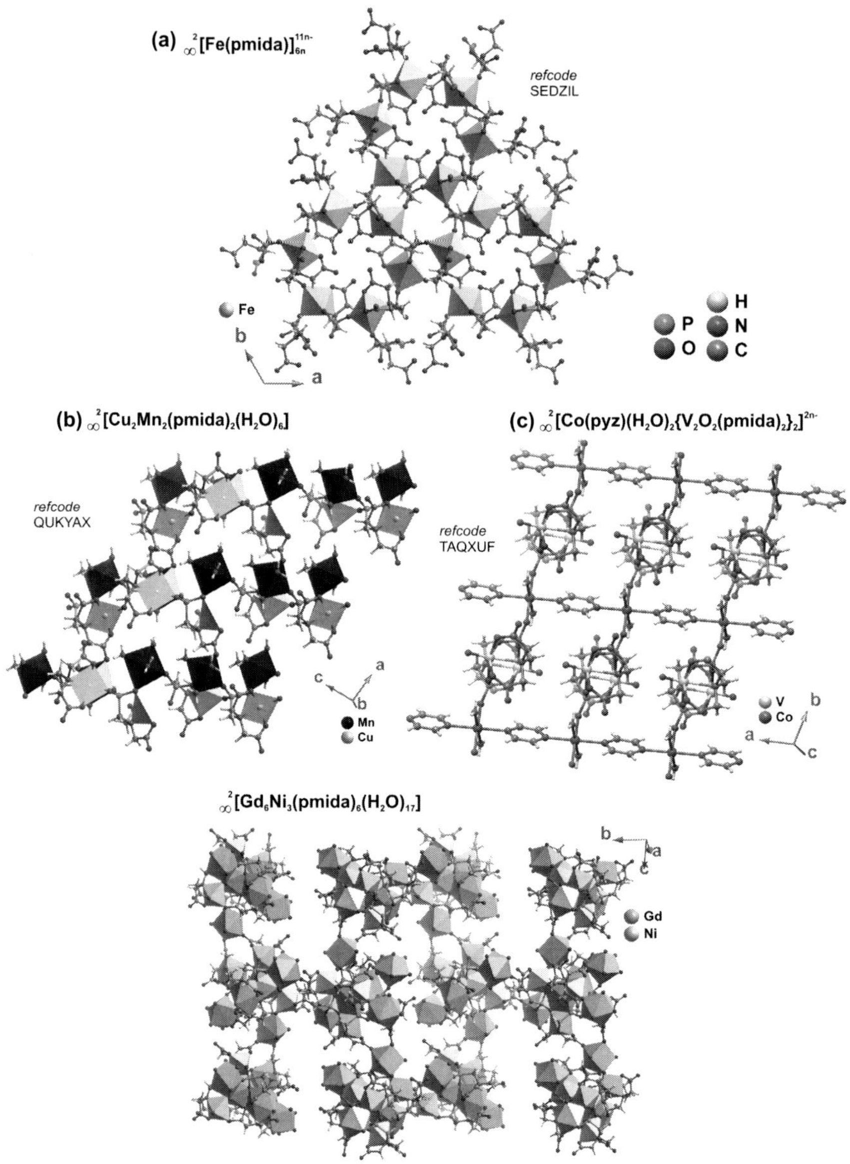

Figure 17.10 Mixed ball-and-stick and polyhedral representations of 2D coordination polymers containing $H_{4-x}pmida^{x-}$ residues and two or more types of metallic centers in their composition (see Scheme 17.2 for a graphical listing).

K^+ and Rb^+ induces a lowering of the overall crystal symmetry (to space groups $P3_1$ and $C2/c$, respectively), at the same time that these alkali sites start to be partially occupied (most certainly a cause-effect occurrence). The authors of this work found that to properly balance the crystal charge some Fe^{3+} centers had to be reduced to Fe^{2+} (also based on Mössbauer spectroscopic data). Nevertheless, the topological and geometrical features of the $_\infty{}^2[Fe(pmida)]_{6n}{}^{11n-}$ anionic plane nets remain unchanged indicating a considerable structural robustness of the coordination polymer. The remaining five layered materials with $H_{4-x}pmida^{x-}$ residues are always constructed from the use of auxiliary molecular unit as explained in detail in the following paragraphs.

The three structures highlighted in Figure 17.9 are constructed from only $H_{4-x}pmida^{x-}$ residues and one type of metal center. Their layered nature arises, however, from the presence of either a second crystallographic independent metal center (with strikingly distinct coordination environment from the center which is directly coordinated to $H_{4-x}pmida^{x-}$) or from the inclusion of a charge-balancing ion which, itself, also participates in the construction of the layer. Three examples are known, all discovered by the research group of Clearfield. Compound $[Zn_2(pmida)(CH_3CO_2H)]\cdot 2H_2O$,[61] containing the plane net depicted in Figure 17.9a, contains a second tetrahedral Zn^{2+} metallic center which bridges adjacent Zn^{2+}-pmida sequestrated moieties. If these bridging tetrahedral Zn^{2+} centers were not present in the structure, only discrete mononuclear complexes, similar to those of Figure 17.1a, would be present having the acetic acid molecule *trans*-coordinated to the nitrogen atom. In the same report Clearfield also described another layered compound: $[Co_2(pmida)(H_2O)_5]\cdot H_2O$ (layer represented in Figure 17.9b).[61] This layer exhibits a remarkable and unprecedented ABA sandwich-type distribution in which the central section is composed of octahedral Co^{2+} cations coordinated to four water molecules. Using a similar rationale, in the absence of these aqua-cations the structure would instead be solely formed by a 1D polymer strongly resembling the $_\infty{}^1[Ni(pmida)(H_2O)]^{2n-}$ one described in the previous section (*syn,anti*-bridge by way of the carboxylate group, with Co···Co separation of about 6.00 Å). The third example is the $_\infty{}^2[Zr_2(PO_4)(H_2pmida)^-(Hpmida)(H_2O)_2]$[42,62] material which can be envisaged as being constructed by 1D undulated tapes (based on Zr^{4+} dimeric units) interconnected by way of phosphate inorganic anions (Figure 17.9c). These inorganic moieties are included in the structure because of the use of phosphoric acid in the reaction medium.

A special emphasis should be further given to $_\infty{}^2[Zr_2(PO_4)(H_2pmida)$-$(Hpmida)(H_2O)_2]$. Besides the unprecedented nature of this layered zirconium(IV) phosphate/phosphonate, this compound was also the subject of further synthetic and application studies from the same research group.[63] Clearfield and co-workers investigated in detail the preparation of the compound by varying, in a systematic fashion, the ligand:phosphoric acid ratio in the reactive mixture. They were able to isolate novel materials highly rich in $HPO_4{}^{2-}$ decorating the inner walls of the layers and having the –OH groups pointing to the interlamellar region. Clearfield developed models for these compounds based on the structure of $_\infty{}^2[Zr_2(PO_4)(H_2pmida)(Hpmida)(H_2O)_2]$ and data from various characterization techniques (mainly powder X-ray diffraction data). They discovered that the increase of $HPO_4{}^{2-}$ anions in the chemical composition promoted the formation of layers typical of the α-phase of zirconium phosphate.[64,65] Compounds intercalate alkylamines ($C_nH_{2n+1}NH_2$ with $n = 3–8$): up to 2 mol of molecules could be intercalated in between $_\infty^2[Zr_2(PO_4)(H_2pmida)(Hpmida)(H_2O)_2]$; the other HPO_4^{2-}-rich phases could take instead about 1.5 mol of these guests.

Gu and Sevov[37] prepared, under relatively mild hydrothermal conditions, mixed-metal (3d–3d or 3d–4f) multi-dimensional coordination polymers. The 3d–3d structures, formulated as $[Cu_2M_2(pmida)_2(H_2O)_6]\cdot 2H_2O$ (where $M^{2+} = Mn^{2+}$, Co^{2+} or Cd^{2+}), are layered coordination polymers with an honeycomb motif. As rationalized above for other compounds, in the absence of the second M^{2+} cation, the structures would only exhibit spiral 1D Cu-pmida coordination polymers promoted by the bridging phosphonate group. In these structures the Cu^{2+} centers appear only coordinated to the sequestering (mode I) $pmida^{4-}$ anions, in a square pyramidal coordination fashion. Figure 17.10b depicts the $_\infty^2[Cu_2Mn_2(pmida)_2(H_2O)_6]$ layer and it is clearly possible to discern that the aforementioned Cu^{2+}-pmida chain is decorated by peripheral Mn^{2+} cations. This promotes the formation of a double-stranded *meso*-helical motif with a pitch of about 19.9 Å. The coalescence of these strands by way of the $[Mn(H_2O)_3]^{2+}$ moieties (placed in alternate locations alongside the chain) leads to the formation of the layered structure. A structural note worth mentioning concerns the fact that the helical nature of the strands induces an orientation of the basal plane of the Cu^{2+} square pyramid toward a sterically-hindered region of space, hence the absence of a sixth coordination moiety to complete the coordination sphere.

Based on the tetranuclear $[V_2O_2(pmida)_2M_2(H_2O)_{10}]$ complexes reported in Section 17.4.1.3 (where M^{2+} are transition metal centers), Paz *et al.*[66] demonstrated how to design a layer-type compound by displacing some coordinated water molecules by rod-like bridging ligands, such as pyrazine (pyz), and neighboring dimeric $[V_2O_2(pmida)_2]^{2-}$ units. The resulting $_\infty^2[Co(pyz)(H_2O)_2\{V_2O_2(pmida)_2\}_2]^{2n-}$ anionic layer (depicted in Figure 17.10c) can thus be envisaged as strings of $[V_2O_2(pmida)_2]^{2-}$ units directly bound to $[Co(H_2O)_2]^{2+}$ units, running parallel to the [010] direction, which are then pillared by the organic pyrazine ligand. This distribution leads to a (4,4) plane net of Co^{2+} metallic centers with two distinct types of bridges: while the pyrazine imposes a $Co\cdots Co$ separation of *ca.* 7.22 Å, the large, bulky V^{4+} dimeric unit separates the same metal centers by *ca.* 10.14 Å. The crystal charge is balanced by hexaaquocobalt(II) complexes which are located in the interstitial spaces of the anionic layers. In Section 17.5 it is further demonstrated how this design strategy, also with the anionic $[V_2O_2(pmida)_2]^{2-}$ dimers, can hierarchically lead to 3D frameworks.

A very recent example of a mixed-metal 3d–4f layered material comes from the research groups of Kong and Long,[67] and is based on a $[Gd_6Ni_3(pmida)_6(H_2O)_{17}]$ high nuclearity complex fragment. Despite almost all nine metallic centers exhibiting strikingly distinct coordination environments, one can discern a couple of common important features: first, all $pmida^{4-}$ anionic ligands coordinate to the metal centers (either Ni^{2+} or Gd^{3+}) *via* coordination mode I; second, the Gd^{3+} centers can be divided into those having coordinated water molecules or those which are completely water-free and located in the inner core of the complex. Two peripheral Gd^{3+} aquacomplexes are responsible for the parallel coalescence of the Gd_6Ni_3 complexes, leading to the undulated plane net (topology 6^3) depicted in Figure 17.10d: $_\infty{}^2[Gd_6Ni_3(pmida)_6(H_2O)_{17}]$. This material was the subject of detailed studies which took advantage of the presence of the nine metallic centers in very close proximity: magnetic measurements indicate the presence of antiferromagnetic interactions; alternating current impedance measurements further showed that the increase of temperature is concomitant with a decrease of the conductance. This last result motivated the authors to further investigate the reasons for such unusual behavior and it was found that even though the crystal structure of the layered material does not change with the increase of temperature, water loss occurs. Thus, the authors speculate that conductance might arise from the transference of protons with the water molecules acting as their suppliers (hence the decrease of conductance with increasing temperature).

17.4.4 3D Frameworks

Framework-type materials based on $H_{4-x}pmida^{x-}$ residues are not very common. This section collects a total of five structures, two mixed-metal d–f frameworks and three homometallic ones. A couple of structures were intentionally omitted from this section as they will be used to exemplify the modular construction of $H_{4-x}pmida^{x-}$-based framework materials in the subsequent section.

Three homometallic 3D frameworks have been prepared and described by Fan *et al.*, $[Mn(H_2O)(H_2pmida)]$,[41] by Tang *et al.*, $[Ln(Hpmida)(H_2O)_2]\cdot H_2O$ (where $Ln^{3+} = Y^{3+}$, Er^{3+}, Lu^{3+}, Gd^{3+}, Tb^{3+}, Dy^{3+} and Yb^{3+}),[38] and Mao *et al.*, $[Pb_2(pmida)]\cdot 1.5H_2O$.[34] The $_\infty{}^3[Mn(H_2O)(H_2pmida)]$ framework contains one of the rare coordination modes of $H_{4-x}pmida^{x-}$ residues (see Section 17.3.4), with the central nitrogen atom being protonated while the remaining functional groups establish the physical bridges between five symmetry related Mn^{2+} cations. As depicted in Figure 17.11a, the compound contains chains (along the [001] direction) of corner-shared alternating $\{MnO_6\}$ octahedra and $\{PO_3H\}$ tetrahedra, which are linked together by the carboxylate groups to form the 3D network. This material was shown to be thermally stable up to *ca.* 300 °C while magnetic susceptibility studies demonstrated weak antiferromagnetic interactions between metallic centers: the closest Mn $\cdots$ Mn proximity is of *ca.* 5.22 Å and occurs along the [100] direction (perpendicular to the plane of Figure 17.11a) with no bridges existing between the metallic centers so to promote an effective magnetic exchange (*i.e.* the closest distance occurs through space).

$_\infty{}^3[Pb_2(pmida)]$ is a remarkable structure discovered in the laboratories of Clearfield (Figure 17.11d).[34] This high-symmetry structure (tetragonal $P4_2/n$ space group) contains a single crystallographic Pb^{2+} cation which is chelated to $pmida^{4-}$ *via* mode I. Because of the lone pair of Pb^{2+}, the coordination geometry is a highly distorted octahedron. Nevertheless, the self-assembly leads to a porous structure with large channels (of about 5 Å and formed by a

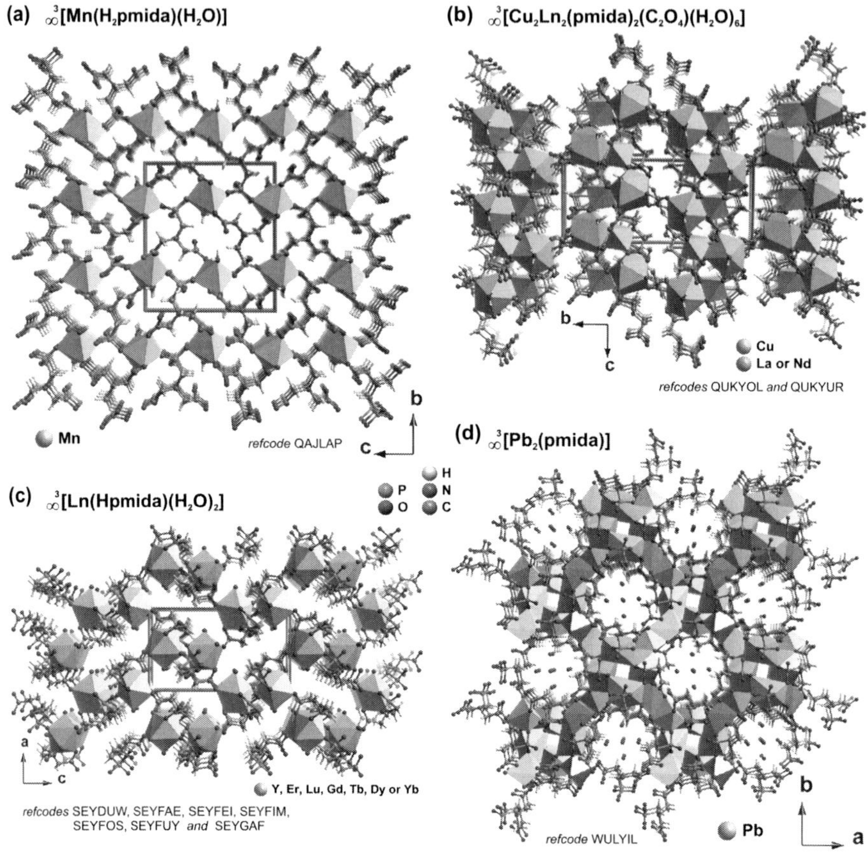

Figure 17.11 Mixed ball-and-stick and polyhedral representations of 3D coordination polymers containing $H_{4-x}pmida^{x-}$ residues.

symmetrical 32-membered ring) running parallel to the *c* axis. Because the structure is truly microporous, the authors showed that all water molecules are readily released below 100 °C, with the framework maintaining its integrity up to approximately 315 °C.

The system formed by the seven isostructural $_{\infty}^{3}$[Ln(Hpmida)(H_2O)$_2$] frameworks (Figure 17.11c) also has the $Hpmida^{3-}$ ligand sequestrating the Ln^{3+} centers *via* coordination mode I. Because these cations have large coordination numbers, they are able to establish four additional bonds (two with water molecules and another two with neighboring $Hpmida^{3-}$ ligands) leading to a complex 3D framework. Both the metal and the ligand are three-connected nodes of the network. Hence the structure can be envisaged as a binodal (10,3) non-interpenetrated network having helical tunnels along the [100] direction (cross-section of about 5.6 × 12.9 Å). It is worth emphasizing two important aspects of these materials: first, all structures have been determined in the chiral $P2_12_12_1$ space group, but crystals were proven to be racemic because of very weak NLO properties; second, the photoluminescence of some materials has been investigated in the solid state, in particular the Tb^{3+} material exhibits strong green emission with a lifetime of about 1 ms (for an excitation wavelength of 380 nm while monitoring the emission at 488 nm).

Figure 17.11b depicts the mixed-metal d–f $_{\infty}^{3}[Cu_2Ln_2(pmida)_2(C_2O_4)(H_2O)_6]$ (where $Ln^{3+} = La^{3+}$ and Nd^{3+}) MOF described by Gu and Sevov[37] in the same report of their layered d–d coordination polymers (see previous section). Remarkably, the molar ratios of the metallic cations and the $pmida^{4-}$ anionic ligand remain the same across these families of compounds but, because of the +3 charge of the lanthanides, an additional oxalate anion *per* two Ln^{3+} was included in the structures. Another similarity between the 2D and 3D materials concerns the fact that Cu^{2+} is chelated by $pmida^{4-}$ by way of coordination mode I. This structural feature led the authors to conclude that because of the employed synthetic strategies (*i.e.* hydrothermal synthesis with all reactants added at the same time), the formation of structures seems to be initiated by the formation of the highly stable $[Cu(pmida)]^{2-}$ monomer. As in the layered structures, also in these 3D materials, Ln^{3+} decorates a 1D chain of Cu-pmida with *anti,anti*-carboxylate bridges promoting the formation of double-decker layers. However, because of the large coordination numbers of lanthanides and the required presence of oxalate, these *bis*-chelating moieties establish further inter-lanthanide linkages (*e.g.* La···La distance of *ca.* 6.53 Å), which are in the core of the formation of the 3D structures.

A very recent family of 3D materials containing $H_{4-x}pmida^{x-}$ residues is based on the centrosymmetric Fe^{2+} units previously mentioned (and depicted in Figure 17.3b), and complements well the structural ideas first described by Wood and co-workers[31] for the anionic $_{\infty}^{2}[Fe(pmida)]_{6n}^{11n-}$ plane net: $Fe^{III}[LnFe^{II}_6(Hpmida)_6]\cdot nH_2O$ (where $Ln^{3+} = Nd^{3+}$, Gd^{3+}, Tb^{3+} and Er^{3+}). This family also contains in its composition an anionic $_{\infty}^{2}[Fe(pmida)]_{6n}^{5n-}$ plane net, for which the dimeric Fe^{2+} units are interlinked by the peripheral functional groups.

While in the structures of Wood and co-workers these connections occur by way of the phosphonate groups, in the new materials prepared by Rocha *et al.*[68] connections are instead established *via* the carboxylate groups (Figure 17.12, bottom). Concomitantly, this type of inter-dimer bridges reduces the Fe···Fe separations to *ca.* 5.16 Å. This structural diversity of the Fe-containing dimers shows that even though both units ($[Fe^{III}_2(pmida)_2]^{2-}$ and $[Fe^{II}_2(Hpmida)_2]^{2-}$, see Figure 17.3) exhibit strikingly similar structural properties, the inability of the phosphonate group to establish further connections because of its protonation in the $_{\infty}^{3}[LnFe^{II}_6(Hpmida)_6]^{n-}$ frameworks can promote a completely new connectivity type for the dimers.

While considering the compounds of Wood,[31] the layers had their charge balanced by alkali cations, in the materials reported by Rocha *et al.*[68] the crystal charge is instead neutralized by the inclusion of rare earth cations (Ln^{3+}). These metallic centers also have the important structural function to "glue together" the layers (Figure 17.12, top), leading to a 3D mixed-metal 3d–4f anionic coordination polymer. In addition, the engineering of optical centers into the hybrid frameworks further allows the isolation of photoluminescent compounds, as demonstrated by the authors for the Tb^{3+}-containing compound. Materials composing this family are porous with charge-balancing Fe^{3+} cations (one per six Fe^{2+} in the structure) and highly disordered water molecules occupying the voids in the framework (not shown in Figure 17.12).

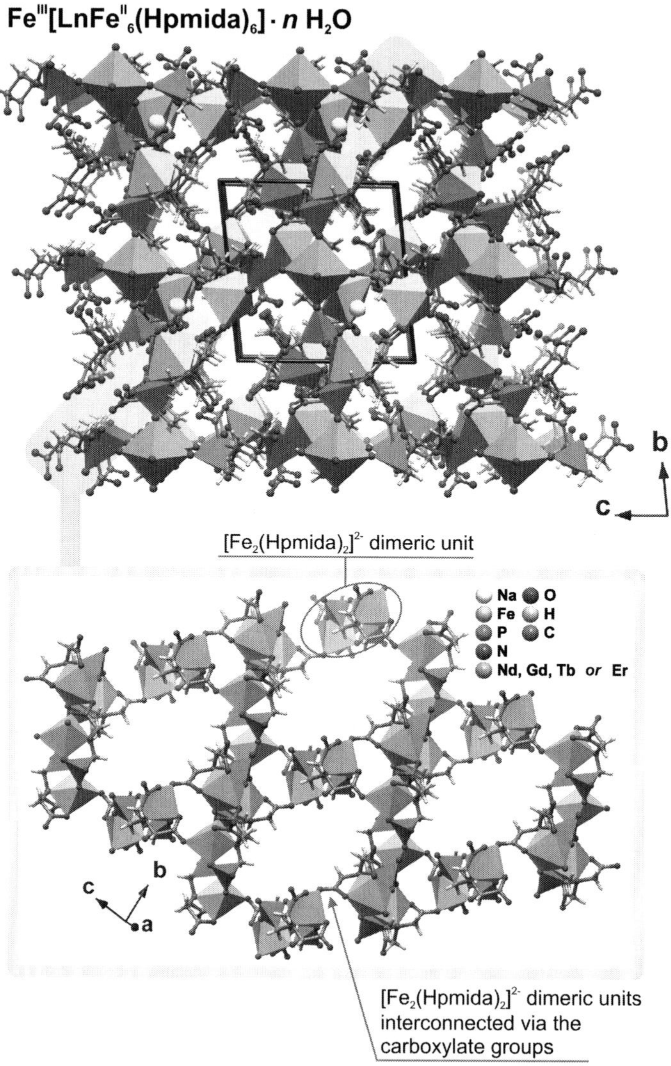

Figure 17.12 Mixed ball-and-stick and polyhedral representation of the (top) 3D anionic coordination polymer $_{\infty}^{3}[LnFe^{II}_6(Hpmida)_6]^{n-}$ (where $Ln^{3+} = Nd^{3+}$, Gd^{3+}, Tb^{3+} and Er^{3+}), assembled by way of Ln^{3+} connections between adjacent (bottom) layers formed by the coalescence of $[Fe_2(Hpmida)_2]^{2-}$ dimeric units through the carboxylate groups.

17.5 Modular Metal–Organic Frameworks

In the introductory sections of this chapter, it was mentioned that the self-assembly of $H_{4-x}pmida^{x-}$ residues with metallic cations could easily generate multi-dimensional coordination polymers (some of which are 3D). These MOF structures belong to the first generation of these materials and for some interesting examples the reader is referred to the previous sections.

While describing the discrete complexes that $H_{4-x}pmida^{x-}$ can form with a wide variety of metallic centers, it was stated that the resulting units (*e.g.* dimers and tetramers) exhibit the required characteristics and properties to be employed as rigid SBUs for the construction of modular MOFs.[48,69,70] In this small, final subsection we summarize the ideas which might be behind this rational design of higher dimensionality structures by the use of rigid building units based of $H_{4-x}pmida^{x-}$ residues.

Attention is focused on two examples of modular 3D networks: one material takes advantage of the large Co_{15} cluster described in Section 17.4.2.3, and the other is based on the stepwise use of the anionic $[V_2O_2(pmida)_2]^{4-}$ units with auxiliary metal cations and bridging ligands.

17.5.1 Example 1: "0D" → 3D

The discrete cationic $^0[Co_{15}(pmida)_6(H_2O)_{24}]^{6+}$ nanocluster discovered by Gu and Sevov was employed as the building unit of a remarkable framework obtained using typical bottom-up approaches (Figure 17.13, right). On one hand, and as mentioned before for the Al^{3+} unit, the existence of several water molecules coordinated to Co^{2+} and decorating the entire surrounding of the 15-cluster makes this unit highly amenable to establish bridges *via* coordination

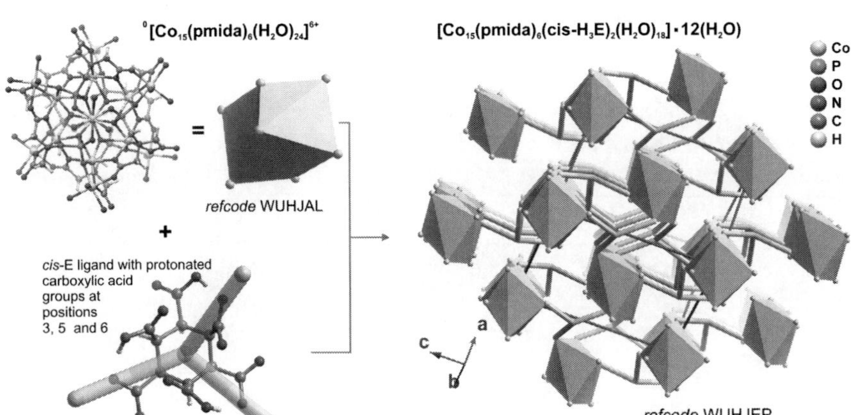

Figure 17.13 Schematic representation of the modular construction of the $[Co_{15}(pmida)_6(cis\text{-}H_3E)_2(H_2O)_{18}]$ framework based on the use of discrete cationic $^0[Co_{15}(pmida)_6(H_2O)_{24}]^{6+}$ nanoclusters interconnected *via* deprotonated *cis*-1,2,3,4,5,6-cyclohexanehexacarboxylic acid residues.

based on ligand replacement strategies. On the other hand, the use of anionic organic ligands (such as phosphonates or carboxylates) is highly convenient as they naturally balance the +6 charge of the cluster while being able to establish, at the same time, bridges between neighboring clusters.

Gu and Sevov[37] employed *cis*-1,2,3,4,5,6-cyclohexanehexacarboxylic acid (*cis*-H_6E) as the auxiliary organic bridging ligand, along with the $^0[Co_{15}(pmida)_6(H_2O)_{24}]^{6+}$ nanocluster to produce an extended structure formulated as $[Co_{15}(pmida)_6(cis\text{-}H_3E)_2(H_2O)_{18}]\cdot 12H_2O$ (Figure 17.13). The authors noted that this organic molecule could only act as a physical bridge between nanoclusters while in its partially deprotonated *cis* enantiomer. The completely deprotonated *trans* form can only act as a counter-ion, as described in Section 17.4.1.4. As well shown by the empirical formulae of the compounds, the two *cis*-H_3E^{3-} ligands displaced as expected a total of six water molecules from the aqua-complexes belonging to the discrete complex, yielding a highly porous 3D structure with helices having a pitch of about 49 Å. It was described that the magnetic properties registered for the individual Co_{15} cluster were straightforwardly transposed into the 3D MOF material, meaning that the bridges *via* *cis*-H_3E^{3-} are unable to potentiate the magnetic interactions among clusters. This occurs, most likely, because the large spatial separation between individual clusters (see Figure 17.13, right) avoids magnetic exchange. This is a notable example of a modular construction in which it is possible to immediately go from a discrete complex (i.e. "0D") to a 3D framework without constructing intermediate coordination polymers.

17.5.2 Example 2: "0D" → "0D" → 2D → 3D

The second selected example of a modular construction of multi-dimensional coordination polymers based on $H_{4-x}pmida^{x-}$ residues takes advantage of the centrosymmetric anionic $[V_2O_2(pmida)_2]^{4-}$ dimer, which was the subject of a detailed description in Section 17.4.1.2 and it was also employed in the design of a layered structure as described in Section 17.4.3.

The first report of such a dimeric unit comes from Crans and co-workers[30] who published, in the late 1990s, a detailed study on the preparation and characterization of two Na^+-containing salts based on this anionic dimer: $Na_8[V_2O_2(pmida)_2]_2\cdot 16H_2O$ and $Na_4[V_2O_2(pmida)_2]\cdot 10H_2O$.[30] Crans synthesized these materials using various different sources of the metallic centers, namely vanadium sulfate, $VO(acac)_2$ or even sodium metavanadate. It was found that this unit can easily be self-assembled in aqueous solution by reacting vanadium(IV) oxide sulfate pentahydrate with H_4pmida, as schematically represented at the top of Figure 17.14.[58] Because of its straightforward preparation and physical robustness we envisaged that this dimer could be, in principle, used as a "building brick" for the construction of higher dimensionality networks. Indeed, in the crystal structures of Crans, $[V_2O_2(pmida)_2]^{4-}$ is completely surrounded by Na^+ cations and it was rationalized[58] that if these alkali metals could be replaced by, for example, other transition metal cations, coordinative connections to the unit could, indeed, be established. The use of

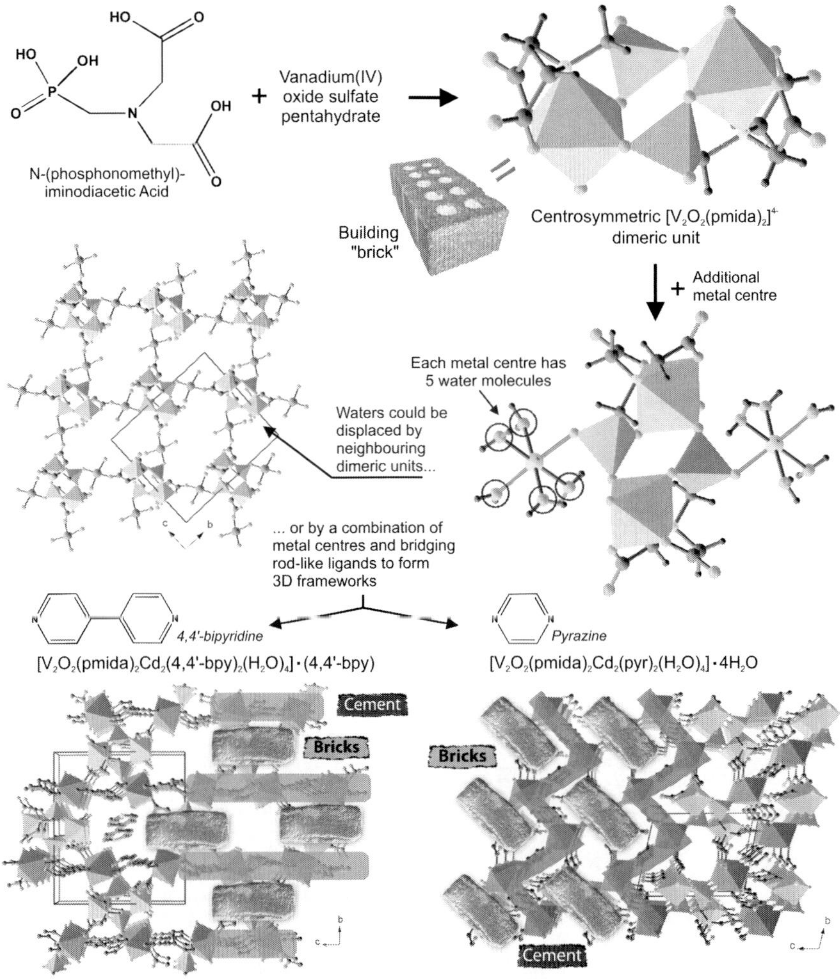

Figure 17.14 Schematic representation of the modular construction of 3D architectures based on the self-assembly of anionic $[V_2O_2(pmida)_2]^{4-}$ dimers with transition metal cations and bridging rod-like organic ligands.

Co^{2+} centers allowed the identification and isolation in the solid state of the tetranuclear neutral $[V_2O_2(pmida)_2Co_2(H_2O)_{10}]$ complex (Figure 17.14, middle right) showing the presence of two $[Co(H_2O)_5]^{2+}$ moieties bound to the phosphonate groups. Immediately after this report similar complexes were described containing Zn^{2+} and Mn^{2+}.[51,56,57] This "0D" → "0D" stepwise construction was of crucial relevance at the design stage because it clearly showed a synthetic way how one could design and isolate a rigid structure having labile water molecules in the periphery of the molecule.

The displacement of the water molecules from $[V_2O_2(pmida)_2M_2(H_2O)_{10}]$ (where $M^{2+} = Zn^{2+}$, Co^{2+}, Mn^{2+} or Cd^{2+}) could occur by way of two different possibilities. First, polymerization might occur by replacing water molecules by phosphonate and/or carboxylate groups from neighboring anionic $[V_2O_2(pmida)_2]^{4-}$ dimers, as depicted in Figure 17.14 (middle left). This leads to a 2D plane net ("0D" → 2D) which is a fragment of a known structure.[58] If water replacement only occurs *via*, for example, the phosphonate group low-dimensional polymers may arise as noted for the cationic $_\infty^2[Co(pyz)(H_2O)_2V_2O_2(pmida)_{22}]^{2n-}$ layer (see Section 17.4.3). The second possibility concerns the use of bridging rod-like organic ligands, of which 4,4-bipyridine is a notable example. Its use allowed us to isolate a 3D framework with empirical formula $[V_2O_2(pmida)_2Cd_2(4,4'\text{-bpy})_2(H_2O)_4]\cdot(4,4'\text{-bpy})$ (Figure 17.14, bottom left; refcodes NACYOG and NACYUM), having large channels running parallel to the *a* axis.[58] Noteworthy, these channels are so large that they can accommodate uncoordinated 4,4′-bipyridine molecules (Figure 17.14, bottom left).

Based on this rational design, we envisaged that the use of a smaller spacer such as pyrazine (pyr), which was already used to design layered networks, could promote the formation of identical framework-type materials but with smaller channels to prevent the trapping of the organic ligand (based solely on geometrical features). A new 3D framework was obtained, $[V_2O_2(pmida)_2Cd_2(pyr)_2(H_2O)_4]\cdot 4(H_2O)$ (Figure 17.14, bottom right),[35] also having 1D channel along the [100] direction but filled instead with highly disordered water molecules.

In short, the results highlighted above provide solid evidence for the reader that the design of porous structures, based on the hierarchically self-assembly of rigid $H_{4-x}pmida^{x-}$-containing building blocks, in particular the anionic $[V_2O_2(pmida)_2]^{4-}$ unit, is possible. Nevertheless, research work based on this strategy is very scarce in the literature and more exploration of this and other anionic units (see, for example, Sections 17.4.1.2 and 17.4.13) is pertinent.

17.6 Conclusions and Outlook

In this chapter we have summarized the structural work available on the self-assembly of $H_{4-x}pmida^{x-}$ residues with metal centers (encompassing the entire periodic table of the elements). Compounds have been organized according to their dimensionalities and also with respect to the way the organic ligand coordinates and participates in the networks.

This review clearly shows that, from the structural standpoint, there is still much room for further investigation in this area. Novel synthetic methodologies and types of reactions should be explored in order to promote the systematic isolation of, for example, highly porous materials. For example, it was shown that by using discrete dimeric units it is possible to design, in a modular fashion, porous compounds. This strategy is still limited to a couple of structures and it needs further investigation, for example by extending it to other anionic units, such as those described here. The use of rare earth cations is very limited and should also be further explored. In this way, it is expected that novel and functional structures based on $H_{4-x}pmida^{x-}$ residues will emerge in following years.

Acknowledgements

Thanks are due to Fundação para a Ciência e a Tecnologia (FCT, Portugal) for supporting our research efforts on coordination polymers by way of their general funding support and the R&D project No. PTDC/QUI-QUI/098098/2008 (FCOMP-01-0124-FEDER-010785).

References

1. C. Janiak, *Dalton Trans.*, 2003, 2781.
2. N. Champness, *Dalton Trans.*, 2006, 877.
3. M. D. Allendorf, C. A. Bauer, R. K. Bhakta and R. J. T. Houk, *Chem. Soc. Rev.*, 2009, **38**, 1330.
4. M. Kurmoo, *Chem. Soc. Rev.*, 2009, **38**, 1353.
5. J. Lee, O. K. Farha, J. Roberts, K. A. Scheidt, S. T. Nguyen and J. T. Hupp, *Chem. Soc. Rev.*, 2009, **38**, 1450.
6. J. R. Li, R. J. Kuppler and H. C. Zhou, *Chem. Soc. Rev.*, 2009, **38**, 1477.
7. L. Q. Ma, C. Abney and W. B. Lin, *Chem. Soc. Rev.*, 2009, **38**, 1248.
8. L. J. Murray, M. Dinca and J. R. Long, *Chem. Soc. Rev.*, 2009, **38**, 1294.
9. G. K. H. Shimizu, R. Vaidhyanathan and J. M. Taylor, *Chem. Soc. Rev.*, 2009, **38**, 1430.
10. D. Zacher, O. Shekhah, C. Woll and R. A. Fischer, *Chem. Soc. Rev.*, 2009, **38**, 1418.
11. N. R. Champness, in *Making Crystals by Design-Methods, Techniques and Applications*, ed. D. Braga, F. Grepioni, Wiley-VCH Verlag GmbH & Co. KGaA, Weinheim, 2007, pp. 193.
12. A. Clearfield, *Curr. Opin. Solid State Mat. Sci.*, 1996, **1**, 268.
13. S. Kitagawa, R. Kitaura and S. Noro, *Angew. Chem., Int. Ed. Engl.*, 2004, **43**, 2334.
14. U. Mueller, M. Schubert, F. Teich, H. Puetter, K. Schierle-Arndt and J. Pastre, *J. Mater. Chem.*, 2005, **16**, 626.
15. S. Horike, S. Shimomura and S. Kitagawa, *Nature Chem.*, 2009, **1**, 695.
16. J. D. Ostergren, G. E. Brown, G. A. Parks and T. N. Tingle, *Environ. Sci. Technol.*, 1999, **33**, 1627.
17. S. Brul and P. Coote, *Int. J. Food Microbiol.*, 1999, **50**, 1.
18. J. J. Chisolm, *J. Toxicol.-Clin. Toxicol.*, 1992, **30**, 493.
19. F. H. Allen, *Acta Crystallogr. B*, 2002, **58**, 380.
20. F. H. Allen and W. D. S. Motherwell, *Acta Crystallogr. B*, 2002, **58**, 407.
21. L. Cunha-Silva, D. Ananias, L. D. Carlos, F. A. A. Paz and J. Rocha, *Z. Kristall.*, 2009, **224**, 261.
22. L. Cunha-Silva, S. Lima, D. Ananias, P. Silva, L. Mafra, L. D. Carlos, M. Pillinger, A. A. Valente, F. A. A. Paz and J. Rocha, *J. Mater. Chem.*, 2009, **19**, 2618.
23. L. Cunha-Silva, L. Mafra, D. Ananias, L. D. Carlos, J. Rocha and F. A. A. Paz, *Chem. Mat.*, 2007, **19**, 3527.

24. N. Chen, S. Wang, C. H. Yao, L. Hui and Y. X. Qu, *J. Chem. Eng. Data*, 2010, **55**, 2613.
25. H. Li, Y. X. Qu and S. Wang, *J. Chem. Eng. Data*, 2010, **55**, 1143.
26. I. J. Bruno, J. C. Cole, P. R. Edgington, M. Kessler, C. F. Macrae, P. McCabe, J. Pearson and R. Taylor, *Acta Crystallogr. B*, 2002, **58**, 389.
27. C. F. Macrae, I. J. Bruno, J. A. Chisholm, P. R. Edgington, P. McCabe, E. Pidcock, L. Rodriguez-Monge, R. Taylor, J. van de Streek and P. A. Wood, *J. Appl. Crystallogr.*, 2008, **41**, 466.
28. K. Branderburg, *Diamond Version 3.2*, Crystal Impact GbR, Bonn, Germany, 2009.
29. *UCSF Chimera package*, The Resource for Biocomputing, Visualization and Informatics at the University of California, San Francisco (supported by NIH P41 RR-01081).
30. D. C. Crans, F. L. Jiang, O. P. Anderson and S. M. Miller, *Inorg. Chem.*, 1998, **37**, 6645.
31. D. Cave, F. C. Coomer, E. Molinos, H. H. Klauss and P. T. Wood, *Angew. Chem., Int. Ed. Engl.*, 2006, **45**, 803.
32. V. Georgantas, N. Kotsakis, C. P. Raptopoulou, A. Terzis, L. Iordanidis, M. Zervou, T. Jakusch, T. Kiss and A. Salifoglou, *J. Inorg. Biochem.*, 2009, **103**, 1530.
33. L. Mafra, F. A. A. Paz, F. N. Shi, J. Rocha, T. Trindade, C. Fernandez, A. Makal, K. Wozniak and J. Klinowski, *Chem.-Eur. J.*, 2006, **12**, 363.
34. J. G. Mao, Z. K. Wang and A. Clearfield, *Inorg. Chem.*, 2002, **41**, 6106.
35. F. A. A. Paz, J. Rocha, J. Klinowski, T. Trindade, F. N. Shi and L. Mafra, *Prog. Solid State Chem.*, 2005, **33**, 113.
36. S. O. H. Gutschke, D. J. Price, A. K. Powell and P. T. Wood, *Angew. Chem., Int. Ed. Engl.*, 1999, **38**, 1088.
37. Z. G. Gu and S. C. Sevov, *J. Mater. Chem.*, 2009, **19**, 8442.
38. S. F. Tang, J. L. Song, X. L. Li and J. G. Mao, *Cryst. Growth Des.*, 2006, **6**, 2322.
39. C. Oldham, in *Comprehensive Coordination Chemistry*, ed. S. G. Wilkinson, Pergamon Press, Oxford, 1987, pp. 435.
40. F. N. Shi, F. A. A. Paz, P. Girginova, J. Rocha, V. S. Amaral, J. Klinowski and T. Trindade, *J. Mol. Struct.*, 2006, **789**, 200.
41. Y. Fan, G. H. Li, Z. Shi, D. Zhang, J. N. Xu, T. Y. Song and S. H. Feng, *J. Solid State Chem.*, 2004, **177**, 4346.
42. D. M. Poojary, B. L. Zhang and A. Clearfield, *Angew. Chem., Int. Ed. Engl.*, 1994, **33**, 2324.
43. B. L. Zhang, D. M. Poojary and A. Clearfield, *Inorg. Chem.*, 1998, **37**, 249.
44. K. R. Ma, Y. L. Zhu and Q. F. Yin, *J. Coord. Chem.*, 2009, **62**, 3243.
45. H. X. Pei, S. M. Lu, Y. X. Ke, J. M. Li, S. B. Qin, S. X. Zhou, X. T. Wu and W. X. Du, *Struct. Chem.*, 2004, **15**, 207.
46. F. A. A. Paz, F. N. Shi, T. Trindade, J. Klinowski and J. Rocha, *Acta Crystallogr. E*, 2005, **61**, m2247.
47. F. N. Shi, F. A. A. Paz, T. Trindade and J. Rocha, *Acta Crystallogr. E*, 2006, **62**, m335.

48. M. Eddaoudi, D. B. Moler, H. L. Li, B. L. Chen, T. M. Reineke, M. O'Keeffe and O. M. Yaghi, *Acc. Chem. Res.*, 2001, **34**, 319.
49. O. M. Yaghi, H. L. Li, C. Davis, D. Richardson and T. L. Groy, *Acc. Chem. Res.*, 1998, **31**, 474.
50. F. N. Shi, F. A. A. Paz, P. I. Girginova, V. S. Amaral, J. Rocha, J. Klinowski and T. Trindade, *Inorg. Chim. Acta*, 2006, **359**, 1147.
51. Y. Q. Gao, W. Gu, X. P. Sun, X. Liu and D. Z. Liao, *Z. Anorg. Allg. Chem.*, 2008, **634**, 1961.
52. J. A. Fernandes, F. N. Shi, J. Rocha and F. A. A. Paz, *Acta Crystallogr. E*, in preparation.
53. F. N. Shi, F. A. A. Paz, P. I. Girginova, L. Mafra, V. S. Amaral, J. Rocha, A. Makal, K. Wozniak, J. Klinowski and T. Trindade, *J. Mol. Struct.*, 2005, **754**, 51.
54. K. R. Ma, J. N. Xu, D. K. Ning, J. Shi, D. J. Zhang, Y. Fan and T. Y. Song, *Inorg. Chem. Commun.*, 2009, **12**, 119.
55. Y. Fan, G. H. Li, W. P. Han, M. Yu, L. Wang, Z. F. Tian, T. Y. Song and S. H. Feng, *J. Solid State Chem.*, 2005, **178**, 2267.
56. L. Cunha-Silva, F. N. Shi, J. Klinowski, T. Trindade, J. Rocha and F. A. A. Paz, *Acta Crystallogr. E*, 2008, **64**, m39.
57. L. Cunha Silva, F. N. Shi, J. Klinowski, T. Trindade, J. Rocha and F. A. A. Paz, *Acta Crystallogr. E*, 2007, **63**, m372.
58. F. A. A. Paz, F. N. Shi, J. Klinowski, J. Rocha and T. Trindade, *Eur. J. Inorg. Chem.*, 2004, 2759.
59. F. A. Cotton and G. Wilkinson, *Advanced Inorganic Chemistry*, 4th edn., Wiley, New York, 1980.
60. J. L. Song, A. V. Prosvirin, H. H. Zhao and J. G. Mao, *Eur. J. Inorg. Chem.*, 2004, 3706.
61. J. G. Mao and A. Clearfield, *Inorg. Chem.*, 2002, **41**, 2319.
62. D. M. Poojary and A. Clearfield, *J. Organomet. Chem.*, 1996, **512**, 237.
63. B. L. Zhang, D. M. Poojary, A. Clearfield and G. Z. Peng, *Chem. Mat.*, 1996, **8**, 1333.
64. A. Clearfield and J. A. Stynes, *J. Inorg. Nucl. Chem.*, 1964, **26**, 117.
65. A. Clearfield, R. H. Blessing and J. A. Stynes, *J. Inorg. Nucl. Chem.*, 1968, **30**, 2249.
66. F. A. A. Paz, F. N. Shi, L. S. Mafra, A. Makal, K. Wozniak, T. Trindade, J. Klinowski and J. Rocha, *Acta Crystallogr. E*, 2005, **61**, m1628.
67. G. L. Zhuang, Y. C. Jin, H. X. Zhao, X. J. Kong, L. S. Long, R. B. Huang and L. S. Zheng, *Dalton Trans.*, 2010, **39**, 5077.
68. J. Rocha, F. A. A. Paz, F. N. Shi, D. Ananias, N. J. O. Silva, L. D. Carlos and T. Trindade, *Eur. J. Inorg. Chem.*, 2011, 2035.
69. M. Eddaoudi, J. Kim, N. Rosi, D. Vodak, J. Wachter, M. O'Keefe and O. M. Yaghi, *Science*, 2002, **295**, 469.
70. O. M. Yaghi, M. O'Keeffe, N. W. Ockwig, H. K. Chae, M. Eddaoudi and J. Kim, *Nature*, 2003, **423**, 705.

CHAPTER 18

Uranyl Phosphonates: A Structural Survey

KARAH E. KNOPE[a,b] AND CHRISTOPHER L. CAHILL[a]

[a] Department of Chemistry, The George Washington University, Washington, DC, 20052, USA; [b] Chemical Sciences and Engineering Division, Argonne National Laboratory, Argonne, IL, 60439, USA

18.1 Introduction

The crystal chemistry of uranyl phosphonates is exceptionally diverse and broad. Despite the maturity of phosphonate chemistry in general (as evidenced by other chapters in this volume), new actinide bearing materials are appearing at an appreciable rate as researchers are applying synthetic methodologies typically reserved for transition metal chemistry (e.g. hydrothermal, self-assembly) with the goal of producing inorganic/organic hybrid materials. One can argue that the interest in these materials is inspired by advances in hybrid materials in general, yet equally influential has been a desire to relate these efforts to various components of the nuclear fuel cycle. While the latter is beyond the scope of this chapter, one can appreciate the link between the local structure of phosphonate (and indeed carboxyphosphonate) complexes and separations and reprocessing issues.[1–4] Further, environmental concerns and stewardship of spent nuclear fuel make fundamental understanding of relevant (and even potential) phases of critical importance.

The goal of this chapter is to provide a comprehensive overview of the structural chemistry of U^{VI} containing phosphonate materials. In addition to tabulating compounds (through early 2010), we provide detailed structural descriptions of a few representative compounds. No materials built from the

Metal Phosphonate Chemistry: From Synthesis to Applications
Edited by Abraham Clearfield and Konstantinos Demadis

Published by the Royal Society of Chemistry, www.rsc.org

other actinide elements are treated herein as these are being presented elsewhere in this volume. Further, we have elected not to review phosphonate chemistry,[5–7] hybrid materials in general,[8–10] or uranium crystal chemistry[11,12] thoroughly, considering the excellent resources already in print. Instead, this presentation will be specific to U^{VI} phosphonates and will be more structural in nature. Unless relevant, we will refrain from detailed discussions of local structural metrics (e.g. bond lengths and angles) and instead focus on extended topologies and synthetic perspective.

18.2 Background: U^{VI} Crystal Chemistry and Hybrid Materials

The crystal chemistry of U^{VI} containing materials has been reviewed extensively and thoroughly. Readers are encouraged to explore Burns' reviews for detailed treatment of this topic.[11–13] Further, uranium containing coordination polymers and metal-organic frameworks with particular relevance to compounds described herein have been reviewed recently as well.[14,15] A brief overview is warranted here, however, to appreciate the relationship between uranium and phosphonate coordination preferences.

Hexavalent uranium exists almost exclusively as the uranyl cation, $[UO_2]^{2+}$, a nearly linear, triatomic species with nominally terminal axial oxo ligands (Figure 18.1). Additional coordination to this ion is therefore in the equatorial plane to promote square, pentagonal, and hexagonal bipyramidal primary building units. In the presence of oxoanions such as iodates, phosphates, carbonates, selenates, etc, edge and point sharing of these polyhedra give rise to a range of layered and framework topologies.[12] Despite the structural diversity and numbers of materials, both synthetic and naturally occurring, the uranyl cation is found almost without exception to be in one of these coordination geometries. Thus one can expect a relatively fixed and limited *local* geometry of the uranyl cation. Diversity thus emerges from variation in packing and coordination to other building units.

A few comments on hybrid materials are in order as well, considering that all of the compounds presented herein are in fact of this type. Broadly speaking, a

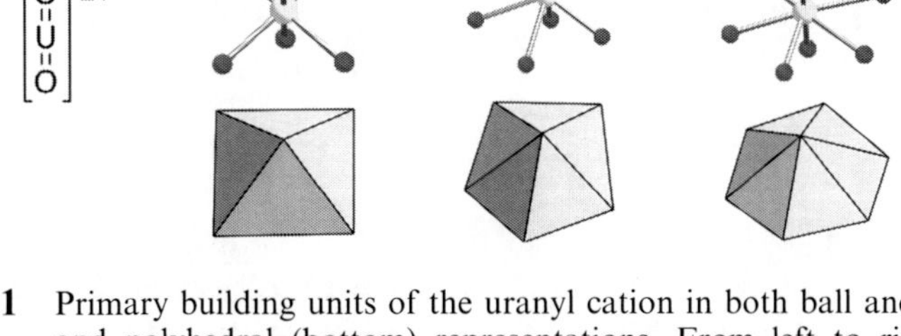

Figure 18.1 Primary building units of the uranyl cation in both ball and stick (top) and polyhedral (bottom) representations. From left to right: square, pentagonal, and hexagonal bipyramids.

hybrid material is one in which an organic component and an inorganic component interact on a molecular level. Specific to these materials, however, we consider the relationship between the uranyl cation (an inorganic building unit) and an organic "linker" possessing appropriate functionality to coordinate favorably and promote extended topologies. This definition is sufficiently broad to cover most coordination polymers and metal-organic framework materials of interest here and elsewhere, yet for this chapter our "linker" species will be restricted to organic molecules of the general formula R-PO_3H (where R = alkyl, phenyl, phosphonate, carboxylate, or sulfonyl). Returning to the comment regarding fixed local geometry of the uranyl cation, similar comments can be made of the functional groups on the organic linker species: the phosphonate (and also carboxylate) entities themselves vary little, yet variation in coordination arrangements such as edge versus point sharing or bridging versus bidentate (respectively) give rise to a range of packing arrangements and assemblies. Further, we will see that flexibility, sterics, and rotational degrees of freedom within the organic linkers themselves contribute to structural geometry.

18.2.1 Uranyl Monophosphonates

Table 18.1 contains a series of uranyl monophosphonates wherein we have defined "mono" as having a single phosphonate functional group and no others of any type. In other words, these are monotopic ligands where the organic components either decorate or are engaged in intermolecular interactions. We have chosen to use the six letter Cambridge Structural Database codes wherever possible and our searches are from Version 5.31 (November 2009).

Figures 18.2 and 18.3 present $[UO_2(HO_3PPh)_2(H_2O)]_2 \cdot 8H_2O$ (α-UPP in Grohol 1996, TOJWEU in the CSD),[16] a 1D material built from UO_2^{2+} pentagonal bipyramids wherein each UO_2^{2+} cation is bound to four phosphonate oxygen atoms (from four distinct phenylphosphonate groups) and one water molecule. Due to the monotopic nature of the phenylphosphonate ligand as well as protonation of one of the phosphonate oxygen atoms, polymerization is limited to one dimension and results in the formation of linear chains, as shown in Figure 18.2. The phenyl groups are arranged nearly perpendicular to the direction of the UO_2^{2+}-phosphonate chains, as shown in Figure 18.3, resulting in a boat shaped conformation.

Interestingly, this non-luminescent (α-UPP) was shown to undergo a phase transition to a luminescent β-UPP, $UO_2(HO_3PPh)_2(H_2O) \cdot 2(H_2O)$ (RUNNET) under conditions of moderate humidity (Figures 18.4 and 18.5).[17] While the local structure of the uranyl cations is similar in both phases, the authors note a bond breaking/reformation mechanism that results in the phenyl groups being in a *trans* configuration in β-UPP as opposed to *cis* in α-UPP (Figure 18.4).

The final monophosphonate to be described is $UO_2(O_3PPh) \cdot 0.7H_2O$ (ZAPRAJ).[18] Like the compounds introduced above, the structure of ZAPRAJ is also built from UO_2^{2+} pentagonal bipyramids (Figure 18.6). There are two

Table 18.1 Uranyl monophosphonates.

Formula	*Crystal system*	*Space group*	*a (Å)*	*b (Å)*	*c (Å)*	*α (°)*	*β (°)*	*γ (°)*	*Reference(s)*	*CSD*
$[C_4N_2H_{12}][UO_2F(PO_3Me)]_2$	monoclinic	$P2_1/n$	6.929	8.117	14.851	90	91.72	90	Doran *et al.*[19] (2003)	BEKREP
$[(C_4N_2H_{12})][UO_2(PO_3Me)(PO_2\text{-}(OH)Me)]_2$	monoclinic	$P2_1/a$	10.068	8.912	12.963	90	90.49	90	Doran *et al.*[19] (2003)	BEKRIT
$[C_4N_2H_{12}][(U_2O_4F_3)(UO_2F(H_2O))\text{-}(PO_3Ph)_2]\cdot 2H_2O$	orthorhombic	Pbcm	6.780	17.282	24.675	90	90	90	Doran *et al.*[19] (2003)	BEKROZ
$[C_4N_2H_{12}]_2[(UO_2)_5(PO_3Bz)_6(PO_2\text{-}(OH)Bz)_2]$	triclinic	P-1	11.473	14.910	15.448	64.23	70.21	84.14	Doran *et al.*[19] (2003)	BEKRUF
$UO_2(O_3PCH_3)$	triclinic	P-1	6.403	6.691	7.098	90.47	99.68	97.33	Grohol (1999)[20]	CAJWIT
$[UO_2(NO_3)_2(camPO_3Me_2)_2]^*$	monoclinic	C2/c	26.739	8.137	19.782	90	127.90	90	Henderson (1998)[21]	CEKFIH
$UO_2(NO_3)_2\{C_3N_3(C_5H_{10}N_2)_2\text{-}[(EtO)_2P(O)\}_2$	triclinic	P-1	10.050	10.628	12.464	73.96	70.25	75.32	Conary (1988)[22]	JADZAP
$UO_2(O_3PCH_2Cl)$	monoclinic	$P2_1/c$	6.742	7.115	13.149	90	98.81	90	Poojary (1995)[23]	REJXOT
$UO_2(HO_3PPh)_2(H_2O)\cdot 2(H_2O)$	monoclinic	$P2_1/c$	7.766	11.255	20.959	90	100.38	90	Grohol (1997)[17]	RUNNET
$[UO_2(HO_3PPh)_2(H_2O)]_2\cdot 8H_2O$	triclinic	P-1	11.724	16.676	11.375	101.61	106.76	102.57	Grohol (1996)[16]	TOJWEU
$UO_2(HO_3PPh)_2\cdot 2(EtOH)$	triclinic	P-1	9.332	11.480	5.672	98.98	92.78	108.54	Grohol (1996)[16]	TOJWIY
$UO_2(O_3PPh)\cdot 0.7H_2O$	hexagonal	P6/mcc	21.827	21.827	7.080	90	90	120	Poojary (1995)[18]	ZAPRAJ
$(UO_2)_3(HO_3PPh)_2(O_3PPh)_2\cdot(H_2O)$	orthorhombic	$P2_12_12_1$	17.197	7.213	27.828	90	90	90	Poojary *et al.* (1996)[24]	ZUQWIR
			16.827	7.167	28.823	90	90	90	Aranda *et al.* (1998)[25]	ZUQWIR01
			17.197	7.213	27.828	90	90	90	Poojary (1996)[26]	ZUQWIR10

*cam = endo-8-camphanyl

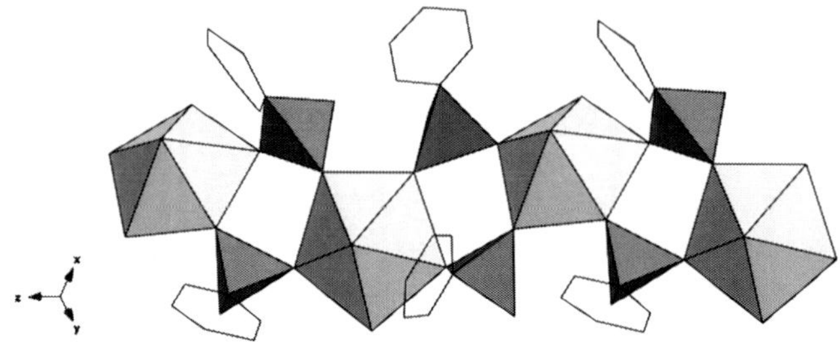

Figure 18.2 A view of 1D chains in structure of $[UO_2(HO_3PPh)_2(H_2O)]_2 \cdot 8H_2O$ (TOJWEU). For this and all figures to follow, uranyl centers are shown as light grey polyhedra, phosphonate groups as dark grey tetrahedra and organic components as black lines.

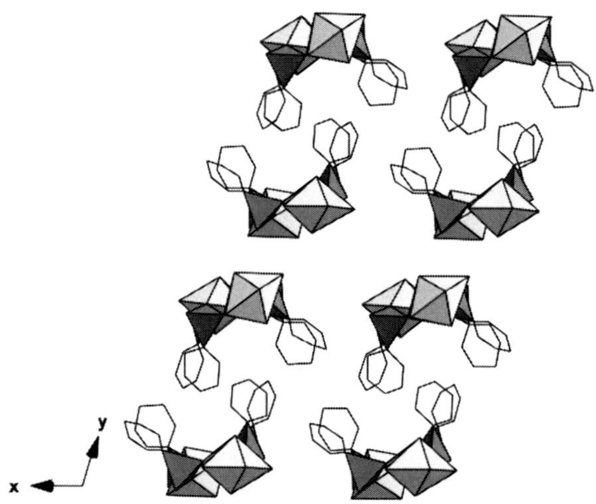

Figure 18.3 Packing diagram of TOJWEU.

crystallographically distinct uranyl sites: one U^{VI} center is bound to five oxygen atoms from four phosphonate ligands while the other is coordinated to four phosphonate oxygen atoms as well as a bound water molecule. Edge sharing of the phosphonate groups with the uranyl polyhedra is present and we note that this is less common than the point sharing observed in many structures. A regular feature that is maintained in this structure, however, is the single coordinated water molecule that completes the pentagonal bipyramidal geometry of the uranyl cation. The beauty of this structure is perhaps best appreciated from a packing diagram (Figure 18.7) wherein one can see molecular "tubes" that pack infinitely down [001]. The circumference of the tube contains six U^{VI} atoms and six phosphorus atoms. The phenyl groups point outward from the tubes. Solvent water occupies the 1D pores that are approximately 12 Å in diameter.

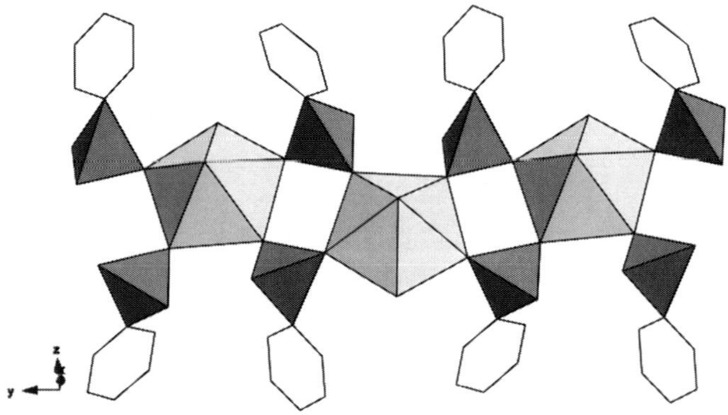

Figure 18.4 View of the 1D chains in structure of $UO_2(HO_3PPh)_2(H_2O) \cdot 2(H_2O)$ (RUNNET).

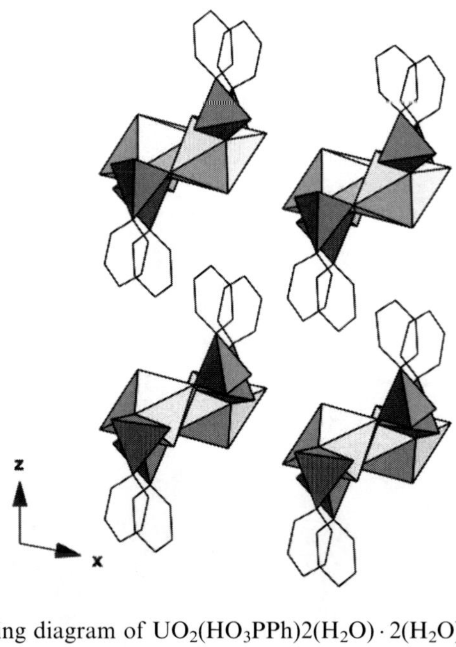

Figure 18.5 Packing diagram of $UO_2(HO_3PPh)2(H_2O) \cdot 2(H_2O)$ (RUNNET).

18.2.2 Uranyl Diphosphonates

Uranyl diphosphonates, at first inspection, are not a radical departure from the monophosphonates introduced above. Coordination motifs between the uranyl cation and phosphonate groups are essentially identical (i.e. point and edge sharing). Diversity of structure types emerges instead for essentially one reason: proximity of the phosphonate groups to one another. Methylene diphosphonates introduce the possibility of a chelation motif through point sharing

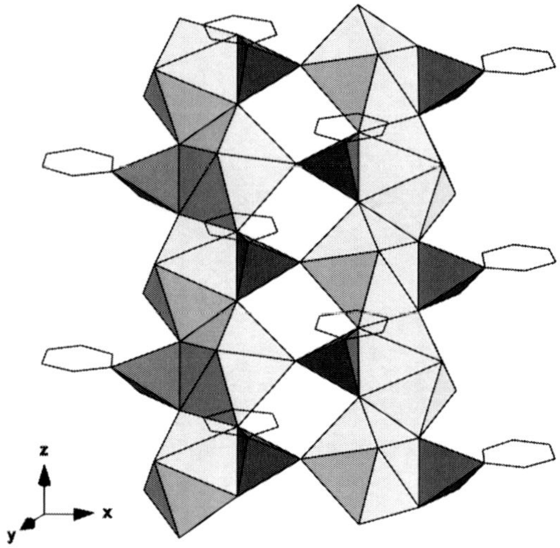

Figure 18.6 Local structure of $UO_2(O_3PPh) \cdot 0.7H_2O$ (ZAPRAJ).

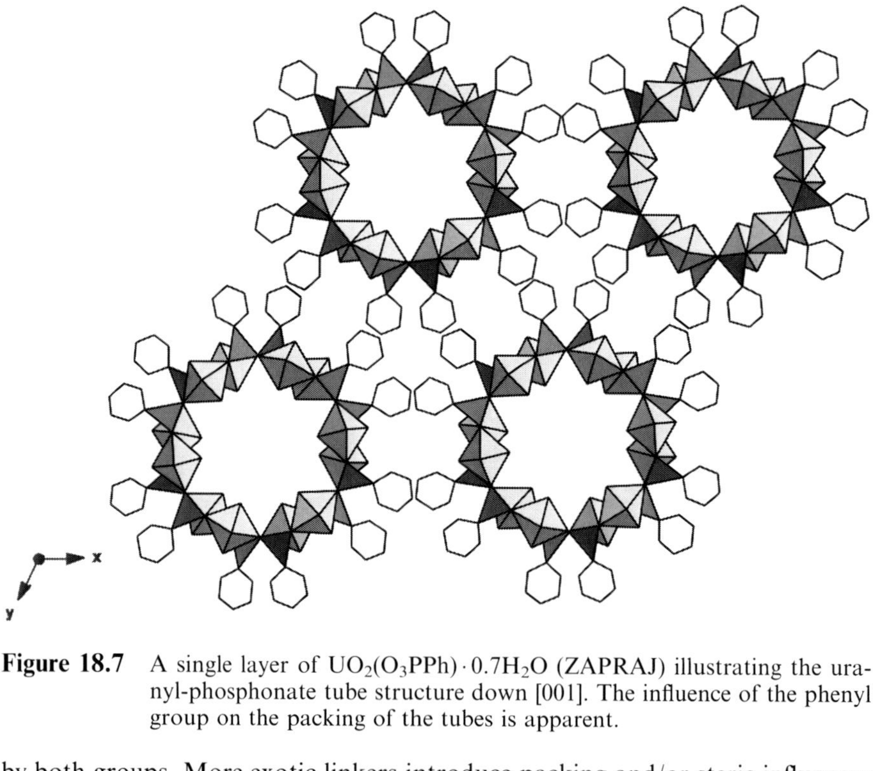

Figure 18.7 A single layer of $UO_2(O_3PPh) \cdot 0.7H_2O$ (ZAPRAJ) illustrating the uranyl-phosphonate tube structure down [001]. The influence of the phenyl group on the packing of the tubes is apparent.

by both groups. More exotic linkers introduce packing and/or steric influences resulting from what is between the phosphonate functional groups. Table 18.2 contains a listing of uranyl diphosphates.

Table 18.2 Uranyl Diphosphonates.

Formula	*Crystal system*	*Space group*	*a (Å)*	*b (Å)*	*c (Å)*	*α (°)*	*β (°)*	*γ (°)*	*Reference(s)*	*CSD*
$(UO_2)_2[CH_2(PO_3)_2](H_2O)_3 \cdot (H_2O)$	monoclinic	$P2_1/c$	8.457	14.442	10.600	90	100.00	90	Nelson (2008)[29]	XOFVEU
$UO_2[CH_2(PO_3H)_2](H_2O)$	orthorhombic	$Pna2_1$	18.759	7.913	5.437	90	90	90	Nelson (2008)[29]	XOFVIY
$[C_{10}H_{10}N_2]_2(C_{24}H_{16}O_{36}P_8U_6) \cdot 2(H_2O)$	triclinic	P-1	9.793	10.659	17.564	83.85	86.32	65.18	Adelani (2009)[30]	GUXZIJ
$[C_4H_{12}N_2]\{(UO_2)_2[C_6H_4(PO_3H)_2]_3\} \cdot 2(H_2O)$	monoclinic	$P2_1/n$	10.241	11.437	15.933	90	95.63	90	Adelani (2009)[30]	GUXZOP
$[C_{10}H_9N_2]_2(C_{18}H_{18}O_{22}P_6U_2) \cdot 2(H_2O)$	triclinic	P-1	8.811	12.008	12.349	63.52	89.23	89.59	Adelani (2009)[30]	GUXZUV
$[C_{12}H_{10}N_2]_2(C_{24}H_{16}O_{36}P_8U_6) \cdot 2(H_2O)$	monoclinic	$P2_1/n$	9.805	19.570	17.339	90	92.02	90	Adelani (2009)[30]	GUYBAE
$(UO_2)[CH_2(PO_3H)_2](H_2O) \cdot 2(H_2O)$	orthorhombic	Pnma	7.401	9.712	14.587	90	90	90	Knope (2010)[27]	NA
$(UO_2)[CH_2(PO_3H)_2](H_2O) \cdot 4(H_2O)$	monoclinic	$P2_1/m$	9.585	9.744	13.026	90	93.39	90	Knope (2010)[27]	NA
$(UO_2)[(O_3PCH_2PO_3H)](H_2O) \cdot 3(H_2O) \cdot (NH_4)$	monoclinic	$P2_1/n$	7.198	13.477	13.365	90	102.93	90	Knope (2010)[27]	NA
$UO_2[CH_2(PO_3H)_2](H_2O) \cdot (H_2O)$	orthorhombic	$Pna2_1$	10.805	8.140	10.582	90	90	90	Knope (2010)[27]	NA
$UO_2(NO_3)_2\{[(EtO)_2P(O)]_2C_5H_3N\}$	monoclinic	C2/c	24.906	6.819	23.213	90	141.26	90	Russell (1995)[31]	ZEVJOZ

A methylenediphosphonate (MDP) that highlights this "chelation" is $(UO_2)[CH_2(PO_3H)_2](H_2O)\cdot 2H_2O$ (Figure 18.8).[27] This material was synthesized at room temperature and is part of a larger study exploring both temperature and counter-ions on the structures of uranyl diphosphonates. There is one unique U^{VI} site and it, like others, is pentagonal bipyramidal and coordinated to four phosphonate oxygen atoms and one water molecule. The bifunctionality of MDP, however, allows the ligand to chelate one U^{VI} center while binding to two other UO_2^{2+} ions in a monodentate fashion. Coordination of the MDP units to three UO_2^{2+} cations as well as protonation of two phosphonate oxygen atoms results in the formation of the 1D chains shown in Figure 18.8. Note that these chains are reminiscent of those found in several $[AnO_2(TO_4)(H_2O)_2\cdot 2H_2O$ (T = S, Se, Mo, Cr) architectures.[28]

The structure of $UO_2[CH_2(PO_3H)_2](H_2O)$ (Figure 18.9; XOFVIY)[29] also contains MDP and adopts a 3D structure. The local structure of each UO_2^{2+} cation is nearly identical to that observed in $(UO_2)[CH_2(PO_3H)_2](H_2O)\cdot 2H_2O$ (see Figure 18.8) wherein each MDP unit coordinates three uranyl cations (one through chelation and two via a single oxygen atom of adjacent $-PO_3H$ groups). In both compounds, each MDP unit is diprotonated. However, whereas the protonation sites of the chelating $-PO_3H$ unit are nearly planar with respect to the bound H_2O molecule in $(UO_2)[CH_2(PO_3H)_2](H_2O)\cdot 2H_2O$, these sites are oriented approximately perpendicularly in $UO_2[CH_2(PO_3H)_2](H_2O)$.

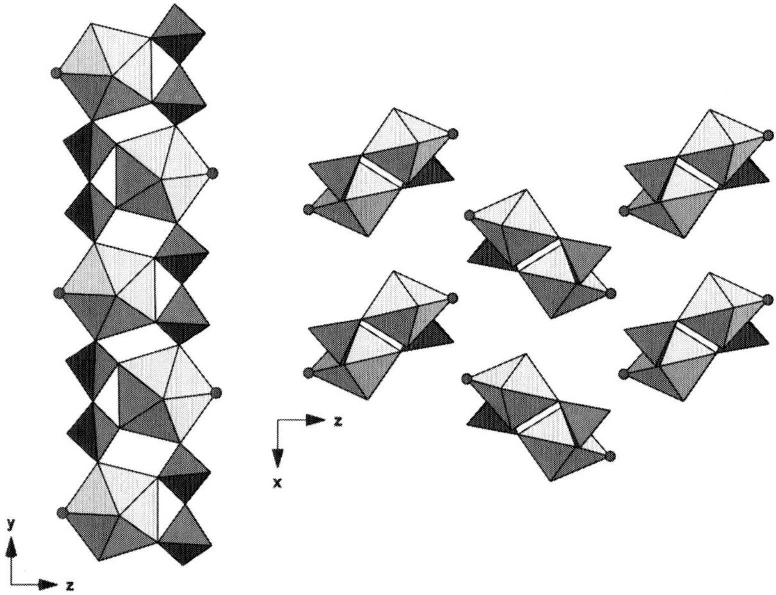

Figure 18.8 Left: polyhedral representation of $[UO_2(CH_2(PO_3H)_2)(H_2O)]\cdot 2H_2O$ illustrating the topology of the 1D chains. Right: packing diagram of $[UO_2(CH_2(PO_3H)_2)(H_2O)]$ chains. Solvent water molecules have been omitted.

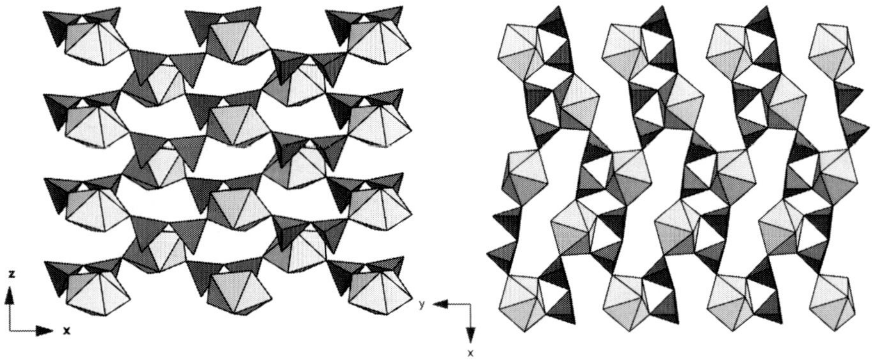

Figure 18.9 Two views of the layered structure of $UO_2[CH_2(PO_3H)_2](H_2O)$ (XOFVIY). Note the layers may be thought of as constructed from chains similar to those in Figure 18.8.

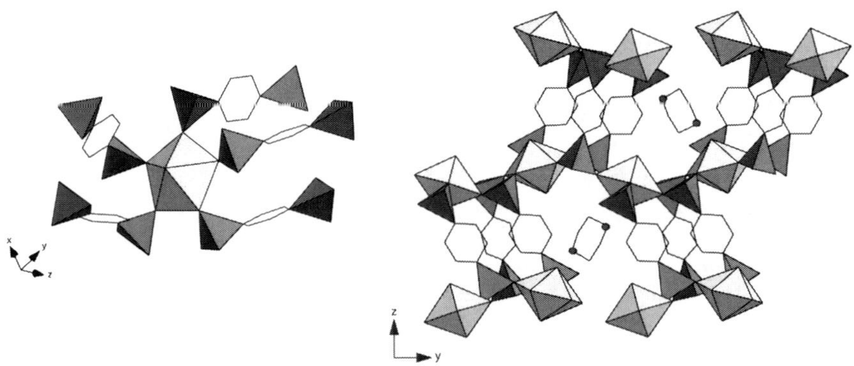

Figure 18.10 The structure of $[C_4H_{12}N_2](UO_2)_2[C_6H_4(PO_3H)_2]_3 \cdot 2(H_2O)$ (GUXZOP). Left: local structure highlighting uranyl point-sharing with five independent phosphonate groups. Right: 3D view of GUXZOP showing piperazine template molecules.

Furthermore, the $-PO_3H$ groups in the latter are oriented along the *c* axis to result in a polar structure.

An interesting series of pillared diphosphonates was reported recently by Adelani and Albrecht-Schmitt.[30] In this study, the authors used 1,4-benzenediphosphonic acid (BDP) as a linker and then explored the influence of noncoordinating template molecules to direct assembly of extended structures. An example where piperazine was used is in the structure of $[C_4H_{12}N_2]\{(UO_2)_2[C_6H_4(PO_3H)_2]_3\} \cdot 2(H_2O)$ (GUXZOP; Figure 18.10). This material adopts a pillared structured wherein layers of UO_2^{2+} pentagonal bipyramids are linked via BDP units. Each UO_2^{2+} cation is bound to five oxygen atoms from five distinct BDP ligands. Coordination is exclusively point sharing and no chelation is observed, as one would anticipate considering the distance between functional groups as well as the rigidity of the ligand. Two

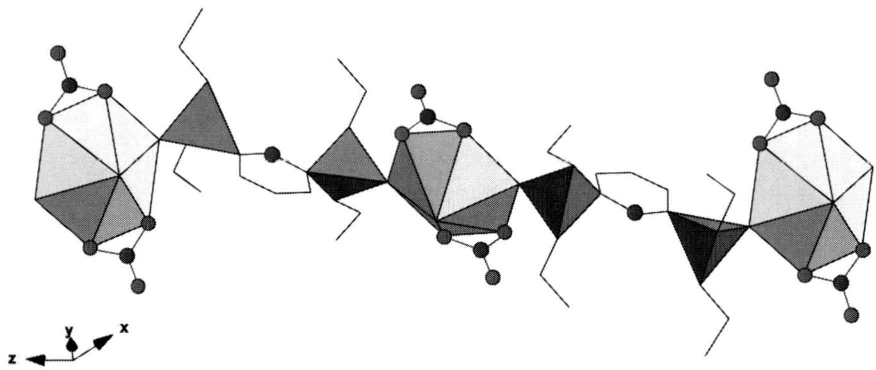

Figure 18.11 A view of the chains found in $UO_2(NO_3)_2[(EtO)_2P(O)]_2C_5H_3N$ (ZEVJOZ).

phosphonate oxygen atoms of each BDP unit bind U^{VI} centers while one oxygen atom remains protonated. The structure also contains protonated piperazine guest molecules that reside in channels measuring 8.0 × 8.3 Å.

A slightly more exotic diphosphonate linker can be found in $UO_2(NO_3)_2\{[(EtO)_2P(O)]_2C_5H_3N\}$ (ZEVJOZ; Figure 18.11) where the phosphonate groups contain ethyl groups on two of the three oxygen sites.[31] Use of this linker was part of a study to explore multidentate chelates for aqueous phase extractants and as the authors mention, resulted in an unanticipated coordination. Rather than adopting a tridentate motif through the two phosphonate oxygens and the N on the pyridine ring, the phosphonate groups instead bind to two distinct U^{VI} sites leaving the N uncoordinated. As a result, the compound adopts a 1D linear chain structure that consists of the UO_2^{2+} cation equatorially bound to six oxygen atoms, including four oxygen atoms from two nitrate groups and two oxygen atoms from two $(EtO)_2P(O)]_2C_5H_3N$ ligands. Two of the phosphonate oxygen atoms of the ligand are "protected," leaving only one P–O site available for metal coordination. The orientation of these sites precludes chelation and limits coordination to one dimension.

18.2.3 Uranyl Carboxyphosphonates

The use of carboxyphosphonate linkers has been quite successful in expanding the catalog of U^{VI} containing hybrid materials. Whereas the phosphonate groups behave as we would anticipate from the above discussions, the introduction of a second functional group, the carboxylate, makes for a true multitopic and heterofunctional linker. From a structural perspective, carboxylate functionalities introduce distinct coordination possibilities as well as the potential for ordered heterometallic materials based on hard/soft acid/base considerations.[15,32] Tables 18.3 and 18.4 include the known examples of both homo- and heterometallic compounds and structural descriptions for two examples of homometallic compounds and three heterometallic examples follow.

Table 18.3 Homometallic uranyl carboxyphosphonates.

Formula	*Crystal system*	*Space group*	*a (Å)*	*b (Å)*	*c (Å)*	*α (°)*	*β (°)*	*γ (°)*	*Reference(s)*	*CSD*
$Na[UO_2(O_3PCH_2CO_2)]\cdot 2(H_2O)$	triclinic	P-1	6.874	7.417	9.635	79.15	78.69	63.89	Alsobrook (2008)[33]	GOBWOK
$UO_2(O_3PCH_2CO_2H)$	monoclinic	$P2_1/c$	6.693	7.135	14.307	90	103.02	90	Knope (2008)[34]	VOLNOA
$(UO_2)_4(HO_3PCH_2CO_2)(O_3PCH_2CO_2)_2(H_2O)_4\cdot 3H_2O$	triclinic	P-1	6.347	15.548	15.731	97.34	91.58	90.89	Knope (2008)[34]	VOLNUG
$(NH_4)[(UO_2)(O_3PCH_2CO_2)]\cdot H_2O$	orthorhombic	Pbca	9.590	13.349	13.450	90	90	90	Knope (2008)[34]	VOLPAO
$(NH_4)_2[(UO_2)_3(O_3PCH(CH_3)CO_2)_2(O_3PCH(CH_3)CO_2H)]\cdot H_2O$	monoclinic	$P2_1/m$	8.754	17.87	9.055	90	104.30	90	Knope (2008)[34]	VOLPES
$(C_{10}H_{10}N_2)[(UO_2)_3(PPA)_2(HPPA)(H_2O)]\cdot 3(H_2O)$	monoclinic	$P2_1/c$	8.980	25.558	13.844	90	100.31	90	Knope (2009)[39]	BUCRAT
$(C_{12}H_{12}N_2)[(UO_2)_3(PPA)_2(HPPA)(H_2O)]\cdot 2(H_2O)$	triclinic	P-1	10.506	10.886	15.646	105.56	95.25	104.02	Knope (2009)[39]	BUCREX
$(C_{12}H_{14}N_2)[(UO_2)_3(PPA)_2(HPPA)(H_2O)]\cdot 2(H_2O)$	triclinic	P-1	10.412	10.906	15.753	105.90	95.65	103.47	Knope (2009)[39]	BUFRAW
$UO_2(HO_3PC_6H_4CO_2H)_2\cdot 2(H_2O)$	triclinic	P-1	5.513	6.766	13.762	90.98	95.00	106.40	Adelani (2010)[40]	NA
$UO_2(O_3PCH_2CO_2H)(H_2O)\cdot 2H_2O$	monoclinic	$P2_1$	8.352	7.001	8.373	90	107.30	90	Knope (2010)[41]	NA
$K[(UO_2)_2(PO_3CH_2CO_2)(PO_3CH_2CO_2H)(H_2O)]\cdot H_2O$	triclinic	P-1	7.892	8.350	12.198	80.14	83.07	85.79	Alsobrook (2009)[42]	NA
$Rb[(UO_2)_2(PO_3CH_2CO_2)(PO_3CH_2CO_2H)(H_2O)]\cdot H_2O$	triclinic	P-1	8.093	8.379	12.321	79.79	81.80	85.37	Alsobrook (2009)[42]	NA
$Cs[(UO_2)(PO_3CH_2CO_2)]$	monoclinic	$P2_1/n$	8.024	9.502	10.522	90	92.99	90	Alsobrook (2009)[42]	NA
$Cs[(UO_2)_2(PO_3CH_2CO_2)(PO_3CH_2CO_2H)](H_2O)\cdot H_2O$	triclinic	P-1	8.469	9.085	11.479	83.59	77.44	84.68	Alsobrook (2009)[42]	NA
$Cs_3[(UO_2)_4(PO_3CH_2O_2)_2(PO_3CH_2CO_2H_{0.5})_2]\cdot nH_2O$	monoclinic	C2/c	27.447	8.790	15.985	90	110.71	90	Alsobrook (2009)[42]	NA
$[C_4N_2H_{14}][(UO_2)_2(PPA)_2]\cdot (H_2O)$	triclinic	P-1	7.076	10.802	14.871	82.893	87.135	85.813	Ramaswamy (2010)[43]	NA

Table 18.4 Heterometallic uranyl carboxyphosphonates.

Formula	*Crystal system*	*Space group*	*a (Å)*	*b (Å)*	*c (Å)*	*α (°)*	*β (°)*	*γ (°)*	*Reference(s)*	*CSD*
$UO_2Cu(PO_3CH_2CO_2)(OH)(H_2O)_2$	monoclinic	$P2_1/c$	6.967	14.231	9.058	90	98.39	90	Alsobrook (2008)[33]	GOBWAW
$(UO_2)_2(O_3PCH_2CO_2)_2Cu(H_2O)_3$	monoclinic	$P2_1/c$	7.636	19.300	11.245	90	100.59	90	Alsobrook (2008)[33]	GOBWEA
$[H_3O][(UO_2)_2Cu_2(PO_3CH_2CO_2)_3(H_2O)_2$	triclinic	P-1	8.813	10.023	14.730	83.44	83.95	64.87	Alsobrook (2008)[33]	GOBWIE
$(UO_2)_2(PPA)_2(H_2O)_2 \cdot Zn(H_2O)_6 \cdot 4H_2O$	triclinic	P-1	7.029	8.300	10.982	81.18	74.14	89.90	Knope (2010)[38]	KUNJIN
$(UO_2)_4(PPA)_2(HPPA)_2 \cdot Zn(H_2O)_6 \cdot 5H_2O$	triclinic	P-1	10.668	11.538	17.214	78.47	87.55	76.31	Knope (2010)[38]	KUNJOT
$(UO_2)_2(H_2O)_2(PPA)_2Zn(H_2O)_4$	triclinic	P-1	6.944	8.371	16.048	97.54	96.43	91.80	Knope (2010)[38]	KUNJUZ
$(UO_2)_2(PPA)_2(HPPA)Zn_2(H_2O)_2 \cdot 3(H_2O)$	triclinic	P-1	9.103	9.728	15.095	91.40	94.50	115.63	Knope (2010)[38]	KUNKAG

18.2.3.1 Homometallic Compounds

The structure of $Na[UO_2(O_3PCH_2CO_2)] \cdot 2H_2O$ (Figure 18.12; GOBWOK)[33] adopts a layered topology wherein monomers of UO_2^{2+} pentagonal bipyramids are linked via phosphonoacetate (PPA) molecules. The PPA ligands bind each UO_2^{2+} cation through both phosphonate and carboxylate oxygen atoms. Bridging of two uranyl sites by the carboxylate groups is common in these materials, as is point sharing of the phosphonate tetrahedra. Interlayer sodium cations charge balance the anionic sheets and are not considered in our definition of "hetero-metallic" (below) as there is no coordination to the carboxyphosphonate linkers.

The structure of $UO_2(O_3PCH_2CO_2H)$ (Figure 18.13; VOLNOA)[34] is built from chains of edge-shared pentagonal bipyramids that propagate along [010]. Coordination of the PPA ligands to the U^{VI} centers in this material occurs exclusively through phosphonate oxygen atoms (via both edge and point sharing) and results in 2D sheets. The protonated and uncoordinated carboxylate units protrude into the interlayer (Figure 18.13) and results in head-to-head O–H · · · O hydrogen bonding interactions. This non-covalent interaction has been discussed in a number of other forums as being a "supramolecular synthon" from which to construct extended topologies.[35]

Interestingly, the sheet topology of this material is related to that found in β-uranophane ($Ca[(UO_2)(SiO_3OH)]_2 \cdot 5H_2O$)[36] where silicate tetrahedra are coordinated similarly to the phosphonates in VOLNOA. There may be a correlation here in that one oxygen atom of the silicate species is a hydroxyl group and essentially terminal. An analogy could be made to the fourth vertex

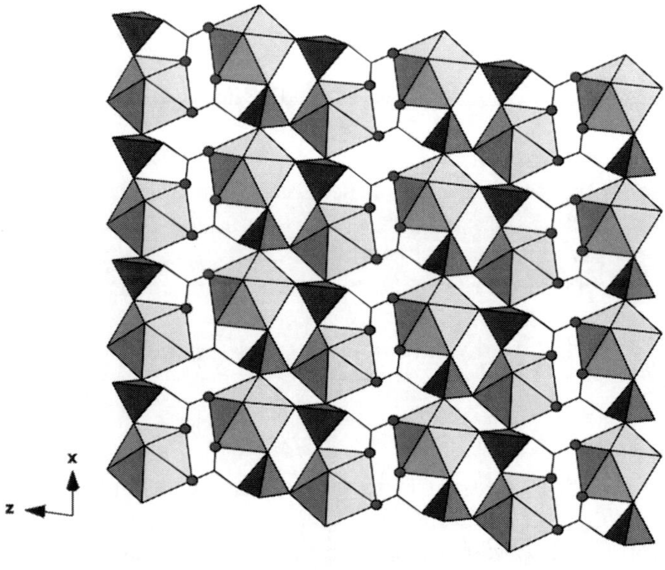

Figure 18.12 The layered structure of $Na[UO_2(O_3PCH_2CO_2)] \cdot 2(H_2O)$ (GOBWOK). Point-sharing by three phosphonate oxygen atoms as well as bridging by carboxylate groups are present.

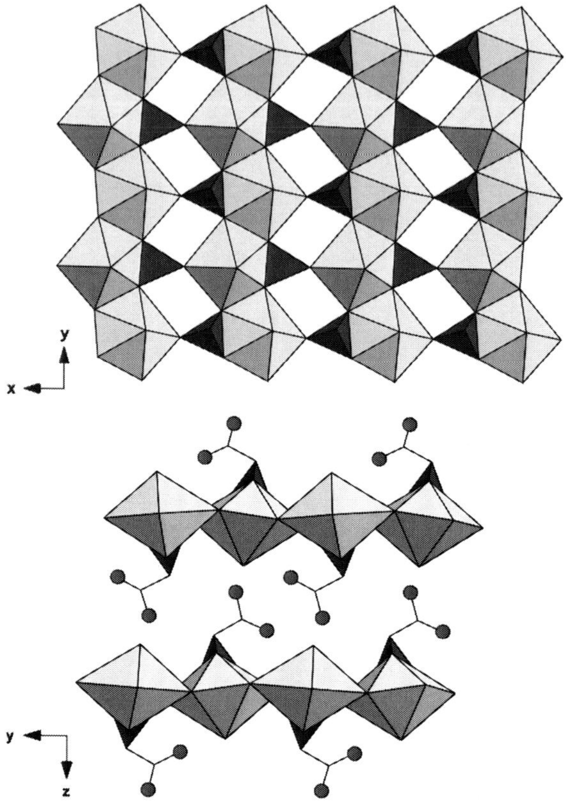

Figure 18.13 The layered structure of $UO_2(O_3PCH_2CO_2H)$ (VOLNOA). Uncoordinated carboxylate groups protrude into the interlayer where they participate in head-to-head H-bonding interactions.

of a phosphonate group being a carbon atom that is not coordinated to any metal centers. In fact, a quick inspection of layered uranyl phosphonate structures in general suggests more similarities to some layered uranyl silicate structures as opposed to uranyl phosphates despite the similarities in composition. Finally, the synthesis of VOLNOA was achieved via direct assembly of U^{VI} with PPA linkers, yet was part of a larger study exploring the *in situ* formation of PPA from trialkylphosphonate starting materials.

18.2.3.2 Heterometallic Compounds

True heterometallic compounds are those in which coordination to both metal centers occurs through the carboxylate and/or phosphonate functional groups. In other words, the second metal center is not strictly charge balancing. The uranyl cation shows an exclusive preference for the phosphonate groups in all compounds reported thus far. Carboxylate groups either bind exclusively to other metal centers, or bridge between the uranyl and other metals.

Figure 18.14 presents the structure of $(UO_2)_2(O_3PCH_2CO_2)_2Cu(H_2O)_3$ (GOBWEA).[33] This compound consists of U^{VI} pentagonal bipyramids linked via PPA ligands. The PPA units bind the UO_2^{2+} centers via both carboxylate and phosphonate oxygen atoms and polymerize the UO_7 units into 2D sheets. Hydrated Cu^{2+} cations are found in the interlayer, yet are coordinated to carboxylate groups. These ions exhibit Jahn–Teller distorted octahedra with two short bonds and two longer bonds. Interaction of the Cu^{2+} cations with axial positions of the UO_2^{2+} serves to link the sheets into a 3D structure. Such coordination is similarly observed in UO_2^{2+}–Cu^{2+} containing minerals such as torbernite and metatorbernite.[37]

$(UO_2)_2(PPA)_2(H_2O)_2 \cdot Zn(H_2O)_6 \cdot 4H_2O$ (KUNJIN; Figure 18.15)[38] consists of discrete UO_2O_5 monomers linked via phosphonoacetate units to form chains

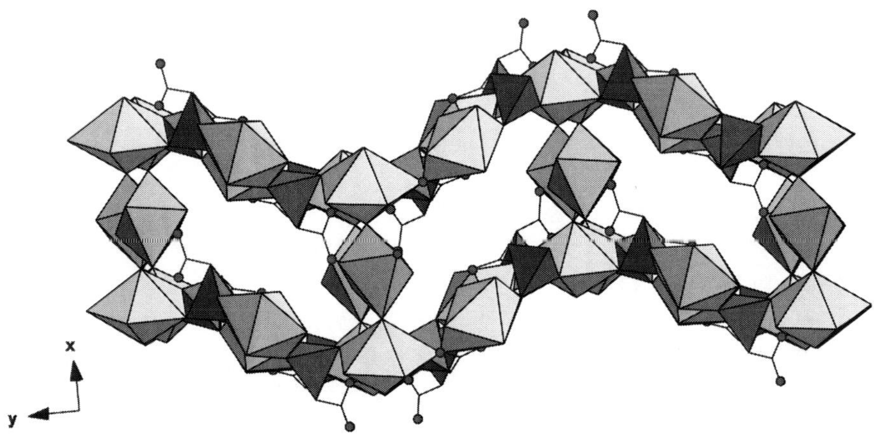

Figure 18.14 The 3D structure of $(UO_2)_2(O_3PCH_2CO_2)_2Cu(H_2O)_3$ (GOBWEA). Jahn–Teller distorted and hydrated Cu^{II} sites are coordinated to the uranyl oxygen atoms of sequential layers.

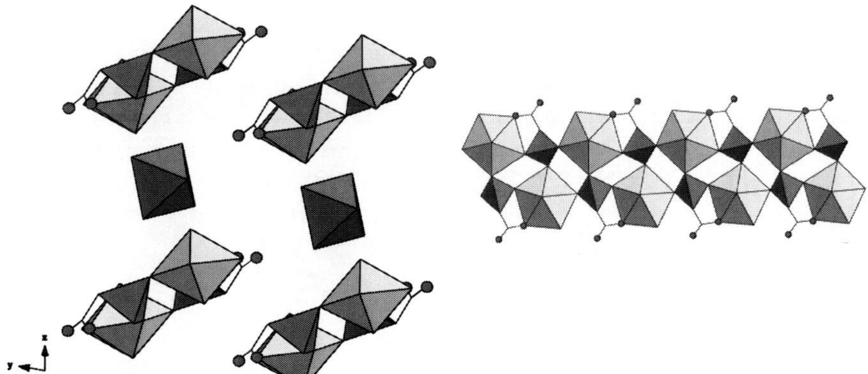

Figure 18.15 The chain structure of $(UO_2)_2(PPA)_2(H_2O)_2 \cdot Zn(H_2O)_6 \cdot 4(H_2O)$ (KUNJIN). Anionic chains (inset) are charge balanced by fully hydrated Zn^{II} ions in the extra-chain region. Solvent water molecules are not shown.

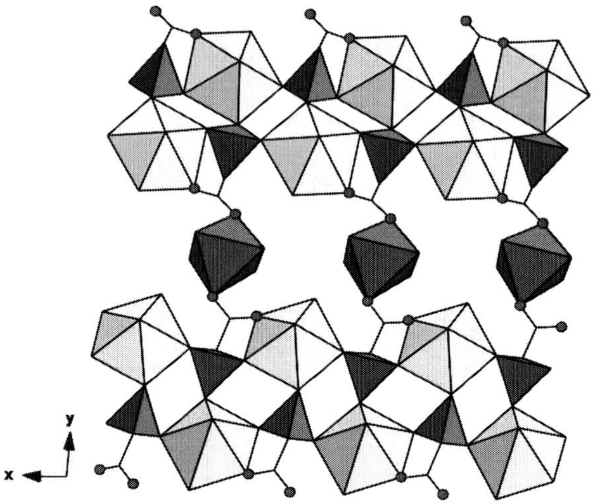

Figure 18.16 The layered structure of $(UO_2)_2(H_2O)_2(PPA)Zn(H_2O)_4$ (KUNJUZ). Partially hydrated Zn^{II} ions bridge $(UO_2)_2(H_2O)_2(PPA)^{2-}$ chains to form layers.

of $[(UO_2)_2(PPA)_2(H_2O)_2]^{2-}$. Coordination of the PPA ligand to the UO_2^{2+} cation occurs via both phosphonate and carboxylate oxygen atoms. Each PPA unit chelates one UO_2^{2+} cation via a phosphonate and carboxylate oxygen atom. The ligand additionally binds two U^{VI} centers in a monodentate fashion via the remaining phosphonate oxygen atoms. One carboxylate oxygen atom remains unbound and protonated. Hydrated $[Zn(H_2O)_6]^{2+}$ cations charge balance the anionic chains and are not coordinated to the PPA linkers.

Technically we should not consider KUNJIN to be a heterometallic compound, yet we have included this material considering its context for the next heterometallic structure to be described, $(UO_2)_2(H_2O)_2(PPA)Zn(H_2O)_4$ (KUNJUZ; Figure 18.16).[38] KUNJUZ forms from similar starting conditions as KUNJIN, yet from higher temperature treatment for a longer duration. The result is a compound with identical uranyl phosphonate chains, yet the Zn^{2+} ions are now less hydrated and coordinated directly to the carboxylate groups of the PPA. The authors of this study note a number of structural changes, largely at the Zn^{2+} site, as a function of time and temperature within this system.

18.2.4 Miscellaneous Compounds

Table 18.5 lists a number of "miscellaneous" uranyl-bearing phosphonate materials. Most of these compounds do not quite fit our definition of a hybrid material in that they are molecular compounds and not polymeric in any dimension. Further, the linker molecules are typically monofunctional or have other features for consideration such as differing substituents or hetero atoms.

Table 18.5 Miscellaneous uranyl phosphonates.

Formula	*Crystal system*	*Space group*	*a (Å)*	*b (Å)*	*c (Å)*	*α (°)*	*β (°)*	*γ (°)*	*Reference(s)*	*CSD*
$UO_2(NO_3)_2[(i\text{-}C_3H_7O)_2P(O)CH_2C(O)N(C_2H_5)_2$	monoclinic	$P2_1/n$	10.520	10.963	20.033	90	103.79	90	Bowen (1983)[45]	BOXPIN
$UO_2(HOCH_2PO_3)\cdot 5H_2O$	monoclinic	$P2_1/c$	7.004	8.579	16.754	90	90.65	90	Britel (1986)[46]	FAFGEY
$UO_2(NO_3)_2\{2\text{-}[(C_2H_5O)_2P(O)]C_5H_4NO\}$	triclinic	P-1	8.074	10.514	12.250	112.65	104.07	92.11	McCabe (1987)[47]	FESHEQ
$UO_2(NO_3)_2\{2\text{-}[(C_6H_5)_2P(O)]C_5H_4NO\}$	triclinic	P-1	9.083	10.644	12.338	78.23	76.65	76.74	McCabe (1987)[47]	NA
$UO_2(NO_3)_2[(i\text{-}C_3H_7O)_2P(O)\text{-}CH_2S(O)_2C_6H_{11}]_2$	monoclinic	$P2_1$	8.374	25.423	10.031	90	100.49	90	Karthikeyan (1989)[48]	JARBOT
$UO_2(NO_3)_2[(C_6H_5)_2P(O)CH_2S(O)_2N\text{-}(CH_3)_2]_2$	monoclinic	$P2_1/c$	10.759	13.811	17.823	90	105.83	90	Cromer (1990) [49]	JISMIH
$UO_2(NO_3)_2[(i\text{-}C_3H_7O)_2P(O)\text{-}CH(CH_2C_6H_4NO_2\text{-}p)C(O)N(C_2H_5)_2]$	triclinic	P-1	9.524	10.111	17.692	70.27	72.48	70.01	Karthikeyan (1988)[50]	SUDPUC
$UO_2\{C_9N_3H_{19}(PO_3H)_3\}\cdot H_2O$	monoclinic	$P2_1/c$	11.818	11.355	14.866	90	112.05	90	Bao (2006)[44]	VEBNOG
$UO_2(NO_3)_2\{[(i\text{-}PrO)_2P(O)][C(O)NEt_2]\text{-}CH\{(CH_2)_3C(O)OEt]\}$	monoclinic	$P2_1/c$	8.855	16.257	20.603	90	92.07	90	Conary (1992)[51]	WAGFIT
$[C_4N_2H_{14}][(UO_2)_2(C_2O_4)(HPPA)_2]$	monoclinic	C2/m	16.079	7.003	10.228	90	104.34	90	Ramaswamy (2010)[43]	NA

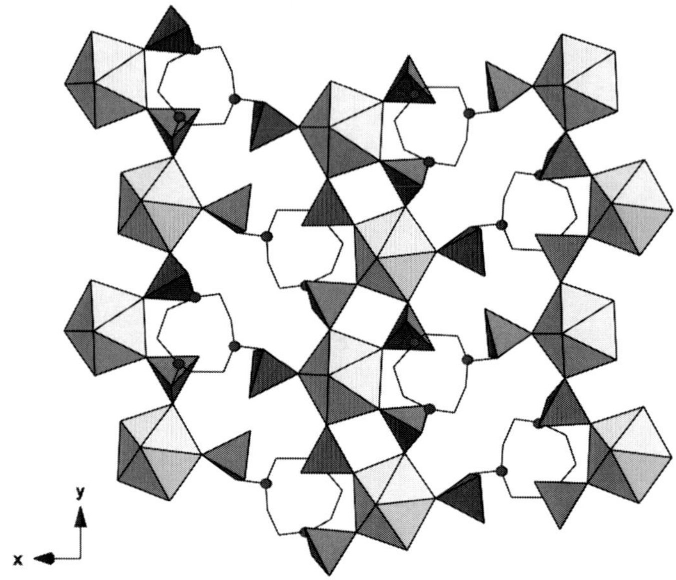

Figure 18.17 The triphosphonate $UO_2C_9N_3H_{19}(PO_3H)_3 \cdot H_2O$ (VEBNOG).

Nonetheless, we have included these for completeness and considering the relevance of the studies during which they were produced.

One compound that does fit nicely with themes developed thus far, however, is $UO_2\{C_9N_3H_{19}(PO_3H)_3\} \cdot H_2O$ (VEBNOG; Figure 18.17)[44] The authors of this study explored phosphonate-functionalized macrocycles as a means for assembling extended topologies, specifically, 1,4,7-triazacyclononane-1,4,7-triyl-tris(methylenephosphonic acid ($notpH_6$). The resulting structure adopts a layered topology wherein uranyl pentagonal bipyramids are formed from five distinct point-shared phosphonate groups. One nitrogen atom and three phosphonate oxygen atoms on each $notpH_6$ unit remain protonated and charge balance an otherwise anionic sheet. This is the only example of a triphosphonate material discussed herein.

18.3 Concluding Remarks

Uranyl phosphonates are, in many respects, still in their infancy in terms of a thorough exploration of their structural potential. A cursory glance at the tables within this chapter indicates that most materials have been prepared within only the last few years. While several earlier examples have been explored as part of systematic efforts to study actinide separations (e.g. JARBOT and SUDPUC, Table 18.5), more recent efforts have utilized a more materials-centric approach. Indeed, exploratory efforts to probe the structural chemistry of the uranyl cation in the presence of phosphonate (and carboxylate) functional groups has led to a respectable catalog of structure types. Moreover, the potential to develop materials with even more complex topologies is present. Phosphonate materials in general are in some ways at an advantage in this regard considering not only the solubility of organic precursors (versus metal phosphates, for example), but also the nearly limitless range of linker geometries and possibilities for substitution and functionalization. We have presented a sampling of representative materials in this contribution and hope to have highlighted a number of recurring structural themes: the relatively fixed local geometry of the uranyl cation and phosphonate groups, the relationship between extended topologies of these materials and their uranyl–transition metal counterparts, as well as the influence of the organic components of the linkers themselves, and the affinity of the uranyl cation for phosphonate groups over carboxylates in heterometallic systems. Considering these features, one may anticipate the diversity of these materials to continue to expand and contribute not only to our general understanding of solid state uranyl chemistry, but also to key separations and environmental challenges for the future.

Acknowledgements

Research efforts at GW were supported by the US Department of Energy during the production of this manuscript.

References

1. M. P. Jensen, J. V. Beitz, R. D. Rogers and K. L. Nash, *J. Chem. Soc., Dalton Trans.*, 2000, 3058.
2. K. L. Nash, *Radiochim. Acta*, 1993, **61**, 147.
3. K. L. Nash, *J. Alloys Compd.*, 1994, **213–214**, 300.
4. K. L. Nash, *J. Alloys Compd.*, 1997, **249**, 33.
5. A. Clearfield, *Curr. Opin. Solid State Mater. Sci.*, 1996, **1**, 268.
6. A. Clearfield, *Progr. Inorg. Chem.*, 1998, **47**, 371.
7. A. Clearfield, *Curr. Opin. Solid State Mater. Sci.*, 2002, **6**, 495.
8. G. Ferey, *Chem. Soc. Rev.*, 2008, **37**, 191.
9. S. L. James, *Chem. Soc. Rev.*, 2003, **32**, 276.
10. S. Kitagawa and R. Matsuda, *Coord. Chem. Rev.*, 2007, **251**, 2490.
11. P. C. Burns, *Rev. Mineral.*, 1999, **38**, 23.
12. P. C. Burns, *Can. Mineral.*, 2005, **43**, 1839.
13. P. C. Burns, M. L. Miller and R. C. Ewing, *Can. Mineral.*, 1996, **34**, 845.
14. C. L. Cahill and L. A. Borkowski, in *Structural Chemistry of Inorganic Actinide Compounds*, ed. S. V. Krivovichev, P. C. Burns, I. G. Tananaev, Elsevier, Amsterdam, 2007, pp. 409–442.
15. C. L. Cahill, D. T. de Lill and M. Frisch, *CrystEngComm*, 2007, **9**, 15.
16. D. Grohol, M. A. Subramanian, D. M. Poojary and A. Clearfield, *Inorg. Chem.*, 1996, **35**, 5264.
17. D. Grohol and A. Clearfield, *J. Am. Chem. Soc.*, 1997, **119**, 4662.
18. D. M. Poojary, D. Grohol and A. Clearfield, *Angew. Chem., Int. Ed. Engl.*, 1995, **34**, 1508.

19. M. B. Doran, A. J. Norquist and D. O'Hare, *Chem. Mater.*, 2003, **15**, 1449.
20. D. Grohol, F. Gingl and A. Clearfield, *Inorg. Chem.*, 1999, **38**, 751.
21. W. Henderson, M. T. Leach, B. K. Nicholson, A. L. Wilkins and t. l. P. A. T. Hoye, *Polyhedron*, 1998, **17**, 3747.
22. G. S. Conary, E. N. Duesler and R. T. Paine, *Inorg. Chim. Acta*, 1988, **145**, 149.
23. D. M. Poojary, D. Grohol and A. Clearfield, *J. Phys. Chem. Solids*, 1995, **56**, 1383.
24. D. M. Poojary, A. Cabeza, M. A. G. Aranda, S. Bruque and A. Clearfield, *Inorg. Chem.*, 1996, **35**, 1468.
25. M. A. G. Aranda, A. Cabeza, S. Bruque, D. M. Poojary and A. Clearfield, *Inorg. Chem.*, 1998, **37**, 1827.
26. D. M. Poojary and A. Clearfield, *J. Organomet. Chem.*, 1996, **512**, 237.
27. K. E. Knope and C. L. Cahill, *Dalton Trans.*, 2010, **39**, 8319.
28. S. V. Krivovichev and P. C. Burns, in *Structural Chemistry of Inorganic Actinide Compounds*, ed. S. V. Krivovichev, P. C. Burns, I. G. Tananaev, Elsevier, Amsterdam, 2007, pp. 95–182.
29. A.-G. D. Nelson, T. H. Bray, W. Zhan, R. G. Haire, T. S. Sayler and T. E. Albrecht-Schmitt, *Inorg. Chem.*, 2008, **47**, 4945.
30. P. O. Adelani and T. E. Albrecht-Schmitt, *Inorg. Chem.*, 2009, **48**, 2732.
31. A. A. Russell, R. L. Meline, E. N. Duesler and R. T. Paine, *Inorg. Chim. Acta*, 1995, **231**, 1.
32. R. G. Pearson, *J. Chem. Educ.*, 1968, **45**, 581.
33. A. N. Alsobrook, W. Zhan and T. E. Albrecht-Schmitt, *Inorg. Chem.*, 2008, **47**, 5177.
34. K. E. Knope and C. L. Cahill, *Inorg. Chem.*, 2008, **47**, 7660.
35. G. R. Desiraju, *Angew. Chem., Int. Ed. Engl.*, 2007, **46**, 8342.
36. K. Viswanathan and O. Harneit, *Am. Mineral.*, 1986, **71**, 1489.
37. A. J. Locock and P. C. Burns, *Can. Mineral.*, 2003, **41**, 489.
38. K. E. Knope and C. L. Cahill, *Eur. J. Inorg. Chem.*, 2010, 1177.
39. K. E. Knope and C. L. Cahill, *Inorg. Chem.*, 2009, **48**, 6845.
40. P. O. Adelani and T. E. Albrecht-Schmitt, *Inorg. Chem.*, 2010, **49**, 5701.
41. K. E. Knope and C. L. Cahill, *Inorg. Chem. Commun.*, 2010, **13**, 1040.
42. A. N. Alsobrook and T. E. Albrecht-Schmitt, *Inorg. Chem.*, 2009, **48**, 11079.
43. P. Ramaswamy, R. Prabhu and S. Natarajan, *Inorg. Chem.*, 2010, **49**, 7927.
44. S.-S. Bao, G.-S. Chen, Y. Wang, Y.-Z. Li, L.-M. Zheng and Q.-H. Luo, *Inorg. Chem.*, 2006, **45**, 1124.
45. S. M. Bowen, E. N. Duesler and R. T. Paine, *Inorg. Chem.*, 1983, **22**, 286.
46. A. Britel, M. Wozniak, J. C. Boivin, G. Nowogrocki and D. Thomas, *Acta Crystallogr., Sect. C*, 1986, **42**, 1502.
47. D. J. McCabe, A. A. Russell, S. Karthikeyan, R. T. Paine, R. R. Ryan and B. Smith, *Inorg. Chem.*, 1987, **26**, 1230.
48. S. Karthikeyan, R. R. Ryan and R. T. Paine, *Inorg. Chem.*, 1989, **28**, 2783.
49. D. T. Cromer, R. R. Ryan, S. Karthikeyan and R. T. Paine, *Inorg. Chim. Acta*, 1990, **172**, 165.
50. S. Karthikeyan, R. T. Paine and R. R. Ryan, *Inorg. Chim. Acta*, 1988, **144**, 135.
51. G. S. Conary, R. L. Meline, R. Schaeffer, E. N. Duesler and R. T. Paine, *Inorg. Chim. Acta*, 1992, **201**, 165.

CHAPTER 19
Structural Chemistry of Transuranium Phosphonates

JUAN DIWU AND THOMAS ALBRECHT-SCHMITT

Department of Civil Engineering and Geological Sciences and Department of Chemistry and Biochemistry, University of Notre Dame, Notre Dame, IN 46556, USA

19.1 Introduction

Phosphonates have played a critical role in lanthanide and actinide chemistry, primarily from the perspective of their use in separation processes.[1] These developments have occurred because of the unusual nature of the P=O–RE (RE = rare earth element) interaction that persists even at low pH. Many phosphonates have been developed for binding actinides in solution, including geminally substituted diphosphonic acids that can be grafted to polymer matrixes, e.g. Diphonix resins.[2] Despite the importance of the formation of actinide phosphonates in nuclear waste treatment, little is known about the solid state structures of these compounds or their physical properties. The only well-developed actinide phosphonate materials are known with thorium[3] and uranium.[4,5] The intricacies of uranium(VI) phosphonates are aptly illustrated by the uranyl phenylphosphonate system that undergoes a transformation from a *cis* (α-UPP) to a *trans* (β-UPP) confirmation at room temperature.[4b,c] Upon exposure to Na^+ or Ca^{2+} cations in an aqueous environment, both of these phases transform from linear chain structures into a hollow nanotubular form, γ-UPP.[4a]

The primary challenge in developing structure–property relationships in actinide phosphonates is that, other than the uranyl compounds, these phases are highly insoluble and generally precipitate as microcrystalline or amorphous

Metal Phosphonate Chemistry: From Synthesis to Applications
Edited by Abraham Clearfield and Konstantinos Demadis

Published by the Royal Society of Chemistry, www.rsc.org

powders. As a result, single crystal structure determinations are rare, thus forcing the use of *ab initio* powder diffraction methods when possible.[4h,5] One solution to this conundrum is to utilize reduction processes that transform actinides from the generally more soluble 6+ oxidation state to the much less soluble 4+ oxidation state. We have discovered that under mild hydrothermal conditions Np^{VI} and Pu^{VI} can be reduced to Np^{IV} and Pu^{IV} even in the presence of certain oxidation agents so long as this process is also driven by the formation of highly insoluble products.[6] In this vain, we review *in situ* reduction reactions of An^{VI} to An^{IV} (An = Np, Pu) that can be used to prepare the first well-characterized examples of transuranium phosphonates.

19.2 Some Transuranium Phosphonates of Interest (Table 19.1)

19.2.1 $Np(CH_3PO_3)(CH_3PO_3H)(NO_3)(H_2O) \cdot H_2O$

19.2.1.1 Synthesis

The hydrothermal reaction of neptunium(VI) nitrate with methylphosphonic acid results in the formation of clusters of dichroic pale pink/green crystals of $Np(CH_3PO_3)(CH_3PO_3H)(NO_3)(H_2O) \cdot H_2O$ (NpMeP-1) as the only solid product.[11]

19.2.1.2 Structural Description

The single crystal structure of this compound shows polar layers that are joined to one another via a hydrogen bonding network involving interlayer water molecules. The layered structure is constructed from corner-sharing NpO_8 distorted dodecahedra that are bridged by $CH_3PO_3^{2-}$ and $CH_3PO_3H^-$ anions. The coordination sphere is completed by a chelating nitrate anion and a coordinating water molecule. These units form layers that extend into the [*ac*] plane. As indicated by the orthorhombic space group $Pca2_1$, the structure is polar. The polarity of $Np(CH_3PO_3)(CH_3PO_3H)(NO_3)(H_2O) \cdot H_2O$ is subtle. A portion of the structure of $Np(CH_3PO_3)(CH_3PO_3H)(NO_3)(H_2O) \cdot H_2O$ is shown in Figure 19.1. The H atoms from the interlayer water molecules as well as those on the $CH_3PO_3H^-$ anions were located from difference maps. The nitrate and $CH_3PO_3H^-$ anions form a hydrogen bonding network that stitches the layers together and provides stability for the overall structure. Both the hydrogen bonding network and the tilt of the nitrate anions can be used to visualize the polarity of the structure along the *c* axis. The structure of $Np(CH_3PO_3)(CH_3PO_3H)(NO_3)(H_2O) \cdot H_2O$ does not parallel that of other M^{IV} phosphonates.

Here the NpO_8 distorted dodecahedra have Np–O bond distances that range from 2.199(3) to 2.589(3) Å, yielding a bond valence sum of 4.07, which is consistent with this compound containing Np^{IV}.[7] The P–O bond distances are normal, and the only one worth mentioning is the longest one of 1.579(4) Å, which corresponds to a P–OH moiety in the $CH_3PO_3H^-$ anions.

Table 19.1 List of transuranium phosphonates compounds.

Formula	*Space group*	*a (Å)*	*b (Å)*	*c (Å)*	*α (°)*	*β (°)*	*γ (°)*	*Ref.*
Neptunium compounds								
$Np(CH_3PO_3)(CH_3PO_3H)(NO_3)(H_2O) \cdot H_2O$	$Pca2_1$	10.607(3)	9.438(3)	11.346(4)	90	90	90	11
$NpO_2(CH_3PO_3)$	$P\bar{1}$	6.2978(10)	6.6804(10)	7.0836(11)	90.614(3)	99.335(3)	96.500(3)	11
$NpO_2(C_2H_5PO_3)$	$P2_1/c$	6.6859(11)	7.0768(11)	13.331(2)	90	98.480(3)	90	n/a
Plutonium compounds								
$Pu(CH_3PO_3)_2$	$P2_1/c$	8.0895(12)	10.7001(10)	9,4541(14)	90	92.331(3)	90	12

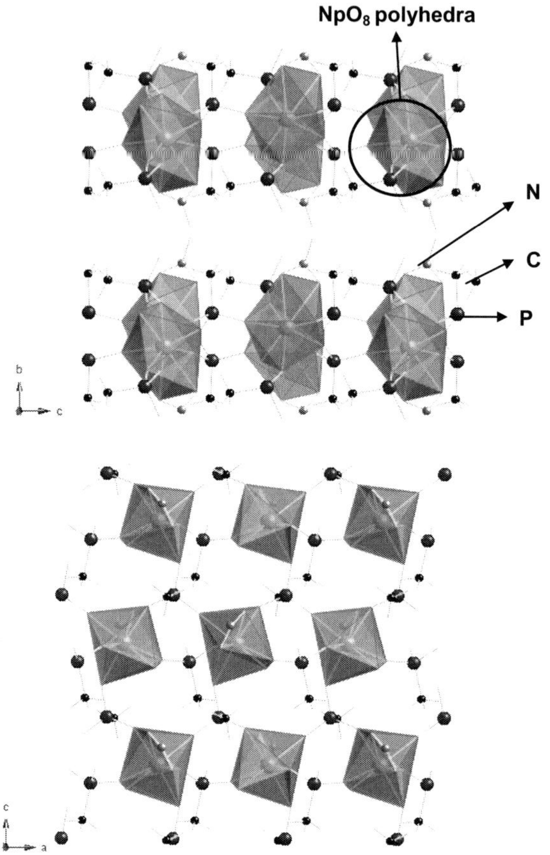

Figure 19.1 Two views of the polar, lamellar structure of $Np(CH_3PO_3)(CH_3PO_3H)(NO_3)(H_2O) \cdot H_2O$.

The isolation of NpMeP-1 as a pure crystalline material and subsequent structural characterization allows us to address potential structure–property correlations. In particular, the magnetic properties of Np^{IV} compounds can be particularly intriguing because of its $5f^3$ electron configuration that has a $^4I_{9/2}$ ground state. The magnetic behavior of Np^{IV} compounds sharply contrasts with that of U^{IV} ($5f^2$, $^3H_4$) compounds that have a tendency of becoming non-magnetic or only exhibiting temperature-independent paramagnetism at low temperatures owing to a singlet ground state. The Np···Np distance is greater than 5.3 Å in NpMeP-1, and there are no μ_3-O groups in the structure, and therefore no opportunity for superexchange. Therefore, the prediction is that the Np^{IV} centers will behave as isolated ions and Curie–Weiss behavior should be observed without long-range magnetic ordering. The inverse magnetic susceptibility is plotted versus the temperature in Figure 19.2. As can be observed from these data, this prediction proves to be true as Curie–Weiss behavior is exhibited at low temperatures. The susceptibility data were least-squares fit

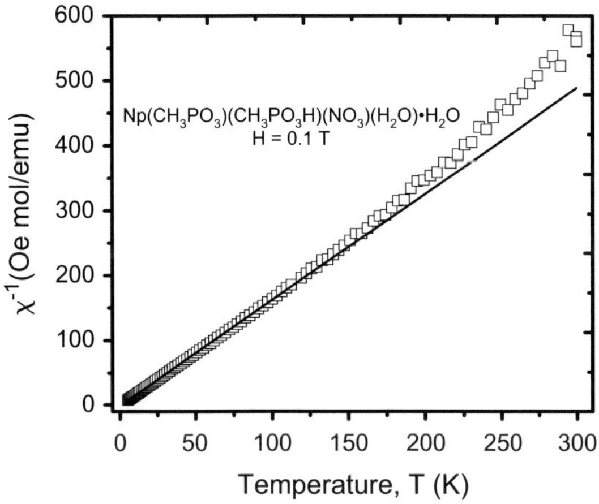

Figure 19.2 Temperature dependence of the reciprocal molar magnetic susceptibility for NpMeP-1 under an applied magnetic field of 0.1 T between 5 and 300 K. The straight line represents the fit to the Curie–Weiss law in the range of 5–100 K.

to $\chi(T) = C/(T + \theta) + \chi_0$, where C is the Curie constant, θ is the Weiss constant, and χ_0 is the temperature-independent susceptibility. We have fit the data from 5 to 100 K to arrive at a θ value of 0.53(6) K, indicating no strong interactions between Np^{IV} ions. The value of μ_{eff} is 2.213(1) μ_B. While this value is considerably lower than that predicted for the free ion (3.62 μ_B), it compares well with other Np^{IV} compounds such as $Np(IO_3)_4$ (2.22 μ_B).[6] This reduced moment is not a reflection of the covalency of the 5f orbitals but rather arises from crystal field effects.[6]

19.2.2 $NpO_2(CH_3PO_3)$

19.2.2.1 Synthesis

The beauty of utilizing neptunium as the core metal ion in phosphonate materials is the ability to tune the oxidation state of neptunium, as demonstrated by the formation of NpMeP-1. By using a large excess of nitrite in the aforementioned reaction, the oxidation state of neptunium can be maintained at 6+,[8] enabling the isolation of the neptunium(VI) methylphosphonate, $NpO_2(CH_3PO_3)$ (NpMeP-2), as orange-brown microcrystals.[11]

19.2.2.2 Structural Description

A view of part of the structure of NpMeP-2 is shown in Figure 19.3. This lamellar compound is isostructural with its uranium(VI) analog.[4h] However, the structure of $UO_2(CH_3PO_3)$ was obtained from high-resolution powder X-ray diffraction data.[4h] The structure of this compound consists of NpO_7

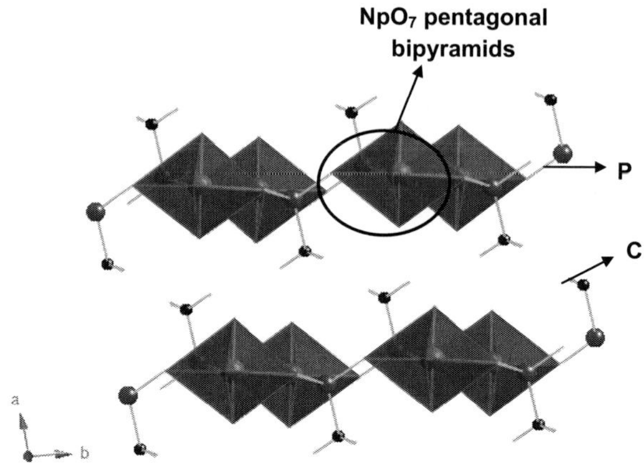

Figure 19.3 View down the *c* axis of the layered structure of the neptunium(VI) methylphosphonate $NpO_2(CH_3PO_3)$.

pentagonal bipyramids containing a nearly linear NpO_2^{2+} neptunyl core. The critical bond distances are the two neptunyl Np–O bonds that average 1.736(7) Å. This distance is statistically shorter than that typically found in uranyl compounds, which average 1.79(3) Å.[9] The NpO_7 pentagonal bipyramids form 1D chains via edge sharing and are, in turn, bridged by $CH_3PO_3^{2-}$ anions to yield layers that extend into the [*bc*] plane. The methyl groups are oriented approximately perpendicular to this plane and point between the layers, as shown in Figure 19.3.

NpMeP-1 and NpMeP-2 were the first reported transuranium phosphonates. We have demonstrated that the redox capabilities of neptunium, which are not paralleled by uranium or other lighter actinides, provide access to new neptunium(IV) and neptunium(VI) phosphonates. Neptunium(V) phosphonates should also be available in the near future. The high coordination numbers of Np in these compounds provide access to structural motifs not available from transition metal phosphonate materials.[10]

19.2.3 $NpO_2(C_2H_5PO_3)$

19.2.3.1 Synthesis

The hydrothermal reaction of neptunium(VI) nitrate with ethylphosphonic acid results in the formation of pale pink tablets of $NpO_2(C_2H_5PO_3)$ as the only solid product.

19.2.3.2 Structural description

The structure of $NpO_2(C_2H_5PO_3)$ is nearly identical with its methylphosphonate analog. The layers extend in the [*ab*] plane. The typology is shown in

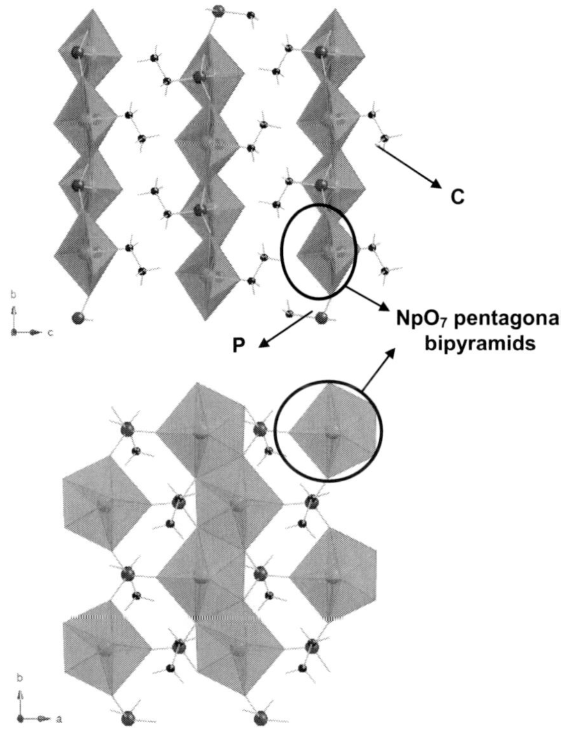

Figure 19.4 Two views of the layered structure of $NpO_2(C_2H_5PO_3)$.

Figure 19.4. The layers are separated by the ethyl groups. The PO_3 here not only chelates to the neptunium centers, but also bridge between the Np chains to form the layers.

19.2.4 $Pu(CH_3PO_3)_2$

19.2.4.1 Synthesis

A stock solution of Pu^{VI} was placed in an autoclave along with $CH_3PO_3H_2$. The autoclave was sealed and heated at 180 °C for 3 days followed by slow cooling to room temperature over a 24 h period. The product consisted of blue prisms of $Pu(CH_3PO_3)_2$.[12]

19.2.4.2 Structural Description

The structure of $Pu(CH_3PO_3)_2$ is surprisingly simple and is formed from Pu^{IV} centers that are bridged by $CH_3PO_3^{2-}$ anions. The binding of the plutonium centers by the methylphosphonate creates an octahedral environment with

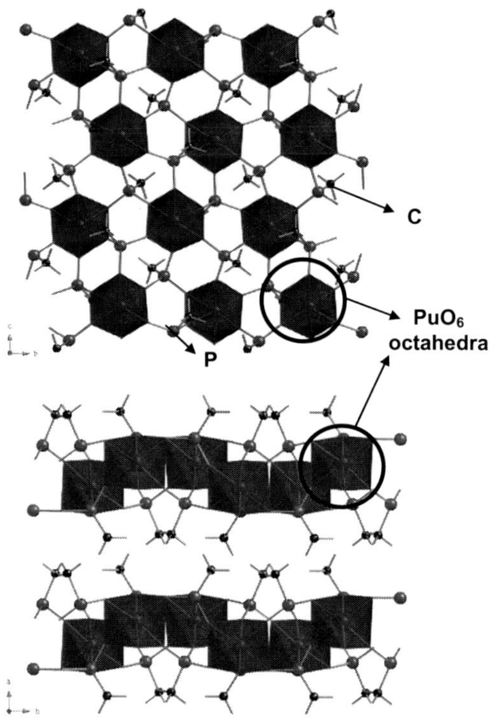

Figure 19.5 Two views of the lamellar structure of $Pu(CH_3PO_3)_2$. PuO_6 octahedra are indicated.

Pu–O bond distances ranging from 2.188(3) to 2.235(3) Å. P–O bonds in the phosphonate are essentially invariant and average 1.523(3) Å. The layered topology that is created is closely related to that of the well-known structure of α-$Zr(HPO_4)_2$, which forms the basis for many zirconium phosphonate materials.[10] Plutonium is not generally found with coordination numbers this low, and this structure clearly represents a case where a non-radioactive surrogate like Zr^{IV} provides a good mimic of Pu^{IV} and vice versa. Different views of the structure are shown in Figure 19.5.

19.3 Some Transuranium Diphosphonates of Interest (Table 19.2)

19.3.1 $Np[CH_2(PO_3)_2](H_2O)_2$ (NpC1P2-1)

19.3.1.1 Synthesis

A stock solution of Np^{VI} was placed in an autoclave along with methylenediphosphonate. The solution immediately changed from pink to green without the formation of a precipitate. The autoclave was sealed and heated at 180 °C

Table 19.2 List of transuranium diphosphonate compounds.

Formula	*Space group*	*a (Å)*	*b (Å)*	*c (Å)*	*α (°)*	*β (°)*	*γ (°)*	*Ref.*
Neptunium compounds								
$Np[CH_2(PO_3)_2](H_2O)_2$	*Pna2₁*	8.8418(15)	9.6381(16)	8.6627(15)	90	90	90	13
$Np[C_6H_4(PO_3H)_2]_2 \cdot 2H_2O$	$P\bar{1}$	7.886(1)	9.863(1)	12.810(2)	100.644(2)	95.617(2)	102.669(1)	36
Plutonium compounds								
α-$Pu[CH_2(PO_3)_2](H_2O)$	$P2_1/m$	6.513(3)	5.753(2)	9.023(4)	90	106.344(7)	90	12
β-$Pu[CH_2(PO_3)_2](H_2O)$	$P\bar{1}$	8.6692(5)	8.7012(5)	9.5318(6)	71.403(4)	72.248(3)	82.859(4)	14
γ-$Pu[CH_2(PO_3)_2](H_2O)$	*Pbca*	8.5990(8)	11.2761(11)	13.7865(13)	90	90	90	14
$Pu[CH_2(PO_3)_2](H_2O) \cdot H_2O$	$P2_1/n$	5.2504(2)	8.2896(3)	16.4629(7)	90	93.4629(7)	90	14
$Pu[C_6H_4(PO_3H)(PO_3H_2)][C_6H_4(PO_3H)(PO_3)] \cdot 2H_2O$	$P\bar{1}$	9.444(3)	10.021(3)	12.670(3)	74.024(2)	78.452(2)	66.042(2)	36
Mixed actinide compounds								
$UO_2Np(H_2O)_2[CH_2(PO_3)(PO_3H)]_2$	$P2_1/m$	5.5279(5)	20.3381(19)	5.9321(6)	90	98.703(2)	90	15
$UO_2Pu(H_2O)_2[CH_2(PO_3)(PO_3H)]_2$	$P2_1/m$	5.5558(4)	20.5994(15)	5.9482(5)	90	99.1140(10)	90	12

for 3 days followed by slow cooling to room temperature over a 24 h period. The product consisted of dichroic pink/green prisms of NpC1P2-1 and colorless mother liquor.[13]

19.3.1.2 Structural Description

$Np[CH_2(PO_3)_2](H_2O)_2$ adopts a 3D network structure where the Np^{IV} centers are coordinated by methylenediphosphonate and water molecules. Various depictions of these structures are shown in Figure 19.6. In both cases the C1P2 ligand binds six different An^{IV} centers. This bonding is quite different from that seen with model complexes containing Ln^{3+} (Ln = lanthanide) cations where one of the C1P2 oxygen atoms on each of the phosphonate moieties is protonated.[1a] The similarities between NpC1P2-1 and uranium counterpart rapidly break down. Foremost among the differences is that the Np^{IV} is eight-coordinate in NpC1P2-1, whereas the U^{IV} in UC1P2-1 is seven-coordinate.

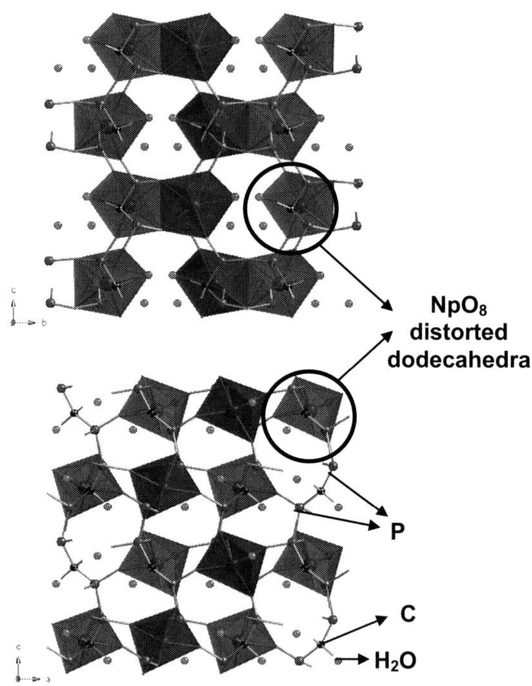

Figure 19.6 Two views of part of the structure of $Np[CH_2(PO_3)_2](H_2O)_2$ showing the polar 3D framework constructed from Np^{IV} centers that are bound by methylenediphosphonate and two water molecules to form isolated NpO_8 distorted dodecahedra. The bottom view shows the channels that the water molecules reside in. From this view it appears that the NpO_8 units share an edge, but this is an illusion. The majority of the methylenediphosphonate ligands have been omitted for clarity.

Therefore, at the most fundamental level the structural chemistry of Np^{IV} is not being mimicked by U^{IV}.

The geometries of both of these high coordination number polyhedra found in NpC1P2-1 and UC1P2-1 deviate strongly from any idealized shape. In the case of NpC1P2-1, the Np^{IV} centers are found as NpO_8 distorted dodecahedra with Np–O bond distances ranging from 2.216(3) to 2.648(3) Å. The two longest bonds of 2.595(3) and 2.648(3) Å are from the coordinating water molecules. These bond distances can be used to arrive at a bond-valence sum of 4.29 for Np(1), which is consistent with this compound containing Np^{IV}.[16] Sharply contrasting with NpC1P2-1, UC1P2-1 contains UO_7 units that are not adequately described by common polyhedral designations such as a capped octahedron or a capped trigonal prism. Instead they most closely resemble square antiprisms with a missing vertex. A UO_7 polyhedron containing U^{IV} was previously recognized in $U(UO_2)(PO_4)_2$.[16] U–O bond distances are found to occur from 2.225(6) to 2.451(6) Å, where again the longest bond length of 2.451(6) Å is to the coordinating water molecule. The bond-valence sum for the U(1) atom is consistent with this compound containing U^{IV}, with a value of 4.18.[7]

The ionic radii of 8-fold coordinated U^{IV} and Np^{IV} are quite similar at 1.00 Å and 0.98 Å, respectively.[17] Both cations are most commonly found in solids in eight- and nine-coordinate environments.[18] There are numerous examples of isostructural U^{IV} and Np^{IV} compounds.[18] Nevertheless, there is an ongoing debate about whether early actinides like $^{232}Th^{IV}$ and $^{238}U^{IV}$, or even non-radioactive Ce^{IV} and Zr^{IV}, can be used as surrogates for more potentially hazardous transuranium elements like $^{237}Np^{IV}$ and $^{239}Pu^{IV}$.[19] The fact is that while NpC1P2-1 and UC1P2-1 were both produced under similar conditions, they do possess different structures and compositions. Thus, here we find another instance where it is clear that if one is going to understand the crystal chemistry of a transuranium element, one must actually work with the element in question and not a surrogate.[20] This is contrary to that often observed for U^{IV} and Np^{IV} phosphate compounds that are heated to higher temperatures.[19,21–24] To further illustrate potential differences, we point to the $Th_4(PO_4)_4P_2O_7$ structure type, which is under consideration as a possible host material for the long-term storage of transuranium elements. It has been shown that this phase can incorporate U^{IV} and Pu^{IV} but not the surrogates Ce^{IV} or Zr^{IV}.[19] The former is reduced to Ce^{III} and the latter is too small to stabilize this lattice. Furthermore, while $M_2O(PO_4)_2$ phases can be prepared when M = Zr, Th, U, and Np, attempts to prepare the Ce^{IV} and Pu^{IV} analogs result in spontaneous reduction to Ce^{III} and Pu^{III}, and monazite-like phases with the composition MPO_4 result.[24]

Another observation of particular interest is the demonstration that the *in situ* hydrothermal reduction of Np^{VI} to Np^{IV} is not unique to the methylphosphonate system but rather is a more general reaction. In fact, our observations are thus far that this reaction is normal under the hydrothermal conditions described and can only be prevented by the incorporation of an oxidant.[11] This reduction chemistry also occurs with U^{VI}, albeit to a very small extent.

19.3.2 $Np[C_6H_4(PO_3H)_2]_2 \cdot 2H_2O$ (NpPhP2)

19.3.2.1 Synthesis

A stock solution of Np^{VI} was placed in an autoclave along with 1,2-phenylenediphosphonic acid (PhP2). After hydrothermal treatment, pale green acicular crystals of NpPhP2 were isolated.[36]

19.3.2.2 Structural Description

The structure of NpPhP2 is a 1D chain structure. Figure 19.7 shows the depiction of the monomer chains that extend down the *b* axis. The NpO_8 units are formed by 1,2-phenylenediphosphonic acid (PhP2) only bridging between two adjacent metal centers. Each NpO_8 is surrounded by 8 PO_3 moieties from 4 PhP2 groups. Figure 19.7 shows the interaction of the benzene rings and the arrangement of the chains in the space with two free water molecules. The average bond distance for NpPhP2 is 2.340 Å, slightly larger than PuPhP2, which is consistent with the actinide contraction.[36]

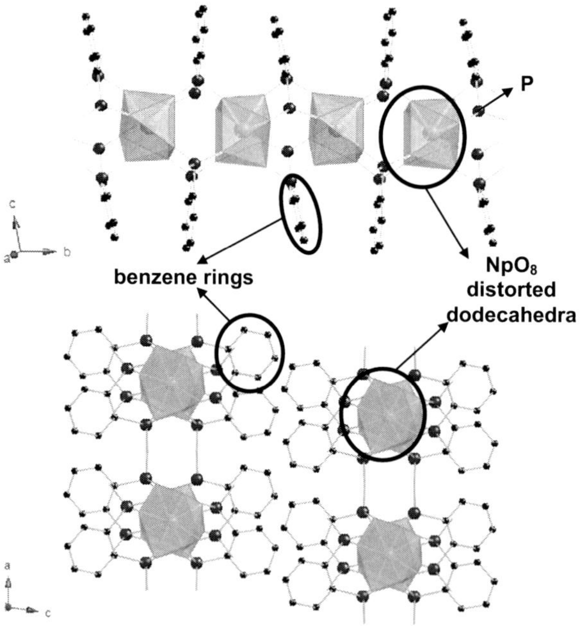

Figure 19.7 Two views of part of the structure of $Np[C_6H_4(PO_3H)_2]_2 \cdot 2H_2O$. The top view shows the 1D chain structure constructed from Np^{IV} centers that are bound by PhP2 and form isolated NpO_8 distorted dodecahedra. The bottom view shows the direction that benzene rings overlap. From this view it appears that the NpO_8 units are monomers.

19.3.3 α-$Pu[CH_2(PO_3)_2](H_2O)$

19.3.3.1 Synthesis

A stock solution of Pu^{VI} was placed in an autoclave along with $CH_2(PO_3H_2)_2$. The autoclave was sealed and heated at 180 °C for 3 days followed by slow cooling to room temperature over a 24 h period. The product consisted of an amorphous red precipitate and blue tablets of $Pu[CH_2(PO_3)_2](H_2O)$.[12]

Our basic conjecture is that even though Pu^{VI} is not as strong of an oxidizing agent as Np^{VI} ($E^\circ = 1.159$ V for Np^{VI}; $E^\circ = 0.936$ V for Pu^{VI}),[25] it will still be reduced from Pu^{VI} to the least soluble oxidation state, Pu^{IV}, under hydrothermal conditions where water is acting as the reductant. These are not homogeneous reactions but rather are driven in part by the free energy of the formation of the highly insoluble Pu^{IV} phosphonates. This hypothesis was tested by starting with a stock solution of PuO_2^{2+} (as the nitrate) to which either methylphosphonic acid or methylenediphosphonic acid were added. The addition of the phosphonic acid does not result in color change or any detectable reduction. However, upon heating, the PuO_2^{2+} is reduced to Pu^{4+} and is isolated in the form of either $Pu(CH_3PO_3)_2$ or α-$Pu[CH_2(PO_3)_2](H_2O)$.

From control studies we were surprised to learn that ThO_2 dissolves under mild hydrothermal conditions in the presence of methylenediphosphonic acid, and this allows for the isolation of $Th[CH_2(PO_3)_2](H_2O)_2$. This implies that we might not need to use PuO_2^{2+} as our source of plutonium, but rather we can simply use brute force methods and digest PuO_2 under acidic conditions in the presence of phosphonates. The reaction of PuO_2 with methylenediphosphonic acid in water at 180 °C under autogenously generated pressure results in the formation of crystals of α-$Pu[CH_2(PO_3)_2](H_2O)$ as we found when PuO_2^{2+} was used as the source of plutonium.

19.3.3.2 Structural Description

Owing to twinning and disorder, the structure of $Pu[CH_2(PO_3)_2](H_2O)$ does not have very low residuals. Nevertheless, given the paucity of structural data on plutonium compounds, its details are important. The monohydrate of Pu^{IV} methylenediphosphonate has the same stoichiometry as that of its U^{IV} analog.[13] Furthermore, it contains similar $An^{IV}O_7$ (An = U, Pu) polyhedra as $U[CH_2(PO_3)_2](H_2O)$. These building units are exceedingly rare in actinide crystal chemistry. The Pu–O bond distances range from 2.18(2) to 2.49(2) Å, with the longest bond being with the terminal water molecule. Both compounds adopt 3D framework structures where the $[CH_2(PO_3)_2]^{4-}$ anions chelate Pu^{IV} and also bridge between Pu^{IV} cations. The compounds are not isotypic, however. $U[CH_2(PO_3)_2](H_2O)$ is actually of higher symmetry and crystallizes in the orthorhombic space group *Pbca*.[13] In contrast, $Pu[CH_2(PO_3)_2](H_2O)$ is monoclinic and crystallizes in $P2_1/m$. A depiction of the structure of $Pu[CH_2(PO_3)_2](H_2O)$ is shown in Figure 19.8.

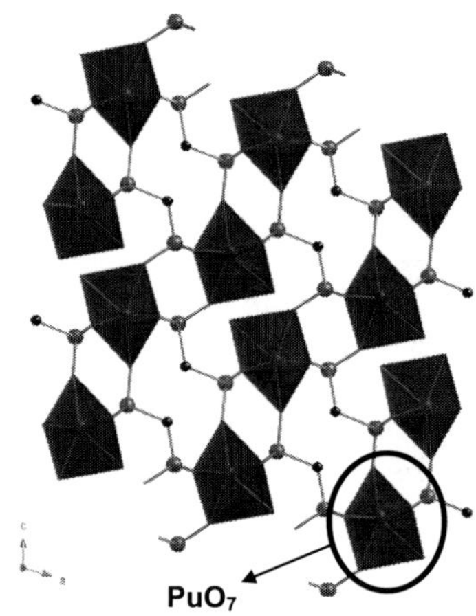

Figure 19.8 Depiction of the structure of α-$Pu[CH_2(PO_3)_2](H_2O)$.

19.3.4 β-$Pu[CH_2(PO_3)_2](H_2O)$ (1), γ-$Pu[CH_2(PO_3)_2](H_2O)$ (2), $Pu[CH_2(PO_3)_2](H_2O) \cdot H_2O$ (3)

19.3.4.1 Synthesis

PuO_2 was combined with $CH_2(PO_3H_2)_2$. Distilled and Millipore filtered water was added to the solids. After hydrothermal treatment the products consisted of a mixture of crystals of different colors. Relatively large green crystals of β-$Pu[CH_2(PO_3)_2](H_2O)$ (**1**) were easily isolated. γ-$Pu[CH_2(PO_3)_2](H_2O)$ (**2**) forms red polycrystalline masses that must be cleaved for the isolation of twinned crystals. $Pu[CH_2(PO_3)_2](H_2O) \cdot H_2O$ (**3**) forms clusters of nearly colorless acicular crystals. A very pale peach coloration is observed if one looks down the length of the crystals.

19.3.4.2 Structural Description

The structures of **1**–**3** are composed of Pu^{4+} centers bound by methylenediphosphonate and water to yield seven-coordinate plutonium in the form of PuO_7 units. While this coordination number is not nearly as common as eight- or nine-coordinate systems, it also occurs in $U[CH_2(PO_3)_2](H_2O)$,[13] as well as several related or isostructural Ce^{4+} phases,[26] and is known more generally from the monoclinic form of zirconia.[27] $[CH_2(PO_3)_2]^{4-}$ can chelate and/or bridge between tetravalent actinides. When it chelates, it uses two oxygen atoms to do this. Each oxo donor is from one of the PO_3 moieties.[1,13,28]

In the process of chelating the plutonium ions with $[CH_2(PO_3)_2]^{4-}$, four of the oxygen atoms are directed away from the first plutonium center, and these serve to bridge between the Pu^{4+} ions to create extended structures. A seventh site is

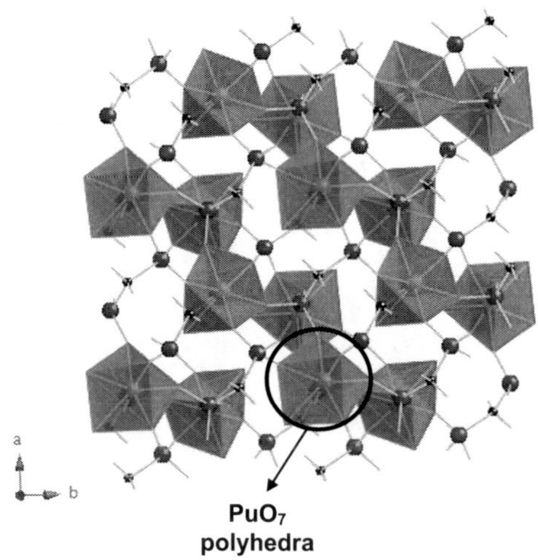

Figure 19.9 Depiction of the 3D framework structure of β-Pu[$CH_2(PO_3)_2$](H_2O) along the *c* axis. This structure is constructed from two seven-coordinate PuO_7 units that are bridged by the methylenediphosphonate.

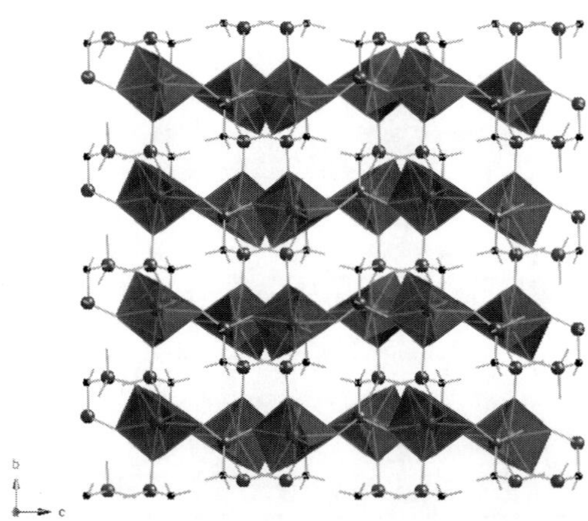

Figure 19.10 View of the 3D framework structure of γ-Pu-[$CH_2(PO_3)_2$](H_2O) along the *a* axis. This structure is constructed from seven-coordinate PuO_7 units that are both bridged and chelated by the methylenediphosphonate.

occupied by water. The PuO_7 coordination environments are shown in Figure 19.12. Views of the extended structures of **1**, **2**, and **3** are shown in Figures. 19.9, 19.10, and 19.11. From these figures it appears in some cases that single PO_3 groups are using two oxygen atoms to chelate plutonium centers. This is an illusion caused by the superposition of the polyhedra. The framework adopted by **3** is more open than that found for the other compounds, and accounts for the occluded water molecule in the structure.

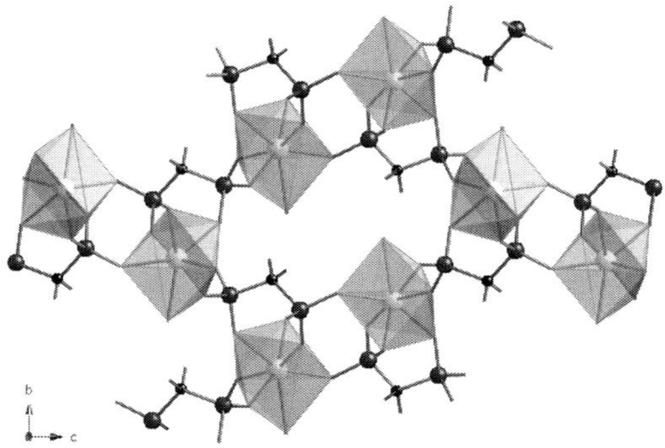

Figure 19.11 View along the *a* axis of the structure of $Pu[CH_2(PO_3)_2](H_2O) \cdot H_2O$. This structure consists of a 3D network constructed from PuO_7 units that are chelated and bridged by methylenediphosphonate. An additional occluded water molecule fills the voids in the structure.

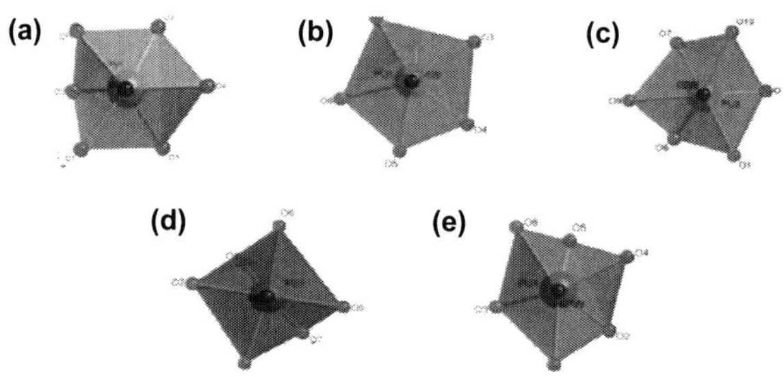

Figure 19.12 PuO_7 coordination environments found in α-$Pu[CH_2(PO_3)_2](H_2O)$ (a); β-$Pu[CH_2(PO_3)_2](H_2O)$ (**1**) (b,c); γ-$Pu[CH_2(PO_3)_2](H_2O)$ (**2**) (d); and $Pu[CH_2(PO_3)_2](H_2O) \cdot H_2O$ (**3**) (e). The description of these units varies from capped octahedra to dodecahedra with a missing vertex.

The PuO_7 units found in these compounds do not conform to idealized shapes that are easily described, and vary from units best described as dodecahedra with a missing vertex to capped octahedra. Of the two PuO_7 units found in β-$Pu[CH_2(PO_3)_2](H_2O)$, the polyhedron containing Pu(2) can be described as a capped octahedron where the capping group is along one of the C axes. The same description applies to the PuO_7 units in $Pu[CH_2(PO_3)_2](H_2O) \cdot H_2O$. The PuO_7 polyhedra in α-$Pu[CH_2(PO_3)_2](H_2O)$ and γ-$Pu[CH_2(PO_3)_2](H_2O)$, as well as the Pu(1) unit in β-$Pu[CH_2(PO_3)_2](H_2O)$, are better described as dodecahedra with a missing vertex. Much like distorted trigonal bipyramids versus square pyramids in transition metal chemistry, there are probably very small differences in energies between these different geometries.

The UV–vis–NIR spectrum of Pu^{IV} is well known, and has been described in several works.[29–32] The spectrum consist of a series of weak, Laporte-forbidden f–f transitions. Normally when these data are acquired from solutions or

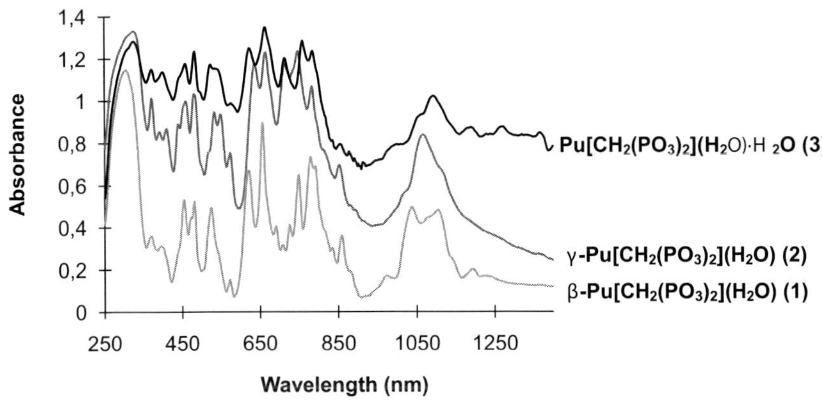

Figure 19.13 UV–vis–NIR spectra of green crystals of β-$Pu[CH_2(PO_3)_2](H_2O)$ (**1**), red crystals of γ-$Pu[CH_2(PO_3)_2](H_2O)$ (**2**), and very pale peach crystals of $Pu[CH_2(PO_3)_2](H_2O){\cdot}H_2O$ (**3**).

powders at room temperature, these transitions are relatively broad for f–f transitions. Data acquired at 4 K yields much sharper features.[33] Here the use of single crystals of **1** and **2**, and a few intergrown crystals for **3**, produces much sharper features than normally obtained at room temperature, and more transitions are clearly resolved. Despite the dramatic color differences, the measurements demonstrate that β-$Pu[CH_2(PO_3)_2](H_2O)$ (**1**), γ-$Pu[CH_2\text{-}(PO_3)_2](H_2O)$ (**2**), and $Pu[CH_2(PO_3)_2](H_2O) \cdot H_2O$ (**3**) are all Pu^{4+} compounds, as shown in Figure 19.12. This indicates that the visual color of transuranium compounds is often not a good indication of oxidation state.[6] The four plutonium compounds can be divided into two groups by similar coloration. The red (**3**) and peach (**2**) compounds have Pu^{4+} chelated by the diphosphonate, and in the blue (α) and green (**1**) compounds the diphosphonate-only bridges. The primary effect on the UV–vis–NIR spectra is that in **1** the peak near 547 nm (green region) is substantially diminished with respect to the intensity of this peak in **2** and **3** (Figure 19.13). This transition is in a region where multiple *J* values contribute, and transitions in this region have not been definitely assigned.[33] The human eye is very sensitive in this region of the visible spectrum, and this subtle change amidst a series of numerous f–f transitions causes the dramatic changes in the observed color even though the differences appear minor in the spectrum.

19.3.4.3 Comparisons of Pu^{IV} and Ce^{IV} Structures

The structure of **1** provides an unusual opportunity to compare Pu–O bond distances not just with Ce–O bond distances from an isostructural compound but also within the same structure because there are two crystallographically unique plutonium centers in **1**. The internal comparison yields an average Pu–O bond distance of 2.248(3) Å for both Pu^{4+} centers. This average compares well with the average Ce–O bond distance of 2.250(4) Å. In fact, if 3σ on the errors is used, then the differences between the distances are not statistically significant. The variations in the Pu–O(H_2O) and Ce–O(H_2O) bond distances are also potentially informative as they might provide different results than found with the Pu–O bonds from the PO_3 moieties. These distances are 2.429(3) and 2.493(3) Å for **1**, and 2.528(4) and 2.472(4) Å for cerium. The averages of 2.461(3) and 2.500(4) Å appear to be different by 0.039(4) Å, but if one looks at the internal variations between the two different sites (0.064(3) and 0.056(4) Å), the difference is larger between sites than it is between compounds. In short, the errors within the models for the structures of **1** and its cerium counterpart are such that the bond distances appear to be essentially the same.

γ-Pu[$CH_2(PO_3)_2$](H_2O) (**2**) is isotypic with β-Ce[$CH_2(PO_3)_2$](H_2O), which provides us an additional opportunity to discern differences in bond distances between Pu^{4+} and Ce^{4+}. The average M–O bond distances to the oxygen atoms from the PO_3 groups are 2.260(8) Å and 2.272(5) Å for plutonium and cerium, respectively. The Pu–O(H_2O) and Ce–O(H_2O) bond distances are 2.419(7) Å and 2.457(5) Å . The average Pu–O bond distances for all seven oxygen donors in **1** and **2** are 2.248(3) Å and 2.260(8) Å, giving a difference of 0.012(8) Å (i.e. they are not statistically different), while the difference between **2** and β-Ce[$CH_2(PO_3)_2$](H_2O) is also 0.012(8) Å . We conclude that there are no measurable differences between Pu–O bonds and Ce–O bonds in this system.

19.3.5 Pu[$C_6H_4(PO_3H)(PO_3H_2)$][$C_6H_4(PO_3H)(PO_3)$] · $2H_2O$

19.3.5.1 Synthesis

A stock solution of Pu^{VI} was placed in an autoclave along with PhP2. Distilled and Millipore filtered water was added to the solids. The closed liner was sealed inside a stainless steel autoclave and then heated to 180 °C for three days in a furnace. The product is pink prisms of Pu[$C_6H_4(PO_3H)(PO_3H_2)$][C_6H_4(PO_3H)(PO_3)] · $2H_2O$ (PuPhP2).[36]

19.3.5.2 Structure Description

PuPhP2 forms 1D chains where the Pu^{IV} centers are coordinated by 1,2-phenylenediphosphonic acid (PhP2). Figure 19.14 shows the depiction of the chains that extend along the *a* axis. The metal centers are eight coordinate with 1,2-phenylenediphosphonate not only bridging, but also chelating the centers via the PO_3 moieties. The phenyl rings extend into the space between the chains. The MO_8 polyhedra share an edge to form dimers within the chains. Six PhP2 moieties are bound to each dimer. Four of these bridge between the dimers to form the infinite chains. The two PO_3 moieties that belong to the same phenyl ring coordinate the metal centers differently. The two ligands that are a part of the dimer itself also chelate the Pu^{IV} centers. When PhP2 chelates, the PO_3 unit of one PhP2 moiety chelates to one metal center while the other PO_3 unit from

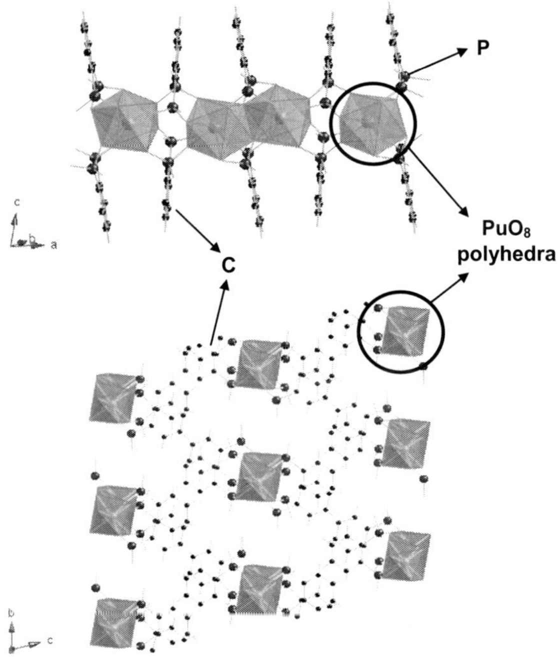

Figure 19.14 Views of $Pu[C_6H_4(PO_3H)(PO_3H_2)][C_6H_4(PO_3H)(PO_3)] \cdot 2H_2O$ 1D chains. The top view shows that the PuO_8 centers form dimers in the chains. The bottom view shows the direction the benzene rings overlap.

the same PhP2 bridges between the two metal centers. The second PhP2 ligand behaves in a similar fashion, but only chelates the metal center that was chelated by the first PhP2 moiety. Figure 19.14 shows the overlap of the phenyl groups and the arrangement of the dimers in space with the two co-crystallized water molecules.

19.3.5.3 Comparison Pu and Np Compounds

The differences between the Pu and Np structures extend beyond the presence of dimers and monomers. The geometries around the MO_8 polyhedra are in fact quite different. Using the algorithm developed by Raymond and coworkers,[19f] we probed the coordination environments around the metal centers. The dihedral angles of the all adjacent faces are calculated and compared with the ideal angles from the three possible distortions of a dodecahedron (D_{4d}, D_{2d}, and C_{2v}). The Kepert ligand repulsion model with $n = 8$ is utilized.[14] The evaluation of the calculation is the standard deviation of the dihedral angles along all edges. The Pu unit is very close to being D_{2d}, which is a trigonal dodecahedron, whereas the Np unit is close to being a D_{4d} square antiprism.[36]

19.3.6 $UO_2Np(H_2O)_2[CH_2(PO_3)(PO_3H)]_2$

19.3.6.1 Synthesis

A solution of Np^{VI} nitrate was heated with UO_3 and methylenediphosphonic acid at 180 °C for three days in a PTFE-lined autoclave, followed by slow cooling to room temperature over 24 h.[15] This results in the isolation of drab crystals of the title compound.

19.3.6.2 Structure Description

Single crystal X-ray diffraction experiments on UNpC1P2–1 reveal a remarkable 3D framework structure constructed from UO_6 tetragonal bipyramids, NpO_8 distorted dodecahedra, and monoprotonated methylenediphosphonate anions (Figure 19.15). The UO_6 tetragonal bipyramids contain a standard uranyl core resting on an inversion center with two U–O bond lengths of 1.776(7) Å. Of the remaining four U–O bonds, two are 2.289(7) Å and two are 2.317(6) Å. From these data a bond-valence sum for the U center can be calculated to be 6.07, which is consistent with this compound containing U^{VI}.[7,34] The NpO_8 units are found as distorted dodecahedra and reside on a single mirror plane. The Np–O bond lengths range from 2.245(7) Å to 2.500(9) Å; with two particularly long bonds of 2.474(11) Å and 2.500(9) Å. Both of these oxygen atoms are terminal and are thought to be from coordinating water molecules, as have been found in the two previous Np^{IV} phosphonates. An additional interesting feature of the structure of UNpC1P2–1 is a single terminal oxygen atom bound to P(2) of the methylenediphosphonate with a P–O bond length of 1.568(8) Å. This distance suggests the oxygen atom is

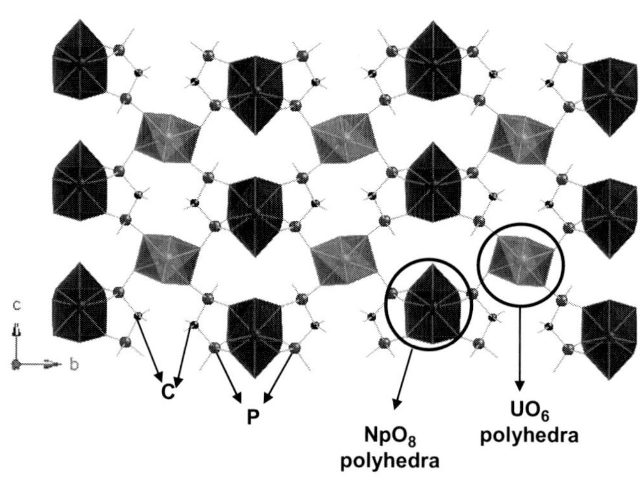

Figure 19.15 A view down the *a* axis (in the [*bc*] plane) of the 3D framework structure of the heterobimetallic U^{VI}/Np^{IV} diphosphonate $UO_2Np(H_2O)_2[CH_2(PO_3)(PO_3H)]_2$ (UNpC1P2-1). UO_6 and NpO_8 coordination polyhedra are shown.

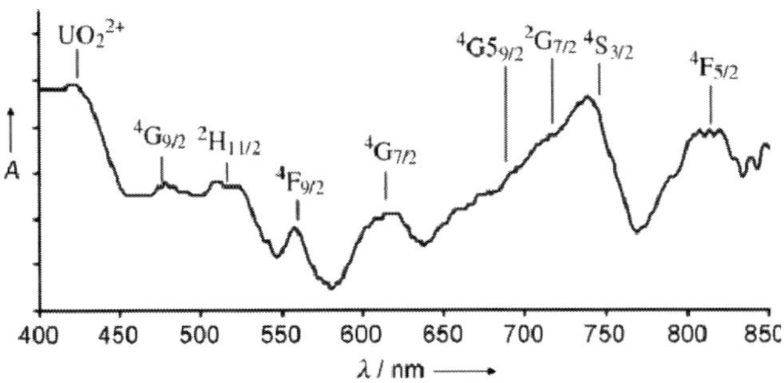

Figure 19.16 Visible diffuse reflectance spectrum of $UO_2Np(H_2O)_2[CH_2(PO_3)(PO_3H)]_2$ (UNpC1P2-1) demonstrating the presence of both U^{VI} and Np^{VI}.

protonated. As can be noted in Figure 19.15, there are small channels extending down the *a* axis that are ideally configured for a series of hydrogen bonding interactions between the P–OH group and coordinating water molecules.

Confirmation of the presence of both U and Np in UNpC1P2–1, while bolstered by reactivity and bonding arguments, cannot be demonstrated unequivocally on the basis of X-ray scattering alone because U and Np differ by only a single electron. Evidence for the inclusion of both elements in the crystals is provided by a variety of spectroscopic techniques starting with energy dispersive X-ray analysis that clearly indicates the presence of both U and Np in the crystals. Further clarity concerning the oxidation states of the actinoids in question is provided by visible diffuse reflectance spectroscopy (Figure 19.16). The spectrum shows characteristic charge-transfer features (centered at 422 nm) for the UO_2^{2+} cation, as well as strong f–f absorption bands (with the most intense feature at 740 nm) consistent with the presence of Np^{4+}.[35] Fluorescence from UNpC1P2–1 is not observed because the paramagnetic Np^{4+} centers quench the emission from the UO_2^{2+} moieties. Using a Raman microscope, vibrational data were collected from UNpC1P2-1. The α1-symmetric stretching mode of the UO_2^{2+} ion was found at 832 cm^{-1}.

19.3.7 $UO_2Pu(H_2O)_2[CH_2(PO_3)(PO_3H)]_2$

19.3.7.1 Synthesis

PuO_2 was reacted with UO_3 and methylenediphosphonic acid in water at 180 °C for 3 days in a PTFE-lined autoclave, followed by slow cooling to room temperature over 24 h. The product mixture consisted of an amorphous red precipitate, blue plates of α-$Pu[CH_2(PO_3)_2](H_2O)$, yellow acicular crystals of $UO_2[CH_2(PO_3H)_2](H_2O)$, and a low yield of orange-pink prisms of the desired product.[12]

Since our discovery of a heterobimetallic mixed-actinide phosphonate, currently represented only by $UO_2Np(H_2O)_2[CH_2(PO_3)(PO_3H)]_2$,[15] we have sought to prepare U^{VI}/Pu^{IV} compounds of this type. Again we started with the *in situ* hydrothermal reduction of PuO_2^{2+} to Pu^{4+} to provide our source of tetravalent plutonium. However, we invariably found that when UO_3 and PuO_2^{2+} are reacted simultaneously with methylenediphosphonic acid that the heterobimetallic compound does not form, but only $Pu[CH_2(PO_3)_2](H_2O)$ and $UO_2[CH_2(PO_3H)_2](H_2O)$ are isolated. However, with our new knowledge concerning the solubility of PuO_2 in the presence of methylenediphosphonic acid, we resorted to the direct reaction of the appropriate oxides, UO_3 and PuO_2, with methylenediphosphonic acid. This reaction afforded crystals of the desired compound, $UO_2Pu(H_2O)_2[CH_2(PO_3)(PO_3H)]_2$, albeit in low yield. Similar reactions were performed by replacing the PuO_2 with ThO_2. These result in the formation of $UO_2Th(H_2O)_2[CH_2(PO_3)(PO_3H)]_2$ in high yield.

19.3.7.2 Structure Description

Given the complex nature of the formula of these compounds, it might be considered surprising that $UO_2An(H_2O)_2[CH_2(PO_3)(PO_3H)]_2$ (An = Th, Np, Pu) form compounds with identical composition that are also isotypic. This is especially true since the compounds are arrived at by different routes. This suggests that this composition and structure might represent an energetic well. The structure of $UO_2An(H_2O)_2[CH_2(PO_3)(PO_3H)]_2$ consists of UO_6 tetragonal bipyramids containing an uranyl, UO_2^{2+}, core, and AnO_8 distorted dodecahedra containing an An^{IV} center that are bridged by partially protonated methylenediphosphonate, as shown in Figure 19.17. We can differentiate between the hexavalent uranium and the tetravalent actinide on the basis of

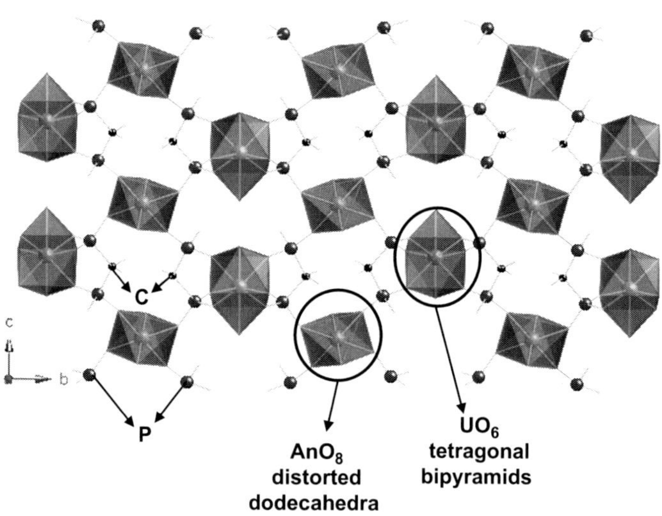

Figure 19.17 Illustration of the structure of $UO_2An(H_2O)_2[CH_2(PO_3)(PO_3H)]_2$ (An = Th, Np, Pu) consisting of UO_6 tetragonal bipyramids containing a uranyl, UO_2^{2+}, core, and AnO_8 distorted dodecahedra, containing an An^{IV} center, that are bridged by partially protonated methylenediphosphonate.

coordination geometry instead of X-ray scattering alone, which would be very similar for the two elements. Reaffirmation of the presence of An^{IV} and U^{VI} is also provided by Raman and visible diffuse reflectance spectroscopic data.[15]

In the case of $UO_2Pu(H_2O)_2[CH_2(PO_3)(PO_3H)]_2$, the uranium atom resides on an inversion center providing one unique U–O bond of 1.753(7) Å, and two unique equatorial bonds of 2.262(7) and 2.293(6) Å. These compare very well with the Np analog.[15] The Pu–O bonds occur from 2.213(7) to 2.475(4) Å. There are in fact six short bonds that range from 2.213(7) to 2.277(7) Å and two long bonds to the coordinating water molecules of 2.465(10) and 2.475(4) Å. Here the Pu–O bonds are slightly shorter than the Np–O bonds as expected from the actinide contraction. The expected difference should be on the order of 0.02 Å.[17] As implied from its larger ionic radius, the Th–O bonds are the longest in this series.

19.4 Conclusions

Transuranium phosphonates still represent a small subset of compounds in the vast family of metal phosphonates. Nevertheless these compounds have demonstrated that neptunium and plutonium do not behave the same as uranium and thorium in many cases. Furthermore, whereas Th^{IV} and U^{VI} have no 5f electrons, all of the common oxidation states of neptunium and plutonium have 5f electrons. This allows for the potential development of magneto-structural correlations. We have also now reached the point where *in situ* hydrothermal reduction of transuranics can be combined with redox-inactive actinides to prepare structurally rich heterobimetallic compounds. Our quest for new transuranium phosphonates continues.

Acknowledgements

We are grateful for support provided by the Chemical Sciences, Geosciences, and Biosciences Division, Office of Basic Energy Sciences, Office of Science, Heavy Elements Program, U.S. Department of Energy, under Grant DE-FG02-01ER16026. This material is based upon work supported as part of the Materials Science of Actinides, an Energy Frontier Research Center funded by the U.S. Department of Energy, Office of Science, Office of Basic Energy Sciences under Award Number DE-SC0001089.

References

1. (a) K. L. Nash, *J. Alloys Compd.*, 1997, **249**, 33; (b) M. P. Jensen, J. V. Beitz, R. D. Rogers and K. L. Nash, *J. Chem. Soc., Dalton Trans.*, 2000, **18**, 3058.
2. R. Chiarizia, E. P. Horwitz, S. D. Alexandratos and M. J. Gula, *Sep. Sci. Technol.*, 1997, **32**, 1.
3. M. B. Dines and P. M. DiGiacomo, *Inorg. Chem.*, 1981, **20**, 92.

4. (a) D. Grohol and A. Clearfield, *J. Am. Chem. Soc.*, 1997, **119**, 9301; (b) D. Grohol and A. Clearfield, *J. Am. Chem. Soc.*, 1997, **119**, 4662; (c) D. Grohol, M. A. Subramanian, D. M. Poojary and A. Clearfield, *Inorg. Chem.*, 1996, **35**, 5264; (d) D. M. Poojary, A. Cabeza, M. A. G. Aranda, S. Bruque and A. Clearfield, *Inorg. Chem.*, 1996, **35**, 5603; (e) D. M. Poojary, D. Grohol and A. Clearfield, *Angew. Chem., Int. Ed. Engl.*, 1995, **34**, 1508; (f) D. M. Poojary, D. Grohol and A. Clearfield, *J. Phys. Chem. Solids*, 1995, **56**, 1383; (g) A. Britel, M. Wozniak, J. C. Boivin, G. Nowogrocki and D. Thomas, *Acta Crystallogr.*, 1986, **C42**, 1502; (h) D. Grohol and A. Clearfield, *Inorg. Chem.*, 1999, **38**, 751; (i) M. B. Doran, A. J. Norquist and D. O'Hare, *Chem. Mater.*, 2003, **15**, 1449.
5. A. Cabeza, M. A. G. Aranda, F. M. Cantero, D. Lozano, M. Martínez-Lara and S. Bruque, *J. Solid State Chem.*, 1996, **121**, 181.
6. T. H. Bray, J. Ling, E.-S. Choi, J. S. Brooks, J. V. Beitz, R. E. Sykora, R. G. Haire, D. M. Stanbury and T. E. Albrecht-Schmitt, *Inorg. Chem.*, 2007, **46**, 3663.
7. (a) N. E. Brese and M. O'Keeffe, *Acta Crystallogr.*, 1991, **B47**, 192; (b) I. D. Brown, *The Chemical Bond in Inorganic Chemistry: The Bond Valence Model*, Oxford University Press, Oxford, UK, 2002.
8. S. Nagasaki, K. Kinoshita, D. S. Wisnubroto, Y. Enokida and A. Suzuki, *J. Nucl. Sci. Technol.*, 1992, **29**, 671.
9. P. C. Burns, *Can. Mineral.*, 2005, **43**, 1839.
10. (a) A. Clearfield, *Prog. Inorg. Chem.*, 1998, **47**, 371; (b) A. Clearfield, *Curr. Opin. Solid State Mater. Sci*, 2003, **6**, 495; (c) J.-G. Mao, *Coord. Chem. Rev.*, 2007, **251**, 1493.
11. T. H. Bray, A.-G. D. Nelson, G. B. Jin, R. G. Haire and T. E. Albrecht-Schmitt, *Inorg. Chem.*, 2007, **46**, 10959.
12. A.-G. D. Nelson, T. H. Bray, A. F. Stanley and T. E. Albrecht-Schmitt, *Inorg. Chem.*, 2009, **48**, 4530.
13. A.-G. D. Nelson, T. H. Bray, W. Zhan, R. G. Haire and T. E. Albrecht-Schmitt, *Inorg. Chem.*, 2008, **47**, 4945.
14. J. Diwu, A.-G. D. Nelson, S. Wang, C. F. Campana and T. E. Albrecht-Schmitt, *Inorg. Chem.*, 2010, **49**, 3337.
15. A.-G. D. Nelson, T. H. Bray and T. E. Albrecht-Schmitt, *Angew. Chem., Int. Ed. Engl.*, 2008, **47**, 6252.
16. P. Benard, D. Louer, N. Dacheux, V. Brandel and M. Genet, *Chem. Mater.*, 1994, **6**, 1049.
17. R. D. Shannon, *Acta Crystallogr.*, 1976, **A32**, 751.
18. (a) R. A. Penneman, R. R. Ryan and A. Rosenzweig, *Struct. Bonding (Berlin)*, 1973, **13**, 1; (b) V. Brandel and N. Dacheux, *J. Solid State Chem.*, 2004, **177**, 4755; (c) V. Brandel and N. Dacheux, *J. Solid State Chem.*, 2004, **177**, 4743.
19. (a) W. Runde, A. C. Bean, T. E. Albrecht-Schmitt and B. L. Scott, *Chem. Commun.*, 2003, **4**, 478; (b) A. C. Bean, B. L. Scott, T. E. Albrecht-Schmitt and W. Runde, *Inorg. Chem.*, 2003, **42**, 5632; (c) F. Weigel and L. W. H. Engelhardt, *J. Less-Common Met.*, 1983, **91**, 339; (d) A. C. Bean, S. M.

Peper and T. E. Albrecht-Schmitt, *Chem. Mater.*, 2001, **13**, 1266; (e) A. E. V. Gorden, D. K. Shuh, B. E. F. Tiedemann, R. E. Wilson, J. Xu and K. N. Raymond, *Chemistry*, 2005, **11**, 2842; (f) A. E. V. Gorden, J. Xu and K. N. Raymond, *Chem. Rev.*, 2003, **103**, 4207; (g) J. Xu, E. Radkov, M. Ziegler and K. N. Raymond, *Inorg. Chem.*, 2000, **39**, 4156.

20. N. Dacheux, R. Podor, V. Brandel and M. J. Genet, *Nucl. Mater.*, 1998, **252**, 179.
21. N. Dacheux, S. Grandjean, J. Rousselle and N. Clavier, *Inorg. Chem.*, 2007, **46**, 10390.
22. N. Dacheux, A. C. Thomas, V. Brandel and M. Genet, *J. Nucl. Mater.*, 1998, **257**, 108.
23. C. Keller and K. H. Walter, *J. Inorg. Nucl. Chem.*, 1965, **27**, 1253.
24. N. Dacheux, N. Clavier, G. Wallez and M. Quarton, *Solid State Sci.*, 2007, **9**, 619.
25. L. R. Morss, N. M. Edelstein, J. Fuger, *The Chemistry of the Actinide and Transactinide Elements*, Springer, Heidelberg, 2006, **4**, Chapter 19, p. 2130.
26. J. Diwu, A.-G. D. Nelson and T. E. Albrecht-Schmitt, *Comments Inorg. Chem.*, 2010, **31**(1&2), 46.
27. J. D. McCullough and K. N. Trueblood, *Acta Crystallogr.*, 1959, **12**, 507.
28. R. Chiarizia, E. P. Horwitz, S. D. Alexandratos and M. Gula, *J. Sep. Sci. Technol.*, 1997, **32**, 1.
29. D. L. Clark, S. D. Conradson, D. W. Keogh, P. D. Palmer, B. L. Scott and C. D. Tait, *Inorg. Chem.*, 1998, **37**, 2893.
30. J. I. Kim, C. Lierse and F. Baumgartner, *ACS Symp. Ser.*, 1983, **216**, 317.
31. M. Eiswirth, J. I. Kim and C. Lierse, *Radiochim. Acta*, 1985, **38**, 1971.
32. H. Capdevila, P. Vitorge, E. Giffaut and L. Delmau, *Radiochim. Acta*, 1996, **74**, 93.
33. W. T. Carnall, G. K. Liu, C. W. Williams and M. F. Reid, *J. Chem. Phys.*, 1991, **95**, 7194.
34. P. C. Burns, R. C. Ewing and F. C. Hawthorne, *Can. Mineral.*, 1997, **35**, 1551.
35. L. P. Varga, M. J. Reisfeld and L. B. Asprey, *J. Chem. Phys.*, 1970, **53**, 250.
36. J. Diwu, S. Wang, Z. Liao and T. E. Albrecht-Schmitt, *Inorg. Chem.*, 2010, **49**, 10074.

Subject Index

Page numbers followed by T refer to a table. Illustrations accompany the text throughout and are not indicated separately in the index.

Entries for individual phosphonic acids may include references to the corresponding phosphonates, and *vice versa*.

Many ligands are referred to by different acronyms and labels by different authors. These are listed in parantheses after the name in the index (and there are cross-references from the acronyms). In case of difficulty in identifying the relevant label on a particular page, the reader is advised to look at the beginning of the chapter concerned for a key to the abbreviations used. For convenience, this information is gathered in the index under "abbreviations and structural formulae".